METAL IONS IN
BIOLOGICAL SYSTEMS

METAL IONS IN
BIOLOGICAL SYSTEMS

Edited by

Astrid Sigel
and **Helmut Sigel**

Institute of Inorganic Chemistry
University of Basel
CH-4056 Basel, Switzerland

VOLUME 39

Molybdenum and Tungsten:
Their Roles in Biological Processes

CRC Press
Taylor & Francis Group
Boca Raton London New York

CRC Press is an imprint of the
Taylor & Francis Group, an **informa** business

CRC Press
Taylor & Francis Group
6000 Broken Sound Parkway NW, Suite 300
Boca Raton, FL 33487-2742

First issued in paperback 2019

© 2002 by Taylor & Francis Group, LLC
CRC Press is an imprint of Taylor & Francis Group, an Informa business

No claim to original U.S. Government works

ISBN-13: 978-0-8247-0765-1 (hbk)
ISBN-13: 978-0-367-39629-9 (pbk)

Visit the Taylor & Francis Web site at
http://www.taylorandfrancis.com

and the CRC Press Web site at
http://www.crcpress.com

Preface to the Series

Recently, the importance of metal ions to the vital functions of living organisms, hence their health and well-being, has become increasingly apparent. As a result, the long-neglected field of "bioinorganic chemistry" is now developing at a rapid pace. The research centers on the synthesis, stability, formation, structure, and reactivity of biological metal ion-containing compounds of low and high molecular weight. The metabolism and transport of metal ions and their complexes is being studied, and new models for complicated natural structures and processes are being devised and tested. The focal point of our attention is the connection between the chemistry of metal ions and their role for life.

No doubt, we are only at the brink of this process. Thus, it is with the intention of linking coordination chemistry and biochemistry in their widest sense that the *Metal Ions in Biological Systems* series reflects the growing field of "bioinorganic chemistry." We hope, also, that this series will help to break down the barriers between the historically separate spheres of chemistry, biochemistry, biology, medicine, and physics, with the expectation that a good deal of future outstanding discoveries will be made in the interdisciplinary areas of science.

Should this series prove a stimulus for new activities in this fascinating "field," it would well serve its purpose and would be a satisfactory result for the efforts spent by the authors.

Fall 1973

Helmut Sigel
Institute of Inorganic Chemistry
University of Basel
CH-4056 Basel, Switzerland

Preface to Volume 39

Molybdenum and tungsten are congeners in group 6 of the Periodic Table and they are the only members of the 4d and 5d metal series with known biological functions. Their importance for biological systems has now been recognized for over 75 and 25 years, respectively. Molybdenum is required by archaea, bacteria, fungi, plants, and animals, including humans, and well over 50 different enzymes are now known. In contrast, tungsten, at least to date, has been identified in only a few organisms, mostly hyperthermophilic archaea that live in volcanic vents on the sea bed at $>100°C$, i.e., in a niche that apparently favors the evolution of tungstoenzymes.

The use of molybdenum by living organisms seems surprising at first glance since its abundance in the Earth's crust is low and many elements that are not biologically essential, like Al, Ti, Zr, La, and Ga, are far more abundant. However, Mo is the most abundant transition metal in the oceans, where it exists, almost exclusively, as the dianionic molybdate ion, MoO_4^{2-}, its average concentration being about 10 µg/L compared with about 0.5 µg/L for Zn or 0.1 µg/L for Cu. The tungsten concentrations are about two orders of magnitude lower, matching those of copper. Only in special environments such as the deep-sea hydrothermal vents are they higher. These and other aspects are covered in Chapter 1, which focuses on the biogeochemistry of molybdenum and tungsten. Chapter 2 deals with the absorption, transport, and homeostasis of the two elements by living systems, i.e., mainly by bacteria, as well as the binding of molybdate and tungstate to proteins.

Molybdenum plays a vital role in the reduction of dinitrogen to ammonia, since it occurs in one type of nitrogenases; these enzymes are used by bacteria, which usually live in a symbiotic association with plants, for the so-called nitrogen fixation which is essential for sustaining life on earth. The properties of molybdenum nitrogenases, which contain a cluster of Mo and seven Fe, are discussed in Chapter 3 whereas the biosynthesis of this iron-molybdenum cofactor of nitrogenases is covered in Chapter 5. Chemical dinitrogen fixation by molybdenum and tungsten complexes is the focus of Chapter 4. Worldwide about 90 million tons of N per year are provided to agriculture through biological nitrogen fixation, and another 90 million tons are supplied via the industrial Haber-Bosch process; without it widespread food shortages would undoubtedly occur. Considering that this latter process uses high temperatures and pressures, typically 350°C and 200–300 bar, and is dependent on fossil fuels generating large quantities of CO_2 as a byproduct, it is evident that achieving nitrogen fixation under ambient conditions corresponding to those at which nitrogenases work is one of the most challenging subjects in chemistry.

The other kind of enzymes important for the nitrogen cycle are the nitrate reductases. They belong to a large class of enzymes with a mononuclear Mo center and catalyze key reactions in the metabolism of C, N, S, etc., in bacteria, plants, animals, and humans. An overview of these molybdenum enzymes containing the pyranopterin cofactor is given in Chapter 6. The crystallographic appearance of this cofactor in enzymes containing Mo or W, as well as the related model chemistry, and its biosynthesis including the molecular biology are covered in the next three chapters, respectively. Chapters 10 through 18 are devoted to individual enzymes containing the pyranopterin cofactor such as nitrate reductase, nitrite oxidoreductase, xanthine oxidoreductase, and several hydroxylases and dehydrogenases.

The tungsten-dependent aldehyde oxidoreductases, which also contain the pterin cofactor, are discussed in Chapter 19, whereas Chapter 20 is devoted to tungsten-substituted molybenum enzymes. In the penultimate chapter, the metabolism of molybdenum and its requirements for humans including deficiency and toxicity are discussed, and the final chapter addresses the toxicity of tungsten for animals and humans.

It is the hope of the authors and editors that this volume, solely devoted to the roles of molybdenum and tungsten in biological systems, will further stimulate this fascinating and rapidly growing research area which owes much to the pioneering work of Bob Bray, to whom the following *In Memoriam*, written by David J. Lowe, pays tribute.

Astrid Sigel
Helmut Sigel

In Memoriam
of Professor Robert (Bob) C. Bray
(4th August 1928–17th August 2001)

The field of molybdenum and tungsten biochemistry notes with sadness the sudden death of Professor Robert C. Bray. Bob was a pioneer of the study of the role of molybdenum in biological processes who continued working in the area up to the day of his death, shortly before this volume went to press. During his entire career he remained at the forefront of the field as a much respected experimentalist, with a deep knowledge of

FIG. 1.   Robert C. Bray, M.A., Ph.D., Sc.D.

molybdenum-containing enzymes, and an active contributor to both written and verbal debates on their structure and function. He will be greatly missed by collaborators and competitors alike as an influential and stimulating scientist, colleague, and friend.

Bob was born in Thornton Heath in the south of London and was educated at Purley Grammar School before studying chemistry at Selwyn College of Cambridge University. He went on to the Chester Beatty Research Institute, the research arm of the Royal Cancer Hospital, where an extensive program of work on the molybdoprotein xanthine oxidase, organized by Professor F. Bergel FRS, had been initiated in 1952. This research arose from the suggestion that reduction of the activity level of this enzyme in tissues might be associated with carcinogenesis, although Bob with his characteristic regard for careful experimental design was skeptical of the evidence for this [1]. Bergel recruited Bob as a member of the team working on xanthine oxidase and became his Ph.D. supervisor. Bob's first publication [2] described the isolation of xanthine oxidase in a crystalline form, although we have had to wait almost 50 years since this achievement for the determination by X-ray crystallography of the structure of the enzyme [3]. In agreement with others, including Green and Beinert [4], his first paper reported the presence of molybdenum in the preparations as well as of flavin, although the authors were less certain of the functionality of the iron that was also present. Thus, the three cornerstones of Bob's scientific interests—molybdenum, flavin, and iron-sulfur in biological systems—were laid at the very start of his career. In his hands xanthine oxidase became a paradigm system for studying the structure and function of multi-center electron transfer within proteins.

Bob was always interested in applying new techniques in order to learn more about his proteins, and as a result xanthine oxidase generally appears at the start of the literature of biological applications of new physical techniques. EPR (electron paramagnetic resonance) is particularly associated with Bob [5], but more recently for example he was one of the first to be aware of the power of using a genetic approach in combination with biophysical methods. His use of EPR dated from visits to Sweden and was strengthened by a long collaboration with John Gibson [6], who was then just down the road from the Chester Beatty in the old Royal School of Science building at Imperial College. The usefulness of EPR was extended by Bob's development of the rapid-freezing method [7], which enabled him to study solutions of enzymes frozen within 5 ms of initiating a reaction and thus to observe spectral changes occurring during a single enzymic turnover. The latest version of this apparatus is still in regular use in my laboratory with many components made and calibrated by Bob, together with a comprehensive and meticulous instruction manual prepared with his usual attention to detail. In the early days he used a combination of EPR and rapid freezing with the Swedish group and also took it to Madison for a productive collaboration with Helmut Beinert and Graham Palmer [8].

In 1970 Bob moved from the Chester Beatty to the University of Sussex in Brighton to take up a readership in the Biochemistry subject group and in the then School of Molecular Sciences. He served as Subject Chair of Biochemistry for many years and was awarded a personal chair in 1990. His work on xanthine oxidase continued and expanded to a number of other enzymes, including nitrogenase [9] and dimethylsulfoxide (DMSO) reductase, the subject of his latest paper [10]; this appeared in the issue of *Biochemistry* dated four days after his death. Several other manuscripts to which he contributed are with journals or still in preparation.

Bob had retired, at least in a technical sense, from the University of Sussex in 1993, but remained actively involved in research. He said recently that he was intending to retire in reality in the near future, but I and others remained unconvinced of this. Bob was in great form at the molybdenum and tungsten enzyme Gordon Conference in July 2001, and was clearly looking forward with enthusiasm to future developments in his chosen field. A key aspect of this meeting was a consideration of the crystal structures of xanthine dehydrogenase and xanthine oxidase. It is particularly pleasing that Bob lived to see the solution of the crystal structure [3] of the molecule he spent his life exploring. As a result of the work of Bob (and others) the environment of the molybdenum was understood before the crystallographic results added much important and additional detail to our knowledge of this fascinating protein. A characteristically outstanding experimental achievement from Bob's early work is shown in Figure 2, with an EPR spectrum of the Very Rapid signal from turning-over ^{95}Mo-enriched bovine milk xanthine oxidase [11]; this showed the world that the molybdenum was "real" and functional, and suggested its likely environment. Indeed, due to Bob's many contributions, and work inspired by them,

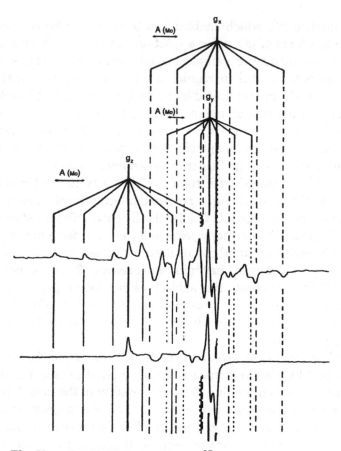

FIG. 2. The Very Rapid EPR signal from ^{95}Mo-substituted xanthine oxidase [11]. (Reproduced with permission from *Nature*.)

we also know much about the nature of the transient intermediates involved in the catalyses effected by these enzymes. These species are not accessible to an X-ray crystallographic study but need to be defined in order to achieve a complete understanding of enzymic function.

I am sure that Bob greatly appreciated the recent significant rise in interest in the field of which he was a pathfinder, as demonstrated by the contents of this volume and by the establishment of a series of Gordon Research Conferences on molybdenum and tungsten enzymes. In fact, his last public scientific appearance was at this Gordon Conference in Oxford only five weeks before he died. Here he talked about DMSO reductase and stayed up well into the early hours in excellent form, helped by only the occasional glass of sustenance, discussing the contents of both his paper and those of others.

The emails received about Bob show that his contributions to science have been an inspiration to many, and will continue to be so. He will be greatly missed by those who knew him well as someone who was intensely loyal, a totally fair, kind and generous person, rigorous and objective in his science, collegial in every sense of the word, and a man of great personal integrity.

ACKNOWLEDGMENTS

I thank many of Bob's friends and colleagues for help and suggestions in the preparation of this contribution and especially Ben Adams, Michael Bray, Pam Bray, John Enemark, Dave Garner, Russ Hille, Peter Knowles, Al McEwan, Ray Richards, Claudio Scazzocchio and Charles Young; they will recognize many of their words in the text.

David J. Lowe
John Innes Centre
Norwich, UK

REFERENCES

1. R. C. Bray, *Ph.D. Thesis: Some Aspects of the Chemistry and Biochemistry of Xanthine Oxidase*, University of London (1956).

2. P. G. Avis, F. Bergel, R. C. Bray, and K. V. Shooter, *Nature, 173*, 1230–1233 (1954).

3. C. Enroth, B. T. Eger, K. Okamoto, T. Nishino, T. Nishino, and E. Pai, *Proc. Natl. Acad. Sci. USA, 97*, 10723–10728 (2000).

4. D. E. Green and H. Beinert, *Biochim. Biophys. Acta, 11*, 599–600 (1953).

5. R. C. Bray, B. G. Malmström, and T. Vänngård, *Biochem. J., 73*, 193–197 (1959).

6. J. F. Gibson and R. C. Bray, *Biochim. Biophys. Acta, 153,* 721–723 (1968).

7. R. C. Bray, *Biochem. J., 81,* 189–193 (1961).

8. R. C. Bray, G. Palmer, and H. Beinert, *J. Biol. Chem., 239,* 2667–2676 (1954).

9. B. E. Smith, D. J. Lowe, and R. C. Bray, *Biochem. J., 130,* 641–643 (1972).

10. R. C. Bray, B. Adams, A. T. Smith, R. L. Richards, D. J. Lowe, and S. Bailey, *Biochemistry 40,* 9810–9820 (August 21, 2001).

11. R. C. Bray and L. Meriwether, *Nature, 212,* 467–469 (1966).

Contents

Contributors

Numbers in parentheses indicate the pages on which the authors' contributions begin.

Dietmar J. Abt Fachbereich Biologie, Universität Konstanz, Postfach M665, D-78457 Konstanz, Germany (369)

Michael W. W. Adams Department of Biochemistry and Molecular Biology, and Center for Metalloenzyme Studies, University of Georgia, Athens, GA 30602-7229, USA <adams@bmb.uga.edu> (673)

Jan R. Andreesen Institute of Microbiology, University of Halle, Kurt-Mothes-Strasse 3, D-06120 Halle, Germany (Fax: +49-345-552-7010) <j.andreesen@mikrobiologie.uni-halle.de> (405)

Carlos D. Brondino Departamento de Química, Centro de Química Fina e Biotecnologia, Faculdade de Ciências e Tecnologia, Universidade Nova de Lisboa, P-2825-114 Caparica, Portugal (539)

Sharon J. Nieter Burgmayer Department of Chemistry, Bryn Mawr College, Bryn Mawr, PA 19010, USA (Fax: +1-610-526-5086) <sburgmay@brynmawr.edu> (265)

Michele Mader Cosper Department of Chemistry, University of Arizona, Tucson, AZ 85721-0041, USA (621)

Carlos A. Cunha Departamento de Química, Centro de Química Fina e Biotecnologia, Faculdade de Ciências e Tecnologia, Universidade Nova de Lisboa, P-2825-114 Caparica, Portugal (539)

Dennis R. Dean Department of Biochemistry, Fralin Biotechnology Center, Virginia Tech, Blacksburg, VA 24061, USA <deandr@vt.edu> (163)

Holger Dobbek Max-Planck-Institut für Biochemie, Abteilung Strukturforschung, Am Klopferspitz 18A, D-82152 Martinsried, Germany <dobbek@biochem.mpg.de> (227)

John H. Enemark Department of Chemistry, University of Arizona, Tucson, AZ 85721-0041, USA (Fax: +1-520-626-8065) <jenemark@u.arizona.edu> (621)

Susanne Fetzner Mikrobiologie, Fachbereich 7 (Biologie), Carl von Ossietzky Universität Oldenburg, D-26111 Oldenburg, Germany <susanne.fetzner@uni-oldenburg.de> (405,481)

Berthold Fischer Lehrstuhl für Analytische Chemie, Ruhr-Universität Bochum, Universitätsstrasse 150, D-44780 Bochum, Germany (Fax: +49-234-3214-420) <fischer@anaserv.anachem.ruhr-uni-bochum.de> (265)

Jeverson Frazzon Department of Food Science, ICTA, Federal University of Rio Grande do Sul, Porto Allegre, RS, 91051-970, Brazil (163)

C. David Garner The Chemistry Department, Nottingham University, Nottingham NG7 2RD, UK (Fax: +44-115-951-3563) <dave.garner@nottingham.ac.uk> (699)

Vadim N. Gladyshev Department of Biochemistry, University of Nebraska, Lincoln, NE 68588-0664, USA <vgladysh@unlnotes.unl.edu> (655)

Masanobu Hidai Science University of Tokyo, Department of Materials Science and Technology, Faculty of Industrial Science and Technology, Noda, Chiba 278-8510, Japan (Fax: +81-471-24-1699) <hidai@rs.noda.sut.ac.jp> (121)

Russ Hille Department of Medical Biochemistry, The Ohio State University, Columbus, OH 43210-1218, USA (<hille.1@osu.edu>) (187)

Robert Huber Max-Planck-Institut für Biochemie, Abteilung Strukturforschung, Am Klopferspitz 18A, D-82152 Martinsried, Germany (FAX: +49-89-8578-3516) <huber@biochem.mpg.de> (227)

Jürgen Hüttermann Fachrichtung Biophysik und Physikalische Grundlagen der Medizin, Universität des Saarlandes, Klinikum Bau 76, D-66421 Homburg (Saar), Germany (Fax: +49-6841-166227) <bpjhue@med-rz.uni-saarland.de> (481)

xxv

Reinhard Kappl Fachrichtung Biophysik und Physikal. Grundlagen der Medizin, Universität des Saarlandes, Klinikum Bau 76, D-66421 Homburg (Saar), Germany <bprkap@med-rz.uni-saarland.de> (481)

Peter M. H. Kroneck Fachbereich Biologie, Universität Konstanz, Postfach M665, D-78457 Konstanz, Germany <peter.kroneck@uni-konstanz.de> (369)

Florence Lagarde Laboratoire de Chimie Analytique et Minérale-UMR 7512, ECPM, 25, rue Becquerel, F-67087 Strasbourg Cédex 02, France (Fax: +33-3-9024-2725) <florence.lagarde@chimie.u-strasbg.fr> (741)

David M. Lawson Department of Biological Chemistry, John Innes Centre, Norwich, NR4 7UH, UK <david.lawson@bbsrc.ac.uk> (31,75)

Maurice Leroy Laboratoire de Chimie Analytique et Minérale-UMR 7512, ECPM, 25, rue Becquerel, F-67087 Strasbourg Cédex 02, France (741)

David J. Lowe Biological Chemistry Department, John Innes Centre, Colney Lane, Norwich, NR4 7UH, UK (Fax: +44-1603-450018) <david.lowe@bbsrc.ac.uk> (*vii*, 455)

Ralf R. Mendel Botanical Institute, Technical University of Braunschweig, D-38023 Braunschweig, Germany (Fax: +49-531-3918-128) <r.mendel@tu-bs.de> (317)

Yasushi Mizobe Institute of Industrial Science, The University of Tokyo, Komaba, Meguro-ku, Tokyo 153-8510, Japan <ymizobe@iis.u-tokyo.ac.jp> (121)

José J. G. Moura Departamento de Química, Centro de Química Fina e Biotecnologia, Faculdade de Ciências e Tecnologia, Universidade Nova de Lisboa, P-2825-114 Caparica, Portugal <jose.moura@dq.fct.unl.pt> (539)

Takeshi Nishino Department of Biochemistry and Molecular Biology, Nippon Medical School, 1-1-5 Sendagi, Bunkyo-ku, Tokyo 113-8602, Japan (431)

Emil F. Pai Departments of Biochemistry, Medical Biophysics, and Molecular and Medical Genetics, University of Toronto, 1 King's College Circle, Toronto, ON, M5S 1A8, Canada, and Division of Molecular and Structural Biology, Ontario Cancer Institute, University Health Network, 610 University Avenue, Toronto, ON,

M5G 2M9, Canada (Fax: +1-416-978-8548)
<pai@hera.med.utoronto.ca> (431)

Richard N. Pau Department of Biochemistry and Molecular Biology,
University of Melbourne, Parkville, VIC 3052, Australia
(Fax: +61-3-9347-7730) <r.pau@unimelb.edu.au> (31)

Maria João Romão Departamento de Química, Centro de Química
Fina e Biotecnologia, Faculdade de Ciências e Tecnologia,
Universidade Nova de Lisboa, P-2825-114 Caparica, Portugal
<mromao@dq.fct.unl.pt> (539)

Roopali Roy Department of Biochemistry and Molecular Biology,
and Center for Metalloenzyme Studies, University of Georgia, Athens,
GA 30602, USA (673)

Günter Schwarz Botanical Institute, Technical University of
Braunschweig, D-38023 Braunschweig, Germany (317)

Barry E. Smith John Innes Centre, Nitrogen Fixation Laboratory,
Norwich, NR4 7UH, UK <barry.smith@bbsrc.ac.uk> (75)

Edward I. Stiefel ExxonMobil Research and Engineering Company,
1545 Route 22 East, Clinton Township, Annandale, NJ 08801-3059,
USA (1)
Present address: Department of Chemistry, Princeton University,
Princeton, NJ 08544-1009, USA <estiefel@princeton.edu>

Lisa J. Stewart The Chemistry Department, Nottingham University,
Nottingham NG7 2RD, UK (699)
Present address: The Daresbury Laboratory, Daresbury, Warrington,
WA4 4AD, UK <l.j.stewart@dl.ac.uk>

Rudolf K. Thauer Max-Planck Institut für terrestrische
Mikrobiologie und Laboratorium für Mikrobiologie, Philipps-
Universität, Karl-von-Frisch-Strasse, D-35043 Marburg, Germany
(Fax: +49-6421-178209) <thauer@mailer.uni-marburg.de> (571)

Judith R. Turnlund USDA/ARS, Western Human Nutrition
Research Center, University of California at Davis, Davis, CA 95616,
USA <jturnlun@whnrc.usda.gov> (727)

Julia Vorholt Max-Planck Institut für terrestrische Mikrobiologie
und Laboratorium für Mikrobiologie, Philipps-Universität, Karl-von-
Frisch-Strasse,D-35043 Marburg, Germany (Fax: +49-6421-178209)
<vorholt@mailer.uni-marburg.de> (571)

Contents of Previous Volumes

---

* Out of print.

---

* Out of print.

**Volume 21. Applications of Nuclear Magnetic Resonance to
            Paramagnetic Species**

Volume 27. Electron Transfer Reactions in Metalloproteins

**Volume 33. Probing of Nucleic Acids by Metal Ion Complexes
of Small Molecules**

Volume 37. Manganese and Its Role in Biological Processes

Comments and suggestions with regard to contents, topics, and the like
for future volumes of the series are welcome.

The following Marcel Dekker, Inc., books are also of interest for any reader involved with bioinorganic chemistry or who is dealing with metals or other inorganic compounds:

Handbook on Toxicity of Inorganic Compounds
edited by Hans G. Seiler and Helmut Sigel, with Astrid Sigel.
In 74 chapters, written by 84 international authorities, this book covers the physiology, toxicity, and levels of tolerance, including prescriptions for detoxification, for all elements of the Periodic Table (up to atomic number 103). The book also contains short summary sections for each element, dealing with the distribution of the elements, their chemistry, technological uses, and ecotoxicity as well as their analytical chemistry.

Handbook on Metals in Clinical and Analytical Chemistry
edited by Hans G. Seiler, Astrid Sigel, and Helmut Sigel.
This book is written by 80 international authorities and covers over 3500 references. The first part (15 chapters) focuses on sample treatment, quality control, etc., and on the detailed description of the analytical procedures relevant for clinical chemistry. The second part (43 chapters) is devoted to a total of 61 metals and metalloids; all these contributions are identically organized covering the clinical relevance and analytical determination of each element as well as, in short summary sections, its chemistry, distribution, and technical uses.

Handbook on Metalloproteins
edited by Ivano Bertini, Astrid Sigel, and Helmut Sigel.
The book consists of 23 chapters written by 43 international authorities. It summarizes a large part of today's knowledge on metalloproteins, emphasizing their structure-function relationships, and it encompasses the metal ions of sodium, potassium, magnesium, calcium, vanadium, chromium, manganese, iron, cobalt, nickel, copper, zinc, molybdenum, and tungsten.

1

The Biogeochemistry of Molybdenum and Tungsten

Edward I. Stiefel*

ExxonMobil Research and Engineering Co., 1545 Rt 22E,
Clinton Township, Annandale, NJ 08801, USA

*Present address: Department of Chemistry, Princeton University, Princeton,
NJ 08544-1009, USA

1. INTRODUCTION

Biogeochemical cycles have been identified for the major elements essential for life on Earth [1]. The cycles of carbon, oxygen, hydrogen, nitrogen, and sulfur are intimately linked and the evolution of a relatively stable set of biogeochemical cycles was likely crucial to the evolution of life on Earth [1].

Many "trace" elements, both essential and nonessential, have biogeochemical cycles as well. Moreover, many of the essential elements including molybdenum and, to a lesser extent, tungsten, play key roles in the cycles of the major elements through their participation in the active sites of metalloenzymes.

The importance of molybdenum and tungsten in biological systems has been known for more than 75 [2] and more than 25 years [3,4], respectively. These two elements, congeners in group 6 of the periodic table, are the only members of the second (Mo) and third (W) transition series with known biological functions. Molybdenum is required by bacteria, archaea, fungi, plants, and animals, while tungsten, at least to date, has only been identified in a limited range of prokaryotes (bacteria and archaea) [5] (see also R. Roy and M. W. W. Adams, Chapter 19). Most organisms, including humans, require molybdenum for their existence. Only a few organisms, mostly hyperthermophilic archaea, require tungsten.

The requirement for Mo and W is due to their presence at the active sites of metalloenzymes that execute key transformations in the metabolism of nitrogen, sulfur, carbon, arsenic, selenium, and chlorine compounds. For the major cycles, molybdenum plays a critical role in the biogeochemistry of nitrogen and sulfur, and both Mo and W participate in anaerobic parts of the carbon cycle.

This chapter addresses two interrelated aspects of the biogeochemistry of molybdenum and tungsten. First, we consider the presence of Mo and W in the environment, starting with their abundances in natural waters and their potential speciation. The possible ancient transformations and speciation of Mo and W may have had profound impacts on the evolution of life on Earth. Second, we consider the cycling and metabolism of major and minor elements and highlight the key roles that Mo or W enzymes play in the modern biogeochemical transformations.

2.   MOLYBDENUM AND TUNGSTEN IN THE ENVIRONMENT

The use of molybdenum by living organisms at first glance appears disproportionate to its abundance in the Earth as a whole or even in Earth's crust [6]. Indeed, many elements that are not biologically essential are far more abundant, including Al, Ti, Zr, La, and Ga. However, if one considers the aqueous phases on our planet [6,7], the abundance of Mo is remarkably high, such that the use molybdenum by myriad organism should not come as a great surprise. Moreover, the unique chemistry of molybdenum may provide competitive advantage to those organisms that learn to exploit it.

2.1.   Natural Abundances and Speciation of Mo and W

Molybdenum is the most abundant transition metal in the modern oceans of Earth. At the slightly alkaline pH (8.3) of the largely oxygen-rich oceans, molybdenum is present, almost exclusively, as the dianionic molybdate ion, MoO_4^{2-}. The average concentration of molybdenum in the oceans is about $10\,\mu g/L$ [6,8] compared to about $0.5\,\mu g/L$ for Zn, 0.1 $\mu g/L$ for Cu, and about $0.003\,\mu g/L$ for Co. In addition to its soluble form, molybdenum is found in colloidal or suspended particles [9] and may be bound in sulfide-rich sedimentary deposits [10].

Molybdate, which resembles sulfate in charge and shape, forms highly soluble salts with sodium, the dominant cation in the oceans. The structural resemblance of the oxyanions sulfate, molybdate, and tungstate, illustrated in Fig. 1, highlights the potential difficulty that biological systems have in differentiating among these ions in recognition, uptake, and assimilation (see also R. Pau and D. Lawson, Chapter 2).

In general, tungsten concentrations at about $0.1\,\mu g/L$ are about two orders of magnitude lower than those of molybdenum [8], except for special environments, such as the deep-sea hydrothermal vents (vide infra) [11] (see also Chapter 19).

A clear distinction can be made between the behavior of molybdenum and tungsten under aerobic vs. anaerobic conditions. Generally, under aerobic conditions, molybdate and tungstate are the principal forms, and these compete with each other for binding sites and uptake systems. Under anaerobic conditions, especially when sulfur levels are high, as in marine environments, molybdenum can be reduced to Mo(V) or Mo(IV) and sulfur coordination becomes prevalent through MoS_2 formation or by binding to organosulfur ligands [10,12]. Such conditions may prevail near deep-sea hydrothermal vents where molybdenum may be unavailable and tungsten plays an important role. In particular, in

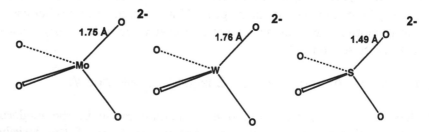

FIG. 1.   The tetrahedral structures of the oxyanions MO_4^{2-} (M=Mo, W, S). The similarities in shape and, for Mo and W, in bond length illustrate the difficulty of distinguishing between the various oxyanions.

such environments the anaerobic interconversions of aldehydes and carboxylic acids are catalyzed by tungsten enzymes (vide infra), which are discussed in Chapter 19.

The form of Mo in soil depends on pH and the nature of other minerals and organic matter present. The average molybdenum concentration in soil is 1–2 mg Mo/kg (i.e., 1–2 ppm) [9,13]. But values from 0.4 to 36 mg/kg have been reported, and local variation with time, lateral position, and depth are often significant. In plants, high phosphorus levels facilitate uptake, whereas high sulfur levels diminish the uptake of Mo [14,15]. Reducing conditions favor molybdenum(IV), where, generally, the compounds are less soluble and hence less available than the molybdenum(VI) species present under oxidizing conditions.

At high concentrations, more typical of the laboratory than of natural soil or aqueous systems, molybdenum forms, through condensation, a series of pH-dependent di- and polynuclear anionic compounds (called polymolybdates or heteropolymolybdates) [16]. While, tungsten forms a related series of poly- and heteropoly salts, these form far more slowly and less reversibly than their Mo analogues. Since the concentration of Mo or W in the environment is usually quite low, such polynuclear species are not generally considered biologically important and are not discussed further. However, in putative molybdenum storage proteins [17–20], such polynuclear species may indeed be present.

At the relatively low concentrations typical of most soils and natural aqueous systems, a very different set of reactions, (1) through (4), occur [21].

$$MoO_4^{2-} + H^+ \rightarrow HMoO_4^- \qquad\qquad (1)$$
$$MoO_4^{2-} + 2H^+ \rightarrow H_2MoO_4 \qquad\qquad (2)$$
$$MoO_4^{2-} + 2H_2O + 3H^+ \rightarrow MoO_2(OH)(OH_2)_3^+ \qquad\qquad (3)$$
$$MoO_4^{2-} + 2H_2O + 4H^+ \rightarrow MoO_2(OH_2)_4^{2+} \qquad\qquad (4)$$

Reactions (1) and (2) produce protonated forms of molybdate (analogous to forming bisulfate and sulfuric acid from sulfate). The molybdenum in $HMoO_4^-$ remains four-coordinate, but an alternative structure has been proposed for H_2MoO_4, involving addition of two water molecules to give the neutral tris(oxo) tris(aqua) species, $MoO_3(H_2O)_3$ [16]. Reactions (3) and (4) clearly involve expansion of the Mo coordination sphere with formation of octahedral six-coordinate cis(dioxo) coordination. Significantly, molybdate and monoprotonated molybdate are anionic,

while the products formed with increasing acidity are neutral or cationic. Since most soils have strong cation binding ability, molybdenum
may be unavailable in acidic soils. Increasing the pH by the addition
of lime (CaO) releases the anionic molybdate ion, which is not strongly
absorbed by the soil matrix and hence is readily available to plants.
However, the high solubility of molybdate can also be a liability, since
its enhanced mobility may, under overly wet conditions, allow extensive
leaching to occur.

Anaerobic waters may contain significant sulfide ion levels, either
from sulfate reducing bacteria or from volcanic sources. Under these
conditions, molybdate is no longer the dominant form present and,
indeed, may be essentially absent. Successive substitution of oxide by
sulfide, as in equation (5), leads, depending on pH and sulfide ion concentration, to formation of various thiomolybdate ions [22,23].

$$MoO_4^{2-} \rightarrow MoO_3S^{2-} \rightarrow MoO_2S_2^{2-} \rightarrow MoOS_3^{2-} \rightarrow MoS_4^{2-} \tag{5}$$

While tungstate can be converted to thiotungstate ions, the corresponding reactions are much slower and not as favorable.

When thiomolybdates are present the possibility of internal redox
reactions must be taken into account [24,25]. These reactions lead to the
formation of oxidized sulfur (disulfide, polysulfide, or elemental sulfur)
and reduced Mo [Mo(V) or Mo(IV)] and form soluble Mo-S species and/or
precipitates of MoS_3 or MoS_2. Tungsten is much less prone to internal
redox compared to molybdenum [25].

Thiomolybdates are strong ligands [24] and may bind elements
such as Cu and Fe. Indeed, the binding of Cu by MoS_4^{2-} is responsible
for the copper molybdenum antagonism that plagues ruminant animals
in molybdenum-rich soils [26].

2.2. Evolutionary Implications of Mo and W
Biogeochemistry

In the history of Earth, a major event was the development of oxygenic
photosynthesis, which may have occurred more than 3.5×10^9 years
before the present (ybp) [27]. Molecular oxygen did not immediately
accumulate in the atmosphere and waters, due largely to the presence
of abundant ferrous iron, which reacted with the oxygen to precipitate

massive banded (ferric) iron formations. Ultimately, as O_2 began to accumulate in the atmosphere and oceans, between 2 and 1×10^9 ybp [27], reaction (6) could occur:

$$2MoS_2 + 7O_2 + 2H_2O \rightarrow 2MoO_4^{2-} + 4SO_2 + 4H^+ \qquad (6)$$

This reaction may have allowed life to proliferate by making the highly soluble molybdate ion available for incorporation into critical metalloenzymes. The catalytic ability imparted to the new metalloenzymes allowed organisms to occupy important new ecological niches. When Mo was less available, the critical enzymes that depend on Mo may have required alternatives. It is possible that the alternative (vanadium and all-iron) nitrogenases, discussed below, are evolutionary precursors of the modern and more efficient molybdenum system.

In carbon metabolism, tungsten enzymes, currently found in hyperthermophilic archaea, could possibly be the evolutionary forerunners of molybdenum enzymes (which do not appear to be phylogenetically related).

3. THE NITROGEN CYCLE

The major reservoir for nitrogen on Earth is the atmosphere, where the dinitrogen molecule, at 78%, is the principal gaseous component. The second most common form of inorganic nitrogen is nitrate ion, found in both aerobic and anaerobic waters and soils. These two major forms of inorganic nitrogen enter the biological nitrogen cycle through catalysis by molybdenum enzymes. From the point of view of life, the nitrogen cycle converts these inorganic forms to organonitrogen species necessary for cellular physiology. The organic compounds contain the reduced form of nitrogen found in proteins, nucleic acids, porphyrins, coenzymes, and numerous other cellular components.

A simplified version of the nitrogen cycle is shown in Fig. 2 [1,28]. The four major processes in the cycle are nitrogen fixation, nitrification, nitrate assimilation, and denitrification. Molybdenum enzymes play a crucial role in each of these pathways.

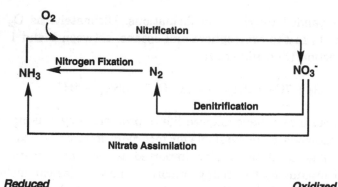

FIG. 2. A simplified schematic of the nitrogen cycle indicating the principal processes of nitrogen fixation, nitrification, nitrate assimilation, and denitrification.

3.1. Nitrogen Fixation

The first point where inorganic nitrogen enters the nitrogen cycle occurs with nitrogen fixation. A diverse, but limited, group of bacteria and archaea convert the N_2 of the atmosphere to the ammonia level. The nitrogenase enzyme involved, discussed in Chapter 3, is present only in prokaryotes, but some of these live symbiotically with plants (e.g., legumes) and animals (e.g., termites), often providing the major input of fixed nitrogen to these systems [29–32]. The nitrogen is said to be "fixed," since it is now accessible to metabolic enzymes for biosynthesis of the diverse set of organonitrogen compounds necessary for life. Economically important symbioses include the anabaena-azolla symbiosis (used as a "green manure" in rice paddies); the rhizobium/bradyrhizobium symbioses with legumes (e.g., peanuts, soybeans, and peas); and the frankia involvement with alder [29–32]. These natural and agriculturally encouraged symbioses reduce or eliminate the need for nitrogenous fertilizer application.

The nitrogenase enzyme system is energy demanding. The active site contains the iron-molybdenum cofactor (FeMoco) [33] as part of a complex protein system that requires low-potential reductant, hydrolyzes ATP, and evolves H_2 [34–36]. The mechanism used to overcome the difficult activation and reduction of the strong nitrogen-nitrogen triple bond has yet to be elucidated at the molecular level, and whether or not molybdenum directly or indirectly interacts with dinitrogen remains a subject of conjecture.

The relationship of molybdenum and nitrogen fixation was first established by Bortels [2] who also found that in some cases vanadium stimulates nitrogen fixation. More recent work [37] led to the identification of two alternative nitrogen fixation systems that resemble the molybdenum system (biochemically and genetically) but contain vanadium or iron in place of molybdenum [38] (see also D. Lawson and B. Smith, Chapter 3). The vanadium and the all-iron nitrogenases are generally less efficient than the molybdenum system and appear to be employed only when molybdenum is limiting. Organisms do not express the alternate versions when molybdenum is sufficient and molybdenum represses the gene for the synthesis of the alternative proteins. Since, to date, no organism has been found that uses only the alternative version, molybdenum appears to be preferred for nitrogen fixation. However, for certain organisms, nitrogen fixation is sufficiently important to warrant the additional investment in the less efficient V and Fe systems as a backup (see Chapter 3).

3.2. Nitrification

The more detailed rendition of the nitrogen cycle shown in Fig. 3 reveals that the other major pathways—nitrification, nitrate assimilation, and denitrification—are multistep processes [39,40]. In nitrification, ammonia is oxidized to nitrate with hydroxylamine and nitrite as intermediates [41,42]. Fertilization with reduced nitrogen compounds (e.g., NH_3, urea) requires nitrification because oxidation to nitrate must occur before most plants can take up the added nitrogen and use it effectively in their metabolism. The microorganisms that carry out nitrification gain energy from the oxidation of NH_3 using O_2 as the oxidant, a thermodynamically very favorable process [41]. The terminal step, involving oxidation of nitrite to nitrate, is catalyzed by the molybdenum enzyme nitrite oxidase [43–45] (see also P. M. H. Kroneck and D. Abt, Chapter 10).

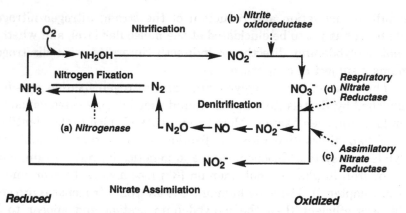

FIG. 3.   The biogeochemical cycle of nitrogen illustrating key molybdenum enzymes: (a) nitrogenase; (b) nitrite oxidoreductase; (c) assimilatory nitrate reductase; (d) respiratory nitrate reductase. The molybdenum enzymes are italicized.

3.3.   Nitrate Assimilation

In addition to nitrogen fixation, a major presence for molybdenum in nitrogen acquisition is in the reduction of nitrate to nitrite catalyzed by nitrate reductase [46]. There are several different types of nitrate reductase [46] (see also P. M. H. Kroneck and D. Abt, Chapter 10 and R. Hille, Chapter 6). One is an assimilatory enzyme, which executes the first step in the pathway used by plants, as well as microorganisms, to convert nitrate (the most abundant oxidized form of aqueous nitrogen) to a reduced level [47,48]. The molybdenum enzyme nitrate reductase catalyzes the conversion of nitrate (NO_3^-) to nitrite (NO_2^-). Many organisms, including all plants, then reduce the nitrite to ammonia using the single enzyme nitrite reductase [49] in the overall process called nitrate assimilation [50].

3.4.   Denitrification

Other (micro)organisms are involved in the process of denitrification [51] [42], where *dissimilatory* nitrate reductase plays the initiating role in this respiratory nitrate reduction pathway [39,40,52] (see also

Chapter 10). The products formed in this pathway are NO, N_2O, or N_2, and the denitrifying organisms harness the energy of nitrate/nitrite reduction to generate transmembrane gradients and potentials for ATP synthesis [39,40,42,53]. Since N_2O is a greenhouse gas and an ozone-depleting gas, considerable attention has been given to N_2O release by denitrifying bacteria in agricultural and nitrate/nitrite runoff regimes [54,55] as well as in wastewater treatment [53].

3.5. Organonitrogen Metabolism

In the chemistry of organonitrogen compounds, a molybdenum enzyme is involved in the reduction of trimethylamine N-oxide to trimethylamine (responsible for the smell of rotten fish) [56] (see also Chapter 6). In addition, across a broad phylogenetic spectrum (plants, animals, fungi, bacteria), molybdenum enzymes are used in the oxidation of a wide variety of N-heterocyclic molecules [5,57] (see also Chapters 11–14).

There is no known role for tungsten in the nitrogen cycle.

4. SULFUR METABOLISM AND THE SULFUR CYCLE

Sulfur plays a crucial role in molybdenum and tungsten enzymes, all of which have sulfur ligands directly bound to Mo or W. In nitrogenase, these take the form of sulfide ions that bridge iron and molybdenum atoms in the iron-molybdenum cofactor (see Chapter 3). Cysteine ligands play an important role in binding the metal sulfide clusters to the protein.

In all other molybdenum enzymes and in all tungsten enzymes, the group 6 metal is bound through sulfur donors to variants of the pyranopterin dithiolene ligand (Mpt to form Moco, see Chapter 6) and in some cases, in addition, by the thiolate sulfur of cysteine.

4.1. The Inorganic Sulfur Cycle

Many microorganisms obtain energy for metabolism by the oxidation or reduction of inorganic compounds of sulfur [58]. Molybdenum enzymes are used in particular steps in these pathways (Chapter 7). The sulfur cycle is complex [58] and not as well defined as the nitrogen cycle. A simplified version is shown in Fig. 4.

Sulfate is the most oxidized form of biological sulfur, and H_2S, HS^-, or sulfide are the most reduced forms. Major classes of bacteria carry out the reactions shown in Fig. 4. The right branch of Fig. 4 involves sulfate-reducing bacteria (SRB), which convert sulfate to sulfide, using sulfate as their terminal electron acceptor under anaerobic conditions and oxidizing organic compounds in the process [59,60]. The left branch of Fig. 4 involves sulfide-oxidizing organisms that are usually aerobes, which use O_2 to oxidize sulfide (or disulfide, polysulfide, thiosulfate, or elemental sulfur) to sulfate. Figure 5 shows some of the many possible intermediate species and highlights where molybdenum enzymes participate.

Molybdenum enzymes of the sulfur cycle are described in detail in Chapter 17. The enzyme sulfite oxidase ($SO_3^{2-} \rightarrow SO_4^{2-}$) is important in human metabolism and in the oxidation of sulfite in aerobic environ-

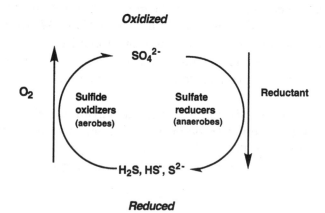

FIG. 4.   A simplified schematic of the sulfur cycle indicating the roles of sulfate-reducing bacteria (right branch) and sulfur-oxidizing bacteria (left branch).

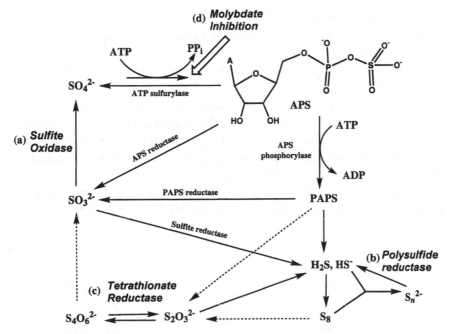

FIG. 5. The biogeochemical sulfur cycle illustrating the key molybdenum enzymes: (a) sulfite oxidase; (b) polysulfide reductase; (c) tetrathionate reductase; and (d) the effect of molybdate as an inhibitor of the ATP sulfurylase system. The molybdenum enzymes are italicized.

ments [61–63]. Under anaerobic conditions, the Mo enzymes tetrathionate reductase ($S_4O_6^{2-} \rightarrow 2S_2O_3^{2-}$) [64,65] and polysulfide reductase ($S_n^{2-} \rightarrow S_{(n-1)}^{2-} + S^{2-}$) [66,67] catalyze inorganic sulfur conversions. The single volatile *inorganic* sulfur molecule is H_2S, which does not have a long atmospheric residence time since it is rapidly oxidized through chemical and photochemical reactions.

4.2. Volatile Organic Sulfur: Dimethyl Sulfide and Dimethyl Sulfoxide

An important part of the sulfur cycle involves formation of the *organic* compound of sulfur, dimethyl sulfide, which is common in the marine environment. Dimethyl sulfide (DMS) is formed in the decomposition of dimethylsulfoniopropionate (DMSP). DMSP is produced by algae [68], possibly as a buoyancy aid, for osmotic balance, or, through its subsequent reaction [69, 70], perhaps as a chemical deterrent or attractant. DMS can be oxidized photochemically and biologically to dimethyl sulfoxide (DMSO) [68].

DMSO serves as a terminal electron acceptor for a variety of bacteria and in the process is reduced to the volatile DMS (Fig. 6). DMSO is widely distributed in marine waters [71] and soil [72]. The molybdenum enzyme DMSO reductase, present in many marine bacteria, catalyzes the reduction of DMSO to DMS [73–77].

Since DMS is relatively insoluble and volatile, it enters the atmosphere, where, as one of the few volatile sulfur compounds, it contri-

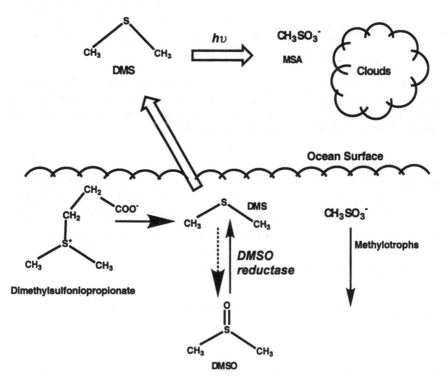

FIG. 6. Organic reactions of the sulfur cycle that lead to the production of DMS and clouds. The molybdenum enzyme DMSO reductase is italicized.

butes to the smell of the sea and attracts sea birds to areas of high algal productivity.

From a global perspective, the key reaction of atmospheric DMS is its rapid chemical and photochemical conversion to methyl sulfonates (see Fig. 6), which can serve as cloud condensation nuclei [68,78]. These sulfur compounds provide one of the few initiators of cloud formation, without the need for land-based input (dust). DMS volatilization and oxidation thus provides a mechanism for cloud formation in open ocean regions. The resultant clouds increase the Earth's albedo (reflection of light), and may contribute to cooling, which could have global climate implications [78]. Although the quantitative significance of the effect remains uncertain, DMS production could be part of an important negative feedback loop that may contribute to global temperature regulation of Earth [79].

Molybdenum enzymes related to DMSO reductase include biotin sulfoxide reductase [80,81] and, possibly, some methionine sulfoxide reductases [82], which also remove oxygen from sulfoxides to generate sulfides.

4.3. The Sulfate Reducers

Sulfate-reducing bacteria use sulfate as their terminal electron acceptor and generate sulfide under anaerobic conditions. These strictly anaerobic organisms are important in sulfide ore formation [83], biocorrosion [83,84], the souring of petroleum under anaerobic conditions [83], and in the Cu-Mo antagonism in ruminants [26,85].

First, the relatively unreactive sulfate is activated to form adenosine phosphosulfate and phosphoadenosine phosphosulfate (see Fig. 5). These "active sulfate" materials can be reduced to sulfite by APS reductase. The sulfite is then reduced through sulfite reductase in a six-electron process to sulfide [60]. Molybdate is a powerful inhibitor of the enzyme ATP sulfurylase, which initiates sulfate activation [86]. This inhibition may be responsible for the effect of high molybdate levels in slowing the growth of SRB [87].

4.4. Sulfide Oxidation Reactions

The oxidative part of the sulfur cycle is the province of a large number of organisms, mostly bacteria, that gain energy from various oxidative conversions of the type shown in Fig. 5 [88–90]. *Thiobacillus* is a well-studied example of an aerobic sulfide-oxidizing organism [89,90]. Thiobacilli are autotrophs that use the energy obtained from oxidation of inorganic sulfur compounds for all of their cellular needs, including the fixation of carbon using the Calvin cycle. Other sulfide oxidizers are important symbionts (vide infra). The organisms use molecular oxygen to oxidize the reduced sulfur species from sulfide to sulfite, with the molybdenum enzyme sulfite oxidase catalyzing the terminal step in the process.

4.5. Deep-Sea Hydrothermal Vents

Macroscopic life at the deep-sea vents is clearly aerobic and uses oxygen in the energy-yielding oxidation of sulfide released by the vents [91,92]. The vent animals harbor symbiotic sulfide-oxidizing bacteria, which are responsible for the carbon fixation that makes the animals (including tubeworms and clams) true chemoautotrophic organisms, albeit dependent on past photosynthesis for the oxygen required in their metabolism.

In contrast to the aerobic nature of the vent animals, there are microniches near the black smokers, where there is little or no molecular oxygen and very high sulfide levels. Here bacterial communities thrive using strictly anaerobic reactions [93]. It is this community of microorganisms, many of which are thermophilic or hyperthermophilic archaea, that uses tungsten enzymes for the metabolism of carbon compounds [94]. If, as has been suggested, life originated at sites similar to modern-day hydrothermal vents, then tungsten enzymes, which are endemic to these habitats, may have been important in the development of early metabolism on Earth (see Chapter 19).

5.   CARBON METABOLISM

5.1.   Aerobic Carbon Metabolism

The major flux in the carbon cycle involves oxygenic photosynthesis, which leads to the reduction of CO_2 through the Calvin cycle and the formation of O_2 using photosystem II and the oxygen-evolving manganese cluster [95]. The carbon cycle is completed by the oxidation of the biomass through respiration or combustion to form CO_2 and H_2O [1]. In contrast to the nitrogen and sulfur cycles discussed above, where Mo enzymes play a key role, Mo and W do not play quantitatively large roles in the aerobic carbon cycle.

However, one molybdenum enzyme that does play an interesting role in aerobic carbon chemistry is CO dehydrogenase (oxidase). This enzyme, isolated from aerobic bacteria (e.g., carboxydo bacteria such as *Oligotropha carboxidovorans*), is capable of oxidizing CO to CO_2 using O_2 as the oxidant [96,97]. The CO can be used as a source of cellular carbon and reducing equivalents for cellular metabolism. The CO dehydrogenase enzyme has an unusual copper-molybdenum site, whose mechanistic implications are under study [98] (see also H. Dobbeck and R. Huber, Chapter 7).

5.2.   Anaerobic Carbon Metabolism

In addition to the highly visible aerobic part of the carbon cycle, there is an anaerobic component of the cycle in which CO, CH_4, CH_3COO^-, and H_2 play key roles. This anaerobic carbon metabolism is largely the realm of prokaryotes, including SRB, methanogenic archaea, clostridia, and many other anaerobic organisms. These organisms are often heterotrophs and utilize a variety of organic substrates to generate the reducing power for their cellular metabolism. Several molybdenum and tungsten enzymes participate in key steps of these processes (Fig. 7) (see also Chapters 16 and 19).

The enzyme formyl methanofuran dehydrogenase (FMDH) is found in methanogens that reduce CO_2 to CH_4 [99] (see also J. Vorholt and R. K. Thauer, Chapter 16 and V. Gladyshev, Chapter 18). This enzyme generally functions in the opposite direction from that

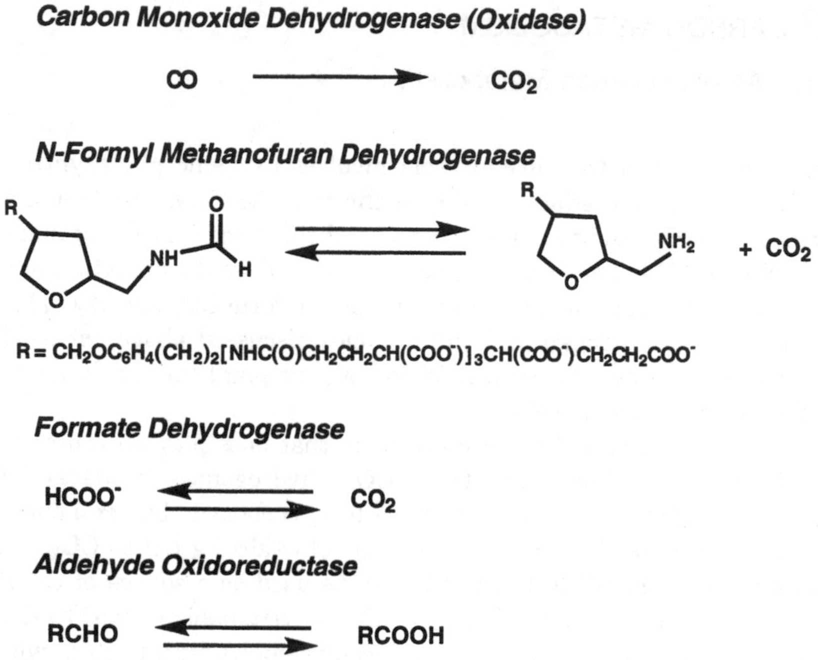

Carbon Monoxide Dehydrogenase (Oxidase)

$$CO \longrightarrow CO_2$$

N-Formyl Methanofuran Dehydrogenase

$R = CH_2OC_6H_4(CH_2)_2[NHC(O)CH_2CH_2CH(COO^-)]_3CH(COO^-)CH_2CH_2COO^-$

Formate Dehydrogenase

$$HCOO^- \rightleftharpoons CO_2$$

Aldehyde Oxidoreductase

$$RCHO \rightleftharpoons RCOOH$$

FIG. 7. Some of the molybdenum and tungsten enzymes involved in carbon metabolism. The aerobic CO dehydrogenase is a Mo enzyme, but FMDH, FDH, and AOR can be either Mo or W enzymes.

implied by its name, i.e., by reducing CO_2 to formyl methanofuran rather than producing CO_2 from the latter [99]. This reaction of CO_2 is the first step in its transformation to CH_4. Interestingly, the FMDH enzymes from mesophilic organisms are generally molybdenum dependent [100], those from thermophilic organisms use Mo or W [101], while those known from hyperthermophiles are tungsten enzymes [102] (see also Chapter 16). The implications of these results are uncertain, but they may indicate that W enzymes are capable of performing at a higher temperature than Mo enzymes. Alternatively, Mo may be less available in hyperthermophilic habitats and have limited availability to thermophilic organisms.

Formate dehydrogenase (FDH) catalyzes the oxidation of formate to give CO_2 and may be either an Mo or a W enzyme. In either case, it is a clear example of a molybdenum enzyme that does *not* involve oxo transfer, even in a stoichiometric sense. The formate dehydrogenase of *Escherichia coli* contains a selenocysteine ligand bound to Mo in its active site [103] (see also Chapter 18). The first tungsten enzyme identified, isolated from the anaerobe *Clostridium thermoaceticum,* was a formate dehydrogenase [3].

A variety of aldehyde-metabolizing enzymes including aldehyde oxidoreductase can be either Mo or W enzymes (see Chapters 7, 15, and 19). The first crystal structure of an Mo enzyme in the xanthine oxidase family was an aldehyde oxidoreductase from *Desulfovibrio gigas* [104,105] (see Chapter 7). The so-called aldehyde oxidoreductase enzymes may in certain organisms function physiologically in the reverse direction, where they serve to reduce a carboxylic acid to the corresponding aldehyde [11,94] (see also Chapter 19).

The W enzymes play a central role in the carbon metabolism of hyperthermophilic archaea [94]. Interestingly, these W enzymes do not appear to have homology with the Mo enzymes that play analogous roles in other organisms. Rather, the W enzymes may have arisen independently for this application. This is an unusual result insofar as the organic part of the prosthetic group is virtually identical in Mo and W enzymes and, moreover, some Mo enzymes (e.g., FMDH) have clear homologies to W enzymes.

6. ARSENIC, SELENIUM, AND CHLORINE METABOLISM

Microorganisms use a variety of molybdenum enzymes in the reactions of oxyanions of main group elements including S (discussed above), Se, As, and Cl. The (mostly) bacteria occupy environmental niches, where the reaction in question is important to the organism's ability to thrive or survive. Thus, molybdenum enzymes are involved in the oxidation of arsenite to arsenate, the reduction of selenate to selenite, and the reduction of perchlorate and chlorate to chlorate and chlorite, respectively. A significant range of bacteria are capable of using arsenate, selenate, or chlorate as their terminal electron acceptor [106–108].

Of the two forms of arsenic available in the environment, arsenite in most cases appears to be significantly more toxic than arsenate. The molybdenum enzyme arsenite oxidase [109], whose crystal structure has recently been solved [110], catalyzes the oxidation of the more toxic As(III) to the less toxic As(V) state. The reaction may be widespread [111,112] and is typical of the Mo enzymes insofar as it formally involves a change in the number of oxo groups in the product [110] (see also Chapter 6).

Selenate can serve as the terminal electron acceptor in several species including *Thauera selenatis* [113,114]. The enzyme from *T. selenatis* has been isolated and shown to reduce selenate (SeO_4^{2-}) to selenite (SeO_3^{2-}) and to contain molybdenum. Importantly, the enzyme will not reduce nitrate, nitrite, chlorate, or sulfate at discernible rates, showing the specificity of the system for the oxidized selenium anion [106].

The ability of Mo enzymes to catalyze perchlorate (ClO_4^-) and chlorate (ClO_3^-) reduction was first recognized through the isolation of chlorate-resistant mutants. While, initially, this resistance was thought to arise from nitrate reductase [115], it is now known that, at least for a significant number of organisms, distinct molybdenum enzymes are involved in the reduction of perchlorate and chlorate [64,116]. This capability is interesting insofar as neither ClO_4^- nor ClO_3^- occurs naturally to a significant extent. Rather, their significant presence in the environment is likely due to the extensive use of oxychlorine compounds in bleaches and explosives over the past century. Apparently, this relatively short time period has been sufficient for this new enzymatic activity to evolve.

The conversions of perchlorate to chlorate and chlorate to chlorite appear to be catalyzed by the same enzyme, designated (per)chlorate reductase. The enzyme is clearly a member of the molybdenum enzyme family [117]. A significant number of organisms can use (per)chlorate as their terminal electron acceptor during the oxidation of organic compounds [107,118–121]. Indeed, such organisms have been suggested as remediation agents for anaerobic hydrocarbon–contaminated soil or water [122].

7. CONCLUSION

Molybdenum and tungsten are available to organisms worldwide, with molybdenum having the far greater abundance and availability, especially in aerobic environments. Nature has exploited these elements in a wide variety of metalloenzymes, which are the *raison d'être* for this book. The molybdenum enzymes are involved in key reactions of the nitrogen and sulfur cycles, whereas both Mo and W enzymes have specialized roles in aerobic and anaerobic aspects of the carbon cycle. In addition, molybdenum enzymes are involved in the metabolism of the oxyanions of selenium, arsenic, and chlorine (in addition to sulfur).

The proliferation of Mo enzymes may have had to await the development of an aerobic atmosphere and the production of molybdate, Mo(VI), as the major soluble and available form of Mo in the largely aerobic waters of the modern world. The evolutionary implications of this transition, from Mo scarcity to abundance, may mean that W or V had been exploited by biological systems before Mo. Tungsten enzymes are prevalent in thermophilic and hyperthermophilic organisms from high-sulfide environments, such as those found around deep-sea hydrothermal vents. Such habitats are sometimes touted as potential sites for the origins of life, giving rise to the notion that tungsten may have been critically important for life long before molybdenum. This conjecture remains a subject for future research and discussion.

ACKNOWLEDGMENT

This chapter was enriched through the author's participation in the Center for Environmental Bioinorganic Chemistry (CEBIC) at Princeton University, supported by the U.S. National Science Foundation and Department of Energy.

ABBREVIATIONS

ADP	adenosine 5'-diphosphate
AOR	aldehyde oxidoreductase
APS	adenosine 5'-phosphosulfate
ATP	adenosine 5'-triphosphate
DMS	dimethyl sulfide
DMSO	dimethyl sulfoxide
DMSP	dimethyl sulfoniopropionate
FDH	formate dehydrogenase
FeMoco	iron-molybdenum cofactor of nitrogenase
FMDH	formyl methanofuran dehydrogenase
Moco	molybdenum cofactor
Mpt	metal-binding pyranopterin dithiolene [5] (also molybdopterin)
MSA	methylsulfonic acid
PAPS	3'-phosphoadenosine-5'-phosphosulfate
SRB	sulfate-reducing bacteria
ybp	years before the present

REFERENCES

1. M. C. Jacobson, R. J. Charlson, H. Rodhe, and G. H. Orians, *Earth System Science: From Biogeochemical Cycles to Global Change*, Academic Press, London, 2000, pp. 1–523.

2. H. Bortels, *Zentralbl. Bakteriol. Parisitenkd. Infektionskr.*, *95*, 193–218 (1936).

3. L. G. Ljungdahl, *Trends Biochem. Sci.*, *1*, 63–65 (1976).

4. L. G. Ljungdahl and J. R. Andreesen, *Meth. Enzymol.*, *53*, 360–372 (1978).

5. R. S. Pilato and E. I. Stiefel, in *Bioinorganic Catalysis*, 2nd ed. (J. Reedijk and E. Bouwman, eds.), Marcel Dekker, New York, 1999, pp. 81–152.

6. P. A. Cox, *The Elements on Earth*, Oxford University Press, Oxford, 1995, pp. 1–287.

7. K. W. Bruland in *Chemical Oceanography*, 2nd ed. (J. P. Riley and R. Chester, eds.), Academic Press, London, 1983, pp. 157–220.

8. J. Brown, A. Colling, D. Park, J. Phillips, D. Rothery, and J. Wright, *Seawater: Its Composition, Properties and Behavior*, Pergamon Press, Oxford, 1989, pp. 1–165.

9. W. M. Jarrell, A. L. Page, and A. A. Elseewi, *Residue Rev.*, 74, 1–43 (1980).

10. G. R. Helz, C. V. Miller, J. M. Charnock, J. F. W. Mosselmans, R. A. D. Pattrick, C. D. Garner, and D. J. Vaughan, *Geochim. Cosmochim. Acta*, 60, 3631–3642 (1996).

11. M. W. W. Adams, *Soc. Appl. Microbiol. Symp. Ser.*, 28, 108S–117S (1999).

12. S. R. Emerson and S. S. Huested, *Mar. Chem.*, 34, 177–196 (1991).

13. D. G. Barceloux, *J. Toxicol. Clin. Toxicol.*, 37, 231–237 (1999).

14. U. C. Gupta (ed.), *Molybdenum in Agriculture*, Cambridge University Press, Cambridge, UK, 1997, pp. 1–276.

15. U. C. Gupta, in *Molybdenum in Agriculture* (U. C. Gupta, ed.), Cambridge University Press, Cambridge, UK, 1997, pp. 71–91.

16. J. J. Cruywagen, *Adv. Inorg. Chem.*, 49, 127–182 (2000).

17. A. M. Grunden and K. T. Shanmugam, *Arch. Microbiol.*, 168, 345–354 (1997).

18. S. M. Hinton and L. E. Mortenson, *J. Bacteriol.*, 162, 477–484 (1985).

19. S. M. Hinton and L. E. Mortenson, *J. Bacteriol.*, 162, 485–493 (1985).

20. P. T. Pienkos and W. J. Brill, *J. Bacteriol.*, 145, 743–751 (1981).

21. K. S. Smith, L. S. Balistrieri, S. M. Smith, and R. C. Severson, in *Molybdenum in Agriculture* (U. C. Gupta, ed.), Cambridge University Press, Cambridge, UK, 1997.

22. M. A. Harmer and A. G. Sykes, *Inorg. Chem.*, 19, 2881–2885 (1980).

23. B. E. Erickson and G. R. Helz, *Geochim. Cosmochim. Acta*, 64, 1149–1158 (2000).

24. E. I. Stiefel, in *Transition Metal Sulfur Chemistry: Biological and Industrial Significance* (E. I. Stiefel and K. Matsumoto, eds.), *ACS Symp. Ser.*, *653*, 1996, pp. 2–38.

25. E. I. Stiefel, *J. Chem. Soc. Dalton Trans.*, 3915–3924 (1997).

26. J. Mason, *Toxicology*, *42*, 99–109 (1986).

27. J. W. Schopf (ed.), *Earth's Earliest Biosphere: Its Origin and Evolution*, Princeton University Press, NJ, Princeton, 1983, pp. 1–523.

28. S. J. N. Burgmayer and E. I. Stiefel, *J. Chem. Educ.*, *62*, 943–953 (1985).

29. A. Quispel, *Encycl. Plant Physiol.*, *New Ser.*, *15A*, 286–329 (1983).

30. P. A. Kaminski, J. Batut, and P. Boistard, *Rhizobiaceae*, 431–460 (1998).

31. R. H. Burris, in *Prokaryotic Nitrogen Fixation* (E. W. Triplett, ed.), Horizon Scientific Press, Wymondham, UK, 2000, pp. 33–41.

32. D. A. Waller, in *Prokaryotic Nitrogen Fixation* (E. W. Triplett, ed.), Horizon Scientific Press, Wymondham, UK, 2000, pp. 225–236.

33. B. K. Burgess, *Chem. Rev.*, *90*, 1377–1406 (1990).

34. B. K. Burgess and D. J. Lowe, *Chem. Rev.*, *96*, 2983–3011 (1996).

35. J. B. Howard and D. C. Rees, in *Prokaryotic Nitrogen Fixation* (E. W. Triplett, ed.), Horizon Scientific Press, Wymondham, UK, 2000, pp. 49–53.

36. B. E. Smith, *Adv. Inorg. Chem.*, *47*, 159–218 (1999).

37. P. E. Bishop and R. Premakumar in *Biological Nitrogen Fixation* (G. S. Stacey, R. H. Burris, and H. J. Evans, eds.), Chapman and Hall, New York, 1992, pp. 736–762.

38. R. R. Eady, *Chem. Rev.*, *96*, 3013–3030 (1996).

39. B. B. Ward, *Microb. Ecol. Oceans*, 427–453 (2000).

40. B. B. Ward, *Microb. Ecol.*, *32*, 247–261 (1996).

41. A. B. Hooper, T. Vannelli, D. J. Bergmann, and D. M. Arciero, *Antonie van Leeuwenhoek*, *71*, 59–67. (1997).

42. H. Bothe, G. Jost, M. Schloter, B. B. Ward, and K. P. Witzel, *FEMS Microbiol. Rev.*, *24*, 673–690 (2000).

43. T. Yamanaka and Y. Fukumori, *FEMS Microbiol. Rev.*, *54*, 259–270 (1988).

44. M. Miencke, E. Bock, D. Kastrau, and P. M. H. Kroneck, *Arch. Microbiol.*, *158*, 127–131 (1992).

45. E. Spieck, S. Ehrich, J. Aamand, and E. Bock, *Arch. Microbiol.*, *169*, 225–230 (1998).

46. J. T. Lin and V. Stewart, *Adv. Microb. Physiol.*, *39*, 1–30 (1998).

47. L. P. Solomonson and M. J. Barber, *Annu. Rev. Plant Physiol. Plant Mol. Biol.*, *41*, 225–253 (1990).

48. C. Moreno-Vivian, P. Cabello, M. Martinez-Luque, R. Blasco, and F. Castillo, *J. Bacteriol.*, *181*, 6573–6584 (1999).

49. J. L. Wray and M. P. Ward, *Nova Acta Leopold.*, *70*, 59–72 (1994).

50. C. Meyer and M. Caboche, *Transgenic Plant Res.*, 125–133 (1998).

51. G. P. Robertson, in *Handbook of Soil Science* (M. E. Sumner, ed.), CRC Press, Boca Raton, FL, 2000, pp. C181–C190.

52. L. Philippot and O. Hojberg, *Biochim. Biophys. Acta*, *1446*, 1–23 (1999).

53. D. Zart, R. Stuven, and E. Bock, in *Biotechnology,* 2nd ed., Vol. 11a (J. Winter, ed.), Wiley-VCH Verlag, Weinheim, 1999, pp. 55–64.

54. H. J. Michel and H. Wozniak, *Agribiol. Res.*, *51*, 3–11 (1998).

55. P. Cellier, J. C. Germon, C. Henault, and S. Genermont, *Colloq. - Inst. Natl. Rech. Agron.*, *83*, 25–37 (1997).

56. M. Czjzek, J.-P. Dos Santos, J. Pommier, G. Giordano, V. Mejean, and R. Haser, *J. Mol. Biol.*, *284*, 435–447 (1998).

57. S. Fetzner, *Naturwissenschaften*, *87*, 59–69 (2000).

58. D. E. Canfield and R. Raiswell, *Am. J. Sci.*, *299*, 697–723 (1999).

59. G. D. Fauque, *Biotechnol. Handb.*, *8*, 217–241 (1995).

60. J. Le Gall and M.-Y. Liu, in *Transition Metals and Microbial Metabolism* (G. Winkelmann and C. J. Carrano, eds.), Harwood, Amsterdam, 1997, pp. 281–310.

61. C. Kisker, H. Schindelin, A. Pacheco, W. A. Wehbi, R. M. Garrett, K. V. Rajagopalan, J. H. Enemark, and D. C. Rees, *Cell*, *91*, 973–983 (1997).

62. J. L. Johnson and K. V. Rajagopalan, *J. Clin. Invest.*, *58*, 551–556 (1976).

63. R. Hille, *Chem. Rev.*, *96*, 2757–2816 (1996).

64. L. F. Oltmann, W. N. M. Reijnders, and A. H. Stouthamer, *Arch. Microbiol.*, *111*, 37–43 (1976).

65. L. F. Oltmann, V. P. Claassen, P. Kastelein, W. N. M. Reijnders, and A. H. Stouthamer, *FEBS Lett.*, *106*, 43–46 (1979).

66. A. Jankielewicz, R. A. Schmitz, O. Klimmek, and A. Kroger, *Arch. Microbiol.*, *162*, 238–242 (1994).

67. I. I. Blumentals, M. Itoh, G. J. Olson, and R. M. Kelly, *Appl. Environ. Microbiol.*, *56*, 1255–1262 (1990).

68. R. P. Kiene, P. T. Visscher, M. D. Keller, and G. O. Kirst, in *Biological and Environmental Chemistry of DMSP and Related Sulfonium Compounds* (R. P. Kiene, P. T. Visscher, M. D. Keller, and G. O. Kirst, eds.), Plenum Press, New York, 1996, pp. 1–430.

69. M. J. E. C. Van der Maarel, M. Jansen, H. M. Jonkers, and T. A. Hansen, *Geomicrobiol. J.*, *15*, 37–44 (1998).

70. M. K. Bacic, S. Y. Newell, and D. C. Yoch, *Appl. Environ. Microbiol.*, *64*, 1484–1489 (1998).

71. A. D. Hatton, G. Malin, S. M. Turner, and P. S. Liss, in *Biological and Environmental Chemistry of DMSP and Releated Sulfonium Compounds* (R. P. Kiene, P. T. Visscher, M. D. Keller, and G. O. Kirst, eds.), Plenum Press, New York, 1996, pp. 405–412.

72. M. Horinouchi, K. Kasuga, H. Nojiri, H. Yamane, and T. Omori, *Recent Res. Dev. Agric. Biol. Chem.*, *2*, 277–282 (1998).

73. E. I. Stiefel, *Science*, *272*, 1599–1600 (1996).

74. F. Schneider, J. Loewe, R. Huber, H. Schindelin, C. Kisker, and J. Knaeblein, *J. Mol. Biol.*, *263*, 53–69 (1996).

75. H. Schindelin, C. Kisker, J. Hilton, K. V. Rajagopalan, and D. C. Rees, *Science*, *272*, 1615–1621 (1996).

76. A. S. McAlpine, A. G. McEwan, and S. Bailey, *J. Mol. Biol.*, *275*, 613–623 (1998).

77. M. K. Johnson, S. D. Garton, and H. Oku, *JBIC, J. Biol. Inorg. Chem.*, *2*, 797–803 (1997).

78. R. J. Charlson, J. E. Lovelock, M. O. Andreae, and S. G. Warren, *Nature*, *326*, 655–661 (1987).

79. A. J. Watson and P. S. Liss, *Philos. Trans. R. Soc. London, Ser. B*, *353*, 41–51 (1998).

80. K. E. Johnson and K. V. Rajagopalan, *J. Biol. Chem.*, *276*, 13178–13185 (2001).

81. V. V. Pollock and M. J. Barber, *Biochemistry*, *40*, 1430–1440 (2001).

82. R. M. Gibson and P. J. Large, *FEMS Microbiol. Lett.*, *26*, 95–99 (1985).

83. W. A. Hamilton, *Biodegradation*, *9*, 201–212 (1998).

84. H. A. Videla, *Biofouling*, *15*, 37–47 (2000).

85. J. Mason, A review of the role of sulfur and sulfur compounds in the molybdenum copper antagonism in ruminants and non-ruminants, in *Proceedings of the Symposium on Sulphur Forages*, An Foras Taluntais, Dublin, 1978, pp. 182–196.

86. Z. Reuveny, *Proc. Natl. Acad. Sci. USA*, *74*, 619–622 (1977).

87. D. R. Ranade, A. S. Dighe, S. S. Bhirangi, V. S. Panhalkar, and T. Y. Yeole, *Bioresour. Technol.*, *68*, 287–291 (1998).

88. D. P. Kelly, J. K. Shergill, W.-P. Lu, and A. P. Wood, *Antonie van Leeuwenhoek*, *71*, 95–107 (1997).

89. I. Suzuki, C. W. Chan, and T. L. Takeuchi, in *Environmental Geochemistry of Sulfur Oxidation* (C. N. Alpers and D. W. Blowes, eds.), *ACS Symp. Ser.*, *550*, 1994, pp. 60–67.

90. J. T. Pronk, R. Meulenberg, W. Hazeu, P. Bos, and J. G. Kuenen, *FEMS Microbiol. Rev.*, *75*, 293–306 (1990).

91. F. Gaill, *FASEB J.*, *7*, 558–565 (1993).

92. H. Felbeck and G. N. Somero, *Trends Biochem. Sci. (Pers. Ed.)*, *7*, 201–204 (1982).

93. C. Jeanthon, *Antonie van Leeuwenhoek*, *77*, 117–133 (2000).

94. A. Kletzin and M. W. W. Adams, *FEMS Microbiol. Rev.*, *18*, 5–63 (1996).

95. D. O. Hall and K. Rao, *Photosynthesis*, 6th ed., 1999, pp. 1–228.

96. O. Meyer and K. Fiebig, Enzymes oxidizing carbon monoxide, in *Proceedings of the Symposium on Gas Enzymology*, Reidel, Dordrecht, 1985, pp. 147–168.

97. O. Meyer and H. G. Schlegel, *Annu. Rev. Microbiol.*, *37*, 277–310 (1983).

98. H. Dobbek, L. Gremer, O. Meyer, and R. Huber, *Proc. Natl. Acad. Sci. USA*, *96*, 8884–8889 (1999).

99. M. Karrasch, G. Boerner, M. Enssle, and R. K. Thauer, *FEBS Lett.*, *253*, 226–230 (1989).

100. P. A. Bertram, M. Karrasch, R. A. Schmitz, R. Boecher, S. P. J. Albracht, and R. K. Thauer, *Eur. J. Biochem.*, *220*, 477–484 (1994).

101. R. A. Schmitz, S. P. J. Albracht, and R. K. Thauer, *Eur. J. Biochem.*, *209*, 1013–1018 (1992).

102. J. A. Vorholt, M. Vaupel, and R. K. Thauer, *Mol. Microbiol..*, *23*, 1033–1042 (1997).

103. J. C. Boyington, V. N. Gladyshev, S. V. Khangulov, T. C. Stadtman, and P. D. Sun, *Science*, *275*, 1305–1308 (1997).

104. J. Rebelo, S. Macieira, J. M. Dias, R. Huber, C. S. Ascenso, F. Rusnak, J. J. G. Moura, I. Moura, and M. J. Romão, *J. Mol. Biol.*, *297*, 135–146 (2000).

105. M. J. Romão, N. Rosch, and R. Huber, *J. Biol. Inorg. Chem.*, *2*, 782–785 (1997).

106. I. Schröder, S. Rech, T. Krafft, and J. M. Macy, *J. Biol. Chem.*, *272*, 23765–23768 (1997).

107. J. D. Coates, U. Michaelidou, R. A. Bruce, S. M. O'Connor, J. N. Crespi, and L. A. Achenbach, *Appl. Environ. Microbiol.*, *65*, 5234–5241 (1999).

108. J. F. Stolz and R. S. Oremland, *FEMS Microbiol. Rev.*, *23*, 615–627 (1999).

109. G. L. Anderson, J. Williams, and R. Hille, *J. Biol. Chem.*, *267*, 23674–23682 (1992).

110. P. J. Ellis, T. Conrads, R. Hille, and P. Kuhn, *Structure*, *9*, 125–132 (2001).

111. G. Ji, S. Silver, E. A. E. Garber, H. Ohtake, C. Cervantes, and P. Corbisier, *Biohydrometall. Technol., Proc. Int. Biohydrometall. Symp.*, 2, 529–539 (1993).

112. J. M. Santini, L. I. Sly, R. D. Schnagl, and J. M. Macy, *Appl. Environ. Microbiol.*, 66, 92–97 (2000).

113. S. A. Rech and J. M. Macy, *J. Bacteriol.*, 174, 7316–7320 (1992).

114. J. M. Macy in *Selenium in the Environment* (W. T. Frankenberger, Jr. and S. Benson, eds.), Marcel Dekker, New York, 1994, pp. 421–444.

115. A. H. Stouthamer, *Antonie van Leeuwenhoek; J. Microbiol. Serol.*, 35, 505–521 (1969).

116. M. D. Roldan, F. Reyes, C. Moreno-Vivian, and F. Castillo, *Curr. Microbiol.*, 29, 241–245 (1994).

117. S. W. M. Kengen, G. B. Rikken, W. R. Hagen, C. G. Van Ginkel, and A. J. M. Stams, *J. Bacteriol.*, 181, 6706–6711 (1999).

118. A. Malmqvist, T. Welander, and L. Gunnarsson, *Appl. Environ. Microbiol.*, 57, 2229–2232 (1991).

119. G. B. Rikken, A. G. M. Kroon, and C. G. van Ginkel, *Appl. Microbiol. Biotechnol.*, 45, 420–426 (1996).

120. W. Wallace, T. Ward, A. Breen, and H. Attaway, *J. Ind. Microbiol.*, 16, 68–72 (1996).

121. J. D. Coates, U. Michaelidou, S. M. O'Connor, R. A. Bruce, and L. A. Achenbach, *Environ. Sci. Res.*, 57, 257–270 (2000).

122. J. D. Coates, R. A. Bruce, J. Patrick, and L. A. Achenbach, *Bioremediation. J.*, 3, 323–334 (1999).

2

Transport, Homeostasis, Regulation, and Binding of Molybdate and Tungstate to Proteins

Richard N. Pau[1] and David M. Lawson[2]

[1]Department of Biochemistry and Molecular Biology,
University of Melbourne,
Parkville 3010, Australia

[2]Department of Biological Chemistry, John Innes Centre,
Norwich NR4 7UH, UK

1. INTRODUCTION

Molybdenum and tungsten are minor components of the Earth, found in lesser amounts than iron, copper, and zinc. However, compared with the biochemically widely used transition metals, both molybdenum and tungsten are more available for biological processes. At a concentration of about 110 nM, molybdenum is the most abundant transition metal in the sea. The concentration of tungsten is much lower, i.e., comparable to that of zinc and copper. The relative abundance of molybdenum and tungsten is, however, dramatically reversed in marine hydrothermal vents, a niche that favors the evolution of tungstoenzymes [1] (see also Chapter 1).

In terrestrial environments, the concentration of bioavailable molybdenum, about 50 nM, is usually lower than in the sea, and orders of magnitude greater than tungsten, though its distribution is uneven. Molybdenum depletion on land is reflected by the evolution of alternative nitrogenases that enable bacteria to fix nitrogen in the absence of molybdenum by substituting vanadium or iron for molybdenum in nitrogenase cofactors [2]. Even when molybdenum is present, the uptake systems in microbial mats and plant rhizospheres may deplete it locally [3]. When grown diazotrophically on solid medium, a strain of *Azotobacter vinelandii* that contains both a molybdenum and molybdenum-independent nitrogenase has a competitive advantage over a strain with only the molybdenum nitrogenase. This is presumably because the uptake system results in local depletion of molybdenum, favoring the growth of cells with a molybdenum-independent nitrogenase [4].

The chemical property that makes molybdenum and tungsten readily accessible is their hydrolysis to water-soluble oxoanionic species. At neutral pH in oxidizing aqueous environments, molybdenum and tungsten occur predominantly as the soluble monomeric oxyanion molybdate ($Mo(VI)O_4^{2-}$) and tungstate ($W(VI)O_4^{2-}$). Molybdate is in rapid equilibrium with other species of molybdenum, particularly the cationic species MoO_2^{2+}, which is at a concentration of 10^{-19} relative to molybdate [5]. Tungstate shows similar speciation [1,6]. Polyoxo species predominate at low pH when the concentrations of monomeric molybdate and tungstate are much lower.

Vanadium must be considered in the context of the biological chemistry of molybdate and tungstate as it has a similar oxygen chemistry [7]. Its concentration in the sea is almost as high as that of molybdenum. At low concentrations and physiological conditions (pH 6–8), the monomeric forms of vanadate are $H_2VO_4^-$ and HVO_4^{2-}. The VO_4^{3-} structure is similar to phosphate, whereas molybdate and tungstate resemble sulfate. The challenge for living organisms is therefore how to distinguish between molybdate, tungstate, and other tetraoxyanions. This chapter considers the special features that enable biological systems to specifically transport and handle molybdate and tungstate.

2. COMPETITION WITH IRON(III)

2.1. Siderophores

Even before entry into cells, molybdenum competes for metal ligands in the environment. In oxidizing aqueous solution at neutral pH, Fe(III) hydrolyzes to form highly insoluble ferric hydroxides and oxides ($K_{sol.} = 4 \times 10^{-38}$ to 4×10^{-44}) [5,8]. As a result, the concentration of free Fe(III) (about 10^{-17} M) is too low for use in biological systems. To overcome this, bacteria synthesize a number of small organic iron chelators, called siderophores, that solubilize Fe(III) for transport into the cell. These iron carriers are based on multidentate nitrogen, enolate, or phenolate oxygen donors. Most siderophores are hexadentate ligands that form highly stable octahedral complexes with iron. One of the most powerful siderophores ($K_{form.} = 10^{52}$) is the hexadentate cyclic catechol enterobactin synthesized by *Escherichia coli* [9]. Siderophores usually have lower affinity for other metals. Competition for molybdenum arises because some siderophores, particular catechols, react readily with molybdate.

2.2. Catecholate Siderophores of *Azotobacter vinelandii*

Catecholate siderophores are best understood for *A. vinelandii*, an aerobic soil-living bacterium that fixes nitrogen. The catecholate siderophores secreted by *A. vinelandii* are shown in Fig. 1. In iron-sufficient conditions, the low-affinity ligand 2,3-dihydroxybenzoic acid is synthesized [10], and at very low iron concentrations (<3 mM), the high-affinity pyoverdin-type siderophore azotobactin is produced [10–13]. At intermediate concentrations the bidentate aminochelin (2,3-dihydroxybenzoylputrecine) and tetradentate azotochelin (*N,N'*-bis-(2,3-dihydroxybenzoyl)-L-lysine) are produced [14,15].

Azotochelin would be expected to form an Fe_2L_3 ferric complex, as in the iron complex of the bidentate siderophore rhodotorulic acid and the gallium complexes of a number of bis(catecholamide) ligands [16,17]. At pH 7, iron is almost entirely present as an insoluble precipitate. It is solubilized by azotochelin, and equilibrium with the bis(catecholamide) iron complexes is reached within a few hours [18]. However, in contrast

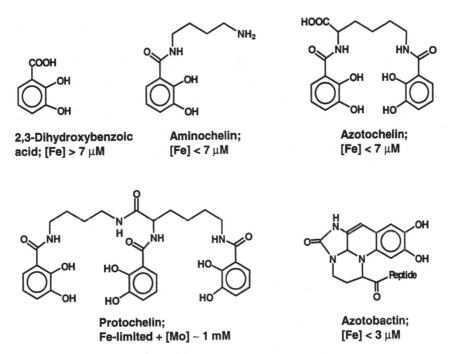

2,3-Dihydroxybenzoic acid; [Fe] > 7 μM

Aminochelin; [Fe] < 7 μM

Azotochelin; [Fe] < 7 μM

Protochelin; Fe-limited + [Mo] ~ 1 mM

Azotobactin; [Fe] < 3 μM

FIG. 1. Siderophores produced by *A. vinelandii*. The metal concentrations required for the synthesis of the siderophores are shown.

to iron, azotochelin is ideally suited for complexation with the cationic species MoO_2^{2+}, which has four vacant coordination sites. As this molybdenum species is in rapid equilibrium with molybdate, azotochelin instantly reacts with molybdate to form a stable soluble complex at neutral pH. Addition of a stoichiometric amount of molybdate to an aqueous solution containing iron dramatically reduces the rate of iron solubilization by azotochelin. The thermodynamically more stable iron complex is only slowly formed and reaches 50% of its final concentration 23 times slower than in the absence of molybdate. Equilibrium is only achieved after several weeks. It is conceivable that heterometallic species, such as $[FeMoO_2(azotochelin)_2]_5$, as well as mono, di, and polymeric species, may be formed. Similar results were obtained under anaerobic conditions [19,20]. The formation of the iron complex would involve a metal exchange reaction with the molybdate complex. The proton independent stability constant for molybdenum azotochelin is

10^{35}, a value comparable with the stability constant of iron(III) bis(catecholamide) complexes [21,22].

The structure of the molybdenum complex of the azotochelin analogue 5-LICAM (N,N'-bis(2,3-dihydroxybenzoyl)-1,5-diaminopentane) (Fig. 2) reveals a dimer in which two metal centers are coordinated in an octohedral environment and bridged by two ligands [19]. This unusual dimeric structure is different from the triply bridged Fe_2L_3 ferric complex expected to be formed. The difference may be important for the cellular recognition of the complex.

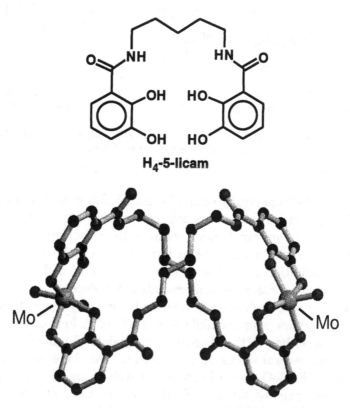

H_4-5-licam

FIG. 2.   Structure of the anion of the 5-LICAM and the molybdate complex $(\Lambda,\Lambda\text{-}[\{MoO_2(\text{5-LICAM})\}_2]^{4-}$. This and all subsequent structural figures were produced using the programs MOLSCRIPT [108] and Raster3D [109].

2.3. A Siderophore Induced by Molybdate

Surprisingly, siderophores in *A. vinelandii* are synthesized not only in response to iron deficiency but also to molybdate excess. The triscatecholamide siderophore protochelin (Fig. 1) is produced under iron-limiting conditions when the concentration of molybdate is greater than about 100 µM [23]. Formally, it is a condensation product of azotochelin and aminochelin, and consequently has three bidentate subunits. In contrast to azotochelin, protochelin is ideally suited for complexation of iron, as its denticity exactly matches the sixfold coordination requirements of iron. Complexation with molybdate would leave one catecholamide subunit free. A 2:3 protochelin-Mo complex is produced at the molybdate concentrations required for its synthesis [20]. Iron hydroxide is solubilized much more rapidly by protochelin than by azotochelin [24]. In the presence of molybdate the solubilization rate of iron is reduced, though by far less than the solubilization of iron by azotochelin in the presence of molybdate. The proton-independent stability constant for ferric protochelin is 10^{47}.

Wild-type *A. vinelandii* secretes azotochelin until its concentration is about the same as the molybdate concentration in the medium (about 100 µM) [21]. Until this point azotochelin is in excess, and almost all the molybdate is complexed. Above about 100 µM molybdate, azotochelin excretion ceases and protochelin is secreted instead. The switch between azotochelin and protochelin production occurs over a limited molybdate concentration range. If it did not occur, the molybdenum azotochelin complex formation would reduce the availability of azotochelin as an iron scavenger. The production of protochelin can be interpreted as a response to the competition of molybdate for azotochelin. The mechanism for the synthesis of protochelin and its induction at high molybdate concentrations is not understood. Molybdate concentration itself does not appear to be the signal as it is formed in an *Azotobacter* strain that does not transport molybdate [20]. It is also not formed in vitro from its components aminochelin and azotochelin in the presence of molybdate.

2.4. Siderophore Uptake

Special protein receptors convey siderophores across the outer membrane of gram-negative bacteria. It is of interest that three major iron-repressible outer-membrane proteins are synthesized by A. *vinelandii* in iron-limited medium [25]. Further, molybdate limitation resulted in the hyperproduction of a 44-kDa outer-membrane protein. Membrane proteins may facilitate the transport of iron and molybdate complexes into the periplasm of A. *vinelandii*, though there is no experimental evidence for this. If present, such transport through the outer membrane might add to the efficiency of molybdate uptake. Both iron and molybdate would be translocated into the periplasm, unless there is a mechanism that discriminates between the molybdenum and iron complexes. Specific transporters then translocate iron siderophore complexes across the inner membrane of gram-negative bacteria. However, this is not essential for the molybdenum-azotochelin complex. Its apparent dissociation constant is $100\,\mu M$ [21]. As the dissociation constant for the periplasmic molybdate-binding protein is much lower (see Sec. 3.3), the binding protein would remove molybdate from the Mo-azotochelin complex in the periplasm. The molybdate-bound form of the binding protein transfers molybdate to the membrane proteins of the high-affinity molybdate transporter for translocation into the cytoplasm (see Sec. 3.2).

2.5. Homocitrate Complexes

In addition to catecholate molybdenum complexes, it is noteworthy that the molybdenum atom in the nitrogenase cofactor is complexed with the hydroxyacid homocitrate. This is secreted by the nitrogen-fixing bacteria *Klebsiella pneumoniae* and A. *vinelandii*, and mutants that are unable to synthesize homocitrate can take it up from the medium [26,27]. It may therefore complex molybdate prior to nitrogenase cofactor formation.

3.   TRANSPORT SYSTEMS

3.1.   High-Affinity Molybdate Transport

The evolution of an uptake system with high affinity and specificity allows prokaryotes to scavenge molybdate in the presence of competing anions. This is particularly critical in the sea, where the concentration of sulfate is 2.6×10^5 times that of molybdate. It has been suggested that sulfate inhibition of molybdate uptake, and hence of nitrogen fixation, may underlie the chronic nitrogen deficiency in marine environments [28]. However, molybdate was not found to be the limiting nutrient for nitrogen fixation in a number of representative marine assemblages [3].

Diazotrophic growth of a strain of *A. vinelandii* that contains a single molybdenum-dependent nitrogenase is not inhibited at sulfate molybdate ratios up to eight times higher than those calculated for seawater [3]. When grown in low molybdate concentrations (10 nM), *E. coli* concentrates molybdate to a level at least 100 times that of the culture medium [29]. *A. vinelandii* is so efficient in scavenging molybdate that it has been used to selectively remove molybdate from growth media [30]. In *K. pneumoniae*, molybdenum uptake is coregulated with nitrogenase synthesis, whereas in *A. vinelandii*, molybdenum uptake is not regulated [31]. Molybdenum is accumulated in *E. coli* and *K. pneumoniae* with a K_m of 20–30 mM [32,33]. Despite the high selectivity for molybdate and tungstate, the specificity of the high-affinity uptake systems may vary in different bacteria. For instance, in *Clostridium pasteurianum*, sulfate is a competitive inhibitor of molybdate uptake [34].

3.2.   The High-Affinity Molybdate Transport Gene Locus

Genes encoding bacterial high-affinity molybdate transporters were identified by mutations that cause pleotropic loss of molybdoenzyme activity [29,35–42] or loss of molybdate-dependent repression of alternative nitrogenases [43,44]. The first molybdate transport mutants were identified in *E. coli* because they affected the activities of both nitrate reductase and formate hydrogen lyase [35]. Furthermore, enzyme activity was restored when the cells were grown in high concentrations of molybdate, indicating that the loss of activity was not due to defective enzymes. Molybdenum transport mutants of *K. pneumoniae,* and of a strain of *E. coli* that expresses *K. pneumoniae* molybdenum nitrogenase, lacked both nitrate reductase and molybdenum nitrogenase activity [45]. Loss of nitrate reductase activity is identified by growth of cells in the presence of chlorate, which the active enzyme converts to the toxic product chlorite. The first molybdenum transport gene identified was called *chlD* [35]. The designation *chl* was subsequently changed to *mod* [46]. *E. coli chlD* is now called *modC.*

Anions are transported to prokaryotes by ATP-binding cassette (ABC) transporters [47]. The high-affinity molybdate transporter consists of three protein components: ModA, ModB, and ModC [38,41,44,48,49]. ModA, the periplasmic binding protein, is the first protein ligand for molybdate. By analogy with other ABC transporters, the oxyanion-bound form of ModA interacts with the homodimeric integral membrane protein ModB. The peripheral membrane protein ModC couples ATP hydrolysis with molybdate translocation through the membrane into the cytoplasm [50]. The organization of genes in the molybdate transport locus is shown in Fig. 3.

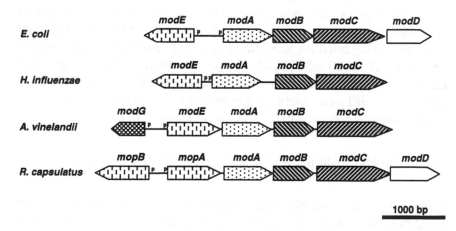

FIG. 3. Organization of genes in the molybdate transport loci of *E. coli, H. influenzae, A. vinelandii,* and *R. capsulatus. modA, modB,* and *modC* encode proteins for the molybdate transporter. *modE* and *R. capsulatus mopA and mopB* encode homologous molybdate-dependent transcriptional regulators. *A. vinelandii modG* encodes a cytoplasmic molybdate-binding protein with a tandem repeat of a Mop-like sequence. The location of promoters are indicated by "p."

The molybdate transporter is genetically one of the simplest ABC transporters because the integral and peripheral membrane proteins are each encoded by a single gene. In contrast to the molybdate transporter, the integral membrane proteins of the high-affinity sulfate and phosphate transporters are both encoded by two separate genes. As in most ABC transporters, the genes of the molybdate transport locus are found very close to each other and form a single transcription unit. However, in *Haemophilus influenzae*, *modA* and *modB* are separated by 171 base pairs [51]. The putative *modA* gene in *Synechocystis* sp., *Bacillus subtilis,* and *Deinococcus radiodurans* is followed by the gene encoding the ATP-binding protein of the transporter. The integral membrane protein is presumably encoded elsewhere.

In *A. vinelandii*, the molybdate transport genes *modA*, *B*, and *C* are duplicated in a distinct operon. ModA2 shows only 50% amino acid identity with ModA1 [52]. The *mod* locus includes genes related to molybdenum metabolism other than those encoding the transporter. The gene *modE*, which encodes the molybdate-dependent regulatory protein, is discussed in Sec. 4.6. In *A. vinelandii* and *E. coli,* the gene *modF* is encoded in the neighboring operon upstream from *modABC*. It is transcribed divergently from the operon containing the genes encoding the transporter. ModF is homologous to ModC. Mutations in *modF* have no detectable phenotype in *E. coli* [53]. An open reading frame, *modD,* is immediately downstream from *modC* in the transport operons of *E. coli, H. influenzae,* and *Rhodobacter capsulatus* [41,44,49]. It encodes a 26-kDa protein in *E. coli* and a 31-kDa protein in *H. influenzae* and *R. capsulatus*. No effect on molybdenum transport was detected in an *E. coli modD* mutant [41]. However, in *R. capsulatus,* a fourfold higher molybdate concentration was needed to repress alternative nitrogenase in a *modD* mutant [44]. The *A. vinelandii* molybdate transport locus also includes ModG that encodes a cytoplasmic molybdate-binding protein (see Sec. 4.1). It is located immediately upstream from, and in the same operon as, ModF.

Open reading frames similar to *modC* are also found outside the transport operon. An open reading frame (orf10) in *A. vinelandii* and *nifC* in *C. pasteurianum* are found in the nitrogenase gene clusters of these bacteria [54,55]. No genes encoding proteins similar to components of the bacterial and *Archaeal* high-affinity molybdate/tungstate transporter have been found in yeast and other eukaryotes. This suggests either that molybdate transport in these organisms may occur via other anion transporters or that a different specific molybdate transporter has been evolved and remains undetected. Recent genetic evidence suggests that *Chlamydomonas reinhardtii* has distinct high- and low-affinity molybdate transport systems, and that the high-affinity system can distinguish between molybdate and tungstate [56].

3.3. The Periplasmic Molybdate-Binding Protein

The high-affinity molybdate-binding protein is free in the periplasm of gram-negative bacteria and tethered to the surface of gram-positive bacteria. Mature ModA is a monomeric 24- to 25-kDa protein (about 230 amino acids), making it one of the smallest periplasmic binding proteins. It is transcribed with a leader peptide for export through the membrane [57]. In gram-positive bacteria, such as *Staphylococcus carnosa*, the N terminus of the mature protein has a hydrophobic binding motif that anchors the protein to the outer side of the cell membrane [58].

When radioactive molybdate is added to the cultures of *A. vinelandii* and *E. coli*, there is a very rapid uptake of ^{99}Mo into the cells, followed by uptake at a slower rate [29,32,40,48]. The rapidly bound ^{99}Mo is exchangeable with nonradioactive molybdate. As this exchangeable molybdate fraction is absent in a *modA* mutant and present in a *modB* mutant of *A. vinelandii*, the initial uptake can be interpreted as the binding of molybdate to ModA [48]. Sulfate and vanadate do not compete for molybdate bound to this fraction.

Mobility on native gel electrophoresis has been used to study the specificity of anion binding to *E. coli* ModA with a wide range of substrates [57]. ModA binds molybdate and tungstate with high-affinity, but not sulfate, phosphate, arsenate, nitrate, chlorate, selenate, metavanadate, perchlorate, and permanganate. The apparent dissociation constants for oxanion binding, determined by UV difference spectroscopy, were estimated to be 3 μM for molybdate and 7 μM for tungstate. A later study, using an isotopic method, revised the K_d for molybdate to 20 nM [59], a value more comparable to the K_m (50 nM) for molybdate transport by whole cells [32]. ModA binds tungstate with approximately the same affinity as molybdate. The stoichiometry of binding is 1 mol of molybdate per mole of protein.

Phylogenetic analysis of amino acid sequences for molybdate and sulfate-binding proteins from different bacteria shows that they are closely related to sulfate-binding proteins. The molybdate-binding proteins can be divided into two groups. The molybdate-binding proteins of *A. vinelandii* and *E. coli* are from different groups and have different hydrogen bonding interactions in their binding pockets (see Sec. 3.5) [52].

3.4. Structure of ModA

Binding proteins of ABC transporters have similar overall structures [60]. *E. coli* ModA and *A. vinelandii* ModA2 consist of two similar globular domains connected by strands at the bottom of a deep cleft between the domains [52,61] (Fig. 4). Each domain has a central, mixed, five-stranded β sheet sandwiched between two layers of α helices. Domain II arises as a single insertion between β strands 4 and 5 of domain I. Thus, both the N and C termini are in domain I, and there are two interdomain connections. The *Salmonella typhimurium* sulfate-binding protein has a very similar structure but contains an additional segment of

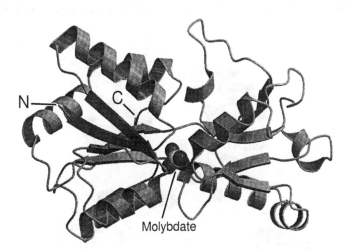

FIG. 4.   Structure of the ModA protein of *A. vinelandii* from Protein Data Bank entry 1ATG. The polypeptide chain is displayed in ribbon representation, with the molybdate in space-filling representation.

50 amino acids at the C terminus, which is in domain II [62]. Its inter-domain hinge therefore consists of three strands.

3.5. The Oxyanion Binding Site

Molybdate and tungstate are bound between the two domains. Closure of the interdomain cleft forms a completely dehydrated pocket around the anion. In *A. vinelandii* ModA2, the nearest water is 6.4 Å away. The anion is held in the binding pocket by seven hydrogen bonds from uncharged residues that lie at the N termini of four α helices that converge on the binding pocket. The α helices provide a high concentration of local dipoles that stabilize the charges on the anion. In *A. vinelandii* ModA2, five hydrogen bonds are donated from domain I by three side chains, Thr9, Asn10, and Ser27, and two backbone amide residues of Asn10 and Ser37. Two hydrogen bonds are donated by the backbone amides of Tyr118 and Val147 from domain II (Fig. 5). Surprisingly, the residues involved in binding molybdate in the *E. coli* are different [61]. Seven hydrogen bonds are donated by the side chains and backbone amides of Ser12 and Ser39, by the backbone amides of Ala125 and Val152, and by the hydroxyl of Tyr170. The predominance of hydrogen bond donors in domain I of both proteins suggests that molybdate binds

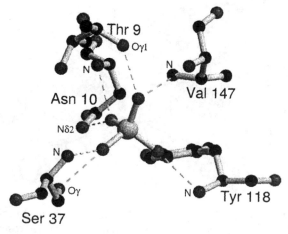

FIG. 5.   The molybdate binding site of *A. vinelandii* ModA.

to this domain first [52]. The binding pockets of the molybdate- and sulfate-binding proteins are very similar, with the anions held by seven hydrogen bonds donated by the protein. Again, the residues involved in binding are different [62]. Five of the seven hydrogen bonds in the sulfate-binding protein involve main-chain interactions.

3.6. Specificity of ModA

The periplasmic phosphate-, sulfate-, and molybdate-binding proteins are highly selective for their respective anions. Each binding site has a high concentration of uncharged polarizable potential hydrogen bond donors. However, in the phosphate-binding protein, a single aspartic acid residue is a hydrogen bond acceptor that plays a key role in the recognition of mono- and dibasic phosphate. It accepts hydrogen bonds from the protons on the phosphate anion and confers specificity by repelling the unprotonated sulfate anions [60,63]. The charges on the anions are dissipated entirely by multiple local dipoles provided at the N termini of four α helices that converge on the site [64].

The nature of the binding surface may also be relevant to the selectivity of the binding proteins. The surfaces of the binding clefts in the phosphate- and sulfate-binding proteins are both negatively charged and thus noncomplementary to their anions. The noncomplementarity may enhance discrimination between similar ligands [65]. By contrast, the surface potentials of the binding pockets of both *A. vinelandii* and *E. coli* molybdate-binding proteins are virtually neutral [52,61]. From simple electrostatic considerations, based on Born charging energies, it is energetically more favorable for an apolar pocket to bind a larger anion with the same overall charge than a smaller one [61]. In other words, this would favor the binding of molybdate over sulfate.

The similarity of the binding pockets indicates that their selectivity does not depend on the differences in hydrogen bonding interactions. Viewed from the protein, molybdate, tungstate, and sulfate all appear as tetrahedrons of oxygen atoms. The average metal to oxygen bond lengths of molybdate and tungstate are very similar at 1.75 Å and 1.76 Å, respectively. By contrast, the sulfur to oxygen bond length is about 0.3 Å smaller in sulfate at 1.47 Å. The smallest sphere that can accommodate a molybdate or a tungstate has a volume of 132 Å^3, whereas sulfate will fit into a sphere of volume 99 Å^3. This difference in anion volume is reflected in the binding pocket volumes. The volumes of the binding pockets of *A. vinelandii* ModA2 and *E. coli* ModA are 74 Å^3 and 72 Å^3, respectively, compared with 59 Å^3 for the sulfate-binding protein. As these volumes are 55–60% of the "smallest sphere" volumes for the respective anions, the binding pockets must fit very closely around their respective bound anions [52]. Thus, binding pocket volume has an important role in the selectivity of the binding pocket of the periplasmic molybdate/tungstate-binding protein. The inability to distinguish between molybdate and tungstate further strengthens this argument [57]. Both anions are therefore transported into the cell, and the uptake of molybdate and tungstate depends solely on their availability. As one or another anion tends to predominate according to the environment, this is not usually a problem.

3.7. The Integral and Peripheral Membrane Proteins ModB and ModC

By analogy with integral membrane proteins of ABC transporters, the 24-kDa ModB protein (about 230 amino acids) is a homodimer that forms a channel through the membrane. Its outer surface interacts with oxyanion-bound ModA. In common with transmembrane proteins of ABC transporters, ModB has a conserved sequence located in a cytoplasmic loop that may interact with the peripheral membrane protein ModC [66-68]. *E. coli, H. influenzae,* and *A. vinelandii* ModB have five transmembrane helices as opposed to the usual six for integral membrane proteins of ABC transporters. However, other putative ModB proteins, such as that of *R. capsulatus,* are predicted to have six transmembrane helices.

The 38- to 39-kDa (about 350 amino acids) peripheral membrane ATP-binding protein ModC is predicted to be closely associated with the cytoplasmic side of the transmembrane protein ModB and to be a homodimer. It contains an ATP-binding domain of about 200 amino acids that is highly conserved among ABC transporters. In addition to this domain, it has a C-terminal domain that is homologous to the cytoplasmic molyb-date-binding proteins (see Fig. 6 in Sec. 4.1).

3.8. High-Affinity Tungstate Transport

Despite the inability of the molybdate transporter to distinguish between molybdate and tungstate, some bacteria can select tungstate in the presence of molybdate. For example, the molybdenum iso-enzyme of formylmethanofuran dehydrogenase isoenzymes in *Methanobacterium wolfei* and *Methanobacterium thermoautotrophicum* is induced by molybdate while the tungsten isoenzyme is synthesized constitutively [69]. That this discrimination takes place at the binding protein of an ABC transporter has recently been demonstrated in the gram-positive aerobic *Eubacterium acidaminophilum* that synthesizes the tungstoenzymes formate dehydrogenase and aldehyde oxidase. Uptake of radioactive tungstate is not affected by molybdate.

Genes encoding an ABC transporter for tungstate (*tup*, *t*ungstate *up*take) are located downstream of the genes for formate dehydrogenase [70]. The purified extracellular binding protein TupA binds tungstate with high affinity, but not molybdate, sulfate, phosphate, chlorate, chro-mate, vanadate, and selenate. The amino acid sequence of the 30.9-kDa binding protein shows only weak similarity with the ModA sequence, though it shows high similarity with putative binding proteins from *Methanobacterium thermoautotrophicum*, *Haloferax volcani*, *Campylobacter jejuni*, and *Vibrio cholerae*, which are therefore pre-sumed to be tungstate specific. Phylogenetic analysis of the amino acid sequence of TupA showed that these proteins form a distinct group of oxyanion-binding proteins. The molecular basis of this extreme case of oxyanion discrimination has yet to be determined.

3.9. Low-Affinity Transport

The high-affinity molybdate transport system concentrates molybdate in cells grown in low concentrations of molybdate. Wild-type *E. coli* cells grown aerobically in 10 nM molybdate contain 1.0 μM of molybdate [29]. When the transport system is inactivated by mutagenesis, the resulting molybdoenzyme deficiency can be restored by high concentrations of molybdate (>100 µM) [33,35,36,48]. This indicates that molybdate can enter the cell by other routes. As the high-affinity uptake system is repressed at high molybdate concentrations, the low-affinity systems are used to transport molybdate in cells grown under these conditions. However, the low-affinity uptake systems do not allow an unregulated rise of the internal molybdate concentration. The concentration of molybdate in wild-type and a *modC* mutant of *E. coli* grown in the presence of 100 µM molybdate is only 5–12 µM [29], lower than the external concentration. So the combined effect of the high- and low-uptake systems is to maintain an internal concentration of molybdenum in the cell within a narrow range.

In *E. coli*, the sulfate transport system may transport molybdate when it is present at high concentrations [71,72]. High sulfate concentration represses the sulfate transport system. The concentration of molybdate required for maximal formate hydrogen lyase activity in high-affinity molybdenum transporter mutants is 10 times higher when *E. coli* cells are grown in 15 mM sulfate compared with low-sulfate medium. Similar results were obtained when formate hydrogen lyase activity of wild-type *E. coli* was compared in a double-mutant strain with impaired high-affinity molybdate and sulfate transport systems. These results suggest that molybdate transport by the sulfate uptake system is 10 times less efficient than that by the molybdate transport system [72].

At high concentrations, molybdate may also be transported into the cell by other systems that have not been fully characterized. In *K. pneumoniae*, a permease of the LacY family of proton/sugar symporters is involved in low-affinity molybdate transport [73].

Tungstate inhibits nitrogenase activity and represses the high-affinity transport system. However, spontaneous mutations in tungsten-tolerant strains of *A. vinelandii* enable them to grow diazotrophically in the presence of high concentrations of tungstate (above 0.1 µM). These strains are likely to have impaired low-affinity molybdate/tungstate uptake systems [74]. They accumulate less molybdenum than the wild-type strain, and a tungstate-tolerant *modC* double mutant synthesizes very little nitrate reductase, even at high molybdate concentrations [48].

4.   CYTOPLASMIC MOLYBDATE-BINDING PROTEINS

4.1.   A Common Molybdate/Tungstate Binding Domain

The early steps in the processing of molybdenum in *Clostridium pasteurianum* were first studied by identifying proteins that bound [99]Mo [75–77]. The major protein that bound molybdenum before the detection of nitrogenase activity was called Mop [78]. It was purified and shown to contain 0.7 mol of Mo per mole of protein. The apparent molecular weight of Mop was estimated to be about 30 kDa, though it dissociated into subunits of approximately 6 kDa.

The Mop protein was associated with fluorescence that was attributed to a pterin and was subsequently referred as a molybdopterin-binding protein. However, the fluorescence spectrum of Mop differed from that of the molybdenum cofactor, and moreover it contained no phosphorus. Furthermore, the putative pterin species did not activate cofactor-free nitrate reductase. It was therefore suggested that Mop may bind a precursor to the molybdenum cofactor. Nevertheless, this assertion was made prior to the full elucidation of the biosynthetic pathway for molybdopterin (see Chapter 9). *C. pasteurianum* synthesizes three distinct Mop proteins encoded by different genes [79]. Each 7-kDa polypeptide consists of 68 amino acids. The three polypeptides have more than 90% amino acid sequence identity and are expressed in different amounts. The amino acid sequences of the Mop proteins are shown in Fig. 6.

Analysis of a high-affinity molybdenum transport locus of *A. vinelandii* first showed that Mop-like sequences are present in very different proteins (Fig. 6) [48]. They are present in (1) proteins that contain only Mop domains, either as a 7-kDa monomer or a 14-kDa tandem Mop repeat; (2) the Mo-responsive regulatory protein ModE; and (3) the C-terminal domain of ModC, the inner-membrane protein of the high-affinity molybdate transporter. Proteins with Mop domains were called molbindins, as they contain a common molybdate-binding domain [80].

Molybdenum is transported to the cytoplasm as molybdate by the high-affinity transporter. One way in which molybdenum could be distinguished from sulfate is by changing from four-coordinate in molybdate to six-coordinate, as found in many of the cofactors that contain molybdenum [5]. Molybdenum K-edge X-ray absorption spectra of

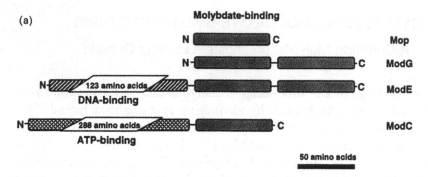

FIG. 6. (a) Organization of Mop-like sequences in *C. pasteurianum* Mop, *A. vinelandii* ModG, and *E. coli* ModE and ModC. (b) Alignments of Mop-like sequences displayed using the program ALSCRIPT [110]. The tandem repeats of ModG, ModE, MopA, and MopB are shown as separate sequences. Residues that show conservation in at least 50% of the sequences are boxed, whereas those showing greater than 75% identity are shaded. The location of secondary structural elements in the ModG and Mop structures is indicated, and the sequence numbering scheme is relative to the first repeat of ModG. The bottom line gives the type of binding site to which the residue belongs in ModG and Mop. The vertical dashed line indicates the domain boundaries. Cp, *Clostridium pasteurianum*; Hi, *Haemophilus influenzae*; So, *Sporomusa ovata*; Av, *Azotobacter vinelandii*; Ec, *Escherichia coli*; Rc, *Rhodobacter capsulatus*.

molybdenum bound in Mop from *H. influenzae*, the Mop-like protein ModG, and the transcriptional regulatory protein ModE from *E. coli* show that they all bind tetrahedral molybdate with an Mo-O distance of 1.76 Å [81]. This is consistent with the value obtained for molybdate from small-molecule X-ray structures. Therefore, the specificity of the cytoplasmic proteins for molybdenum does not depend on a change of coordination number or geometry.

4.2. Mop-like Proteins: The Structures of Mop and ModG

The X-ray crystal structures of two Mop-like proteins have been solved, namely, those of *Sporomusa ovata* Mop [82] and *A. vinelandii* ModG [83]. They are highly symmetrical, multimeric proteins with the basic repeating structural unit being the 7-kDa Mop domain. These domains

(b)

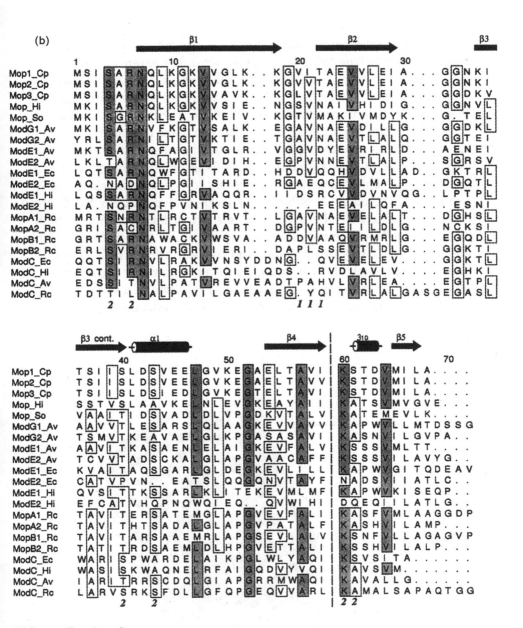

FIG. 6.   Continued

adopt an OB (oligonucleotide/oligosaccharide-binding) fold motif consisting of a five-stranded β barrel in a Greek key arrangement that is
capped by a two-turn, amphipathic α helix [84]. Each protein contains
six Mop units in the oligomer and are closely superposable, although the
Mop protein is hexameric, whereas ModG is a trimer (Fig. 7). In the
latter, the subunits contain a pair of Mop domains that share 41% and
are related to one another by pseudo-twofold symmetry. In Mop, pairs of
twofold-related subunits exchange their C-terminal β strands with one
another [82]. This places the C terminus of one subunit close the N
terminus of the other. It is here that pairs of Mop units in ModG are
fused into a single 14-kDa subunit via a short surface loop [83]. In both
proteins the Mop domains pack closely together, and in ModG each
subunit makes 34 hydrogen bonds and two salt bridges with its neigh-
bors. Approximately 40% of potential solvent accessible surface of the
ModG subunits is buried at the interfaces, resulting in a very compact,
roughly spherical molecule approximately 50 Å in diameter [83].

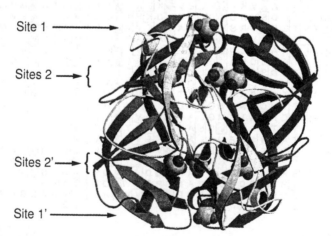

FIG. 7.   Structure of the fully loaded ModG protein of *A. vinelandii* from PDB
entry 1H9M. The three subunits of the trimer are shown in ribbon representation as varying shades of gray. The molybdates are depicted as space-filling
models.

4.3. Oxyanion Binding Sites

Both Mop and ModG have a total of eight potential anion binding sites that belong to two novel and distinct types. In Mop, two equivalent sites (type 1) lie on the threefold axis at opposite poles of the molecule. The six remaining sites (type 2) are located between pairs of subunit interfaces and are arranged as two three-membered rings, one ring in each half of the molecule [82]. The situation in ModG is slightly more complex because the two types of site are further subdivided into two similar but not equivalent sites associated with the different domains. The sites in one half of the molecule are referred to as 1 and 2, whereas those in the other half are designated 1' and 2' [83].

The type 1 binding sites of ModG are essentially identical and involve three contiguous residues from symmetry-equivalent loops of the three subunits: Ala92, Val93, Asn94 in site 1, and Ala20, Val21, and Asn22 in site 1' (Fig. 8). The ligand lies on the trimer threefold axis, which passes through the molybdenum atom and one of the oxygens. Each nonaxial oxygen receives hydrogen bonds from the backbone amides of each of these three residues. In addition, the side chain of each symmetry-equivalent Asn donates a hydrogen bond to the single axial oxygen that is pointing toward the center of the protein [83]. Thus, the molybdate is held by 12 hydrogen bonds. As in the periplasmic molybdate-binding protein ModA [52,61], no charged residues are involved.

In Mop, the oxyanion (actually tungstate) bound in site 1 interacts only with backbone amides, and each of the nonaxial oxygens makes only two such contacts with the protein. Moreover, the "axial" oxygen faces outward and is tilted so that it accepts a hydrogen bond from the amide of only one of the three Thr20 residues [82]. The equivalent of Asn22 in ModG is modeled as Met in Mop, and its side chain does not interact with the anion. The result is that molybdate in Mop is bound by 7 hydrogen bonds as opposed to 12 in ModG [83].

The type 2 sites of Mop-like proteins lie at the interfaces between pairs of subunits and use a mixture of charged and uncharged interactions to bind ligands (Fig. 9). Binding site 2' of ModG most closely resembles the type 2 site of Mop and will therefore be described in more detail. The ligand is bound through favorable contacts with both

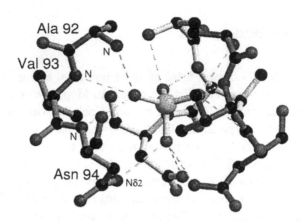

FIG. 8. The type 1 molybdate binding site of *A. vinelandii* ModG. This is structurally equivalent to the type 1′ site.

domains of one subunit. In domain I the side chain of Lys132 salt bridges to the molybdate, whereas the main-chain amide of Ala133 donates a hydrogen bond. From domain II, the side chain hydroxyl of Ser76 and the backbone amide of Arg78 both donate hydrogen bonds. Interactions with the other subunit involve just two residues from domain I. Thr40 donates hydrogen bonds from both its backbone amide and side chain hydroxyl to the anion, whereas it is only the side chain of Ser43 that interacts with the anion. The binding site 2 equivalent of the latter is Ala115, and thus there is no interaction with this residue. Otherwise binding sites 2 and 2′ are essentially the same. Another important feature of the binding site is a water molecule that is associated with the molybdate [83].

The type 2 sites of Mop use residues structurally equivalent to those of ModG binding site 2′ to bind the ligand. Moreover, in a super-position of the two protein structures, the oxyanions in these sites virtually eclipse one another. However, no solvent was reported in the binding pocket of the Mop structure [82]. Although the Arg (residue 78) is conserved in binding site 2 of ModG, in Mop, and many other Mop-like sequences, in all the crystal structures solved to date, the side chain does not interact with the bound ligand. It was therefore suggested to have a role in ligand capture [83].

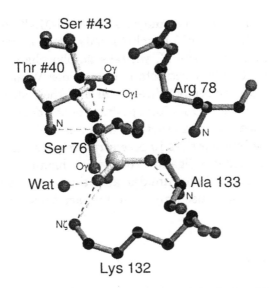

FIG. 9. The type 2' molybdate binding site of *A. vinelandii* ModG. The hash symbols denote residues from one subunit, which also have bonds given in lighter gray. This is closely similar to the type 2 binding site, although the equivalent of Ser43 is an Ala, and thus there is no interaction with this residue.

4.4.   Affinity and Specificity

The crystal structures of ModA have shown that ligand size is a major determinant of specificity in the oxyanion binding proteins [52,61]. This would indicate that perhaps the type 1 binding sites of ModG are the most selective, since they have an average cavity volume of 83 Å^3 that is comparable to the average volume of 73 Å^3 reported for the two ModA structures. By contrast, the type 2 sites are considerably larger with an average volume of 170 Å^3, suggesting a much looser fit [83]. Another feature of the ModA binding pockets is their electrostatic neutrality. This is also reflected in the ModG and Mop type 1 binding sites.

Conversely, the type 2 sites make use of a positively charged Lys side chain, which most likely increases the affinity of these sites for oxyanions but at the same time reduces their selectivity for molybdate (and tungstate). Indeed, when molybdate-loaded ModG is crystallized from high concentrations of phosphate, this latter oxyanion replaces the molybdate in binding site 2′, whereas no phosphate is seen in the type 1 sites [83]. Unlike electrostatic or van der Waals interactions, hydrogen bonds are extremely directional and their strength is critically dependent on the correct juxtaposition of the donor and acceptor atoms. Through the use of interactions with relatively rigid main-chain NH groups, these binding sites impose strict requirements on the size and shape of the cognate ligand. The type 1 binding sites of ModG make use of nine backbone amides and thus are highly selective, whereas the type 2 sites use only three and are thus less discriminating [83].

4.5. Cooperativity

Within each half of the ModG trimer, the type 1 and type 2 sites are separated by approximately 10 Å [83]. Upon closer inspection, it is apparent that they are linked by a short hydrogen bond network (Fig. 10). This involves only three hydrogen bonds: the ligand in the type 2 binding site is hydrogen-bonded to a water molecule that in turn is hydrogen-bonded to the side chain of Asn22/94. The latter forms part of the type 1 binding site and hydrogen-bonds to the axial oxygen of the molybdate. In fact, the axial oxygen is also linked to two additional molybdates by symmetry equivalent networks. This observation raises the possibility of cooperative ligand binding effects in ModG. Given the presence of a positively charged Lys in the type 2 binding sites, it seems likely that these become occupied first. The concomitant binding of a water molecule then may be responsible for the establishment of the network as far as the side chain of the Asn. Indeed, this link to the Asn may correctly align the side chain for ligand incorporation in the type 1 binding site. Thus the binding of molybdate in the type 2 sites may increase the affinity of the type 1 sites for molybdate. In the *S. ovata* Mop structure the equivalent of the Asn in the type 1 binding site is modeled as a Met that effectively precludes the existence of such a network [82]. However, sequence alignments suggest that this network could be present in other Mop proteins.

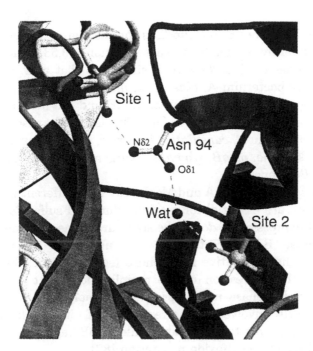

FIG. 10.   The hydrogen-bonding network linking sites 1 and 2 in ModG. An equivalent network is also observed between sites 1' and 2'.

4.6.   The Molybdate-Dependent Transcriptional Factor ModE

The transcriptional factor ModE senses molybdate concentration and regulates the transcription of operons involved in molybdenum uptake and utilization. In the presence of high concentrations of molybdate, it represses transcription of the molybdenum transport operon, thus limiting uptake of molybdate by the high-affinity transporter [85]. ModE may regulate the low-affinity molybdate transport in *A. vinelandii* [48]. It also acts as a positive regulator for genes involved in molybdoenzyme and cofactor biosynthesis.

ModE has only been identified in a few bacteria in which it is found in the molybdate transport locus. The position of the *modE* gene in the transport locus varies (Fig. 3). In *E. coli*, *modE* is located in a separate operon that is immediately upstream and divergently transcribed from the *modABCD* operon. *E. coli modE* is constitutivly expressed at a low level independently of molybdate concentration [49,53]. The organization of the genes in *H. influenzae* is similar to that of *E. coli*. In the nitrogen-fixing bacterium *A. vinelandii, modE* is the first gene in the *modEABC* operon. The expression of ModE and the molybdate transporter are therefore autoregulated [86]. This is also the case with *modE* homologue, *mopA*, in *R. capsulatus*. However, *R. capsulatus* has a second *modE*-like gene, *mopB*, which is divergently transcribed from the *mopAmodABCD* operon.

In *R. capsulatus*, MopA and MopB may have different roles. MopB and molybdate are essential for maximal dimethyl sulfoxide reductase activity and expression of *dorA*, the structural gene for dimethyl sulfoxide reductase. By contrast, both *mopA* and *mopB* are involved in regulation of the alternative nitrogenase (see Sec. 4.8). They are not required for the expression of the periplasmic nitrate reductase or xanthine dehydrogenase in the presence of molybdate. Thus, MopA may be primarily involved in the regulation of nitrogen fixation gene expression in response to molybdate, while MopB has a role in nitrogen fixation and dimethyl sulfoxide respiration [87].

4.7. Regulation of Molybdate Transport

The *mod* operon of *E. coli* is well expressed when wild-type strains are grown in molybdenum-deficient minimal medium or in molybdate transport mutants [42,72,85]. However, even 100 mM molybdate does not repress expression of the *modABCD* transcription in a *modA* mutant strain that also has a mutation in *modE* [42]. *A. vinelandii modE* mutants similarly transcribe the transport operon independently of molybdate concentration [86]. There is no evidence for molybdate-dependent regulation of the transporter in *S. carnosus* [58].

4.8. Regulation of *R. capsulatus* Alternative Nitrogenase

The alternative nitrogenase of *R. capsulatus* contains a cofactor that lacks molybdenum. The structural genes for the alternative nitrogenase, *anfHDK*, are only transcribed in the absence of molybdate. Molybdate-dependent regulation of alternative nitrogenase is indirect [88]. In the presence of molybdate, the ModE homologues MopA and MopB repress transcription of *anfA*, the gene for the activator of *anfA* transcription. Transcription of *anfA* is also regulated by transcriptional factor NtrC, making expression of the alternative nitrogenase dependent on both nitrogen and molybdenum. The situation is different in *A. vinelandii* in which both the vanadium and iron nitrogenases are still repressed by molybdate in a *modE* mutant strain [86].

4.9. Transcriptional Activation

In the presence of molybdate, ModE activates transcription of genes for dimethyl sulfoxide reductase (*dmsA*), fumarate reductase [89], molybdenum cofactor biosynthesis (*moaABC*) [90], formate hydrogenlyase (*hyc* and *fdhF*), and nitrate reductase (*narGHJI*) [91].

4.9.1. The E. coli dms Operon

Oxygen, molybdate, and nitrate regulate transcription of the *dms* operon for dimethyl sulfoxide reductase. This multisignal regulation is complex [89]. It involves two transcription start sites. Transcription is impaired in *modE* mutant strains. Anaerobic induction of *dmsA* transcription depends on the transcription factor Fnr, and a Fnr binding site is located in a distal promoter P1. A ModE binding site in the proximal P2 promoter overlaps a putative −10 RNA polymerase binding site, a location similar to the ModE binding site in the *modABCD* transport operon that is repressed by ModE. Expression of the P2 promoter alone is repressed in a molybdate-dependent manner by ModE and derepressed in a ModE mutant strain. The impaired transcription of the *dms* operon from the full-length promoter in a *modE* mutant strain is not yet understood. It may involve the participation of intermediate host factor that binds at a site overlapping the ModE binding site and may induce a conformation change in the DNA.

NarL mediates nitrate-dependent repression of *dms* transcription and molybdate abolishes nitrate repression. A NarL binding site overlaps the −10 consensus sequence of the P1 promoter. Again, as ModE does not bind to the P1 promoter, other factors, possibly the NarX/NarQ nitrate sensor, may be involved in the molybdate requirement for nitrate-dependent repression.

4.9.2. The E. coli moa Operon

In *E. coli*, the first step of molybdenum cofactor biosynthesis is the synthesis of an intermediate, precursor Z, from a guanosine nucleotide or related compound. This committed step requires the first three of five genes (*moaABCDE*) in the *moa* operon. Its expression requires anaerobic conditions and molybdate. Furthermore, it is repressed when the bacterium synthesizes active molybdenum cofactor, so the positive regulation by molybdate is only observed in strains incapable of synthesizing molybdenum cofactor [90]. Like the *dms* operon, the *moa* operon has two promoters and the Fnr-dependent distal promoter is the site of anaerobic regulation. Molybdate-dependent activation depends on ModE binding to a site in the proximal promoter. Tungstate can substitute for molybdate for ModE-dependent activation. Molybdenum cofactor repression acts at the proximal promoter, though its mechanism has not been established. Cofactor-dependent repression of *moa* is lost in the presence of tungstate, presumably because it forms a nonfunctional molybdenum cofactor.

4.9.3. The E. coli fdhF, hyc, and nar Operons

Molybdate and ModE are required for optimal transcription of the operons that encode *E. coli* formate hydrogen lyase (*fdhF* and *hyc*) and respiratory nitrate reductase (*nar*) [92]. Transcription of both operons depends on ModE as well as the presence of the molybdenum cofactor. Transcription of the *fdhF* and *hyc* operons is also activated by the regulatory protein FhlA. Several *fhlA* mutants activate *hyc* transcription in the absence of molybdate or ModE. Activation by the molybdenum cofactor is also mediated by *fhlA* [93].

4.10. ModE Binding to DNA

The binding of ModE to promoters has been demonstrated by gel shift assays and the binding sites identified by DNAse I footprinting.

Molybdenum-dependent binding of ModE has been demonstrated for the *E. coli modA*, dmsA, and *moaA* promoters. Molybdate is essential for high-affinity binding of ModE to a *modA* promoter fragment [94]. ModE binds molybdate with a dissociation constant of 0.8 μM. The apparent dissociation constant of ModE for the binding site in the *modA* promoter was estimated to be 30 nM in the presence of 0.1 M molybdate, whereas it is 10 times this value in the absence of molybdate.

Similar binding affinities have been shown for binding to the *E. coli dmsA* (28 nM) [89], *modA* (45 nM) [95], and *moaA* (24 nM) promoter fragments [90]. A much lower apparent K_d of 0.3 nM for the binding of ModE to *E. coli modA* promoter has also been reported [96].

DNAse I footprinting experiments have located the ModE binding sites in the promoters of the *E. coli modA*, dms, and *moa* operons [89,94,95] (Fig. 11). Putative ModE binding sites were also identified by comparison of the promoters of genes regulated by molybdenum [88]. The binding sites determined by DNAse I footprinting have the palindromic sequence CGnTnTnT-8n-AnAnAnCG (where n is a variable nucleotide). The ModE binding site in the promoter of the *E. coli modABCD* operon overlaps the start site of transcription and the −10 polymerase binding sites [89,94].

A putative ModE binding site is present at a similar position in the promoter of the *R. capsulatus mopAmodABCD* operon. ModE therefore probably represses expression of the molybdate transport operon by preventing RNA polymerase binding to the promoters. In contrast to promoters that are repressed by ModE, the ModE binding site in the *E. coli* proximal *moaA* promoter, which is activated by ModE, is adjacent to the −35 RNA polymerase binding site. It therefore binds in a position where it could interact with RNA polymerase.

E. coli *modA*	agtCGtTATATtgtcgccTAcATAACGttacat
E. coli *dmsA*	attCGaTgTATAcaagccTATATAgCGaactgc
E. coli *moaA*	tgaCGcTATATacatgatTAcATAgCGaaagtg
Identical nucleotides	**CG.T.TAT.......TA.ATA.CG**

FIG. 11. DNA binding sites of ModE determined in the promoters of *E. coli modA, dmsA,* and *moaA*.2. Invariant nucleotides are shown in capital letters and palindromic nucleotides are shown in bold letters.

4.11. Structure of ModE

E. coli ModE is a 54-kDa homodimeric protein of 262 amino acids. Molybdate is required for ModE to bind to DNA, and the K_d for DNA binding in the presence of molybdate is about 30 nM. Titration of intrinsic protein fluorescence shows that *E. coli* ModE binds two molecules of molybdate with a K_d of 0.8 µM. Molybdate binding is accompanied by a large quench (50%) in fluorescence [94]. The X-ray crystal structure of full-length *E. coli* ModE has been determined in the absence of molybdate (Fig. 12) [97]. It consists of distinct N-terminal and C-terminal domains separated by a linker region of 11 amino acids. The majority of the interactions between the two subunits are in the N-terminal domain, which consists of 121 residues (five α helices and four β strands). It contains a winged helix-turn-helix motif. Dimerization provides the twofold related residues required for binding to the palindromic DNA binding site. Putative residues implicated in DNA binding have been identified. The C-terminal domain comprises residues 122–262. Like ModG it consists of a tandem repeat of Mop-like sequences and thus ModE has four Mop domains per dimer. These closely correspond in structure to four subunits of Mop and two subunits of ModG [83]. Nevertheless, the subunit interface is noticeably more open in ModE, reflecting its ligand-free status.

The oligomerization state of ModE is inconsistent with the presence of type 1 binding sites because these require a trimeric arrangement of Mop domains [82,83]. Indeed, both loops, corresponding to those that bind type 1 molybdates in Mop and ModG, bear acidic side chains in ModE, suggesting a tendency to repel oxyanions. Furthermore, sequence analysis of *E. coli* ModE indicates that the second Mop repeat lacks all the side-chain groups implicated in molybdate binding at type 2 sites in ModG and Mop [80] (Fig. 6). Inspection of the structure reveals that these substitutions effectively render one face of the subunit incapable of binding molybdate. However, unlike ModG the di-Mop units of ModE are inverted with respect to one another such that these faces are on the outside, whereas the faces bearing the conserved type 2 binding site residues come together [97] (Fig. 13). This gives rise to two equivalent type 2 sites comprising Thr163 and Ser166 from one subunit and Ser126, Arg128, Lys183, and Ala184 from the other. A superposition of a

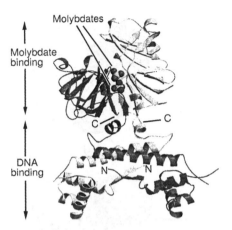

FIG. 12.   Structure of apo-ModE from *E. coli* depicted in ribbon representation from Protein Data Bank entry 1B9N. The predicted locations of the two molybdate binding sites derived from a superposition with a ModG "dimer" are indicated by the space-filling models.

hypothetical ModG dimer containing type 2 molybdates on the C-terminal domains of ModE also places the oxyanions in the vicinity of these two sites [83] (Fig. 12).

Very recently, a truncated form of ModE consisting of only the C-terminal Mop domains was crystallized in the presence of molybdate [98]. While adopting essentially the same structure as seen for the apo ModE, two molecules of molybdate were bound per dimer and the subunits were more closely associated. The oxyanions were bound using the residues described above in an analogous fashion to the type 2 molybdates of both ModG and Mop. Indeed, even the bound water molecule seen in ModG was structurally conserved. Curiously, this water is hydrogen-bonded to Gln144 that is structurally equivalent to the Asn in the type 1 binding sites of ModG, although, for reasons described above, ModE is unlikely to support type 1 binding sites. In agreement with ModG and Mop, only the backbone amide of the conserved Arg (residue 128) interacted with the bound ligand, whereas the side chain was involved in stabilizing the dimer interface through a salt bridge to Glu281 in the neighboring subunit. The conformational changes observed in the C-terminal domains of ModE upon molybdate binding could be propagated to the N-terminal DNA binding domains, thereby triggering recognition of their target DNA sequences.

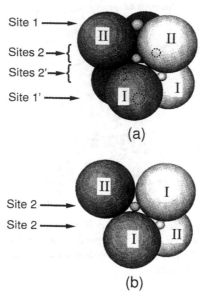

Site 1

Sites 2

Sites 2'

Site 1'

(a)

Site 2

Site 2

(b)

FIG. 13. Schematic diagrams indicating the quaternary structures of ModG and the C-terminal domains of *E. coli* ModE, together with the locations of their binding sites.

4.12. The Mop Domain of ModC

A number of the peripheral membrane proteins of ABC transporters have distinct C-terminal domains that extend beyond the highly conserved ATP binding domain. The important regulatory roles of these C-terminal domains have recently been recognized [99]. The C terminus of ModC, the peripheral membrane protein of the high-affinity molybdate transporter, resembles a single Mop domain and therefore could potentially bind molybdate immediately after it enters the cytoplasm (Figs. 1 and 2). As ModC is predicted to be dimeric, molybdate could bind at the interface of the Mop dimer. However, even the type 2 sites of Mop, ModG, and ModE require the participation of three Mop domains. Therefore, if ModC can bind molybdate, it must do so in a slightly different fashion. Although there have been no studies on the role of the C-terminal domain of ModC, we can imagine that molybdate binding to these domains might result in a conformational change in the conserved ATP binding domain of the protein. It may thus directly regulate the activity of the transporter, effecting a rapid response to excess molybdate as opposed to a slower response effected by transcriptional regulation. Alternatively, the Mop domain may interact with other proteins, transferring molybdate directly to them. In this context, it is of interest that a strain of *A. vinelandii* that lacks the regulatory protein ModE, and therefore should have a fully derepressed transporter, shows very impaired molybdate uptake compared to the wild-type strain [86].

4.13. Storage Proteins of *Azotobacter* and *Chlamydomonas*

An early study of showed that *A. vinelandii* accumulates much higher levels of molybdenum than *K. pneumoniae*. In *A. vinelandii*, molybdenum was bound to a tetrameric storage protein with pairs of 21- and 24-kDa subunits. It binds approximately 15 atoms of molybdenum per tetramer [31]. It is notable that in this study the molybdate-binding protein ModG was not isolated. Nucleotide, either ATP or ADP, is required for the accumulation of molybdenum on this protein [100].

An interesting carrier protein that binds a molybdenum cofactor (MoCo) is found in *Chlamydomonas reinhardtii* [101]. It is a protein of 66 kDa with subunits of 16.5 kDa. In contrast to free MoCo, MoCo bound to the carrier protein is remarkably protected against inactivation by aerobic conditions and basic pH. The carrier protein transferred active MoCo to aponitrate reductase in vitro without addition of molybdate.

5. CONCLUSIONS

Although molybdate forms complexes with iron siderophores, it is captured and transported into bacteria and *Archaea* by highly specific binding proteins. They mostly bind both molybdate and tungstate with high affinity [52,61]. Anion size plays a major role in the binding proteins' ability to differentiate molybdate and tungstate from other oxyanions.

Recently, a transport system that is highly specific for tungstate has been described in prokaryotes that synthesizes tungstoenzymes [70].

Within the cytoplasm, molybdate binds to a number of proteins with domains characterized by Mop-like sequences and an oligosaccharide/oligonucleotide fold. This fold associates in three different ways in different proteins. The C-terminal Mop-like domain of the transporter protein ModC is predicted to have the simplest organization, comprising a dimer of Mop units. Four Mop units are found in ModE, which binds two moles of molybdate in a single type of binding site (Fig. 13). Both the Mop protein and ModG have six Mop domains and bind up to 8 mol of molybdate in two types of binding site. Because *C. pasteurianum* Mop and ModG incorporate different Mop polypeptides or sequences into the two types of binding site, they contain a greater variety of binding sites than that found in a simple Mop protein, such as that of *S. ovata*. Thus, *C. pasteurianum* Mop proteins may contain three different polypeptides and ModG has two tandem Mop-like sequences [83]. Furthermore, it is not inconceivable that Mop polypeptides may associate with the Mop-like domains of ModC and ModE. Parallels may exist with the iron storage protein ferritin. The functional unit is a highly symmetrical molecule composed of 24 subunits [102,103]. They exist as a range of multimers through the admixture of two (or three in some species) similar subunits. In higher organisms the proportions were also shown to be tissue dependent [104], suggesting functional differences. Subsequently it was demonstrated that these "isoferritins" display a gradation in properties, including iron uptake rates, iron capacity [105], and protein stability [106], in accordance with their subunit composition.

The multiple binding sites and polypeptide combinations of Mop-like proteins enable them to bind molybdate over a wide concentration range. As the highest affinity site becomes saturated, the lower affinity sites continue to bind the ligand. The only evidence for the wholly molybdate-binding role of Mop-like proteins has been obtained in *A. vinelandii* [48]. Very little nitrate reductase activity is observed in wild-type cells until they are grown in culture medium that contains more than about 10 nM molybdate. A strain lacking ModG has a similar response of nitrate reductase activity to increasing molybdate concentration. However, a strain with deletions of both ModG and ModE shows nitrate reductase activity at much lower molybdate concentrations than both the wild-type or *modG* mutant. This suggests that the Mop-like domains of both ModG and ModE both affect the internal molybdate concentration by binding molybdate with high affinity. Furthermore, ModG and ModE are not required for molybdenum cofactor synthesis. Many transition metals bind to specific cytoplasmic proteins that act as metal chaperones. This involves highly specific transfer of the metal between donor and target proteins that involves changes in metal coordination geometries [107]. The above experiments suggest that ModG and ModE in *A. vinelandii* do not act as molybdate chaperones. It is notable that *E. coli* does not have a Mop protein. The details of the conversion of molybdate to the nitrogenase and molybdenum nitrogenase cofactors have not been fully resolved. The biosynthesis of these cofactors is reviewed in Chapters 5 and 9.

ACKNOWLEDGMENTS

We thank A. Duhme-Keir for valuable discussions and for providing Figs. 1 and 2. RNP is also grateful to Prof. I. Bertini at the Magnetic Resonance Center (CERM) of the University of Florence. DML acknowledges support from the Biotechnology and Biological Sciences Research Council.

ABBREVIATIONS AND DEFINITIONS

ABC transporter	ATP-binding cassette transporter
ADP	adenosine diphosphate
anf	an alternative nitrogenase gene
ATP	adenosine 5'-triphosphate
di-Mop unit	two 7-kDa Mop-like units fused in a single polypeptide
dms	a dimethyl sulfoxide reductase gene
FhlA	regulatory protein controlling transcription of formate hydrogenlyase
hyc/fdh	a formate hydrogenlyase gene
5-LICAM	*N,N'*-bis(2,3-dihydroxybenzoyl)-1,5-diaminopentane

moa	a molybdenum cofactor biosynthesis gene
MoCo	molybdenum cofactor
mod	a gene from the *mod* operon associated with molybdenum transport and processing
ModA	periplasmic molybdate-binding protein
ModA1/ModA2	two isoforms of ModA from *A. vinelandii*
ModB	transmembrane component of ABC transporter for molybdate
ModC	peripheral membrane component of ABC transporter for molybdate with ATPase activity
ModD/ModF	proteins derived from the *mod* operon of no known function
ModE	molybdenum responsive transcriptional regulator
ModG	homotrimeric cytoplasmic molybdate-binding protein made up of 14-kDa subunits
Mop domain/unit	7-kDa Mop-like unit
Mop	homohexameric cytoplasmic molybdate-binding protein made up of 7-kDa subunits
MopA/MopB	homologues of ModE
nar	a nitrate reductase gene
nif	a molybdenum nitrogenase gene
OB-fold	oligonucleotide/oligosaccharide-binding fold
PDB	Protein Data Bank
tup	a tungstate uptake gene
TupA	periplasmic tungstate-binding protein

REFERENCES

1. A. Kletzin and M. W. W. Adams, FEMS Microbiol. Rev., *18*, 5–63 (1996).

2. R. R. Eady in *Vanadium and Its Role in Life* (H. Sigel and A. Sigel, eds.), Vol. 31 of *Metal Ions in Biological Systems*, Marcel Dekker, New York, 1995, pp. 363–405.

3. D. M. Paulsen, H. W. Paerl, and P. E. Bishop, *Limnol. Oceanog.*, *36*, 1325–1334 (1991).

4. R. H. Maynard, R. Premakumar, and P. E. Bishop, *J. Bacteriol.*, *176*, 5583–5586 (1994).

5. J. J. R. Frausto da Silva and R. J. P. Williams, *The Biological Chemistry of the Elements*, Clarendon Press, Oxford, 1993.

6. W. R. Hagen and A. F. Arendsen, *Struct. Bond.*, *90*, 161–192 (1998).

7. G. R. Willsky, in *Vanadium in Biological Systems* (N. D. Chasteen, ed.), Kluwer Academic, Amsterdam, 1990, pp. 1–24.

8. U. Schwertman, *Plant Soil*, *130*, 1–25 (1991).

9. A. Bagg and J. B. Neilands, *Microbiol. Rev.*, *51*, 509–518 (1987).

10. W. J. Page and M. Huyer, *J. Bacteriol.*, *158*, 496–502 (1984).

11. J. L. Corbin, I. L. Karle, and J. Karle, *Chem. Commun.*, 186–187 (1970).

12. P. Demange, A. Bateman, A. Dell, and M. A. Abdallah, *Biochemistry*, *27*, 2745–2752 (1988).

13. W. J. Page, K. S. Collinson, P. Demange, A. Dell, and M. A. Abdallah, *BioMetals*, *4*, 217–222 (1991).

14. J. L. Corbin and W. A. Bulen, *Biochemistry*, *27*, 2745–2752 (1969).

15. W. J. Page and M. von Tigerstrom, *J. Gen. Microbiol.*, *134*, 453–460 (1988).

16. R. C. Scarrow, D. L. White, and K. N. Raymond, *J. Am. Chem. Soc.*, *107*, 654 (1985).

17. E. J. Enemark and T. D. P. Stack, *Angew. Chem.*, *34* 996–998 (1995).

18. A.-K. Duhme, R. C. Hider, and H. Khodr, *BioMetals*, *9*, 245–248 (1996).

19. A.-K. Duhme, Z. Dauter, R. C. Hider, and S. Pohl, *Inorg. Chem.*, *1996*, 3059–3061 (1996).

20. A. S. Cornish and W. J. Page, *Appl. Environ. Microbiol.*, *66*, 1580–1586 (2000).

21. A.-K. Duhme, R. C. Hider, M. J. Naldrett, and R. N. Pau, *J. Biol. Inorg. Chem.*, *3*, 520–526 (1998).

22. R. C. Scarrow, D. J. Ecker, C. Ng, S. Liu, and K. N. Raymond, *J. Am. Chem. Soc.*, *101*, 6097–6104 (1991).

23. A. S. Cornish and W. J. Page, *BioMetals*, *8*, 332–338 (1995).

24. A. K. Duhme, R. C. Hider, and H. H. Khodr, *Chem. Ber.*, *130*, 969–973 (1997).

25. W. J. Page and M. von Tigerstrom, *J. Bacteriol.*, *151*, 237–242 (1982).

26. T. R. Hoover, A. D. Robertson, R. L. Cerny, R. N. Hayes, V. K. Imperial, V. K. Shah, and P. W. Ludden, *Nature*, *329*, 855–857 (1987).

27. M. S. Madden, T. D. Paustian, P. W. Ludden, and V. K. Shah, *J. Bacteriol.*, *173*, 5403–5405 (1991).

28. R. W. Howarth and J. J. Cole, *Science*, *229*, 653–655 (1985).

29. D. Scott and N. K. Amy, *J. Bacteriol.*, *171*, 1284–1287 (1989).

30. K. Schneider, A. Müller, K.-U. Johannes, E. Diemann, and J. Kottmann, *Anal. Biochem.*, *193*, 292–298 (1991).

31. P. T. Pienkos and W. T. Brill, *J. Bacteriol.*, *145*, 743–751 (1981).

32. G. L. Corcuera, M. Bastidas, and M. Dubourdieu, *J. Gen. Microbiol.*, *139*, 1869–1875 (1993).

33. J. Imperial, R. A. Ugalde, V. K. Shah, and W. J. Brill, *J. Bacteriol.*, *163*, 1285–1287 (1985).

34. B. B. Elliot and L. E. Mortenson, *J. Bacteriol.*, *124*, 1295–1301 (1975).

35. J. H. Glaser and J. A. DeMoss, *J. Bacteriol.*, *108*, 854–860 (1971).

36. G. T. Sperl and J. A. DeMoss, *J. Bacteriol.*, *122*, 1230–1238 (1975).

37. M. Debourdieu, E. Andrade, and J. Puig, *Biochem. Biophys. Res. Commun.*, *70*, 766–773 (1976).

38. S. Johann and S. M. Hinton, *J. Bacteriol.*, *169*, 1911–1916 (1987).

39. R. J. Maier, L. Graham, R. G. Keefe, T. Pihl, and E. Smith, *J. Bacteriol.*, *169*, 2548–2554 (1987).

40. S. Hemschemeier, M. Grund, B. Keuntje, and R. Eichenlaub, *J. Bacteriol.*, *173*, 6499–6506 (1991).

41. J. A. Maupin-Furlow, J. K. Rosentel, J. H. Lee, U. Deppenmeier, R. P. Gunsalus, and K. T. Shanmugam, *J. Bacteriol.*, *177*, 4851–4856 (1995).

42. S. Rech, U. Deppenmeier, and R. P. Gunsalus, *J. Bacteriol.*, *177*, 1023–1029 (1995).

43. F. Luque, L. A. Mitchenall, M. Chapman, R. Christine, and R. N. Pau, *Mol. Microbiol.*, *7*, 447–459 (1993).

44. G. Wang, S. Angermüller, and W. Klipp, *J. Bacteriol.*, *175*, 3031–3042 (1993).

45. C. Kennedy and J. R. Postgate, *J. Gen. Microbiol.*, *98*, 551–557 (1977).

46. T. Shanmugam, V. Stewart, R. P. Gunsalus, D. H. Boxer, J. A. Cole, and M. Chippaux, *Mol. Microbiol.*, *22*, 3452–3454 (1992).

47. W. Boos and J. M. Lucht, in *Escherichia coli and Salmonella typhimurium*, Vol. 1 (F. C. Neidhardt, ed.), ASM Press, Washington, DC, 1996, pp. 1175–1209.

48. N. J. Mouncey, L. A. Mitchenall, and R. N. Pau, *J. Bacteriol.*, *177*, 5294–5302 (1995).

49. H. M. Walkenhorst, S. K. Hemschemeier, and R. Eichenlaub, *Microbiol. Res.*, *150*, 347–361 (1995).

50. J. Chen, S. Sharma, F. A. Quiocho, and A. L. Davidson, *Proc. Natl. Acad. Sci. USA*, *98*, 1525–1530 (2001).

51. R. D. Fleishmann, M. D. Adams, O. White, R. A. Clayton, E. F. Kirkness, and A. R. Kerlavge, *Science*, *169*, 496–512 (1995).

52. D. M. Lawson, C. E. Williams, L. A. Mitchenall, and R. N. Pau, *Structure*, *6*, 1529–1539 (1998).

53. A. M. Grunden, R. M. Ray, J. K. Rosentel, F. G. Healy, and K. T. Shanmugam, *J. Bacteriol.*, *178*, 735–744 (1996).

54. M. R. Jacobson, K. E. Brigle, L. T. Bennett, R. A. Setterquist, M. S. Wilson, V. L. Cash, J. Beynon, W. E. Newton, and D. R. Dean, *J. Bacteriol.*, *171*, 1017–1027 (1989).

55. S. Z. Wang, J. S. Chen, and J. L. Johnson, *Biochem. Biophys. Res. Commun.*, *169*, 1122–1128 (1990).

56. A. Llamas, K. L. Kalakoutskii, and E. Fernández, *Plant Cell Environ.*, *23*, 1247–1255 (2000).

57. S. Rech, C. Wolin, and R. P. Gunsalus, *J. Biol. Chem.*, *271*, 2557–2562 (1996).

58. H. Neubauer, I. Pantel, P. E. Lindgren, and F. Gotz, *Arch. Microbiol.*, *172*, 109–115 (1999).

59. J. Imperial, M. Hadi, and N. K. Amy, *Biochim. Biophys. Acta*, *1370*, 337–346 (1998).

60. F. A. Quiocho and P. S. Ledvinda, *Mol. Microbiol.*, *20*, 17–25 (1996).

61. Y. Hu, S. Rech, R. P. Gunsalus, and D. C. Rees, *Nature Struct. Biol.*, *4*, 703–707 (1997).

62. J. W. Pflugarth and F. A. Quiocho, *Nature*, *314*, 257–260 (1985).

63. H. Luecke and F. A. Quiocho, *Nature*, *347*, 402–406 (1990).

64. P. S. Ledvina, A. L. Tsai, Z. Wang, E. Koehl, and F. A. Quiocho, *Protein Sci.*, *7*, 2550–2559 (1998).

65. P. S. Ledvina, N. Yao, A. Choudhary, and F. A. Quiocho, *Proc. Natl. Acad. Sci. USA*, *93*, 6786–6791 (1996).

66. E. Dassa, W. Saurin, and W. Koster, *Mol. Microbiol.*, *12*, 993–1004 (1994).

67. M. Mourez, M. Hofnung, and E. Dassa, *EMBO J.*, *16*, 3066–3077 (1997).

68. W. Saurin and E. Dassa, *Protein Sci.*, *3*, 325–344 (1994).

69. A. Hochheimer, R. Hedderich, and R. K. Thauer, *Arch. Microbiol.*, *170*, 389–393 (1998).

70. K. Makdressi, J. R. Andreesen, and A. Pich, *J. Biol. Chem. 276*, 24557–24564 (2001).

71. J. H. Lee, J. C. Wendt, and K. T. Shanmugam, *J. Bacteriol.*, *172*, 2079–2087 (1990).

72. J. K. Rosentel, F. Healy, J. A. Maupin-Furlow, J. H. Lee, and K. T. Shanmugam, *J. Bacteriol.*, *177*, 4857–4864 (1995).

73. M. Grund, Ph.D. thesis, University of Bielefeld, Germany, 1994.

74. R. Premakumar, S. Jacobitz, S. C. Ricke, and P. E. Bishop, *J. Bacteriol.*, *178*, 691–696 (1996).

75. B. B. Elliott and L. E. Mortenson, *J. Bacteriol.*, *127*, 770-779 (1976).

76. S. M. Hinton and L. E. Mortenson, *J. Bacteriol.*, *162*, 477–484 (1985).

77. S. M. Hinton and L. E. Mortenson, *J. Bacteriol.*, *162*, 485–493 (1985).

78. S. M. Hinton and B. Merritt, *J. Bacteriol.*, *168,* 688–693 (1986).

79. S. M. Hinton, C. Slaughter, W. Eisner, and T. Fisher, *Gene*, *54*, 211–219 (1987).

80. D. M. Lawson, C. E. Williams, D. J. White, A. P. Choay, L. A. Mitchenall, and R. N. Pau, *J. Chem. Soc., Dalton Trans.*, 3981–3984 (1997).

81. A. K. Duhme, W. Meyer-Klaucke, D. J. White, L. Delarbre, L. A. Mitchenall, and R. N. Pau, *J. Biol. Inorg. Chem.*, *4*, 588–592 (1999).

82. U. G. Wagner, E. Stupperich, and C. Kratky, *Structure*, *8*, 1127–1136 (2000).

83. L. Delarbre, C. E. M. Stevenson, D. J. White, L. A. Mitchenall, R. N. Pau, and D. M. Lawson, *J. Mol. Biol.*, *308*, 1063–1079 (2001).

84. A. G. Murzin, *EMBO J.*, *12*, 861–867 (1993).

85. J. B. Miller, D. J. Scott, and N. K. Amy, *J. Bacteriol.*, *169*, 1853–1860 (1987).

86. N. J. Mouncey, L. A. Mitchenall, and R. N. Pau, *Microbiology*, *142*, 1997–2004 (1996).

87. P. S. Solomon, A. L. Shaw, M. D. Young, S. Leimkühler, G. R. Hanson, W. Klipp, and A. G. McEwan, *FEMS Microbiol. Lett.*, *190*, 203–208 (2000).

88. M. Kutsche, S. Leimkühler, and W. Klipp, *J. Bacteriol.*, *178*, 2010-2017 (1996).

89. P. M. McNicholas, R. C. Chiang, and R. P. Gunsalus, *Mol. Microbiol.*, *27*, 197–208 (1998).

90. L. A. Anderson, E. McNairn, T. Leubke, R. N. Pau, and D. H. Boxer, *J. Bacteriol.*, *182*, 7035–7043 (2000).

91. W. T. Self, A. M. Grunden, A. Hasona, and K. T. Shanmugam, *Microbiology*, *145*, 41–55 (1999).

92. A. Hasona, W. T. Self, R. M. Ray, and K. T. Shanmugam, *FEMS Microbiol. Lett.*, *169*, 111–116 (1998).

93. W. T. Self and K. T. Shanmugam, *FEMS Microbiol. Lett.*, *184*, 47–52 (2000).

94. L. A. Anderson, T. Palmer, N. C. Price, S. Bornemann, D. H. Boxer, and R. N. Pau, *Eur. J. Biochem.*, *246*, 119–126 (1997).

95. P. M. McNicholas, S. A. Rech, and R. P. Gunsalus, *Mol. Microbiol.*, *23*, 515–524 (1997).

96. A. M. Grunden, W. T. Self, M. Villain, J. E. Blalock, and K. T. Shanmugam, *J. Biol. Chem.*, *274*, 24308–24315 (1999).

97. D. R. Hall, D. G. Gourley, G. A. Leonard, E. M. Duke, L. A. Anderson, D. H. Boxer, and W. N. Hunter, *EMBO J.*, *18*, 1435–1446 (1999).

98. D. G. Gourley, A. W. Schuttelkopf, L. A. Anderson, N. C. Price, D. H. Boxer, and W. N. Hunter, *J. Biol. Chem.*, *276*, 20641–20647 (2001).

99. W. Boos and A. Böhm, *Trends Genet.*, *16*, 404–409 (2000).

100. R. M. Allen, J. T. Roll, P. Rangaraj, V. K. Shah, G. P. Roberts, and P. W. Ludden, *J. Biol. Chem.*, *274*, 15869–15874 (1999).

101. C. P. Witte, M. I. Igeno, R. Mendel, G. Schwarz, and E. Fernandez, *FEBS Lett.*, *431*, 205–209 (1998).

102. R. J. Hoare, P. M. Harrison, and T. G. Hoy, *Nature*, *255*, 653–654 (1975).

103. D. M. Lawson, P. J. Artymiuk, S. J. Yewdall, J. M. A. Smith, J. C. Livingstone, A. Treffry, A. Luzzago, S. Levi, P. Arosio, G. Cesareni, C. D. Thomas, W. V. Shaw, and P. M. Harrison, *Nature*, *349*, 541–544 (1991).

104. T. G. Adelman, P. Arosio, and J. W. Drysdale, *Biochem. Biophys. Res. Commun.*, *63*, 1056–1062 (1975).

105. D. Boyd, C. Vecoli, D. M. Belcher, S. K. Jain, and J. W. Drysdale, *J. Biol. Chem.*, *260*, 11755–11761 (1985).

106. A. Bomford, C. Conlon-Hollingshead, and H. N. Munro, *J. Biol. Chem.*, *256*, 948–955 (1981).

107. T. V. O'Halloran and V. C. Culotta, *J. Biol. Chem.*, *275*, 25057–25060 (2000).

108. P. J. Kraulis, *J. Appl. Crystallogr.*, *24*, 946–950 (1991).

109. E. A. Merritt and D. J. Bacon, *Meth. Enzymol.*, *277*, 505–524 (1997).

110. G. J. Barton, *Protein Eng.*, *6*, 37–40 (1993).

3
Molybdenum Nitrogenases: A Crystallographic and Mechanistic View

David M. Lawson and Barry E. Smith

Department of Biological Chemistry, John Innes Centre,
Norwich NR4 7UH, UK

1. INTRODUCTION

1.1. Nitrogen Fixation

If a plant is healthy and has enough water, then in most agricultural soils its growth is limited by the supply of nitrogen. In many agriculture systems the nitrogen available in soils is supplemented by nitrogenous fertilizer generated by the industrial Haber-Bosch process through the catalytic hydrogenation of atmospheric N_2. Worldwide about 90 million tonnes of N per year is supplied to agriculture in this way, and without it widespread food shortages would undoubtedly occur. However, an additional 90 million tonnes of N is provided to worldwide agriculture through the process of biological nitrogen fixation. This process is mediated solely by bacteria, usually in symbiotic association with plants, which are mainly legumes, but the fern *Azolla* and some shrubs and trees are also important.

The Haber-Bosch process uses high temperatures and pressures, typically 350°C and 200–300 atm, and is dependent on fossil fuels, generating large quantities of CO_2 as a byproduct. In contrast, the biological process functions at field temperatures and 0.8 atm N_2 pressure. Both processes use the abundant atmospheric N_2 as feedstock and both have NH_3 as primary product but bacteria use nitrogenases, the complex metalloenzymes that are the subject of this chapter, to carry out this important and apparently difficult chemistry under very mild conditions.

2. THE NITROGENASES

2.1. Conventional Molybdenum Nitrogenase

The most well-characterized form of nitrogenase is undoubtedly the conventional molybdenum nitrogenase. Indeed, until the mid-1980s it was widely thought to be the only form of the enzyme (but see Sec. 2.2). Most work has been done on the enzymes from the free-living bacteria *Azotobacter vinelandii*, *Clostridium pasteurianum*, and *Klebsiella pneumoniae*, but enough is known of the enzyme from other species, including the symbiotic, agriculturally important *Rhizobia*, to know that the following description is common to all.

The enzyme consists of two oxygen-sensitive metallosulfur proteins, i.e., the molybdenum-iron (MoFe) protein or Component 1 and the iron (Fe) protein or Component 2, both of which are essential for activity (for recent reviews, see [1–3]). The nitrogenase proteins from various organisms are designated by a capital letter to indicate the genus, a lower case letter to indicate the species, and a number to indicate the component; thus, Av1 is the MoFe protein from *A. vinelandii* and Kp2 is the Fe protein from *K. pneumoniae*, etc. For consistency, an amino acid sequence numbering scheme relative to *Azotobacter* proteins will be adopted throughout.

In addition to N_2 the enzyme can reduce a range of other small substrates, notably the proton to H_2, and acetylene to ethylene. Mo nitrogenase proteins are encoded by *nif* genes of which there are 20 arranged contiguously in 7 operons on the chromosome of *K. pneumoniae*. Only 3 genes encode the structural polypeptides of Kp1 and Kp2; the others encode proteins for electron transfer, regulation, and biosynthesis (see Chapter 5). The Fe and MoFe proteins have been characterized by a wide range of spectroscopies and their 3-D X-ray crystallographic structures have been solved (see Secs 3 and 4). The Fe protein is a homodimer (encoded by *nifH*) of M_r ~60,000 Da, with an Fe_4S_4 cluster bound between the subunits (see Sec. 4). There is a remarkable degree of sequence homology (~90%) between the Mo nitrogenase Fe proteins from a wide range of bacteria. Early work [4–7] demonstrated that the Fe protein acts as a very specific, MgATP-activated, electron donor to the MoFe protein in a reaction where at least two MgATP molecules are hydrolyzed to MgADP and P_i for every electron transferred to substrate (Fig. 1). The intricacy of the interactions between the MoFe and Fe proteins and the stability of the complex formed between them during turnover (see Sec. 6) suggests that the Fe protein may not simply be an electron donor but could have additional, as yet unrecognized, roles in the enzymic reaction.

The MoFe protein is an $\alpha_2\beta_2$ tetramer (encoded by *nif K* and *D)* of M_r ~ 220,000 Da containing two each of two types of unique metallosulfur clusters, i.e., the Fe_8S_7 P clusters and the $MoFe_7S_9$·homocitrate iron-molybdenum cofactor (FeMoco) centers. The P clusters are bound between the α and β subunits and the FeMoco centers are bound in the α subunit. The structures (Sec. 3) and functions (Secs 4 and 7) of these

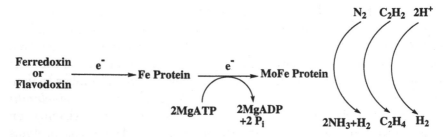

FIG. 1.   The electron transfer path through molybdenum nitrogenase.

clusters will be discussed below. Neither of them has been found elsewhere in biology nor have they been successfully synthesized chemically.

2.2. Alternative Nitrogenases

In addition to the Mo nitrogenase three alternative forms of the enzyme
are known: two are very similar to Mo nitrogenase with the Mo replaced
by V or Fe and encoded by *vnf* or *anf* genes, respectively. These enzymes
have MoFe protein equivalents in the VFe and FeFe proteins and also Fe
proteins encoded by *vnfH* and *anfH* genes which show considerable
homology to Fe proteins from the Mo enzyme. The third alternative
nitrogenase is completely different and seems to be more closely related
to the nonnitrogenase molybdenum enzymes with a probable Mo pterin
active site rather than one similar to FeMoco.

2.2.1. *Vanadium Nitrogenases*

The existence of alternative nitrogenases was initially controversial but
was established beyond doubt when strains of *A. vinelandii*, deleted for
the Mo nitrogenase structural genes *nifKD* and *H*, were shown to fix
nitrogen [8,9]. Furthermore, growth was enhanced by adding V. Genetic
analysis revealed that the V nitrogenase in *Azotobacter* is encoded by
different structural genes from the Mo enzyme but shares some of the
biosynthetic genes [10,11]. The VFe protein has an additional small
subunit that is essential for activity and is encoded by the *vnfDK* and
G genes [12,13]. No X-ray crystallographic data are available but the V
EXAFS spectra revealed that the environment of the V atom is very
similar to that of the Mo in FeMoco [14,15]. Fe EXAFS of the extracted
cofactor, FeVaco, showed long-range Fe-Fe interactions consistent with
those seen with FeMoco [16,17]. Mössbauer [18] and electron paramagnetic resonance (EPR) [19] data are fully consistent with the presence of
P clusters, and the VnfD and K amino acid sequences are closely homologous to those of the NifD and K proteins [10,11]. These data make it
probable that the VFe protein has a structure very similar to the MoFe
protein. However, the substrate-reducing profiles of the two enzymes
differ in detail. Specifically, V nitrogenase fixes N_2 less efficiently than
the Mo enzyme at 30°C although this relationship is reversed at low

temperatures [13]. In contrast to the Mo enzyme a small amount of hydrazine is also a product of N_2 reduction [20]. In addition, the V enzyme reduces acetylene to ethylene and a minor product, ethane [13]. FeVaco, extracted from the VFe protein, activated the MoFe protein polypeptides from a *nifB* mutant which is unable to biosynthesize FeMoco [21]. The resultant hybrid enzyme could reduce acetylene to ethylene and ethane but was unable to reduce N_2. These data imply that the chemistry of acetylene reduction by the two enzymes is largely defined by the respective cofactors but that the reduction of N_2 depends also on important interactions with the polypeptides (see also Sec. 7).

2.2.2.   *Iron-Only Nitrogenases*

There is clear genetic evidence for a third nitrogenase in *A. vinelandii* [22]. The enzyme has specific genes encoding the structural polypeptides, *anfHDG* and *K*, but relies on the Mo or V systems for biosynthetic genes. Early preparations of the FeFe protein (an $(\alpha_2\beta_2\gamma_2$ hexamer) had relatively   low   activity,   but   more   recent   preparations   from *Rhodospirillum rubrum* [23] and *Rhodobacter capsulatus* [24] have activities comparable to the VFe protein and contained essentially no Mo or V but only substantial levels of Fe. Substrate-reducing activities were similar to those of V nitrogenase. Some ethane as well as ethylene was produced from acetylene, which was a relatively poor substrate. N_2 reduction was also poor, with much of the electron flux going into H_2 evolution.

2.2.3.   Streptomyces thermoautotrophicus *Nitrogenase*

Streptomyces thermoautotrophicus was isolated from the soil covering burning charcoal piles and uses gases as sources of energy [25]. With CO, or H_2 with CO_2, as growth substrates the organism can fix N_2 aerobically in a reaction that appears to require less ATP than other nitrogenases. Electrons for the reaction are provided by oxidation of CO by an Mo-containing CO dehydrogenase that reduces O_2 to superoxide anion radicals, which act as the enzyme reductant. Thus, the nitrogenase is not damaged by O_2. It consists of two proteins, ST1 and ST2. The former is a heterotrimer containing Mo, Fe and S^{2-}, and the latter is a homodimer containing substoichiometric levels of Mn and Zn. The authors concluded that ST2 is a superoxide oxidoreductase and suggested that ST1 might contain an Mo-pterin active site. The N_2 reduction activity of the enzyme, as prepared, was very low and it did not reduce acetylene or ethylene.

3. THE MOLYBDENUM-IRON PROTEIN

3.1. Protein Structure

The crystal structures of Av1 [26] and Cp1 [27] were solved independently almost a decade ago at resolutions of 2.7 and 2.2 Å, respectively, although the resolution of the former was subsequently extended to 2.05 Å [28]. More recently, the Kp1 structure was also determined at 1.6 Å resolution [29]. All three models are closely superposable although the agreement between Av1 and Kp1 is closest. The $\alpha\beta$ halves of the molecule are arranged around a noncrystallographic twofold axis that coincides with a long, solvent-filled channel passing through the center of the tetramer, which is around 8 Å across at its narrowest point (Fig. 2). To date, no function has been ascribed to this channel. The $\alpha\beta$ halves are held together largely by interactions between neighboring β subunits, although there are some interactions of the type α/β. In all structures reported to date a cation, Ca^{2+} or Mg^{2+}, which appears to have a purely structural role, is found at each β/β interface. In Av1 and Kp1, both subunits are around 500 residues in length, have similar sequences, and share roughly the same fold. Indeed, the subunits within each $\alpha\beta$ half are related to one another by pseudo-twofold symmetry. However, the pseudo-twofold axes for each $\alpha\beta$ half are not colinear but roughly parallel, passing either side of the noncrystallographic twofold axis, and thus the tetramer does not exhibit 222 symmetry. The subunits can be subdivided into three domains each comprising a central parallel β sheet sandwiched between α helices. In the α subunit, an elongated pocket, which is occupied by the FeMoco, is formed at the interface of the three domains. The equivalent region in the β subunit is filled with bulky aromatic side chains. The P cluster is bound pseudosymmetrically between α and β subunits.

A nonglycine *cis*-peptide was observed between Trp253 and Ser254 (Av1 numbering) in the α subunit of Kp1 within 10 Å of the cofactor. It

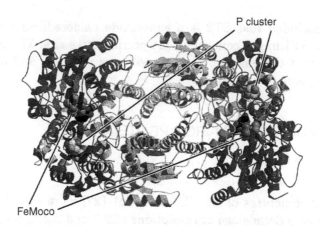

FIG. 2. Structure of the tetrameric MoFe protein of *Klebsiella pneumoniae* from Protein Data Bank (PDB) entry 1QGU. The polypeptide chain is displayed in ribbon representation, with the α subunits in darker grey, and the metal-sulfur clusters depicted as space-filling models. This and all subsequent structural figures were produced using the programs MOLSCRIPT [131] and Raster3D [132].

was speculated that this may undergo cis/trans isomerization and thereby act as a conformational switch, perhaps gating proton and/or electron transfer [29]. However, this *cis*-peptide has not been reported elsewhere although the deposited data are consistent with its presence.

A description of the extensive analysis by EXAFS spectroscopy of the metal-sulfur clusters is beyond the scope of this chapter and moreover has been recently reviewed [3,29]. However, it is sufficient to state that the interatomic distances obtained from these studies are generally consistent with the current models derived from crystallography that are described below.

3.2. P Cluster

The P cluster is found at the interface between the α and β subunits. It is located 14 Å from the FeMoco and 10 Å below the surface of the protein, which places it in a favorable position to facilitate electron transfer to the FeMoco [28,30], a contention that is also supported by kinetic, mutagenic, and spectroscopic data [31–33]. In the original structure at 2.7 Å resolution, it was proposed that this was an Fe_8S_8 cluster consisting of two Fe_4S_4 clusters bridged by two cysteine thiol ligands, Cysα88 and Cysβ95, and a long disulfide bond between two of the cluster sulfurs [26]. A further four cysteine residues, two from each subunit, serve as additional ligands to the cluster (Cysα62, Cysα154, Cysβ70, and Cysβ153). These cysteine ligands had been correctly identified by mutagenesis studies [34,35]. An additional residue, Serβ188, was also implicated as a cluster ligand.

Shortly afterward a slightly different model was proposed for the Cp1 P cluster structure at 2.2 Å resolution [27]. This structure was an Fe_8S_7 cluster, having the appearance of two Fe_4S_4 cubanes fused at one corner, giving rise to a central sulfur atom with distorted octahedral coordination to six iron atoms. This model was subsequently ratified by the higher resolution structures of the dithionite-reduced states of Av1 [28] and Kp1 [29] and is designated P^N (see Sec. 3.4 and Fig. 3a). However, upon oxidation through the loss of two electrons, one half of the P cluster adopts a more open structure; two of the iron atoms lose their coordination to the central sulfur and form new interactions with the protein, one with the main-chain nitrogen of Cysα88 and the other with the side chain of Serβ188 to give the P^{OX} structure (Fig. 3b). In the case of Av1, the oxidation states of the crystals were also confirmed by single-crystal EPR spectroscopy [28]. It seems likely that the original Fe_8S_8 model was a misinterpretation of low-resolution data collected from crystals of mixed oxidation state. Indeed, in the 1.6-Å resolution study of Kp1, such a "mixed" crystal was analyzed. Fortunately, at this resolution it is possible to resolve the individual atoms of the cluster, with the "mobile" iron atoms being associated with elongated electron density peaks [29]. The structural changes observed around the P cluster as a result of cycling between reduced and oxidized states imply that one or more protein groups are becoming alternately protonated and deprotonated [28]. Since dinitrogen reduction requires both electrons and protons this may provide a mechanism whereby the transfer of these two substrates is coupled through the P cluster.

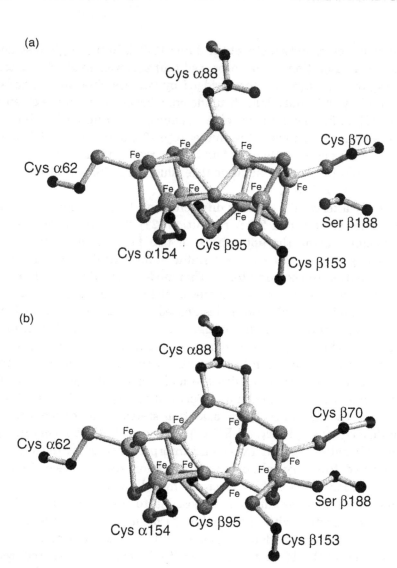

FIG. 3.   Structure of the P cluster from Kp1 shown with coordinating residues.
(a) The dithionite-reduced P^N state (PDB entry 1QGU). (b) The two electron
oxidized P^{OX} state (PDB entry 1QH1). The atomic radii increase in the order
C/N/O < S < Fe and the atoms are coloured C = black, S/N/O = dark grey,
Fe = light grey.

3.3.   Iron-Molybdenum Cofactor

The iron-molybdenum cofactor, or FeMoco, is contained entirely within the α subunit of the MoFe protein, has the empirical formula $MoFe_7S_9$·homocitrate, and is now generally accepted as the site of substrate reduction (see Sec. 7.1). The metal sulfur core of FeMoco is constructed of two partial cubanes of $MoFe_3S_3$ and Fe_4S_3 that are joined by three bridging sulfurs (Fig. 4). Although in the initial Av1 model the identity of one of these bridging atoms was in doubt [26], in the light of higher resolution structures all have been confirmed as sulfurs [28,29]. The cofactor has only two covalent linkages to the protein, through the Mo at one end to the $N^{\delta 1}$ of Hisα442, and through an iron at the opposite end to the thiol group of Cysα275. These two residues had previously been identified as probable ligands through directed mutagenesis [36] or analysis of Nif$^-$ mutants [37]. The essential organic moiety (R)-homocitrate is also coordinated to the Mo atom through its 2-hydroxy and 2-carboxyl groups [38]. Thus, the molybdenum is octahedrally coordinated by three sulfurs from the cofactor itself, a hydroxyl and carboxyl from the homocitrate and an amide from the histidine (Fig. 4).

There are approximately 30 buried water molecules, many of which are structurally conserved between Av1 and Kp1, adjacent to

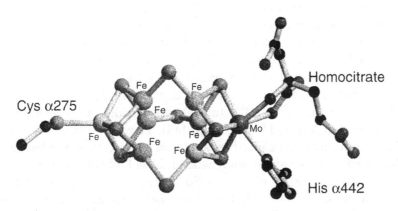

FIG. 4.   Structure of the FeMoco from Kp1. Color scheme as for Fig. 3, with the single Mo in dark grey and labeled.

the homocitrate and almost on a direct path between the P cluster and the FeMoco. It is possible that they form an efficient network for electron tunneling between the two metal centers [39], and may also provide a source of protons for substrate reduction. By contrast, the environment at the cysteine-ligated end of the FeMoco is virtually free of solvent. The cofactor itself is surrounded by a number of hydrophobic residues, including $Val\alpha70$, $Ile\alpha231$, $Leu\alpha358$, and $Phe\alpha381$, whilst hydrogen bonds between sulfur atoms of the FeMoco in Av1 and the side chains of $Arg\alpha96$, $His\alpha195$, $Arg\alpha359$, and the backbone amide groups from $Gly\alpha356$ and $Gly\alpha357$ have been indicated (see Sec. 7.4.1 and Fig. 11). These electrostatic interactions could have a role in the stabilization of negatively charged intermediates generated during substrate reduction [30].

3.4. Redox States of the MoFe Protein

The redox chemistry of the MoFe protein is complex because spectroscopic signals arise from both the FeMoco and the P cluster metal centers. In the dithionite-reduced state, FeMoco is paramagnetic, exhibiting an $S = 3/2$ EPR signal with g values near 4.3, 3.7, and 2, whereas Mössbauer and EPR spectroscopy show the P cluster to be a diamagnetic, all-ferrous cluster (P^N), with a spin state of $S = 0$ [4,7,40]. On oxidation, the P cluster is oxidized first at −340 mV [41] by two electrons to an $S = 3$ EPR spin state [42–44]. Careful EPR studies using parallel and perpendicular mode EPR techniques identified an intermediate one-electron oxidation of the P cluster to states that exhibited $S = 5/2$ and $S = 1/2$ spin signals [44]. Further oxidation of the protein results in the removal of an electron from the FeMoco centers at a potential that is both pH and species dependent. At pH 8, the E_m of the FeMoco for the $M^N \rightarrow M^{OX}$ (one-electron oxidation) is −180 mV in Kp1, −47 mV in Av1 and 0 mV in Cp1 [45]. To reach this M^{OX} state the MoFe protein must be oxidized by a total of six electrons, rendering the FeMoco EPR inactive and diamagnetic and the P clusters paramagnetic [44]. Further oxidation of the MoFe protein to an E_m of +90 mV removes at least one more electron from each P cluster (and eight electrons total from the MoFe protein) and results in a P^{OX2} state associated with $S = 1/2$ and $S = 7/2$ EPR signals [44]. Oxidation of Kp1 above

+200 mV results in oxidative damage [41] and with Av1 oxidation above +100 mV results in an unstable $S = 9/2$ EPR signal [46]. Of these redox states, only that associated with the initial oxidation of the P clusters is likely to be involved in enzyme turnover.

4. THE IRON PROTEIN

4.1. Protein Structure

The highest resolution crystal structure for an iron protein is that of Cp2 at 1.93 Å resolution [47]; however, we shall restrict our discussion to the Av2 structure as this has been the most extensively studied by crystallography. The latter has recently been determined to a resolution of 2.15 Å with MgADP bound [48]. It comprises a homodimer of ~30 kDa subunits that are bridged by a solvent-exposed Fe_4S_4 cluster (Fig. 5). It is the obligate electron donor to the MoFe protein, a process that requires the consumption of MgATP. In line with its function as the ATPase, each subunit of the iron protein has a nucleotide binding site and these sites lie some 18 Å apart in the dimer. The protein is folded into a central eight-stranded β sheet flanked by nine α helices, terminating in an extended chain of approximately 20 residues that wraps around the neighboring subunit. The Fe_4S_4 cluster forms part of the dimer interface, which involves, in addition, numerous van der Waals and polar interactions. These favorable interactions are sufficiently strong to maintain the integrity of the dimer even when the cluster is removed [49]. The nucleotide binding sites lie around 16 Å from the Fe_4S_4 cluster.

4.2. Fe_4S_4 Cluster

The Fe protein contains a single Fe_4S_4 cluster that is symmetrically coordinated to both subunits through the side chains of Cys97 and Cys132. It is perhaps the most solvent exposed of all known Fe_4S_4 clusters, which may give rise to its unique properties whereby it can exist in three different oxidation states [50] (see Sec. 5). Both of the cluster ligands are located near the N termini of α helices that are directed

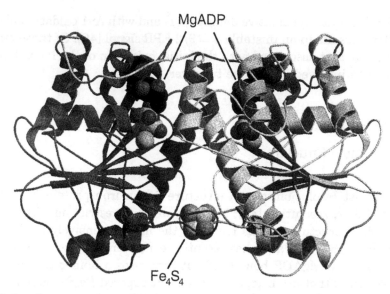

FIG. 5. Structure of the Av2 homodimer (PDB entry 1FP6) in ribbon representation. Dark and light grey shading are used to distinguish between the two
subunits. The two bound nucleotides and the Fe_4S_4 cluster are shown as space-
filling representations.

toward the cluster, permitting favorable electrostatic interactions
between the terminal amide groups of the helices and the anionic cluster.

4.3. Nucleotide Binding Site

In the first iron protein crystal structure to be solved, that of Av2 at
2.9 Å resolution, only a single ADP molecule was observed. This was
bound with low occupancy in a cross-subunit arrangement, with the
adenosine bound to one subunit (partially overlapping one binding
site) and the phosphates to the other [51]. This binding mode is
undoubtedly nonproductive and most likely an artefact. Indeed, the
recently determined 2.15-Å resolution structure of Av2 showed both
nucleotide binding sites fully occupied with MgADP [48]. In this structure, the two nucleotides are close to and aligned roughly parallel to the

dimer interface with the β-phosphates pointing toward the Fe_4S_4 cluster. In addition to having a similar secondary structure, the iron protein has a nucleotide binding site that shares a number of structural features with other nucleotide-binding proteins, such as the G-proteins and P21 *ras* [47]. These are the Walker A motif and two switch regions denoted switches I and II.

The Walker A motif has the sequence GKGGIGKS (where consensus residues are underlined) and corresponds to residues 9–16 in the Fe protein (Fig. 6). It forms a loop between the end of the N-terminal β strand and the beginning of an α helix that wraps around the phosphate groups of the nucleotide, and for this reason is also referred to as the phosphate-binding or P loop. The invariant lysine forms a salt bridge with the β phosphate, whereas the hydroxyl of residue 16 coordinates the Mg^{2+} ion.

Switch II includes residues 125–129, with the sequence DVLGD, and lies just upstream of one of the Fe_4S_4 cluster ligands, Cys132.

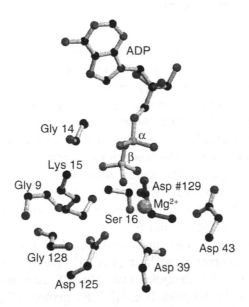

FIG. 6.   The nucleotide binding site of Av2 with MgADP bound (PDB entry 1FP6). Only the consensus residues of the P-loop and those implicated in ATP hydrolysis and signal transduction are displayed. Asp129 is labelled with a hash sign (#) to indicate that it resides in the neighboring subunit.

Asp125 and Gly128 from one subunit and Asp129 from the neighboring subunit interact with either the Mg^{2+} ion or the phosphate groups via water molecules. However, the cross-subunit interaction with Asp129 is only present in the structure of the transition state (see Sec. 6.1). This region provides a direct communication link between the cluster and the nucleotide binding site.

Switch I involves residues 38–44 that include Asp39, the side chain of which is structurally equivalent to a critical main-chain carbonyl oxygen seen in other nucleotide switch proteins [52]. This is suitably placed to activate a water molecule for nucleophilic attack on the γ-phosphate of the nucleotide and is also involved in coordinating the Mg^{2+} ion. Alterations in the coordination of the Mg^{2+} ion during turn-over are transduced via switch I through the peptide backbone to a surface helix involved in docking to the MoFe protein comprising residues 59–69 (see Sec. 6).

5. MECHANISTIC STUDIES

Figure 1 outlines the electron transfer pathway through nitrogenase to substrates and indicates the biological electron donors. However, because of its oxygen sensitivity, nitrogenase is routinely isolated in the presence of dithionite and almost all in vitro experiments on nitrogenase have used dithionite as electron donor. This is the case for the Lowe-Thorneley model for nitrogenase action [53], which for more than 15 years has been used as the basis of descriptions and explanations of the enzyme mechanism. The model was developed from stopped-flow spectrometric data coupled with a series of rapid-quench experiments measuring the time courses of product formation and is described by the two cycles shown in Figs. 7 and 8.

Figure 7 shows the Fe protein cycle in which one electron is transferred from the Fe protein, with an $[Fe_4S_4]^{1+}$ cluster, to the MoFe protein with concomitant hydrolysis of two MgATP molecules. In the model MgATP binds to the reduced $[Fe_4S_4]^{1+}$ Fe protein (F_{red}), which then complexes with the MoFe protein (here M represents only one $\alpha\beta$ dimer of the tetrameric MoFe protein). Rapid protein-protein electron transfer then occurs, leaving a complex with an oxidized $[Fe_4S_4]^{2+}$ Fe

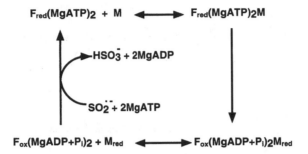

FIG. 7.   The Fe protein cycle of Mo nitrogenase (after Ref. [53]) describing the transfer of a single electron from the Fe protein (F) to one $\alpha\beta$ half of the MoFe protein (M) with the accompanying hydrolysis of MgATP. The rate determining step in the cycle is proposed to be the dissociation of $F_{ox}(MgADP + P_i)_2$ from M_{red}. Subscript red = reduced and ox = oxidized (reproduced with permission from Ref. 3).

protein and a one-electron-reduced MoFe protein. Opinions differ as to whether MgATP hydrolysis follows or coincides with electron transfer, but there is agreement that this overall reaction step is complex and includes partial reactions [2,54,55]. The protein-protein complex then dissociates in what is proposed as the rate-limiting step of the cycle. The cycle is completed by the dissociation of MgADP, reduction of the Fe protein, and its combination with fresh MgATP.

The MoFe protein cycle (Fig. 8), consists of eight sequential one-electron steps, each of which corresponds to an Fe protein cycle [53]. As in Fig. 7, M depicts an $\alpha\beta$ dimer half of the MoFe protein. It is plausibly assumed that each transferred electron is neutralized by addition of a proton. This could result in the formation of hydrides. The chemical reactions shown in Fig. 8 are consistent with the observed stoichiometry of enzymic N_2 reduction:

$$N_2 + 8H^+ + 8e^- \rightarrow 2NH_3 + H_2$$

and with well-established model chemistry of N_2 and N-NH$_2$, etc., complexes that have been fully characterized spectroscopically and crystallographically [56,57]. A transient species, releasing hydrazine (N_2H_4) upon quenching of the enzyme reaction with acid or alkali, has been detected and is consistent with the presence of an intermediate M N-NH$_2$ species as shown [53]. The enzyme reaction has not been charac-

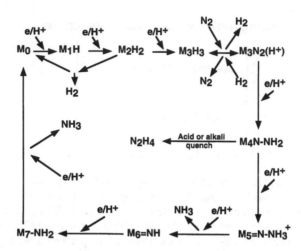

FIG. 8. The MoFe protein cycle of Mo nitrogenase depicting the proposed sequence of events in the reduction of N_2 to $2NH_3 + H_2$ based on well established model chemistry [56,57] and the pre-steady state kinetics of product formation by the enzyme [53]. As in Fig. 7, M represents an $\alpha\beta$ half of the MoFe protein. Subscripts 0-7 indicate the number of electrons transferred to M from the Fe protein through the Fe protein cycle in Fig. 7 (reproduced with permission from Ref. 3).

terized beyond species M_5 since the chemical agents used to quench the reaction would react with the proposed intermediates to yield NH_3. The reactions shown in the two cycles have been computer simulated on the assumption that the rates of reaction in the Fe protein cycle are independent of the redox state of the MoFe protein. This assumption has been experimentally justified for the first two steps in Fig. 8 [58]. Using the computer simulations, Lowe and Thorneley were able to rationalize all of the data then available on H_2 evolution and N_2 reduction and have since been able to extend the model to include acetylene reduction and more recent data. In particular, some partial reactions within Figs. 7 and 8 have been investigated and successfully incorporated into the overall model [55].

However, some recent experimental observations have indicated that the mechanism may be more complex, at least under some conditions. Specifically, it has been shown that some reductants can reduce the

Fe protein [Fe$_4$S$_4$] cluster completely, i.e., to the [Fe$_4$S$_4$]^0 state [59]. Furthermore, when in this state the Fe protein could pass two electrons onto the MoFe protein [60], but the number of ATP molecules consumed stayed the same and so the ATP hydrolyzed/2e$^-$ ratio dropped from ~5 to ~2, i.e., the enzyme became much more energy efficient. This effect was more marked when the enzyme was reducing substrates relatively slowly. To reach the [Fe$_4$S$_4$]^0 state of the Fe protein most workers have used Ti(III) citrate as reductant [59,61], but this state can also be reached with methyl viologen [59], by photoreduction with eosin/NADH [62], or, most importantly, with flavodoxin hydroquinone [54], a natural electron donor, as reductant. It has been suggested that the [Fe$_4$S$_4$]^0 state may be important in whole cells under some physiological conditions. A further departure from the Lowe-Thorneley model is the observation that flavodoxin hydroquinone [54] and the eosin/NADH photoreductant [63] can pass electrons to the oxidized [Fe$_4$S$_4$]$^{2+}$ Fe protein while it is complexed to the MoFe protein. Furthermore, recent work has shown that dissociation of the complex need not occur for MgADP to be replaced by MgATP [54]. These last two reactions appear to bypass the proposed rate-determining step of the enzymic mechanism.

In addition to the above complications, a reexamination of Fe protein-MoFe protein titration curves has indicated that they may be sigmoidal at very low Fe protein levels [64,65]. This is not a novel observation; others had reported similar data on several occasions over the past 30 years, but the observation remains controversial. The authors interpreted their data in terms of two Fe protein molecules interacting with each half of the MoFe protein. This suggestion may be consistent with the observation (above) that it is possible to reduce the oxidized Fe protein while it is complexed with the MoFe protein.

An additional complication is the very recent report that in a complex between Cp2 and Kp1 the two halves of the Kp1 have different binding affinities for Cp2 [66]. The authors proposed a long-range conformational change across the Kp1 to explain their data and further suggested that the data implied a shuttle mechanism for nitrogenase action whereby each half of the MoFe protein reacts in turn.

All of the above observations, if substantiated, must be incorporated into an overall nitrogenase mechanism. However, it is possible that there may be more than one competent reaction scheme, with the predominant scheme depending on the precise conditions. The Lowe-

Thorneley model has been so successful in explaining a wide range of data that it is likely to form the core of any mechanism.

6. THE NITROGENASE COMPLEX

6.1. Transition State Structure

The initial model of the Av1/Av2 complex was produced by simply docking the individual crystal structures together to minimize the separation between the Fe_4S_4 cluster of the Fe protein and the P cluster of the MoFe protein [26]. The resultant structure was essentially consistent with a body of cross-linking and mutagenesis data that implicated specific residues as being at the Av1/Av2 interface [1]. While useful for speculation and correct in outline, the proposed docking model proved to be no substitute for the crystal structure. Through the use of $ADP \cdot AlF_4^-$ as an analogue for the transition state of ATP hydrolysis it is possible to lock the two component proteins together in a 2:1 complex [67,68]. The resultant 360-kDa complex for Av was subsequently crystallized and its structure determined at 3.0 Å resolution [69].

There is one Fe protein dimer (referred to as γ_2) per $\alpha\beta$ half of the MoFe protein, giving an overall subunit composition of $(\alpha\beta\gamma_2)_2$. This arrangement was independently confirmed by small-angle X-ray scattering for the Kp enzyme [70]. The two $\alpha\beta\gamma_2$ halves of the complex are related by a noncrystallographic twofold axis. Moreover, the molecular twofold axis of the Av2 dimer coincides with the pseudo-twofold axis relating the α and β subunits of Av1 (Fig. 9).

Upon complex formation, major conformational changes occur in the Fe protein. In global terms there is a ~13° rotation of the Av2 subunits toward the dimer axis, giving rise to a more compact structure that can make more intimate contact with the surface of the MoFe protein. As an additional consequence, the Fe_4S_4 cluster is projected ~4 Å toward the protein surface, such that in the complex the Fe_4S_4 cluster is only 14 Å distant from the P cluster as opposed to the separation of around 18 Å derived from the simple docking model (Fig. 10). Now the P cluster is roughly equidistant between the Fe_4S_4 cluster of the Fe protein and the FeMoco, thereby consistent with its probable role in mediating electron transfer between these two centers. By contrast,

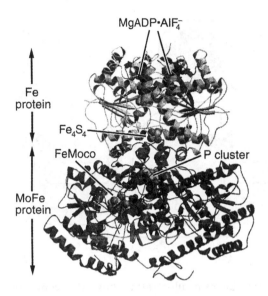

FIG. 9.   Structure of the putative transition state complex that is formed when Fe protein (light grey) is added to the MoFe protein (dark grey) in the presence of MgADP·AlF$_4^-$ (PDB entry 1N2C). In the interests of clarity only one $\alpha\beta$ half of the MoFe protein is depicted.

no large structural changes were noted in the MoFe protein upon formation of the complex.

In the complex, the Fe$_4$S$_4$ cluster becomes completely buried in the Av2/Av1 interface region, where it is in van der Waals contact with loop regions α157–159 and β156–158 of the MoFe protein. Furthermore, the carbonyl oxygens of Ileα123 and Valα124 in helix α5 of the MoFe protein α subunit, and those of their pseudosymmetry-related residues Alaβ123 and Valβ124 on helix α7 of the β subunit, form hydrogen bonds to the main-chain amides of both Cysγ97 residues of the Fe protein. Thus, the Fe$_4$S$_4$ cluster is linked directly to the MoFe protein in this putative transition state complex [69].

Deletion of Leuγ127 in Av2 locks it into a conformation resembling the MgATP-bound state of the Av1/Av2 complex. This mutant forms a nondissociating complex with Av1 that displays a slow but detectable rate of electron transfer in the absence of MgATP [71,72]. Recently, the crystal structures of this complex with and without ATP have been determined [73]. In these structures, the conformations of the Fe pro-

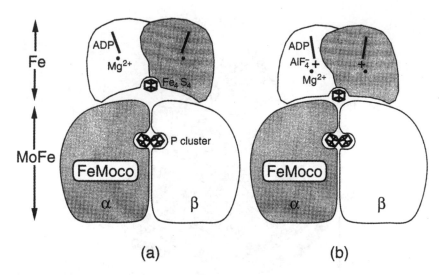

FIG. 10. Schematic diagram illustrating how the conformational change observed in the Fe protein upon formation of the transition state complex allows a closer approach of the Fe_4S_4 cluster to the P cluster of the MoFe protein and thus facilitates electron transfer. (a) Docking model based on independent crystal structures of the two component proteins. (b) Transition state complex.

tein are intermediate between that observed for the free Fe protein and that seen in the $ADP \cdot AlF_4^-$ complex.

6.2. Implications for Turnover

In isolation, the Fe protein exhibits low levels of ATPase activity [2]. It is only when the protein-protein complex is formed that the rate of ATP hydrolysis becomes relevant to the enzyme mechanism since rapid electron transfer within the nitrogenase enzyme is an ATP-dependent process. This rate enhancement is most likely due to the movement of a key residue into the correct position to catalyze MgATP hydrolysis or, alternatively, the stabilization of a Fe protein-nucleotide intermediate that is not attainable by the Fe protein alone [69].

All the crystal structures seen so far for free Fe proteins (i.e., not complexed to MoFe protein) have similar conformations. The binding of MgATP initiates a conformational change in the reduced Fe protein that results in a more compact structure. The conformation seen in the

ADP·AlF$_4^-$ complex may be partially caused by nucleotide binding but must be largely stabilized through protein-protein interactions upon complex formation. The outward projection of the Fe$_4$S$_4$ cluster when MgATP binds to the Fe protein would favor electron transfer from the cluster to the MoFe protein and is concomitant with a decrease in its midpoint potential from -300 mV to -420 mV.

The Ala157Gly mutant of Av2 was impaired in its ability to reduce substrates because of its diminished ability to undergo the MgATP-induced conformational change [74,75]. The substrate reduction rate of this mutant was 20% that of the wild-type protein for hydrogen evolution and ethylene production. Analysis of the crystal structures of Av2, Cp2, and the Av1/Av2 transition state complex indicated that changes in steric interactions were responsible for the slower conformational change. Nevertheless, after adopting the wild-type MgATP-bound conformation, this mutant was still capable of complex formation, electron transfer, and MgATP hydrolysis.

The role of ATP hydrolysis remains the subject of some debate, partly because it is not known whether it follows or coincides with electron transfer. Although both ATP hydrolysis and phosphate release occur more rapidly than turnover [55], it is possible that they initiate complex dissociation and thus determine the lifetime of the complex. Whether or not ATP hydrolysis (and phosphate release) initiates complex dissociation, it seems likely that it is reversion of the Fe protein to the more open conformation seen in the free structures that ultimately causes the component proteins to separate. Then the shape complementarity between the respective docking surfaces is reduced and thereby promotes complex dissociation. Furthermore, opening of the twofold axis facilitates the exchange of MgATP for MgADP in preparation for a subsequent Fe protein cycle [69].

If ATP hydrolysis follows electron transfer, then the conformational change induced may "gate" some process within the MoFe protein, such as electron transfer from the P cluster to the FeMoco. It is clear that protein-protein interactions between the two component proteins are vitally important and that some of the energy derived from the hydrolysis of two MgATP for each electron transferred is likely to be required for some critical step within the MoFe protein. The idea of expending so much energy on MgATP hydrolysis for complex dissociation alone seems improbable.

7. THE ACTIVE SITE

7.1. Evidence for FeMoco as the Active Site

In 1977, Shah and Brill reported the extraction of the FeMoco center from the MoFe protein and showed that it could combine with extracts of certain mutants to generate an active MoFe protein [76]. These observations, though extremely important, did not demonstrate that FeMoco was the enzyme's active site because active enzyme clearly needed the presence of both FeMoco and P clusters. More convincing evidence came from studies on the MoFe protein from *nif V* mutants Nif V⁻ Kp1. These mutants were poor N_2 fixers but reduced protons and acetylene as effectively as the wild type. Critically, their H_2 evolution activity, unlike that of the wild type, was inhibited by CO [77]. In 1984, Hawkes et al. demonstrated that when the FeMoco from Nif V⁻ Kp1 was extracted and combined with extracts of a mutant deficient in FeMoco biosynthesis, the *nif V* mutant phenotype was transferred with the FeMoco to the resultant MoFe protein [78]. Comparable experiments with wild-type FeMoco yielded wild-type MoFe protein. These data showed clearly that FeMoco constitutes at least part of the enzyme's active site. (As will be discussed below in Secs. 7.2 and 7.3, it is evident that the environment provided by the protein is also critical for activity.) The *nif V* gene encodes a homocitrate synthase [79] and in Nif V⁻ Kp1 homocitrate is thought to be replaced by citrate [80]. Alternative substitutions for homocitrate were made possible by the development of an in vitro biosynthetic system for FeMoco (see Chapter 5). All of these substitutions altered the substrate specificity of the resultant enzymes, thus confirming the importance of homocitrate and the fact that FeMoco forms the substrate reduction site [81].

7.2. Substrate and Inhibitor Binding to the MoFe Protein

Early EPR studies [5] noted a pH-dependent change in the g values of the $S = 3/2$ EPR signal from FeMoco in the MoFe protein and that binding acetylene perturbed the pK_a of this change. Nevertheless no sign of a direct interaction between the substrate and FeMoco could be detected. Analysis of the pH dependence of nitrogenase activity indicated that a group with a $pK_a = 6.3$ must be deprotonated and another with a $pK_a = 9$ must be protonated for activity [82]. The latter is consistent with the earlier EPR findings. Several EPR signals have been reported [83] from nitrogenase turning over in the presence of a range of substrates or products. In none of these cases is there yet definitive proof that FeMoco is the site of substrate binding. For example, an EPR signal observed when the enzyme was turning over under the product ethylene was shown, by ^{57}Fe labeling of the FeMoco, to derive from FeMoco, but again no direct bonding of ethylene to FeMoco could be detected by EPR [84].

However, CO, a potent, and noncompetitive inhibitor of N_2 reduction, has been shown to bind to the cluster. During enzyme turnover under CO two distinct EPR signals can be observed, one (lo-CO) at low and the other (hi-CO) at high CO partial pressures [83,85,86]. ^{13}C ENDOR spectroscopy [87,88] facilitated detection of interactions between these signals and ^{13}CO, and ^{57}Fe ENDOR spectroscopy showed that the signals arise from FeMoco [89]. The authors interpreted their data as indicating that in the lo-CO species one CO was bound to FeMoco, possibly in a bridging mode between two of the central Fe atoms, and that the hi-CO species had two CO molecules bound as terminal ligands to adjacent Fe atoms.

CO binding to the enzyme during turnover has also been studied using stopped-flow Fourier transform infrared spectroscopy (SF-FTIR) [90]. The data were more complex than those observed with EPR/ENDOR, presumably because the latter monitors only paramagnetic states, whereas SF-FTIR can also monitor diamagnetic states. When stoichiometric levels of CO were used, a band at $1904 \, cm^{-1}$ appeared in the IR spectrum. The intensity of this peaked after about 7 s and then slowly declined to be replaced by a band at $1715 \, cm^{-1}$ [91]. The time course for the appearance of the 1904-cm^{-1} band correlated with the onset of inhibition of the disappearance of the substrate azide during turnover. Under a 10-fold excess of CO, the 1904-cm^{-1} band had a lower intensity and bands at $1936 \, cm^{-1}$, $1880 \, cm^{-1}$, and $1958 \, cm^{-1}$ appeared. The first two of these exhibited the same time course whereas the 1958-cm^{-1} band appeared more slowly. As these bands decayed, the 1715-cm^{-1} band was formed, but at a significantly lower rate than found with stoichiometric CO. Isotopic labeling experiments were consistent with the 1936-cm^{-1} and 1880-cm^{-1} bands arising from the symmetrical and asymmetrical coupled vibrations of two CO molecules bound trans to each other on the same metal atom and the 1904-cm^{-1} and 1958-cm^{-1} bands arising from distinct, uncoupled, terminally bound CO species. The authors concluded that the 1715-cm^{-1} band could arise from a CO molecule bound to a face of FeMoco. This may correspond to the hi-CO species observed by EPR spectroscopy.

7.3. Reactivity of Isolated FeMoco

7.3.1. Substrate and Inhibitor Binding to Isolated FeMoco

There have been numerous studies on the interactions between extracted FeMoco and substrates and inhibitors. Relating these studies to the enzymic behavior is complicated because FeMoco when isolated has two additional binding sites, probably initially occupied by solvent, where it was bound to the polypeptide through Cysα275 to the tetrahedral Fe and Hisα442 to Mo. FeMoco binds one mol thiol/mol and in doing so its EPR spectrum is sharpened toward that observed from the MoFe protein [92]. Fe and Se EXAFS data show that one selenophenol binds to an Fe atom of FeMoco. By analogy, thiols are likely to do the same, most probably to the tetrahedral Fe [93].

The electrochemistry of isolated FeMoco is perturbed under an atmosphere of CO compared with that under N_2 or Ar and indicates that two sequential two-electron reduced intermediates can be accessed [94]. Under elevated CO pressures FTIR spectroscopy in a thin-layer cell revealed CO stretch bands at 1883 and 1922 cm^{-1}, probably respectively associated with these two intermediates. Addition of thiophenol had no effect on these two bands but imidazole split the 1922 band into two bands at 1910 and 1929 cm^{-1} whereas the 1883 band was shifted to 1870 cm^{-1}. All of these bands were assigned to terminally bound CO, and the evidence suggested that at least two CO molecules could be bound at the same time. At low CO pressures a band at 1808 cm^{-1} was observed to precede formation of the other features and might be from a bridging CO species. These data are not dissimilar to those observed from the enzyme during turnover (see Sec. 7.2). Cyanide binding, probably to two sites with different binding constants, results in a marked sharpening toward an axial form of the EPR spectrum of

FeMoco [95]. On one of these sites, most probably the tetrahedral Fe, cyanide appears to be competitive with thiol. In addition, there is EXAFS evidence for cyanide binding to Mo in FeMoco [96]. ^{19}F NMR using p-CF$_3$C$_6$H$_4$S$^-$ as a reporter ligand has also provided evidence of cyanide and methyl isocyanide binding [97], although these authors could find no evidence from Mo K-edge EXAFS that methyl isocyanide binds to Mo.

When the reaction between thiophenolate and FeMoco is monitored at 450 nm by stopped-flow spectrophotometry, the rate of reaction is independent of the concentration of thiol [98]. This is consistent with rate-limiting dissociation of a ligand, presumably the N-methyl formamide (NMF) anion, from FeMoco and its subsequent replacement with thiophenolate. Other compounds that interact with FeMoco affect the electron distribution within FeMoco and thus the strength of the FeMoco-NMF bond. This in turn affects the dissociation rate of the NMF and consequently the rate of reaction with thiophenolate. The extent of perturbation indicates not only which ligands bind FeMoco but also the propinquity of the substrate binding site to the dissociating bond. Acetylene or CO had no affect on the rate of reaction with thiophenolate and thus did not bind to FeMoco in its dithionite reduced $S = 3/2$ state. However, azide, $tert$-butyl isocyanide, cyanide, imidazole, and protons all perturbed the rate of reaction with thiophenolate.

The reaction with cyanide was the most complex, the data being consistent with cyanide binding to the tetrahedral iron atom and being displaced by thiophenolate via parallel dissociative and associative pathways. Excess cyanide produced additional small effects indicating binding of an additional cyanide remote from the tetrahedral iron and probably at Mo. Thus, these data are consistent with the EPR data described above. Azide accelerates and $tert$-butyl isocyanide inhibits the rate of reaction with thiophenolate, presumably as a consequence of their relative electron releasing and withdrawal properties, but the effects were relatively small indicating binding at the Mo end of cofactor. Electrochemical and EPR studies have shown that both oxidized and semireduced FeMoco can be protonated [99,100]. Protonation of FeMoco by the weak acid [NHEt$_3$]BPh$_4$ resulted in an increased rate of its reaction with thiophenolate and, by analogy with the simpler iron-sulfur clusters, this was interpreted as indicating protonation of bridging sulfur [98].

R homocitrate is an essential ligand for FeMoco, and when it is replaced by citrate, as in the *nif V* mutant of *K. pneumoniae*, the resultant enzyme is a poor nitrogen fixer. FeMoco from the MoFe protein from a *nif V* mutant showed the same reactivity with thiophenolate as the wild type when bound to cyanide, azide, or the proton [101]. However, when imidazole was bound, the kinetics of the reactions of thiophenolate with the two cofactors were very different. The only difference between the two cofactors is the existence of an extra CH_2 group in the pendant carboxyl arm of R homocitrate compared with citrate. Molecular mechanics calculations showed that this pendant $CH_2CH_2CO_2$ arm of R homocitrate is sufficiently long and flexible that it can form a hydrogen bond with the NH group of the imidazole ligand. This would impose imidazolate character on the ligand, which in turn would release electron density into the cluster core and perturb the Fe-NMF bond. However, the pendant arm of citrate, $CH_2CO_2^-$, in the Nif V$^-$ cofactor is shorter and cannot interact with the imidazole ligand. Although the X-ray structure of the MoFe protein indicates that homocitrate cannot hydrogen-bond to Hisα442, these observations are relevant to enzyme function if we assume that during turnover, reduction and protonation could activate the Mo atom by dissociation of the carboxyl group, as has been established with model compounds [102]. Molecular modeling reportedly shows that monodentate R homocitrate in the protein can establish a hydrogen bond between the pendant carboxyl group and the NH group of Hisα442 without rotation of the histidine ligand while maintaining stabilizing hydrogen bonding with the polypeptide [101]. If homocitrate were to become monodentate, a vacant site would be opened up on the Mo, which would be suitable for binding N_2. The binding of N_2 to a metal site and the ability of that N_2 ligand to be protonated are favored by electron-rich sites. Thus, the above process would favor N_2 binding and its reduction and would explain the need for homocitrate as part of FeMoco.

In summary, the proposal is that during enzyme turnover electrons passed to FeMoco induce the carboxyl group of R homocitrate to dissociate from Mo. The monodentate homocitrate then hydrogen bonds to Hisα442, which causes an increase in the electron richness of FeMoco, particularly at the Mo site, and allows binding of N_2. Additional electron transfer to FeMoco followed by protonation would result in the formation of ammonia.

7.3.2. Substrate Reduction by Isolated FeMoco

Early reports of acetylene reduction to ethylene by FeMoco using sodium borohydride as the reductant were controversial, but it was later established that isolated FeMoco combined with a very strong reductant is an active catalyst for acetylene reduction [3]. However, this activity is common to a number of metal complexes and thus is perhaps not surprising.

More recently, the catalytic reduction of acetylene to ethylene by isolated FeMoco using zinc or europium amalgams as the reductants, DMF as the solvent, and thiophenol or citric acid as the source of protons has been reported [103]. Reactivity increased more than an order of magnitude on going from zinc to europium amalgam, i.e., with the decrease (more negative) in the redox potential of the reductant. With the latter, FeMoco catalyzed C_2H_4 and C_2H_6 production from C_2H_2, showing sigmoidal kinetics consistent with substrate-induced cooperativity between at least two sites for C_2H_2 reduction. CO inhibited acetylene reduction with both reductants. The different reducing agents apparently access different redox levels of FeMoco, and therefore different numbers and types of binding sites for substrates and inhibitors were detected. The enzyme reconstitution activity of FeMoco before and after the reaction was the same, so during the reactions the structural integrity of FeMoco was retained.

At low, unsaturating, C_2H_2 pressure, N_2 inhibited acetylene reduction catalyzed by FeMoco in DMF solution with europium amalgam as a reducing agent [104]. The inhibition was competitive and reversible; $K_i = 0.45$ atm N_2. The authors concluded that it is possible to obtain a state of the isolated cofactor that is capable of binding the N_2 molecule, but the products of N_2 reduction were not detected.

Recently, the electrochemical behavior of FeMoco in NMF has been reinvestigated [100] and, in common with earlier work by Schultz et al. [105], two redox waves at -0.32 V and at close to -1.0 V relative to NHE were observed and were interpreted in terms of the equilibria:

$$(FeMoco)_{ox} \rightleftharpoons (FeMoco)_{sred} \rightleftharpoons (FeMoco)_{red}$$

where $(FeMoco)_{sred}$ corresponds to the $S = 3/2$ state.

The potentials were modified slightly by binding thiols to FeMoco. If a relatively acidic thiol, pentafluorothiophenol, was used in large

excess (75-fold), a catalytic process was observed. It was concluded that
the thiol was acting both as a ligand and as an acid, and providing
protons for the catalytic evolution of H_2 [100]. In a coulometric experiment in a sealed vessel, at least 85% of the electrons passed were recov-
ered as H_2; thus, isolated FeMoco can be induced to evolve H_2 at the
relatively high potential of $-0.28\,V$. Reduction of $(FeMoco)_{ox}$ to the
semireduced (sred) state with concomitant protonation must allow
further electron transfer, accessing a $(FeMoco)_{red}$ state at high potential.
Presumably attack of a proton on this species would liberate H_2 and
regenerate $(FeMoco)_{ox}$. However, it is possible that lower (more
reduced) redox states could be accessed on protonation/reduction.
Importantly, at fast cyclic voltammetry scan rates, the catalysis step is
circumvented and the system approaches reversibility. This confirms
the role of the $(FeMoco)_{ox/sred}$ system in the turnover and the fact that
the integrity of the cluster is maintained. Clearly, by coupling electron
and proton addition, access to more reduced states of FeMoco is
achieved at high potential and facilitates electrocatalytic H_2 evolution.
There are distinct parallels with the proposed mechanism of H_2 evolution by the enzyme.

7.4. Where on FeMoco Do Substrates and Inhibitors Bind?

7.4.1. Evidence from Mutagenesis of Residues Around FeMoco

Crystallography (Fig. 4) has confirmed that FeMoco is covalently
bound to the polypeptide only through the residues Cysα275 to the
tetrahedral Fe atom and Hisα442 to the Mo atom. However, the inability to induce isolated FeMoco to reduce N_2 indicates that other interactions with the polypeptides are also probably important.
Consequently every residue in the immediate environment of FeMoco
(Fig. 11) has been subjected to directed mutagenesis and the effects of
these alterations on the enzyme's substrate-reducing activities
assessed. Several of these mutants had interesting properties that
demonstrated the fine control exerted by its environment on the properties of FeMoco. For example, if Argα277, which is close to the ligand

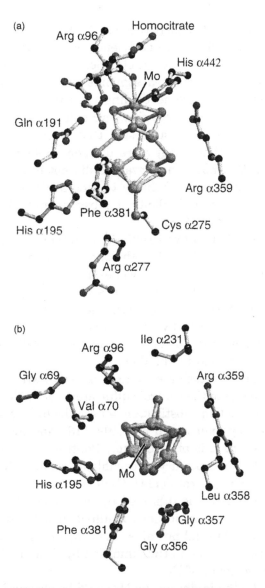

FIG. 11.   The environment of the FeMoco in Kp1 (PDB entry 1QGU). (a) Side view. (b) End view produced by rotation through approximately 80° about the horizontal axis. The homocitrate, Hisα442 and Cysα275 have been removed for clarity.

Cysα275, is altered to His, then the resultant enzyme is incapable of reducing N_2 but can still reduce acetylene, cyanide, azide, and protons [106]. Under nonsaturating CO, acetylene, but not cyanide or azide, reduction by the mutant showed sigmoidal kinetics, indicating cooperativity between two acetylene binding sites in the presence of CO, i.e., the presence of at least three sites on FeMoco that can be occupied simultaneously. The authors suggested that since cyanide and azide did not show cooperativity, they were unlikely to share the acetylene binding sites. Furthermore, since cyanide did not induce cooperativity in acetylene reduction it was unlikely to bind at the same site as CO. Overall the data imply a multiplicity of substrate and inhibitor binding sites on protein-bound FeMoco.

Mutating Glnα191 to Lys results in a mutant that cannot reduce N_2 but can reduce acetylene to ethylene and also to some ethane [107,108]. N_2 does not inhibit acetylene or proton reduction, but CO inhibits H_2 evolution by this mutant in a manner reminiscent of *nifV* mutants. As can be seen in Fig. 11a the highly conserved residue Glnα191 approaches and is hydrogen-bonded to one of the carboxyl groups of the homocitrate. This would imply a relationship between this region of FeMoco and CO binding.

A particularly interesting residue is Hisα195, which is thought to hydrogen-bond to one of the central sulfur atoms of FeMoco. Most work has been done on the Gln substitution, where the hydrogen bond to the central S is probably maintained [107,109–113], and the Asn substitution [109,110], where it is probably not. Contrary to an earlier report [114], the Hisα195Gln mutant was capable of reducing N_2, but only at about 2% of the wild-type rate [111]. C_2H_2 was a good substrate but at low levels its reduction was inhibited by N_2, which also inhibited H_2 evolution. Both of these inhibitions resulted in lowered electron flux through the enzyme but did not affect ATP hydrolysis and thus raised the ATP/2e$^-$ ratio. H_2 relieved the inhibition by N_2 and HD was formed under D_2 in the presence of N_2.

Interestingly, the inhibition by N_2 could be observed when the Fe protein to MoFe protein ratio was only 2.5:1, indicating that it was not necessary to access very reduced states in the MoFe protein cycle (Fig. 8) for N_2 to bind. Three new EPR signals in the $g = 2$ region were detected from the Hisα195Gln mutant during turnover under C_2H_2 [112]. A Q-band EPR and ^{13}C and ^{1}H ENDOR study [113] of the most prominent of

these was interpreted as evidence that it arose from an intermediate in
C_2H_2 reduction. Coupling to three distinct ^{13}C atoms was detected, indicating the presence of at least two distinct C_2H_2-derived species. A
strongly coupled nonexchangeable ^{1}H was also detected. The authors
concluded that these with other data were indicative of the binding of
C_2H_2 to at least two sites in a side-on manner, probably on the central
4Fe faces of FeMoco. The observations with the Hisα195Asn mutant
were also striking. In this case, no N_2 reduction could be detected
although N_2 still inhibited H^+ and C_2H_2 reduction. Also, no HD was
formed under an N_2/D_2 atmosphere, indicating that simple N_2 binding
was not sufficient for HD formation [110].

Dean and colleagues have developed an alternative method of identifying residues in the vicinity of FeMoco that are important in substrate
reduction. They started from observations in the literature that there
are two sites for C_2H_2 binding and reduction on the MoFe protein. They
then selected a strain that was supersensitive to C_2H_2 inhibition of
growth and sought mutants that escaped this inhibition. Some of the
mutants were simply revertants, but two mapped to Glyα69 which had
been substituted by Ser [115]. This substitution eliminated a high-affinity site for C_2H_2 binding, leaving a second lower affinity site intact. N_2
reduction activity was unaffected. In contrast to the wild type, where
both CO and C_2H_2 are noncompetitive inhibitors of N_2 reduction, with
the Glyα69Ser mutant both exhibited competitive inhibition. CO also
became a competitive inhibitor of C_2H_2, azide, and nitrous oxide reduc-
tions [116]. The reciprocity of substrate and inhibitor interactions in
this mutant implies that this form of nitrogenase is acting as a classic
single-site enzyme. These data were interpreted in terms of a two-site
model: one high-affinity site for C_2H_2 where CO but not N_2, azide, or
nitrous oxide binds, and the other site where C_2H_2, N_2, azide, nitrous
oxide, and CO can all bind. Glyα69 is immediately adjacent to Valα70,
which approaches one 4Fe face of FeMoco (Fig. 11b), and the authors
concluded that both of the C_2H_2 binding sites are in close proximity to
this face.

7.4.2. Theoretical Studies

Following the discovery of the structure of FeMoco, there have been
several attempts to use theory to calculate the most favorable modes

of the binding and reduction of N_2 on it. Here we shall not attempt to describe the calculations in detail but will simply outline the main findings [3,117]. Most of these calculations have focused on binding at the central six Fe sites. This bias is based on several observations: first, Fe is common to the cofactors from the Mo, V, and Fe-only nitrogenase systems; secondly, the six trigonal Fe sites appear to be coordinatively unsaturated and therefore available for substrate binding; and thirdly, the Mo is six-coordinate and so appears to be coordinatively saturated. However, some Mo complexes are seven- or eight-coordinate, so this alone is not sufficient reason for excluding the possibility that Mo is involved in substrate binding [56,57]. Furthermore, as noted above, there is sound chemical precedent [102] for carboxylate ligands dissociating from Mo sites on reduction, thus leaving room for other ligands, such as N_2.

Extended Hückel (EH) [118], complete (CNDO) [119], or intermediate (INDO) [120–122] neglect of differential overlap theory and density functional theory (DFT) [117,123–125] have all been applied to the problem. Since there is insufficient symmetry in the FeMoco structure, all have had to simplify it in some way to make computation practical. Most of the calculations detect considerable metal-metal bonding between the central six Fe atoms, which presumably would at least partially satisfy their apparent coordinative unsaturation. The trigonal Fe atoms from each half of the cluster form two three-membered rings, the rings being separated by only 2.6 Å. However, in 1.6-Å resolution Kp1 electron density maps contoured at the 1σ level, the electron densities for the two rings remain separated by a well-defined cleft whereas the electron densities for the atoms within a ring are merged. This suggests some overlapping of orbitals between atoms in a ring, despite the slightly longer average interatomic distance of 2.7 Å [29]. Most calculations support the conclusion that N_2 bonding is weak; however, the most favorable mode of binding is contentious. Some authors prefer end-on binding to a single Fe atom [123–127], others suggest binding to a four-iron face [120–125], either linearly across the face, or with one N bound to all four Fe atoms [118,128] and the other interacting with just two. Others suggest that the N_2 molecule could be bound within the 6Fe cage [120–122]. Most agree that the S atoms are important both as a repository for electrons during turnover and also as part of a proton delivery system. The inherent flexibility of FeMoco, being tethered to the protein only at its extremes, is also considered important in many schemes because it would allow the Fe atoms to move to positions that are more accessible to substrate. Moreover, when the N_2 molecule is bound between metal atoms, the flexibility could be an important contributor to weakening of the N–N bond.

Very recently, two papers concentrating on the Mo atom as the site of binding were published. Both consider the effect of dissociation of the carboxyl group of homocitrate to generate a site for N_2 binding and also carried out calculations on model compounds. From EH calculations [129] it was concluded that although N_2 could bind to such a site, its activation for reduction was unlikely. DFT calculations [130] indicated that binding was energetically more favorable to such a Mo (or V) site than to trigonal Fe. After binding of N_2 to the Mo atom and its partial reduction to NNH_2 (an intermediate that has been detected with the enzyme), the energetically favored reaction path was for the NNH_2 to form a bridge between the Mo and a nearby Fe atom. This would facilitate cleavage of the N–N bond.

8.   CONCLUSIONS

Determinations of the three-dimensional structures of the MoFe and Fe proteins and their putative transition state complex with Mg ADP·AlF$_4^-$ have considerably illuminated our understanding of nitrogenase structure-function relationships, although many important questions remain.

It is clear that the Fe protein can be fully reduced to the $[Fe_4S_4]^0$ state and that this state can pass two electrons to the MoFe protein, resulting in a diminished requirement for ATP hydrolysis. Furthermore, MgATP/MgADP exchange can occur on the MoFe:Fe protein complex, and flavodoxin hydroquinone can apparently reduce the oxidized $[Fe_4S_4]^{2+}$ state of the Fe protein within this complex, implying that complex dissociation may not always be the rate-determining step in turnover. These observations require further investigation to determine how important they might be in vivo since they have major implications for the energetics and detailed mechanism of nitrogenase turnover. In addition, there is now good evidence that both the Fe and MoFe proteins can pass through a series of conformational states within the complex, and the relevance of these to turnover and in particular to the role(s) of ATP hydrolysis requires further study.

No definitive answer on the site of N_2 binding and reduction on FeMoco has yet been obtained. Theoretical calculations implicate Mo or Fe sites or both. Results with mutated MoFe proteins indicate the importance of hydrogen bonding to the central S atoms and implicate one of the 4Fe faces in providing the high-affinity binding site for acetylene, but N_2 is not competitive at this site. CO binding is to at least two sites, probably on separate metals, at least one of which must then be Fe. However, R-homocitrate bound to Mo is clearly essential for N_2 reduction, and this strongly implies the involvement of Mo in the process. To counter this, mutation of Argα277, close to the tetrahedral Fe end of FeMoco, also destroys N_2 reduction activity. The jury is clearly still out on where N_2 binds and will probably remain so until direct interactions between N_2 and/or its reduction products and specific metal sites are observed.

ACKNOWLEDGMENTS

The authors gratefully acknowledge useful discussions with Dr. Suzanne Mayer, financial support from the Biotechnology and Biological Sciences Research Council, and the award of an Emeritus Fellowship from the John Innes Foundation to B.E.S.

ABBREVIATIONS AND DEFINITIONS

anf	an iron-only nitrogenase gene
CNDO	complete neglect of differential overlap
DFT	density functional theory
DMF	dimethylformamide
EH	extended Hückel
ENDOR	electron nuclear double resonance
EPR	electron paramagnetic resonance
EXAFS	extended x-ray absorption fine structure spectroscopy
FeMoco	the iron molybdenum cofactor of nitrogenase

Fe protein	the iron protein of molybdenum nitrogenase
INDO	intermediate neglect of differential overlap
MgADP	magnesium complex of adenosine 5'-diphosphate
MgATP	magnesium complex of adenosine 5'-triphosphate
MoFe protein	the molybdenum iron protein of molybdenum nitro-genase
NHE	normal hydrogen electrode
nif	a molybdenum nitrogenase gene
Nif	a molybdenum nitrogenase gene product
NMF	N-methyl formamide
P cluster	the Fe_8S_7 cluster of nitrogenase
P_i	inorganic phosphate
PDB	Protein Data Bank
SF-FTIR	stopped-flow Fourier transform infrared spectroscopy
VFe protein	the vanadium iron protein of vanadium nitrogenase
vnf	a vanadium nitrogenase gene
Vnf	a vanadium nitrogenase gene product

The nitrogenase proteins from various organisms are designated by a capital letter to indicate the genus, a lower case letter to indicate the species and a number to indicate the component. Thus Av1 is the MoFe protein (Component 1) from *A. vinelandii*, Kp2 is the Fe protein (Component 2) from *K. pneumoniae*, etc.

REFERENCES

1. J. B. Howard and D. C. Rees, *Chem. Rev.*, *96*, 2965–2982 (1996).

2. B. K. Burgess and D. J. Lowe, *Chem. Rev.*, *96*, 2983–3011 (1996).

3. B. E. Smith, in *Advances in Inorganic Chemistry* (A. G. Sykes and R. Cammack, eds.), Vol. 47, Academic Press, New York, 1999, pp. 159–218.

4. B. E. Smith, D. J. Lowe, and R. C. Bray, *Biochem. J.*, *130*, 641–643 (1972).

5. B. E. Smith, D. J. Lowe, and R. C. Bray, *Biochem. J.*, *135*, 331–341 (1973).

6. W. H. Orme-Johnson, W. D. Hamilton, T. L. Jones, M. Y. Tso, R. H. Burris, V. K. Shah, and W. J. Brill, *Proc. Natl. Acad. Sci. USA*, *69*, 3142–3145 (1972).

7. B. E. Smith and G. Lang, *Biochem. J.*, *137*, 169–180 (1974).

8. P. E. Bishop, R. Premakumar, D. R. Dean, M. R. Jacobson, J. R. Chisnell, T. M. Rizzo, and J. Kopczynski, *Science*, *232*, 92–94 (1986).

9. P. E. Bishop, M. Hawkins, and R. R. Eady, *Biochem. J.*, *238*, 437–442 (1986).

10. R. R. Eady, *Chem. Rev.*, *96*, 3013–3030 (1996).

11. R. R. Eady, in *Vanadium and Its Role in Life* (H. Sigel and A. Sigel, eds.), Vol. 31 of *Metal Ions in Biological Systems*, Marcel Dekker, New York, 1995, pp. 363–405.

12. S. I. Waugh, D. M. Paulsen, P. V. Mylona, R. H. Maynard, R. Premakumar, and P. E. Bishop, *J. Bacteriol.*, *177*, 1505–1510 (1995).

13. R. R. Eady, R. L. Robson, T. H. Richardson, R. W. Miller, and M. Hawkins, *Biochem. J.*, *244*, 197–207 (1987).

14. J. M. Arber, B. R. Dobson, R. R. Eady, P. Stevens, S. S. Hasnain, C. D. Garner, and B. E. Smith, *Nature*, *325*, 372–374 (1987).

15. G. N. George, C. L. Coyle, B. J. Hales, and S. P. Cramer, *J. Am. Chem. Soc.*, *110*, 4057–4059 (1988).

16. J. Chen, J. Christiansen, R. C. Tittsworth, B. J. Hales, S. J. George, D. Coucouvanis, and S. P. Cramer, *J. Am. Chem. Soc.*, *115*, 5509–5515 (1993).

17. J. M. Arber, B. R. Dobson, R. R. Eady, S. S. Hasnain, C. D. Garner, T. Matsushita, M. Nomura, and B. E. Smith, *Biochem. J.*, *258*, 733–737 (1989).

18. N. Ravi, V. Moore, S. G. Lloyd, B. J. Hales, and B. H. Huynh, *J. Biol. Chem.*, *269*, 20920–20924 (1994).

19. B. J. Hales, E. E. Case, J. E. Morningstar, M. F. Dzeda, and L. A. Mauterer, *Biochemistry*, *25*, 7251–7255 (1986).

20. M. J. Dilworth and R. R. Eady, *Biochem. J.*, *277*, 465–468 (1991).

21. B. E. Smith, R. R. Eady, D. J. Lowe, and C. Gormal, Biochem. J., 250, *299–302 (1988).*

22. J. R. Chisnell, R. Premakumar, and P. E. Bishop, *J. Bacteriol.*, *170*, 27–33 (1988).

23. R. Davis, L. Lehman, R. Petrovich, V. K. Shah, G. P. Roberts, and P. W. Ludden, *J. Bacteriol.*, *178*, 1445–1450 (1996).

24. K. Schneider, U. Gollan, M. Drottboom, S. Selsemeier Voigt, and A. Muller, *Eur. J. Biochem.*, *244*, 789–800 (1997).

25. M. Ribbe, D. Gadkari, and O. Meyer, *J. Biol. Chem.*, *272*, 26627–26633 (1997).

26. J. Kim and D. C. Rees, *Nature*, *360*, 553–560 (1992).

27. J. T. Bolin, N. Campobasso, S. W. Muchmore, T. V. Morgan, and L. E. Mortenson in *Molybdenum Enzymes, Cofactors and Model Systems* (E. I. Stiefel, D. Coucouvanis, and W. E. Newton, eds.) of *ACS Symposium Series 535*, American Chemical Society, Washington, D.C., 1993, pp. 186–195.

28. J. W. Peters, M. H. B. Stowell, S. M. Soltis, M. G. Finnegan, M. K. Johnson, and D. C. Rees, *Biochemistry*, *36*, 1181–1187 (1997).

29. S. M. Mayer, D. M. Lawson, C. A. Gormal, S. M. Roe, and B. E. Smith, *J. Mol. Biol.*, *292*, 871–891 (1999).

30. J. Kim and D. C. Rees, *Biochemistry*, *33*, 389–397 (1994).

31. D. J. Lowe, K. Fisher, and R. N. F. Thorneley, *Biochem. J.*, *292*, 93–98 (1993).

32. W. N. Lanzilotta and L. C. Seefeldt, *Biochemistry*, *35*, 16770–16776 (1996).

33. L. Ma, M. A. Brosius, and B. K. Burgess, *J. Biol. Chem.*, *271*, 10528–10532 (1996).

34. D. R. Dean, R. A. Setterquist, K. E. Brigle, D. J. Scott, N. F. Laird, and W. E. Newton, *Mol. Microbiol.*, *4*, 1505–1512 (1990).

35. H. M. Kent, M. Baines, C. Gormal, B. E. Smith, and M. Buck, *Mol. Microbiol.*, *4*, 1497–1504 (1990).

36. H. M. Kent, I. Ioannidis, C. Gormal, B. E. Smith, and M. Buck, *Biochem. J.*, *264*, 257–264 (1989).

37. D. Govezensky and A. Zamir, *J. Bacteriol.*, *171*, 5729–5735 (1989).

38. T. R. Hoover, J. Imperial, P. W. Ludden, and V. K. Shah, *Biochemistry*, *28*, 2768–2771 (1989).

39. H. B. Gray and J. R. Winkler, *Annu. Rev. Biochem.*, *65*, 537–561 (1996).

40. E. Münck, H. Rhodes, W. H. Orme-Johnson, L. C. Davis, W. J. Brill, and V. K. Shah, *Biochim. Biophys. Acta.*, *400*, 32–53 (1975).

41. B. E. Smith, M. J. O'Donnell, G. Lang, and K. Spartalian, *Biochem. J.*, *191*, 449–455 (1980).

42. R. C. Tittsworth and B. J. Hales, *J. Am. Chem. Soc.*, *115*, 9763–9767 (1993).

43. J. H. Spee, A. F. Arendsen, H. Wassink, S. J. Marritt, W. R. Hagen, and H. Haaker, *FEBS Lett.*, *432*, 55–58 (1998).

44. A. J. Pierik, H. Wassink, H. Haaker, and W. R. Hagen, *Eur. J. Biochem.*, *212*, 51–61 (1993).

45. M. J. O'Donnell and B. E. Smith, *Biochem. J.*, *173*, 831–839 (1978).

46. W. R. Hagen, H. Wassink, R. R. Eady, B. E. Smith, and H. Haaker, *Eur. J. Biochem.*, *169*, 457–465 (1987).

47. J. L. Schlessman, D. Woo, L. Joshua Tor, J. B. Howard, and D. C. Rees, *J. Mol. Biol.*, *280*, 669–685 (1998).

48. S. B. Jang, L. C. Seefeldt, and J. W. Peters, *Biochemistry*, *39*, 14745–14752 (2000).

49. G. L. Anderson and J. B. Howard, *Biochemistry*, *23*, 2118–2122 (1984).

50. P. Strop, P. M. Takahara, H. J. Chiu, H. C. Angove, B. K. Burgess, and D. C. Rees, *Biochemistry*, *40*, 651–656 (2001).

51. M. M. Georgiadis, H. Komiya, P. Chakrabarti, D. Woo, J. J. Kornuc, and D. C. Rees, *Science*, *257*, 1653–1659 (1992).

52. E. F. Pai, U. Krengel, G. A. Petsko, R. S. Goody, W. Kabsch, and A. Wittinghofer, *EMBO J.*, *9*, 2351–2359 (1990).

53. R. N. F. Thorneley and D. J. Lowe, in *Molybdenum Enzymes* (T. G. Spiro, ed.), Wiley, New York, 1985, pp. 221–284.

54. M. G. Duyvis, H. Wassink, and H. Haaker, *Biochemistry*, *37*, 17345–17354. (1998).

55. D. J. Lowe, in *Nitrogen Fixation: Fundamentals and Applications* (I. A. Tikhonovich, N. A. Provorov, V. I. Romanov, and W. E. Newton, eds.), Kluwer Academic, Dordrecht, 1995, pp. 103–108.

56. C. J. Pickett, *J. Bioinorg. Chem.*, *1*, 601–606 (1996).

57. D. J. Evans, R. A. Henderson, and B. E. Smith, in *Bioinorganic Catalysis* (J. Reedijk and E. Bouwman, eds.), Marcel Dekker, New York, 1999, pp. 153–207.

58. K. Fisher, D. J. Lowe, and R. N. F. Thorneley, *Biochem. J.*, *279*, 81–85 (1991).

59. G. D. Watt and K. R. N. Reddy, *J. Inorg. Biochem.*, *53*, 281–294 (1994).

60. J. A. Erickson, A. C. Nyborg, J. L. Johnson, S. M. Truscott, A. Gunn, F. R. Nordmeyer, and G. D. Watt, *Biochemistry*, *38*, 14279–14285 (1999).

61. S. J. Yoo, H. C. Angove, B. K. Burgess, E. Munck, and J. Peterson, *J. Am. Chem. Soc.*, *120*, 9704–9705 (1998).

62. S. Y. Druzhinin, L. A. Syrtsova, A. M. Uzenskaja, and G. I. Likhtenstein, *Biochem. J.*, *290*, 627–631 (1993).

63. L. A. Syrtsova, V. A. Nadtochenko, N. N. Denisov, E. A. Timofeeva, N. I. Shkondina, and V. Y. Gak, *Biochemistry (Moscow)*, *65*, 1145–1152 (2000).

64. J. L. Johnson, A. C. Nyborg, P. E. Wilson, A. M. Tolley, F. R. Nordmeyer, and G. D. Watt, *Biochim. Biophys. Acta*, 1543, *36–46 (2000)*.

65. J. L. Johnson, A. C. Nyborg, P. E. Wilson, A. M. Tolley, F. R. Nordmeyer, and G. D. Watt, *Biochim. Biophys. Acta*, *1543*, 24–35 (2000).

66. S. Maritano, S. A. Fairhurst, and R. R. Eady, *J. Bioinorg. Chem.* *6*, 590–600 (2001).

67. K. A. Renner and J. B. Howard, *Biochemistry*, *35*, 5353–5358 (1996).

68. M. G. Duyvis, H. Wassink, and H. Haaker, *FEBS Lett.*, *380*, 233–236 (1996).

69. H. Schindelin, C. Kisker, J. L. Schlessman, J. B. Howard, and D. C. Rees, *Nature*, *387*, 370–376 (1997).

70. J. G. Grossman, S. S. Hasnain, F. K. Yousafzai, B. E. Smith, and R. R. Eady, *J. Mol. Biol.*, *266*, 642–648 (1997).

71. M. J. Ryle and L. C. Seefeldt, *Biochemistry*, *35*, 4766–4775 (1996).

72. W. N. Lanzilotta, K. Fisher, and L. C. Seefeldt, *Biochemistry*, *35*, 7188–7196 (1996).

73. H. J. Chiu, J. W. Peters, W. N. Lanzilotta, M. J. Ryle, L. C. Seefeldt, J. B. Howard, and D. C. Rees, *Biochemistry*, *40*, 641–650 (2001).

74. N. Gavini and B. K. Burgess, *J. Biol. Chem.*, *267*, 21179–21186 (1992).

75. E. H. Bursey and B. K. Burgess, *J. Biol. Chem.*, *273*, 29678–29685 (1998).

76. V. K. Shah and W. J. Brill, *Proc. Natl. Acad. Sci. USA*, *74*, 3249–3253 (1977).

77. P. A. McLean and R. A. Dixon, *Nature*, *292*, 655–656 (1981).

78. T. R. Hawkes, P. A. McLean, and B. E. Smith, *Biochem. J.*, *217*, 317–321 (1984).

79. L. M. Zheng, R. H. White, and D. R. Dean, *J. Bacteriol.*, *179*, 5963–5966 (1997).

80. J. Liang, M. Madden, V. K. Shah, and R. H. Burris, *Biochemistry*, *29*, 8577–8581 (1990).

81. R. M. Allen, R. Chatterjee, M. S. Madden, P. W. Ludden, and V. K. Shah, *Crit. Rev. Biotechnol.*, *14*, 225–249 (1994).

82. D. N. Pham and B. K. Burgess, *Biochemistry*, *32*, 13725–13731 (1993).

83. D. J. Lowe, R. R. Eady, and R. N. F. Thorneley, *Biochem. J.*, *173*, 277–290 (1978).

84. T. R. Hawkes, D. J. Lowe, and B. E. Smith, *Biochem. J.*, *211*, 495–497 (1983).

85. M. G. Yates and D. J. Lowe, *FEBS Lett.*, *72*, 121–126 (1976).

86. L. C. Davis, M. T. Henzl, R. H. Burris, and W. H. Orme-Johnson, *Biochemistry*, *18*, 4860–4869 (1979).

87. R. C. Pollock, H. I. Lee, L. M. Cameron, V. J. Derose, B. J. Hales, W. H. Orme-Johnson, and B. M. Hoffman, *J. Am. Chem. Soc.*, *117*, 8686–8687 (1995).

88. H. I. Lee, L. M. Cameron, B. J. Hales, and B. M. Hoffman, *J. Am. Chem. Soc.*, *119*, 10121–10126 (1997).

89. P. D. Christie, H. I. Lee, L. M. Cameron, B. J. Hales, W. H. Orme-Johnson, and B. M. Hoffman, *J. Am. Chem. Soc.*, *118*, 8707–8709 (1996).

90. S. J. George, G. A. Ashby, C. W. Wharton, and R. N. F. Thorneley, *J. Am. Chem. Soc.*, *119*, 6450–6451 (1997).

91. R. N. F. Thorneley, G. A. Ashby, and S. J. George, in *Nitrogen Fixation: From Molecules to Crop Productivity* (F. O. Pedrosa, M. Hungria, G. Yates, and W. E. Newton, eds.), Kluwer Academic, Dordrecht, The Netherlands, 2000, pp. 39–40.

92. J. Rawlings, V. K. Shah, J. R. Chisnell, W. J. Brill, R. Zimmermann, E. Münck, and W. H. Orme-Johnson, *J. Biol. Chem.*, *253*, 1001–1004 (1978).

93. I. Harvey, R. W. Strange, R. Schneider, C. A. Gormal, C. D. Garner, S. S. Hasnain, R. L. Richards, and B. E. Smith, *Inorg. Chim. Acta*, *276*, 150–158 (1998).

94. S. K. Ibrahim, K. Vincent, C. A. Gormal, B. E. Smith, S. P. Best, and C. J. Pickett, *Chem. Commun.*, 1019–1020 (1999).

95. A. J. M. Richards, D. J. Lowe, R. L. Richards, A. J. Thomson, and B. E. Smith, *Biochem. J.*, *297*, 373–378 (1994).

96. H. B. I. Liu, A. Filipponi, N. Gavini, B. K. Burgess, B. Hedman, A. Dicicco, C. R. Natoli, and K. O. Hodgson, *J. Am. Chem. Soc.*, *116*, 2418–2423 (1994).

97. S. D. Conradson, B. K. Burgess, S. A. Vaughn, A. L. Roe, B. Hedman, K. O. Hodgson, and R. H. Holm, *J. Biol. Chem.*, *264*, 15967–15974 (1989).

98. K. L. C. Gronberg, C. A. Gormal, B. E. Smith, and R. A. Henderson, *Chem. Commun.*, 713–714 (1997).

99. W. E. Newton, S. F. Gheller, B. J. Feldman, W. R. Dunham, and F. A. Schultz, *J. Biol. Chem.*, *264*, 1924–1927 (1989).

100. T. LeGall, S. K. Ibrahim, C. A. Gormal, B. E. Smith, and C. J. Pickett, *Chem. Commun.*, 773–774 (1999).

101. K. L. C. Gronberg, C. A. Gormal, M. C. Durrant, B. E. Smith, and R. A. Henderson, *J. Am. Chem. Soc.*, *120*, 10613–10621 (1998).

102. D. L. Hughes, S. K. Ibrahim, C. J. Pickett, G. Querne, A. Laouenan, J. Talarmin, A. Queiros, and A. Fonseca, *Polyhedron*, *13*, 3341–3348 (1994).

103. T. A. Bazhenova, M. A. Bazhenova, G. N. Petrova, S. A. Mironova, and V. V. Strelets, *Kinet. Catal.*, *41*, 499–510 (2000).

104. T. A. Bazhenova, M. A. Bazhenova, G. N. Petrova, and A. E. Shilov, *Kinet. Catal.*, *40*, 851–852 (1999).

105. F. A. Schultz, S. F. Gheller, B. K. Burgess, S. Lough, and W. E. Newton, *J. Am. Chem. Soc.*, *107*, 5364–5368 (1985).

106. J. Shen, D. R. Dean, and W. E. Newton, *Biochemistry*, *36*, 4884–4894 (1997).

107. D. J. Scott, H. D. May, W. E. Newton, K. E. Brigle, and D. R. Dean, *Nature*, *343*, 188–190 (1990).

108. D. J. Scott, D. R. Dean, and W. E. Newton, *J. Biol. Chem.*, *267*, 20002–20010 (1992).

109. K. Fisher, M. J. Dilworth, C. H. Kim, and W. E. Newton, *Biochemistry*, *39*, 10855–10865 (2000).

110. K. Fisher, M. J. Dilworth, and W. E. Newton, *Biochemistry*, *39*, 15570–15577 (2000).

111. M. J. Dilworth, K. Fisher, C. H. Kim, and W. E. Newton, *Biochemistry*, *37*, 17495–17505 (1998).

112. M. Sorlie, J. Christiansen, D. R. Dean, and B. J. Hales, *J. Am. Chem. Soc.*, *121*, 9457–9458 (1999).

113. H. I. Lee, M. Sorlie, J. Christiansen, R. T. Song, D. R. Dean, B. J. Hales, and B. M. Hoffman, *J. Am. Chem. Soc.*, *122*, 5582–5587 (2000).

114. C. H. Kim, W. E. Newton, and D. R. Dean, *Biochemistry*, *34*, 2798–2808 (1995).

115. J. Christiansen, V. L. Cash, L. C. Seefeldt, and D. R. Dean, *J. Biol. Chem.*, *275*, 11459–11464 (2000).

116. J. Christiansen, L. C. Seefeldt, and D. R. Dean, *J. Biol. Chem.*, *275*, 36104–36107 (2000).

117. T. H. Rod and J. K. Norskov, *J. Am. Chem. Soc.*, *122*, 12751–12763 (2000).

118. H. B. Deng and R. Hoffmann, *Angew. Chem.*, *32*, 1062–1065 (1993).

119. S. J. Zhong and C. W. Liu, *Polyhedron*, *16*, 653–661 (1997).

120. K. K. Stavrev and M. C. Zerner, *Chem. A Eur. J.*, *2*, 83–87 (1996).

121. K. K. Stavrev and M. C. Zerner, *Theoret. Chem. Acc.*, *96*, 141–145 (1997).

122. K. K. Stavrev and M. C. Zerner, *Int. J. Quantum Chem.*, *70*, 1159–1168 (1998).

123. I. G. Dance, *Austral. J. Chem.*, *47*, 979–990 (1994).

124. I. Dance, *Chem. Commun.*, 165–166 (1997).

125. I. Dance, *Chem. Commun.*, 523–530 (1998).

126. T. H. Rod, B. Hammer, and J. K. Norskov, *Phys. Rev. Lett.*, *82*, 4054–4057 (1999).

127. T. H. Rod, A. Logadottir, and J. K. Norskov, *J. Chem. Phys.*, *112*, 5343–5347 (2000).

128. P. E. M. Siegbahn, J. Westerberg, M. Svensson, and R. H. Crabtree, *J. Phys. Chem. B*, *102*, 1615–1623 (1998).

129. F. Barriere, C. J. Pickett, and J. Talarmin, *Polyhedron*, *20*, 27–36 (2001).

130. M. C. Durrant, *Inorg. Chem. Commun.*, *4*, 60–62 (2001).

131. P. J. Kraulis, *J. Appl. Cryst.*, *24*, 946–950 (1991).

132. E. A. Merritt and D. J. Bacon, *Meth. Enzymol.*, *277*, 505–524 (1997).

4

Chemical Dinitrogen Fixation by Molybdenum and Tungsten Complexes: Insights from Coordination Chemistry

Masanobu Hidai[1] *and Yasushi Mizobe*[2]

[1]Department of Materials Science and Technology, Faculty of Industrial Science and Technology, Science University of Tokyo, Noda, Chiba 278–8510, Japan

[2]Institute of Industrial Science, University of Tokyo, Komaba, Meguro-ku, Tokyo 153–8505, Japan

1. INTRODUCTION

Industrial N_2 fixation producing ammonia from the reaction with H_2 gas by the use of a heterogeneous Fe catalyst (equation 1) started in September 1913. Ammonia is now the second largest synthetic chemical product, mainly because of the rapidly increasing demand of ammonium salts as fertilizer. However, in spite of extensive studies to explore more efficient N_2-fixing systems from the early 20th century until now, there have been no fundamental changes in the process for the synthesis of ammonia. The catalysis still requires extremely drastic conditions [1]. Development of a new type of catalysts by which N_2 is transformed into ammonia and other nitrogen-containing compounds under mild conditions is one of the most challenging subjects in chemistry.

$$N_2 + 3H_2 \quad \xrightarrow[\text{400−500 °C, >100 bar}]{\text{Fe catalyst}} \quad 2NH_3 \qquad (1)$$

Chemical nitrogen fixation at room temperature under atmospheric pressure was first observed in 1964 by Vol'pin and Shur, who detected the formation of ammonia after acidolysis of the reaction mixtures of certain transition metal salts with reducing agents under N_2 [2]. Coordination of the N_2 molecule to a metal center was then confirmed in 1965 by Allen and Senoff, who isolated the first stable N_2 complex $[Ru(NH_3)_5(N_2)]^{2+}$, although its N_2 ligand had its origin not in gaseous

N_2 but in hydrazine [3]. After these two pioneering findings, the chemistry of N_2 fixation has expanded dramatically; a number of N_2 complexes have been prepared from gaseous N_2 and the bonding nature of the N_2 ligands bound to one or more metals have been characterized unambiguously by X-ray crystallography.

From the early stages of these studies, N_2 complexes of Mo have been attracting particular interest because of their relevance to the natural enzyme nitrogenase, which has long been known to contain Mo as a key metal in addition to Fe, although it is now recognized that nitrogenases containing V instead of Mo, as well as those with only Fe as a metal component, are also present. It was in 1969 that the first fully characterizable N_2 complex of Mo, *trans*-[Mo(N_2)$_2$(dppe)$_2$] (**1a**), was prepared in this laboratory [4]. Subsequently, related Mo and W complexes of this type with tertiary phosphine co-ligands have been prepared numerously by Chatt and his coworkers. It is noteworthy that although almost all d-block transition metals along with some rare-earth metals and Li are now known to give N_2 complexes if appropriate ligands are bound to the metals, intriguing reactivities of the coordinated N_2 are observed only in a limited number of complexes. In this line, Mo and W complexes are quite outstanding, since their N_2 ligands show remarkably diversified reactivities. Not only the nitrogen hydrides, i.e., ammonia and hydrazine, but also a variety of organonitrogenous compounds are now available from Mo and W dinitrogen complexes under mild conditions. Furthermore, the mechanisms for the formation of these nitrogen-containing compounds have been clarified in detail by successful isolation of a significant number of important intermediate stages [5].

This chapter summarizes the chemistry of N_2 complexes of Mo and W, emphasizing, as far as possible, the recent findings on the reactivities of coordinated N_2 ligands. The reader can also refer to several excellent reviews described previously, which cover the related or wider area of chemical dinitrogen fixation by using soluble metal complexes [6].

2. PREPARATION AND STRUCTURES OF DINITROGEN COMPLEXES

2.1. Mononuclear Complexes with End-on N_2 Ligand

Most diversely prepared Mo and W complexes are zero-valent bis(dini-trogen) complexes of the type $[M(N_2)_2(L)_4]$ (L = tertiary phosphine) including *trans*- or *cis*-$[M(N_2)_2\{R_2P(CH_2)_nPR_2\}_2]$ (R = alkyl, aryl; n = 1-3) and *cis*- or *trans*-$[M(N_2)_2(PR_nPh_{3-n})_4]$ (R = alkyl, n = 1-3). Most of these compounds are now available from the reactions of Mo or W halides with reducing agents such as Mg and Na-Hg under 1 atm of N_2. Among these complexes, the chemistry of *trans*-$[M(N_2)_2(dppe)_2]$ (**1**) and *cis*-$[M(N_2)_2(PMe_2Ph)_4]$ (**2**) shown in equation 2 has been studied most extensively, whose structures were determined in detail by X-ray analyses for **1a** [7], **1b** (M = W) [8], and **2b** (M = W) [9]. A number of related zero-valent complexes are known, which include mono-, bis-, and tris(dinitrogen) complexes with various monodentate to tetradentate co-ligands containing not only P but also other donor atoms. The tetra-thioether complex *trans*-$[Mo(N_2)_2(Me_8[16]aneS_4)]$ (**3**) [10] might be noteworthy as a N_2 complex having only S-donor co-ligands. Mononuclear N_2 complexes with a higher valent metal center are less common, although the Mo(II) complex $[Cp*MoCl(N_2)(PMe_3)_2]$ [11] and the Mo(III) complex $[Mo(N_2)[N_3N]]$ [12] are known.

$$\text{MoCl}_5 \xrightarrow[\text{Mg or Na-Hg}]{4\text{PMe}_2\text{Ph or 2dppe}} \quad\text{or}\quad \tag{2}$$

P⌣P = dppe   **1a**: M = Mo   **2a**: M = Mo
                  1b: M = W   **2b**: M = W

In all of these mononuclear complexes, the N_2 ligands are bonded to the metal center in an end-on fashion. Single-crystal X-ray diffraction studies have disclosed that the M−N−N linkages are essentially linear, in which the N−N distances ($\sim$ 1.09-1.14 Å) are almost the same as or slightly elongated from that of free N_2 (1.0976(2) Å). In contrast, the $\nu(N{\equiv}N)$ frequencies ($\sim$1890-2110 cm^{-1}) observed in their infrared (IR) spectra are significantly lower than that of free N_2 (2331 cm^{-1}; Raman) and vary considerably depending on the nature of the metal center and ancillary ligands.

2.2. Homobimetallic N_2 Complexes

In multinuclear metal complexes, an N_2 ligand is known to exhibit several coordination modes to metals as illustrated in Fig. 1. However, with respect to homometallic multinuclear Mo and W complexes, only type **i** containing essentially a linear $M-N-N-M$ array has been observed for both the zero-valent and higher valent M. In dinuclear N_2 complexes with two Mo(0) and W(0) centers which can be prepared in an analogous manner to that for $[M(N_2)_2(L)_4]$, the bridging N_2 ligand can be described as $M-N{\equiv}N-M$. Figure 2 depicts the structures of $[\{W(N_2)_2(PEt_2Ph)_3\}_2(\mu\text{-}N_2)]$ [13] and $[\{Mo(\eta^6\text{-}C_6H_3Me_3)(dmpe)\}_2(\mu\text{-}N_2)]$ [14] as the typical N_2 complexes of this type, and, indeed, the $N-N$ bond distance of 1.145(7) Å in the latter [15] is comparable to those in mononuclear zero-valent N_2 complexes.

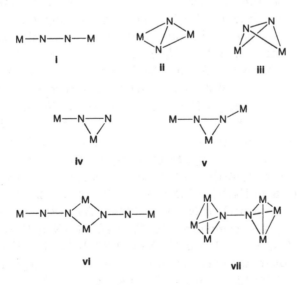

FIG. 1. Coordination modes demonstrated for the N_2 ligand bound to two or more metals.

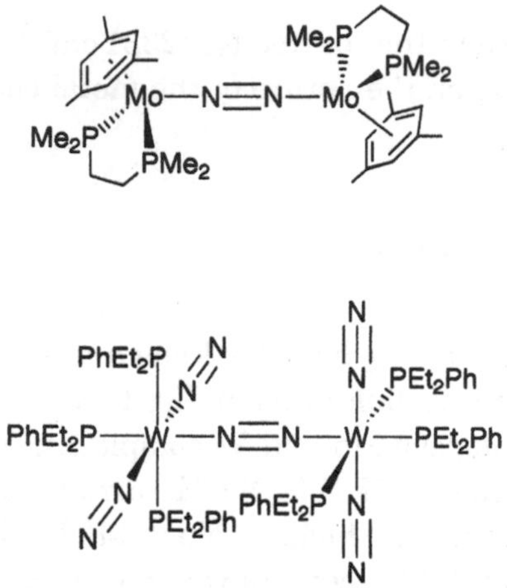

FIG. 2.   Structures of zero-valent dinuclear complexes with bridging N_2 ligand.

In contrast, the N_2 bridges between high-valent metal centers generally have much longer $N-N$ bond lengths along with the shorter $M-N$ bond distances, indicating that these N_2 units are rather interpreted as diazenido(2−) (N_2^{2-}) or hydrazido(4−) (N_2^{4-}) ligands represented as M=N=N=M or M≡N−N≡M, respectively. The N_2 ligands in the formal Mo(VI) and W(IV) hydrazido(4−) complexes, such as $[(Cp^*MoMe_3)_2(\mu-N_2)]$ ($d(N-N) = 1.236(3)$ Å) [16] and $[\{WCl_2(PhCCPh)(DME)\}_2(\mu-N_2)]$ ($d(N-N) = 1.292(16)$ Å) (Scheme 1) [17], are derived from deprotonation of hydrazine, while there also exist the compounds available straightforwardly by the use of atmospheric N_2, e.g., the formal M(IV) diazenido(2−) complexes $[\{[N_3N]M\}_2(\mu-N_2)]$ (M = Mo, R in [N$_3$N] = SiMe$_2$Bu^t: $d(N-N) = 1.20(2)$ Å; M = W, R in [N$_3$N] = neopentyl: $d(N-N) = 1.39(2)$ Å) [18] and the formal W(VI) hydrazido(4−) complexes $[(Cp^*WMe_2X)_2(\mu-N_2)]$ (X = Me, thiolates, phenolates; X = SMes: $d(N-N) = 1.27(2)$ Å) [19]. For comparison, the $N-N$ distances in cis- or trans-RN=NR (R = alkyl or aryl) and NH_2-NH_2 are 1.22-1.26 Å and 1.451(5) Å, respectively.

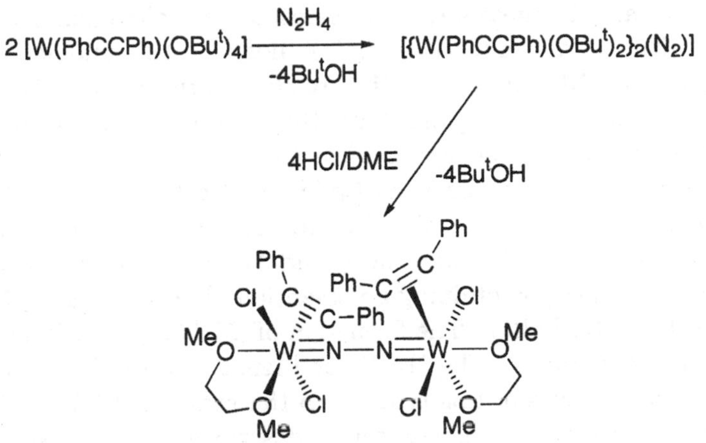

$$2 \text{ [W(PhCCPh)(OBu}^t)_4] \xrightarrow[\text{-4Bu}^t\text{OH}]{\text{N}_2\text{H}_4} \text{[\{W(PhCCPh)(OBu}^t)_2\}_2(\text{N}_2)]}$$

4HCl/DME $\diagup$ -4Bu^{t}OH

Scheme 1

2.3. Heterobimetallic N_2 Complexes

Numerous heterobimetallic complexes are known which contain an N_2 ligand bridging between Mo or W and another transition metal or a main-group metal. First examples of this type, e.g., *trans*-[Mo{(μ-NN)AlEt$_3$}$_2$(dppe)$_2$] [20] and [ReCl(PMe$_2$Ph)$_4$(μ-NN)MoCl$_4$(OMe)] [21], were synthesized through simple binding of Lewis acidic metal species to the remote basic N atom in the coordinated N_2 ligand in **1a** and *trans*-[ReCl(N$_2$)(PMe$_2$Ph)$_4$], respectively. More recently, [Mo(NBu$^t_2$Ph)$_3$(μ-N$_2$)U(NRAr)$_3$] (R = C(CD$_3$)$_2$CH$_3$, Ar = 3,5-C$_6$H$_3$Me$_2$) was prepared similarly from the reaction of in situ-generated [Mo(N$_2$)(NBu$^t_2$Ph)$_3$] with [U(NRAr)$_3$(THF)] (R = C(CD$_3$)$_2$Me, Ar = 3,5-C$_6$H$_3$Me$_2$) [22] as shown in equation 3.

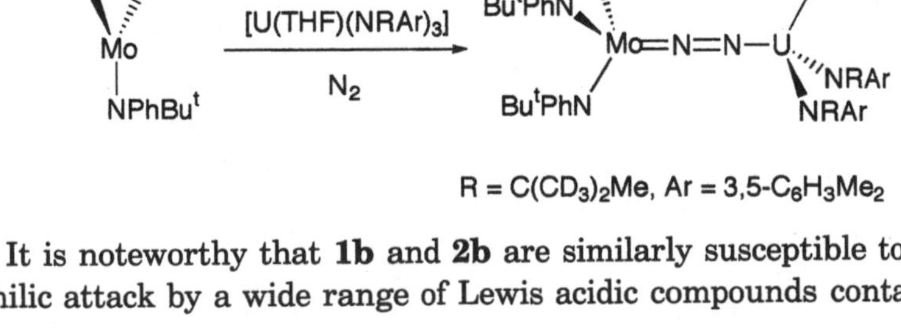

$$\text{(3)}$$

R = C(CD$_3$)$_2$Me, Ar = 3,5-C$_6$H$_3$Me$_2$

It is noteworthy that **1b** and **2b** are similarly susceptible to electrophilic attack by a wide range of Lewis acidic compounds containing group 4, 5, and 13 metals to give corresponding bimetallic μ-N_2 complexes with a diazenido(2−) (N_2^{2-}) ligand. An anionic complex [NBu^n_4][$W(N_2)(NCS)(dppe)_2$] (4b), readily derived from 1b, has a more electron-rich N_2 ligand than the parent neutral complex 1b and often shows higher reactivities. Complexes 1b, 2b, and/or 4b react with electrophiles such as $Cp'M'Cl_3$, $Cp'_2M'Cl_2$, $Cp^*_2M'Cl_2$/NaI, and $Cp_2ZrMeCl$/NaI (M = Ti, Zr, Hf, Cp' = Cp, Cp*) as well as $Cp'M'Cl_4$ (M′ = Nb, Ta), Me_3TaCl_2, and Me_2NbCl_3 to give a series of dinuclear complexes containing an essentially linear $XW(\mu$-$N_2)M'$ moiety (X = Cl, I, NCS) [23]. The formation of [$MCl(dppe)_2(\mu$-$N_2)TiCpCl_2$] is shown in equation 4. The N−N bond lengths in the fully characterized seven complexes of this type fall in the range 1.21-1.28 Å, indicating that the bridging N_2 moiety can be described as M=N=N=M. With respect to group 13 metal compounds, the reaction of 4b with $HCMe_2CMe_2BH_2$ and that of 2b with [9-BBN][OTf] give rise to the formation of the boryldiazenido(1−) complexes trans-[$W(NCS)\{NNB(H)CMe_2CMe_2H\}$ $(dppe)_2$] having a bent N−N−B linkage (133.6(6)°) and trans-[$W(OTf)(NN$-9-BBN)$(dppe)_2$] with an almost linear N−N−B array (168.3(7)°), respectively [24]. As for the reaction with Al compounds, it was demonstrated previously that from the reaction of 2b with $AlCl_3$ in the presence of py [$\{WCl(py)(PMe_2Ph)_3(\mu_3$-$N_2)\}_2(AlCl_2)_2$] was obtained, whose atom-connecting scheme was confirmed by a preliminary X-ray result [25]. A more recent study has disclosed that 1b reacts with GaX_3 (X = Cl, Br) or Ga_2Cl_4 to yield the analogous compounds [$\{WX(PMe_2Ph)_4$ $(\mu_3$-$N_2)\}_2(GaX_2)_2$] (equation 5), the X-ray analysis of the compound with X = Cl revealing considerably long N−N bonds (1.31(2) and 1.32(2) Å) [26]. The N−N distances determined by the X-ray analyses and the ν(NN) values observed in the IR spectra of these complexes fall in the range of N−N double bonds, where the W center has a formal oxidation state of +2.

$$(4)$$

$$M = W, Mo$$

$$(5)$$

2b $X = Cl, Br; \mathbf{P} = PMe_2Ph$

Heterobimetallic N_2 complexes containing higher valent Mo or W centers are also known widely. These complexes are derived from the metathetical replacement of group 1 and 2 metal ions attached to the remote N atom of the end-on N_2 ligand [or rather the diazenido(2−) or hydrazido(4−) ligand] by transition metal cations. By the use of this method, multinuclear diazenido(2−) type complexes [{[N$_3$N]Mo(μ-N$_2$)}$_n$**M**] ($n = 3$: **M** = Fe; $n = 2$: **M** = VCl(THF), ZrCl$_2$, etc.) [27] and the hydrazido(4−) type complex [Cp*WMe$_3$(μ-N$_2$)TaCp*Me$_2$] [28] were obtained. Synthesis of the Fe-Mo dinitrogen complex [{[N$_3$N]Mo(μ-N$_2$)}$_3$Fe], which might have some relevance to the active site cluster in MoFe nitrogenase, is shown in equation 6, while the stoichiometry for the formation of the latter W-Ta N_2 complex is depicted in Scheme 2.

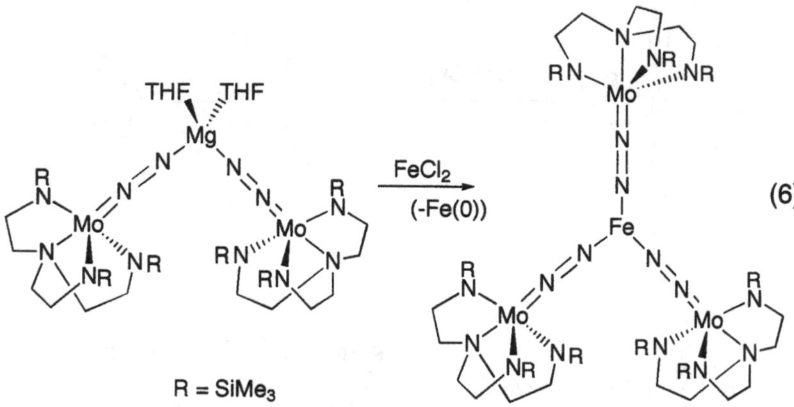

$$R = SiMe_3 \qquad (6)$$

$$[Cp^*WMe_4][PF_6] \xrightarrow[-[N_2H_5][PF_6]]{+2N_2H_4} [Cp^*WMe_4(NHNH_2)] \xrightarrow{-CH_4} [Cp^*WMe_3(NNH_2)]$$

$$[Cp^*WMe_3(NNH_2)] \xrightarrow[-2C_4H_{10}]{2Bu^nLi} [Cp^*WMe_3(NNLi_2)]_x$$

$$[Cp^*WMe_3(NNLi_2)]_x + 2Cp^*TaMe_3Cl$$

$$\xrightarrow[-2LiCl]{} [Cp^*WMe_3(\mu\text{-}N_2)TaCp^*Me_2] + [Cp^*TaMe_4]$$

Scheme 2

3.   REACTIONS OF COORDINATED DINITROGEN

3.1.   Direct N−N Bond Scission of a Bridging N₂ Ligand Forming Metal Nitrides

An interesting reaction causing direct cleavage of the N−N triple bond of a bridging N_2 ligand between two Mo atoms has been discovered by Cummins and his coworkers (Scheme 3). An Mo(III) complex with sterically encumbered amido ligands $[Mo(NRAr)_3]$ $[R = C(CD_3)_2Me,$

Ar = 3,5-$C_6H_3Me_2$] reacts with N_2 (1 atm) at $-35°C$ to give a spectro-scopically characterized dinuclear complex [{Mo(NRAr)$_3$}$_2$(μ-N_2)], which is readily converted into 2 equivalents of the Mo nitrido complex [N≡Mo(NRAr)$_3$] upon warming to 25°C [29]. The N−N bond order of about 2 for the intermediate μ-N_2 complex with a linear Mo=N=N=Mo array has been inferred from the Raman data as well as the x-ray absorption fine structure results. The N−N bond cleavage of this inter-mediate μ-N_2 complex is presumed to proceed through a transition state with a zig-zag Mo=N−N=Mo core, where a strong π-donating ability of the Mo(NRAr)$_3$ unit might make the N−N bond weaker to enhance the N−N bond scission [29b,30]. Unfortunately, the nitrido complex pro-duced is so stable that further conversion of the nitride ligand to other nitrogenous ligands or nitrogen-containing compounds has not been achieved. It is to be noted that direct scission of the N−N bond of a μ-N_2 ligand forming bridging nitride ligands has been observed more recently for a V diamidoamine complex [31] and an Nb calix[4]arene complex [32].

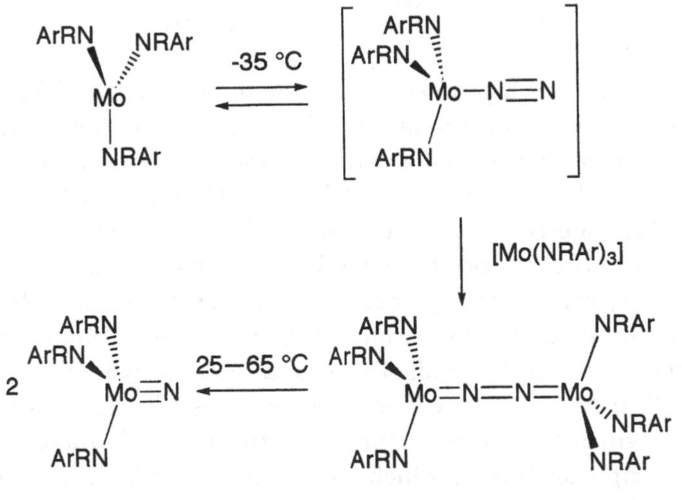

R = C(CD$_3$)$_2$Me, Ar = 3,5-$C_6H_3Me_2$

Scheme 3

3.2. Nucleophilic Reactions of the Remote Nitrogen Atom in a Terminal N_2 Ligand

3.2.1. Reactions with Inorganic Acids

In relation to the biological N_2 fixation, forming ammonia through coupled protonation and electron transfer (equation 7), reactivities of coordinated N_2 toward Brønsted acids have been investigated for a variety of well-defined N_2 complexes. Studies on reactions of 1 and 2 with protic acids have been carried out particularly because Mo has been believed to be a key metal in nitrogenase. Interestingly, protonation proceeds in a stepwise manner to give ammonia and/or hydrazine in significant yields, for which key intermediates are isolable and well characterizable. Although a considerable number of other transition metal N_2 complexes are known to yield nitrogen hydrides upon protonation, intermediate complexes containing a protonated N_2 ligand have never been isolated in these reactions.

$$N_2 + 8H^+ + 8e^- \xrightarrow{\text{nitrogenase}} 2NH_3 + H_2 \qquad (7)$$

Protonation of 2 proceeds quite readily, where the yields of nitrogen hydrides depend on the nature of the acid, solvent, and metal. The reaction courses for producing ammonia and hydrazine are summarized in Scheme 4. Thus, the reaction of 2b with H_2SO_4 in MeOH results in nearly stoichiometric conversion of one N_2 ligand into ammonia through successive $N-H$ bond formation with six protons accompanied by transfer of six electrons from W, whereas similar protonation of 2a leads to the formation of ammonia in much lower yield since it follows predominantly the route involving disproportionation of a hydrazido(2−) intermediate [33]. The reaction of the latter type has been demonstrated in a more convincing manner for a triphosphine complex, trans-$[Mo(N_2)_2(triphos)(PPh_3)]$, which, upon treatment with HBr in THF, produces about 0.72 mol of ammonia and 1.5 mol of N_2 per mole of the N_2 complex together with an almost quantitative amount of $[MoBr_3(triphos)]$ [34]. Differences in the reactivities between the complexes containing W and Mo as the central metal presumably arise from the more reducing nature of the former, which favors the oxidation reaction up to W(VI), while the latter undergoes oxidation only to

Mo(III). In contrast, reactions of the Mo and W complexes **2** with HCl gas in DME produce hydrazine as a major product and are believed to proceed via a hydrido-hydrazido(2−) intermediate [35]. Several hydrido-hydrazido(2−) complexes have been isolated from related reactions and characterized unequivocally by spectroscopy and X-ray analysis [35,36]. For the N_2 complexes with dppe ligands **1**, protonation with H_2SO_4 or HX aq. (**1a**: X = Br, I; **1b**: X = Cl, Br, I) in MeOH smoothly gives hydrazido(2−) complexes. However, further protonation does not occur under mild conditions [33a,37].

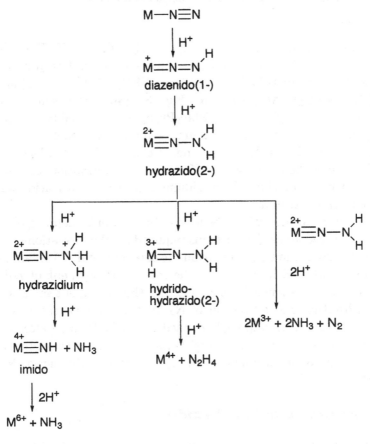

Scheme 4

Numerous compounds corresponding to the intermediate stages shown in Scheme 4 have been isolated to date which include dia-

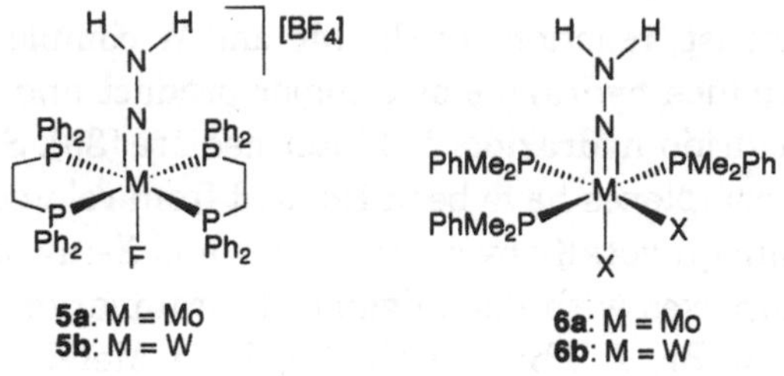

FIG. 3.   Structures of hydrazido(2−) complexes **5** and **6**.

zenido(1−) complexes with dppe ligands [MX(NNH)(dppe)$_2$] (X = halogen) [38], hydrazido(2−) complexes with PMe$_2$Ph or dppe ligands such as *trans*-[MF(NNH$_2$)(dppe)$_2$][BF$_4$] (**5**; Fig. 3) [39] and *cis*,-*mer*-[MX$_2$(NNH$_2$)(PMe$_2$Ph)$_3$] (**6**; Fig. 3) [40], and some hydrido-hydrazido(2−) complexes with PMe$_2$Ph ligands, e.g. [WClBr(NNH$_2$)-(PMe$_2$Ph)$_3$]Br [35]. Hydrazidium complexes such as *trans*-[WCl(NNH$_3$)(PMe$_3$)$_4$]Cl$_2$ [41] and *trans*-[WF(NNH$_3$)(depe)$_2$][BF$_4$]$_2$ [42] have also been prepared. However, the formation of ammonia upon treatment of these hydrazidium complexes with acids has not yet been demonstrated.

In these protonation reactions, high-valent metal species are finally produced since the electrons required for the formation of ammonia and hydrazine are supplied from the central metal. To obtain ammonia catalytically with respect to the metal, employment of reductive conditions is indispensable to regenerate the low-valent metal centers for coordination and activation of N$_2$. In this context, the reaction of *trans*-[W(OTs)(NNH$_2$)(dppe)$_2$]$^+$ carried out in THF under electroreductive conditions in the presence of 1 atm of N$_2$ is noteworthy, since it produces ammonia accompanied by regeneration of the parent N$_2$ complex **1b** [43].

3.2.2.   Reactions with Metal Hydrides

Complex **2b** also reacts with acidic transition metal hydrides such as [HCo(CO)$_4$], [H$_2$Fe(CO)$_4$], and [HFeCo$_3$(CO)$_{12}$] to afford ammonia in moderate yields, when the reaction mixtures are subjected to base distillation [44]. It is noteworthy that not only these acidic metal hydrides but also the hydridic metal compounds such as [Cp$_2$ZrHCl] and Na[AlH$_2$(OCH$_2$CH$_2$OMe)$_2$] are employed to obtain ammonia [44a]. Since no intermediates have been characterized for these reactions with metal hydrides, the mechanism operating in these systems is still unknown. However, a series of hydrazido(2−) complexes trans-[W(OR)(NNH$_2$)(dppe)$_2$][A] [A = Co(CO)$_4$, FeCo$_3$(CO)$_{12}$] were isolated from the related system' **1b** with [HCo(CO)$_4$] or [HFeCo$_3$(CO)$_{12}$] in various ROH [44c].

3.2.3. Reactions with an Acidic H_2 Ligand

It seems quite difficult to carry out effectively both protonation and electroreduction or chemical reduction in a one-pot manner. Hence, reduction of N_2 using H_2 gas as both a proton source and a reducing agent is expected to be more promising. However, it is commonly recognized that coordinated N_2 dissociates readily from a metal without N−H bond formation under H_2. Interestingly, the H_2 complex [CpRu(dtfpe)H$_2$][BF$_4$] has been found to react with **1b**, giving the hydrazido(2−) complex **5b** and a Ru hydride as shown in equation 8. The H_2 complex exists in solution as a mixture of slowly interconverting η^2-H$_2$ and dihydride species, the pK_a values of which are 4.3 and 4.4, respectively. However, it is to be noted that this H_2 complex is not available directly from gaseous hydrogen [45].

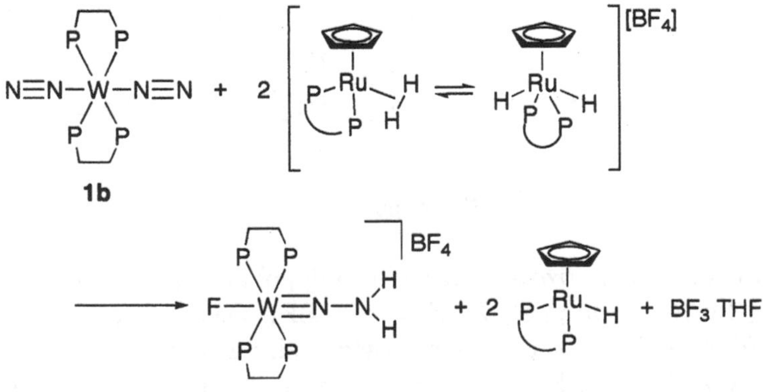

(8)

1b

5b   P⌣P = dppe, P⌣P = dtfpe

This has led to the recent finding [46] that ammonia is formed by Ru-assisted protonation of N_2 coordinated to W with H_2. The complex $[RuCl(dppp)_2][PF_6]$ is partially converted to $trans$-$[RuCl(\eta^2$-$H_2)(dppp)_2][PF_6]$ upon treatment with H_2 (1 atm) at room temperature in benzene-dichloromethane, the ratio of the former to the latter complex being about 9:1 [47]. The pK_a value of this H_2 complex is estimated to be 4.4. When 2b is allowed to react with this mixture containing totally 10 equivalents of these Ru complexes at 55°C under 1 atm of H_2, 0.55 mol NH_3/W atom is produced after 24 h, for which 0.10 mol NH_3 is present in a free form in the reaction mixture, while the remaining 0.45 mol NH_3 is released after base distillation of the evaporated reaction mixture residue [46]. The yield of hydrazine is negligible. After the reaction, the formation of $trans$-$[RuHCl(dppp)_2]$ (1.50 mol/mol 2b charged) is observed. Although 2b is completeley consumed with concurrent liberation of small amounts of PMe_2Ph and $[HPMe_2Ph]^+$, the fate of W is uncertain. Based on these observations, the ideal stoichiometry of the reaction between 2b and the H_2 complex $trans$-$[RuCl(\eta^2$-$H_2)(dppp)_2][PF_6]$ may be described by equation 9.

$$
\textbf{2b} \qquad\qquad (9)
$$

$$
\xrightarrow[-N_2]{H_2\ (1\ atm),\ 55\ °C}\quad 2\,NH_3\ +\ 6\left(\ \right)\ +\ W^{6+}\ \text{species}
$$

$$
\overset{P\ \frown\ P}{} = dppp
$$

In contrast to $trans$-$[RuCl(\eta^2$-$H_2)(dppp)_2][PF_6]$, the H_2 complexes $trans$-$[RuCl(\eta^2$-$H_2)(dppe)_2]X$ (X = PF_6, BF_4, OTf) are generated from $[RuCl(dppe)_2]X$ in a quantitative yield under 1 atm of H_2, and the H_2 complex with X = PF_6 has a pK_a value of 6.0. The reactions with 2b under similar conditions result in the formation of ammonia in the yields of 0.55-0.79 mol/W atom [46b]. On the other hand, the yield of

NH$_3$ decreases markedly when H$_2$ complexes with pK_a values larger than 10 are employed. The presence of free ammonia in the reaction mixture clearly indicates that conversion of the coordinated N$_2$ into ammonia proceeds by the use of H$_2$ gas under these mild conditions.

The N−H bond formation is presumed to proceed by direct nucleophilic attack of the remote N atom in the ligating N$_2$ on the coordinated H$_2$. This leads to the heterolytic cleavage of H$_2$ into proton and hydride accompanied by the formation of a N−H bond and the hydride complex trans-[RuHCl(dppe)$_2$], respectively. It should be noted that the yield of ammonia also depends markedly on the nature of the dinitrogen complexes. Thus, the reaction of trans-[W(N$_2$)$_2$(PPh$_2$Me)$_4$] (7) with 10 equivalents of trans-[RuCl(η^2-H$_2$)(dppe)$_2$][OTf] results in the formation of only 0.05 mol NH$_3$/W atom, while the analogous reaction of 2a gives no ammonia. In contrast, both of the reactions of 2b and 7 with 10 equivalents of HOTf afford ammonia in 1.0-1.2 mol/W atom, while protonation of 2a with H$_2$SO$_4$ produces 0.68 mol NH$_3$/Mo atom.

When 2b is treated with 2 equivalents of trans-[RuCl(η^2-H$_2$)(dppe)$_2$][OTf], the reaction proceeds very rapidly at 55°C or even at room temperature to form a fully characterized hydrazido(2−) intermediate complex trans-[W(OTf)(NNH$_2$)(PMe$_2$Ph)$_4$][OTf] together with trans-[RuHCl(dppe)$_2$] [46b]. At 55°C, this hydrazido(2−) complex is slowly transformed into another hydrazido(2−) complex trans-[WCl(NNH$_2$)(PMe$_2$Ph)$_4$][OTf] by reacting with trans-[RuHCl(dppe)$_2$], and these two hydrazido(2−) complexes afford ammonia by further treatment with trans-[RuCl(η^2-H$_2$)(dppe)$_2$][OTf] (Scheme 5). The W-N$_2$ complex 7 is also rapidly converted to the hydrazido(2−) stage trans-[W(OTf)(NNH$_2$)(PPh$_2$Me)$_4$][OTf] by treatment with trans-[RuCl(η^2-H$_2$)(dppe)$_2$][OTf]. However, in contrast to the corresponding PMe$_2$Ph complex, this hydrazido(2−) complex is presumably not susceptible to further electrophilic attack by the H$_2$ ligand, which leads to the poor yield of ammonia (vide supra). The Mo-N$_2$ complex 2a does not give even the corresponding hydrazido(2−) complex under these conditions. The differences observed might be ascribable to the weaker electron-donating ability of PPh$_2$Me than PMe$_2$Ph and of Mo than W.

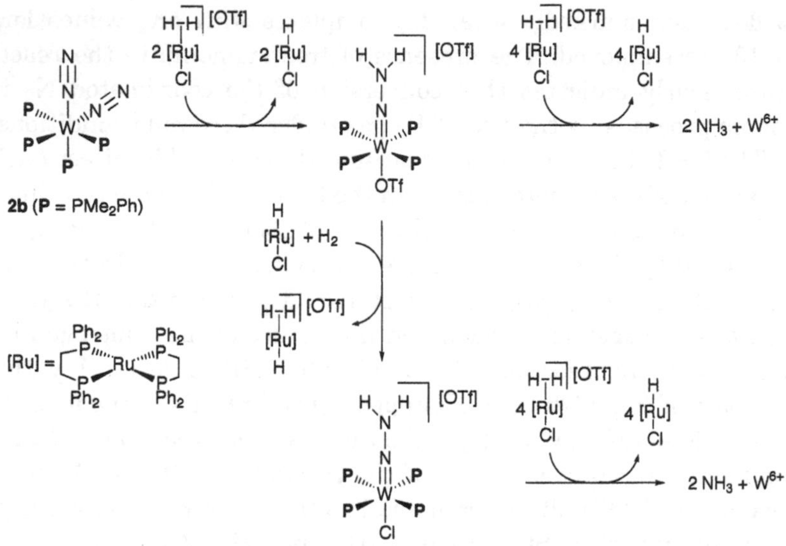

Scheme 5

3.2.4. Reactions with Metal Hydrosulfides or H_2 Activated on a Metal-Sulfido Site

The ligating N_2 in **1** and **2** is not protonated upon treatment with organic thiols and H_2S. From these reactions certain thiolato complexes or a sulfido-hydrosulfido cluster are obtained, e.g. [Mo(SR)$_2$(dppe)$_2$] (R = Et, Pr^n, Bu^n, Ph) [48], [W$_4$(μ_3-S)$_2$(μ-S)$_4$(SH)$_2$(PMe$_2$Ph)$_6$] [49], and [{Mo(SBu^t)$_2$(PMe$_2$Ph)}$_2$(μ-S)$_2$] [50] with concomitant liberation of N_2 gas. Interestingly, the reactions of **2b** with 10 equivalents of hydrosulfido-bridged dinuclear complexes such as [Cp*Ir(μ-SH)$_3$IrCp*]Cl and [Fe(triphos)(μ-SH)$_3$Fe(triphos)][BF$_4$] at 55°C under N_2 for 24 h afford a small amount of free ammonia, and the formation of further ammonia is observed after base distillation of the evaporated reaction mixture residue (see, e.g., equation 10). The combined yields of ammonia reach 0.78 and 0.38 mol/mol **2b**, respectively [51]. Other μ-hydrosulfido complexes, such as [Cp*Rh(η-SH)$_3$RhCp*]Cl, [Cp*Ir(μ-SH)$_2$IrCp*Cl], and [Cp*RuCl(μ-SH)$_2$TiCp$_2$], also react with **2b** to give the mixtures which liberate ammonia after base distillation, but the yields are quite low (<0.09 mol/mol **2b**). These observations are of particular interest since several groups have claimed that protonation of N_2 in the FeMo cofactor proceeds with the aid of the bridging hydrosulfido ligands [52]; protons first bind to the bridging sulfides in the MoFe$_7$S$_9$ cluster and then migrate to the adjacent N$_2$ molecule coordinated to the multi-iron site [52a].

$$\mathbf{2b}$$

The reaction of **1b** with [Cp*Ir(μ-SH)$_3$IrCp*]Cl or [Fe(triphos)(μ-SH)$_3$Fe(triphos)][BF$_4$] at 55°C affords *trans*-[WCl(NNH$_2$)(dppe)$_2$]Cl and **5b**, respectively, in moderate yields, while treatment of **1b** with [Fe(triphos)(μ-SD)$_3$Fe(triphos)][BF$_4$] results in the formation of a significant amount of the NND$_2$ analogue of **5b**. These findings suggest that the above protonation of **2b** to give ammonia also proceeds via a hydrazido(2−) intermediate formed by the transfer of protons from the μ-SH ligands to the coordinated N$_2$ [51].

As an extension, reactions of W dinitrogen complexes with the sulfido-bridged dimolybdenum complexes [Cp$'$Mo(μ-S$_2$CH$_2$)(μ-S)(μ-SR)MoCp$'$][OTf] (Cp$'$ = Cp or η^5-C$_5$H$_4$Me, R = H or Me) under H$_2$ have been investigated, since the latter complexes are known to facilitate the cleavage of the C≡N bond in nitriles under H$_2$ [53] and heterolytic cleavage of H$_2$ by reaction with pyridine under H$_2$ [54].

When **2b** and **7** are treated with 10 equivalents of these dimolybdenum complexes (Cp$'$ = Cp: R = H, Me; Cp$'$ = η^5-C$_5$H$_4$Me: R = H) at 25°C and then 55°C under H$_2$, the formation of 0.33-0.52 mol NH$_3$/W atom is observed [55]. The yield of free ammonia is ca. 0.05 mol/W atom or less, while most of ammonia produced is detected as NH$_4^+$ in the water extract of the reaction mixture. The NMR spectrum of the reaction mixture of **2b** with 2 equivalents of the dimolybdenum complex (Cp$'$ = Cp, R = H) at 25°C under H$_2$ has manifested the formation of the hydrazido(2−) complex *trans*-[W(OTf)(NNH$_2$)(PMe$_2$Ph)$_4$][OTf] in 39% yield. Furthermore, it has turned out that treatment of this hydrazido(2−) complex with 10 equivalents of the dimolybdenum complex at 25°C and then at 55°C under H_2 gives ammonia in the yield of 0.99 mol/W atom.

When the dppe-N_2 complex **1b** is used instead of **2b**, conversion of the ligating N_2 into a hydrazido(2−) ligand has been demonstrated in a well-defined manner. For example, equation 11 shows the stoichiometry clarified for the reaction of **1b** with 2 equivalents of the dimolybdenum complex (Cp$' = \eta^5$-C_5H_4Me, R = Me) at 25°C under H_2 [55]. These reactions are noteworthy in that heterolytic cleavage of H_2 is accomplished on the metal-sulfido site, leading to the N−H bond formation of coordinated N_2.

$$(11)$$

3.2.5. Direct Conversion of Coordinated N_2 into Organonitrogenous Ligands

Conversion of ligating N_2 into organonitrogenous compounds is of particular interest. If high-value-added nitrogenous compounds can be synthesized directly from molecular nitrogen, such reactions, being fundamentally different from the present biological and industrial N_2-fixing processes, are presumably of great industrial importance. Results of the studies on the C−N bond forming reactions conducted in this context are summarized below in Secs. 3.2.5, 3.2.6, and 3.3. It might be noteworthy that conversion of coordinated N_2 into organonitrogenous ligands and compounds is observed quite rarely for N_2 complexes of transition metals other than Mo and W; the well-defined reactions are, to our knowledge, limited to the formation of acyl- or aroyldiazenido(1−) complexes from [ReCl(N$_2$)(L)(PMe$_2$Ph)$_3$] (L = py, PMe$_2$Ph) and RCOCl [56] and that of [CpMn(CO)$_2$(MeN=N−)Li$^+$] from [CpMn(CO)$_2$(N$_2$)] and MeLi [57]. The latter is quite different from the former reactions in that the nitrogen adjacent to the metal is attacked by a nucleophile, methyl carbanion. When the latter reaction product is treated with [Me$_3$O][BF$_4$] and then with pressurized N$_2$, MeN=NMe is liberated from the metal, accompanied by regeneration of the parent N$_2$ complex in low yield.

The first unambiguous C−N bond formation at ligating N$_2$ was found in the acylation and aroylation of 1 with RCOCl [58]. These reactions are considered to proceed via nucleophilic attack of the remote N atom in the N$_2$ ligand on the electron-deficient acyl and aroyl carbons [6f]. Thus, 1 reacts with RCOCl to give acyl- or aroyldiazenido(1−) complexes trans-[MCl(NNCOR)(dppe)$_2$] together with the hydrazido(2−) derivatives trans-[MCl(NNHCOR)(dppe)$_2$]Cl. The latter arises from the former through protonation by in situ-generated HCl impurity [56]. The benzoyldiazenido(1−) complex [Mo(NRAr)$_3$(NNCOPh)] is also derived from [Mo(NRAr)$_3$(μ-N$_2$)Na(THF)$_x$] and PhCOCl (R = 1-adamantyl, Ar = 3,5-C$_6$H$_3$Me$_2$) [59]. The other C−N bond formation clarified in detail is the alkylation of N$_2$ ligand in 1 by alkyl halides, which generally proceeds under photolytic conditions through the attack of alkyl radicals (see Sec. 3.3).

The remote N atom in coordinated N$_2$ undergoes electrophilic attack by silyl iodides. Thus, 1 and 2 react with Me$_3$SiI at 50°C in benzene under rigorously dry conditions to afford the silyldiazenido(1−) complexes [MI(NNSiMe$_3$)(dppe)$_2$] and trans-[MI(NNSiMe$_3$)(PMe$_2$Ph)$_4$] (equation 12). In the presence of in situ-generated HI from Me$_3$SiI and adventitious moisture, the silylhydrazido(2−) complexes [MI(NNHSiMe$_3$)(dppe)$_2$]I and cis,mer-[MI$_2$(NNHSiMe$_3$)(PMe$_2$Ph)$_3$] become the major products [60]. Numerous silyldiazenido(1−) complexes can be prepared more conveniently by treatment of 2b with a series of easily available R$_3$SiCl in the presence of excess NaI, and furthermore this method is applicable to the synthesis of the germyldiazenido(1−) complexes trans-[MI(NNGeR$_3$)(PMe$_2$Ph)$_4$] (R = Me, Ph) (equation 12) [61]. Other silyl compounds, such as Me$_3$SiOTf and [Ph$_2$MeSiCo(CO)$_4$], also serve as the silylating reagents for 1 and/or 2, yielding, e.g., trans-[W(OTf)(NNSiMe$_3$)(PMe$_2$Ph)$_4$] [62]

and *trans*-[{M(NNSiPh$_2$Me)(dppe)$_2$}(μ-OC)Co(CO)$_3$], respectively [63]. Disilylation of the remote N atom in ligating N$_2$ has been attained by treating **2b** with ClSiMe$_2$CH$_2$CH$_2$SiMe$_2$Cl/NaI to give the disilylhydrazido(2−) complex *cis,mer*-[WI$_2$(NNSiMe$_2$CH$_2$CH$_2$SiMe$_2$)(PMe$_2$Ph)$_3$] [64]. Other silylated complexes derived more recently from N$_2$ complexes and silyl chlorides include [Mo(NRAr)$_3$(NNSiMe$_3$)] [44] and [Mo(N$_3$N)(NNSiPr$_3^i$)] [18a].

$$
\begin{array}{c}
\underset{\textbf{2}}{
\begin{array}{c}
\text{N} \\ \text{|||} \\ \text{N} \\
\text{PhMe}_2\text{P}_{\prime\prime\prime\prime\prime\prime}\text{M}_{\prime\prime\prime}\text{N}{\equiv}\text{N} \\
\text{PhMe}_2\text{P} \qquad \text{PMe}_2\text{Ph} \\
\text{PMe}_2\text{Ph}
\end{array}}
\quad
\xrightarrow[\text{R}_3\text{ECl/NaI}]{\text{Me}_3\text{SiI or}}
\quad
\begin{array}{c}
\text{N}{-}\text{ER}_3 \\ \text{||} \\ \text{N} \\
\text{PhMe}_2\text{P}_{\prime\prime\prime\prime\prime\prime}\text{M}_{\prime\prime\prime}\text{PMe}_2\text{Ph} \\
\text{PhMe}_2\text{P} \qquad \text{PMe}_2\text{Ph} \\
\text{I}
\end{array}
\qquad (12)
$$

$$\text{E = Si, Ge; R = alkyl, aryl}$$

Detailed investigations of the reactivities of the silyldiazenido(1−) complexes *trans*-[MI(NNSiMe$_3$)(PMe$_2$Ph)$_4$] (M = Mo, W) have finally led to the exploitation of a catalytic N$_2$-fixing system to produce silylamines. Thus, treatment of Me$_3$SiCl with an equimolar amount of Na in THF under N$_2$ (1 atm) in the presence of **1** or **2** results in the formation of N(SiMe$_3$)$_3$ together with a small amount of HN(SiMe$_3$)$_2$ (equation 13) [65]. The optimum yield of 37% ($\sim$24 mol silylamines/Mo atom) is attained from the reaction at 30°C for 15 h using 0.5 mol% of **2a**. Among the complexes **1** and **2**, the catalytic activity decreases in the order **2a** $\gg$ **1a** $\gg$ **1b**, **2b**. The major byproduct is Me$_3$SiSiMe$_3$, the formation of which is also enhanced by employment of **1** and **2**. In addition to Me$_3$SiCl, a range of chlorosilanes can be used similarly for the catalytic conversion of N$_2$ to silylamines, although the turnover numbers are much lower [61].

$$
\text{N}_2 + \text{Me}_3\text{SiCl} + \text{Na} \xrightarrow{\textbf{1 or 2}} \tag{13}
$$
$$
\text{N(SiMe}_3)_3 + \text{HN(SiMe}_3)_2 + \text{Me}_3\text{SiSiMe}_3
$$

The mechanism for this catalytic reaction is not clear. However, it is likely that the silyl radical generated by homolytic cleavage of R$_3$SiCl within the coordination sphere of Mo or W attacks the N$_2$ ligand in the initial step to give a silyldiazenido(1−) species. Attack of alkyl radicals on coordinated N$_2$ has been unambiguously demonstrated in the reactions of **1** with alkyl halides to form alkyldiazenido(1−) complexes, which provide another important route to the C−N bond formation at coordinated N$_2$ (Sec. 3.3).

3.2.6. Formation of Organonitrogenous Ligands via a Hydrazido(2−) Intermediate

Detailed studies on the reactivities of the hydrazido(2−) complexes **5** and **6**, which are readily available from **1** and **2** as intermediates in the protonation reactions (vide supra), have disclosed that these can also serve as versatile precursors to form organonitrogenous ligands and compounds. In the typical reactions, **5** and **6** undergo electrophilic attack by a range of organic carbonyl compounds owing to the substantial nucleophilicity of the remote nitrogen of the hydrazido(2−) ligand.

Thus, **5b** undergoes reactions with succinyl chloride to give a diacylhydrazido(2−) complex containing the $\overline{\text{NNCOCH}_2\text{CH}_2\text{CO}}$ ligand [66], while **6b** (X = Cl) reacts with a ketene Ph$_2$C=C=O to afford *cis,mer*-[MCl$_2$(NNHCOCHPh$_2$)(PMe$_2$Ph)$_3$] [67].

More versatile reactions of the hydrazido(2−) complexes **5** and **6** are the condensation with ketones and aldehydes to form diazoalkane complexes (equations 14 and 15) [68]. Interestingly, the N$_2$ complex **1b** reacts with acetylacetone to give an alkenyldiazenido(1−) complex *mer*-[W(acac)(NNCMe=CHCOMe)(PMe$_2$Ph)$_3$]. This reaction probably proceeds through the initial formation of the hydrazido(2−) intermediate *mer*-[W(acac)(NNH$_2$)(PMe$_2$Ph)$_3$][acac]. This is followed by condensation with acetylacetone to give a diazoalkane species, and subsequent elimination of one acetylacetone from the latter gives rise to the final product [69].

$$(14)$$

$$(15)$$

Direct conversion of the ligating N_2 in **2b** into acetone azine is accomplished by treating **2b** with an MeOH/acetone mixture at 50°C. It is likely that this reaction also involves hydrazido(2−) and diazoalkane species as intermediates [70]. Furthermore, treatment of **2b** in benzene-acetone with [RuCl(dppp)$_2$]X (X = PF$_6$, BF$_4$, OTf) under H$_2$ at 55°C [46a] gives acetone azine in good yield. The reaction is also believed to proceed via the initial formation of a hydrazido(2−) complex followed by the condensation with acetone to form a diazoalkane complex. Acetone azine is released by the reaction of the latter complex with the H$_2$ complex trans-[RuCl(η^2-H$_2$)(dppp)$_2$]X. Indeed, similar treatment of **1b** with [RuCl(dppp)$_2$]X (X = PF$_6$, BF$_4$) under H$_2$ results in the formation of the diazoalkane complexes trans-[WF(NN=CMe$_2$)(dppe)$_2$]X.

The condensation reactions of the hydrazido(2−) complexes **5** and **6** have been extended to the synthesis of nitrogen heterocycles. Thus, when these complexes are allowed to react with dialdehydes or their equivalents, the complexes containing heterocyclic ligands are obtained as summarized in Scheme 6, from which heterocyclic compounds can be liberated under certain reaction conditions.

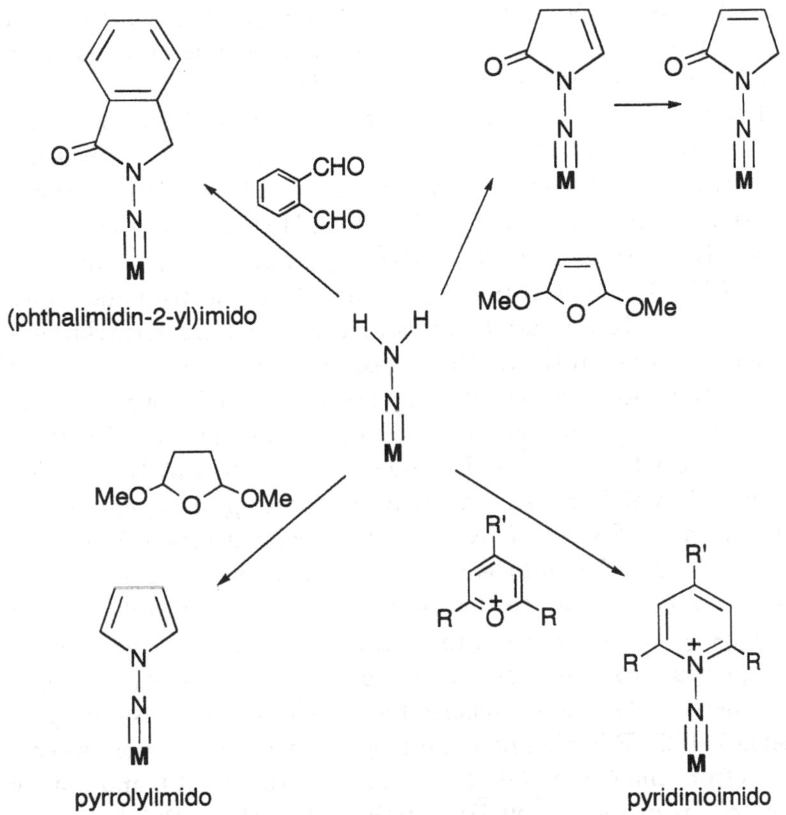

Scheme 6

Reaction of **6b** (X = Cl, Br) with phthalaldehyde in the presence of a catalytic amount of HCl or HBr affords the (phthalimidin-2-yl)imido complexes cis,mer-[WX$_2$(NNCH$_2$C$_6$H$_4$CO)(PMe$_2$Ph)$_3$]. Although the reaction of **5b** under similar conditions results in the formation of only the monocondensation product $trans$-[WF(NN=CHC$_6$H$_4$CHO)(dppe)$_2$][BF$_4$], this diazoalkane complex is transformed into the desired (phthalimidin-2-yl)imido complex $trans$-[WF(NNCH$_2$C$_6$H$_4$CO)(dppe)$_2$][BF$_4$] by further reaction with AlCl$_3$ in refluxing THF. The reaction of cis,mer-[WCl$_2$(NNCH$_2$C$_6$H$_4$CO)(PMe$_2$Ph)$_3$] with HBr gas in CH$_2$Cl$_2$ liberates 2-aminophthalimidine, while treatment with KOH in methanol, ethanol, or THF produces phthalimidine and ammonia [71]. In an analogous manner, **5** reacts with 2,5-dimethoxy-2,5-dihydrofuran to give initially $trans$-[MF(NNCH=CHCH$_2$CO)(dppe)$_2$][BF$_4$], which is gradually converted to a thermodynamically more stable compound $trans$-[MF(NNCH$_2$CH=CHCO)(dppe)$_2$][BF$_4$] [71].

Employment of 2,5-dimethoxytetrahydrofuran led to the synthesis of pyrrolylimido ligands. Thus, the cyclic acetal of succinaldehyde reacts with **5** and **6b** to form $trans$-[MF(NNCH=CHCH=CH)(dppe)$_2$] [BF$_4$] and cis,mer-[WX$_2$(NNCH=CHCH=CH)(PMe$_2$Ph)$_3$] (X = Cl, Br), respectively. Interestingly, the pyrrole ring in $trans$-[MF(NNCH=CHCH=CH)(dppe)$_2$]$^+$ undergoes electrophilic substitutions, including bromination, cyanation, sulfonation, formylation, and acylation predominantly at the β position owing to the steric effect caused by the dppe ligands, although free pyrrole is known to be susceptible to electrophilic attack at the α position exclusively. Treatment of $trans$-[MF(NNCH=CHCH=CH)(dppe)$_2$][BF$_4$] with LiAlH$_4$ followed by workup with MeOH results in the formation of pyrrole, ammonia, and N-aminopyrrole. For the W complex, the reaction proceeds more selectively via the N−N bond cleavage to give pyrrole and ammonia as the major products. It is to be noted that a significant amount of [MH$_4$(dppe)$_2$] is isolated from these reaction mixtures. Since the parent N$_2$ complexes **1** can be regenerated from these tetrahydrido complexes, the synthetic cycle is accomplished for pyrroles from the N$_2$ complexes **1** (Scheme 7) [72]. Pyrrole and N-aminopyrrole are much more selectively released from the pyrrolylimido complexes with PMe$_2$Ph and π-acceptor ligands $cis,trans$-[WX$_2$(NNCH=CHCH=CH)(L)(PMe$_2$Ph)$_2$]. For example, the reaction of the complex (X = Br, L = Bu^{t}NC) with LiAlH$_4$ followed by methanolysis gives pyrrole and ammonia almost quantitatively, whereas treatment of the complexes (X = Cl, Br; L = CO, PhC≡CH, PhCHO) with KOH in alcohol results in the quantitative formation of N-aminopyrrole [73].

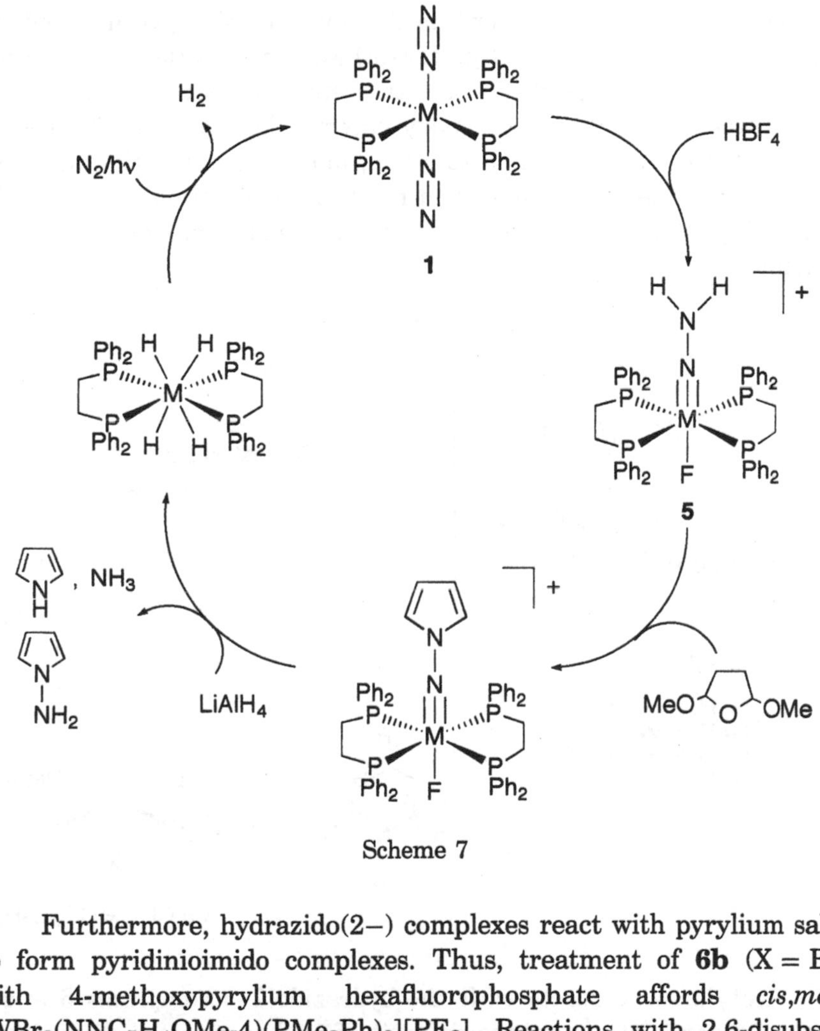

Scheme 7

Furthermore, hydrazido(2−) complexes react with pyrylium salts to form pyridinioimido complexes. Thus, treatment of **6b** (X = Br) with 4-methoxypyrylium hexafluorophosphate affords *cis,mer-*[WBr$_2$(NNC$_5$H$_4$OMe-4)(PMe$_2$Ph)$_3$][PF$_6$]. Reactions with 2,6-disubstituted pyrylium salts hardly proceed in a similar manner. However, the sterically less hindered hydrazido(2−) complexes *cis,trans-*[WX$_2$(NNH$_2$)(L)(PMe$_2$Ph)$_2$] (X = Cl, Br; L = CO, C$_2$H$_4$), readily available from **6b**, smoothly react not only with 4-methoxypyrylium salt but also with 2,4,6-trimethylpyrylium and 2,6-diethoxycarbonyl-4-phenylpyrylium salts, forming the corresponding pyridinioimido complexes [74]. These pyridinioimido complexes with a π-acceptor ligand undergo facile N−N bond cleavage upon reduction with cobaltocene in the presence of a proton source, affording the corresponding pyridine and the imido complex $cis,trans$-[WCl$_2$(NH)(L)(PMe$_2$Ph)$_2$] (equation 16). Since the pyridinioimido ligand can be interpreted as a substituted hydrazidium ligand, the $N-N$ bond cleavage observed here, which is induced by the coupled reduction and protonation, might represent the first well-defined example for demonstrating the degradation step of the $MN-NH_3^+$ stage into the imido species and ammonia shown in Scheme 4.

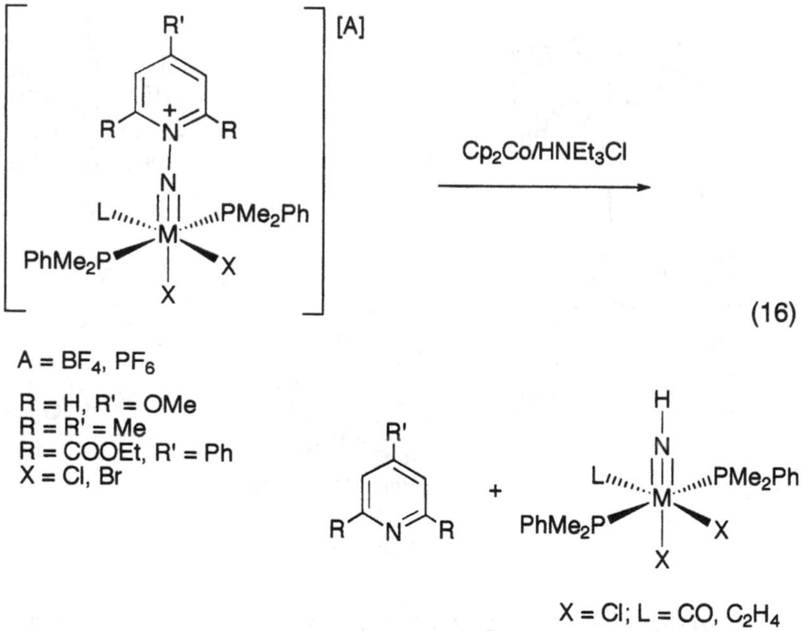

$$\text{(16)}$$

A = BF$_4$, PF$_6$

R = H, R' = OMe
R = R' = Me
R = COOEt, R' = Ph
X = Cl, Br

X = Cl; L = CO, C$_2$H$_4$

Acetylacetone reacts with the hydrazido(2−) complexes **5** and **6** in the presence of a catalytic amount of acid to give the diazoalkane complexes $trans$-[MF(NNC=MeCH$_2$COMe)(dppe)$_2$][BF$_4$] and cis,mer-[MX$_2$(NN=CMeCH$_2$COMe)(PMe$_2$Ph)$_3$] (X = Cl, Br). The NMR studies show that the diazoalkane ligand in the latter PMe$_2$Ph complexes is present as a mixture of keto and enol forms in solution [68a–c,69]. Similarly, a series of diazoalkane complexes [WCl$_2$(NN=CRCHR'COR'')(L)(PMe$_2$Ph)$_2$] are obtained from [WCl$_2$(NNH$_2$)(L)(PMe$_2$Ph)$_2$] (L = PMe$_2$Ph, CO) and β-diketones RCOCHR'COR'' (R, R'' = alkyl, Ph; R' = H, Me), which give rise to the formation of pyrazoles by treatment with a KOH/EtOH mixture (Scheme 8). Reaction of **2b** with a PhCOCH$_2$COMe/EtOH/KOH mixture also resulted in the formation of 5-methyl-3-phenylpyrazole but in lower yield [75].

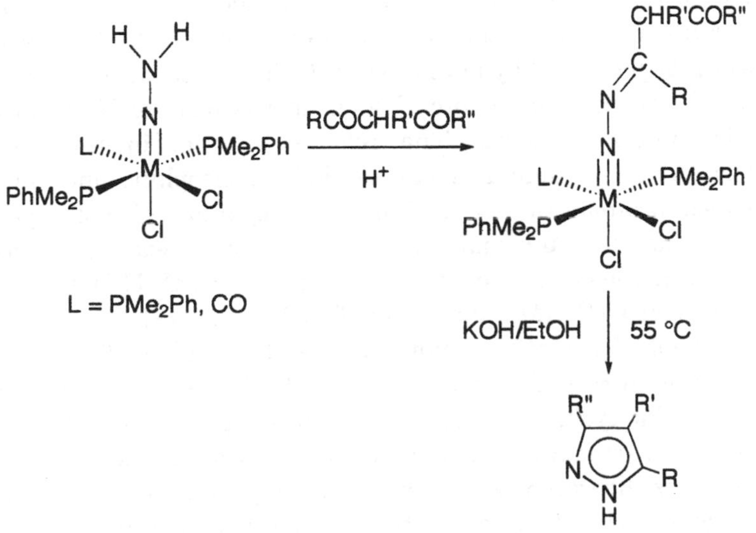

Scheme 8

The remote N atom in a diazenido(1−) moiety also undergoes electrophilic attack by organic halides. Conversion of hydrazido(2−) complexes into alkenyldiazenido(1−) complexes is also successful by reacting the diazenido(1−) complexes *trans*-[WX(NNH)(dppe)$_2$], generated from **5b** (X = F, Br) through deprotonation, with strongly activated cyanoalkenes bearing leaving groups such as Cl and OEt. When **5b** is treated with an equimolar amount of the alkene RR′C=CR″Z in the presence of a slight molar excess of NEt$_3$, *trans*-[WX(NNCR″=CRR′)(dppe)$_2$] is obtained (e.g., R = R′ = CN, R″ = H or Cl, Z = Cl or OEt) [76]. Alkenyldiazenido(1−) complexes of this type without electron-withdrawing substituents have later been derived from the reactions of the diazoalkane complexes *trans*-[MF{NN=C(R″)CHRR′}(dppe)$_2$][BF$_4$] (M = Mo, W) with strong bases such as LiNPr$_2^i$ and NaN(SiMe$_3$)$_2$ [77].

3.3. Attack of Carbon Radicals on the Remote Nitrogen Atom in a Terminal N_2 Ligand

Reactions of 1 with a variety of alkyl halides RX under irradiation of a W lamp afford the corresponding alkyldiazenido(1−) complexes trans-[MX(NNR)(dppe)$_2$] through a radical mechanism [56,78]; the alkyl radical, which is generated by homolysis of RX in the coordination sphere of Mo or W, attacks the remote N atom in the ligating N_2 to give the alkyldiazenido(1−) ligand. The remote N atom in the resulting alkyldiazenido(1−) ligand is substantially nucleophilic and therefore undergoes protonation by acid or the alkylation of S_N2-type. The alkylhydrazido(2−) and dialkylhydrazido(2−) complexes thus obtained include numerous compounds of the types trans-[MX(NNHR)(dppe)$_2$]$^+$ and trans-[MBr(NNRMe)(dppe)$_2$]Br (R = Me, Et) [79].

Dialkylhydrazido(2−) complexes are also available from the reactions of 1 with α,ω-dibromoalkane Br(CH$_2$)$_n$Br ($n = 2$–4) through the intramolecular cyclization [80]. It is interesting to note that this reaction has been extended to accomplish the cyclic system where N-amino-piperidine is liberated with concurrent regeneration of the parent N_2 complex 1a (Scheme 9) by electroreduction of trans-[MoBr{N$\overline{\text{N(CH}_2\text{)}_4\text{CH}_2}$}(dppe)$_2$]Br in THF [81]. The intermediate [Mo{N$\overline{\text{N(CH}_2\text{)}_4\text{CH}_2}$}(dppe)$_2$] has been isolated later by chemical reduction of the cationic dialkylhydrazido(2−) complex. It is to be noted that protonation of the former neutral dialkylhydrazido(2−) complex by acid leads to the N−N bond cleavage, giving piperidine and trans-[MoBr(NH)(dppe)$_2$]Br [82]. On the other hand, treatment of 1b with gem-dibromides RCHBr$_2$ (R = Me, H) gives rise to the diazoalkane complexes trans-[WBr(NN=CHR)(dppe)$_2$]Br [83].

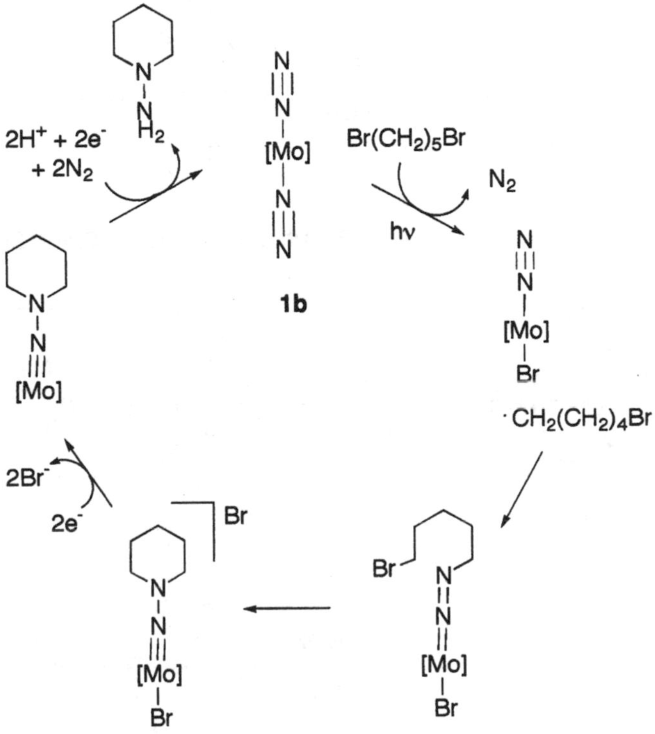

Scheme 9

Interestingly, the N_2 complex **3** containing a thiocrown ether ligand reacts with organic halides ranging from MeI and PhX (X = Br, I) to give organodiazenido(1−) complexes at room temperature without irradiation (Scheme 10) [84]. It is noteworthy that arylation of coordinated N_2 occurs in these reactions; however, the mechanism is not clear at present.

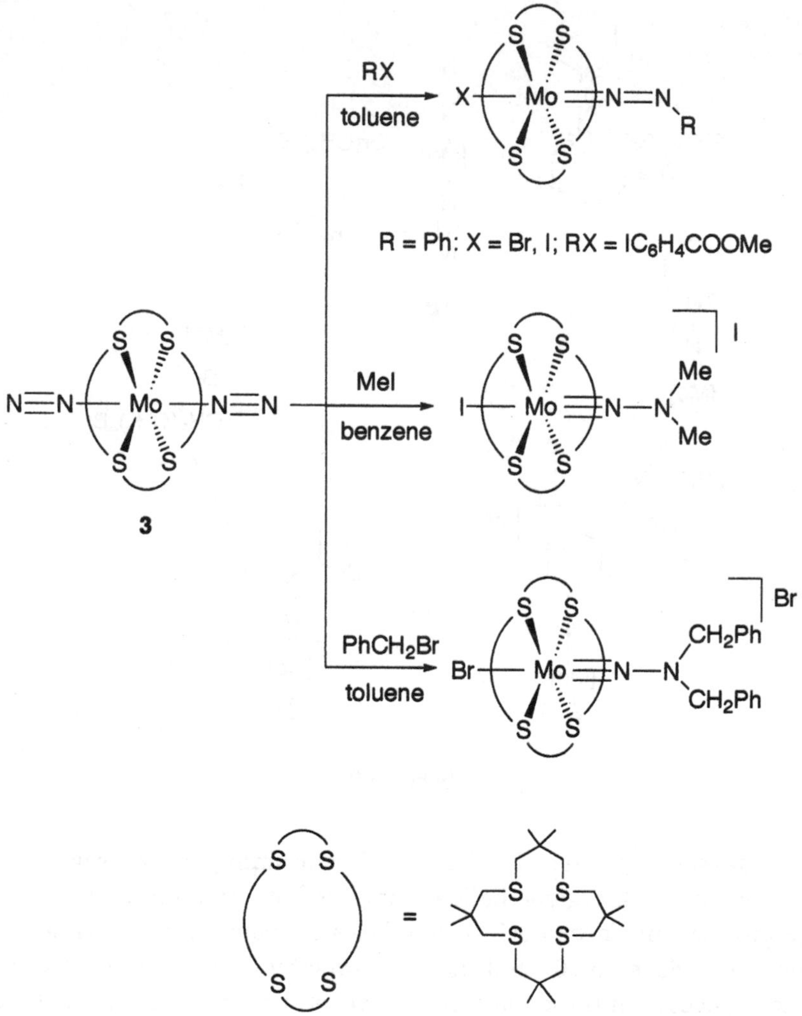

Scheme 10

Arylation of the N_2 ligand bound to the $W(dppe)_2$ moiety is observed in some cases. Thus, the isolated or in situ-generated anionic N_2 complex **4b** reacts with the activated aryl fluorides such as $[(\eta^6\text{-}FC_6H_4COOMe)Cr(CO)_3]$ and $[(\eta^6\text{-}FC_6H_4R)RuCp][PF_6]$ (R = H, Me, OMe, COOMe) to form the bimetallic aryldiazenido(1−) complexes shown in Scheme 11 [85]. The reaction is presumed to proceed via nucleophilic attack of the remote nitrogen of the N_2 ligand on the activated aryl fluorides.

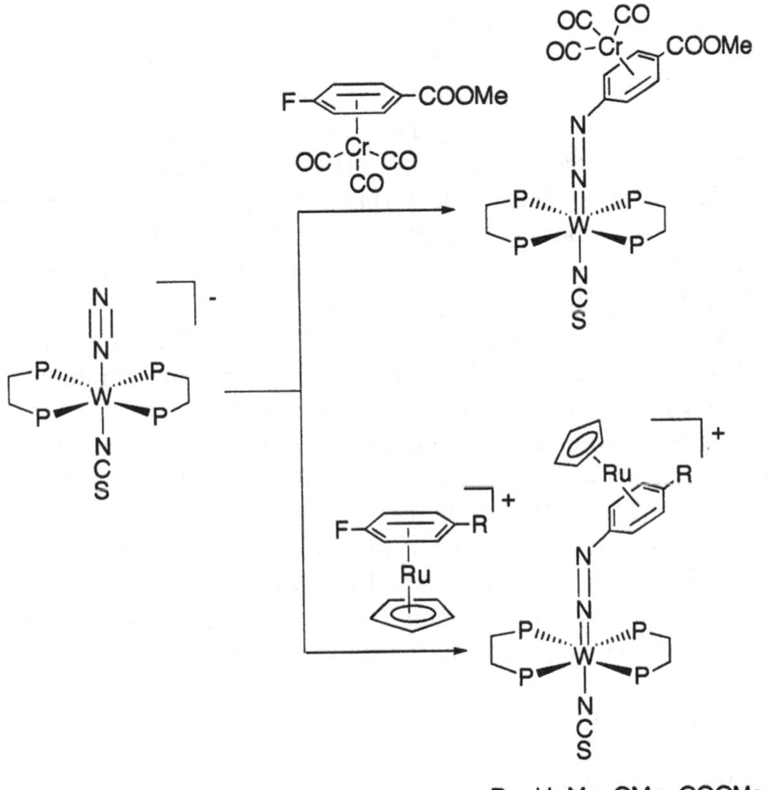

R = H, Me, OMe, COOMe

Scheme 11

3.4. Reactions of a Bridging N_2 Ligand with Inorganic Acids

In some cases μ-N_2 complexes react with acids but protonation does not occur at the N_2 ligand. Thus, $[\{Mo(\eta^6\text{-}C_6H_3Me_3)(dmpe)\}_2(\mu\text{-}N_2)]$ reacts with HBF_4 to give $[\{MoH(\eta^6\text{-}C_6H_3Me_3)(dmpe)\}_2(\mu\text{-}N_2)]^{2+}$ [14], while treatment of $[(Cp^*WMe_3)_2(\mu\text{-}N_2)]$ with 2 equivalents of HX (X = Cl or HOTf) results in the formation of $[(Cp^*WMe_2X)_2(\mu\text{-}N_2)]$

in high yields with concurrent liberation of methane [16]. However, when $[(Cp^*WMe_3)_2(\mu\text{-}N_2)]$ or its Mo analogue is treated with excess amount of HCl and the resulting reaction mixture is subjected to base distillation, 0.34 and 0.32 mol NH_3 per mole of the W_2 or Mo_2 complex, respectively, are produced. On the other hand, the reaction of the complex tentatively formulated as $[\{W(PhCCPh)(OBu^t)_2\}_2(\mu\text{-}N_2)]$ with 4 equivalents of HCl in the presence of DME affords the fully characterizable compound $[\{WCl_2(PhCCPh)(DME)\}_2(\mu\text{-}N_2)]$ (vide supra). However, hydrazine is formed in high yield upon treatment with excess HCl followed by hydrolysis with water [17].

Dinuclear N_2 complexes derived from **2b** such as $[WX(PMe_2Ph)_4(\mu\text{-}N_2)MCp_2Cl]$ (M = Ti, X = Cl; M = Zr, X = I), $[WCl(PMe_2Ph)_4(\mu\text{-}N_2)TiCpCl_2]$, and $[WCl(PMe_2Ph)_4(\mu\text{-}N_2)MCp'Cl_3]$ (M = Nb, Ta), give 0.34-0.88 mol NH_3/W atom and, commonly, less hydrazine by treatment with H_2SO_4 in MeOH [23].

For these protonation reactions of the bridging N_2 complexes containing Mo or W, the mechanisms are still elusive because of the poor nitrogen balance and the ambiguity of the metal products. In contrast, the mechanism for the quantitative formation of hydrazine from $[\{M(S_2CNEt_2)_3\}_2(\mu\text{-}N_2)]$ (M = Nb, Ta) with HX (X = Cl, Br) has been established unequivocally, which involves a series of dinuclear complexes with a bridging protonated dinitrogen ligand as intermediates [86].

4. CONCLUSIONS

The chemistry of dinitrogen complexes has advanced remarkably during the last three decades, and most transition metal ions are now known to form N_2 complexes. It should be emphasized that Mo and W dinitrogen complexes are especially promising for further syntheses because the coordinated N_2 displays versatile reactivities resulting in the formation of a variety of nitrogenous ligands and compounds.

Furthermore, future studies on Mo and W dinitrogen complexes may lead to the discovery of novel and efficient homogeneous catalysts for N_2 fixation.

ABBREVIATIONS

acac	acetylacetonate
9-BBN	9-borabicyclo[3.3.1]nonyl
Bu^n	n-butyl group
Bu^t	t-butyl group
Cp	η^5-C_5H_5
Cp*	η^5-C_5Me_5
depe	$Et_2PCH_2CH_2PEt_2$
DME	$MeOCH_2CH_2OMe$
dmpe	$Me_2PCH_2CH_2PMe_2$
dppe	$Ph_2PCH_2CH_2PPh_2$
dppp	$Ph_2PCH_2CH_2CH_2PPh_2$
dtfpe	$(4\text{-}CF_3C_6H_4)_2PCH_2CH_2P(4\text{-}C_6H_4CF_3)_2$
Et	ethyl group
Me	methyl group
$Me_8[16]aneS_4$	3,3,7,7,11,11,15,15-octamethyl-1,5,9,13-tetrathiacyclohexadecane
Mes	$2,4,6\text{-}C_6H_2Me_3$
$[N_3N]^{3-}$	η^4-$[N(CH_2CH_2NR)_3]^{3-}$
OTf	OSO_2CF_3
OTs	$OSO_2\text{-}4\text{-}C_6H_4Me$
Ph	phenyl group
Pr^i	i-propyl group
Pr^n	n-propyl group
py	pyridine
THF	tetrahydrofuran
triphos	$PhP(CH_2CH_2PPh_2)_2$

REFERENCES

1. M. Appl, *Ammonia*, Wiley-VCH, Weinheim, 1999.
2. (a) M. E. Vol'pin and V. B. Shur, *Dokl. Akad. Nauk. SSSR*, *156*, 1102 (1964); (b) M. E. Vol'pin, *Pure Appl. Chem.*, *30*, 607 (1972).

3. A. D. Allen and C. V. Senoff, *J. Chem. Soc., Chem. Commun.*, 621 (1965).

4. (a) M. Hidai, K. Tominari, Y. Uchida, and A. Misono, *J. Chem. Soc., Chem. Commun.*, 1392 (1969); (b) M. Hidai, K. Tominari, and Y. Uchida, *J. Am. Chem. Soc.*, *94*, 110 (1972).

5. (a) M. Hidai and Y. Mizobe, in *Activation of Unreactive Bonds and Organic Synthesis* (S. Murai, ed.), Springer-Verlag, Berlin, 1999, p. 227; (b) M. Hidai, *Coord. Chem. Rev.*, *185–186*, 99 (1999); (c) M. Hidai and Y. Ishii, *Bull. Chem. Soc. Jpn.*, *69*, 819 (1996); (d) M. Hidai and Y. Ishii, *J. Mol. Cat.*, *107*, 105 (1996); (e) M. Hidai and Y. Mizobe, *ACS Symp. Ser.*, *535*, 346 (1993); (f) T. A. George and J. R. D. DeBord, *ACS Symp. Ser.*, *535*, 346 (1993).

6. (a) M. Hidai and Y. Mizobe, *Chem. Rev.*, *95*, 1115 (1995); (b) R. R. Schrock, *Pure Appl. Chem.*, *69*, 2197 (1997); (c) R. L. Richards, *Coord. Chem. Rev.*, *154*, 83 (1996); (d) S. Gambarotta, *J. Organomet. Chem.*, *500*, 117 (1995); (e) T. A. Bazhenova and A. E. Shilov, *Coord. Chem. Rev.*, *144*, 69 (1995); (f) G. J. Leigh, *Acc. Chem. Res.*, *25*, 177 (1992); (g) R. A. Henderson, *Transition Met. Chem.*, *15*, 330 (1990); (h) J. R. Dilworth and R. L. Richards, in *Comprehensive Organometallic Chemistry* (G. Wilkinson, F. G. A. Stone, and E. Abel, eds.), Pergamon Press, Oxford, 1982, Vol. 8, Chap. 60; (i) A. E. Shilov, *Pure Appl. Chem.*, *64*, 1409 (1992).

7. (a) T. Uchida, Y. Uchida, M. Hidai, and T. Kodama, *Bull. Chem. Soc. Jpn.*, *43*, 2883 (1971); (b) T. Uchida, Y. Uchida, M. Hidai, and T. Kodama, *Acta Crystallogr.*, *B31*, 1197 (1975).

8. C. Hu, W. C. Hodgeman, and D. W. Bennett, *Inorg. Chem.*, *35*, 1621 (1996).

9. H. Dadkhah, J. R. Dilworth, K. Fairman, C. T. Kan, R. L. Richards, and D. L. Hughes, *J. Chem. Soc., Dalton Trans.*, 1523 (1985).

10. T. Yoshida, T. Adachi, M. Kaminaka, and T. Ueda, *J. Am. Chem. Soc.*, *110*, 4872 (1988).

11. R. T. Baker, J. C. Calabrese, R. L. Harlow, and I. D. Williams, *Organometallics*, *12*, 830 (1993).

12. M. B. O'Donoghue, N. C. Zanetti, W. M. Davis, and R. R. Schrock, *J. Am. Chem. Soc.*, *119*, 2753 (1997).

13. S. N. Anderson, R. L. Richards, and D. L. Hughes, *J. Chem. Soc., Dalton Trans.*, 245 (1986).

14. M. L. H. Green and W. E. Silverthorn, *J. Chem. Soc, Dalton*, 2164 (1974).

15. R. A. Forder and K. Prout, *Acta Crystallogr., B30*, 2778 (1974).

16. R. R. Schrock, R. M. Kolodziej, A. H. Liu, W. M. Davis, and M. G. Gale, *J. Am. Chem. Soc., 112*, 4338 (1990).

17. M. R. Churchill, Y.-J. Li, K. H. Theopold, and R. R. Schrock, *Inorg. Chem., 23*, 4472 (1984).

18. (a) M. Kol, R. R. Schrock, R. Kempe, and W. M. Davis, *J. Am. Chem. Soc., 116*, 4382 (1994); (b) K. Shih, R. R. Schrock, and R. Kempe, *J. Am. Chem. Soc., 116*, 8804 (1994); (c) M. Scheer, J. Müller, M. Schiffer, G. Baum, R. Winter, *Chem. Eur. J.*, 6, 1252 (2000).

19. M. B. O'Regan, A. H. Liu, W. C. Finch, R. R. Schrock, and W. M. Davis, *J. Am. Chem. Soc., 112*, 4331 (1990).

20. M. Aresta, *Gazz. Chim. Ital., 102*, 781 (1972).

21. M. Mercer, R. H. Crabtree, and R. L. Richards, *J. Chem. Soc., Chem. Commun.*, 808 (1973).

22. A. L. Odem, P. L. Arnold, and C. C. Cummins, *J. Am. Chem. Soc., 120*, 5836 (1998).

23. (a) H. Ishino, T. Nagano, S. Kuwata, Y. Yokobayashi, Y. Ishii, M. Hidai, and Y. Mizobe, *Organometallics, 20*, 188 (2001); (b) Y. Mizobe, Y. Yokobayashi, Y. Oshita, T. Takahashi, and M. Hidai, *Organometallics, 13*, 3764 (1994).

24. H. Ishino, Y. Ishii, and M. Hidai, *Chem. Lett.*, 677 (1998).

25. T. Takahashi, T. Kodama, A. Watakabe, Y. Uchida, and M. Hidai, *J. Am. Chem. Soc., 105*, 1680 (1983).

26. K. Takagahara, H. Ishino, Y. Ishii, and M. Hidai, *Chem. Lett.*, 897 (1998).

27. M. B. O'Donoghue, W. M. Davis, R. R. Schrock, and W. M. Reiff, *Inorg. Chem., 38*, 243 (1999).

28. T. M. Glassman, A. H. Liu, and R. R. Schrock, *Inorg. Chem., 30*, 4723 (1991).

29. (a) C. E. Laplaza and C. C. Cummins, *Science, 268*, 861 (1995); (b) C. E. Laplaza, M. J. A. Johnson, J. C. Peters, A. L. Odem,

E. Kim, C. C. Cummins, G. N. George, and I. J. Pickering, *J. Am. Chem. Soc.*, *118*, 8623 (1996).

30. Q. Cui, D. G. Musaev, M. Svensson, S. Sieber, and K. Morokuma, *J. Am. Chem. Soc.*, *117*, 12366 (1995).

31. G. K. B. Clentsmith, V. M. B. Bates, P. B. Hitchcock, and F. G. N. Cloke, *J. Am. Chem. Soc.*, *121*, 10444 (1999).

32. A. Caselli, E. Solari, R. Scopelliti, C. Floriani, N. Re, C. Rizzoli, and A. Chiesi-Villa, *J. Am. Chem. Soc.*, *122*, 3652 (2000).

33. (a) J. Chatt, A. J. Pearman, and R. L. Richards, *Nature*, *253*, 39 (1975); (b) J. Chatt, A. J. Pearman, and R. L. Richards, *J. Chem. Soc. Dalton Trans.*, 1852 (1977); (c) S. N. Anderson, M. E. Fakley, R. L. Richards, and J. Chatt, *J. Chem. Soc. Dalton Trans.*, 1973 (1981).

34. J. A. Baumann, G. E. Bossard, T. A. George, D. B. Howell, L. M. Koczon, R. K. Lester, and C. M. Noddings, *Inorg. Chem.*, *24*, 3568 (1985).

35. T. Takahashi, Y. Mizobe, M. Sato, Y. Uchida, and M. Hidai, *J. Am. Chem. Soc.*, *102*, 7461 (1980).

36. J. Chatt, M. E. Fakley, P. B. Hitchcock, R. L. Richards, and N. T. Luong-Thi, *J. Chem. Soc. Dalton Trans.*, 345 (1982).

37. (a) J. Chatt, G. A. Heath, and R. L. Richards, *J. Chem. Soc. Chem. Commun.*, 1010 (1972); (b) J. Chatt, G. A. Heath, and R. L. Richards, *J. Chem. Soc. Dalton Trans.*, 2074 (1974).

38. J. Chatt, A. J. Pearman, and R. L. Richards, *J. Chem. Soc. Dalton Trans.*, 1520 (1976).

39. M. Hidai, T. Kodama, M. Harakawa, and Y. Uchida, *Inorg. Chem.*, *15*, 2694 (1976).

40. J. Chatt, A. J. Pearman, and R. L. Richards, *J. Chem. Soc. Dalton Trans.*, 1766 (1978).

41. A. Galindo, A. Hills, D. L. Hughes, R. L. Richards, M. Hughes, and J. Mason, *J. Chem. Soc. Dalton Trans.*, 283 (1990).

42. J. E. Barklay, A. Hills, D. L. Hughes, G. J. Leigh, C. J. Macdonald, M. Abur-Bakar, and H. Mohd-Ali, *J. Chem. Soc. Dalton Trans.*, 2503 (1990).

43. (a) C. J. Pickett and J. Talarmin, *Nature*, 317, 652 (1985); (b) C. J. Pickett, K. S. Ryder, and J. Talarmin, *J. Chem. Soc. Dalton Trans.*, 1453 (1986).

44. (a) M. Hidai, T. Takahashi, I. Yokotake, and Y. Uchida, *Chem. Lett.*, 645 (1980); (b) H. Nishihara, T. Mori, T. Saito, and Y. Sasaki, *Chem. Lett.*, 667 (1980); (c) H. Nishihara, T. Mori, Y. Tsurita, K. Nakano, T. Saito, and Y. Sasaki, *J. Am. Chem. Soc.*, *104*, 4367 (1982).

45. G. Jia, R. H. Morris, and C. T. Schweitzer, *Inorg. Chem.*, *30*, 593 (1991).

46. (a) Y. Nishibayashi, S. Iwai, and M. Hidai, *Science*, *279*, 540 (1998); (b) Y. Nishibayashi, S. Takemoto, S. Iwai, and M. Hidai, *Inorg. Chem.*, *39*, 5946 (2000).

47. E. Rocchini, A. Mezzetti, H. Rüegger, U. Burckhardt, V. Gramlich, A. D. Zotto, P. Martinuzzi, and P. Rigo, *Inorg. Chem.*, *36*, 711 (1977).

48. J. Chatt, J. P. Lloyd, and R. L. Richards, *J. Chem. Soc. Dalton Trans.*, 565 (1976).

49. S. Kuwata, Y. Mizobe, and M. Hidai, *J. Chem. Soc. Dalton Trans.*, 1753 (1997).

50. P. Dahlstron, J. R. Dilworth, J. Hutchinson, S. Kumar, R. L. Richards, and J. Zubieta, *J. Chem. Soc. Dalton Trans.*, 1489 (1983).

51. Y. Nishibayashi, S. Iwai, and M. Hidai, *J. Am. Chem. Soc.*, *120*, 10559 (1998).

52. (a) I. Dance, *Chem. Commun.*, 165 (1997) and 523 (1998); (b) S.-J. Zhong and C.-W. Liu, *Polyhedron*, *16*, 653 (1997); (c) D. Sellmann and J. Sutter, *Acc. Chem. Res.*, *30*, 460 (1997); (d) R. L. Richards, *Coord. Chem. Rev.*, *154*, 83 (1996).

53. P. Bernatis, J. C. V. Laurie, and M. Rakowski DuBois, *Organometallics*, *9*, 1607 (1990).

54. J. C. V. Laurie, L. Duncan, R. C. Haltiwanger, R.-T. Weberg, and M. Rakowski DuBois, *J. Am. Chem. Soc.*, *108*, 6234 (1986).

55. Y. Nishibayashi, I. Wakiji, K. Hirata, M. Rakowski DuBois, and M. Hidai, *Inorg. Chem.*, *40*, 578 (2001).

56. J. Chatt, A. A. Diamantis, G. A. Heath, N. E. Hooper, and G. J. Leigh, *J. Chem. Soc. Dalton Trans.*, 688 (1977).

57. D. Sellmann and W. Weiss, *J. Organomet. Chem.*, *160*, 183 (1978).

58. J. Chatt, G. A. Heath, and G. J. Leigh, *J. Chem. Soc. Chem. Commun.*, 444 (1972).

59. J. C. Peters, J.-P. F. Cherry, J. C. Thomas, L. Baraldo, D. J. Mindiola, W. M. Davis, and C. C. Cummins, *J. Am. Chem. Soc.*, *121*, 10053 (1999).

60. M. Hidai, K. Komori, T. Kodama, D.-M. Jin, T. Takahashi, S. Sugiura, Y. Uchida, and Y. Mizobe, *J. Organomet. Chem.*, *272*, 155 (1984).

61. H. Oshita, Y. Mizobe, and M. Hidai, *J. Organomet. Chem.*, *456*, 213 (1993).

62. K. Komori, S. Sugiura, Y. Mizobe, M. Yamada, and M. Hidai, *Bull. Chem. Soc. Jpn.*, *62*, 2953 (1989).

63. A. C. Street, Y. Mizobe, F. Gotoh, I. Mega, H. Oshita, and M. Hidai, *Chem. Lett.*, 383 (1991).

64. H. Oshita, Y. Mizobe, and M. Hidai, *Organometallics*, *11*, 4116 (1992).

65. K. Komori, H. Oshita, Y. Mizobe, and M. Hidai, *J. Am. Chem. Soc.*, *111*, 1939 (1989).

66. K. Iwanami, Y. Mizobe, T. Takahashi, T. Kodama, Y. Uchida, and M. Hidai, *Bull. Chem. Soc. Jpn.*, *54*, 1773 (1981).

67. T. Aoshima, Y. Mizobe, M. Hidai, and J. Tsuchiya, *J. Organomet. Chem.*, *423*, 39 (1992).

68. (a) M. Hidai, Y. Mizobe, M. Sato, T. Kodama, and Y. Uchida, *J. Am. Chem. Soc.*, *100*, 5740 (1978); (b) P. C. Bevan, J. Chatt, M. Hidai, and G. J. Leigh, *J. Organomet. Chem.*, *160*, 165 (1978); (c) Y. Mizobe, Y. Uchida, and M. Hidai, *Bull. Chem. Soc. Jpn.*, *53*, 1781 (1980); (d) Y. Mizobe, R. Ono, Y. Uchida, M. Hidai, M. Tezuka, S. Moue, and A. Tsuchiya, *J. Organomet. Chem.*, *204*, 377 (1981).

69. M. Hidai, S. Aramaki, K. Yoshida, T. Kodama, T. Takahashi, Y. Uchida, and Y. Mizobe, *J. Am. Chem. Soc.*, *108*, 1562 (1986).

70. (a) A. Watakabe, T. Takahashi, D.-M. Jin, I. Yokotake, and Y. Uchida, *J. Organomet. Chem.*, *254*, 75 (1983); (b) M. Hidai, M. Kurano, and Y. Mizobe, *Bull. Chem. Soc. Jpn.*, *58*, 2719 (1985).

71. H. Seino, Y. Ishii, and M. Hidai, *Inorg. Chem.*, *36*, 161 (1997).

72. H. Seino, Y. Ishii, T. Sasagawa, and M. Hidai, *J. Am. Chem. Soc.*, *117*, 12181 (1995).

73. T. Sasagawa, H. Seino, Y. Ishii, Y. Mizobe, and M. Hidai, *Bull. Chem. Soc. Jpn.*, *72*, 425 (1999).

74. H. Ishino, S. Tokunaga, H. Seino, Y. Ishii, and M. Hidai, *Inorg. Chem.*, *38*, 2489 (1999).

75. Y. Harada, Y. Mizobe, Y. Ishii, and M. Hidai, *Bull. Chem. Soc. Jpn.*, *71*, 2701 (1998).

76. H. M. Colquhoun, A. E. Crease, S. A. Taylor, and D. J. Williams, *J. Chem. Soc. Dalton Trans.*, 2781 (1988).

77. (a) Y. Ishii, H. Miyagi, S. Jitsukuni, H. Seino, B. S. Harkness, and M. Hidai, *J. Am. Chem. Soc.*, *114*, 9890 (1992); (b) H. Seino, Y. Ishii, and M. Hidai, *Organometallics*, *13*, 364 (1994).

78. G. E. Bossard, D. C. Busby, M. Chang, T. A. George, and S. D. A. Iske, Jr., *J. Am. Chem. Soc.*, *102*, 1001 (1980).

79. W. Hussain, G. J. Leigh, H. Mohd-Ali, and C. J. Pickett, *J. Chem. Soc. Dalton Trans.*, 1473 (1986).

80. (a) M. Abu Bakar, D. L. Hughes, and G. J. Leigh, *J. Chem. Soc. Dalton Trans.*, 2525 (1988); (b) J. Chatt, W. Hussain, G. J. Leigh, and F. P. Terreros, *J. Chem. Soc. Dalton Trans.*, 1408 (1980).

81. C. J. Pickett and G. J. Leigh, *J. Chem. Soc., Chem. Commun.*, 1033 (1981).

82. R. A. Henderson, G. J. Leigh, and C. J. Pickett, *J. Chem. Soc. Dalton Trans.*, 425 (1989).

83. R. Ben-Shoshan, J. Chatt, G. J. Leigh, and W. Hussain, *J. Chem. Soc. Dalton Trans.*, 771 (1980).

84. T. Yoshida, T. Adachi, T. Ueda, M. Kaminaka, N. Sasaki, T. Higuchi, T. Aoshima, I. Mega, Y. Mizobe, and M. Hidai, *Angew. Chem. Int. Ed.*, *28*, 1040 (1989).

85. (a) Y. Ishii, M. Kawaguchi, Y. Ishino, T. Aoki, and M. Hidai, *Organometallics*, *13*, 5062 (1994); (b) Y. Ishii, Y. Ishino, T. Aoki, and M. Hidai, *J. Am. Chem. Soc.*, *114*, 5429 (1992).

86. (a) R. A. Henderson, S. H. Morgan, and A. N. Stephens, *J. Chem. Soc. Dalton Trans.*, 1101 (1990); (b) R. A. Henderson and S. H. Morgan, *J. Chem. Soc. Dalton Trans.*, 1107 (1990).

5

Biosynthesis of the Nitrogenase Iron-Molybdenum-Cofactor from *Azotobacter vinelandii*

Jeverson Frazzon[1] *and Dennis R. Dean*[2]

[1]Department of Food Science, ICTA, Federal University of Rio Grande do Sul, Porto Alegre, RS, 91051–970, Brazil

[2]Department of Biochemistry, Fralin Biotechnology Center, Virginia Tech, Blacksburg, VA 24061, USA

1. INTRODUCTION

The reduction of dinitrogen gas to ammonia is called *nitrogen fixation*. Nitrogen fixation occurs by natural combustion events, such as by lightning, through an industrial process, called the Haber-Bosch process, and biologically through the action of a select group of microorganisms, called diazotrophs. Biological nitrogen fixation is the most significant contributor to the global nitrogen cycle [1] and is necessary for sustaining life on Earth. One reason for this is that pools of fixed forms of nitrogen, e.g., nitrate and ammonia, are continuously sequestered into sediments and are thereby made unavailable for life processes. The second major reason is that ammonia and nitrate are continuously converted to dinitrogen through the combined microbiological processes of nitrification and denitrification. All known Mo enzymes, except some of those involved in biological nitrogen fixation, contain Mo within a group of structurally related cofactors generically referred to as *molybdopterins*. Molybdopterin-containing enzymes are the topic of other chapters in this volume.

Enzymes that catalyze biological nitrogen fixation are referred to as *nitrogenases*. There are, in fact, four different kinds of nitrogenases that have been characterized. Three of these are structurally and functionally very similar to each other, with the major differentiating feature being the nature of the heterometal (Mo, V, or Fe) contained in their active site metal cluster [2]. The fourth class is a superoxide-dependent enzyme identified in the bacterium *Streptomyces thermoautotrophicus* [3]. This fourth type of nitrogenase contains a molybdopterin cofactor. It is difficult to know with any certainty the specific contributions of the individual nitrogenase types to the biogeochemical nitrogen cycle. However, the enzyme that contains Mo, but not in the form of molybdopterin, is the one best studied and is the most important from the agronomic perspective. This enzyme is referred to as the Mo-dependent nitrogenase and is composed of two separately purifiable components, usually called the Fe protein and the MoFe protein (see Chapter 3 and [4] and [5] for recent reviews).

The Fe protein is composed of two identical subunits (64 kDa) and contains a single [4Fe-4S] cluster bridged between the two subunits. The Fe protein also contains two nucleotide binding sites, one on each subunit. The MoFe protein is an $\alpha_2\beta_2$ heterotetramer (250 kDa) that contains two pairs of metalloclusters called the P cluster [8Fe-7S] and the iron-molybdenum cofactor (FeMo cofactor). Each $\alpha\beta$ dimer of the MoFe protein contains one P cluster and one FeMo cofactor (7Fe-9S-Mo-homocitrate, see Fig. 1). For many years it has been assumed that each $\alpha\beta$ dimer of the MoFe protein functions as an independent unit together with one Fe protein. However, there is emerging evidence that communication might occur between the two "halves" of the MoFe protein during catalysis [6].

The overall reaction catalyzed by the Mo-dependent nitrogenase is usually indicated as:

$$N_2 + 8H^+ + 16MgATP + 8e^- \Rightarrow 2NH_3 + H_2 + 16MgADP + 16Pi$$

Detailed aspects of the catalytic process are described in Chapter 3, so only several salient features are described here. During catalysis the Fe protein serves as a specific MgATP-dependent reductant of the MoFe protein, which in turn provides the substrate reduction site. For each electron transfer cycle an Fe protein with two MgATP bound (one on each subunit) transiently associates with one $\alpha\beta$ unit of the MoFe protein. During this association, the two MgATP molecules are hydrolyzed to MgADP and Pi in an event coupled to the transfer of one electron from the Fe protein [4Fe-4S]$^+$ cluster to the MoFe protein. The P clusters are thought to act as intermediate electron acceptors within the MoFe protein before passing the electrons on to the FeMo cofactor where substrate becomes bound and is reduced. Following electron transfer, the oxidized Fe protein, with MgADP bound, dissociates from the MoFe protein. Because dinitrogen reduction requires multiple reducing equivalents and only one electron is transferred from the Fe protein to the MoFe protein, multiple rounds of component protein association and dissociation must occur to accomplish substrate reduction.

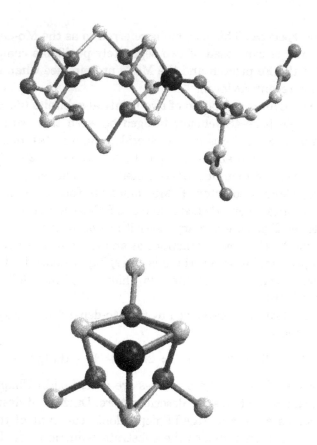

FIG. 1. FeMo cofactor structure. Top view shows the entire FeMo cofactor structure and the bottom view shows an end-on view of FeMo cofactor without homocitrate attached to the Mo. Within the metal-sulfur core the darkest shaded atom is Mo, light gray atoms correspond to S, and medium shaded atoms correspond to Fe. In the upper view, homocitrate is attached to the Mo atom.

There are several lines of evidence that FeMo cofactor provides the substrate binding and reduction site. First, an FeMo cofactorless MoFe protein that contains intact P clusters is designated apo-MoFe protein. Apo-MoFe protein is incapable of catalyzing any substrate reduction but can be activated by addition of FeMo cofactor extracted from intact MoFe protein [7]. Second, an altered MoFe protein that produces an FeMo cofactor that contains citrate rather than homocitrate as its organic constituent has altered substrate reduction properties [8]. For example, MoFe protein that contains citrate within its cofactor is able to effectively reduce protons and acetylene but only has a poor capacity to reduce dinitrogen. Finally, altered forms of MoFe protein that have amino acid substitutions for residues that provide the first shell of non-covalent interactions with FeMo cofactor also exhibit alterations in substrate reduction properties [9]. For example, substitution of the α subunit His[195] residue by Gln[195] results in an enzyme that retains the ability to effectively reduce protons and acetylene but cannot reduce dinitrogen.

It was not until 1992 with the availability of a crystallographically determined structure of the MoFe protein that a structure of FeMo cofactor also became available [10,11]. Structural models have now been obtained for MoFe proteins from three different sources—*Azotobacter vinelandii* [10], *Clostridium pasteurianum* [12], and *Klebsiella pneumoniae* [13]—and each of these MoFe proteins contains an FeMo cofactor having the same structure and composition. FeMo cofactor is constructed from S-bridged [4Fe-3S] and [3Fe-3S-Mo] sub-fragments that are geometrically analogous to cuboidal clusters. An organic constituent, homocitrate, is coordinated to the Mo atom through its 2-hydroxy and 2-carboxyl groups (Fig. 1). Opposite ends of the FeMo cofactor are covalently attached to the MoFe protein α subunit through α-Cys[275], S-coordinated to an Fe atom at one end of the molecule, and α-His[442], N-coordinated to the Mo atom at the other end of the molecule. The unusual structure of FeMo cofactor has attracted the attention of bioinorganic chemists interested in determining where and how substrates interact with FeMo cofactor during catalysis (discussed in Chapter 3), and biochemists and geneticists interested in how this complex entity is constructed biologically (for another review, see [14]). In this chapter we review what is known about FeMo cofactor biosynthesis using *Azotobacter vinelandii* as a model organism.

2. BIOCHEMICAL GENETIC ANALYSIS OF FeMo COFACTOR BIOSYNTHESIS

Genes whose products are involved in the formation of an active nitrogenase are indicated as *"nif"* genes, referring to *ni*trogen *f*ixation.

There are two general biochemical-genetic approaches that have been used to analyze those *nif* genes whose products are involved in the biosynthesis of FeMo cofactor. The first of these was to inactivate individual *nif* genes and ask if such lesions resulted in a specific defect in MoFe protein activity [15] and, if so, could such a defect be correlated to an FeMo cofactor-deficient MoFe protein or one containing an altered FeMo cofactor. The second approach has involved mixing extracts prepared from strains having complementary defects in FeMo cofactor assembly or insertion in an attempt to reconstitute activity [16].

The first approach led to the identification of four genes (*nifH*, *nifE*, *nifN*, and *nifB*) that are absolutely required for FeMo cofactor biosynthesis. However, as will be discussed below, a number of other *nif* gene products also participate in FeMo cofactor formation or insertion, although their individual functions are either partially or completely dispensable. The second approach led to the important observation that FeMo cofactor can be assembled in the absence of the α and β subunits of the MoFe protein. In other words, FeMo cofactor accumulates in extracts of strains that are deleted for the structural genes (*nifD* and *nifK*), respectively, encoding the MoFe protein α and β subunits. This means that FeMo cofactor is not assembled directly into the MoFe protein in a stepwise fashion but rather is separately formed and then inserted into an apo form of the MoFe protein. Thus, understanding how FeMo cofactor is assembled for the maturation of the nitrogenase MoFe protein presents the following challenges. How are the Fe and S needed to form the [Fe-S] core of FeMo cofactor activated and targeted for FeMo cofactor formation? Where and how is the [Fe-S] core formed? How is the Mo necessary for FeMo cofactor formation acquired, mobilized, and inserted into an FeMo cofactor precursor? How is homocitrate formed and how does it become attached to the Mo site within FeMo cofactor? What is the sequence of these events and how are they coordinated together? Where does FeMo cofactor accumulate prior to its insertion into the MoFe protein and how is the insertion process accomplished. Although it appears that most or all of the players involved in these processes are now known, the molecular details of the events involved in FeMo cofactor biosynthesis are far from being understood.

3. A GENERAL MODEL FOR FeMo COFACTOR BIOSYNTHESIS

Here we itemize general aspects of the formation of FeMo cofactor. In subsequent sections some specific aspects of this process are described. (a) FeMo cofactor is separately synthesized from the MoFe protein subunits and then inserted into an apo form of the MoFe protein that contains intact P clusters but no FeMo cofactor. (b) Certain small proteins might serve as molecular props that hold the apo-MoFe protein in the proper conformation for FeMo cofactor insertion. (c) These small proteins also appear to serve a dual function in trafficking FeMo cofactor or its intermediates between assembly sites. FeMo cofactor is sequentially assembled upon a series of different scaffolds. Thus, intermediates in the assembly process appear to be transferred from one scaffold to the next.

The first stage of FeMo cofactor formation probably involves the participation of the *nifU* and *nifS* gene products. These proteins form a complex upon which simple [2Fe-2S] and or [4Fe-4S] units are assembled. These units are then transferred to the product of the *nifB* gene on which an entity called B cofactor is assembled. NifB cofactor is composed only of Fe and S and probably represents the [Fe-S] framework necessary for FeMo cofactor formation. NifB cofactor is then transferred to an $\alpha_2\beta_2$ complex composed of the products of the *nifE* and *nifN* gene products. The NifEN complex bears strong sequence similarity when compared with the MoFe protein. It is therefore possible that the [Fe-S] core delivered from NifB activity to the NifEN complex subsequently rearranges to assume a structure that has three-dimensional features similar to FeMo cofactor.

However, neither Mo nor homocitrate appears to be attached to the FeMo cofactor intermediate contained on the NifEN complex. The incorporation of Mo and homocitrate into FeMo cofactor therefore probably occurs at a late stage in FeMo cofactor biosynthesis in processes that might be facilitated by small carrier proteins described above. The other catalytic component of nitrogenase, the Fe protein, is also required for both the formation of FeMo cofactor and its insertion into the apo-MoFe protein. The specific roles of the Fe protein in the formation and insertion of FeMo cofactor is not known.

4. MOBILIZATION OF IRON AND SULFUR

Because the FeMo cofactor metal-sulfur core is constructed from two
partial cubes connected by three inorganic sulfides, two possible routes
for its construction are reasonable. One possibility is that the two sub-
fragments are separately formed and then joined to form the [Fe-S] core
of FeMo cofactor, a model that we prefer, or that the core is assembled
intact as one species. In any event, intracellular Fe and S destined to
form the [Fe-S] core must in some way be targeted for that purpose.

There is now evidence that the products of two *nif*-specific genes,
nifS and *nifU*, participate in the initial stages of [Fe-S] cluster core
formation. NifS, product of the *nifS* gene, is a pyridoxal phosphate containing homodimer [17]. It is likely that NifS is the ultimate S source for
all of the nitrogenase metalloclusters, including FeMo cofactor. This
enzyme uses pyridoxal phosphate chemistry and L-cysteine as substrate
to form an enzyme-bound persulfide, which is the activated form of S
destined for metallocluster formation [18]. The product of a second *nif*-
specific gene, *nifU*, provides a site for initial formation of an [Fe-S]
cluster unit that is probably later donated to another protein, NifB, to
form the [Fe-S] core of FeMo cofactor [19,20]. The physiological source
of Fe that is delivered to the NifU scaffold is not yet known, but some
details about how the [Fe-S] cluster core is initially formed on NifU are
beginning to emerge.

Examination of the NifU primary sequence shows that it is a modular protein composed of three separate domains [21]. One of the
domains is located in the N-terminal third of NifU and provides a scaffold for assembly of a transient [Fe-S] cluster destined for formation of
FeMo cofactor and other nitrogenase metalloclusters. The second
domain is located in the central third of NifU and contains one [2Fe-
2S] cluster in each subunit of the as-isolated homodimeric NifU protein
[19,20]. These [2Fe-2S] clusters have been designated "permanent"
clusters as they are not released from the protein by chelating reagents
and they are capable of being reversibly oxidized and reduced. The third
domain is located within the C-terminal third of the protein. This
domain contains two conserved cysteine residues separated by two
other amino acids. This arrangement is reminiscent of a thioredoxin
motif, although the function of the C-terminal NifU domain is not

known and has been shown by mutagenesis to be unnecessary for nitro-genase metallocluster formation.

In vitro studies using purified preparations of NifS and the oxi-dized form of NifU have shown that it is possible to achieve the sequen-tial catalytic formation of [2Fe-2S] and [4Fe-4S] units within the N-terminal domains of the NifU homodimer ([22], our unpublished results). It is not yet known whether the [2Fe-2S] form or the [4Fe-4S] form assembled within the transient cluster site is the one mechan-istically relevant to FeMo cofactor biosynthesis. Both species are extre-mely redox-sensitive and are immediately and quantitatively released from the assembly site upon reduction by dithionite or other reducing agents. This feature leads to speculation that one role of the permanent [2Fe-2S] clusters could be the redox-dependent release of the transient cluster following transfer of an electron from a reduced form of the permanent cluster. One attractive possibility is that such intramolecular electron transfer and cluster release could be triggered by specific inter-action between a form of NifU that contains transient clusters and the target protein to which the cluster is delivered.

5. NifB COFACTOR

Ludden and coworkers have pioneered attempts to develop a cell-free system for FeMo cofactor biosynthesis involving the use of only pur-ified components, a sulfur source, Fe, Mo, and homocitrate [14]. Although this effort has not yet come to complete fruition, the approach has provided considerable insight toward the biosynthetic pathway. In its present stage of development the in vitro biosynthetic system is not strictly catalytic. In other words, some of the proteins necessary for FeMo cofactor formation in vitro, for example NifB (pro-duct of the *nifB* gene), do not appear to serve as catalysts but rather are consumed as reagents. There could be a trivial explanation for this phenomenon because dithionite is added to the in vitro biosynthetic system in order to maintain an anoxic environment. Thus, the sensi-tivity of [Fe-S] cluster precursors formed on NifU to reducing reagents, as discussed above, could preclude the correct donation of FeMo cofac-tor building blocks to the assembly site in the current system. The L-cysteine desulfurization reaction catalyzed by NifS is also extremely sensitive to dithionite.

Although the exact function of NifB is not yet known, it seems likely that it provides a scaffold for the formation of the [Fe-S] core necessary for FeMo cofactor biosynthesis. Evidence that this is so, emerged from development of the in vitro FeMo cofactor biosynthetic system. Early in the development of the in vitro FeMo cofactor biosynthetic system it was found that deletion of the *nifE*, *nifN*, or *nifB* gene results in the production of apo-MoFe protein that can be activated by the simple addition of isolated FeMo cofactor. These are the only *nif*-specific genes whose inactivation gives this phenotype. By mixing extracts of *nifE*- or *nifN*-deficient mutants with an extract of a *nifB*-deficient mutant, all of which produce inactive apo-MoFe protein, the inactive MoFe protein becomes activated [16]. This observation permitted attempts to purify components that contribute to the complementing activities. By using this approach it was possible to purify a protein complex contained in *nifB*-deficient or other extracts that reconstitutes MoFe protein activity when added to *nifE*- or *nifN*-defective extracts. This protein turned out to be an $\alpha_2\beta_2$ heterotetramer composed of the *nifE* and *nifN* gene products [23]. The NifEN complex is discussed in more detail in the next section.

The same approach used for purification of the NifEN complex did not work for isolation of NifB. However, this approach did lead to the purification of a novel [Fe-S] cluster that is apparently the product of NifB activity. Shah and coworkers found that addition of a detergent-soluble membrane fraction prepared from *nifEN*-deficient extracts could be used to activate FeMo cofactorless MoFe protein present in *nifB*-deficient extracts [24]. This fraction was designated NifB cofactor, and elemental analysis indicated that it is composed of Fe and S but does not contain Mo or homocitrate. Evidence that NifB cofactor is an FeMo cofactor precursor that is assembled and then donated to the NifEN complex was obtained by showing that NifB cofactor binds to the NifEN complex. This was shown in two ways [25,26]. First, addition of NifB cofactor to isolated NifEN complex results in a conformational change in the NifEN complex as monitored by native gel electrophoresis. Second, incorporation of ^{55}Fe- or ^{35}S-enriched NifB cofactor into either the isolated NifEN complex, or ultimately into FeMo cofactor, could be monitored.

Questions that emerge then are: what is the structure of NifB cofactor and what is the role of NifB in its formation? The answers to these questions are not known, but the roles of NifU and NifS in formation of [Fe-S] cluster intermediates, as well as the primary sequence of NifB, permit some speculation on these points. If it is assumed that NifB cofactor contains the complete complement of Fe and S necessary for formation of the FeMo cofactor [Fe-S] core, then a minimum of 7Fe and 9S atoms is necessary. Also, if we consider that NifU is capable of donating [2Fe-2S] or [4Fe-4S] units to the NifB scaffold then multiple rounds of donation from NifU to NifB could result in a complement of 8Fe and 9S for core formation. This is one Fe more than needed and one S less than needed for FeMo cofactor formation. It is easy to imagine that the "extra" Fe could be removed and then replaced by Mo at a subsequent step during FeMo cofactor assembly. Thus, it seems reasonable to expect that NifB function could involve the fusion of two [4Fe-4S] clusters, ultimately donated from NifU, through insertion of a S bridge. The subsequent rearrangement of such a cluster, either on NifB or at a subsequent step in FeMo cofactor biosynthesis, could occur to yield a [8Fe-9S] core having the same geometrical shape as FeMo cofactor but without homocitrate and with the position usually occupied by Mo in the mature cofactor substituted by Fe.

The primary sequence of NifB [27] lends some credence to this hypothesis. Analysis of the primary sequence of NifB has recently indicated that it is a member of a large family of proteins that generate a radical species by reductive cleavage of S-adenosylmethionine. Such proteins catalyze a diverse number of reactions including methylation, isomerization, sulfur insertion, ring formation, anaerobic oxidation, and radical formation [28]. The sequence signature of this protein family is a characteristic Cys-X-X-X-Cys-X-X-Cys motif involved in coordinating an unusual [Fe-S] cluster and two other separate primary sequence motifs that are spatially conserved in the NifB primary sequence (Fig. 2). Included in this family are proteins known to function in catalyzing sulfur insertion into prosthetic groups, including MiaB, BioB, and LipA. In the case of BioB (biotin synthase), there is evidence that the [Fe-S] cluster contained within BioB itself serves as the direct S donor [29]. If NifB serves a similar function in providing the proposed "extra" S for formation of NifB cofactor, then this feature could also explain why

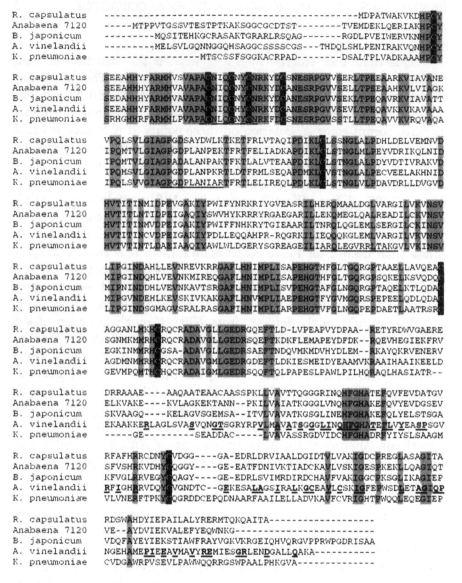

```
R. capsulatus    ------------------------------------------------MDPATWAKVKDHPGY
Anabaena 7120    -----MTPPVTGSSVTESTPTKAKSGGCGCDTST--------TVEMDEKLQERIAKHPGY
B. japonicum     ---------MQSITEHKGCRASAKTGRARLRSQAG-------RGDLPVEIWERVKNHPGY
A. vinelandii    ---------MELSVLGQNNGGQHSAGGCSSSSCGS---THDQLSHLPENIRAKVQNHPGY
K. pneumoniae    ----------------MTSCSSFSGGKACRPAD--------DSALTPLVADKAAAHPGY

R. capsulatus    SEEAHHYFARMHVSVAPACNIQCNYCNRKYDCSNESRPGVVSERLTPEEAARKVIAVANE
Anabaena 7120    SEEAHHHYARMHVAVAPACNIQCNYCNRKYDCANESRPGVVSELLTPEEAAHKVLVIAGK
B. japonicum     SEDAHHHYARMHVAVAPACNIQCNYCNRKYDCANESRPGVVSEKLTPEQAVRKVIAVATT
A. vinelandii    SEEAHHYFARMHVAVAPACNIQCHVCNRKYDCANESRPGVVSEVLTPEQAVKKVKAVAAA
K. pneumoniae    SRHGHHRFARMHLPVAPACNLQCNYCNRKFDCSNESRPGVSSTLLTPEQAVVKVRQVAQA

R. capsulatus    VPQLSVLGIAGPGDSAYDWLKTKETFRLVTAQIPDIKLGLSSNGLALPDHLDELVEMNVD
Anabaena 7120    IPQMTVLGIAGPGDPLANPEKTFRTFELIADKAPDIKLGLSTNGLMLPEYVDRIKQLNID
B. japonicum     IPQMTVLGIAGPADALANPAKTFKTLALVTEAAPDIKLGLSTNGLALPDYVDTIVRAKVD
A. vinelandii    IPQMSVLGIAGPGDPLANPKRTLDTFRMLSEQAPDMKLGVSTNGLALPECVEELAKHNID
K. pneumoniae    IPQLSVVGIAGPGDPLANIARTFRTLELIREQLPDLKLGLSTNGLVLPDAVDRLLDVGVD

R. capsulatus    HVTITINMIDPEVGAKIYPWIFYNRKRIYGVEASRILHERQMAALDGLVARGILVKVNSV
Anabaena 7120    HVTITLNTIDPEIGAQIYSWVHYKRRRYRGAEGARILLEKQMEGLQALREADILCKVNSV
B. japonicum     HVTITINMVDPEIGAKIYPWIFFNHKRYTGIEAARILTNRQLQGLEMLSERGILCKINSV
A. vinelandii    HVTITINCVDPEIGAKIYPDLLEQQAHPR-RQGRKILIEQQQKGLEMLVARGILVKVNSV
K. pneumoniae    HVTVTINTLDAEIAAQIYAWLWLDGERYSGREAGEILIARQLEGVRRLTAKGVLVKINSV

R. capsulatus    LIPGINDAHLLEVNREVKRRGAFLHNIMPLISAPEHGTHFGLTGQRGPTAAELLAVQEAG
Anabaena 7120    MIPGINDQHLVEVNKMIREQGAFLHNIMPLISAPEHGTHFGLTGQRGPSQKELKSVQDQG
B. japonicum     MIPNINDDHLVEVNKAVTSRGAFLHNIMPLISVPEHGTAFGLNGQRGPTAQELKTLQDAG
A. vinelandii    MIPGVNDEHLKEVSKIVKAKGAFLHNVMPLIAEPEHGTFYGVMGQRGPSEPEELQDLQDAG
K. pneumoniae    LIPGINDSGMAGVSRALRASGAFIHNIMPLIARPEHGTVFGLNGQPEPDAETLAATRSRG

R. capsulatus    AGGANLMKHCRQCRADAVGLLGEDRGQEFTLD-LVPEAPVYDPAA--RETYRDVVGAERE
Anabaena 7120    SGNMKMMRHCRQCRADAVGLLGEDRSQEFTKDKFLEMAPEYDFDK--RQEVHEGIEKFRV
B. japonicum     EGKINMMRHCGSA-ADAVGLLGEDRSAEFTNDQVMKMDVHYDLEM--RKAYQKRVENERV
A. vinelandii    AGDMNMMRHCRQCRADAVGMLGEDRGDEFTLDKIESMEIDYEAAMVKRAAIHAAIKEELD
K. pneumoniae    GEVMPQMTCHQCRADAIGMLGEDRSQQFTQLPAPESLPAWLPILHQRAQLHASIATR--

R. capsulatus    DRRAAAE----AAQAATEAACAASSPKLLVAVTTQGGGRINQHFGHATEFQVFEVDATGV
Anabaena 7120    ELKVAKE----KVLAGKEKTANN--PKILVAIATKGGGLVNQHFGHATEPQVYEVDGSEV
B. japonicum     SKVAAGQ----KELAGVSGEMSA--ITVLVAVATKGSGLINEHFGHAKEPLYELSTSGA
A. vinelandii    EKAAKKERLAGLSVASVQNGTSGRYRPVLMAVATSGGGLINQHFQHATEFLVYEASPSGV
K. pneumoniae    -----GE---------SEADDAC-----LVAVASSRGDVIDCHFGHADRFYIYSLSAAGM

R. capsulatus    RFAFHRRCDNYCVDGG----GA-EDRLDRVIAALDGIDTVLVAKIGDCPREGLASAGITA
Anabaena 7120    SFVSHRKVDHYGQGGY----GE-EATFDNIVKTIADCKAVLVSKIGESPKEKLLQAGIQT
B. japonicum     KFVGLRRVEGYCQAGY----GE-EDRLSVIMRDIRDCHAVFVAKIGGCPKSGLIKAGIEP
A. vinelandii    RFIGHRRVDQYCVGNDTC--GEKESALAGSIRALKGCEAVLCSKIGFEFWSDLETAGIQP
K. pneumoniae    VLVNERFTPKYTQGRDDCEPQDNAARFAAILELLADVKAVFCVRIGHTPWQQLEQEGIEP

R. capsulatus    RDSWAHDYIEPAILALYRERMTQKQAITA------------------
Anabaena 7120    VE--AYDVIEKVALEFYEQWNKG------------------
B. japonicum     VDQFAYEYIEKSTIAWFRAYVGKVKRGEIQHVQRGVPPRWPGDRISAA
A. vinelandii    NGEHAMEPIEEAVMAVYREMIESGRLENDGALLQAKA-----------
K. pneumoniae    CVDGAWRPVSEVLPAWWQQRRGSWPAALPHKGVA-------------
```

FIG. 2. Comparison of the primary sequences of *nifB* gene products from *Rhodobacter capsulatus*, *Anabaena* 7120, *Bradyrhizobium japonicum*, *Azotobacter vinelandii*, and *Klebsiella pneumoniae*. Conserved residues other than conserved cysteine residues are shaded in gray. Conserved cysteine residues are shaded in black. The three primary sequence signatures characteristic of *S*-adenosylmethionine-dependent enzymes are underlined. Residues within the *A. vinelandii* sequence that have corresponding residues in the *A. vinelandii nifX* gene product sequence (see also Fig. 5) are indicated in bold and are underlined.

a source of NifB serves as a reagent rather than as a catalyst in FeMo cofactor formation in the in vitro biosynthetic system.

6. THE NifEN COMPLEX

Prior to the purification of the NifEN complex, primary sequence analysis revealed that NifE bears significant sequence identity when compared with the α subunit of the MoFe protein and that NifN bears sequence identity when compared with the MoFe protein β subunit [30]. These primary sequence conservations and the fact that FeMo cofactor is synthesized separately from the MoFe protein subunits led to the proposal that NifEN forms an $\alpha_2\beta_2$ scaffold on which FeMo cofactor is formed. The original idea of the NifEN scaffold was that FeMo cofactor would be formed stepwise within the NifEN complex and then transferred intact to the apo-MoFe protein. Although NifE and NifN form an $\alpha_2\beta_2$ complex and provide an intermediate site for FeMo cofactor formation, the scaffold concept in its original form was not entirely correct. First, at least two other scaffolds for [Fe-S] core formation, NifU and NifB, appear to be necessary prior to the participation of the NifEN complex. Second, FeMo cofactor does not appear to be completed on the NifEN complex because no form of the intermediate attached to NifEN during FeMo cofactor formation has yet been found to contain Mo or homocitrate.

Examination of the NifE and NifN primary sequences reveals several interesting features [30]. For example, there is strong primary sequence conservation between the known FeMo cofactor binding pocket contained in the MoFe protein α-subunit and the predicted FeMo cofactor intermediate binding site in NifE. Of particular relevance is that the α-275^{Cys} residue providing the thiolate ligand to FeMo cofactor in the MoFe protein is also conserved in a corresponding region of the NifE sequence (residue 250^{Cys} in the A. vinelandii NifE sequence). In contrast, the MoFe protein α-442^{His} residue providing the N ligand to the FeMo cofactor in the MoFe protein is replaced by an asparagine residue in the corresponding NifE sequence. A second interesting feature is that, of the six P-cluster-coordinating Cys residues contained within the MoFe protein, only four of them are conserved in the corresponding NifEN primary sequences. This feature led to the suggestion that a bridging [4Fe-4S] cluster might also be contained in the NifEN complex in a position corresponding to the MoFe protein P cluster.

Purification of large amounts of the NifEN complex, made possible through the application of gene fusion technology to increase expression and the development of an affinity purification system, revealed that there are two identical redox-active [4Fe-4S] clusters contained in each NifEN tetramer [31]. These [4Fe-4S] clusters are almost certainly bridged between NifE and NifN subunits, with the NifE subunit providing three Cys ligands and NifN one Cys ligand. The function of the [4Fe-4S] clusters contained in the NifEN complex is not known. However, by analogy to the role of the P clusters as intermediate electron transfer agents to the FeMo cofactor during catalysis, the role of the NifEN [4Fe-4S] clusters could be to supply reducing equivalents necessary either to drive rearrangement of the FeMo cofactor precursor or its release from the NifEN complex.

7. HOMOCITRATE FORMATION AND MOLYBDENUM INSERTION

Identification of homocitrate as a constituent of FeMo cofactor occupies an interesting position in the history of nitrogenase research. It was the analysis of mutants unable to catalyze homocitrate formation, defective in the *nifV* gene, which provided the first clear indication that FeMo cofactor provides the site of substrate reduction. Early analysis of *nifV* mutants revealed that they produce an altered MoFe protein able to effectively reduce acetylene or protons as substrates, whereas the natural substrate dinitrogen is a very poor substrate [32]. Although it was not known at the time that NifV catalyzes homocitrate formation, Smith and coworkers were able to extract FeMo cofactor from MoFe protein produced by a NifV-defective strain and use it to reconstitute apo-MoFe protein produced by a NifB-deficient mutant [8]. Such reconstituted MoFe protein also exhibited the *nifV* phenotype, effective proton, and acetylene reduction but poor nitrogen reduction.

Some years later during their development of the in vitro FeMo cofactor biosynthetic system, Ludden and coworkers discovered that a low molecular weight factor was required for FeMo cofactor biosynthesis and that an intact *nifV* gene was required to produce that factor [33]. Chemical analysis of NifV factor revealed that it is homocitrate [34]. Purification of NifV from *A. vinelandii* heterologously produced in *Escherichia coli* revealed that it is a homodimer that catalyzes the condensation of acetyl-CoA and α-ketoglutarate to form homocitrate [35]. Conclusive evidence that homocitrate is a constituent of FeMo cofactor was obtained using radiolabel tracer studies [36] and the fact that homocitrate can be recognized in the crystallographically determined structure of MoFe protein [10]. That citrate replaces homocitrate in FeMo cofactor produced in *nifV* mutants has been established both by analysis of FeMo cofactor extracted from MoFe protein produced by a *nifV* mutant [37] and by X-ray crystallographic analysis of MoFe protein produced by a *Klebsiella pneumoniae nifV* mutant (S. M. Mayer, personal communication).

A great deal is known about how Mo is acquired by microbial cells and this topic is discussed in detail in Chapter 2. It appears that the initial steps for the acquisition of Mo for both molybdopterin cofactor and FeMo cofactor are the same [38,39]. However, very little is known about how Mo becomes specifically targeted for FeMo cofactor formation and nothing is known about how Mo is inserted into FeMo cofactor. The product of the *nifQ* gene is the only *nif*-specific gene product known to have some involvement in specifically targeting Mo for FeMo cofactor formation. Inactivation of the *nifQ* gene leads to an increased requirement for Mo to sustain normal diazotrophic growth [27,40].

This phenotype can be suppressed by the addition of more Mo than normally required for diazotrophic growth or by addition of cysteine to the growth medium. Comparison of NifQ primary sequences from *R. capsulatus*, *K. pneumoniae*, and *A. vinelandii* (Fig. 3) shows that conserved sequences are confined to the C-terminal third of the respective proteins and that there are five conserved cysteine residues. Taken together this information indicates that a Mo-containing metal-sulfur cluster could be assembled on NifQ and that this entity is the Mo donor to the FeMo cofactor precursor after it leaves the NifEN complex. It seems unlikely that an [Fe-S-Mo] cluster intermediate is assembled on NifQ and then transferred intact to some form of NifB cofactor because NifQ is dispensable under certain conditions. Another completely unknown aspect of FeMo cofactor formation concerns when and how

FIG. 3. Comparison of the primary sequences of *nifQ* gene products from *Rhodobacter capsulatus*, *Klebsiella pneumoniae*, and *Azotobacter vinelandii*. Conserved residues other than conserved cysteine residues are shaded in gray. Conserved cysteine residues are shaded in black.

homocitrate becomes attached to the Mo atom. For example, it is not yet known if homocitrate becomes attached to Mo before or after Mo is inserted into the [Fe-S] core of the FeMo cofactor precursor. There is some speculation that the products of the *nifW* and *nifZ* genes could be involved in some aspect of homocitrate attachment to the Mo atom [41], but there is not yet any definitive information on that point.

Another interesting aspect concerning the mobilization and insertion of Mo into the FeMo cofactor is the oxidation state of Mo prior to its insertion. Again, there is no direct evidence that addresses this issue but it is intriguing that a gene cotranscribed with the *A. vinelandii nifQ* gene encodes a putative product that bears significant sequence identity when compared with the *ArsC* gene of *E. coli* [27,42]. The ArsC protein is believed to be an arsenate binding protein that is involved in the reduction of arsenate to facilitate its efflux from the cell [42] (Fig. 4). The same *A. vinelandii* transcription unit that includes the *ArsC*-like encoding gene and the *nifQ* gene also encodes a ferredoxin [27]. It is therefore possible that the ArsC-like protein is a specific Mo-binding protein involved in targeting Mo for FeMo cofactor formation and perhaps participates in placing Mo in the correct redox state for that purpose.

```
MTSIVFYEKPGCATNRLQKQLLRSAGLHLEVRNLLSEPWTPERLRPFFGDR--PVVEWFNRSAPAIKYGELDPTQ
MSNITIYHNPACGTSRNTLEMIRNSGTEPTIIHYLETPPTRDELVKLIADMGISVRALLRKNVEPYEELGLAEDK

LSASQALELMLAHPILIRRPLIHYGSQYMAGFDLVKLCTHLPAKNDLPDAGEDLEKRSVGQNPIGCLAAGNQ
FIDDRLIDFMLQHPILINRPIVVTPLGTRLCRPSEVVLEILPDAQKGAFSKEDGEKVVDEAGKRLK
```

FIG. 4.   Comparison of the primary sequence the product of a gene contained in the same transcription unit as *nifB* and *nifQ* in *Azotobacter vinelandii* (upper sequence) and the product of the *Escherichia coli ArsC* gene product (lower sequence).

8.   ROLE OF INTERMEDIATE CARRIERS IN FeMo COFACTOR BIOSYNTHESIS

As we have already mentioned, the FeMo cofactor is separately synthesized and then inserted into the apo-MoFe protein. Also, FeMo cofactor precursors are sequentially formed on a series of molecular scaffolds. Thus, FeMo cofactor and its precursors must have some way to make it from one site to the next. This could occur by direct transfer from one assembly site to the next or through the action of intermediate carriers. At least one intermediate FeMo cofactor carrier, designated gamma, has been identified for *A. vinelandii*. Gamma was first discovered when it was found to be attached to apo-MoFe protein isolated from a *nifB*-mutant. Apo-MoFe protein isolated in this form was found to have an $\alpha_2\beta_2\gamma_2$ organization [43]. The gamma protein was also identified in extracts lacking the MoFe protein subunits. Free gamma protein appears to exist in one of two forms: a dimer that does not contain FeMo cofactor or a monomer to which one molecule of FeMo cofactor is also attached [44]. The model that has emerged is that gamma is a chaperone insertase that is able to bind the FeMo cofactor and deliver it to the apo-MoFe protein. The transfer of FeMo cofactor from gamma to the apo-MoFe protein also requires MgATP and the Fe protein.

Other proteins that may serve as intermediate carriers, as well as sites for certain aspects of FeMo cofactor assembly, are the products of the *nifY* gene, the *nifX* gene, and the Fe protein. For example, recent studies have shown that radiolabeled NifB cofactor can be found attached to either the Fe protein or NifX when added to an in vitro biosynthetic system that lacks the MoFe protein subunits [45]. These and related experiments have been interpreted to indicate that once the FeMo cofactor precursor leaves the NifEN complex it becomes attached first to the Fe protein and then to NifX. Evidence that a mature FeMo cofactor might be formed on NifX was obtained by showing that radiolabel accumulated on NifX could be transferred to apo-MoFe protein with a corresponding activation in catalytic activity. The current working model that Ludden and coworkers [45] have suggested is that the [Fe-S] core formed on the NifEN complex is first transferred to the Fe protein, where Mo might become inserted, and then transferred to NifX where FeMo cofactor assembly is completed by attachment of homocitrate.

One complication in the analysis of possible carrier proteins involved in FeMo cofactor biosynthesis is that there appears to be a variety of proteins that might participate in this process and these proteins exhibit a conservation in primary sequence when compared with each other. For example, the C-terminal domain of NifB has primary sequence conservation when compared with either NifX or NifY, both of which have primary sequence conservation when compared with each other [46,47] (Fig. 5). It has recently been found that the primary sequence of gamma also has primary sequence conservation when compared with NifX and NifY (P. W. Ludden, personal communication). Such sequence conservations are not too surprising when one considers

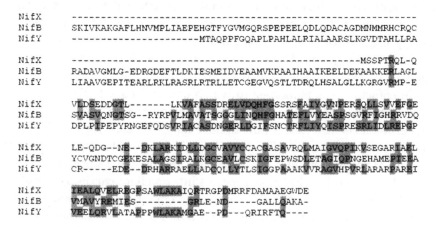

FIG. 5. Comparison of the primary sequences of the *Azotobacter vinelandii* *nifX* and *nifY* gene products with the C-terminal portion of the *nifB* gene product (also see Fig. 2). Conserved regions are shaded in gray.

that these proteins are likely to be involved in binding structurally related FeMo cofactor precursors. However, this situation makes it difficult to assign specific functions as these different players might exhibit a certain level of functional cross-talk. The possibility that such cross-talk does occur is indicated by the fact that neither *nifX* nor *nifY* mutants have obvious phenotypes [47].

9. ROLE OF THE IRON PROTEIN IN COFACTOR ASSEMBLY

The complexity of FeMo cofactor biosynthesis is perhaps best illustrated by the many roles of the Fe protein in the process. So far the Fe protein has been shown to interact with the NifEN complex [48], bind a FeMo cofactor precursor [45], and be required for the attachment of gamma to the apo-MoFe protein subunits [44]. These results are all in line with a variety of studies that have shown that the Fe protein is required for both the formation of FeMo cofactor and its insertion into the apo-MoFe protein [49–54]. At first glance some of this makes some sense. For example, it is known that binding of MgATP to the Fe protein induces a conformational change within the Fe protein, so that it is competent for interaction with the MoFe protein. Thus, one could imagine that the involvement of MgATP and Fe protein for FeMo cofactor insertion could involve the nucleotide-dependent interaction between the Fe protein and the apo-MoFe protein, so that the apo-MoFe protein assumes a conformation that is amenable for cofactor insertion. However, inactive forms of the Fe protein that have an amino acid substitution within the nucleotide binding domain remain capable of promoting FeMo cofactor insertion [50,55]. One could also reasonably explain a similar role for the Fe protein in interaction with the NifEN complex to promote the attachment or release of an FeMo cofactor intermediate from the NifEN complex.

Such an explanation finds support in the observation that the NifEN complex is structurally similar to the MoFe protein. This possibility is further supported by the observation that an altered form of the Fe protein, one that is trapped in the MgATP bound conformation [56], is also able to form a non-dissociating complex with the NifEN heterotetramer and is an inhibitor of in vitro FeMo cofactor biosynthesis [48].

However, as mentioned above, an active form of Fe protein is not necessary for FeMo cofactor formation or insertion. Also, even though a reductant is necessary for in vitro FeMo cofactor formation [57], the Fe protein is probably not the required source of such reducing equivalents. This conclusion is based on the observation that an apo form of the Fe protein lacking the [4Fe-4S] cluster is sufficient for in vitro FeMo cofactor formation [58]. Finally, data that an FeMo cofactor precursor can bind to the Fe protein is convincing [45], yet there is no obvious explanation as to how and why such interaction occurs.

10. SUMMARY AND FUTURE PROSPECTS

So far the products of nine *nif*-specific genes—*nifQ*, *nifV*, *nifB*, *nifX*, *nifU*, *nifS*, *nifX*, *nifN* and *nifE*—have been directly implicated in the biosynthesis of FeMo cofactor. Two themes have emerged from the biochemical and genetic analysis of these gene products that could be relevant to the general process of biological metallocluster formation. The first is that FeMo cofactor is synthesized separately and then inserted into an apo-MoFe protein. The second is that FeMo cofactor biosynthesis does not occur stepwise at a single site, but occurs sequentially with precursors being transferred from one intermediate assembly site to the next. Although the proteins that participate in this process and how they generally fit into a temporal scheme for FeMo cofactor biosynthesis are known, the molecular details of the process are largely not understood.

In our view, two issues have been particularly daunting. First, neither the structures nor the exact chemical composition of any FeMo cofactor precursor is yet known. Recent work showing that FeMo cofactor precursors accumulate on the NifEN complex, the Fe protein, and the NifX protein provides some encouragement that X-ray crystallographic analyses of certain of these proteins, both with and without precursor bound, will be profitable. The challenge here will be to isolate stable forms of these proteins that are sufficiently loaded with the appropriate precursor so that crystallization trials are feasible. The second major problem is that two of the key players in FeMo cofactor biosynthesis, NifQ and NifB, have so far proven intractable to purification despite enormous efforts by several laboratories. We believe that the most reasonable approach to solving this problem will involve the expression of distinct domains of the individual proteins, which can be easily discerned by primary sequence comparisons, and that such protein fragments might well be more amenable to purification. The challenge to this approach will be to establish that biochemical activities are associated with specific protein fragments.

ACKNOWLEDGMENT

Our work is supported by the National Science Foundation.

ABBREVIATIONS

acetyl-CoA	acetyl coenzyme A
ADP	adenosine 5′-diphosphate
ATP	adenosine 5′-triphosphate
Pi	inorganic phosphate

REFERENCES

1. J. A. Cole and S. Ferguson (eds.), *The Nitrogen and Sulfur Cycles*, University Press, Cambridge, MA, 1988.

2. R. R. Eady, *Chem. Revs, 96*, 3013–3030 (1996).

3. M. Ribbe, D. Gadkari, and O. Meyer, *J. Biol. Chem, 272*, 26627–26633 (1997).

4. J. B. Howard and D. C. Rees, *Chem Reviews, 96*, 2965–2982 (1997).

5. B. K. Burgess and D. J. Lowe, *Chem Reviews, 96*, 2983–3011 (1997).

6. T. A. Clarke, F. K. Yousafzai , and R. R. Eady, *Biochemistry, 38*, 9906–9913 (1999).

7. V. K. Shah and W. J. Brill, *Proc. Natl. Acad. Sci. USA*, *74*, 3249–3253 (1977).

8. T. R. Hawkes, P. A. McLean, and B. E. Smith, *Biochem. J*, *217*, 317–321 (1984).

9. C.-H. Kim, W. E. Newton, and D. R. Dean, *Biochemistry*, *34*, 2798–2808 (1995).

10. J. Kim and D. C. Rees, *Nature*, *360*, 553–560 (1992).

11. M. K. Chan and D. C. Rees, *Science*, *260*, 792–794 (1992).

12. J. Kim, D. Woo, and D. C. Rees, *Biochemistry*, *32*, 7104–7115 (1993).

13. S. M. Mayer, D. M. Lawson, C. A. Gormal, S. M. Roe, and B. E. Smith, *J. Mol. Biol.* *292*, 871–891 (1999).

14. P. Rangaraj, C. Ruttiman-Johnson, V. K. Shah, and P. W. Ludden, in *Prokaryotic Nitrogen Fixation: A Model System for Analysis of a Biological Process* (E. Triplett, ed.), Horizon Scientific Press, Wymondham, UK, 2000, pp. 55–79.

15. G. P. Roberts and W. J. Brill, *J. Bacteriol.* *144*, 210–221 (1980).

16. V. K. Shah, J. Imperial, R. A. Ugalde, P. W. Ludden, and W. J. Brill, *Proc. Natl. Acad. Sci. USA*, *83*, 1636–1640 (1986).

17. L. Zheng, R. H. White, V. L. Cash, R. F. Jack, and D. R. Dean, *Proc. Natl. Acad. Sci. USA*, *90*, 2754–2758 (1993).

18. L. Zheng, R. H. White, and D. R. Dean, *Biochemistry*, *33*, 4714–4720 (1994).

19. W. Fu, R. F. Jack, V. Morgan, D. R. Dean, and M. K. Johnson, *Biochemistry*, *33*, 13455–13463 (1994).

20. P. Yuvaniyama, J. N. Agar, V. L. Cash, M. K. Johnson, and D. R. Dean, *Proc. Natl. Acad. Sci. USA*, *97*, 599–604 (2000).

21. C. Ouzounis, P. Bork, and C. Sander, *Trends Biochem. Sci.*, *19*, 199–200 (1994).

22. J. N. Agar, C. Krebs, J. Frazzon, B. H. Huynh, D. R. Dean, and M. K. Johnson, *Biochemistry*, *39*, 7856–7862 (2000).

23. T. D. Paustian, V. K. Shah, and G. P. Roberts, *Proc. Natl. Acad, Sci. USA*, *86*, 6082–6086 (1989).

24. R. Chatterjee, R. M. Allen, P. W. Ludden, and V. K. Shah, *J. Biol. Chem.*, *271*, 6819–6826 (1996).

25. R. M. Allen, R. Chatterjee, P. W. Ludden, and V. K. Shah, *J. Biol. Chem.*, *270*, 26890–26896 (1995).

26. J. T. Roll, V. K. Shah, D. R. Dean, and G. P. Roberts, *J. Biol. Chem.*, *270*, 4432–4437 (1995).

27. R. D. Joerger and P. E. Bishop, *J. Bacteriol.*, *170*, 1475–1487 (1988).

28. H. J. Sofia, G. Chen, B. G. Hetzler, J. F. Reyes-Spindola, and N. E. Miller, *Nucleic Acids Res.* *29*, 1097–1106 (2001).

29. B. T. S. Bui, D. Florentin, F. Fournier, O. Ploux, A. Mejean, and A. Marquet, *FEBS Lett.*, *440*, 226–230 (1998).

30. K. E. Brigle, M. C. Weiss, W. E. Newton, and D. R. Dean, *J. Bacteriol.*, *169*, 1547–1553 (1987).

31. P. J. Goodwin, J. N. Agar, J. T. Roll, G. P. Roberts, M. K. Johnson, and D. R. Dean, *Biochemistry*, *37*, 10420–10428 (1998).

32. P. A. McLean and R. A. Dixon, *Nature*, *292*, 655–656 (1981).

33. T. R. Hoover, V. K. Shah, G. P. Roberts, and P. W. Ludden, *J. Bacteriol.*, *167*, 999–1003 (1986).

34. T. R. Hoover, A. D. Robertson, R. L. Cerny, R. N. Hayes, J. Imperial, V. K. Shah, and P. W. Ludden, *Nature*, *329*, 855–857 (1987).

35. L. Zheng, R. H. White, and D. R. Dean, *J. Bacteriol.*, *179*, 5963–5966 (1997).

36. T. R. Hoover, J. Imperial, P. W. Ludden, and V. K. Shah, *Biochemistry*, *28*, 2768–2771 (1990).

37. J. Liang, M. Madden, V. K. Shah, and R. H. Burris, *Biochemistry*, *29*, 8577–8581 (1990).

38. R. A. Ugalde, J. Imperial, V. K. Shah, and W. J Brill, *J. Bacteriol.*, *164*, 1081–1087 (1985).

39. J. Imperial, R. A. Ugalde, V. K. Shah, and W. J. Brill, *J. Bacteriol.*, *163*, 1285–1287 (1985).

40. J. Imperial, R. A. Ugalde, V. K. Shah, and W. J. Brill, *J. Bacteriol.*, *158*, 187–194 (1984).

41. B. Masepohl, S. Angermuller, S. Hennecke, P. Hubner, C. Moreno-Vivian, and W. Klipp, *Mol. Gen. Genet.*, *238*, 369–382 (1993).

42. C.-M. Chen, T. K. Misra, S. Silver, and B. P. Rosen, *J. Biol. Chem.*, *261*, 15030–15038 (1986).

43. T. D. Paustian, V. K. Shah, and G. P. Roberts, *Biochemistry*, *29*, 3515–3522 (1990).

44. M. J. Homer, D. R. Dean, and G. P. Roberts, *J. Biol. Chem.*, *270*, 24745–24752 (1995).

45. P. Rangaraj, C. Ruttiman-Johnson, V. K. Shah, and P. W. Ludden, *J. Biol. Chem.*, *276*, *15968–15974 (2001)*.

46. C. Moreno-Vivian, M. Schmehl, B. Masepohl, W. Arnold, and W. Klipp, *Mol. Gen. Genet.*, *216*, 353–363 (1989).

47. M. R. Jacobson, K. E. Brigle, L. T. Bennett, R. A. Setterquist, M. S. Wilson, V. L. Cash, J. Beynon, W. E. Newton, and D. R. Dean, *J. Bacteriol.*, *171*, 1017–1027 (1989).

48. P. Rangaraj, M. J. Ryle, W. N. Lanzilotta, P. J. Goodwin, D. R. Dean, V. K. Shah, and P. W. Ludden, *J. Biol. Chem.*, *274*, 29413–19419 (1999).

49. A. C. Robinson, T. W. Chun, J.-G. Li, and B. K. Burgess, *J. Biol. Chem.*, *264*, 10088–10095 (1989).

50. N. Gavini and B. K. Burgess, *J. Biol. Chem.*, *267*, 21179–21186 (1992).

51. S. Tal, T. W. Chun, N. Gavini, and B. K. Burgess, *J. Biol. Chem.*, *266*, 10654–10657 (1991).

52. M. W. Ribbe, E. Bursey, and B. K. Burgess, *J. Biol. Chem.*, *275*, 17631–17638 (2000).

53. W. A. Filler, R. M. Kemp, J. C. Ng, T. R. Hawkes, and B. E. Smith, *Eur. J. Biochem.*, *160*, 371–377 (1986).

54. A. C. Robinson, D. R. Dean, and B. K. Burgess, *J. Biol. Chem.*, *262*, 14327–14332 (1987).

55. D. Wolle, D. R. Dean, and J. B. Howard, *Science*, *258*, 992–995 (1992).

56. W. N. Lanzilotta, K. Fisher, and L. C. Seefeldt, *Biochemistry*, *35*, 7188–7196 (1996).

57. R. M. Allen, R. Chatterjee, P. W. Ludden, and V. K. Shah, *J. Biol. Chem.*, *271*, 4256–4260 (1996).

58. P. Rangaraj, V. K. Shah, and P. W. Ludden, *Proc. Natl. Acad. Sci. USA*, *94*, 11250–11255 (1997).

6

Molybdenum Enzymes Containing the Pyranopterin Cofactor: An Overview

Russ Hille

Department of Molecular and Cellular Biochemistry,
The Ohio State University, Columbus, OH 43210-1218 USA

1. INTRODUCTION

1.1. Introductory Comments

Molybdenum is present in the enzyme nitrogenase as part of a multi-nuclear cluster with seven iron atoms, but in most if not all other molybdoenzymes it is found in a mononuclear center [1–3]. In these mononuclear enzymes the metal of the active site is coordinated by one or two equivalents of a unique pyranopterin cofactor [4,5] found only in these and the closely related tungsten-containing enzymes [6]. Indeed, as is reflected in other chapters in this volume, over the past several years it has become evident that there is a close functional and structural relationship between the molybdenum- and tungsten-containing enzymes. This chapter provides an overview of those molybdenum enzymes that possess a mononuclear center, as well as a more extended discussion of several of these enzymes that are not considered specifically in greater detail in other chapters.

1.2. Biological Distribution of Molybdenum Enzymes

The mononuclear molybdenum enzymes are widely distributed in the biosphere and catalyze a variety of important reactions in the metabolism of nitrogen- and sulfur-containing compounds, and also of various carbonyl compounds (e.g., aldehydes, formate, CO, and CO_2). In higher plants, for example, nitrate reductase catalyzes the first and rate-limiting step in the assimilation of nitrate from the soil [7], and an indole-3-acetaldehyde oxidase is responsible for the final step in the biosynthesis of the hormone indole-3 acetate [8]. In bacteria and archaea, there are a number of molybdenum-containing enzymes expressed when grown on aromatic, heterocyclic compounds as sole carbon source (e.g., nicotinic acid); these enzymes are typically found to catalyze the hydroxylation of a carbon center as the first step in the breakdown of these compounds [9–12]. In addition, bacterial enzymes, such as trimethylamine-N-oxide (TMAO) reductase and dimethyl sulfoxide (DMSO) reductase, catalyze terminal oxidation reactions under a variety of conditions [13]. Interestingly, the substrates for both enzymes are osmolytes from marine organisms, TMAO from fish and DMSO from algae, that counter osmotic pressure due to the salinity of the environment. DMSO reduction to dimethyl sulfide (DMS) is an especially important process environmentally, as the reaction product plays a key role in atmospheric cloud formation; due to its role in increasing global albedo, it is an anti-greenhouse gas. DMS was one of five compounds specifically traced in the course of the iron seeding experiment that took place in the equatorial Pacific Ocean in 1996 [14].

In humans, xanthine oxidoreductase (in its dehydrogenase or oxidase form) catalyzes the final two steps in purine metabolism: sequential hydroxylation of hypoxanthine to xanthine and then uric acid. It is the target of the antihyperuricemic drug allopurinol, which is oxidized by the enzyme to alloxanthine and then able to form a tight complex with the reduced form of the enzyme [15]. Historically, allopurinol was developed as an inhibitor of the enzyme to prevent metabolism of the chemotherapeutic agent 6-mercaptopurine, which was oxidized by the enzyme in vivo, thereby compromising its effectiveness. Treatment with allopurinol and 6-mercaptopurine (or thioguanine) together made it possible to work with much lower levels of the latter and constituted the first rationally designed tandem drug treatment strategy (work for which Gertrude Elion and George Hitchings shared the 1988 Nobel Prize in Physiology or Medicine). Conversion of the dehydrogenase to the oxidase form of xanthine oxidoreductase continues to be implicated in tissue damage associated with oxidative stress, in particular that observed in hypoxia/reperfusion injury [16]. The issue remains controversial, however, and at present it is unresolved as to whether the enzyme plays a role in the pathophysiology of the process [17]. In addition to xanthine oxidoreductase, the closely related mammalian aldehyde oxidases are important in the metabolism of a variety of aldehyde compounds and have been implicated specifically in the biosynthesis of retinoic acid, both in the retina and in the developing nervous system [18]. Finally, sulfite oxidase is responsible for the final step of sulfur metabolism in humans: oxidation of sulfite to sulfate [19]. Loss of sulfite oxidase activity, due to either a mutation in the gene encoding the enzyme or an inability to synthesize the pyranopterin cofactor required by the enzyme, is an invariably fatal genetic lesion [20]. It is significant that the principal effects of loss of function are on the central nervous system, where sulfatides and related compounds constitute major components of the membranes of the myelin sheath. The clinical manifestations of the genetic lesion are thus likely the result of a dysfunction in lipid metabolism, not protein/amino acid metabolism as is frequently inferred in the literature.

1.3.   Reactions Catalyzed

The largest group of mononuclear molybdenum enzymes catalyze the hydroxylation of carbon centers, usually of relatively activated sites such as those found in aromatic heterocyclic compounds and aldehydes. These enzymes catalyze reactions that involve $C-H$ bond cleavage, formally involving the removal of hydride and replacement with hydroxide. CO dehydrogenase from the aerobe *Oligotropha carboxidovorans* is atypical of this group of enzymes in that it catalyzes a reaction that does not involve $C-H$ bond cleavage [21]. (This enzyme is not to be confused with the nickel-containing CO dehydrogenase from organisms such as *Methanosarcina thermophila* [22].) Another enzyme with an unusual reaction stoichiometry that appears to belong to this family is pyrogallol transhydroxylase from *Pelobacter acidigallici* [23], which catalyzes the transfer of a hydroxyl group from one molecule of pyrogallol to another.

Most other molybdenum-containing enzymes catalyze oxygen atom transfer to (for oxidases) or from (for reductases) a substrate lone pair of electrons and, as indicated in Sec. 2, fall into two different families, depending on whether the source organism is pro- or eukaryotic. This generalization notwithstanding, it is to be noted that a growing number of the prokaryotic enzymes catalyze reactions other than oxygen atom transfers. Thus, polysulfide reductase, selenate reductase, and formate dehydrogenase H catalyze dehydrogenation reactions: the reductive cleavage of sulfide from polysulfide [24], the reduction of selenate to selenite [117], and the oxidation of formate to carbonate [25], respectively. The first two reactions involve an (apparently) straightforward reduction, while the latter represents a dehydrogenation, with a hydride transferred to enzyme in the course of the reductive half-reaction of the catalytic cycle. Another type of reaction is catalyzed by formylmethanofuran dehydrogenase, which functions physiologically to reductively amidate a terminal amine, using CO_2 as source of the added carbonyl group [26]. Finally, and quite interestingly, an ethylbenzene dehydrogenase from an isolate of the denitrifying bacterium *Azoarcus* (EbN1) has recently been reported that hydroxylates the substrate to (S)-1-phenylethanol [118]. This reaction represents the sole example of hydroxylation of an unactivated carbon center without utilization of O_2. The enzyme also possesses a unique composition compared with other enzymes of this family (see below), with molybdenum, Fe/S, and *b*-type cytochrome prosthetic groups in an $(\alpha\beta\gamma)_n$ subunit stoichiometry.

1.4. The Pyranopterin Cofactor

The pyranopterin cofactor common to all the mononuclear molybdenum and tungsten enzymes has the structure shown in Fig. 1 [4,5], and coordinates to the metal via the enedithiolate side chain. As discussed elsewhere in this volume, biosynthesis of this cofactor is via a dedicated metabolic pathway that involves no fewer than 12 distinct gene products. In eukaryotes, the cofactor possesses the structure as shown, but in prokaryotes is most often conjugated to nucleosides—usually

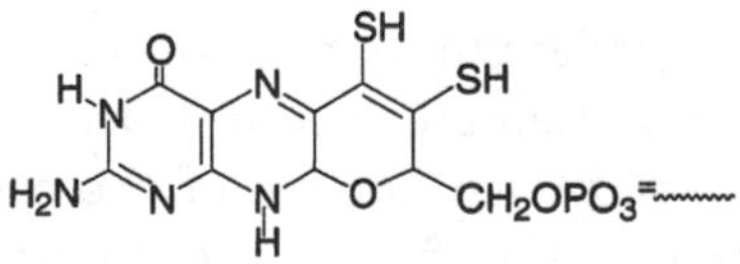

FIG. 1. The pyranopterin cofactor. In eukaryotes the cofactor is as shown, but in prokaryotes it is usually found as the dinucleotide of cytosine, guanosine, adenosine, or inosine, as indicated by the wavy line.

guanosine or cytosine, but occasionally adenosine or inosine. The cofactor is frequently referred to in the literature as molybdopterin, but this is awkward given that the identical cofactor is found in tungsten-containing enzymes and it will be referred to here as pyranopterin.

The pyranopterin cofactor represents an integral component of the molybdenum centers of these enzymes, but does not participate directly in catalysis. Hydrogen bonding interactions with the polypeptide are typically extensive, and particularly in the case of the molybdenum hydroxylases, where there is no protein ligand to the metal, the pyranopterin constitutes the principal means by which the polypeptide binds and retains the molybdenum center. In addition, the pyranopterin appears to be involved in electron transfer. The clearest case is with the molybdenum hydroxylases for which crystal structures exist [27–30], where the distal amino group of the cofactor is hydrogen-bonded to a cysteine residue of a nearby iron-sulfur center. It is evident from the structures that the cofactor represents the path of electron transfer out of the molybdenum center once it has been reduced by substrate. For xanthine oxidase, a specific mechanism by which the conformation of the pyranopterin modulates electron egress has been proposed, involving alternate conformations of the molybdenum center: one with the oxo group of the LMoOS core cis to the enedithiolate group (here designated as L), and the other with the sulfido group cis [31]. The former configuration is that observed in the crystal structure of the resting enzymes [27–30], while the latter is the configuration expected on the basis of the nearly exact similarity of the magnetic circular dichroism (CD) spectrum exhibited by a catalytic intermediate [that giving rise to the "very rapid" electron paramagnetic resonance (EPR) signal; see Sec. 4.1] to that of well-characterized Mo(V) model compounds [31]. It is proposed that in the sulfido-*cis* configuration, the active site possesses an electronic structure appropriate for reaction with substrate, while in the oxo-*cis* configuration the redox-active d_{xy} orbital of the (at least partially reduced) molybdenum is positioned to interact with ligand orbitals in a σ bonding fashion. The σ nature of this interaction is essential, given that the pterin portion of the cofactor is present in a dihydro form that lacks significant π conjugation. The cofactor in essence "directs" reducing equivalents out of the molybdenum center and does not necessarily provide a uniquely effective pathway for rapid electron transfer. Electron transfer out of the molybdenum center has been determined to occur with a rate constant of approximately 8500 s^{-1} [32] over a distance of about about 14.5 Å (to the nearer iron-sulfur center in the protein structure; see Sec. 2.2), a rate constant comparable to that expected for electron transfer over such a distance, assuming no uniquely efficient pathway [33].

The above mechanism for modulation of electron transfer out of the molybdenum center has been referred to as the "oxo-gate" hypothesis since it is the disposition of the Mo=O group relative to the enedithiolate of the pyranopterin that dictates whether electron transfer can occur. A central premise is that a high degree of covalency in the enedithiolate Mo−S bonds (which significantly enhances orbital overlap of donor and acceptor in the electron transfer event) dictates the direction of electron egress from the molybdenum center [34]. The extent to which this is a general principle among the molybdenum enzymes is not known at present. In the cases of the *E. coli* formate dehydrogenase H [35], *Paracoccus denitrificans* nitrate reductase [36], and *Alcaligenes faecalis* arsenite oxidoreductase [37], molybdenum oxidation involves electron transfer to an iron-sulfur center in the same polypeptide (a 4Fe/4S cluster in the first two cases, a 3Fe/4S cluster in the last). Each of these enzymes possesses two equivalents of the pyranopterin cofactor (see Sec. 2.4), designated P and Q in the basis of their disposition in the polypeptide [38]. In all three cases, the iron-sulfur clusters are disposed on the same side of the molybdenum center as the pyranopterin cofactor designated Q, although not at its distal end. Rather, they lie to one side of the cofactor with a coordinating cysteine near the pyrazine ring of the cofactor. In each enzyme, a hydrogen-bonding network exists between the pyrazine ring nitrogen and the cysteine residue. On the basis of the homology with the DMSO and TMAO reductases, members of the same family of molybdenum enzyme that lack any redox-active center other than the molybdenum, it may be assumed that the Q pyranopterin represents the site of electron entry into the molybdenum center. In both these proteins, the distal end of the Q pterin lies close to the surface of the protein, near the interface of three structural domains and opposite the substrate access channel. It appears likely that this represents the docking site for the physiological reductant for these proteins.

By contrast, in sulfite oxidase oxidation of the molybdenum involves electron transfer to a *b*-type cytochrome that resides in a separate domain of the protein [39]. In the crystal structure, the heme domain does not lie at the distal end of the pyranopterin, and in fact is well removed from the molybdenum center with a Mo−Fe distance of 32 Å. This distance is implausibly long for a single electron transfer event in a biological system [33]. However, the heme domain is connected to the molybdenum-binding portion of the protein by a tether of approximately 11 amino acid residues (minimally, residues 85–95) that has no obvious secondary structure, and it is possible that in solution the heme domain is able to assume a position with the heme group closer to the molybdenum, and specifically near the distal end of the pyranopterin cofactor. Significantly, the heme domains of the two subunits of the dimer are not identically disposed relative to their respective molybdenum domains in the asymmetric unit of the crystal, suggesting that there is some flexibility regarding the orientation of the two redox-active centers [39]. If this is the case, however, it is surprising that the docking site for electron transfer would exhibit such weak binding for the heme domain that it is not occupied in the crystal. Nevertheless, it may be that a low intrinsic affinity is the reason why the two domains are tethered together rather than occurring as separate proteins. It remains to be determined whether this is the case.

A final function of the pyranopterin cofactor involves modulating the electronic structure of the molybdenum center, in particular the reduction potential of the cluster. It has been pointed out that the greater the covalency in the Mo−S bonds, the lower the effective charge on the metal and the more difficult it will be to formally reduce the molybdenum from the (VI) to the (IV) valence state [40]. Thus, changes in covalency from one protein to the next undoubtedly contribute significantly to differences in reduction potential. Furthermore, model compound studies have demonstrated a significant "electronic buffer" effect in which sulfur ligands to molybdenum moderate change in electronic structure that attend changes in oxidation states [41]. These two phenomena represent two specific mechanisms by which the pyranopterin cofactor can influence the structure and reactivity of the molybdenum center, contributing to the considerable diversity in electrochemical, spectroscopic, and chemical properties manifested by these enzymes.

2. CLASSIFICATION OF THE MONONUCLEAR MOLYBDENUM ENZYMES

2.1. Rationale for the Classification Scheme Presented

Over the past five or six years, as the number of molybdenum enzymes whose X-ray crystal structures have been determined has grown, it has become increasingly evident that a classification based on structural homology of the active sites is appropriate [1]. The enzymes thus can be grouped into three families, exemplified by xanthine oxidase, sulfite oxidase, and DMSO reductase. The active sites of these enzymes are shown in Fig. 2, and each group is discussed in turn below.

2.2. The Molybdenum Hydroxylases

As indicated above, the molybdenum hydroxylases constitute the largest group of mononuclear molybdenum enzymes, with more than 20 enzymes characterized to varying degree from a wide range of sources.

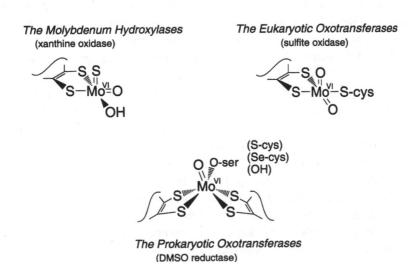

FIG. 2. Active site structures for the three principal families of molybdenum enzymes.

These enzymes range from the well-characterized xanthine and aldehyde oxidoreductases from higher organisms to bacterial enzymes
responsible for the hydroxylation of a diverse range of aromatic heterocycles [1]. Even some eukaryotes express enzymes of this latter type, in
particular the pyridoxal oxidase of *Drosophila melanogaster* [119].

   The aldehyde oxidoreductase from *Desulfovibrio gigas* was the first
mononuclear enzyme for which an X-ray crystal structure was reported,
at 2.25 Å resolution [27] (later refined to 1.8 Å resolution [28]), and
subsequently the structures for CO dehydrogenase from *Oligotropha
carboxidovorans* [29] and bovine xanthine oxidoreductase [30] have
been determined at 2.2 Å and 2.1 Å resolution, respectively (see Fig.
3). As indicated in Fig. 2, the canonical active site for this group of
enzymes consists of a single equivalent of the pyranopterin cofactor
coordinated to an LMoOS(OH) unit, with an overall square pyramidal
coordination geometry. In the crystal structure of the *D. gigas* aldehyde
oxidoreductase [28] the Mo=S group is apical, with the Mo=O group
occupying one of the axial positions, approximately trans to the ene-
dithiolate. The Mo−OH group (considered to be a Mo−OH$_2$ group on
the basis of the apparent Mo−O distance of 2.2 Å in the crystal structure, but this is likely to significantly overestimate the true distance [42]
and the group is more likely present as the Mo−OH depicted here)
points to the solvent access channel to the active site. As discussed
further in Sec. 4.1, it is the Mo−OH group that represents the oxygen
transferred to substrate in the course of reaction. In the two bacterial
systems the cofactor is present as the dinucleotide of cytosine, while in
the bovine enzyme it is present as the mononucleotide. Given that the
pyranopterin cofactor interacts principally with the second of the two
domains that together constitute the molybdenum-binding portion of
the protein, it is not surprising that the bovine sequence for this second
domain diverges from those for the other two enzymes [30]; still, the
overall polypeptide folds are quite similar.

   Interestingly, the molybdenum center of the CO dehydrogenase
represents a significant departure from the structure shown in Fig. 2
in that the active site core is an LMoO$_2$ rather than an LMoOS unit [29].
A second important structural difference is the presence of a selanylcysteine residue, for which a specific catalytic role has been proposed (see
Sec. 4.1). Again, the chemistry catalyzed by this enzyme is also distinct
from that of other members of the family. Nevertheless, it is evident on

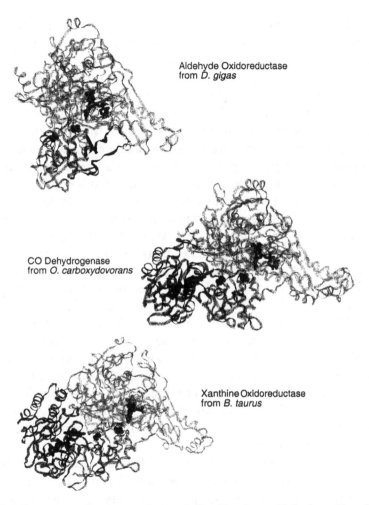

Aldehyde Oxidoreductase
from *D. gigas*

CO Dehydrogenase
from *O. carboxydovorans*

Xanthine Oxidoreductase
from *B. taurus*

FIG. 3. Structures for (top to bottom) the *D. gigas* aldehyde oxidoreductase
[27], the *O. carboxydovorans* CO dehydrogenase [29], and the bovine xanthine
oxidoreductase [30]. The Fe/S centers are rendered in space-filling mode, the
FAD and pterin cofactors in wireframe. The iron-sulfur domains are in medium
gray (*lower left* in the aldehyde oxidoreductase structure, *lower center* in the CO
dehydrogenase and xanthine oxidoreductase structures). The flavin domains of
the two latter proteins are in dark gray (*lower left* in both cases). The molybdenum-binding portions of all three proteins are in light gray. The region of the
aldehyde oxidoreductase that, in the absence of a flavin domain in this protein,
connects the Fe/S- and Mo-binding portions of the protein is indicated by the
dark ribbon.

the basis of the overall structural homology to the aldehyde and xanthine oxidoreductases that the CO dehydrogenase is properly assigned to the molybdenum hydroxylase group.

Enzymes of the hydroxylase family without exception possess multiple redox-active centers in addition to the molybdenum. In the *D. gigas* aldehyde oxidoreductase [27], a pair of 2Fe/2S iron-sulfur centers is present in the N-terminal portion of the enzyme: the first is in a domain possessing an overall fold consisting principally of β sheet and bearing considerable homology to spinach ferredoxin; the second is in a domain having a unique fold, with two long α helices flanked by two shorter ones. The iron-sulfur cluster lies at one end of the two longer helices, and there is a pseudo-twofold axis of symmetry in the domain that bisects the iron-sulfur cluster. Two types of EPR signal have long been observed in enzymes from the molybdenum hydroxylase family, designated Fe/S I and Fe/S II, and it has recently been established that the Fe/S I signal, with EPR properties more typical of ferredoxin-like centers, arises from the cluster in the second (α helical) domain, and the Fe/S II signal from the cluster in the ferredoxin-like domain [43–45]. In the aldehyde oxidoreductase, the polypeptide chain passes from the second Fe/S domain to the molybdenum-binding portion of the enzyme, which consists of two large, roughly peanut-shaped domains that lie across one another with the molybdenum center at the domain–domain interface. In xanthine oxidoreductase an FAD-binding domain intervenes [30] that is homologous to the vanillyl-alcohol oxidase from *Penicillum simplicissimum* [46].

In both aldehyde and xanthine oxidoreductases the polypeptide makes only a single pass from one domain to the next, meaning that the domains are encoded by contiguous stretches in the genes for these proteins [47]. Evidently, these rather complex proteins have been built up over the course of evolution from modular units encoding ancestral proteins possessing each individual redox-active centers. Indeed, in the case of the CO dehydrogenase, the iron-sulfur-, flavin-, and molybdenum-binding portions of the enzyme are in separate subunits rather than domains within a single polypeptide [29]. Approximately half of the enzymes falling into the molybdenum hydroxylase family (all from bacterial sources) are so organized. A majority of the molybdenum hydroxylases possess FAD, as is found in xanthine oxidoreductase and CO dehydrogenase.

The layout of redox-active centers in xanthine oxidoreductase is shown in Fig. 4, with an unambiguous electron transfer sequence: Mo → Fe/S I → Fe/S II → FAD. This is consistent with the long known observation that the oxidizing substrate (oxygen in the case of the oxidase form of the oxidoreductase, NAD^+ in the case of the dehydrogenase form) reacts with the enzyme flavin, at the distal end of the intramolecular electron transfer pathway. Intramolecular electron transfer is thus an integral aspect of catalysis. Electron transfer out of the molybdenum center to Fe/S I occurs via the pyranopterin with a rate

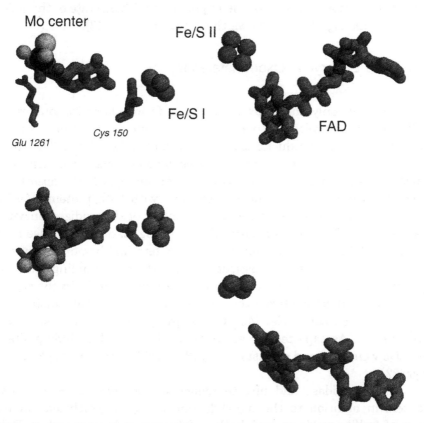

FIG. 4. The disposition of the redox-active centers in xanthine oxidoreductase. From left to right: the molybdenum center, Fe/S I (with the intervening Cys150 that is hydrogen-bonded to the amino terminus of the pterin cofactor), Fe/S II, and FAD. The lower representation is rotated 90° about the horizontal axis relative to the upper one.

constant of about 8500 s^{-1} at pH 8.5 [32]. Electron transfer onto the flavin adenine dinucleotide (FAD) takes place with much smaller rate constants of 150–330 s^{-1} over the pH range 6.0–10.0 [48]. Given the identical UV/visible and CD spectral signatures associated with the reduction of the two iron-sulfur centers [49], it has not yet proven possible to determine a rate constant for electron transfer between the two explicitly, but there is indirect evidence to suggest that it is at least as rapid as the Mo $\rightarrow$ Fe/S I process [32]. It has been shown that electron transfer to the flavin center is impeded by the requirement to take up a proton concomitant with flavin reduction (the enzyme does not accommodate the anionic form of the semiquinone oxidation state of the cofactor), and this significantly slows the overall process [48].

2.3.   The Eukaryotic Oxotransferases

At present the only known oxotransferases from eukaryotic sources are sulfite oxidase (typically isolated from chicken or rat liver) and the assimilatory nitrate reductases from algae and higher plants. The structure of the chicken sulfite oxidase has been determined crystallographically [39], and in conjunction with the considerable X-ray absorption spectroscopic studies that have been done with both proteins [50,51] has led to a consensus that the active centers of the oxidized enzymes have the $LMoO_2$(O-ser) structure shown in Fig. 2. Sulfite oxidase is an α_2 dimer of 110 kDa, and consists of an N-terminal heme domain and two larger domains that constitute the molybdenum-binding and subunit interface portions of the protein, respectively (Fig. 5). In the crystal structure, a sulfate ion (from the mother liquor) is found at a distance of 2.4 Å from the equatorial Mo$-$O, occupying the presumed substrate binding site. As expected for an anionic substrate, this binding site is positively charged and consists of Arg138, Arg190, Arg450, Trp204, and Tyr322.

   Sulfite oxidase and nitrate reductase both possess redox-active centers in addition to the molybdenum: a b-type cytochrome in the case of sulfite oxidase, and both a b-type cytochrome and FAD in nitrate reductase. The molybdenum and heme domains of the two proteins are homologous, but the layout of the domains differs: the heme is N-terminal to the molybdenum-binding portion of sulfite oxi-

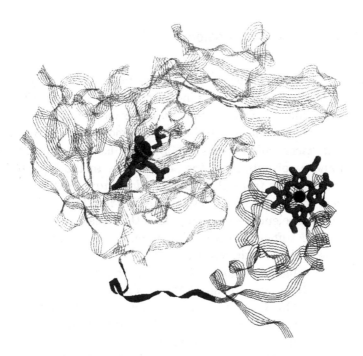

FIG. 5. The structure of sulfite oxidase [39]. The heme domain is in dark gray (*lower right*) and the molybdenum domain in light gray. The tether connecting the two is indicated by the dark ribbon.

dase, and C-terminal in nitrate reductase [52]. The flavin-binding domain lies C-terminal to the heme domain in nitrate reductase, and the structure of this fragment of the protein has been determined crystallographically [53]; it is a member of the ferredoxin reductase family of flavoproteins [54]. A recent X-ray absorption spectroscopy (XAS) analysis of oxidized and reduced *A. thaliana* nitrate reductase has concluded that the as-isolated recombinant protein (from a *Pichia pastoris* expression system) has weakened Mo$-$S coordination to one of the enedithiolate sulfurs which is restored in the course of oxidation by nitrate to give an $LMoO_2$(S-cys) structure consistent with that shown in Fig. 2 [55]. Suggestive evidence is reported for a change in molybdenum coordination geometry in the course of catalysis comparable to that proposed for xanthine oxidase.

2.4. The Prokaryotic Oxotransferases

The prokaryotic oxotransferases represent the most structurally diverse family of molybdenum enzymes. As indicated in Fig. 2, the active sites include two equivalents of the pyranopterin cofactor with two other ligands in a trigonal prismatic coordination geometry. In DMSO reductase from *Rhodobacter sphaeroides* [38,56] these last two ligands are a terminal oxo and a serinate from the polypeptide. In the NAP dissimilatory nitrate reductase from *Desulfovibrio desulfuricans* [36] the serine is replaced by a cysteine and in the *Escherichia coli* formate dehydrogenase by a selenocysteine [35]. In both of these latter proteins, the Mo=O bond is sufficiently long that it may be protonated to give Mo−OH. In the arsenite oxidoreductase from *Alcaligenes faecalis* [37] there is no protein ligand to the metal, the sixth ligand position being occupied by an −OH group; the short Mo−O distance of the fifth ligand is consistent with a Mo=O, as in the DMSO reductases (G. George et al., unpublished XAS results). In addition to the variation in the nature of the sixth (protein) ligand to the molybdenum, there is evidence that at least some enzymes of this class possess a Mo=S rather than a Mo=O group [57]. Structures for representatives of each subgroup of this family are shown in Fig. 6.

This family of enzymes also exhibits considerable variation in the number and organization of additional redox-active centers in the enzyme. Enzymes such as the *Rhodobacter* DMSO reductases, TMAO reductase and biotin-S-oxide reductase have no redox-active center other than the molybdenum, while all the dissimilatory nitrate reductases, formate dehydrogenase H and arsenite oxidoreductase possess iron-sulfur centers in the same polypeptide as the molybdenum; the arsenite oxidoreductase also has a second subunit with a Rieske-type 2Fe/2S center in it [58]. Furthermore, by contrast to the *Rhodobacter* DMSO reductases, the DMSO reductase from *E. coli* is a multi-subunit protein [13]. The molybdenum subunit is comparable in structure to the *Rhodobacter* proteins, but it also possesses a cysteine motif that appears to represent a vestigial 4Fe/4S center that has been lost in the course of evolution. A second subunit of the *E. coli* enzyme possesses four 4Fe/4S iron-sulfur clusters, and a third is the membrane anchor, which lacks redox-active centers but has a quinol

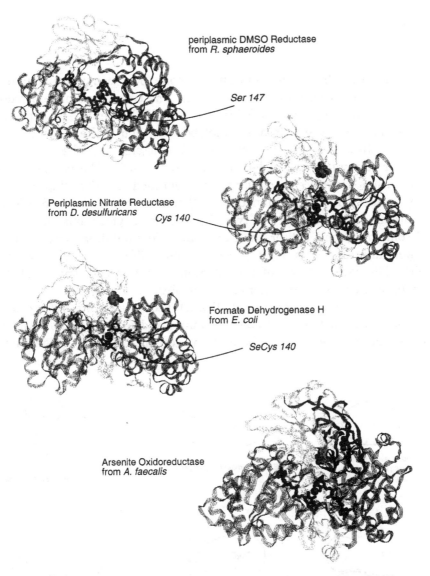

periplasmic DMSO Reductase
from *R. sphaeroides*

Ser 147

Periplasmic Nitrate Reductase
from *D. desulfuricans*   Cys 140

Formate Dehydrogenase H
from *E. coli*

SeCys 140

Arsenite Oxidoreductase
from *A. faecalis*

FIG. 6.   The structures of (top to bottom) the *R. sphaeroides* DMSO reductase
[38], *D. desulfuricans* nitrate reductase [36], *E. coli* formate dehydrogenase H
[35], and *A. faecalis* arsenite oxidoreductase [37]. All orientations view the active
site down the broad solvent access channel; domains II and III, which are
responsible for binding the P and Q pterins, respectively, are indicated in medium gray (*lower left*) and dark gray (*lower right*). Domain I (including the Fe/S
centers of the last three proteins) is in light gray (*top*) and domain IV in light
gray (*bottom*). The Rieske subunit of arsenite oxidoreductase is represented as
dark gray ribbon, which in the view shown sits behind the Fe/S-containing
domain 1 of the Mo subunit.

binding site (the intramembrane quinone pool being the source of reducing equivalents for the reductase). The two *E. coli* nitrate reductases (encoded by the *narGHJI* and *narZYWV* operons) also possess three subunits: the first has a molybdenum and a 4Fe/4S center, the second three 4Fe/4S and one 3Fe/4S clusters, and the membrane anchor subunit a pair of *b*-type cytochromes [59]. The nitrate reductase from *Paracoccus denitrificans* (encoded by the Nap operon) represents a final variation on the theme: it has a Mo- and 4Fe/4S-containing subunit like the two *E. coli* Nar enzymes, but has a cytochrome *b* subunit rather than an iron-sulfur-containing one, and the membrane anchor subunit possesses four different cytochrome sites [60]. The several types of overall protein architecture and prosthetic group composition are summarized in Fig. 7.

The crystal structure of DMSO reductase from *Rhodobacter sphaeroides* was the first to be determined and represents the paradigm in this family of enzymes. The protein is organized into four domains in

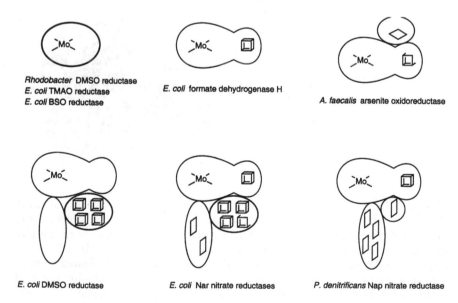

FIG. 7.   The overall protein architectures of members of the DMSO reductase group of molybdenum enzymes. Representative members of each architecture are indicated.

a compound α/β architecture, with the second and third domains responsible for many of the interactions with the two equivalents of the pterin cofactor. These two domains are related by a pseudo-twofold axis of symmetry, and the cofactors associated with them are designated P and Q, respectively. In the *E. coli* formate dehydrogenase [35] and *D. desulfuricans* nitrate reductase [36], the first domain includes a β-sheet region with a 4Fe/4S iron-sulfur cluster that occupies a position in the protein approximately 12–14 Å from the molybdenum. The Q pterin intervenes between the two centers.

3. CONSIDERATION OF SELECTED ENZYMES NOT COVERED IN OTHER CHAPTERS

3.1. Carbon Monoxide Dehydrogenase

An air-stable, molybdenum-containing CO dehydrogenase has been characterized from several organisms, where it catalyzes the key first step of carbon assimilation under conditions of chemo-lithotrophic growth on CO. In *O. carboxidovorans*, the enzyme is encoded by the *coxMSL* operon [61], which encodes the flavin-, iron-sulfur-, and molybdenum-containing subunits, respectively, of an $\alpha_2\beta_2\gamma_2$ hexamer.

The crystal structure of CO dehydrogenase from *O. carboxidovorans* has been obtained at a resolution of a 2.2 Å [29], with the overall structure resembling that of other members of the molybdenum hydro-xylase enzymes as discussed above. Mechanistically, the most remark-able aspect of the structure is that the active site has been modeled as including a selanylcysteine residue, $R-S-Se^-$, as shown in Fig. 8. A reaction mechanism has been proposed in which this selanylcysteine plays a central catalytic role [29] (Fig. 8). CO reacts with the selanylcys-teine residue to give a transient $Se=C=O$ intermediate, which then reacts with the molybdenum center as indicated to give CO_2 and regen-erate the selanylcysteine residue. The four-coordinate Mo(IV) species generated at the end of the reaction is presumably made five-coordinate by the (possibly concomitant) addition of hydroxide from solvent to give the reduced, resting form of the enzyme. This mechanism has recently been called into question, however, as a more detailed crystallographic

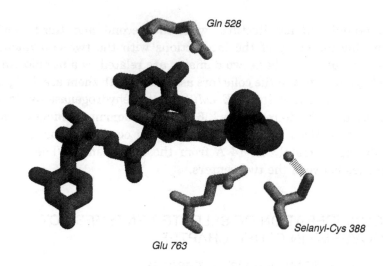

FIG. 8. The active site of carbon monoxide dehydrogenase. *Upper*, the active site structure, including a selanylcysteine residue as proposed in Ref. 29. *Lower*, a reaction mechanism proposed on the basis of the structure [29].

analysis which suggests strongly that the selenium of the putative catalytic selanylcysteine is in fact a partially occupied copper site (H. Dobbek, personal communication). Thus, it is evident that much work remains to be done with this system.

3.2. Assimilatory Nitrate Reductase

Assimilatory nitrate reductase catalyzes the first and rate-limiting step in nitrogen assimilation in higher plants [7], a metabolic role different from that of the dissimilatory nitrate reductases of bacteria, which are terminal oxidases. As indicated above, the enzyme has a large molybdenum binding domain at the N terminus, followed by a central b-type cytochrome domain and a C-terminal domain containing FAD, and is an α_2 dimer in solution. Effective expression systems have been developed for each of the domains from various organisms [62–64], as well as the holoenzyme from A. thaliana [65]. The recombinant flavin- and flavin-heme fragments react with the physiological reductant NADH, which is known to react at the flavin site of the holoenzyme [66,67]. Interestingly, electron transfer from the flavin to the heme of the flavin-heme fragment does not occur until a $FADH_2 \cdot NAD^+$ charge transfer intermediate in the reductive half-reaction decays, presumably by dissociation of the NAD^+ [67].

Assimilatory nitrate reductase is unique among known molybdenum enzymes in that its activity is tightly regulated by posttranslational modification [7]. The enzyme is inactivated at night, apparently to control the concentration of the product nitrite, which would otherwise accumulate in the absence of photosynthetically generated reducing equivalents. This inactivation involves phosphorylation at a specific serine residue in the linker region between the molybdenum and heme binding domains of the protein (Ser534 in the A. thaliana enzyme [68]), catalyzed by a specific calmodulin-dependent nitrate reductase kinase [69]. Other kinases, mediating responses to low levels of CO_2 (which inactivate nitrate reductase) or O_2 (which activates the enzyme), may also be involved [70,71]. Phosphorylation per se does not inactivate the enzyme but rather creates a recognition site for a 14-3-3 protein. It is binding of the 14-3-3 protein to a consensus recognition sequence (LKKSVpSTP in A. thaliana; [72]) that is responsible for inhibition, possibly by preventing effective electron transfer to the molybdenum center rather than by directly inhibiting the chemistry of nitrate reduction. Such a mechanism could effectively inhibit the enzyme were mobility between the molybdenum and heme domains of nitrate reductase prove as important as appears to be the case for sulfite oxidase.

3.3. TMAO Reductase

Trimethylamine-N-oxide is the principal osmolyte of marine fish, and as such it is abundant in marine ecosystems; a variety of microorganisms possess the ability to utilize this compound as a terminal electron acceptor under anaerobic growth conditions. $E.$ $coli$ possesses three enzymes that reduce TMAO: the TorA [73] and TorZ [74] (formerly BisZ) enzymes that are unable to act on S-oxides; and the DmsA-encoded DMSO reductase (see above) that is able to act on both N- and S-oxides. TorZ and DmsA are constitutively expressed, but TorA is inducible via the TorR/TorS/TorT signal transduction cascade [75]. TMAO reductase from $Shewanella$ $massilia$ and similar species is a TMAO-inducible, periplasmic protein encoded by the $torA$ gene of the polycistronic $torECAD$ operon [76]; it is homologous to the $E.$ $coli$ $torCAD$ system. $torC$ encodes an integral membrane pentaheme protein that represents the physiological reductant for the TorA TMAO reductase itself, while $torD$ encodes a cytosolic protein that has been ascribed a TorA-specific chaperonin function [76]. $torE$ encodes a putative membrane protein of unknown function, but which is homologous to the NapE protein of the $P.$ $denitrificans$ nitrate reductase system. No comparable gene is found in the $E.$ $coli$ or $R.$ $sphaeroides$ systems.

The structure of the $S.$ $massilia$ TMAO reductase has been determined crystallographically at a resolution of 2.5 Å [77]. The polypeptide is found to be organized into four domains in the same way as the DMSO reductases from $Rhodobacter$ [38,78]. The Q pyranopterin, which on the basis of the above discussion is likely to be involved in electron transfer into the molybdenum center, is oriented in the polypeptide such that its distal amino group points to the interface junction of domains I, III, and IV. This area on the surface of the protein thus represents a candidate region for interaction of TMAO reductase with its physiological reductant (the membrane-integral pentaheme TorC protein). The principal differences in the active site structure that presumably are important in dictating N-oxide rather than S-oxide reduction for the enzyme include [77] (1) the absence of an active site tyrosine residue (Y114 in the $R.$ $sphaeroides$ DMSO reductase), and (2) a more positively charged surface for the funnel providing access to the active site. The hydrophobic amino acid residues of the substrate-binding pocket, however, are conserved in the TMAO and DMSO reductases.

The active site molybdenum center of TMAO reductase was originally modeled as L_2MoO_2(O-ser) [77], as had been seen for the *R. capsulatus* DMSO reductase [78], but significantly different than the monooxo L_2MoO(O-ser) for the *R. sphaeroides* enzyme [38]. However, it has recently been shown that the second oxygen is an artifact due to the presence in the crystal of a second form of the enzyme in which one of the pyranopterins has dissociated from the molybdenum, being replaced by a second oxo groups [56]. It is the Mo=O that is hydrogen-bonded to Trp118 (in the TMAO reductase structure) that is the genuine Mo=O of the oxidized enzyme, whose active site is represented as a six-coordinate L_2MoO(O-ser) with a trigonal prismatic coordination geometry. Thus, like the DMSO reductases, TMAO reductase presumably functions catalytically by cycling between monooxo Mo(VI) and *des*oxo Mo(IV) species (see Sec. 4.3).

Interestingly, a tungsten-substituted form of TMAO reductase, has been reported [79], and is found to exhibit only modestly reduced catalytic power (a factor of 2 decrease in k_{cat}/K_m) relative to the molybdenum form of the enzyme. The tungsten form is able to utilize DMSO as substrate, and is even able to support bacterial growth on DMSO. It is more oxygen-sensitive and thermostable, and of lower reduction potential than the native molybdenum-containing form of the enzyme. These characteristics are consistent with the idea that tungsten-containing enzymes arose originally in obligate anaerobic, hyperthermophiles in the early biosphere, with the tungsten gradually replaced by molybdenum as the environment cooled and became more aerobic [4]. Its properties are consistent with the behavior of bis(enedithiolate) tungsten compounds, relative to their molybdenum-containing congeners [80].

3.4. Arsenite Oxidase

Bacteria metabolize a variety of arsenicals, most of which are toxic. Arsenite in particular binds to sulfhydryl groups of proteins, particularly to dithiols such as glutathione, and disrupts intracellular redox homeostasis; it is more than an order of magnitude more toxic than arsenate and several hundred times more toxic than methylated arsenicals [81]. As a result, microbial oxidation to arsenate represents an effective detoxification strategy.

In *Alcaligenes faecalis*, arsenite induces a dimeric arsenite oxido-reductase, a molybdenum-containing enzyme of the DMSO reductase family [58]. The X-ray crystal structure of this enzyme has been determined [37], demonstrating that the enzyme possesses two subunits (Fig. 6). The larger subunit is 822 amino acid residues long and incorporates the molybdenum center as well as a 3Fe/4S cluster. As with other members of this family of enzyme, the molybdenum center lies at the bottom of a broad, funnel-shaped depression in the protein surface that provides solvent access to the active site. Unlike other enzymes of the DMSO reductase group, however, there is no protein ligand to the metal of arsenite oxidoreductase: the amino acid residue corresponding to the cysteine, selenocysteine, or serine coordinating the molybdenum in related proteins is Ala199 in arsenite oxidoreductase. The enzyme thus belongs to a new subcategory of the DMSO reductase family whose active site structures more closely resemble that seen in the tungsten-containing aldehyde:ferredoxin oxidoreductases [4], which also possess two equivalents of the pyranopterin cofactor and lack a protein ligand to the metal. The enzyme sample used in the X-ray crystallography was apparently reduced in the X-ray beam, and had a L_2MoO active site with a square pyramidal coordination geometry with a $Mo-O$ distance of 1.6 Å. In conjunction with XAS studies of both oxidized and reduced arsenite oxidoreductase (G. George et al., unpublished), the results are consistent with uptake of hydroxide from solvent upon reoxidation of the molybdenum, giving a six-coordinate $L_2Mo^{VI}O(OH)$ species in the oxidized enzyme.

Substrate docking studies suggest that His195, Glu203, Arg419, and His423 are responsible for substrate binding at the active site. Histidine residues had previously been implicated in substrate binding from chemical modification studies [82]. Binding of arsenite positions it ideally for nucleophilic attack of the arsenite lone pair on the $Mo^{VI}=O$ group, as seen in the reaction of model complexes with phosphines [83,84]. The N-terminal portion of the larger subunit possesses the 3Fe/4S cluster of the protein and exhibits significant structural homology to formate dehydrogenase H and the Nap nitrate reductase, other iron-sulfur-possessing members of the DMSO reductase family. Both of these latter proteins have a 4Fe/4S cluster, and the fourth cysteine ligand seen in these enzymes is replaced by Ser99 in arsenite oxido-reductase.

The smaller subunit of arsenite oxidoreductase, which consists of 134 amino acid residues, has a Rieske-type 2Fe/2S cluster and an overall fold similar to other proteins possessing such a center [85–88]. The Cys60 $-$ X $-$ His62 $-$ X$_{15}$ $-$ Cys78 $-$ X$_2$ $-$ His81 sequence binding the Rieske center in arsenite oxidoreductase conforms to the consensus Cys $-$ X $-$ His $-$ X$_{15\text{-}17}$ $-$ Cys $-$ X$_2$ $-$ His motif for Rieske centers. The (redox-active) iron coordinated by His62 and His81 lies nearer the sur-face of the subunit; both histidine residues lie at the surface of the subunit, with His62 buried within the subunit-subunit interface and close to the 3Fe/4S cluster of the large subunit, with His81 exposed to solvent.

The overall layout of the redox-active centers strongly suggests a linear sequence of electron transfer from the molybdenum center to the 3Fe/4S cluster in the large subunit, then on to the Rieske center of the small subunit. The 3Fe/4S cluster is approximately 14 Å from the molyb-denum atom and sits adjacent to the Q pyranopterin. The two iron-sulfur clusters are 12 Å apart, with the shortest pathway for electron transfer between them via Ser99 of the large subunit (the residue whose mutation from cysteine is responsible for formation of a 3Fe/4S rather than a 4Fe/4S cluster in the protein) and His62 of the small subunit, one of the ligands to the Rieske cluster. Electron transfer out of arsenite oxidoreductase to its physiological oxidants, either a copper-containing azurin or a c-type cytochrome [58], is likely via His81, which is the sole solvent-exposed ligand to the Rieske center. This residue is located in a large concave portion of the protein surface that seems well suited for binding small globular redox partners.

4.　MECHANISTIC CONSIDERATIONS

4.1.　The Molybdenum Hydroxylases

The best-studied molybdenum hydroxylase is the bovine xanthine oxi-doreductase (studied in the oxidase form as it is typically isolated from cow's milk). In the course of substrate hydroxylation, the oxygen atom incorporated into product is ultimately derived from solvent [89], but the proximal donor is a catalytically labile site on the enzyme that is regenerated in the course of each catalytic sequence by oxygen from solvent [90]. These enzymes thus represent a fundamentally different solution to the hydroxylation of a carbon center from that taken by monooxygenase enzymes [1], in which the oxygen incorporated into product is derived from O_2 rather than water, and reducing equivalents are consumed rather than generated, as is the case with the molybdenum-containing enzymes.

The catalytic sequence for xanthine oxidase takes place as indicated in Fig. 9 [91–93], with the enzyme forming first a Mo(IV)·P complex which, depending on the specific substrate and reaction conditions,

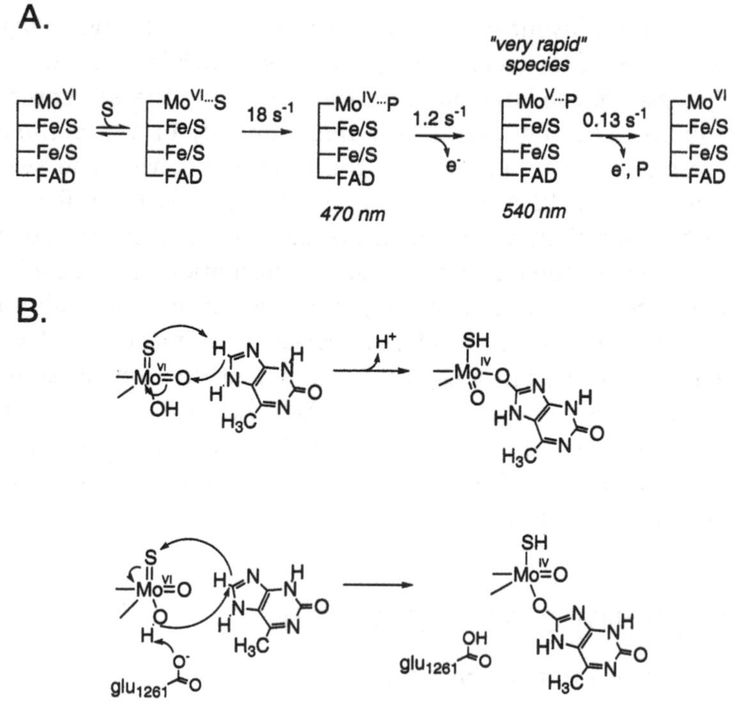

A.

B.

FIG. 9.   Kinetic (A) and chemical (B) mechanisms for xanthine oxidoreductase. In B, alternate chemical mechanisms in which catalysis is initiated by either attack of a substrate cation on the Mo=O group (*upper*) or nucleophilic attack of a Mo−OH on C8 of substrate (*lower*) are indicated

oxidizes by intramolecular electron transfer to a corresponding Mo(V)·P species (that giving rise to the "very rapid" EPR signal, so named on the basis of the kinetics of its formation and decay with xanthine as substrate [94]) or undergoes product dissociation followed by reoxidation, circumventing a Mo(V)·P species. With most substrates (including xanthine) and under most conditions, the reaction proceeds via initial product dissociation rather than one-electron oxidation, but at high pH the Mo(V)·P species can accumulate to 10–15% of the total molybdenum present in the enzyme [95].

The chemistry of the first step of the kinetic mechanism shown in Fig. 9A can be envisaged as occurring in one of two ways, with either the Mo=O or the Mo−OH as the catalytically labile oxygen of the active site (Fig. 9B). The two oxygen sites can be distinguished on the basis of the strength of ^{17}O hyperfine coupling (I = 5/2) to the unpaired electron spin in the Mo(V) state: model compound studies have convincingly demonstrated that Mo=O groups are weakly and isotropically coupled whereas Mo−OH groups are more strongly and anisotropically coupled [96,97]. The physical basis for this observation is that the strong-field Mo=O will define the molecular z axis, and the paramagnetic d_{xy} orbital will thus be less able to interact with the oxo group than a hydroxyl group in the xy plane. In an electron-nuclear double resonance (ENDOR) study of the "very rapid" species, Bray and coworkers [98] were unable to identify a weakly coupled Mo=O underneath the strongly coupled Mo−OR oxygen of the "very rapid" species, and concluded that a "buried" water—possibly coordinated to the molybdenum—rather than the Mo=O oxygen was incorporated into product in the course of the reaction. The presence of a strongly coupled oxygen in the signal-giving species complicates the interpretation, however, and it remains possible a priori even in its absence that a weakly coupled oxygen may have been present but gone undetected. More recently, exchange of ^{17}O from solvent into the active site of unlabeled xanthine oxidase under strictly single-turnover conditions has demonstrated unambiguously that oxygen is incorporated into a strongly and anisotropically coupled site consistent with label having been exchanged quantitatively into the Mo−OH site in a single turnover [99]. There is thus general agreement that it is Mo−OH rather than Mo=O that represents the catalytically labile oxygen of the active site.

The initial ENDOR study above also examined ^{13}C coupling in the "very rapid" species [98]. On the basis of the observed anisotropy in the hyperfine coupling that is attributable to the through-space interaction between the magnetic moments of the unpaired electron and ^{13}C nucleus, a Mo−C distance no greater than 2.4 Å was inferred. This was taken to reflect a side-on η_2 mode of bonding for the C(8)=O group of product, and on the basis of this conclusion a new reaction mechanism was proposed that involved insertion of the C−H bond of substrate across the Mo=S of the molybdenum center, yielding a species with a Mo−C bond [98]. The ^{13}C ENDOR of the "very rapid" species has since been revisited, however, with a specific eye to ascertaining the strength of the evidence in favor of a Mo−C bond in the signal-giving species [100]. A nominal Mo−C distance of 2.24 Å was obtained, not inconsistent with the above study. This distance is arrived at, however, by assuming that all of the anisotropy seen in the ^{13}C coupling is dipolar in nature, yet the large isotropic component to the hyperfine tensor (A = 7.9 MHz) indicates a degree of spin density localized onto the C8 carbon which must also contribute to the anisotropy of the coupling. Correction for the spin density on carbon is complicated by the fact that the anisotropic component of this "local" coupling is dependent on the hybridization of the carbon atom. However, assuming sp^2 hybridization, the corrected distance estimate becomes about 2.8 Å; to the extent that the p-hybridization increases to sp^3 in the signal-giving species, the distance estimate increases to more than 3.0 Å. The ENDOR data thus provide no evidence for an unusually short Mo−C distance in the "very rapid" species. Indeed, it is most likely that the C8 is at a non-bonding distance from the molybdenum.

This conclusion is supported by a computational study [101], in which it is found that a pathway leading to product via an intermediate with a Mo−C bond lies about 20 kcal/mol to higher energy than one proceeding via metal-assisted nucleophilic attack of a hydroxyl group on C8. Furthermore, a nucleophilic attack mechanism is consistent with the known pH dependence of the reaction [92], in which the kinetic parameter k_{cat}/K_m (which follows the reaction of free substrate with free enzyme through the first irreversible step of the reaction) exhibits a bell-shaped curve. The higher pK_a of 7.4 agrees well with that of substrate, with neutral substrate rather than the monoanion being acted on by the enzyme. The pK_a of 6.6 must be due to an ionizable group on the protein, acting as a general base in the course of catalysis. This latter pK_a has been assigned to Glu1271 (in the bovine sequence) [28, 92], which is highly conserved among the molybdenum hydroxylases, and it has been suggested that this residue functions by deprotonating the Mo–OH group to make it a more effective nucleophile in initiating the reaction.

The role of xanthine tautomerization in the course of the reaction has also been investigated by computational methods [102]. The upshot from this study is that in the course of the reaction a proton moves from N3 to N9 as negative charge accumulates on the imidazole subnucleus of substrate during the nucleophilic attack. To the extent that the enzyme active site accommodates or facilitates this tautomerization, it will accelerate the reaction as this translates directly into transition state stabilization. Substrate tautomerization provides a rationale as to why methylation of N3 results in loss of reactivity, an observation difficult to rationalize on electronic or steric grounds.

4.2. The Eukaryotic Oxotransferases

The catalytic mechanisms of sulfite oxidase and nitrate reductase are best understood in the context of the now well-established catalytic power of mononuclear MoO_2 complexes to catalyze oxygen atom tranfer reactions. Several robust systems have been developed that are able to catalyze oxygen atom transfer from a suitable donor (e.g., DMSO) to an acceptor (typically a phosphine), cycling between $Mo(VI)O_2$ and $Mo(IV)O$ species [103–106]. In each case, bulky ligands to the metal prevent the formation of μ-oxo dimers, which are otherwise very stable. In one system, addition of an oxidant permits water to serve as the source of the oxygen used to regenerate the second oxo group in the oxidative limb of the catalytic cycle [106]. The catalytic competence of various oxo donor/acceptor pairs can be understood on thermodynamic grounds [83], and it is for this reason that the sulfite-utilizing enzyme is an oxidase, whereas the nitrate utilizing enzyme is a reductase: the enthalpy for formation of the S=O bond is greater than that for the Mo=O bond, whereas the reverse is true for the N=O bond.

The reactivity of a $Mo(VI)O_2$ complex with phosphine has been examined computationally [107], and the reactivity of the MoO_2 unit understood in the context of a "spectator oxo" effect [108] in which a portion of the enthalpic cost of cleaving the one Mo=O bond is regained by a significant strengthening of the remaining one. This is due to the fact that the two oxo groups of the Mo(VI) species compete in back-bonding into the d_{xy}, d_{xz}, d_{yz} orbital set in the bonding interaction (i.e., those orbitals able to form π-bonding interactions with both oxo groups), a competition that is lost on removal of one of the oxo groups. From this computational study, the key element of the reaction can be understood as the nucleophilic attack of a phosphine lone pair on a π^* orbital of the Mo=O group, with concomitant formation of the P−O bond of product and loss of one Mo−O bond to give what is in essence product coordinated to a now reduced, monooxo Mo(IV) species.

Although it has not been unambiguously demonstrated for either sulfite oxidase or nitrate reductase that the enzyme-catalyzed reaction proceeds in a manner analogous to that seen for the model complexes, with enzyme alternating between dioxo Mo(VI) and monooxo Mo(IV), other mechanisms, involving transient coordination of sulfite to a vacant coordination position or the molybdenum via one of the substrate oxyanion groups, are not likely. Specifically, it has been shown that methylation of the oxyanion groups of sulfite has only a negligible effect on the limiting value of the rate constant for reduction of enzyme in a rapid reaction experiment, k_{red} decreasing from 194 s^{-1} with sulfite to 170 s^{-1} with dimethyl sulfite [109]. The K_d for the methylated substrate is increased, consistent with the crystal structure for sulfite oxidase which shows that the substrate binding site (occupied by sulfate from the mother liquor in the crystal structure) includes three lysine residues interacting with substrate electrostatically. Overall, the available data are consistent with a reaction mechanism for sulfite oxidase involving initial lone-pair attack to give molybdenum reduction and formation of the S−O bond of product by analogy to the reaction of the Mo(VI)O$_2$ complexes with phosphine. In the case of the enzyme, the reaction is plausibly completed by displacement of product from the molybdenum coordination sphere by hydroxide from solvent, which upon reoxidation of the metal deprotonates to give the starting Mo(VI)O$_2$ species. The catalytic mechanism of nitrate reductase presumably proceeds in the reverse of that for sulfite oxidase, with nitrate binding to a Mo(IV)O active site and transferring an oxo group to it.

4.3. The Prokaryotic Oxotransferases

As indicated above, generalizations regarding reaction mechanism for this family of enzymes are difficult because of the diverse reactions catalyzed, and attention here will focus on the *Rhodobacter* DMSO reductases, which have been the subject of the greatest attention. DMSO reductase appears to cycle between monooxo Mo(VI) and *des*oxo Mo(IV) in the course of catalysis [1,40,110,111]. It was originally determined that the enzyme was able to quantitatively transfer oxygen from ^{18}O-labeled DMSO to a water-soluble phosphine under single-turnover conditions [110], and subsequently shown by resonance Raman spectroscopy that the oxygen became incorporated into a single Mo=O group when reduced enzyme was reacted with labeled DMSO [111].

The kinetics of the *R. capsulatus* DMSO reductase has been examined, and spectroscopic evidence for a complex of reduced enzyme with DMSO as a catalytic intermediate has been obtained [112]. This complex can also be formed by treatment of oxidized enzyme with DMS, and a crystal structure of the complex has been reported [113]. The nature of the reaction is such that there can be no spectator oxo effect, and the question arises as to how the bisdithiolene coordination of the metal labilizes a sole Mo=O group. Certainly one factor is the hydrogen-bonding interaction between Trp116 (in the *R. sphaeroides* structure), which would be expected to weaken the Mo=O bond, giving it more Mo−OH character. A recent magnetic circular dichroism study of model Mo(V)OS$_4$ complexes [114] has indicated that the Mo−S bonds in these systems exhibit a high degree of covalency which, along with pseudo-σ and π donation of sulfur lone-pair electron density into the nearly isoenergetic molybdenum d orbitals weaken the Mo=O bond, labilizing it. Thus, the intrinsic electronic structure of the molybdenum center, as one might well expect, may have as great (or even greater) influence on the intrinsic lability of the Mo=O bond as specific hydrogen-bonding interactions with the polypeptide chain.

That this is the case is reflected in recent studies of model bisdithiolene-molybdenum complexes by Holm and coworkers [115,116]. The L$_2$Mo(IV)(OAr) (L = S$_2$C$_2$Me$_2$ or S$_2$C$_2$Ph$_2$) can be prepared by reaction of the dicarbonyl parent with phenols. These compounds react with sulfoxides to generate the oxidized L$_2$Mo(VI)O(OAr) species, which has the long-wavelength absorbance characteristic of all members of the DMSO reductase family of enzymes. Unfortunately, this oxidized compound is not intrinsically stable and to date has not been crystallographically characterized. On the basis of structural and spectroscopic homologies to the cognate tungsten compounds, however, it is likely that the coordination geometry of the oxidized species is distorted octahedral rather than trigonal prismatic as is found in the enzymes [116]. Importantly, this work demonstrates that bisdithiolene-molybdenum complexes are intrinsically capable of catalyzing oxygen atom transfer, cycling between monooxo Mo(V) and *des*oxo Mo(IV) states.

5. CONCLUDING REMARKS

During the past five years, an increasing number of protein crystal structures for enzymes possessing a mononuclear molybdenum center in their active sites has been reported, bringing to nine the number of proteins that have been crystallographically characterized. The structure of at least one member of each family and subfamily of enzymes has been reported. These structures have provided important insight into the physiological electron transfer pathways operating in these enzymes and provided a structural context for mechanistic studies of these enzymes. In work described in detail in other chapters of this volume, the development of expression systems for recombinant proteins holds considerable promise for understanding the role of specific amino acid residues in governing their spectroscopic and catalytic properties. At the same time, the structures that emerge from the active sites of these enzymes have provided new impetus for the synthesis of increasingly accurate structural and functional inorganic models for these enzymes. Future work with these model systems will undoubtedly indicate new lines of investigation for studies of the enzymes themselves.

ACKNOWLEDGMENT

Work in the author's laboratory is supported by grants from the National Institutes of Health (GM58481 and GM59953).

ABBREVIATIONS

BSO	biotin-S-oxide
CD	circular dichroism
DMS	dimethyl sulfide
DMSO	dimethyl sulfoxide
ENDOR	electron-nuclear double resonance
EPR	electron paramagnetic resonance
FAD	flavin adenine dinucleotide (oxidized)
$FADH_2$	flavin adenine dinucleotide (reduced)
NAD^+	nicotinamide pyridine dinucleotide (oxidized)
NADH	nicotinamide pyridine dinucleotide (reduced)
NAP	dissimilatory nitrate reductase from *Paracoccus denitrificans*
TMAO	trimethylamine-N-oxide
XAS	X-ray absorption spectroscopy

REFERENCES

1. R. Hille, *Chem. Rev.*, *96*, 2757–2816 (1996).

2. C. Kisker, H. Schindelin, and D. C. Rees, *Annu. Rev. Biochem.*, *66*, 233–268 (1997).

3. R. S. Pilato and E. I. Stiefel, *Bioinorg. Catal.*, 81–152 (1999).

4. R. R. Mendel and G. Schwarz, *Crit. Rev. Plant Sci.*, *18*, 33–69 (1999).

5. K. V. Rajagopalan and J. L. Johnson, *J. Biol. Chem.*, *267*, 10199–10202 (1992).

6. K. K. Johnson, D. C. Rees, and M. W. W. Adams, *Chem. Rev.*, *96*, 2817–2839 (1996).

7. W. H. Campbell, *Annu. Rev. Plant Physiol.*, *50*, 277–303 (1999).

8. H. Sekimoto, M. Seo, N. Dohmae, K. Takio, Y. Kamiya, and T. Koshiba, *J. Biol. Chem.*, *272*, 15280–15285 (1997).

9. V. N. Gladyshev, S. V. Khangulov, and T. C. Stadtman, *Biochemistry*, *35*, 212–223 (1996).

10. M. Lehmann, B. Tshisuaka, S. Fetzner, and F. Lingens, *J. Biol. Chem.*, *270*, 14420–14429 (1995).

11. M. Bläse, C. Bruntner, B. Tshisuaka, S. Fetzner, and F. Lingens, *J. Biol. Chem.*, *271*, 23068–23079 (1996).

12. C. Canne, I. Stephan, J. Finsterbusch, F. Lingens, R. Kappl, S. Fetzner, and J. Hüttermann, *Biochemistry*, *36*, 9780–9790 (1997).

13. J. H. Weiner, R. A. Rothery, D. Sambasivarao, and C. A. Trieber, *Biochim. Biophys. Acta*, *1102*, 1–18 (1992).

14. S. M. Turner, P. D. Nightingale, L. J. Spokes, M. I. Liddicoat, and P. S. Liss, *Nature*, *383*, 513–517 (1996).

15. V. Massey, H. Komai, G. Palmer, and G. B. Elion, *J. Biol. Chem.*, *245*, 2837–2844 (1970).

16. T. Nishino, *J. Biochem.*, *116*, 1–6 (1994).

17. R. Hille and T. Nishino, *FASEB J.*, *9*, 995–1003 (1995).

18. D.-Y. Huang, A. Furukawa, and Y. Ichikawa, *Arc. Biochem. Biophys.*, *364*, 264–272 (1999).

19. K. V. Rajagopalan, in *Molybdenum and Molybdenum-Containing Enzymes* (M. P. Coughlan, ed.), Pergamon Press, New York, 1980, pp. 241–272.

20. J. L. Johnson and S. K. Wadman, in *Metabolic Basis of Inherited Disease*, 6th ed., Vol. 1 (C. R. Scriver, A. L. Bandet, W. S. Sly, and D. Valle, eds.), McGraw-Hill, New York, 1989, pp. 1463–1475.

21. O. Meyer, K. Frunzke, and G. Mörsdorf, in *Microbial Growth on C_1 Compounds* (J. C. Murrell and D. P. Kelly, eds.) Intercept, Ltd., Andover, UK, 1993, pp. 433–459.

22. R. I. L. Eggen, R. van Kranenburg, A. J. M. Vriesema, A. C. M. Geerling, M. F. J. M. Verhagen, W. R. Hagen, and W. M. de Vos, *J. Biol. Chem.*, *271*, 14256–14263 (1996).

23. S. Sommer, W. Reichenbacher, B. Schink, and P. M. H. Kroneck, *J. Inorg. Biochem.*, *59*, 729 (1995).

24. T. Krafft, M. Bokranz, O. Klimmek, I. Schröder, F. Fahrenholz, E. Kojro, and A. Köger, *Eur. J. Biochem.*, *206*, 503–510 (1992).

25. S. V. Khangulov, V. N. Gladyshev, G. C. Dismukes, and T. C. Stadtman, *Biochemistry*, *37*, 3518–3528 (1998).

26. R. K. Thauer, *Microbiology*, *144*, 2377–2406 (1998).

27. M. J. Romão, M. Archer, I. Moura, J. J. G. Moura, J. LeGall, R. Engh, M. Schneider, P. Hof, and R. Huber, *Science*, *270*, 1170–1176 (1995).

28. R. Huber, P. Hof, R. O. Duarte, J. J. G. Moura, I. Moura, J. LeGall, R. Hille, M. Archer, and M. Romão, *Proc. Natl. Acad. Sci. USA*, *93*, 8846–8851 (1996).

29. H. Dobbek, L. Gremer, O. Meyer, and R. Huber, *Proc. Natl. Acad. Sci. USA*, *96*, 8884–8889 (1999).

30. C. Enroth, B. T. Eger, K. Okamoto, T. Nishino, T. Nishino, and E. F. Pai, *Proc. Natl. Acad. Sci. USA*, *97*, 10723–10728 (2000).

31. R. M. Jones, F. E. Inscore, R. Hille, and M. L. Kirk, *Inorg. Chem.*, *38*, 4963–4970 (1999).

32. R. Hille and R. F. Anderson, *J. Biol. Chem.*, *266*, 5608–5615 (1991).

33. C. C. Moser, C. C. Page, X. Chen, and P. L. Dutton, in *Enzyme Catalyzed Electron and Radical Transfer* (A. Holzenberg and N. S. Scrutton, eds.), Plenum Press, New York, 2000, pp. 1–28.

34. R. L. McNaughton, M. E. Helton, N. D. Rubie, and M. L. Kirk, *Inorg. Chem.*, *39*, 4386–4387 (2000).

35. J. C. Boyington, V. N. Gladyshev, S. V. Khangulov, T. C. Stadtman, and P. D. Sun, *Science*, *275*, 1305–1308 (1997).

36. J. M. Dias, M. E. Than, A. Humm, R. Huber, G. P. Bourenkov, H. D. Bartunik, S. Bursakov, J. Calvete, J. Caldiera, C. Carniero, J. J. G. Moura, I. Moura, and M. J. Romão, *Structure Fold Res.*, *7*, 65–79 (1999).

37. P. Ellis, T. Conrads, R. Hille, and P. Kuhn, *Structure*, *9*, 125–132 (2001).

38. H. Schindelin, C. Kisker, J. Hilton, K. V. Rajagopalan, and D. C. Rees, *Science*, *272*, 1615–1621 (1996).

39. C. Kisker, H. Schindelin, A. Pacheco, W. A. Wehbi, R. M. Garrett, K. V. Rajagopalan, J. H. Enemark, and D. C. Rees, *Cell*, *91*, 973–983 (1997).

40. R. H. Holm, P. Koppenol, and E. I. Solomon, *Chem. Rev.*, *96*, 2239–2314 (1996).

41. B. L. Westcott, N. E. Gruhn, and J. H. Enemark, *J. Am. Chem. Soc.*, *120*, 3382–3386 (1998).

42. H. Schindelin, C. Kisker, and D. C. Rees, *J. Biol. Inorg. Chem.*, *2*, 773–781 (1997).

43. L. Gremer, S. Kellner, H. Dobbek, R. Huber, and O. Meyer, *J. Biol. Chem.*, *275*, 1864–1872 (2000).

44. J. Caldiera, V. Belle, M. Asso, B. Guigliarelli, I. Moura, J. J. G. Moura, and P. Bertrand, *Biochemistry*, *39*, 2700–2707 (2000).

45. C. Canne, D. J. Lowe, S. Fetzner, B. Adams, A. T. Smith, R. Kappl, R. C. Bray, and J. Hüttermann, *Biochemistry*, *38*, 14077–14087 (1999).

46. M. W. Fraaije, R. H. H. van den Heuve, W. J. H. van Berkel, and A. Mattevi, A., *J. Biol. Chem.*, *275*, 38654–38658 (2000).

47. K. Ichida, Y. Amaya, K. Noda, S. Minoshima, T. Hosoya, O. Sakai, N. Shimizu, and T. Nishino, *Gene*, *133*, 279–284 (1993).

48. R. Hille, *Biochemistry*, *30*, 8522–8529 (1991).

49. R. Hille, W. R. Hagen, and W. R. Dunham, *J. Biol. Chem.*, *260*, 10569–10575 (1985).

50. J. M. Berg, K. O. Hodgson, S. P. Cramer, J. L. Corbin, A. Elsberry, N. Periyadath, and E. I. Stiefel, *J. Am. Chem. Soc.*, *101*, 2774–2776 (1979).

51. G. N. George, C. A. Kipke, R. C. Prince, R. A. Sunde, J. H. Enemark, and S. P. Cramer, *Biochemistry*, *28*, 5075–5080 (1989).

52. M. J. Barber, and P. J. Neame, *J. Biol. Chem.*, *265*, 20912–20915 (1990).

53. G. Lu, W. H. Campbell, G. Schneider, and Y. Lindqvist, *Structure*, *2*, 809–821 (1994).

54. P. A. Karplus, M. J. Daniels, and J. R. Herriott, *Science*, *251*, 60–66 (1991).

55. G. N. George, J. A. Mertens, and W. H. Campbell, *J. Am. Chem. Soc.*, *121*, 9730–9731 (1999).

56. H.-K. Li, C. Temple, K. V. Rajagopalan, and H. Schindelin, *J. Am. Chem. Soc.*, *122*, 7673–7680 (2000).

57. M. J. Barber, H. D. May, and J. G. Ferry, *Biochemistry*, *25*, 8150–8155 (1986)

58. G. L. Anderson, J. Williams, and R. Hille, *J. Biol. Chem.*, *267*, 23674–23682 (1992).

59. F. Blasco, C. Iobbi, G. Giordano, M. Chippaux, and V. Bonnefoy, *Mol. Gen. Genet.*, *218*, 249–256 (1989).

60. B. C. Berks, D. J. Richardson, A. Reilly, A. C. Willis, and S. J. Ferguson, *Biochem. J.*, *309*, 983–992 (1995).

61. U. Schübel, M. Kraut, G. Mördorf, and O. Meyer, *J. Bacteriol.*, *177*, 2197–2203 (1995).

62. A. C. Cannons, N. Iida, and L. P. Solomonson, *Biochem. J.*, *278*, 203 200 (1001).

63. G. E. Hyde and W. H. Campbell, *Biochem. Biophys. Res. Commun.*, *168*, 1285–1291 (1990).

64. W. H. Campbell, *Plant Physiol.*, *99*, 693–699 (1992).

65. W. Su, J. A. Mertens, K. Kanamaru, W. H. Campbell, and N. M. Crawford, *Plant Physiol.*, *115*, 1135–1143 (1999).

66. K. Ratnam, N. Shiraishi, W. H. Campbell, and R. Hille, *J. Biol. Chem.*, *270*, 24067–24072 (1995).

67. K. Ratnam, N. Shiraishi, W. H. Campbell, and R. Hille, *J. Biol. Chem.*, *272*, 2122–2128 (1997).

68. P. Douglas, N. Morrice, and C. MacKintosh, *FEBS Lett.*, *177*, 113–117 (1995).

69. P. Douglas, G. Moorhead, Y. Hong, N. Morrice, and C. MacKintosh, *Planta*, *206*, 435–442 (1998).

70. N. G. Halford and D. G. Hardie, *Plant Mol. Biol*, *37*, 735–748 (1998).

71. R. J. Ferl, *Ann Rev. Plant Physiol. Plant Mol. Biol.*, *47*, 49–73 (1996).

72. S. C. Huber, M. Bachmann, and J. L. Huber, *Trends Plant Sci.*, *1*, 432–438 (1996).

73. V. Méjean, M. Iobbi-Nivol, M. Lepelletier, G. Giordano, M. Chippaux, and M. C. Pascal, *Mol. Microbiol.*, *11*, 1169–1179 (1994).

74. S. Gon, J.-C. Patte, V. Méjean, and C. Iobbi-Nivol, *J. Bacteriol.*, *182*, 5779–5786 (2000).

75. C. Jourlin, A. Bengrine, M. Chippaux, and V. Méjean, *Mol. Microbiol*, *20*, 1297–1306 (1996).

76. J. P. Dos Santos, C. Iobbi-Nivol, C. Couillault, G. Giordano, and V. Méjean, *J. Mol. Biol.*, *284*, 421–433 (1998).

77. M. Czjzek, J. P. Dos Santos, J. Pommier, G. Giordano, and V. Méjean, *J. Mol. Biol.*, *284*, 435–447 (1998).

78. A. S. McAlpine, A. G. McEwan, A. L. Shaw, and S. Bailey, *J. Mol. Biol.*, *263*, 53–63 (1996).

79. J. Buc, C.-L. Santini, R. Giordani, M. Czjcek, L.-F. Wu, and G. Giordano, *J. Biol. Inorg. Chem*, *2*, 690–701 (1997).

80. K.-M. Sung and R. H. Holm, *J. Am. Chem. Soc.*, *123*, 1931–1943 (2001).

81. N. Proust, J. Guery, and A. Picot, Andre, Actual. Chim., 3–11 (2000).

82. L. McNellis and G. L. Anderson, *J. Inorg. Biochem.*, *69*, 253–257 (1998).

83. R. H. Holm and J. P. Donahue, *Polyhedron*, *12*, 571–589 (1993).

84. J. H. Enemark and C. G. Young, *Adv. Inorg. Chem 40*, 1–88 (1994).

85. C. J. Carrell, H. Zhang, W. A. Cramer, and J. L. Smith, *Structure*, *5*, 1613–1625 (1997).

86. S. Iwata, M. Saynovits, T. A. Link and H. Michel, *Structure*, *4*, 567–79 (1996).

87. S. Iwata, J. W. Lee, K. Okada, J. K. Lee, M. Iwata, B. Rasmussen, T. A. Link, and S. Ramaswamy B. K. Jap. *Science*, *281*, 64–71 (1998).

88. B. Kauppi, K. Lee, E. Carredano, R. E. Parales, D. T. Gibson, H. Eklund, and S. Ramaswamy, *Structure*, *6*, 571–586 (1998).

89. K. N. Murray, J. G. Watson, and S. Chaykin, *J. Biol. Chem.*, *241*, 4798–4801 (1966).

90. R. Hille and H. Sprecher, *J. Biol. Chem.*, *262*, 10914–10917 (1987).

91. R. B. McWhirter and R. Hille, *J. Biol. Chem.*, *266*, 23724–23731 (1991).

92. J. H. Kim, M. G. Ryan, H. Knaut, and R. Hille, *J. Biol. Chem.*, *271*, 6771–6780 (1996).

93. M. Mondal and S. Mitra, *Biochemistry*, *33*, 10305–10312 (1994).

94. R. C. Bray and T. Vånngård, *Biochem. J.*, *114*, 725–734 (1969).

95. D. E. Edmondon, D. P. Ballou, A. van Heuvelen, G. Palmer, and V. Massey, *J. Biol. Chem.*, *248*, 6135–6144 (1973).

96. D. Dowerah, J. T. Spence, R. Singh, A. G. Wedd, G. Wilson, F. Farchione, J. H. Enemark, J. Kristofski, and M. Bruck, *J. Am. Chem. Soc.*, *109*, 5655–5665 (1987).

97. R. J. Greenwood, G. L. Wilson, J. R. Pilbrow, and A. G. Wedd, *J. Am. Chem. Soc.*, *115*, 5385–5392 (1993).

98. B. D. Howes, R. C. Bray, R. L. Richards, N. A. Turner, B. Bennett, and D. J. Lowe, *Biochemistry*, *35*, 1432–1443 (1996).

99. M. Xia, R. Dempski, and R. Hille, *J. Biol. Chem.*, *274*, 3323–3330 (1999).

100. P. Manikandan, E.-Y. Choi, R. Hille, and B. M. Hoffman, *J. Am. Chem. Soc.*, *123*, 2658–2663 (2001).

101. P. Ilich and R. Hille, *J. Phys. Chem. (B)*, *103*, 5406–5412 (1999).

102. P. Ilich and R. Hille, *Inorg. Chim. Acta*, *263*, 87–94 (1997).

103. J. M. Berg and R. H. Holm, *J. Am. Chem. Soc.*, *107*, 925–932 (1985).

104. J. M. Berg and R. H. Holm, *J. Am. Chem. Soc.*, *107*, 917–924 (1985).

105. B. E. Schultz, S. F. Gheller, M. C. Muetterties, M. J. Scott, and R. H. Holm, *J. Am. Chem. Soc.* *115*, 2714–2722 (1993).

106. Z. Xiao, C. G. Young, J. H. Enemark, and A. G. Wedd, *J. Am. Chem. Soc.*, *114*, 9194–9195 (1992).

107. M. A. Pietsch and M. B. Hall, *Inorg. Chem.* *35*, 1273–1278 (1996).

108. A. K. Rappé and W. A. Goddard, III, *Nature*, *285*, 311–312 (1980).

109. M. S. Brody and R. Hille, *Biochim. Biophys. Acta*, *1253*, 133–135 (1995).

110. B. E. Schultz, R. Hille, and R. H. Holm, *J. Am. Chem. Soc.*, *117*, 827–828 (1995).

111. S. D. Garton, J. Hilton, H. Oku, B. R. Crouse, K. V. Rajagopalan, and M. K. Johnson, *J. Am. Chem. Soc.*, *119*, 12906–12916 (1997).

112. B. Adams, A. T. Smith, S. Bailey, A. G. McEwan, and R. C. Bray, *Biochemistry*, *38*, 8501–8511 (1999).

113. A. S. McAlpine, A. G. McEwan, and S. Bailey, *J. Mol. Biol.*, *275*, 613–623 (1998).

114. J. McMaster, M. D. Carducci, Y.-S. Yang, E. I. Solomon, and J. H. Enemark, *Inorg. Chem.*, *40*, 687–702 (2001).

115. B. S. Lim, J. P. Donahue, and R. H. Holm, *Inorg. Chem.*, *39*, 263–273 (2000).

116. B. S. Lim and R. H. Holm, *J. Am. Chem. Soc.*, *123*, 1920–1930 (2001).

117. I. Schröder, S. Rech, T. Krafft, and J. M. Macy, *J. Biol. Chem.*, *272*, 23765–23768 (1997).

118. O. Kneimeyer and J. Heider, *J. Biol. Chem.*, *276*, 21381–21386 (2001).

119. C. K. Warner, D. T. Watts, and V. Finnerty, *Mol. Gen. Genet.*, *180*, 449–453 (1980).

7
The Molybdenum and Tungsten Cofactors: A Crystallographic View

Holger Dobbek and Robert Huber

Max-Planck-Institut für Biochemie, Abteilung Strukturforschung,
Am Klopferspitz 18a, D-82152 Martinsried, Germany

1. INTRODUCTION

In a wide set of mostly redox-active enzymes, molybdenum and tungsten fulfill crucial catalytic tasks in the global metabolism of nitrogen, sulfur, and carbon (see Chapter 1 in this volume). With the exception of a multinuclear complex iron-molybdenum-sulfur center found in nitrogenases (see Chapter 3 in this volume) [1], the two metals are associated with an organic pterin derivative called molybdopterin or its derivatives, building the molybdenum or tungsten cofactor (moco). The term *moco* is used in this context to designate the cofactors of both the Mo- and the W-containing enzymes. In the past several years, the number of enzymes utilizing moco increased and lies now well over 50 different members [2] isolated from microorganisms, plants, and animals.

Our current understanding of the moco has been largely enriched by high-resolution X-ray crystal structures of Mo- and W-containing enzymes, which started with the elucidation of the structures of the W-containing aldehyde oxidoreductase (AOR) of *Pyrococcus furiosus* [3] and the Mo-containing AOR of *Desulfovibrio gigas* [4]. Following their example several crystal structures of Mo- and W-containing enzymes have been solved, proving the general versatility of the moco. Based on structural and spectroscopic characteristics, different families of moco-containing enzymes can be distinguished, of which at present at least one example per family has been characterized by X-ray crystallography. This chapter will focus on the information about the structure and role of cofactors gained by crystallography in the four moco-containing families.

2.   STRUCTURE OF THE Mo/W COFACTOR

2.1.   Identification of Molybdopterin

Although the tight association of Mo with molybdoproteins had been early recognized, the identification of the organic components coordinating Mo has a history of several decades. A key event was the detection of a dissociable entity capable of reconstituting inactive apoprotein of nitrate reductase (nit-1), shown by Nason and coworkers [5–7] in the early 1970s. As this cofactor was not amenable to normal procedures of isolation, structural studies had to focus on inactive derivatives of molybdopterin, like dicarboxyamidomethyl molybdopterin [8], form A [9] and form B [10]. Based on these studies, Rajagopalan and coworkers proposed models for the moco [8] containing a bicyclic pterin derivative with the pterin ring substituted at position 6 by a phosphorylated 3′,4′-dihydroxybutyl side chain containing a 1′,2′-*cis*-dithiolene bond. The sulfur atoms of the dithiolene group were proposed to coordinate the Mo atom, with a stoichiometry of one molybdopterin per Mo atom (for review, see [11]). Elucidation of the crystal structures of the AORs from *P. furiosus* [3] and *D. gigas* [4] showed the general validity of the proposed model, but with the unanticipated findings that molybdopterin is a tricyclic pyranopterin structure in the protein interior. This has been confirmed by all moco-containing structures determined so far. The pyran ring is formed by the dihydroxybutyl side chain through an attack of its 3'-hydroxyl group on C7 of a dihydropterin. It has been suggested that pyran adduct formation stabilizes the pterin ring against destructive oxidative modifications of the cofactor [12].

2.2. Structure

The pyranopterin ring has so far only been detected in the protein environment, and its opening seems to occur when the cofactor gets released from this stabilizing environment. Elucidation of several crystal structures in the past 5 years supplied additional unquestionable proof that both Mo and W are bound by the dithiolene group of the tricyclic pyranopterin system, thereby ruling out other proposed alternative binding modes. The molybdopterin observed in all known crystal structures has the basis structure shown in Fig. 1.

The molybdopterin is distinctly nonplanar in its pyran and central pyrazine ring. The pterin and pyran rings in the tricycle enclose an angle of approximatey 40°, the pyrazine ring is twisted with C6 exo and C7 endo. The three chiral centers at C6, C7, and C3' are in the R conformation. The pyran ring adopts a half chair conformation with O3' endo. This conformation seems not to be fixed as flexibility between the neighboring rings is obvious from a least-squares superimposition of the pterins [13] (Fig. 2).

Additional conformational flexibility in the cofactor can be observed in the hydroxymethyl side chain (Fig. 2). The oxidation state of the pyrano-pterin system is structurally equivalent to the fully

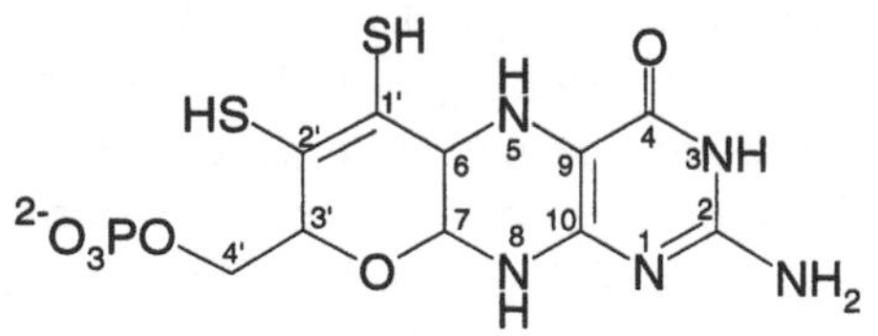

FIG. 1.   Structure of molybdopterin.

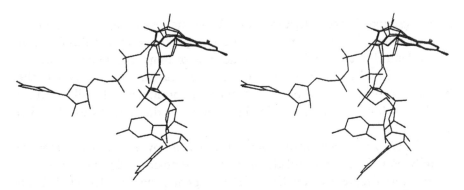

FIG. 2.   Stereoview of the superimposition of the pterin moiety of four different moco conformations as found for the molybdopterin-cystosine dinucleotide cofactor of CO dehydrogenase [20], the Q- and P- molybdopterin-guanine dinucleotide cofactor of DMSO reductase [33], and the molybdopterin cofactor of formaldehyde ferredoxin oxidoreductase [48]. The angle between the best planes defined by the pyran ring and the pterin system are between 30° and 60°. Further flexibility is observed for the hydroxymethyl side chain and the connected nucleotides.

reduced tetrahydropterin oxidation state with a protonation at N5 and N8, although its redox state is equivalent to the two-electron more oxidized dihydropterin, probably 5,6-dihydropterin, which could be formed after opening of the pyran ring. According to the nomenclature used for pterins, the enzyme-bound molybdopterin is perhaps best described as a tetrahydropterin-pyran system.

2.3.   Coordination of Molybdenum

The Mo ion can be coordinated by three types of ligands: the dithiolene-sulfurs of molybdopterin, protein side chains, and nonprotein groups such as single oxygen and sulfur atoms. Mo binding by the cofactor is achieved through coordination by its thiolate groups with apparent Mo−S bond length of approximately 2.4 Å. The ligation is usually accompanied by a significant displacement of Mo from the plane of the dithiolene group. The binding of Mo can result in a 1 : 1 or 2 : 1 molybdopterin-to-Mo ratio and involves different numbers of oxygen and/or sulfur groups. Protein ligands such as serine, cysteine, and selenocysteine are found in some moco-containing proteins. Oxygen and sulfur atoms bound to the Mo ion can include one or more oxo, hydroxo, water, sulfido and hydrosulfido groups. In addition, the number and type of the ligands depends on the oxidation state of the Mo ion. In all enzymes studied so far the Mo ion cycles between the oxidation states Mo(+VI) and Mo(+IV).

In enzymes from prokaryotic sources the pyrano-pterin system is usually bound to a nucleotide of guanine, cytosine, adenine, or hypoxanthine. As first shown for the AOR of *D. gigas*, the dinucleotide is in an extended conformation [4] (Figs. 2 and 3).

3. MOCO-CONTAINING ENZYME FAMILIES

3.1. Introduction

Moco-containing enzymes can be distinguished according to their amino acid sequences, spectroscopic properties, active site structures, and catalyzed reactions. Considering this, the molybdenum-containing enzymes fall into three distinct families plus a family for the tungsten-containing enzymes such as AOR and formaldehyde ferredoxin oxidoreductase from *P. furiosus* (Fig. 4) [2].

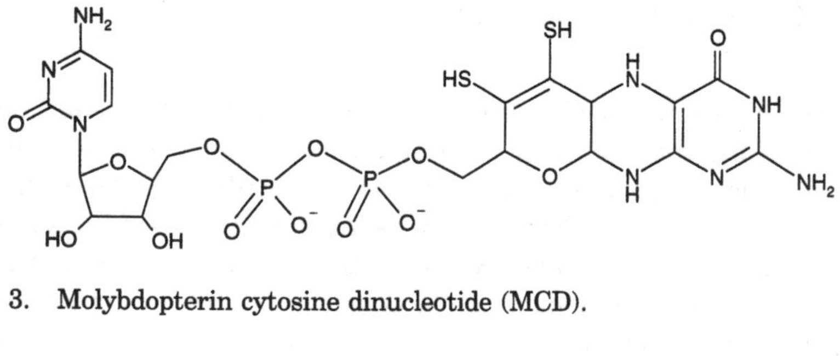

FIG. 3. Molybdopterin cytosine dinucleotide (MCD).

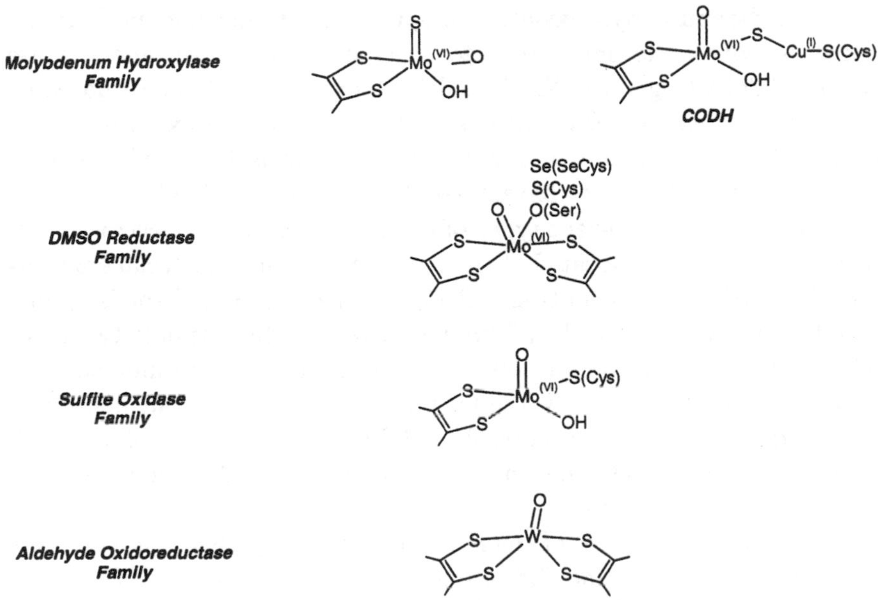

Molybdenum Hydroxylase Family

CODH

DMSO Reductase Family

Se(SeCys)
S(Cys)
O(Ser)

Sulfite Oxidase Family

Aldehyde Oxidoreductase Family

FIG. 4.   The four families of moco-containing enzymes.

3.2.   Mono- and Binuclear Active Sites of Molybdenum Hydroxylases

3.2.1.   Introduction

The members of the large family of molybdenum hydroxylases generally catalyze the oxidative hydroxylation of a diverse range of aldehydes, aromatic heterocycles, and carbon monoxide. While the oxidation of aldehydes and aromatic heterocycles involves the cleavage of a $C-H$ bond during catalysis [14], such is not the case for the substrate CO [15–17]. The catalyzed reaction follows the general equation:

$$SH + H_2O \rightarrow SOH + 2H^+ + 2e^-$$

Molybdenum hydroxylases are properly considered as hydroxylases that use water as their source of oxygen becoming incorporated into the product. Electrons generated by this reaction are transferred to external acceptors using an intramolecular path.

Molybdenum hydroxylases comprise eukaryotic and prokaryotic enzymes with different substrate specificities like xanthine oxidase (XO) and dehydrogenase (XDH), aldehyde oxidase (AO) and oxidoreductase (AOR), formate dehydrogenase (FDH), quinoline-2-oxidoreductase, isoquinoline 1-oxidoreductase, nicotinic acid hydroxylase, and CO dehydrogenase (CODH). The characterized enzymes are all rather complex and use up to five different types of cofactors to transfer electrons from or to their active site located at the moco. While some bacterial enzymes like the AOR of *D. gigas* [4] and *D. desulfuricans* [18] and the isoquinoline 1-oxidoreductase of *Pseudomonas diminuta* [19] use only two types of [2Fe-2S] clusters in addition to the moco, most of the molybdenum hydroxylases, like XDH, XO, AO, and CODH, contain an additional FAD-carrying domain or subunit [14,20,21]. An additional [4Fe-4S] cluster has been detected in 4-hydroxybenzoyl-CoA reductase from *Thauera aromatica* [22].

Crystal structures of molybdenum hydroxylases are currently available for the molybdo-iron-sulfur flavoproteins CODH of *Oligotropha carboxidovorans* [17,20,23] and *Hydrogenophaga pseudoflava* [24], and XO and XDH from bovine milk [21] (Fig. 5). The AORs of *D. gigas* [4,25,26] and *D. desulfuricans* [18] are examples of Mo hydroxylases in which an FAD-containing domain/subunit is absent (Fig. 5).

All enzymes described so far are dimeric in solution and crystallized in their dimeric form. While the dimer is often found in one asymmetrical unit (a.u.) of the crystal, one monomer is found per a.u. in the case of the AORs and the dimer is formed by a crystallographic dyad in both crystal packings [4,18].

General fold of all structures determined so far is very similar (Fig. 5) and the molybdo-iron-sulphur flavoproteins resemble the shape of butterflies. The structure of the FAD-containing enzymes can be divided into three functional subunits/domains. Each monomer is composed of a molybdoprotein, carrying the moco, an iron-sulfur protein, carrying two types of [2Fe-2S] clusters, and a flavoprotein, harboring a non-covalently bound FAD molecule. The molybdoprotein has a heart-like shape with dimensions of approximately $60 \times 80 \times 70$ Å and can be divided into two domains, which both are interacting with the cofactor. A tight network of hydrogen bonds holds the cofactor using

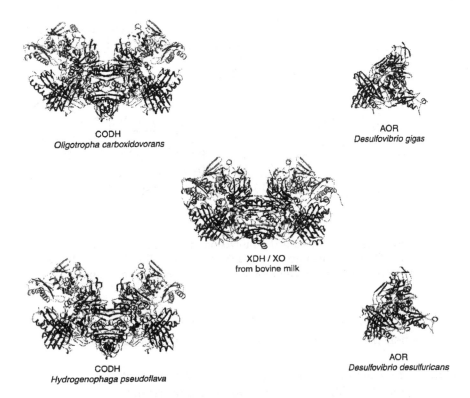

FIG. 5.   Crystal structures of molybdenum hydroxylases. Coordinates are taken from the protein data base (PDB) files 1ALO (AOR of *D. gigas*) [4,25], 1DGJ (AOR of *D. desulfuricans*) [18], 1FFU (CODH of *H. pseudoflava*) [24], 1FO4 (XDH from bovine milk) [21], 1QJ2 (CODH of *O. carboxidovorans*) [20,23].

residues that are mostly conserved in the molybdenum hydroxylase family (Fig. 5).

    Five conserved moco binding motifs have been identified in molybdenum hydroxylases [24]. They are localized in distinct protein segments of the molybdoprotein and interact with parts of the moco (Fig. 6).

    In addition, molybdenum hydroxylases can be divided in subfamilies employing different structurally related substrates. Each of these subfamilies has a characteristic motif at a position where an active site loop has been found in CODHs [24].

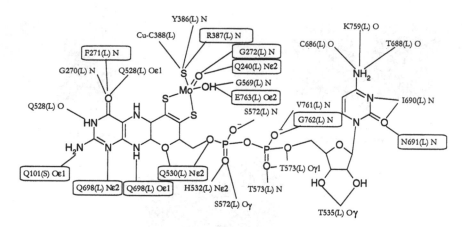

FIG. 6.   Residues coordinating the molybdopterin cytosine dinucleotide (MCD) cofactor in CO dehydrogenase of *O. carboxidovorans*. Residues which are conserved in more than 90% of the molybdenum hydroxylases are boxed.

An additional distinction of the molybdenum hydroxylases can be made when the direct molybdenum environment is inspected.

3.2.2.   Molybdenum-containing Aldehyde Oxidoreductases from Desulfovibrio gigas and D. desulfuricans

The active site of AOR (Fig. 7) contains a pentacoordinated Mo(+VI) center in the air-oxidized enzyme that exhibits a distorted square pyramidal coordination geometry. The equatorial plane is formed by the two dithiolene sulfurs (S1′ and S2′) and two oxygen ligands interpreted as a water (O2) and an oxogroup (O3). While the as-isolated enzyme contains a third oxygen ligand, modeled as an oxogroup (O1), in the apical position of the pyramid, a cyanolysable sulfido group is expected in the functional sulfo form of the enzyme in analogy to xanthine oxidase [25]. No direct interactions of the surrounding polypeptide chain with the metal were detected, although the side chain of Glu869 (Fig. 7) is in 3.5 Å distance to the metal ion and forms a hydrogen bond with one of the equatorial oxygen ligands. Following crystallographic investigations resolved the desulfo, sulfo, oxidized, reduced, and alcohol-bound forms at 1.8 Å [25]. In the sulfo form, formed upon resulfuration with Na_2S, the moco has a dithiolene-bound *fac*-$[Mo, =O, =S, -(OH_2)]$ sub-

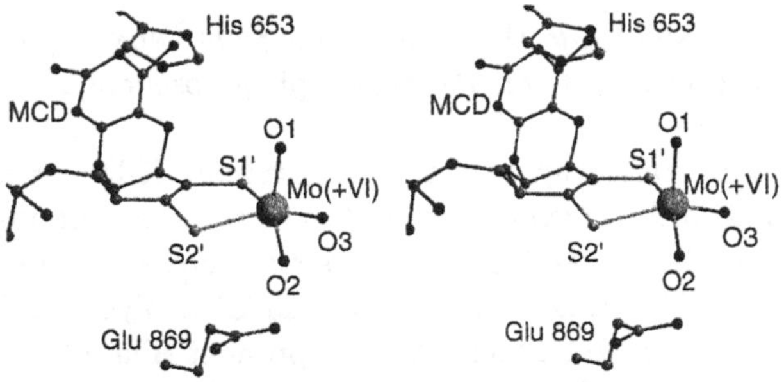

FIG. 7. Stereoview of the active site of AOR from *Desulfovibrio gigas*. Coordinates are taken from the PDB file 1ALO. Distances taken from Huber et al. [25]: *ox*.: Mo−O1, 1.6 Å; Mo−O2, 1.5 Å; Mo−O3, 2.5 Å. *red*.: Mo−O1, 1.9 Å; Mo−O2, 2.5 Å; Mo−O3, 2.1 Å.

strate [25], with the sulfo ligand at the former position of the apical oxo group (O1). The short distance between the dithiolene sulfur atoms of oxidized AOR may indicate a partial disulfide bond character. Reduction of the enzyme leads to an enlargement of the S−S distance and increased dithiolene puckering [25].

3.2.3. Xanthine Oxidase/Dehydrogenase from Bovine Milk

Xanthine oxidoreductases of mammals catalyze the last two steps in the degradation of purines to urate. Conversion of XDH to XO can be achieved by the oxidation of sulfhydryl groups or by limited proteolysis [27].

The crystal structures of bovine milk XDH and XO formed by proteolysis have been solved to 2.1 and 2.5 Å resolution, respectively [21]. The overall structure of both enzyme forms is very similar, with the major difference being a loop which in the XO form blocks the access of larger molecules, such as NAD^+ or FAD.

Three ligands are bound to Mo in addition to the dithiolene sulfurs. As their identity could not be determined unambiguously [21], they have been modeled in analogy to the sulfo species of AOR of *D. gigas* [25] (Fig. 7).

3.2.4. Carbon Monoxide Dehydrogenases from Oligotropha carboxidovorans and Hydrogenophaga pseudoflava

While a first crystallographic characterization of CODH from *O. carboxidovorans* led to the suggestion that the active site contains a Se atom bound to a cysteine residue [20], investigations conducted at atomic resolution of a 23 U·mg^{-1} enzyme and MAD analysis [23] revealed a different, unanticipated active site environment (Fig. 8).

The active site of as-isolated CODH has been found to be built up by an unprecedented binuclear metal cluster, formed on one side by a Cu ion bound to the Sγ-atom of Cys388 and on the other side by a Mo-oxo/hydroxo group ligated by the dithiolene sulfurs and a bridging μ-sulfido ligand between Mo and Cu(I), thereby forming the [CuSMo] cluster (Fig. 8) [23]. The ligands around Mo form a tetrahedral geometry, with the dithiolene group straddling over one vertex of the tetrahedron (Fig. 8). Residues in the second coordination sphere are Gln240, establishing a hydrogen bond to the oxo ligand and Glu763 (Fig. 8). The two oxo atoms of Glu763 show a tight hydrogen-bonding interaction with Cys388 of the active site loop and Oε2 interacts with Mo in the oxidized state. The μ-sulfido ligand and the Sγ atom of Cys388 ligate the Cu(I) ion in a distorted linear geometry.

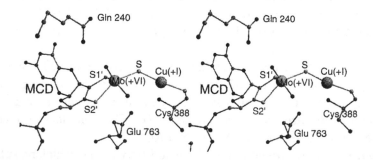

FIG. 8. Stereoview of the active site of CODH from *O. carboxidovorans*. Distances: *ox*.: Mo−OH, 1.85 Å; Mo−Oxo, 1.77 Å; Mo−S, 2.28 Å; Mo−S1′, 2.45 Å; Mo−S2′, 2.49 Å: Mo−Cu(I), 3.74 Å; S−Cu, 2.21 Å; γS−Cu, 2.22 Å; Oε2−Mo, 3.16 Å. *red*.: Mo−OH, 2.03 Å; Mo−Oxo, 1.74 Å; Mo−S, 2.32 Å; Mo−S1′, 2.36 Å; Mo−S2′, 2.44 Å; Mo−Cu(I), 3.93 Å; S−Cu, 2.18 Å; γS−Cu, 2.20 Å; Oε2−Mo, 3.32 Å.

An inactive species of CODH is formed upon treatment of the enzyme with cyanide (CN), leading to the formation of thiocyanate. Crystals that were soaked with 10 mM KCN for more than 24 h show a disrupted [CuSMo] cluster. Cu(I) and the μ-sulfido ligand are cleaved off, probably upon forming SCN$^-$ and [Cu(CN)$_2$]$^-$, leaving an additional oxo ligand at Mo at the former position of the μ-sulfido ligand. The direct Mo environment is similar to the one found for AOR of *D. gigas* [4,25] (Fig. 7).

The equatorial oxygen ligands in the CN-inactivated enzyme have equal bond lengths to Mo, with values between the typical bond lengths of oxo and hydroxo ligands. The Sγ atom of Cys388 moves away from Mo by 0.8 Å, indicating the potentially tensed conformation of the residue in the active state (Fig. 8). The distance between Oε2 of Glu763 and Mo increased by more than 0.3 Å.

Further insight into the binding, activation, and hydroxylation of substrates by molybdenum hydroxylases has been provided by structural studies with a substrate analogue bound to the [CuSMo] cluster of CODH.

An exchange of the oxygen atom of CO by the group NR is the characteristic of isocyanides (CNRs). CO and CNRs share the presence of a nonbonding pair of electrons in the sp-hybridized orbital of the terminal carbon atom.

n-Butylisocyanide (nBIC) has extensively been used as a CO analogue in studies of heme proteins and has been found to resemble CO in the acetyl-CoA synthase/CODH of *Clostridium thermoaceticum* [28]. nBIC also acts as a strong inhibitor of active oxidized CODH of *O. carboxidovorans*, while reduction of the enzyme with CO or sodium dithionite protects it from inhibition by nBIC (L. Gremer, unpublished results). After soaking with nBIC and subsequent reduction with sodium dithionite, Mo(+V) "sulfo" EPR signals are no longer detectable and only 5–10% of "desulfo" EPR signals are recorded.

The isocyanide group of nBIC binds in the active site at the [CuSMo] cluster, forming covalent bonds to the μ-sulfido ligand, the hydroxo ligand, and the Cu atom (Fig. 9), while the alkyl chain extends to the hydrophobic interior of the substrate channel. This observation marks the former hydroxo group at the Mo ion as the oxygen donor of the hydroxylation. A resulting planar four-membered metallocycle exhibits short distances between not directly bonded ring members (Fig. 8),

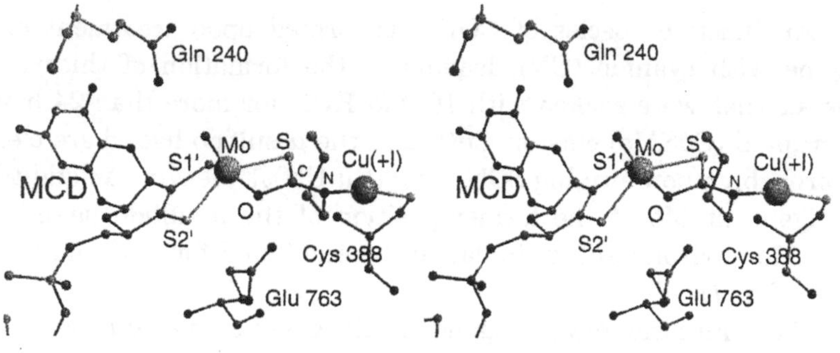

FIG. 9. Active site of CODH from *O. carboxidovorans* after nBIC soaking. Distances: Mo−O, 2.00 Å; Mo−Oxo, 1.67 Å; Mo−S, 2.35 Å; Mo−S1', 2.31 Å; Mo−S2', 2.36 Å: Mo−Cu(I), 5.07 Å; S−Cu, 3.13 Å, γS−Cu, 2.02 Å; Oε2−Mo, 3.44 Å; Cu−N(nBIC), 1.81 Å; N(nBIC)−C(nBIC), 1.22 Å; Mo−N(nBIC), 2.67 Å.

including the C atom of nBIC and Mo. Binding of the inhibitor results in larger structural changes for all atoms of the [CuSMo] cluster and leads to the "open state" of the active site cluster, in which the distance between Cu and Mo increased by more than 1 Å. Opening of the cluster also caused conformational changes at the active site loop. These changes are obviously transmitted to Glu736, resulting in an increase of the distance between Oε2 of Glu763 and Mo [23].

CODH can be reduced by its substrates CO and H_2 or by the addition of reducing agents, such as sodium dithionite. All reduced structures show a very similar active site geometry. Although structural changes for all cofactors of the enzyme upon reduction were observed, no changes agreeing with a reversible reduction of the organic portion of molybdopterin, as found for the AOR of *D. gigas* [25], have been detected.

Reduction results in an increase of the distance between Cu and Mo from 3.74 to 3.93 Å, an increase in bond length between the hydroxyl group and Mo to 2.03 Å, and a change in the bond length between Mo and the dithiolene group between the oxidized and the nBIC-bound species. The reduced [CuSMo] cluster is slightly more open than the oxidized cluster and shows larger distances for Mo and the carboxyl group of Glu763. The presence of bound substrates or reductants (CO, H_2, and dithionite) at the reduced active site has not been observed.

Although oxidized and reduced CO dehydrogenase are structurally very similar, the reduced enzyme is no longer capable of binding CO or nBIC firmly, as demonstrated by the substrate-free CO-reduced species and the inability of nBIC to inhibit reduced CODH, which emphasizes the parallels between CO and nBIC interaction with the [CuSMo] cluster.

Mo(+VI) is regenerated through a chain of redox-active cofactors (type I [2Fe-2S], type II [2Fe-2S], FAD), facilitating reoxidation of the reduced Mo ion (+V or +IV) through electron transfer from the active site Mo to external redox partners like cytochrome b_{561} in the case of CODH [29]. In molybdenum hydroxylases the moco is adjacent to the N-terminal type I [2Fe-2S] cluster [20,21,30], with the molybdopterin ring positioned between the molybdenum and the [2Fe-2S] cluster supporting electron transfer through its conjugated double bonds. The N2' amino group of the fused pyrimidine ring is at a distance of 5.5 Å from the Fe1 atom of the C-terminal cluster. From the C-terminal type I [2Fe-2S] cluster the electrons are transferred to the N-terminal type II [2Fe-2S] cluster and further to the FAD [20,30], where the electrons are carried to an external electron acceptors. The electron transfer between FAD and external redox partners is a general feature of molybdenum hydroxylases, although the type of electron acceptors used differs among the enzymes of the family.

3.3.   Molybdenum Coordinated by Two Mocos in the DMSO Reductase Family

3.3.1.   Introduction

Members of the DMSO reductase family typically catalyze a proper oxygen atom transfer reaction to or from an available electron pair on the substrate. In this way they are similar to enzymes from the sulfite oxidase family, discussed later.

The DMSO reductase family comprises enzymes from bacteria and archaea like the dimethyl sulfoxide reductase (DMSOR), biotin-S-oxide reductase, trimethylamine oxidoreductase (TMAOR), dissimilatory nitrate reductase (NAR), formate dehydrogenases (FDH), polysulfide reductases, arsenite oxidase (AOX), and formylmethanofuran dehydrogenase.

Prosthetic groups found in members of the enzyme family are, in addition to a bis(MGD) group at the active site, [2Fe-2S], [3Fe-4S], and [4Fe-4S] clusters, b-type heme groups, or flavin groups. Several members of the DMSOR family have been structurally characterized by X-ray crystallography: DMSOR of *Rhodobacter sphaeroides* [31,32] and *Rhodobacter capsulatus* [33–36], FDH-H of *Escherichia coli* [37], NAR of *D. desulfuricans* [38], TMAOR of *Shewanella massilia* [39], and AOX of *Alcaligenes faecalis* [40] (Fig. 10).

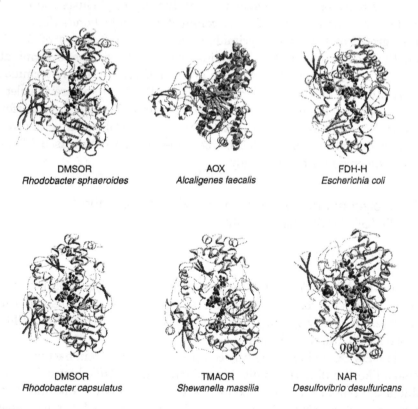

DMSOR
Rhodobacter sphaeroides

AOX
Alcaligenes faecalis

FDH-H
Escherichia coli

DMSOR
Rhodobacter capsulatus

TMAOR
Shewanella massilia

NAR
Desulfovibrio desulfuricans

FIG. 10.   Crystal structures of the members of the DMSO reductase family. Coordinates were taken from the PDB files: 1AA6 (FDH-H) [37], 1DMS (DMSOR of *R. capsulatus*) [33–36], 1EU1 (DMSOR of *R. sphaeroides*) [31, 32], 1TMO (TMAOR) [39], 2NAP (NAR) [38], and 1G8J (AOX) [40].

3.3.2. *DMSO Reductase from* Rhodobacter sphaeroides *and* R. capsulatus

DMSOR serves as a terminal electron acceptor during the anaerobic growth in the presence of DMSO. The enzymes from *R. sphaeroides and R. capsulatus* are monomeric, composed of a single subunit. No additional cofactors except for the bis(MGD) are found in this enzyme. The monomer is built up by four domains, grouped around the cofactor. The two MGDs, designated P-MGD and Q-MGD, form an elongated structure with a maximum extension of about 35 Å between the N2 atoms of the two guanine moieties [31]. The active site is located at the bottom of a large depression in the protein surface, which makes it accessible [31,33]. A protein ligand at the Mo is provided by the hydroxyl group of Ser147. Despite a very similar overall fold found for all DMSOR structures (Fig. 10), different Mo environments have been described. Schindelin et al. [31] proposed for the oxidized DMSOR structure of *R. sphaeroides* a Mo(+VI) ion coordinated by five ligands, S1′ and S2′ of P-MGD, S1′ of Q-MGD, Oγ-Ser147 and one oxo group, plus an additional weak ligand, S2′ of Q-MGD, arranged in a distorted trigonal prismatic geometry (Fig. 11).

Schneider et al. [33] found a pentacoordinated Mo site in the DMSOR structure of *R. capsulatus*. Mo is coordinated here only by the dithiolene sulfurs of P-MGD (Fig. 12). In addition, not one but two oxo groups have been found to build up the square pyramidal

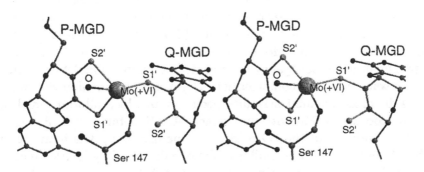

FIG. 11.   Stereoview of the active site of DMSOR from *R. sphaeroides* [31]. Coordinates were taken from the PDB file 1CXS.

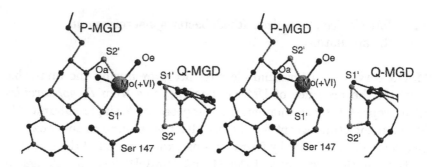

FIG. 12. Stereoview of the active site of DMSOR from *R. capsulatus* [33]. Coordinates were taken from the PDB file 1DMS. Distances: S1′–S2′ (P-MGD), 3.2 Å; S1′(P-MGD)−Mo, 2.4 Å; S2′(P-MGD)−Mo, 2.6 Å; O(Ser147)−Mo, 2.0 Å; Mo−Oa, 1.7 Å; Mo−Oe, 1.7 Å; S1′−S2′ (Q-MGD), 2.5 Å; S1′(Q-MGD)−Mo, 3.5 Å; S2′(Q-MGD)−Mo, 3.9 Å.

arrangement of the ligands. The sulfurs of Q-MGD are not coordinating the central ion but interact with each other at a distance of 2.5 Å.

A third variant for the oxidized active site was provided by McAlpine et al. [34], who described a heptacoordinated Mo ion at the active site of *R. capsulatus* DMSOR (Fig. 13). Seven ligands are provided by the four dithiolene sulfurs (P-MGD and Q-MGD), Oγ of Ser147, and two oxo groups. Measurements conducted from crystals harvested in buffers of different pH showed that the deviation of this model from the two previously studied ones [31,33] are not due to pH effects.

A recently published high-resolution structure [32] revealed that the active site of *R. sphaeroides* DMSOR contained two distinct molybdenum coordination environments that could not be resolved in the earlier 2.2-Å structure of the enzyme [31] (Fig. 11). Two alternative conformations for the Mo were found in a distance of 1.6 Å, with approximately 0.6 occupancy for a pentacoordinated and 0.4 occupancy for a hexacoordinated conformation. The hexacoordinated environment has a distorted trigonal prismatic geometry in which all four dihiolene sulfurs coordinate Mo with an average Mo−S distance of 2.43 Å (Fig. 14b). In addition to the Oγ of Ser147, one oxo group has been found. This Mo environment is similar to the one described by McAlpine et al. [34] and comprises four Mo−S bonds (Fig. 13). However, it lacks the additional oxo group whose position is occupied by the Mo conformation

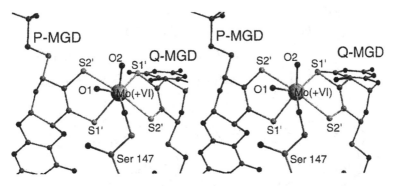

FIG. 13. Stereoview of the active site of DMSOR from *R. capsulatus* [34]. Coordinates were taken from the PDB files 1DMR (ox.) and 2DMR (red.). *Ox.*: Distances: S1'−S2' (P-MGD), 3.1 Å; S1'(P-MGD)−Mo, 2.5 Å; S2'(P-MGD)−Mo, 2.5 Å; O(Ser147)−Mo, 1.9 Å; Mo−O1, 1.6 Å; Mo−O2, 1.8 Å; S1'−S2' (Q-MGD), 3.0 Å; S1'(Q-MGD)−Mo, 2.4 Å; S2'(Q-MGD)−Mo, 2.6 Å. *Red.*: Distances: S1'−S2' (P-MGD), 3.2 Å; S1'(P-MGD)−Mo, 2.5 Å; S2'(P-MGD)−Mo, 2.3 Å; O(Ser147)−Mo, 1.8 Å; Mo−O1, 2.0 Å; S1'−S2' (Q-MGD), 2.6 Å; S1'(Q-MGD)−Mo, 2.5 Å; S2'(Q-MGD)−Mo, 2.8 Å.

of the pentacoordinated environment. In the pentacoordinated conformation (Fig. 14a), Mo is coordinated by the dithiolene sulfurs of P-MGD, as well as two oxo groups and the side chain of Ser147. This square pyramidal geometry is very similar to the structure described by Schneider et al. [33].

The high-resolution structure [32] revealed also a Hepes buffer molecule bound in the active site pointing with its hydroxyethyl side chain toward Mo. The temperature factors of this molecule indicate only a partial occupancy. An exchange of Hepes against cacodylic acid as buffer substance resulted in the complete disappearance of the pentacoordinated state, suggesting that the presence of Hepes within the active site causes the pentacoordinated state [32]. As Hepes can be seen as a crystallization artifact, the hexacoordinated mono-oxo environment was interpreted as the catalytical active Mo coordination [32]. Investigations by Bray et al. who studied the effects of reversible thiolate dissociation from Mo by crystallography, UV/Vis spectroscopy, and enzyme assays [36] proved that the Mo environment found in the crystal depends largely on the types of buffer used. DMSOR crystallized from citrate shows that all four thiolates of the bis(MGD) group are in bond-

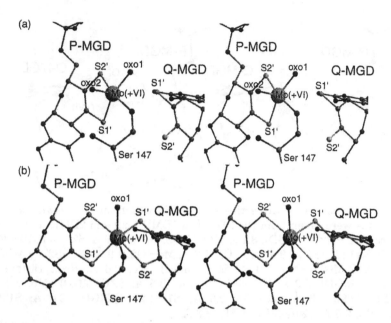

FIG. 14. The active site of DMSOR from *R. sphaeroides* [32] resolved a mixture of two coordination geometries. Coordinates were taken from the PDB file 1EU1. (a) Pentacoordinated geometry: Distances: S1′−S2′ (P-MGD), 3.1 Å; S1′(P-MGD)−Mo, 2.5 Å; S2′(P-MGD)−Mo, 2.4 Å; O(Ser147)−Mo, 1.9 Å; Mo−oxo1, 1.7 Å; Mo−oxo2, 1.7 Å; S1′−S2′ (Q-MGD), 3.1 Å; S1′(Q-MGD)−Mo, 3.6 Å; S2′(Q-MGD)−Mo, 4.5 Å. (b) Hexacoordinated geometry: Distances: S1′−S2′ (P-MGD), 3.1 Å; S1(P-MGD)−Mo, 2.5 Å; S2′(P-MGD)−Mo, 2.4 Å; O(Ser147)−Mo, 1.8 Å; Mo−oxo1, 2.1 Å; S1′−S2′ (Q-MGD), 3.1 Å; S1′(Q-MGD)−Mo, 2.5 Å; S2′(Q-MGD)−Mo, 2.4 Å.

ing distance to Mo, while in the presence of Na-Hepes one pair of thiolates dissociates from the metal. The activity of DMSOR in a backward assay decreased upon exposure with Na-Hepes under aerobic, but not under anaerobic, conditions, suggesting an oxygen-dependent inactivation. The inactivation of DMSOR together with the structural loss of a pair of thiolates provided evidence that sulfur ligand dissociation is an artifact and not part of the catalytic cycle.

In summary, the work of Li et al. [32] and Bray et al. [36] indicates that DMSOR cycles between a mono-oxo and des-oxo state, in which at least the oxidized state is coordinated by four sulfur atoms. Evidence that the common ligand, oxo1 in the work of Li et al. [32] and O2 in the work of McAlpine et al. [34], respectively, is the catalytically labile oxo ligand has been demonstrated by soaking oxidized DMSOR with dimethyl sulfide, which yielded a structure in which DMSO replaces this oxo ligand [35].

The careful revision of all the DMSOR crystal structures solved so far implies that the dithiolene group is binding either with both thiolate groups, as is found for most enzymes, or does not bind at all, as found for the Q-MGD in the DMSOR of Schneider et al. [33] or in the pentacoordinated DMSOR structure reported by Li et al. [32]. This binding mode suggests the description of the dithiolene group as a bidentate ligand since a coordination of Mo through one thiolate group alone has not been observed. When we count the dithiolene group as a single ligand both described ligand geometries (pentacoordinated and hexacoordinated) found for the oxidized enzymes may be described as tetrahedrally distorted (Fig. 15), in which the dithiolene groups are straddling over one vertex of the tetrahedron.

3.3.3. Formate Dehydrogenase H from Escherichia coli and Nitrate Reductase from Desulfovibrio desulfuricans

FDH catalyzes the oxidation of formate to carbon dioxide concomitant with the release of a proton and two electrons. The crystal structure of monomeric FDH-H from E. coli [37] is built up by four domains that are

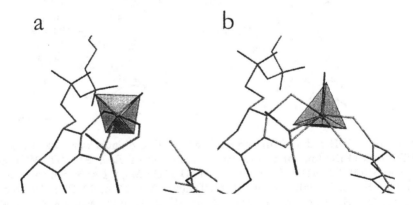

FIG. 15.   Superimposition of the (a) penta- and (b) hexacoordinated active site of DMSOR (32) with the tetrahedral MoO_4^{2-} ion.

holding the bis(MGD) group through an extensive hydrogen-bonding network between the second and third domain. Additional prosthetic groups of FDH-H are a [4Fe-4S] cluster and selenocysteine (SeCys).

The active site of FDH-H shows a similar geometry as found for the active site of DMSOR. The coordination geometry of the formate-reduced FDH-H is approximated by a square pyramid in which the four dithiolene sulfurs of the bis(MGD) group donate the equatorial ligands and the γSe of SeCys140 provides the axial ligand (Fig. 16a). As in the reduced state, all four dithiolene sulfurs coordinate Mo in the oxidized Mo(+VI) structure, displaying a trigonal prismatic geometry (Fig. 16b). A sixth ligand at Mo was modeled as a hydroxyl/water group in 2.2 Å distance from Mo. Oxidation of the enzyme results in a

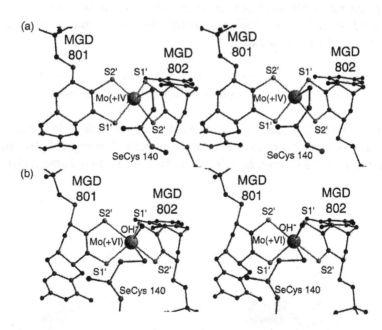

FIG. 16.   FDH reduced (a) and oxidized (b). Coordinates were taken from the PDB files 1FDO (ox.) and 1AA6 (red.). Distances (a) Red.: S1'−S2' (MGD 801), 3.1 Å; S1'(MGD 801)−Mo, 2.3 Å; S2'(MGD 801)−Mo, 2.1 Å; Se(SeCys)−Mo, 2.5 Å; S1'−S2' (MGD 802), 3.2 Å; S1'(MGD 802)−Mo, 2.3 Å; S2'(P-MGD)−Mo, 2.5 Å. (b) Ox: S1'−S2' (MGD 801), 3.1 Å; S1'(MGD 801)−Mo, 2.4 Å; S2'(MGD 801)−Mo, 2.5 Å; Se(SeCys)−Mo, 2.6 Å; Mo−OH, 2.1 Å; S1'−S2' (MGD 802), 3.2 Å; S1'(MGD 802)−Mo, 2.4 Å; S2'(P-MGD)−Mo, 2.4 Å.

rotation of the pterin portion of MGD 801 by 27° away from the equatorial plane and 16° along its own long axis [37], whereas little movement can be found in MGD 802 or in the guanine nucleotide portion of MGD 801 (Fig. 16). A putative site for substrate binding has been identified in the structure with a nitrite molecule (inhibitor) bound species [37]. Nitrite displaces the OH group, while its oxygen atom is bound to Mo. A very similar active site geometry, as described for the active site of oxidized FDH-H (Fig. 16b) [37], has been found in NAR [38], where a Cys residue replaces the active site SeCys found in FDH-H.

Electron transfer from the reduced Mo(+IV) site to regain the Mo(+VI) environment involves a [4Fe-4S] cluster situated near the moco in FDH-H [37], and in NAR [38], with a nearest Mo-Fe distance of about 12 Å. The most direct path between Mo and the [4Fe-4S] cluster involves the partially conjugated ring system of MGD 802, exiting through the N-1 position (Fig. 1) at the pterin.

3.3.4. Arsenite Oxidase from Alcaligenes faecalis

A different varient of the active site environment at Mo was provided recently by the solution of the crystal structure of AOX from A. faecalis [40]. AOX catalyzes the oxidation of arsenite to the less toxic arsenate. It is structured into two subunits, where the small subunit contains a Rieske-type [2Fe-2S] cluster and the large subunit carries the Mo site and a [3Fe-4S] cluster (Fig. 17).

The Mo site is similar to the other members of the family in that it contains a bis(MGD) group coordinating with all four dithiolene sulfurs to Mo. It is unique in the family in having no protein ligand bound to Mo. As an additional ligand a single O atom in a distance of 1.6 Å has been found and was modeled as an oxo group [40]. The five ligands form a square pyramidal geometry. X-ray absorption studies on AOX suggested a five-coordinated Mo site only for the reduced state of the enzyme, which has been explained by photoreduction through synchrotron radiation.

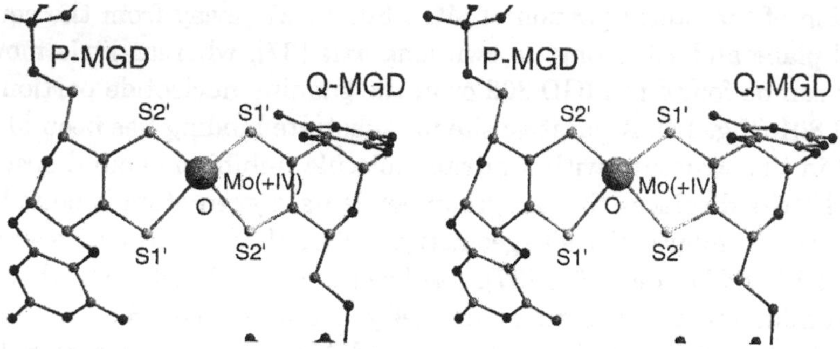

FIG. 17. The active site of AOX from *A. faecalis* [40], determined at 1.6 Å resolution. Coordinates were taken from the PDB file 1G8J.

3.4. The Sulfite Oxidase Family

3.4.1. Introduction

The sulfite oxidase family comprises the eukaryotic molybdoproteins sulfite oxidase (SO), which is ubiquitous among animals, and the assimilatory nitrate reductase (NR) as found in algae, fungi, and higher plants.

Both enzymes catalyze the net transfer of an oxygen atom between the substrate and water in a two-electron oxidation-reduction reaction. SO catalyzes the physiological vital oxidation of sulfite to sulfate, the terminal reaction in the oxidative degradation of the sulfur-containing amino acids cysteine and methionine. NR catalyzes the reduction of nitrate to nitrite in the assimilatory utilization of nitrate.

NR harbors additionally to the moco and b-type heme-containing domains (also found in SO), a C-terminal domain of approximately 30 kDa, containing FAD and a NAD(P)$^+$ binding site [41]. SO forms homodimers with a molecular mass of 101–110 kDA.

3.4.2. Sulfite Oxidase from Chicken Liver

Each monomer of SO is built up by three domains (Fig. 18). The N-terminal domain carries the heme group and has been found to be similar to bovine cytochrome b_5. The second domain exhibits a new fold and harbors the moco. The C-terminal domain contains two anti-parallel β sheets with a Greek key motif, as found in the C2 subtype of

SO
from chicken liver

FIG. 18.   Crystal structure of sulfite oxidase from chicken liver. Coordinates were taken from the PDB file 1SOX [42].

the immunoglobulin superfamily. A head-to-head arrangement of the C–terminal domain forms the main dimer interaction [42]. SO contains a single pyrano-pterin system binding the Mo ion. The moco in SO is not bound to an additional nucleotide and is deeply buried inside the protein matrix. Residues forming side chain contacts with the cofactor are usually highly conserved within the protein family.

Mo is coordinated by five ligands with approximately square pyr-amidal coordination geometry (Fig. 19). The equatorial plane is occupied by three sulfur ligands, one water/hydroxo ligand and an oxo group in the axial position. Two of the sulfur ligands in the equatorial plane originate from the dithiolene sulfurs while the third is the Sγ atom of Cys185 in a distance of 2.5 Å. Cys185 is conserved throughout all SOs

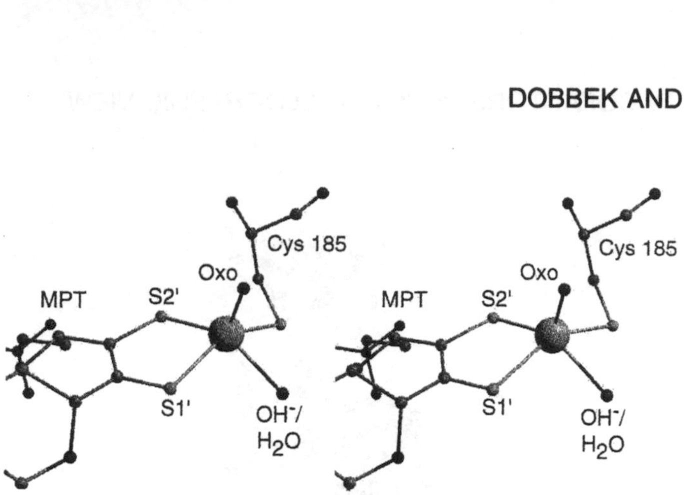

FIG. 19. The active site of SO [42]. Coordinates were taken from the PDB file 1SOX. Distances: S1′–S2′, 3.2 Å; S1′–Mo, 2.4 Å; S2′–Mo, 2.4 Å; Mo–oxo, 1.7 Å; Mo–OH/H$_2$O, 2.2 Å; Mo–S(Cys 185), 2.5 Å.

and NRs, indicating its S-donor function for Mo in all of these proteins. The importance of Cys185 for catalysis is also shown by site-directed mutagenesis, where substitution with Ser yields an inactive enzyme [43]. The observation of a hydroxo/water ligand instead of an oxo ligand in the equatorial plane disagrees with EXAFS experiments [44], which indicated the presence of two oxo groups in the oxidized form of SO. This has been explained by partial reduction of the active site, which might have arisen from traces of sulfite in the precipitant lithium sulfate, photoreduction by the X-ray beam or the reducing properties of the cryoprotectant glycerol [42].

A sulfate ion in the active site has been found near the Mo with a Mo–S (sulfate) distance of 5.2 Å. The presence of a weak oxo atom of the sulfate ion indicates a mixture of sulfate and the reduced sulfite. The weak oxygen group is in close vicinity to the water/hydroxo ligand of the Mo ion (2.4 Å distance). This finding has been taken as an indication that the equatorial oxo ligand is the catalytically labile oxygen, whereas the axial ligand is the spectator oxygen that is not exchanged during catalysis [45].

After substrate oxidation two electrons have to be transferred from the active site to the b-type heme group from where they are carried to the external electron acceptor cytochrome c. A distance of 32 Å between Mo and the heme Fe make a direct transfer unlikely [46]. Furthermore, the pterin ring is not orientated to the heme group and therefore seems not to be part of the electron transfer pathway [42]. This raises the possibility that in solution the heme-containing domain might adopt a different conformation, placing the heme Fe nearer to the Mo ion. Interestingly, all active enzyme states of molybdenum enzymes reported so far contained at least three sulfur ligands of different origin bound to the Mo. The presence of three sulfur ligands in the equatorial plane and a spectator oxo ligand in an axial position are common both to SO [42] and CODH [17,23].

3.5. The Tungsten-containing Enzymes of the Aldehyde Oxidoreductase Family

3.5.1. Introduction

The biological function of W was discovered in the early 1970s [47], and the first purification of a W-containing enzyme, an FDH, was reported in 1983. At present 14 W-containing enzymes have been characterized from anaerobic organisms, which are mostly thermophilic [48]. These enzymes have been grouped in three general classes: called aldehyde oxidoreductase (AOR), formate dehydrogenase (FDH), and the acetylene hydratase (for a review, see [49]). While crystal structures for two members of the AOR family are available, no structural information is published for the other two families. Sequence homologies between W- and Mo-containing FDHs, together with the identification of Se in both enzyme classes suggest similar structures [49].

The majority of W-containing enzymes fall into the AOR class and catalyze the oxidation of aldehydes to the corresponding carboxylic acids, using a ferredoxin as the physiological electron acceptor. Selected members of the AOR family are the aldehyde ferredoxin oxidoreductase (AOR), the formaldehyde ferredoxin oxidoreductase (FOR), and the glyceraldehyde-3-phosphate ferredoxin oxidoreductase of the hyperthermophilic archaeon *Pyrococcus furiosus*. Typically, in addition to W coordinated by bis(MGD), these enzymes also contain [4Fe-4S] clusters. Crystal structures have been determined for AOR [3] and FOR [48] of *P. furiosus*, both showing a similar overall fold (Fig. 20). No homology has been detected between the W- and Mo-containing AORs.

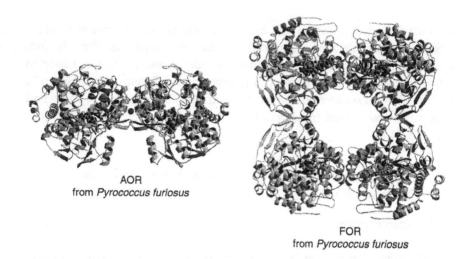

FIG. 20. Overview of the AOR family with solved crystal structures. Coordinates were taken from the PDB files: 1AOR (AOR from *P. furiosus*) [3] and 1B25 (FOR from *P. furiosus*) [48].

3.5.2. *Aldehyde Oxidoreductase from* P. furiosus

AOR from *P. furiosus* forms a homodimer, in which each 66-kDa subunit is composed by three domains [3]. Three different types of metals sites have been found in the structure: the pyranopterin tungsten cofactor, bound between the three domains, a [4Fe-4S] cluster, and a single metal site, tetrahedrally coordinated in the dimer interface. While the [4Fe-4S] cluster and the W ion are approximately 10 Å apart, the single metal site is about 25 Å away from these groups and is not likely to play a role in electron transfer or catalysis. The dimer positions the two [4Fe-4S] clusters in a distance of about 50 Å.

The W ion is coordinated by both pairs of dithiolene sulfurs arranged in a distorted square pyramid, with an angle between the two planes built by the dithiolene groups and the W ion of about 97° [3] (Fig. 21). No protein residues have been found to coordinate the W, but residual electron density at the W ion is likely to originate from glycerol and/ or an oxo group [3]. A linkage of the two MPT molecules is accomplished through their phosphate groups, which coordinate the axial sites of a

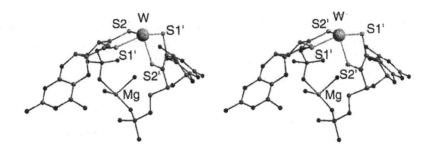

FIG. 21.   Active site of AOR from *P. furiosus*. Coordinates were taken from the PDB file 1AOR.

central magnesium ion (Fig. 21). Both MPT ligands are approximately related by a twofold rotation axis passing through the W and Mg sites [3].

In contrast to the dimeric AOR, FOR forms a tetramer in solution, which is found in the asymmetrical unit of the crystal [48] (Fig. 20). The arrangement of the homotetramer creates W–W distances of about 40 Å. As in AOR, W is bound by all four dithiolene sulfurs of the bis(MPT) with an average W–S distance of 2.49 Å. In addition, W is probably coordinated by one oxygen atom in 2.1 Å distance (Fig. 22).

The [4Fe-4S] cluster is in a similar position as in AOR. A complex between FOR and its physiological electron acceptor positions the [4Fe-4S] clusters of AOR and the ferredoxin approximately 15 Å apart [48]. The glutamate residue Glu308 near the W site is reminiscent of the Mo near glutamate found in molybdenum hydroxylases [48]. However, the

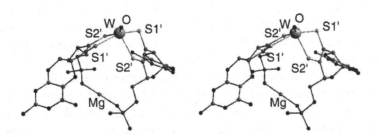

FIG. 22.   Active site of W-containing FOR from *P. furiosus*. Coordinates have been taken from the PDB file 1B25. Distances: average W–S distance 2.49 Å; W–O, 2.1 Å.

distance of 4.9 Å between W and the nearest oxygen from Glu308 certainly does not allow a similar mode of action as found in CODH [23], where the Mo−O distance alternates between 3.1 and 3.5 Å in the course of the catalytic cycle.

4. LIMITATIONS OF A CRYSTALLOGRAPHIC MODEL

X-ray crystallography of macromolecules allows the study of structures of large complexity, while still revealing atomic details of the studied probe. Crystallographic analysis of moco-containing enzymes provided insight into the direct Mo/W environment, which largely corroborated previous results obtained by spectroscopic methods. In addition, insight was provided on the structure of the moco, its binding and stabilization of the protein, access of the substrate and water to the active site, geometry of substrate bound states, architecture of complex active sites like that of CODH [23], as dynamic aspects of the enzyme encountered upon substrate binding.

Still, macromolecular structures can be misleading, and care must be taken in their interpretation. First, in comparison with crystals of small molecules, macromolecular crystals are somewhat flexible and have a high solvent content. This generally results into larger thermal vibrations and local disorder within the crystal, which weakens the diffraction intensity especially at higher scattering angles.

The resulting lower resolution of macromolecular diffraction data decreases the ratio of observations to parameters. To overcome this, additional information from stereochemical information of small-molecule structures [50] is usually incorporated into macromolecular refinements through restrains, which enlarge the rate of convergence of the least-squares refinement to the final value. For metalloproteins with unknown active site geometries the use of restrains has the disadvantage that the metal coordination is biased toward the imposed model [51]. On the other hand, not using restrains may result in incorrect geometries. Problems caused by the incorporation of restrains can be largely overcome when data to true atomic resolution ($d < 1.2$ Å) are available. These not only allow the localization of small atoms near electron-rich scatterers, as Fourier synthesis at atomic resolution are

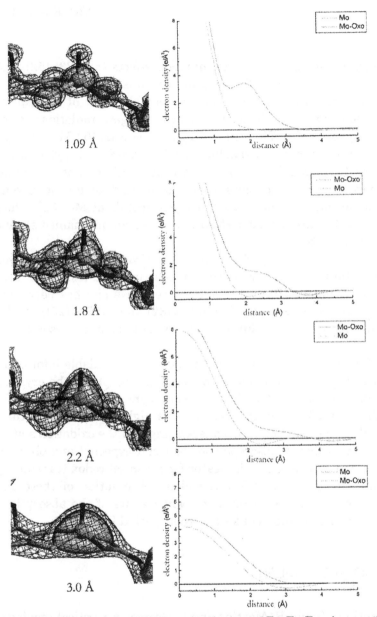

FIG. 23. Resolution and series termination on $2F_o$–F_c Fourier synthesis. Densities are calculated at the resolution given and contoured at 1σ (net) and at 2.5σ (surface). The diagrams show the effect of series termination errors on a calculated Fourier synthesis (F_{calc}/φ_{calc} Fourier synthesis) around Mo alone, respectively, around a Mo−oxo group with a bonding distance of 1.7 Å. Temperature factors of 10 Å^2 were assigned to both atoms

largely unaffected by series termination effects (Fig. 23) [52], but also allow a refinement with fewer or no restrains.

Second, an unintended and unavoidable effect of the exposure of macromolecular crystals to ionizing high energetic radiation is a damage of the crystalline probe, initiated by free radicals [53,54]. Among reactions like the breakage of disulfide bonds or the decarboxylation of acidic residues [54], metals can be reduced or oxidized. For Mo-containing enzymes this redox processes can easily lead to misinterpretations, e.g., of the number and type of oxygen ligands at Mo [52], which are of prime relevance for the postulation of reaction mechanisms relying on a crystal structure.

Third, a general problem of biochemical and structural investigations is the unnoticed presence of two or more states of the inspected enzyme. A mixture of two states in the active site has been found in the first characterizations of DMSOR [32] and CODH [23], which went undetected as the resolution was not sufficient to resolve the two states.

The need for high-resolution data to gain reliable information on the active site environment of metalloproteins places the crystallographer in a quandary. On the one hand, synchrotron sources offer the possibility of collecting excellent data to high resolutions; on the other hand, their intense radiation causes changes of the oxidation state of the metal ion. Although discrepancies between expected and observed Mo ligands can be used as an indicator for potential redox transitions, synchrotron facilities can allow the direct observation of these changes during or after the measurement, when shifts of the absorption edge of redox-sensitive metals like Fe are inspected [40].

ACKNOWLEDGMENT

The authors thank B. Martins and L. Gremer for critical reading of the manuscript.

ABBREVIATIONS AND DEFINITIONS

AO	aldehyde oxidase
AOR	aldehyde oxidoreductase
AOX	arsenite oxidase
CN	cyanide
CNR	isocyanide
CO	carbon monoxide
CoA	coenzyme A
CODH	carbon monoxide dehydrogenase
Cys	cysteine
DMSOR	dimethyl sulfoxide reductase
EPR	electron paramagnetic resonance
EXAFS	extended X-ray absorption fine structure
FAD	flavin adenine dinucleotide
FDH	formate dehydrogenase
FOR	formaldehyde ferredoxin oxidoreductase
Gln	glutamine
Glu	glutamate
Hepes	N-2-hydroxyethylpiperazine-N'-2-ethanesulfonic acid
MAD	multiple wavelength anomalous dispersion
MCD	molybdopterin cytosine dinucleotide
MGD	molybdopterin guanosine dinucleotide
moco	molybdenum cofactor
MPT	molybdopterin
NAD	nicotinamide adenine dinucleotide
NAR	dissimilatory nitrate reductase
nBIC	n-butylisonitrile
NR	assimilatory nitrate reductase
PDB	Protein Data Base
SeCys	selenocysteine
Ser	serine
SCN	thiocyanate (rhodanide) anion
SO	sulfite oxidase
TMAOR	trimethylamine oxidoreductase
UV/Vis	ultraviolet/visible
XDH	xanthine dehydrogenase
XO	xanthine oxidase

REFERENCES

1. J. Kim and D. C. Rees, *Science*, *257*, 1677–1682 (1992).
2. R. Hille, in *Essays in Biochemistry*, Vol. 34 (D. P. Ballou, ed.), Portland Press, 1999.
3. M. K. Chan, S. Mukund, A. Kletzin, M. W. Adams, and D. C. Rees, *Science*, *267*, 1463–1469 (1995).
4. M. J. Romão, M. Archer, I. Moura, J. J. Moura, J. LeGall, R. Engh, M. Schneider, P. Hof, and R. Huber, *Science*, *270*, 1170–1176 (1995).
5. P. A. Ketchum, H. Y. Cambier, W. A. Frazier, C. H. Madansky, and A. Nason, *Proc. Natl. Acad. Sci. USA*, *66*, 1016–1023 (1970).
6. A. Nason, A. D. Antoine, P. A. Ketchum, W. A. d. Frazier, and D. K. Lee, *Proc. Natl. Acad. Sci. USA*, *65*, 137–144 (1970).
7. A. Nason, K. Y. Lee, S. S. Pan, P. A. Ketchum, A. Lamberti, and J. DeVries, *Proc. Natl. Acad. Sci. USA*, *68*, 3242–3246 (1971).
8. S. P. Kramer, J. L. Johnson, A. A. Ribeiro, D. S. Millington, and K. V. Rajagopalan, *J. Biol. Chem.*, *262*, 16357–16363 (1987).
9. J. L. Johnson, B. E. Hainline, K. V. Rajagopalan, and B. H. Arison, *J. Biol. Chem.*, *259*, 5414–5422 (1984).
10. J. L. Johnson and K. V. Rajagopalan, *Proc. Natl. Acad. Sci. USA*, *79*, 6856–6860 (1982).
11. K. V. Rajagopalan and J. L. Johnson, *J. Biol. Chem.*, *267*, 10199–10202 (1992).
12. K. V. Rajagopalan, *J. Biol. Inorg. Chem.*, *2*, 786–789 (1997).
13. D. C. Rees, Y. Hu, C. Kisker, and H. Schindelin, *J. Chem. Soc., Dalton Trans.*, 3909–3914 (1997).
14. R. Hille, *Chem. Rev.*, *96*, 2757–2816 (1996).
15. O. Meyer and H. G. Schlegel, *J. Bacteriol.*, *141*, 74–80 (1980).
16. O. Meyer, *J. Biol. Chem.*, *257*, 1333–1341 (1982).
17. O. Meyer, L. Gremer, R. Ferner, M. Ferner, H. Dobbek, M. Gnida, W. Meyer-Klaucke, and R. Huber, *Biol. Chem.*, *381*, 865–876 (2000).

18. J. Rebelo, S. Maciera, J. M. Dias, R. Huber, C. S. Ascenso, F. Rusnak, J. J. G. Moura, I. Moura, and M. J. Romão, *J. Mol. Biol.*, *297*, 135–146 (2000).

19. M. Lehmann, B. Tshisuaka, S. Fetzner, P. Roger, and F. Lingens, *J. Biol. Chem.*, *269*, 11254–11260 (1994).

20. H. Dobbek, L. Gremer, O. Meyer, and R. Huber, *Proc. Natl. Acad. Sci. USA*, *96*, 8884–8889 (1999).

21. C. Enroth, B. T. Eger, K. Okamoto, T. Nishino, T. Nishino, and E. F. Pai, *Proc. Natl. Acad. Sci. USA.*, *97*, 10723–10728 (2000).

22. K. Breese and G. Fuchs, *Eur. J. Biochem.*, *251*, 916–923 (1998).

23. H. Dobbek, L. Gremer, R. Kiefersauer, R. Huber, and O. Meyer (submitted).

24. P. Hänzelmann, H. Dobbek, L. Gremer, R. Huber, and O. Meyer, *J. Mol. Biol.*, *301*, 1221–1235 (2000).

25. R. Huber, P. Hof, R. O. Duarte, J. J. Moura, I. Moura, M. Y. Liu, J. LeGall, R. Hille, M. Archer, and M. J. Romão, *Proc. Natl. Acad. Sci. USA*, *93*, 8846–8851 (1996).

26. M. J. Romão, N. Rösch, and R. Huber, *J. Biol. Inorg. Chem.*, *2*, 782–785 (1997).

27. W. R. Waud and K. V. Rajagopalan, *Arch. Biochem. Biophys.*, *172*, 365–379 (1976).

28. M. Kumar and S. W. J. Ragsdale, *J. Am. Chem. Soc.*, *117*, 11604–11605 (1995).

29. O. Meyer, K. Frunzke, and G. Mörsdorf, in *Microbial Growth on C1 Compounds* (J. C. Murrell, and D. P. Kelly, eds.), Intercept, Andover, 1993.

30. L. Gremer, S. Kellner, H. Dobbek, R. Huber, and O. Meyer, *J. Biol. Chem.*, *275*, 1864–1872 (2000).

31. H. Schindelin, C. Kisker, J. Hilton, K. V. Rajagopalan, and D. C. Rees, *Science*, *272*, 1615–1621 (1996).

32. H. K. Li, C. Temple, K. V. Rajagopalan, and H. Schindelin, *J. Am. Chem. Soc.*, *122*, 7673–7680 (2000).

33. F. Schneider, J. Löwe, R. Huber, H. Schindelin, C. Kisker, and J. Knäblein, *J. Mol. Biol.*, *263*, 53–69 (1996).

34. A. S. McAlpine, A. G. McEwan, A. L. Shaw, and S. Bailey, *J. Biol. Inorg. Chem.*, *2*, 690–701 (1997).

35. A. S. McAlpine, A. G. McEwan, and S. Bailey, *J. Mol. Biol.*, *275*, 613–623 (1998).

36. R. C. Bray, B. Adams, A. T. Smith, B. Bennett, and S. Bailey, *Biochemistry*, *39*, 11258–11269 (2000).

37. J. C. Boyington, V. N. Gladyshev, S. Khangulov, T. C. Stadtman, and P. D. Sun, *Science*, *275*, 1305–1308 (1997).

38. M. J. Dias, M. E. Than, A. Humm, R. Huber, G. P. Bourenkov, H. D. Bartunik, S. Bursakov, J. Calvete, J. Caldeira, C. Carneiro, J. J. Moura, I. Moura, and M. J. Romão, *Structure Fold. Des.*, *7*, 65–79 (1999).

39. M. Czjzek, J. P. Dos Santos, J. Pommier, G. Giordano, V. Mejean, and R. Haser, *J. Mol. Biol.*, *284*, 435–447 (1998).

40. P. J. Ellis, T. Conrads, R. Hille, and P. Kuhn, *Structure*, *9*, 125–132 (2001).

41. Y. Kubo, N. Ogura, and H. Nakagawa, *J. Biol. Chem.*, *263*, 1968–9 (1988).

42. C. Kisker, H. Schindelin, A. Pacheco, W. A. Wehbi, R. M. Garrett, K. V. Rajagopalan, J. H. Enemark, and D. C. Rees, *Cell*, *91*, 973–988 (1997).

43. R. M. Garrett and K. V. Rajagopalan, *J. Biol. Chem.*, *271*, 7387–7391 (1996).

44. G. N. George, C. A. Kipke, R. C. Prince, R. A. Sunde, J. H. Enemark, and S. P. Cramer, *Biochem. 28*, 5075–5080 (1989).

45. A. K. Rappe and W. A. Goddard III, *Nature*, *285*, 311–312 (1980).

46. C. C. Page, C. C. Moser, X. Chen, and P. L. Dutton, *Nature*, *402*, 47–52 (1999).

47. J. R. Andreesen and L. G. Ljungdahl, *J. Bacteriol.*, *116*, 867–873 (1973).

48. Y. Hu, S. Faham, R. Roy, M. W. Adams, and D. C. Rees, *J. Mol. Biol.*, *286*, 899–914 (1999).

49. M. K. Johnson, D. C. Rees, and M. W. W. Adams, *Chem. Rev.*, *96*, 2817–2839 (1996).

50. R. A. Engh and R. Huber, *Acta Crystallogr. D*, *4*, 392–400 (1991).

51. A. Hodel, S. H. Kim, and A. T. Brünger, *Acta Crystallogr. A48*, 851–858 (1992).

52. H. Schindelin, C. Kisker, and D. C. Rees, *J. Biol. Inorg. Chem.*, 2, 773–781 (1997).

53. C. Nave, *Radiation Phys. Chem.*, 45, 483–490 (1995).

54. R. B. G. Ravelli and S. M. McSweeney, *Structure*, 8, 315–328 (2000).

8

Models for the Pyranopterin-Containing Molybdenum and Tungsten Cofactors

Berthold Fischer[1] *and Sharon J. Nieter Burgmayer*[2]

[1]Lehrstuhl für Analytische Chemie, Ruhr-Universität Bochum, Universitätsstrasse, 150 D-44780 Bochum, Germany

[2]Department of Chemistry, Bryn Mawr College, Bryn Mawr, PA 19010, USA

1. INTRODUCTION

Molybdenum enzymes have long fascinated synthetic chemists with the challenge of creating viable model complexes for duplication of structural, spectroscopic, and reaction chemistry of these enzymes, which are necessary to all forms of life [1–37]. The discovery of tungsten enzymes in hyperthermophilic bacteria means that both group 16 metals have the exceptional status of being the only metals of the second and third transition series that are essential as micronutrients [21]. To date, more than 37 distinct enzymes having Mo or W at the catalytic reaction site are known. Extensive current reviews describing these enzymes are available [21,25,36,37].

The focus of this chapter is the recent model chemistry directly related to the active centers of Mo and W enzymes. Since the advent of the first X-ray structures for Mo and W enzymes 6 years ago, the challenge to chemists modeling the catalytic site has greatly escalated. There is more information to direct syntheses, but the relevant chemical units are increasingly difficult to manipulate. The context for this chapter is as a sequel for two reviews appearing in the last decade that concentrated on model compounds for the Mo and W enzymes [20,36]. Also available are several reviews with a reduced scope [15,19,24,27,29,32], including two that dealt mainly with metal-pterin chemistry [38,39]. We apologize to all authors of works we could not cite due to scope or space limitations or of works inadvertently overlooked.

1.1. The Mission of Synthetic Models

The mission of model chemists is to learn something from small molecules that can be applied to the enzymes, revealing their intricate mechanisms and function. Chemists seeking to develop synthetic models for the pyranopterin-dependent Mo and W enzymes may adopt one of several strategies. Synthetic compounds may be intended for use (1) as structural models for the coordination pattern of the relevant metal ion, or the ligand, or the complete system; (2) as spectroscopic models mimicking certain or all features of natural systems; (3) as probes for the possible electronic interaction either between the relevant redox parts of the system or with other natural electron transfer centers not directly in contact with the pyranopterin-metal active site; (4) as reactivity models mimicking single functional features or complete cycles of the natural enzyme systems. As a result, this mission may generate (1) new synthetic approaches to the natural system; (2) complexes with precise geometries and spectroscopic data that can be used as references

for measurements in natural systems or as an input for better theoretical work; (3) compounds for binding, inhibitor and activity testing
of natural systems; and (4) systems for the testing of mechanistic models
and (5) new drugs or biocatalysts for industrial use.

1.2.   The Synthetic Problem of Building Three Redox-
Active Sites into a Highly Reactive System

The basic knowledge of the monomeric nature of the active metal site
was gathered by Rajagopalan and coworkers [16,18] in the 1980s
through a series of remarkable experiments using biological sources
[4]. Their research revealed the three fundamental features that com-
pose the "common molybdenum cofactor" (Moco): a metal ion (Mo), an
ene-dithiolate group, and a hydrogenated pterin unit (1). It took another
decade and the enormous effort of numerous microbiologists and protein
X-ray crystallographers [40] to identify the detailed structure of the
unique ligand universally associated with Moco. During this period
tungsten enzymes were discovered and found to require a ligand with
identical characteristics to that seen in Mo enzymes. This ligand features an unprecedented combination of three redox-active sites that
includes the metal atom (Mo or W) coordinated by an ene-dithiolate
ligating unit, and a pyranopterin. This ligand chelated to Mo is illustrated as 1.

The special ligand for Mo was coined "molybdopterin" and abbreviated MPT. Note that molybdopterin contains no molybdenum atom
and will also chelate tungsten in tungsten enzymes, so that this name
can lead to confusion for researchers not familiar with the area. For the
sake of compatibility with older literature, we will use in this review the

abbreviation MPT but comprehend it as *metal*-binding *pyranopterin* ene-di*t*hiolate **2**, similar to a proposal by Pilato and Stiefel [36].

There has been some consideration of an equilibrium between the tricyclic pyranopterin structures found in X-ray crystallography of several enzymes and ring-opened bicyclic pterins like **3** which may be found in solution and in the course of the enzyme-catalyzed reaction ([36] and references therein). However, there is no experimental evidence for such a reversible ring opening of the *N,O*-acetal function to form the 5,6-dihydro-MPT intermediate **3**. Such evidence has also not been found in synthetic chemistry related to pyranopteridines [41,42]. In metal-pyranopterin chemistry [42], in any spectroscopic results of chemical biology related to pyranopterin derivatives of MPT [44,45] or biosynthetic MPT precursors [46,47]. The easy formation and reversibility of such adducts is only known for the basic pteridine system [48–50] and simplified model heterocycles such as quinoxalines. Neither of these is found in nature and both react quite differently from the 2-amino-4-oxo-pteridine, *pterin* (for numbering, see **3**), that is prevalent in nature [51,52].

Therefore, in this chapter we will use the numbering scheme of pyranopteridines given in **2** for MPT, which is compatible with IUPAC recommendations and is also used in the few structural reports of isolated pyranopterins [41b,c,d]. We will not elaborate a detailed discussion about applicable oxidation states of pyranopteridines, which may be important to their function in the enzymes, but rather refer to an extensive consideration in [53]. Interested readers can also find in this reference an extended discussion of nomenclature considerations. Before 1990 it was assumed that MPT bound to Mo constituted a *common molybdenum cofactor, Moco*. Now it is evident that molybdenum enzymes do not contain a single, universal Mo complex as a cofactor; rather, there exists a general class of related cofactors [26,30,35,36,54,55]. For some enzymes from bacterial sources, the phosphate group of MPT is linked to a nucleotide derived from bases such as guanine, cytosine, and so forth, whose role, as suggested by X-ray structures, is to help bind the cofactor to the proteins [30,54]. As shown in Fig. 1 (4), five distinct MPT-containing pyranopterin ligands, including MPT itself, have been identified. These derivatives incorporating nucleic acids have been termed dinucleotides, although the pterin part of the "dinucleotide" does not actually have a cyclic sugar attached to the "base" pyranopterin. The nucleotides are bound to MPT via a pyrophosphate linkage [56–59]. The abbreviations of the various MPT derivatives follow according to the nomenclature for pyrophosphates, as MPTpC, MPTpG, MPTpA, and MPTpH for MPT derivatives containing the purine bases cytosine, guanine, adenine, and hypoxanthine, respectively [36].

The variation among Mo and W enzymes extends beyond the different forms of molybdopterin mentioned above. Subclasses composed of MPT-containing proteins with Mo or W differ in metal-MPT stoichiometry, ancillary ligands, coordination numbers, and geometries (Fig. 2). The key differences are (1) either one or two bound MPT ligands; (2) additional ligation from protein residues such as deprotonated hydroxo (serine), deprotonated thiolate (cysteine), or selenate (selenocysteine) groups; and (3) terminal oxido and/or sulfido ligands, in addition to water, hydroxido, or hydrosulfido ligands [17,21,22,25].

None of the cofactors has been isolated and completely structurally characterized in a fully intact state due to their extreme instability [16,18,60,61]. Deistung and Bray reported an anaerobic protocol for the isolation of active "Moco" from xanthine oxidase (XO) but gave no spectroscopic or other data besides an estimate from gel filtration of a

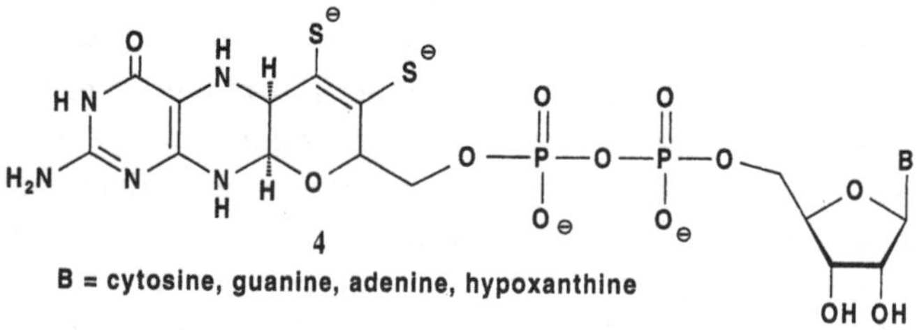

B = cytosine, guanine, adenine, hypoxanthine

FIG. 1. MPT nucleotide variants: MPTpC, MPT-cytosine dinucleotide; MPTpG, MPT-guanosine dinucleotide; MPTpA, MPT-adenosine dinucleotide; MPTpH, MPT-inosine dinucleotide [36].

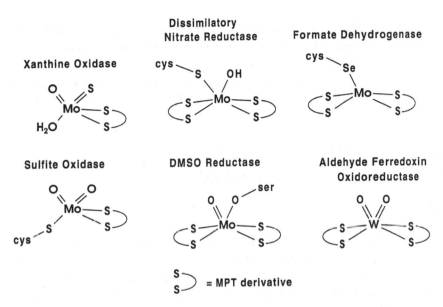

FIG. 2. Overview of the metal center variations in the subclasses of molybdenum and tungsten enzymes relevant to the scope of model compounds. Metal centers are shown in the most stable oxidation state, i.e., +VI; resulting net ionic charges of the centers are not taken into account. One example of an enzyme corresponding to the respective drawn schematic metal center is given in the figure.

mass of about 600 [62]. Due to the facile metal ion dissociation from molybdopterin in conjunction with the well-known oxygen sensitivity of reduced pterins [51,52,62], only fully oxidized pterin decomposition products lacking the dihydropyrano ring have been isolated [36]. A hydrogenated pterin that is a precursor to Moco in the biosynthetic pathway of molybdenum and tungsten enzymes has been isolated [46].

Given the extreme reactivity of isolated Moco, it is not surprising that the synthetic problem of incorporating the three redox-active pieces of an ene-dithiolene, a reduced pterin, and a metal ion into a model system expected to be equally highly reactive and air-sensitive is yet unsolved. As is a usual approach, such complex systems are often broken down into smaller, more easily handled pieces as model systems to facilitate the study of individual properties. In our case, this is (1) the metal ion coordinated by different additional ligand atoms (oxo, thioxo, etc.), (2) the M center coordinated by the ene-dithiolate ligand(s), and (3) the hydrogenated (pyrano)-pterin with or without an incorporated ene-dithiolate component and with or without coordination to a metal ion. Accordingly, this chapter addresses the efforts made by researchers working on one or more of these pieces of the complete Moco structure.

2. EARLY MODELS

2.1. Complexes as Structural Models

2.1.1. Evolution from Dimeric Compounds

From the earliest efforts to model the molybdenum site in the enzymes, it was recognized that dimeric molybdenum species were prevalent in the emerging molybdenum-oxo and -thioxo chemistry of the relevant higher oxidation states (+4 to +6). EPR evidence, and later the experiments of Rajagopalan and coworkers, supported a monomeric Mo center. Model synthesis of mononuclear complexes was inherently hampered by the thermodynamic stability of the dimeric Mo(+V)-cores $[Mo_2O_3]^{4+}$ (**5**) and $[Mo_2O_4]^{2+}$ (**6**).

One of the tasks of the protein in the enzymes and of ligands in model chemistry is to prevent this dimerization of the Mo−oxo sites. Significant advances in coordination chemistry connected with this area are associated with the research groups of Enemark, Holm, Spence, Stiefel, and Wedd [6,8,15,20,63,64]. Their clever ligand design led to the successful preparation of mononuclear Mo(IV)O and Mo(VI)O_2 model compounds, and some systems were shown to participate in oxygen atom transfer (OAT) reactions. Much of this model chemistry aimed to mimic the structure, spectroscopy, and reactivity of the enzyme without any knowledge of the critical molybdopterin ligand required by the molybdenum enzymes. Nonetheless, this chemistry had enormous impact on developing trends in coordination structure and reactivity as well as on identifying those features required for producing the characteristic spectroscopic properties of the Mo sites of the enzymes.

2.1.2. Mononuclear Structural Models

With the discovery of the monomeric nature of the enzyme metal center, numerous mononuclear models were synthesized as structural probes for the different enzyme centers (Fig. 2). Large, multidentate ligands have been devised to prevent dimerization of the Mo−oxo monomers, for which some important examples with S- and N-donor sets are shown in Fig. 3. These were constructed with S-, O-, and N-donor ligands since this donor atom set was consistent with the inner coordination environment determined by EXAFS studies [15,20]. The chemistry of the Mo(VI) ion is dominated by complexes containing the cis-$[MoO_2]^{2+}$ fragment (6,20,65). Incorporation of the latter unit into model complexes is generally straightforward via reaction of standard starting compounds, such as $MoO_2(acac)_2$ with the free acid of appropriate ligands (e.g., Fig. 3). In contrast, no standard procedures are available for preparing the thioxo unit $[MoSO]^{2+}$ in model compounds and accordingly, very few examples of model compounds with this structural characteristic of XO exist (see also Sec. 3). An excellent coverage of the model chemistry of these structural types up to 1994 can be found in the reviews of Enemark and Young [20] and Garner and Bristow [65].

The most valuable compounds from these early studies are those that mimicked enzyme substrate reactions, those that existed in several oxidation states of the molybdenum center, and those having spectroscopic properties similar to the enzymes. These issues will be covered in Secs. 2.2. and 2.3.

2.2. Complexes as Spectroscopic Models

Developing a model compound having spectroscopic properties that match those of the enzyme centers has been a major incentive for chemists in the model business. Relatively easy methods, such as electronic spectroscopy, yielded little useful data because of interference from other chromophores. Most of the pterin-containing Mo and W enzymes

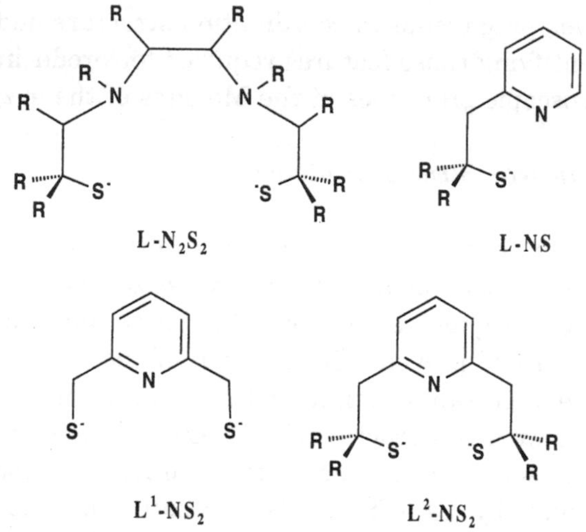

L-N$_2$S$_2$

L-NS

L^1-NS$_2$

L^2-NS$_2$

R = H, Me, Ph, 4-t-Bu-Ph

FIG. 3.   Examples of multidentate S- and N-donor ligands used in model chemistry and their abbreviations.

possess additional prosthetic groups (Fe-S, flavin, heme) whose electronic transitions dominate the electronic spectra of the enzymes [1–5]. To this point the only useful UV/Vis spectra concerning the Mo center have been obtained from two bacterial DMSO reductases (DMSOR), which contain no other highly absorbing prosthetic groups [66,67]. Absorptions at 350 nm in the oxidized enzymes have been assigned to thiolate-Mo(VI) charge transfer by comparison to model compounds [66]. Assignments of some long-wavelength absorptions in the enzyme were uncertain [66].

Resonance Raman spectroscopy, while not very widespread, is also a promising tool that has been used to probe the metal coordination in a variety of metalloproteins [68]. However, again other strongly absorbing prosthetic groups can dominate the resonance Raman spectra in most pterin-containing molybdenum enzymes allowing only the low-frequency vibrational modes around 350 cm^{-1} to be clearly assigned to Mo−S vibrations. This feature is related to metal-ene-dithiolate complexes and will therefore be described in more detail in Sec. 3.

EPR spectroscopy, which is only applicable to paramagnetic Mo(V) centers, is an extremely sensitive method for tracking coordination changes at the molybdenum center in enzymes and model compounds [14]. The molybdenum and ligand hyperfine splittings can offer additional information about the coordination environment of the Mo(V) species and the chemical reactions at the molybdenum center. The first EPR spectra for XO were reported in 1959 by Bray et al. [69], and Bray and coworkers have continued to develop the application of EPR spectroscopy to molybdenum enzymes [14]. Soon thereafter, model complexes with EPR spectra comparable to the ones from enzymes were discovered [20]. In particular, spectroscopic analysis performed on a series of Mo(V) complexes—[MoOXL] (L is the tetradentate ligand N,N'-dimethyl-N,N'-bis(2-mercaptophenyl)ethylenediamine (see type L-N_2S_2 in Fig. 3) and X is an oxido, hydroxido, sulfido, hydrosulfido, or chloro ligand—revealed a close relationship with the spectra of different reduction states of XO [70–74]. Extensive isotopic labeling with ^{95,97,98}Mo, ^{17}O, and ^{33}S was utilized, and comparison of the data with the rapid and slow signals of XO showed a clear connection [74]. In particular, the "very rapid" signal of [MoOSL]$^-$ changes upon protonation of the sulfido ligand to form [MoO(SH)L] to the spectrum of the "rapid" signal observed in XO. In the rapid and slow signals of XO [74], as well as in [MoO(OH)L] and [MoO(SH)L], the ^{1}H hyperfine splitting is also consistent with the proposed hydroxyl or hydrosulfido ligands [70–73]. This L-N_2S_2 ligand was used in a more recent EPR study where it served as part of a model for sulfite oxidase (SO). The complex MoO(SR)(L-N_2S_2) was shown to exhibit EPR parameters that best matched those of SO, as well as the "very rapid" form of XO, due to the arrangement of the three thiolate donors [75].

The technique of magnetic circular dichroism (MCD) is more frequently being applied to study model complexes in addition to the enzymes. Reports available include studies on complexes having MoOS$_4$ inner coordination sphere of mono- and bidentate thiolates [76,77] and of mono- and bis-ene-dithiolene model complexes (Sec. 3) [78,79]. Discussion of specialized spectroscopic methods for early mononuclear model compounds have been exhaustively reviewed by Enemark and Young [20].

2.3. Complexes as Functional Models

It is helpful to define the fundamental reaction types involved in the enzyme substrate reaction before considering how synthetic models duplicate these. Generally, the overall reaction involved in enzyme catalysis may be separated into two half-reactions. As an example, the oxidation of substrate X in Eq. (3) can be described by half-reactions in Eqs. (1) and (2), where M is the metal ion (Mo, W) and X the substrate [15,20,27,36]:

$$M^{VI}O + X \rightleftharpoons M^{IV} + OX \tag{1}$$
$$M^{IV} + H_2O \rightleftharpoons M^{VI}O + 2H^+ + 2e^- \tag{2}$$

Therefore

$$X^{n+} + H_2O \rightleftharpoons OX^{(n+2)} + 2H^+ + 2e^- \tag{3}$$

Most molybdenum and tungsten enzymes catalyze a reaction identical to Eq. (1), which has been called oxygen atom transfer. The reaction in Eq. (2) regenerates the active metal site and is called a coupled electron–proton transfer (CEPT) [20,27,36]. The net reaction (3) represents formally the enzyme-catalyzed oxygen atom exchange between substrate X and water.

2.3.1. Oxygen Atom Transfer

One of the first successful examples for OAT reactivity with a model Mo compound was in the $Mo(VI)O_2(L-NS_2)$ system (Fig. 3) [80]. This model system is quite robust and able to turn over repeatedly, albeit slowly, in the presence of a suitable oxygen atom donor/acceptor pair (Fig. 4a). Sulfoxides are often used as donors and phosphines (cf. triphenylphosphine) as acceptors in the OAT model systems. Subsequently, other similar systems have been developed [15,20,27,36,81] and have helped to understand the basic OAT half-reactions. The OAT reaction under catalytic conditions is rate-limited by the transfer of the oxygen atom from the $Mo(VI)O_2$ complex to the oxo acceptor. Another model system based on the trispyrazolylborate ligands (L-N$_3$) has been successful due to the special ability of L-N$_3$ to stabilize octahedral Mo complexes in oxidation states from +4 to +6. [20,27,36]. One example is a catalytic

system based on $(L-N_3)MoO_2(SPh)$, where many of the species participating in the catalysis have been spectroscopically and structurally characterized. This system functions in a catalytic sulfoxide/phosphine OAT
cycle analogous to Fig. 4a, but additionally can use water as an oxygen
atom source (in the presence of a suitable oxidant), better mimicking the
enzyme reactions [82,83] (Fig. 4b).

The next generation of systems to model enzyme analogue OAT
chemistry under development are based on complexes bearing ene-
dithiolate sulfur-donating ligands more similar to Moco and tungsten
cofactors, and this work is addressed in Sec. 3.

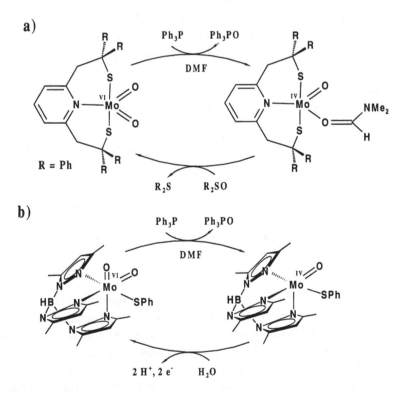

FIG. 4. Examples of oxygen atom transfer.

2.3.2.   Coupled Electron–Proton Transfer

The CEPT reaction [see Eq. (2)] is described as the transfer of one proton and two electrons which formally corresponds to a "hydride" transfer [36]. This concept and the consequences for biological catalysis with metal centers was initially presented by Stiefel [84], and this representation is quite helpful for understanding (1) why Mo(V) states might be very important in the enzyme reaction, (2) what mechanisms activate terminal M=O bonds, (3) why the transfer of protons and electrons must occur concomitantly, and (4) why water activation is a part of a CEPT mechanism in an enzyme. A very detailed explanation is beyond the scope of this chapter but can be found in the review of Pilato and Stiefel [36].

CEPT describes the dramatic modulation of the pK_a of a metal-coordinated E atom (E=O, S) when the oxidation state of a metal center M changes. This effect is due to π bonding in Mo=E systems which, on formal reduction of a cis-Mo(VI)EO center, are populated in the π^* orbitals having significant oxo or E character. This shift of electron density to the oxygen or E atom on formal Mo reduction results in increased O/E basicity. A simple example from model chemistry is shown schematically in Eq. (4) [85,86], where $L = L-N_2S_2$ or $(L-N_3)(SPh)$.

$$LMo^{VI}O_2 + e^- \longrightarrow [LMo^VO_2]^- + H^+ \longrightarrow LMo^VO(OH) \tag{4}$$

3.   MODEL COMPLEXES OF 1,2-ENE-DITHIOLATES

3.1.   Types of Complexes Containing 1,2-Ene-dithiolate Ligands

Recent modeling efforts have concentrated on preparing and characterizing Mo- and W-1,2-ene-dithiolate complexes, in accordance with precise knowledge of the metal environment from protein X-ray structures. Like in the enzyme systems that are differentiated by having one vs. two 1,2-ene-dithiolate MPT ligands (Fig. 2), so can the model systems be discussed as well according to the number of 1,2-ene-dithiolate ligands in the complex.

3.1.1. Metal–Tris-1,2-Ene-dithiolate Complexes

The first studies on Mo- and W-ene-dithiolate complexes dates back to the end of the 1960s when the unusual properties of their tris complexes were revealed. These complexes aroused great interest due to their reversible redox activity, which can be ligand- and/or metal-centered. Most frequently the ligands $(CN)_2C_2S_2^{2-}$ (mnt^{2-} from "maleonitrile") or $(CF_3)_2C_2S_2^{2-}$ were used [87–89]. Some important work was also done with the "$quasi$-ene-dithiolates", $C_6H_4S_2^{2-}$ (bdt^{2-}) and $(CH_3)C_6H_3S_2^{2-}$ (tdt^{2-}), as ligands to Mo and W [90]. Within the last decade, ene-dithiolates bearing quinoxaline substituents have served as models where quinoxaline served as a pterin analogue [20,24,36,91–101]. The quinoxalyl-ene-dithiolates exhibit a reactivity and redox pattern that is somewhat different from that of pterins. All of these tris-ene-dithiolate compounds are thermodynamically stable in one or more oxidation states [87–89]. One instructive example is the use of an activated acetylene in the side chain of a quinoxaline **7** reacting with a molybdenum polysulfide complex **8** to yield the tris-ene-dithiolate complex **9** [97].

One important result from an investigation of the Mo complexes of quinoxalyl-ene-dithiolates demonstrated that these undergo dithiolene dissociation and oxidation to form a thiophene product. This is an exact mimic of the natural metabolic oxidation of Moco to the pterinylthiophene form B (see Sec. 5.2).

3.1.2. Metal–Bis-1,2-Ene-dithiolate Complexes

Metal–bis-1,2-ene-dithiolate complexes have lately received the greatest attention as model compounds due both to their occurrence in natural enzyme centers (see Fig. 2, Sec. 1.2) and to their somewhat easier synthetic availability as compared with metal–mono-1,2-ene-dithiolate complexes. $MoO(mnt)_2^{2-}$ was initially reported in 1969–1970 [89,102] as a side product in the synthesis of the tris-1,2-ene-dithiolate complex. The most important initial work on Mo- and W-bis-1,2-ene-dithiolate complexes was done in the early and middle 1990s by the research groups of Nakamura, Ueyama [103–114], and Sarkar [115–120], and has yielded a large number of oxidobis-1,2-ene-dithiolate complexes. These complexes have the composition $[Mo^{VI}O_2L_2]^{2-}$ or $[Mo^{IV}OL_2]^{2-}$, where M = Mo or W and L is a 1,2-ene-dithiolate ligand, mostly mnt^{2-} or bdt^{2-}, formed from reaction with Na_2MoO_4 under carefully controlled pH conditions. The interesting complexes $Mo^{IV}O[S_2C_2(COOCH_3)_2]_2^{2-}$ and an acid amide derivative ($Mo^{IV}O[S_2C_2(CONH_2)_2]_2^{2-}$) have been reported for studying the relative case of OAT [106,113]. Analogous tungsten compounds have also been described and structurally characterized [111,112,117].

A different approach for Mo- and W-1,2-ene-dithiolate complexes was chosen by the groups of Garner and Joule to gain greater diversity of the 1,2-ene-dithiolate substituents. Ene-dithiol groups attached to heterocyclic systems (often quinoxalines), were masked as 1,3-dithiole-2-thiones or -2-ones [95,98–101,121,122] and were then brought to reaction with $[MoO_2(CN)_4]^{4-}$ and $[WO_2)(CN)_4]^{4-}$ to obtain the corresponding MoO- and WO-bis-1,2-ene-dithiolate dianions respectively [100,101]. One example is given in Scheme 1. The advantage of this strategy is manifest in analogous pterin chemistry (Sec. 5).

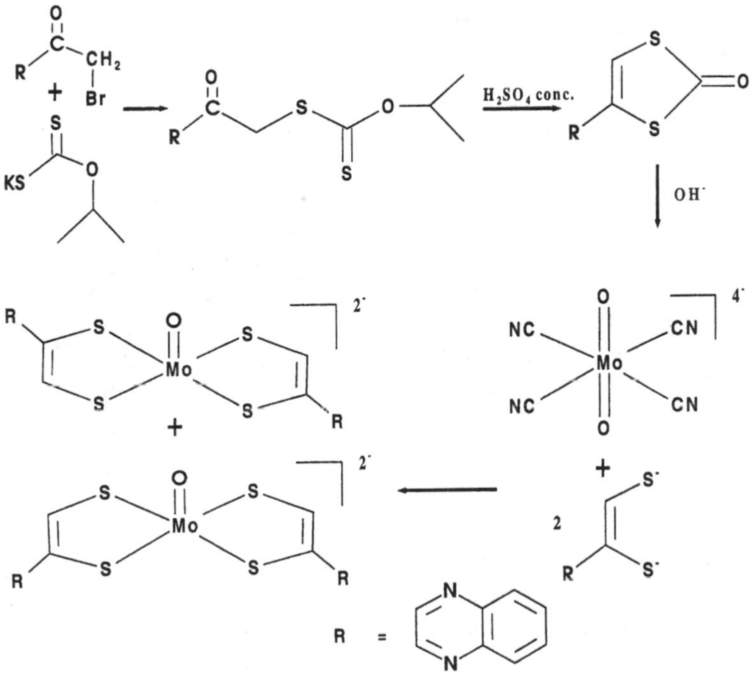

Scheme 1

A recent important addition to this area of ene-dithiolate models have been provided by Holm's group, which has produced an enormous number of new Mo- and W-bis-1,2-ene-dithiolate complexes characterized structurally, spectroscopically and in terms of chemical reactivity [123–135]. Several new Mo-ene-dithiolate complexes were prepared using different Mo reagents and synthetic routes [125]. One new route uses the reaction of 1,2-ene-dithiolate salts with $[Mo^{IV}OCl(MeNC)_4]^+$ and $[Mo^{IV}Cl_4(MeCN)_2]$ to conveniently prepare square pyramidal $[Mo^{IV}O(S_2C_2R_2)_2]^{2-}$ complexes. This system also yielded the new isonitrile complexes $[Mo(S_2C_2R_2)_2(RNC)_2]$ lacking additional terminal oxo groups and possessing an idealized trigonal prismatic C_{2V} stereochemistry [125]. A different yet more promising route to new Mo- and W-bis-1,2-ene-dithiolate complexes uses simple ligand exchange of $Mo(MeCN)_3(CO)_3$ or $W(MeCN)_3(CO)_3$ and $Ni(S_2C_2R_2)_2$, leading to the bis(1,2-ene-dithiolate)dicarbonyl complexes $Mo(CO)_2(S_2C_2R_2)_2$ or $W(CO)_2(S_2C_2R_2)_2$ (R = Me, Ph) [128,129,131,133,134] (Fig. 5). These

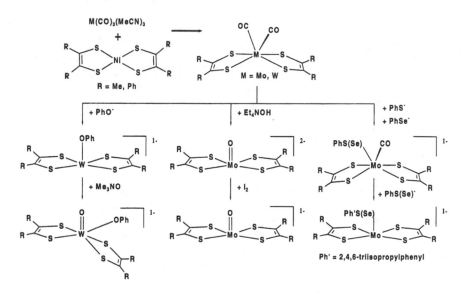

FIG. 5.   Reaction of M(MeCN)$_3$(CO)$_3$(M = Mo, W) and Ni(S$_2$C$_2$R$_2$)$_2$ leading to bis(1,2-ene-dithiolate)dicarbonyl complexes M(CO)$_2$(S$_2$C$_2$R$_2$)$_2$ (R = Me, Ph) and some examples of follow-up products [128,129,131,133,134].

M(CO)$_2$(S$_2$C$_2$R$_2$)$_2$ (M = Mo, W) complexes are extremely versatile precursors to a large number of useful models for several pertinent forms of members of the DMSOR family, which includes aldehyde oxidoreductases and formate dehydrogenases (see Fig. 2) by carbonyl substitution reactions (Fig. 5).

3.1.3.   Metal–Mono-1,2-Ene-dithiolate Complexes

A single 1,2-ene-dithiolate ligand on a metal is a complex difficult to obtain. Use of bulky ligands like cyclopentadienylato ligands to occupy residual coordination sites has been successful in limiting the number of 1,2-ene-dithiolate ligands, but these are biologically not relevent [96,136].

   One of the earliest characterized mono-1,2-ene-dithiolate complexes was reported from the reaction of MoO$_2$Cl$_2$ with Na$_2$(S$_2$C$_2$(CN)$_2$) in tetrahydrofuran to give an orange species that is stable below $-45°C$ and that reacts with phosphines to produce isolable derivatives such as **10** [137].

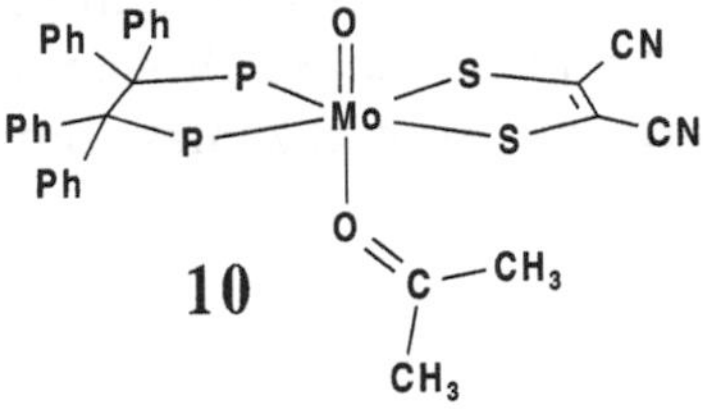

Another type of metal–mono-1,2-ene-dithiolate model complex is based on the trispyrazolylborate ligands (L–N$_3$) and has yielded compounds like (L–N$_3$)MoO(tdt) where tdt is toluene-3,4-dithiolate [138,139]. In addition, the group of Young presented a remarkable reaction where (L–N$_3$)W^{VI}S$_2$X (X = monoanion like PhO$^-$, PhS$^-$, or PhSe$^-$) reacted with alkynes to yield 1,2-ene-dithiolate tungsten(IV) complexes [140,141] (Fig. 6).

The same principle has been employed to synthesize the complex (L–N$_3$)W(SePh){S$_2$C$_2$(Ph)(2-quinoxalinyl)} from (L–N$_3$)W^{VI}S$_2$X and 2-(phenylethynyl)quinoxaline [141]. This molecule is a prototype for a whole series of metal–mono-1,2-ene-dithiolate model complexes containing nitrogen heterocycles, most of which are quinoxalines. The first example for this type of quinoxaline-containing compound was the complex Cp$_2$Mo{S$_2$C$_2$(2-quinoxaline)(C(O)Me)} **13**, which has been prepared according to the sequence shown in Scheme 2 [96,136]. In this

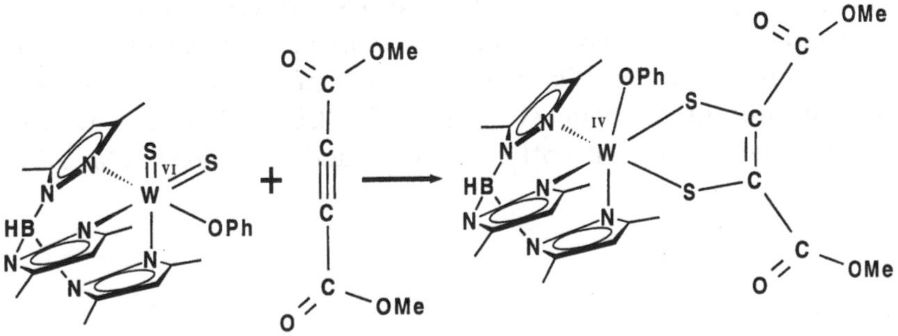

FIG. 6. Reaction of (L-N$_3$)W^{VI}S$_2$X (X = monoanion such as PhO$^-$, PhS$^-$, or PhSe$^-$) with alkynes to yield 1,2-ene-dithiolate tungsten(IV) complexes [140,141].

reaction sequence the interesting intermediate 1-persulfido-2-ene-thio-late (trithiolene) complex, $Cp_2Mo\{S_3C_2(2\text{-quinoxaline})(C(O)Me)\}$ **12**, is generated as a precursor to the corresponding 1,2-ene-dithiolate by a reaction of **7** with **11**. Compounds like this are discussed by Stiefel as possible turnover intermediates in enzymes containing Mo-thioxo centers [32].

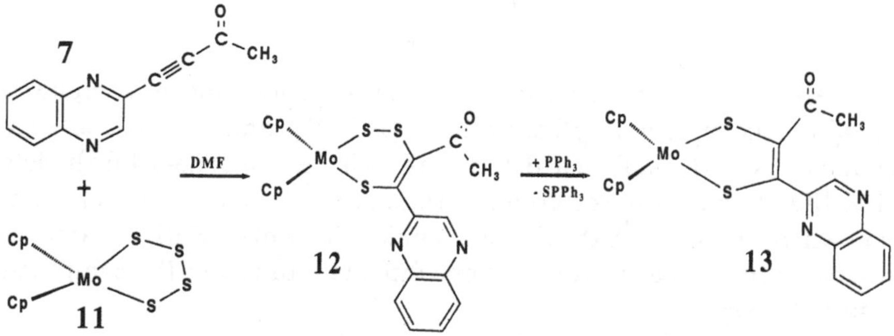

Scheme 2

A similar type of Cp_2Mo-ene-dithiolate-quinoxaline complex has also been prepared by Pilato's group via a different route. They demonstrated that an activated α-bromoketoquinoxaline **15** when reacted with $Cp_2Mo(SH)_2$ **14** results in the formation of a Cp_2Mo-ene-dithiolate-quinoxaline complex **16** [142]. This route has striking similarities to the biosynthetic formation of Moco in its various forms, which proceeds through an α-phosphorylated ketone precursor to MPT [46]. Other simple analogues to this biological phosphate precursor have also been prepared by the group of Pilato and provide important precedents for understanding the biochemical syntheses [143,144]. Similar conversions have also been shown with other reagents and metal ions [145–148].

3.1.4. Metal–Thioxo-1,2-Ene-dithiolate Complexes

Mo- and W-thioxo-1,2-ene-dithiolate complexes are somewhat difficult to obtain. A few examples are known for Cp*MS(S$_2$C$_2$R$_2$) (M = Mo, W) [149,150]. Interesting silylated precursors with the composition MOS(OSiPh$_3$)$_2$(Me$_4$phen), which may soon allow access to metal–thioxo-1,2-ene-dithiolate complexes, were reported lately by Holm's group [126,132]. The complex WS$_2$(OSiPh)$_2$(Me$_4$phen) reacts immediately with two activated alkynes, resulting in dithiolene ring closure and formation of W^{IV}(OSiPh)$_2$(Me$_4$phen)(S$_2$C$_2$R$_2$)] (R = CO$_2$Me, Ph) [132].

3.2. Complexes as Spectroscopic Models

In terms of overall spectroscopic features, many Mo- and W-1,2-ene-dithiolate complexes differ only slightly, if at all, from their respective M-thiolate-counterparts.

The UV/Vis data available for the MoO-bis-1,2-ene-dithiolate complex dianion in Scheme 1 (Sec. 3.1.2) fail to correlate with those reported for DMSOR, but the features have not yet been clearly assigned to specific transitions [100]. Similarly, EPR spectral parameters were characteristic of a standard MoO(ene-dithiolate)$_2$ center comparable to complexes like MoO(mnt)$_2$, where the ene-dithiolate is asymmetric, but again the parameters obtained from the model do not match those reported for DMSOR [100].

Resonance Raman spectroscopy provides one of the few specific features for the ene-dithiolate unit. The complex $Cp_2Mo\{S_2C_2(2$-quinoxaline)(C(O)Me)\}$ **13** was used as a spectroscopic model in resonance Raman studies, where the major, sulfur-sensitive band at 349 cm^{-1} shifts to 341 cm^{-1} upon ^{34}S enrichment [96]. This result confirms results from DMSOR from *Rhodobacter sphaeroides*, where a band at 350 cm^{-1} shifts to 341 cm^{-1} upon enrichment with ^{34}S [151].

MCD is now frequently being applied to the study of model complexes in addition to the enzymes they represent. This powerful technique can examine individual orbital contributions to observed differences in ionization and reduction potentials at high resolution. For example, recent results showed that the intensity of the $S_{ip} \rightarrow Mo\ d_{xy}$ transition can reveal the relative covalency contributions of the Mo d_{xy} redox orbital and the dithiolate S_{ip} orbitals [79]. MCD has also shown large differences in the $S_{op} \rightarrow Mo\ d_{xz,yz}$ transition between $(L-N_3)MoO(tdt)$ and $(L-N_3)MoO(qdt)$ (qdt = quinoxaline-2,3-dithiolate), which indicate an increase in the effective nuclear charge on Mo in $(L-N3)MoO(qdt)$. The qdt^{2-} ligand possesses considerable contributions from the dithione resonance structure, limiting its ability to donate charge to the Mo atom. The results for the tdt^{2-} ligand, however, parallel photoelectron spectroscopy (PES) studies that have demonstrated this ligand's remarkable ability to poise the Mo center at nearly constant ionization potential. With the exception of charge-deficient dithiolates, such as qdt^{2-} [79], this "electronic buffer" effect has been suggested to be a general property of dithiolate ligands. It has therefore been hypothesized that a primary role of the pyranopterin-1,2-ene-dithiolate in molybdenum enzymes is to maintain a nearly constant effective nuclear charge on Mo during the course of catalysis [152].

3.3. Complexes as Functional Models

Mo- and W-1,2-ene-dithiolate complexes are quite similar in terms of overall functional features of their respective M-thiolate counterparts and to other systems already mentioned in Sec. 2.3. As reported by the group of Nakamura and Ueyama, $Mo^{IV}O(bdt)_2^{2-}$ can be converted to the dioxo molybdenum(VI) complex in good yield by the following reaction:

$$MoO(bdt)_2^{2-} + Me_3NO \rightarrow MoO_2(bdt)_2^{2-} + Me_3N \qquad (5)$$

This conversion is a clean reactivity model for the Mo enzyme trimethylamine N-oxide reductase (TMAOR) [107]. The interesting complex $Mo^{IV}O[S_2C_2(CONH_2)_2]_2^{2-}$ has been reported to accelerate the rate of reduction of Me_3NO in OAT considerably [113]. It appears that NH···S hydrogen bonding is modulating the system.

Lately, Holm's group has presented a series of Mo- and W-ene-dithiolate complexes of the general type $[M^{IV}(QR')(S_2C_2Me_2)_2]^{1-}$ as reactivity analogues of the reduced sites of the enzymes DMSOR and TMAOR [$QR' = PhO^-$, 2-AdO^- ($Ad = $2-adamantyl), Pr^iO^-], dissimilatory nitrate reductase ($QR' = $2-$AdS^-$), and formate dehydrogenase ($QR' = $2-$AdSe^-$) [133,134]. These complexes engaged in direct OAT with N-, As-, S-, and Se-oxides to afford the respective M^{VI} complexes in clean transformations. Direct atom transfer was demonstrated with labeled $Ph_2Se^{18}O$. Comparison of relative reactivities of Mo/W isoenzymes and pairs of isostructural Mo/W complexes at parity of ligation reveals that OAT from substrate to metal is faster with tungsten and from metal to substrate is faster with molybdenum. While the OAT reactions in analogue systems are substantially slower than in enzyme systems, this observation demonstrates the usefulness of these models in revealing a kinetic metal effect in OAT reactions for isostructural isoenzymes.

4.   METAL COMPLEXES OF PTERINS

4.1.   Interactions of Pterins with Metals

Before X-ray crystallography provided definitive structures for the metal sites in the Mo and W enzymes, several groups began investigating how the pterin portion of the molybdopterin structure might interact with molybdenum and with other metals. It was observed that metals bind pterins preferentially at the O, N chelate site created by the carbonyl oxygen atom O4 and pyrazine nitrogen N5. A highly delocalized, covalent bond is formed when metal and pterin oxidation states are optimally matched. This section gives an overview of the significant results of this work. In light of the indisputable evidence for molybdopterin binding at the ene-dithiolate fragment, these pterin complexes have little value as structural models for Moco. However, the insight gained into the electronic redistribution that occurs as a result of metals binding oxidized and reduced pterins does have some relevance to the unique character of the pterin piece of molybdopterin and hints at its purpose at the enzyme catalytic site. Molybdenum and pterin ligands, in particular, comprise an unusual reaction system because each has a wide range of oxidation states available and, since pterins and metals individually are redox-active species, one predicts that their reactions will favor electron transfer and metal-pterin redox events [38]. This is the premise that spurred initial explorations Mo(+6) compounds and reduced tetrahydropterin in order to understand the potential reactivity available to oxidized Mo(+6, +5, +4) and the reduced pterin unit of molybdopterin at the catalytic center of the enzymes [153,154].

It is clear that pterins join the other chelating ligands [155,156], notably ene-dithiolates [157], that are well known for "noninnocent" behavior when coordinated to transition metals. Noninnocent ligands have reduced and oxidized forms and can behave as electronic buffers [152] to molybdenum. While the examples used here focus primarily on molybdenum, we and others have previously reported such reactivity for other transition metals, such as copper [158], cobalt [159], and iron [160], in addition to molybdenum [153,154,161–166] systems.

4.2. Metal Complexes of Oxidized and Reduced Pterins and Pteridines

The antecedents of metal-pterin studies are rooted in the work of Hemmerich [167] dating nearly half a century ago. To a large extent, results from more recent studies have merely reiterated Hemmerich's conclusions. Fully saturated ring systems of oxidized alloxazines and pterins have little affinity for high-valent metals or for many redox-active divalent metals but prefer to bind to low-valent, soft metals, such as Mo(0), Ag(I), and Cu(I). Thus, there is only one example of an oxidized pterin coordinated to Mo(+6) [168] or to W(+6) [169], in contrast to several examples of pterins and pteridines bound to low-valent and other metals. Selected examples are depicted in Fig. 7.

When one investigates the metal binding preferences of electron-rich reduced pterins, i.e., tetrahydropterins, it is observed that these molecules utilize the same O4, N5 chelation site for donation to

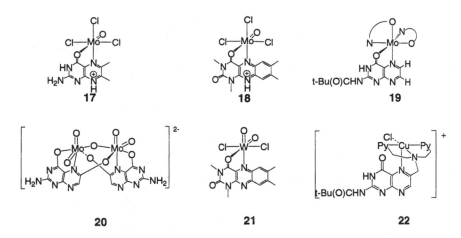

FIG. 7.   Examples of molybdenum and other transition metal pterin complexes from oxidized pterin reagents. **17**, MoOCl₃ (dimethylpterinH) [165]; **18**, MoOCl₃-(tetramethylalloxazineH) [165]; **19**, Mo(pivaloylpterin)₃ [166]; **20**, Na₂(Mo₂O₅-(xanthopterinate)₂] [168]; **21**, WO₂Cl₂(tetramethylalloxazine) [169]; **22**, CuCl-(bis(ethylpyridyl)pivaloylpterin)(PF₆) [170].

high-valent, electropositive metals, notably Mo(+6). Numerous examples [160–164] exist that demonstrate the reaction of Mo(+6) complexes with tetrahydropterins to produce isolable Mo-pterin compounds, and a few are illustrated in Fig. 8. Note that one of the two oxo ligands from a dioxo-Mo(+6) reagent is lost, presumably as water, in parallel with deprotonation of the pterin at N5 in the pyrazine ring.

Recent work shows that the O, N chelation pattern has been indicated in the reaction of pyranopteridines with Mo(+6) complexes [43]. UV/Vis- and NMR spectroscopic results showed close similarities to properties of the isolated Mo-pterin compounds shown in Fig. 8, but X-ray structures were not yet obtained.

4.3. Structural Features

Pterins chelate metals preferentially using the O, N chelate site and frequently the carbonyl O4 is binding as an amidate following deprotonation at N3. A few examples also exist where the metal uses ligation

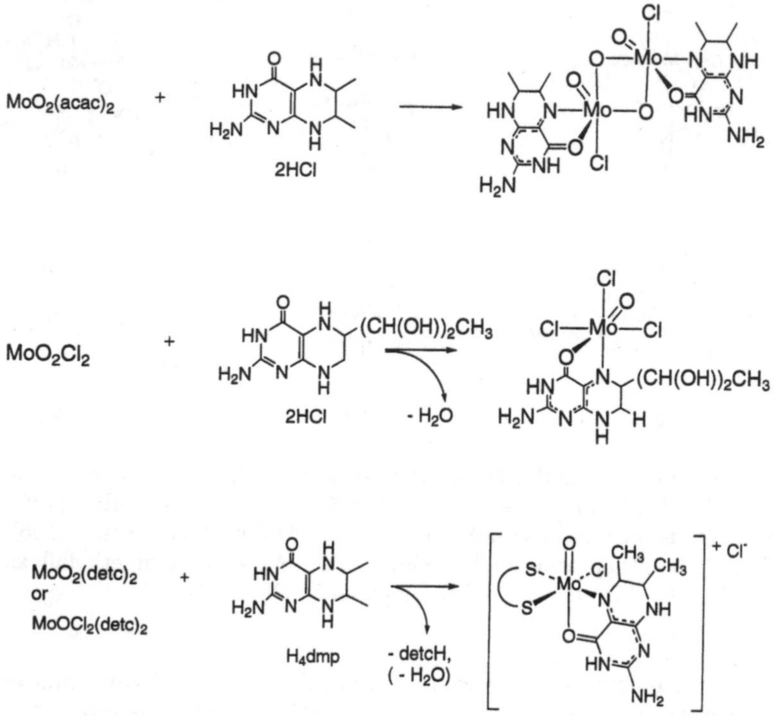

FIG. 8. Examples of reactions of reduced pterins with molybdenum reagents
[160–164].

sites substituted elsewhere on the pterin core in addition to or in place of
the O4/N5 donor atoms (see **22** in Fig. 7). In the case of complexes of
oxidized pterins (or pteridines) with oxidized molybdenum(+6) [or
tungsten(+6)], the M—O4 bond is the shorter and stronger interaction
(Mo—O): 2.084(5) Å [168]; W—O: 2.232(3) Å [169]) as compared with a
modest or weak M—N5 bond (Mo—N: 2.234(6) Å [168]; W—N; 2.462(3)
Å [169]). The C—C and C—N distance of the pterin rings in these cases
is usually minimally changed from the distances in the free ligand, and
they retain the same alternation of single and double bonds as indicated
by longer and shorter distances. Tetrahydropterins bound to Mo (and
other metals) exhibit markedly different intramolecular distances and
these data are entirely consistent with the view that the pterin has been
partially oxidized and the Mo partially reduced. In these complexes it is

now the Mo$-$N5 (cf. 2.027(7) Å [153]; 2.013(4) Å [165]) interaction that is the shorter and stronger as compared with the Mo$-$O4 bond (cf. 2.290(6) Å [163], 2.238(3) Å [165]). The C$-$C and C$-$N distances are altered in ways expected for the partial oxidation of the pyrazine ring. This is depicted in Fig. 8 using dashed lines to show those bonds intermediate between single and double order [160–164].

4.4.   Spectroscopic Features

Model work has proved that a high degree of electronic delocalization occurs in these systems, but oxidation state assignments for the metal and pterin pieces of the product have remained ambiguous and difficult to make. One can find in the literature different oxidation state assignments made for Mo and the pterin ligand [153–154]. One approach to resolve this problem of ambiguous oxidation states successfully used X-ray photoelectron spectroscopy (XPS) to measure Mo3d binding energies to indicate the effective charge on Mo in these complexes [166]. A comparison of Mo3d binding energies with standard molybdenum complexes served to bracket appropriate Mo oxidation states and subsequently the pterin oxidation level. From this study it was concluded that complexes obtained from reaction of Mo(VI) complexes with tetrahydropterins are best described using Mo(V)-trihydropterin assignments. Similarly, it was determined that complexes prepared from Mo(IV) and fully oxidized pterins exhibit Mo3d binding energies that correspond to Mo(V)-pterin radical anion species.

The physical appearance of pterin complexes of transition metals is a good indicator of the extent of delocalization in the pterin-metal unit. For example, complexes formed from Mo(+6) and tetrahydropterins [160–164] are characterized by deep red-purple colors, observed as intense electronic absorptions ($\varepsilon \sim 10\,000$ M^{-1} cm^{-1}) in the visible spectrum. These absorptions may be considered charge transfer transitions from the frontier orbital of the metal and pterin which are well matched in energies and overlap. This interpretation is entirely consistent with the Mo(V)-trihydropterin formulation from XPS results. Similarly, several tris-pteridine complexes of Mo(0) [166] have been described which are diamagnetic, neutral molecules having very intense ($\varepsilon \sim 40\,000$ M^{-1} cm^{-1}) MLCT absorptions near 500 nm. The electronic spectroscopic properties are in accord with XPS results which suggest that the pterin ligands in these tris complexes behave as strong π acids for Mo(0). Conversely, when such intense absorptions are not observed in the visible region of the electron spectrum, little delocalization is indicated and the pterin ligand simply behaving as a Lewis base. Such is the case with complexes of Mo(+6) and W(+6) with oxidized pterin and alloxazine (see **20** and **21** Fig. 7) [168, 169] whose yellow color is entirely due to the ligand. A similar correlation has been observed with complexes of first-row transition metals [38].

4.5. Functional Features

The reactivity of molybdenum-pterin compounds includes ligand substitution, OAT, and intermolecular electron transfer where the reactivity exhibited is determined by subtle variations at the pterin as well as in the Mo coordination sphere. The chemical behavior of molybdenum-tetrahydropterin complexes is sometimes typical of a tetrahydropterin complex of Mo having a formal oxidation state +6, but in other reactions the observed outcomes are typical of an oxo-Mo(+4) center.

4.5.1. Oxo Transfer

Oxygen atom transfer has been observed for several Mo-pterin systems. Scheme 3 illustrates a variety of outcomes when a typical OAT substrate, DMSO, is allowed to react with tetrahydropterin complexes of Mo. **23** undergoes only ligand substitution without OAT. **26** shows that the presence of sulfur donor ligands at Mo enables the OAT reaction, as would be expected from earlier model chemistry, if there is an open coordination site for DMSO binding prior to its reduction, otherwise no reaction occurs, as observed for **24**. Peculiarly, in another system **25** that lacks S donors, the simple exchange of methyl groups for hydrogen atoms on the pterin imparts OAT reactivity. Note that for **26**, OAT proceeds to produce dimethyl sulfide from DMSO reduction in addition to Mo and pterin and dithiocarbamate ligand oxidation, resulting in an overall five-electron redox reaction.

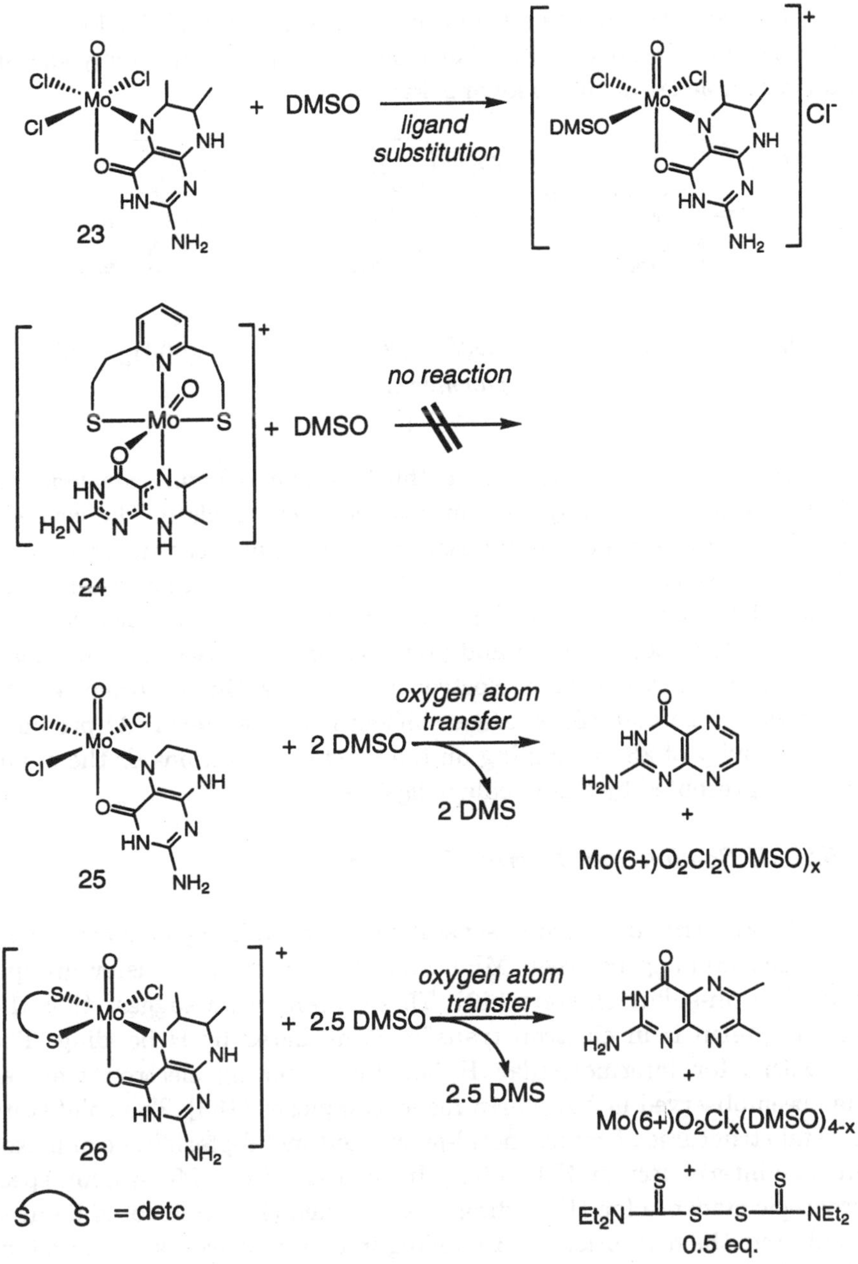

Scheme 3

Consideration of the resonance structures A, B, and C below can be used to understand the variable reactivity types of the molybdenum pterin complexes in the following way.

Mo(VI)(H$_4$pterin$^-$)	Mo(V)(H$_3$pterin$^\bullet$)	Mo(IV)(H$_2$pterin$^+$)
	preferred structure	
A	**B**	**C**

In the preferred structure B this lone pair is highly covalent and shared between the molybdenum and the pterin chelate. The covalent pair in structure B also emphasizes the strong antiferromagnetic coupling in these complexes instead of the diradical view that is suggested by the Mo(V)-trihydropterin formulation. In all cases, this electron pair is accessible to both the Mo and pterin ligand in subsequent reactions. The shifting of the shared electron pair to the Mo (C) produces the reactivity of **25** and **26**, whereas shifting it to the pterin (A) produces the reactivity of **23**, depending on the reaction conditions or the environment presented by the ancillary ligands.

4.5.2. H Bonding and Electron Transfer

The X-ray structures from several pterin-containing enzymes show hydrogen bonding between MPT and other electron transfer groups, notably iron-sulfur clusters [40]. These interactions suggest that the role of pterin is in electron transfer as mediated by H bonding. The precedent for intermolecular H bonding initiating electron transfer has been observed in Mo-pterin model complexes [163]. The solid-state crystal structures of many metal-pterin and metal-pteridine complexes include intermolecular H bonding. In the case of one Mo system, spectroscopic evidence for H bonding was obtained [163] and could be used to observe the disruption of H bonding in certain solvents or the retention of H bonding in other media. Complex decomposition was observed for an H-bonded complex in DMF that yielded both semioxidized and

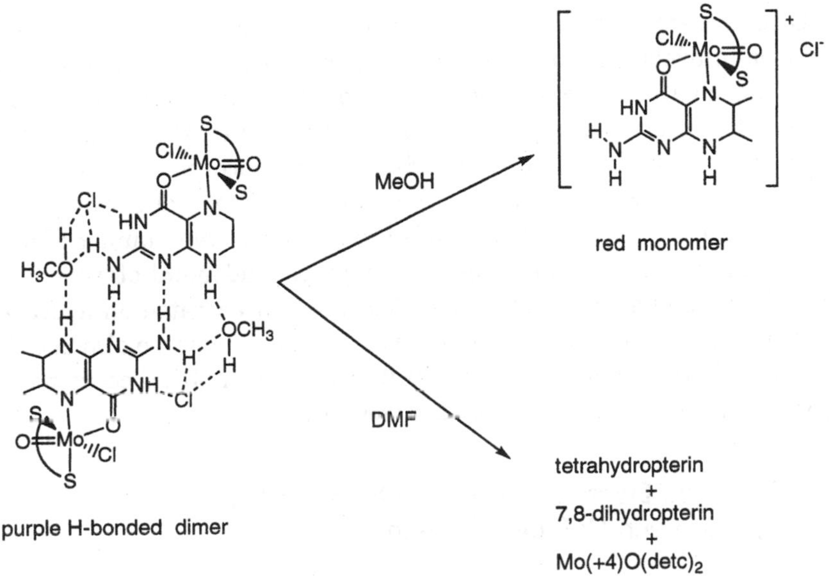

FIG. 9. H bonding preceding disproportion of a Mo(+V)-trihydropterin complex to dihydro- and tetrahydropterins [163].

fully reduced forms of the pterin ligand (Fig. 9). The apparent pterin disproportionation was interpreted as proton transfer accompanying electron transfer between Mo-pterin species in solution.

The correlation between electron transfer and proton transfer also finds an example in the complexes formed from oxidized pterins coordinating to Mo(+4). The products, viewed as Mo(V) species with one-electron reduced pterins, illustrate how H transfer must accompany partial pterin reduction [165].

5.   MODELS WITH ALL THREE REDOX SITES

5.1.   Synthetic Approaches to Mo and W Complexes of Pyranopterin-1,2-Ene-dithiolate

No complete model containing all three redox sites—the molybdenum or tungsten ion, the ene-dithiolate, and the hydrogenated pyranopterin—

has been presented up to now. In addition, there have not been any fully functional models that could duplicate an enzyme process under natural conditions. Synthetic approaches to Mo- or W-pyranopterin-ene-dithiolate complexes have been mainly guided by two principal strategies. The first general route described below (Sec. 5.3) has incorporated ideas and compounds from the oxidative degradation of the natural pyranopterin cofactor system (Sec. 5.2). This has led to an impressive number of new compounds and new synthetic methods in pterin and metal-pterin chemistry. The second general route (Sec. 5.5) has concentrated on hydrogenated pterins and their interactions with the metal ions (see also Sec. 4). Very recently this approach led to the first metal complexes coordinated to pyranopterins via the pyrano ring (see Sec. 5.5).

5.2. Oxidative Degradation Products of the Natural Pyranopterin Cofactor System

Moco can be released from enzymes via treatment with heat or SDS [62], typically with loss of Mo in addition to MPT degradation (see also Sec. 1.2). It can then be selectively oxidized with iodine or Fe^{3+} under oxygen following protocols described first by Rajagopalan's group [44,60,171]. These reactions yield oxidized pterins called form A and form B (see Fig. 10). Form B is structurally similar to urothione, a metabolic degradation product of MPT [61] first isolated in 1940 from human urine [172]. The structure of urothione was identified via synthesis by Sakurai and Goto [173]. The structures of form A and B have been confirmed by direct synthesis [24,174–176], an effort that was to be the foundation of further successful pterin model chemistry (see Sec. 5.3). MPT has been isolated in an oxidized form and also as a dialkylated derivative bearing carboxymethylamide groups at the two ene-dithiolate sulfurs (camMPT) [177] (Fig. 10). This technique of cam alkylation of dithiolene sulfur atoms is now a widely used method to identify MPT in its dinucleotide forms [16,45,58]. Note that the much used abbreviation camMPT is misleading because the reduced pyranopterin structure has been lost.

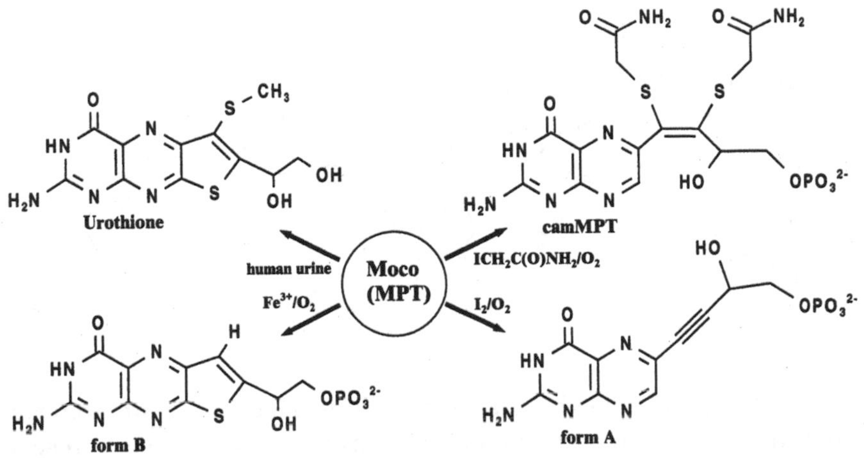

FIG. 10. Moco- and MPT-derived products, urothione, form A, form B, and camMPT.

5.3. Pterin-Ene-Dithiolate Ligands from Form A Precursors

Form A was exploited for use in a well-known inorganic preparation of ene-dithiolates from condensation of activated alkynes and metal tetrasulfides. This strategy is shown in the reaction of **27** and Cp_2MoS_4 **11** yielding the complex $Cp_2Mo\{S_2C_2(6\text{-}N(2\text{-}pivaloylpterin})(C(O)Me)\}$ **29** [96,136]. The reaction proceeds via the 1-persulfido-2-ene-thiolate (trithiolene) intermediate **28** as observed for the analogous quinoxaline derivative described in Sec. 3.1.3. While the bis-Cp ligand environment in **29** is abiological, the system proves the feasibility of the method.

This approach has been successfully extended to obtain a pterin-ene-dithiolate within a coordination environment more closely resembling the catalytic center. Burgmayer's lab has recently reacted a bistetrasulfide oxo-Mo reagent with a solubilized pterin derivative to form complex **30** [178].

Different approaches to pterin precursors were chosen by the groups of Garner and Joule. They built ene-dithiols masked as 1,3-dithiole-2-thiones or -2-ones into pterin side chains, based on the successful strategy with quinoxaline proligands [95,98–101] (see Sec. 3). Utilizing this route they were able to isolate a MoO-bis-1,2-ene-dithiolate-pterin complex **31** at its bis-PPh$_4$ salt [100].

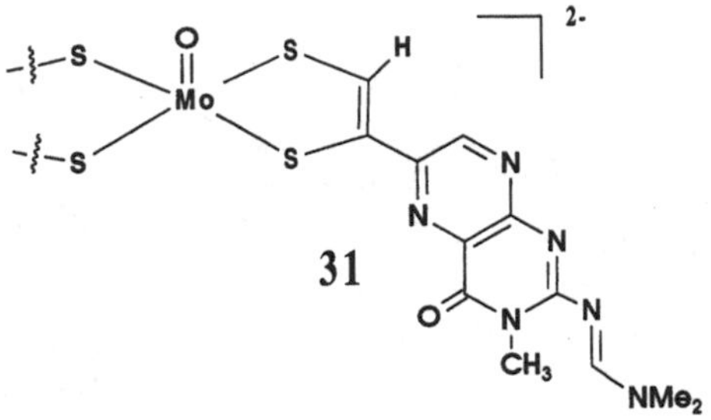

31

The above examples demonstrate the feasibility of synthesizing a pterin-substituted ene-dithiolate. The remaining challenge is to transform the oxidized pterin into the three-ring system of a pyranopterin. The most recent progress toward synthesizing the elusive pyranopterin-ene-dithiolate ligand has come from the labs of Garner and Joule [179]. The synthetic route built on a procedure developed for synthesis of a pyrano[2,3-b]quinoxaline-3,4-dithiolate, following its release from the corresponding 1,3-dithiole-2-one via treatment with CsOH and trapped as its cobalt complex **32** [101].

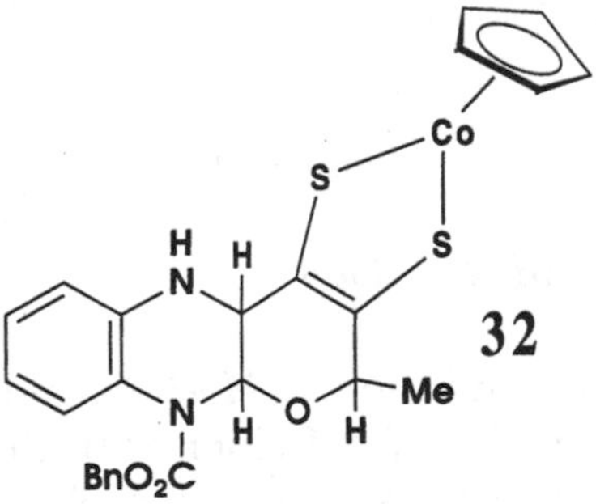

32

Scheme 4 outlines the steps in the synthesis of the pyranopterin. In this sequence the closing of the pyranopterin ring is an astounding reaction, which has no precedent in synthetic pterin chemistry [179].

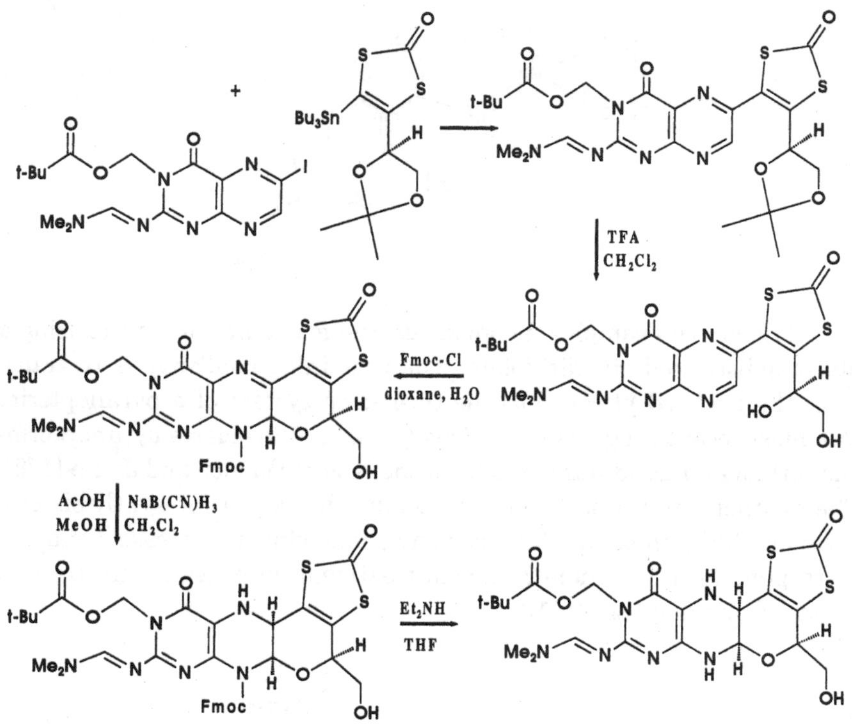

Scheme 4

5.3.1. Spectroscopy and Reactivity

The relatively recent acquisition of Mo-1,2-ene-dithiolate-pterin complexes has not yet come into full fruition with their complete spectroscopic characterization and how closely it duplicates that of the active enzyme centers. The $Cp_2Mo\{S_2C_2(6\text{-}N(2\text{-pivaloylpterin})(C(O)Me)\}$ complex **29** [96,136] has been used as a model in resonance Raman studies, exhibiting a major, sulfur-sensitive band at 349 cm^{-1} shifts to 341 cm^{-1} upon ^{34}S enrichment. This result confirms results from DMSOR from *Rhodobacter sphaeroides*, where a band at 350 cm^{-1} shifts to 341 cm^{-1} upon enrichment with ^{34}S [151]. The UV/Vis data available for the MoO-bis-1,2-ene-dithiolate-pterin complex **31** has not been clearly assigned to specific transitions, although the lack of correlation between the UV/Vis data reported and those for DMSOR was noted [100]. A problem emerged from analysis of its EPR properties, which served to characterize a standard MoO(ene-dithiolate)$_2$ center comparable to complexes such as MoO(mnt)$_2$, in which the ene-dithiolate is symmetrical. It was noted, however, that the measured model parameters and those reported for DMSOR did not match [100].

To the best of our knowledge, no reactivity data has been reported of any of the Mo- or W-1,2-ene-dithiolate-pterin complexes.

5.4. The Biological Precursor to Moco

A hydrogenated pterin precursor mentioned in Sec. 3.1.3 has been identified from the common biosynthetic pathway of molybdenum and tungsten enzymes [46]. This precursor can be obtained both from the molybdenum-deprived organism *N. crassa* and from oxidase-deficient children, neither of which produces active Moco [180]. This MPT precursor, called precursor Z, has been initially formulated as a quinonoid, 5,6- or 5,8-dihydropterin [36,44,46], an assignment that is unlikely to be correct. All known quinonoid compounds are extremely short lived [51,52,181], whereas the reported half-life of precursor Z is much longer and is similar to those reported for synthetic pyranopteridines [41b,c,43,53]. Recent experiments have shown a one-step air oxidation of the precursor to an unambiguously identified product, form Z, **33** as shown by UV/Vis [46,182]. NMR studies with multilabeled precursor derivatives to precursor Z [183] also provided evidence to substantiate the formulation of precursor Z not as a dihydropterin but instead as a pyranopterin **34**. Neither oxidized form Z nor the chemically very interesting precursor has been synthetically prepared.

5.5.   Pyranopterin Model Systems

The pioneering work on isolated and characterized pyranopteridines was reported by Pfleiderer's group in 1990 [41b,c]. Since then there have been very few papers on general pyranopteridine chemistry [41], with the exception of the above-mentioned short communication by the groups of Garner and Joule [179]. Recently, Fischer's group reported that the behavior of pyranopteridines is analogous to that of tetrahydrogenated pterins in the reaction with $Mo^{VI}O_2$ compounds (see Sec. 4), resulting in the same N,O-chelating coordination [43]. They have therefore investigated new routes to substituted pyranopterins that may prevent this N,O chelation while leaving open potential coordinating units in the pyrano ring. A protocol has been devised for the synthesis and characterization of a series of diastereomerically pure pyranopterins with differing numbers of protecting acetyl groups [42]. Selective deacetylation, starting from a tetraacetyl derivative, produces the monoacetylated compound **35**. This is an air-stable product that can coordinate via the *cis*-diol functionality to various metal centers [184]. A beneficial outcome of this work is overcoming the notorious oxygen sensitivity of hydrogenated pterins and pyranopterins through a protective alkylation on N5 in the pyrazine ring.

35

In the course of this work it was also possible to obtain the first crystal structure of a synthetic pyranopterin -*4aS,10aR*-tetraacetylpyranopterin (Fig. 11), which exhibits the same steric configuration at C4a and C10a as the natural cofactors in the active centers of molybdenum and tungsten enzymes [42].

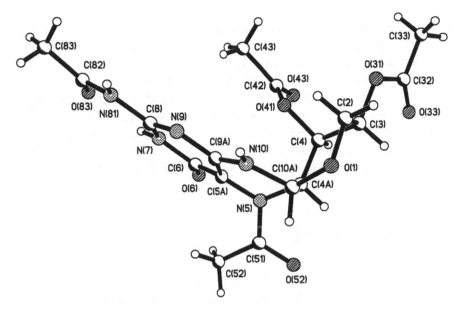

FIG. 11. X-ray structure of the tetraacetyl derivative of 8-amino-3S,4R-dihydroxy-3,4,4a,5,10,10a-hexahydro-2H-pyrano-(3,2-g)-pteridine-6(7H)-one [(4aS,10aR)] [42].

6.   CONCLUSION

The challenging chemical structures of the Mo and W cofactors have been an inspiration for many bioinorganic chemists in the field of model chemistry for more than three decades. Up to this point, an enormous wealth of information has been gathered on the individual components of this unique cofactor system—the metal atom and its environment, the 1,2-ene-dithiolate ligating group, and the pyranopterin heterocycle—each of which possesses its own redox activity. A good deal of model chemistry reaffirms the conviction that the metal atom with its cycle of oxidation states is the center of substrate processing and that the 1,2-ene-dithiolate group is most likely responsible for adjustment and fine tuning of the redox potentials at the metal center. Numerous proposals have been suggested for the pyranopterin's purpose, such as acting as an antenna for directed electron transfer [40b], helping to "store" one electron as a radical [185] or supporting the 1,2-ene-dithiolate group in its potential-tuning role [152]. However, the function of this piece will likely be revealed much more slowly due to the difficulties of synthesizing and handling this reactive system.

The evolving model systems presented in this chapter are ample evidence that close synthetic analogues of the Mo and W catalytic sites, including the pyranopterin portion, will soon be in hand. With such complete models, chemists will be able to focus more closely on finding answers to mechanistic questions and the significance of the puzzling pyranopterin unit.

ACKNOWLEDGMENTS

We acknowledge many years of stimulating collaborative research interactions with Professors W. Pfleiderer and Ed Stiefel, and we thank them for helpful discussions during the preparation of this chapter. We are grateful to W. Belliston for critically reading the manuscript.

ABBREVIATIONS

acac	2,4-pentanedionate (acetylacetonate)
bdt	benzene-1,2-dithiolate
Bn	benzyl
Bu	n-butyl
cam	carboxyamidomethyl
CEPT	coupled electron-proton transfer
Cp	η^5-cyclopentadienyl
Cp*	η^5-pentamethylcyclopentadienyl
detc	diethyldithiocarbamate
DMF	dimethylformamide
DMSO	dimethyl sulfoxide
DMSOR	dimethyl sulfoxide reductase
EPR	electron paramagnetic resonance
EXAFS	extended X-ray absorption fine structure
Fmoc	fluoren-9-ylmethoxycarbonyl

L–N$_3$	tris-pyrazolylborate
MCD	magnetic circular dichroism
Me	methyl residue
Me$_4$phen	3,4,7,8-tetramethyl-1,10-phenanthroline
MLCT	metal to ligand change transfer
mnt	1,2-dicyanoethylene-1,2-dithiolato (maleonitrile)
Moco	molybdenum cofactor
MPT	*m*etal-binding *p*yranopterin ene-di*t*hiolate
NMR	nuclear magnetic resonance
OAT	oxygen atom transfer
PES	photoelectron spectroscopy
Ph	phenyl residue
phen	1,10-phenanthroline
Pr^i	*iso*propyl residue
Py	pyridine
qdt	quinoxaline-2,3-dithiolate
SDS	sodium dodecyl sulfate
SO	sulfite oxidase
t-Bu	*tert*-butyl
tdt	toluene-3,4-dithiolate
TFA	trifluoroacetic acid
THF	tetrahydrofuran
TMAOR	trimethylamine *N*-oxide reductase
XO	xanthine oxidase
XPS	X-ray photoelectron spectroscopy

REFERENCES

1. M. P. Coughlan (ed.), *Molybdenum and Molybdenum-Containing Enzymes*, Pergamon Press, Oxford, 1980.

2. W. E. Newton and S. Otsuka (eds.), *Molybdenum Chemistry of Biological Significance*, Plenum Press, New York, 1980.

3. T. G. Spiro (ed.), *Molybdenum Enzymes*, Wiley, New York, 1985.

4. E. I. Stiefel, D. Coucouvanis, and W. E. Newton (eds.), *Molybdenum Enzymes, Cofactors, and Model Systems.* ACS Symp. Ser., Vol. 535, ACS, Washington, DC, 1983, pp. 1–142.

5. R. C. Bray, in *The Enzymes*, 3rd ed., Vol. 12, Part B (P. D. Boyer, ed.), Academic Press, New York, 1975, p. 299.

6. E. I. Stiefel, *Prog. Inorg. Chem.*, *22*, 1–223 (1977).

7. R. C. Bray, *Adv. Enzymol. Relat. Areas Mol. Biol.*, *51*, 107 (1980).

8. J. T. Spence, *Coord. Chem. Rev.*, *48*, 59 (1983).

9. R. Colton, *Coord. Chem. Rev.*, *62*, 145 (1985).

10. S. J. N. Burgmayer and E. I. Stiefel, *J. Chem. Educ.*, *62*, 943–953 (1985).

11. C. D. Garner and J. M. Charnock, *Compre. Coord. Chem.*, *3*, 1329 (1987).

12. E. I. Stiefel, *Compre. Coord. Chem.*, *3*, 1375 (1987).

13. C. D. Garner, *Compre. Coord. Chem.*, *3*, 1421 (1987).

14. R. C. Bray, *Q. Rev. Biophys.*, *21*, 299 (1988).

15. R. H. Holm, *Coord. Chem. Rev.*, *100*, 183–221 (1990).

16. K. V. Rajagopalan, *Adv. Enzymol. Relat. Areas Mol. Biol.*, *64*, 215–290 (1991).

17. J. C. Wootton, R. E. Nicolson, J. M. Cock, D. E. Walters, J. F. Burke, W. A. Doyle, and R. C. Bray, *Biochim. Biophys. Acta*, *1057*, 157–185 (1991).

18. K. V. Rajagopalan and J. L. Johnson, *J. Biol. Chem.*, *267*, 10199–10202 (1992).

19. S. Goswami, *Heterocycles.*, *35*, 1551–1570 (1993).

20. J. H. Enemark and C. G. Young, *Adv. Inorg. Chem.*, *40*, 1–88 (1994).

21. M. K. Johnson, D. C. Rees, and M. W. W. Adams, *Chem. Rev.*, *96*, 2817–2839 (1996).

22. A. Kletzin and M. W. W. Adams, *FEMS Microbiol. Rev.*, **18**, 5–63 (1996).

23. R. Hille, *J. Biol. Inorg. Chem.*, *1*, 397–404 (1996).

24. D. Collison, C. D. Garner, and J. A. Joule, *Chem. Soc. Rev.*, *25*, 25–32 (1996).

25. R. Hille, *Chem Rev.*, *96*, 2757–2816 (1996).

26. J. L. Johnson, *Handbook of Nutritionally Essential Elements*, Marcel Dekker, New York, 1997.

27. C. G. Young and A. G. Wedd, *J. Chem. Soc., Chem. Commun.*, 1251–1257 (1997).

28. C. G. Young, *J. Biol. Inorg. Chem.*, 2, 810–816 (1997).

29. J. H. Enemark and C. D. Garner, *J. Biol. Inorg. Chem.*, 2, 817–822 (1997).

30. C. Kisker, H. Schindelin, and D. C. Rees, *Annu. Rev. Biochem.*, 66, 233–267 (1997).

31. Y. Moriwaki, T. Yamamoto, and K. Higashino, *Histol. Histopathol.*, 12, 513–524 (1997).

32. E. I. Stiefel, *J. Chem. Soc. Dalton Trans.* 3915–3923 (1997).

33. R. R. Mendel, *Planta*, 203, 399–405 (1997).

34. R. R. Mendel and G. Schwarz, *Crit. Rev. Plant Sci.*, 18, 33–69 (1999).

35. C. Kisker, H. Schindelin, D. Baas, J. Rétey, R. U. Meckenstock, and P. M. H. Kroneck, *FEMS Microbiol. Rev.*, 22, 503–521 (1999).

36. R. S. Pilato and E. I. Stiefel, in *Bioinorganic Catalysis* (J. Reedijk and E. Bouwman, eds.), Marcel Dekker, New York, 1999, pp. 81–152.

37. R. Hille, J. Rétey, U. Bartlewski-Hof, W. Reichenbecher, and B. Schink, *FEMS Microbiol. Rev.*, 22, 489–501 (1999).

38. S. J. N. Burgmayer, *Struct. Bonding* (Berlin) 92, 67–119 (1998).

39. W. Kaim, B. Schwederski, O. Heilmann, and F. M. Hornung, *Coord. Chem. Rev..*, 182, 323–342 (1999).

40. (a) M. K. Chan, S. Mukund, A. Kletzin, M. W. W. Adams, and D. C. Rees, *Science*, 267, 1463–1469 (1995); (b) M. J. Romão, M. Archer, I. Moura, J. J. G. Moura, J. LeGall, R. Engh, M. Schneider, P. Hof, and R. Huber, *Science*, 270, 1170–1176 (1995); (c) H. Schindelin, C. Kisker, J. Hilton, K. V. Rajagopalan, and D. C. Rees, *Science*, 272, 1615–1621 (1996); (d) R. Huber, P. Hof, R. O. Duarte, J. J. G. Moura, I. Moura, M. Y. Liu, J. LeGall, R. Hille, M. Archer, and M. J. Romão, *Proc. Natl. Acad. Sci. USA*, 93, 8846–8851 (1996); (e) F. Schneider, R. Huber, H. Schindelin, C. Kisker, and J. Knaeblein, *J. Mol. Biol.*, 263, 53–69 (1996); (f) J. C. Boyington,

V. N. Gladyshev, S. V. Khangulov, T. C. Stadtman, and P. D. Sun, *Science*, *275*, 1305–1308 (1997); (g) C. Kisker, H. Schindelin, A. Pacheco, W. A. Wehbi, R. M. Garrent, K. V. Rajagopalan, J. H. Enemark, and D. C. Rees, *Cell*, *92*, 973–983 (1997); (h) A. S. McAlpine, A. G. McEwan, A. L. Shaw, and S. Bailey, *J. Inorg. Biol. Chem.*, *2*, 690 (1997); (i) J. M. Dias, M. E. Than, A. Humm, R. Huber, G. P. Bourenkov, H. D. Bartunik, S. Bursakov, J. Calvete, J. Caldeira, C. Carneiro, J. J. G. Moura, I. Moura, and M. J. Romão, *Structure*, *7*, 65–79 (1999); (j) H. Dobbek, L. Gremer, O. Meyer, and R. Huber, *Proc. Natl. Acad. Sci. USA*, *96*, 8884–8889 (1999); (k) C. Enroth, B. T. Eger, K. Okamoto, T. Nishino, T. Nishino, and E. F. Pai, *Proc. Natl. Acad. Sci. USA*, *97*, 10723–10728 (2000).

41. (a) S. Matsura, H. Traub, and L. F. Armarego, *Pteridines 1*, 73–82 (1989); (b) R. Soyka, W. Pfleiderer, and R. Prewo, *Helv. Chim. Acta 73*, 808–826 (1990); (c) R Soyka and W. Pfleiderer, *Pteridines 2*, 63–74 (1990); (d) J. N. Low, E. Cadoret, G. Ferguson, M. D. Lopez, M. L. Quijano, A. Sanchez, and M. Nogueras, *Acta Crystallogr.*, *C51*, 2141–2143 (1995).

42. D. Guschin, W. Belliston, I. M. Müller, and B. Fischer (submitted).

43. B. Fischer, M. vom Orde, K. Leidenberger, A. Pacheco, and L. Bigler, in *Chem. Biol. Pteridines, Proc. 11th Int. Symp. Pteridines and Folates* (W. Pfleiderer and H. Rokos, eds.), Blackwell Scientific, Berlin, 1997, pp. 23–28.

44. K. V. Rajagopalan, *Biochem. Soc. Trans.*, *25*, 757–761 (1997).

45. J. Tachil and O. Meyer, *FEMS Microbiol. Lett.*, *148*, 203–208 (1997).

46. M. M. Wuebbens and K. V. Rajagopalan, *J. Biol. Chem.*, *268*, 13493–13498 (1993).

47. B. Fischer, unpublished results.

48. D. T. Hurst, in *Chemistry and Biochemistry of Pyrimidines, Purines, Pteridines*, Wiley, New York, 1980.

49. A. Albert and H. Mizuno, *J. Chem. Soc.*, *B*, 2423–2427 (1971).

50. A. Albert and H. Mizuno, *J. Chem. Soc., Perkins Trans.*, 1615–1619 (1973).

51. W. Pfleiderer, in *Second Supplements to the 2nd Edition of Rodd's Chemistry of Carbon Compounds*, Vol. 4, K(partial)/L, (M. Sainsbury, ed.), Elsevier, Amsterdam, 1999, pp. 269–330.

52. W. Pfleiderer, in *Comprehensive Heterocyclic Chemistry*, Vol. 7 (A. R. Katritzky, C. W. Reese, and E. F. Scriven II, eds.), Pergamon Press, Oxford, 1996, p. 679.

53. B. Fischer, J. H. Enemark, and P. Basu, *J. Inorg. Biochem.*, *72*, 13–21 (1998).

54. M. J. Romão, *Rev. Port. Quim.*, *3*, 11–22 (1996).

55. M. J. Romão, J. Knäblein, R. Huber, and J. J. G. Moura, *Prog. Biophys. Mol Biol.*, *68*, 121–144 (1997).

56. B. Krüger and O. Meyer, *Eur. J. Biochem.*, *157*, 121–128 (1986).

57. J. L. Johnson, K. V. Rajagopalan, and O. Meyer, *Arch. Biochem. Biophys.*, *283*, 542–545 (1990).

58. J. L. Johnson, N. R. Bastian, and K. V. Rajagopalan, *Proc. Natl. Acad. Sci. USA*, *87*, 3190–3194 (1990).

59. M. Karrasch, G. Boerner, and R. K. Thauer, *FEBS Lett.*, *274*, 48–52 (1990).

60. J. L. Johnson, B. E. Hainline, and K. V. Rajagopalan, *J. Biol. Chem.*, *255*, 1783 (1980).

61. J. L. Johnson and K. V. Rajagopalan, *Proc. Natl. Acad. Sci. USA*, *79*, 6856–6860 (1982).

62. J. Deistung and R. C. Bray, *Biochem. J.*, *263*, 477–483 (1989).

63. A. G. Wedd, *Coord. Chem. Rev.*, *154*, 5–11 (1996).

64. M. T. Pope, *Prog. Inorg. Chem.*, *39*, 181–257 (1977).

65. C. D. Garner and S. Bristow, in *Molybdenum Enzymes* (T. G. Spiro, ed.), Wiley, New York, 1985, pp. 343–409.

66. N. R. Bastian, C. J. Kay, M. J. Barber, and K. V. Rajagopalan, *J. Biol. Chem. 266*, 45–51 (1991).

67. A. G. McEwan, S. J. Ferguson, and J. B. Jackson, *Biochem. J.*, *274*, 305 (1991).

68. T. G. Spiro (ed.), *Biological Applications of Raman Spectroscopy*, Vol. 3, Wiley, New York, 1988.

69. R. C. Bray, B. G. Malmström, and T. Vänngård, *Biochem J.*, *73*, 193 (1959).

70. F. Farchione, G. R. Hanson, C. G. Rodrigues, T. D. Bailey, R. N. Bagchi, A. M. Bond, J. R. Pilbrow, and A. G. Wedd, *J. Am. Chem. Soc.*, *108*, 831–832 (1986).

71. G. L. Wilson, M. Kony, E. R. T. Tiekink, J. R. Pilbrow, J. T. Spence, and A. G. Wedd, *J. Am. Chem. Soc.*, *110*, 6923–6925 (1988).

72. G. L. Wilson, R. J. Greenwood, J. R. Pilbrow, J. T. Spence, and A. G. Wedd, *J. Am. Chem. Soc.*, *113*, 6803–6812 (1991).

73. D. Dowerah, J. T. Spence, R. Singh, A. G. Wedd, G. L. Wilson, F. Farchione, J. H. Enemark, J. Kristofzski, and M. Bruck, *J. Am. Chem. Soc.*, *109*, 5655–5665 (1987).

74. R. C. Bray, B. D. Howes, G. L. Bennett, and J. David, in *New Trends in Biological Chemistry* (T. Ozawa and K. Yagi, eds.), Japan Sci. Soc. Press, Tokyo, 1991, pp. 47–58.

75. M. Mader, M. D. Carducci, and J. H. Enemark, *Inorg Chem.*, *39*, 525–531 (2000).

76. J. McMaster, M. D. Carducci, Y.-S. Yang, E. I. Solomon, and J. H. Enemark, *Inorg Chem.*, *40*, 687–702 (2001).

77. F. E. Inscore, R. McNaughton, B. L. Westcott, M. E. Helton, R. Jones, I. K. Dhawan, J. H. Enemark, and M. L. Kirk, *Inorg Chem.*, *38*, 1401–1410 (1999).

78. R. L. McNaughton, M. E. Helton, N. D. Rubie, and M. L. Kirk, *Inorg. Chem.*, *39*, 4386–4387 (2000).

79. M. E. Helton, N. E. Gruhn, R. L. McNaughton, and M. L. Kirk, *Inorg. Chem.*, *39*, 2273–2278 (2000).

80. J. M. Berg and R. H. Holm, *J. Am. Chem. Soc.*, *106*, 3035–3036 (1984).

81. R. H. Holm and J. M. Berg, *Acc. Chem. Res.*, *19*, 363–370 (1986).

82. S. A. Roberts, C. G. Young, C. E. Kipke, W. E. Cleland, Jr., K. Yamanouchi, M. D. Carducci, and J. H. Enemark, *Inorg. Chem.*, *29*, 3650–3656 (1990).

83. Z. Xiao, C. G. Young, J. H. Enemark, and A. G. Wedd, *J. Am. Chem. Soc.*, *114*, 9194–9195 (1992).

84. E. I. Stiefel, *Proc. Natl. Acad. Sci. USA*, *70*, 988–992 (1973).

85. D. Dowerah, J. T. Spence, R. Singh, A. G. Wedd, G. L. Wilson, F. Farchione, J. H. Enemark, J. Kristofzski, and M. Bruck, *J. Am. Chem. Soc.*, *109*, 5655–5665 (1987).

86. Z. Xiao, R. W. Gable, A. G. Wedd, and C. G. Young, *J. Am. Chem. Soc.*, *118*, 2912–2921 (1996).

87. J. A. McCleverty, *Proc. Inorg. Chem.*, *10*, 49–221 (1968).

88. R. Eisenberg, *Prog. Inorg. Chem.*, *12*, 295–369 (1970).

89. E. I. Stiefel, L. E. Bennett, Z. Dori, T. H. Crawford, C. Simo, and H. B. Gray, *Inorg. Chem.*, *9*, 281–286 (1970).

90. P. I. Clemenson, *Coord. Chem. Rev.*, *106*, 171–203 (1990).

91. L. Larsen, D. J. Rowe, C. D. Garner, and J. A. Joule, *Tetrahedron Lett.*, *28*, 1453–1456 (1988).

92. L. Larsen, D. J. Rowe, C. D. Garner, and J. A. Joule, *J. Chem. Soc., Perkin Trans. 1*, 2317–2327 (1989).

93. J. R. Russell, C. D. Garner, and J. A. Roule, *Tetrahedron Lett.*, *33*, 3371–3374 (1992).

94. J. R. Russell, C. D. Garner, and J. A. Joule, *J. Chem. Soc., Perkin Trans. 1*, 1245–1249 (1992).

95. C. D. Garner, E. M. Armstrong, M. J. Ashcroft, M. S. Austerberry, J. H. Birks, D. Collison, A. J. Goodwin, J. A. Joule, L. Larsen, D. J. Rowe, and J. R. Russell, in *Molybdenum Enzymes, Cofactors, and Model Systems.* (E. I. Stiefel, D. Coucouvanis, and W. E. Newton, eds.), ACS Symp. Ser., Vol. 535, Washington, DC, 1993, pp. 98–113.

96. R. S. Pilato, K. A. Eriksen, M. A. Greaney, E. I. Stiefel, S. Goswami, L. Kilpatrick, T. G. Spiro, E. C. Taylor, and A. L. Rheingold, *J. Am. Chem. Soc.*, *113*, 9372–9374 (1991).

97. C. L. Soricelli, V. A. Szalai, and S. J. N. Burgmayer, *J. Am. Chem. Soc.*, *113*, 9877–9878 (1991).

98. A. Dinsmore, J. H. Birks, C. D. Garner, and J. A. Joule, *J. Chem. Soc., Perkin Trans. 1*, 801–807 (1997).

99. E. S. Davies, R. L. Beddoes, D. Collison, A. Dinsmore, A. Docrat, J. A. Joule, C. R. Wilson, and C. D. Garner, *J. Chem. Soc., Dalton Trans.*, 3985–3995 (1997).

100. E. S. Davies, G. M. Aston, R. L. Beddoes, D. Collison, A. Dinsmore, A. Docrat, J. A. Joule, C. R. Wilson, and C. D. Garner, *J. Chem. Soc., Dalton Trans.*, 3647–3656 (1998).

101. B. Bradshaw, A. Dinsmore, C. D. Garner, and J. A. Joule, *Chem. Commun.*, 417–418 (1998).

102. J. A. McCleverty, J. Locke, B. Ratcliff, and E. J. Wharton, *Inorg. Chim. Acta, 3,* 283–286 (1969).

103. N. Ueyama, N. Yoshinaga, T. Okamura, H. Zaima, and A. Nakamura, *J. Mol. Catal., 64,* 247–256 (1991).

104. N. Ueyama, H. Oku, and A. Nakamura, *J. Am. Chem. Soc., 114,* 7310–7311 (1992).

105. M. Kondo, N. Ueyama, K. Fukuyama, and A. Nakamura, *Bull. Chem. Soc. Jpn., 66,* 1391–1396 (1993).

106. H. Oku, N. Ueyama, A. Nakamura, Y. Kai, and N. Kanehisa, *Chem. Lett.,* 607–610 (1994).

107. H. Oku, N. Ueyama, M. Kondo, and A. Nakamura, *Inorg. Chem., 33,* 209–216 (1994).

108. H. Oku, N. Ueyama, and A. Nakamura, *Chem. Lett.,* 621–622 (1995).

109. H. Oku, N. Ueyama, and A. Nakamura, *Chem. Lett.,* 1131–1132 (1996).

110. N. Ueyama, H. Oku, M. Kondo, T. A. Okamura, N. Yoshinaga, and A. Nakamura, *Inorg. Chem., 35,* 643–650 (1996).

111. H. Oku, N. Ueyama, and A. Nakamura, *Bull. Chem. Soc. Jpn., 69,* 3139–3150 (1996).

112. H. Oku, N. Ueyama, and A. Nakamura, *Chem. Lett.,* 31–32 (1996).

113. H. Oku, N. Ueyama, and A. Nakamura, *Inorg. Chem., 36,* 1504–1516 (1997).

114. H. Oku, B. P. Koehler, N. Ueyama, A. Nakamura, and M. K. Johnson, *J. Inorg. Biochem., 67,* 14 (1997).

115. J. Yadav, S. K. Das, and S. Sarkar, *J. Am. Chem. Soc., 119,* 4315–4316 (1997).

116. P. K. Chaudhury, S. K. Das, and S. Sarkar, *J. Biochem., 319,* 953–959 (1996).

117. S. K. Das, D. Biswas, R. Maiti, and S. Sarkar, *J. Am. Chem. Soc., 118,* 1387–1397 (1996).

118. S. K. Das, P. K. Chaudhury, D. Biswas, and S. Sarkar, *J. Am. Chem. Soc., 116,* 9061–9070 (1994).

119. S. Sarkar and S. K. Das, *Proc. Ind. Acad. Sci., Chem. Sci., 104,* 533–534 (1992).

120. S. Sarkar and S. K. Das, *Proc. Ind. Acad. Sci.*, *Chem. Sci.*, *104*, 437–441 (1992).

121. A. Dinsmore, C. D. Garner, and J. A. Joule, *Tetrahedron 54*, 9559–9568 (1998).

122. A. Dinsmore, C. D. Garner, and J. A. Joule, *Tetrahedron 54*, 3291–3302 (1998).

123. G. C. Tucci, J. P. Donahue, and R. H. Holm, *Inorg. Chem.*, *37*, 1602–1608 (1998).

124. C. Lorber, J. P. Donahue, C. A. Goddard, E. Nordlander, and R. H. Holm, *J. Am. Chem. Soc.*, *120*, 8102–8112 (1998).

125. J. P. Donahue, C. R. Goldsmith, U. Nadiminti, and R. H. Holm, *J. Am. Chem. Soc.*, *120*, 12869–12881 (1998).

126. A. Thapper, J. P. Donahue, K. B. Musgrave, M. W. Willer, E. Nordlander, B. Hedman, K. O. Hodgson, and R. H. Holm, *Inorg. Chem.*, *38*, 4104–4114 (1999).

127. K. B. Musgrave, J. P. Donahue, C. Lorber, R. H. Holm, B. Hedman, and K. O. Hodgson, *J. Am. Chem. Soc.*, *121*, 10297–10307 (1999).

128. B. S. Lim, J. P. Donahue, and R. H. Holm, *Inorg. Chem.*, *39*, 263–273 (2000).

129. K.-M. Sung and R. H. Holm, *Inorg. Chem.*, *39*, 1275–1281 (2000).

130. K. B. Musgrave, B. S. Lim, K.-M. Sung, R. H. Holm, B. Hedman, and K. O. Hodgson, *Inorg. Chem.*, *39*, 5238–5247 (2000).

131. B. S. Lim, K.-M. Sung, and R. H. Holm, *J. Am. Chem. Soc.*, *122*, 7410–7411 (2000).

132. M. Miao, M. W. Willer, and R. H. Holm, *Inorg. Chem.*, *39*, 2843–2849 (2000).

133. B. S. Lim and R. H. Holm, *J. Am. Chem. Soc.*, *123*, 1920–1930 (2001).

134. K.-M. Sung and R. H. Holm, *J. Am. Chem. Soc.*, *123*, 1931–1943 (2001).

135. B. S. Lim and R. H. Holm, *Inorg. Chem.*, *40*, 645–654 (2001).

136. R. S. Pilato, Y. Gea, K. A. Eriksen, M. A. Greaney, E. I. Stiefel, S. Goswami, L. Kilpatrick, T. G. Spiro, E. C. Taylor, and A. L. Rheingold, in *Molybdenum Enzymes, Cofactors, and Model*

Systems. (E. I. Stiefel, D. Coucouvanis, and W. E. Newton, eds.), ACS Symp. Ser., Vol. 535, Washington, DC., 1993, pp. 83–97.

137. K. M. Nicholas and M. A. Khan, *Inorg. Chem.*, *26*, 1633–1636 (1987).

138. U. Küsthardt and J. H. Enemark, *J. Am. Chem. Soc.*, *109*, 7926 (1987).

139. U. Küsthardt, M. J. LaBarre, and J. H. Enemark, *Inorg. Chem.*, *29*, 3182 (1990).

140. A. A. Eagle, S. M. Harben, E. R. T. Tiekink, and C. G. Young, *J. Am. Chem. Soc.*, *116*, 9749–9750 (1994).

141. A. A. Eagle, G. N. George, E. R. T. Tiekink, and C. G. Young, *J. Inorg. Biochem.*, *76*, 39–45 (1999).

142. J. K. Hsu, C. J. Bonangelino, S. P. Kaiwar, C. M. Boggs, J. C. Fettinger, and R. S. Pilato, *Inorg. Chem.*, *35*, 4743–4751 (1996).

143. K. A. van Houten, C. Boggs, and R. S. Pilato, *Tetrahedron*, *54*, 10973–10986 (1998).

144. S. P. Kaiwar, J. K. Hsu, L. M. Liable-Sands, A. L. Rheingold, and R. S. Pilato, *Inorg. Chem.*, *36*, 4234–4240 (1997).

145. D. J. Rowe, C. D. Garner, and J. A. Joule, *J. Chem. Soc. Perkin Trans. 1*, 1907–1910 (1985).

146. G. N. Schrauzer, *Acc. Chem. Res.*, *2*, 72–80 (1969).

147. G. N. Schrauzer and V. P. Mayweg, *J. Am. Chem. Soc.*, *87*, 1483–1489 (1965).

148. G. N. Schrauzer, V. P. Mayweg, and W. Heinrich, *Inorg. Chem.*, *4*, 1615 (1965).

149. H. Kawaguchi and K. Tatsumi, *J. Am. Chem. Soc.*, *117*, 3885–3886 (1995).

150. H. Kawaguchi, K. Yamada, J.-P. Lang, and K. Tatsumi, *J. Am. Chem. Soc.*, *119*, 10346–10358 (1997).

151. L. Kilpatrick, K. V. Rajagopalan, J. Hilton, N. R. Bastian, E. I. Stiefel, R. S. Pilato, and T. G. Spiro, *Biochemistry*, *34*, 3032–3039 (1995).

152. B. L. Westcott, N. E. Gruhn, and J. H. Enemark, *J. Am. Chem. Soc.*, *120*, 3382–3386 (1998).

153. B. Fischer, J. Strähle, and M. Viscontini, *Helv. Chim. Acta*, *74*, 1544–1554 (1991).

154. S. J. N. Burgmayer, A. Baruch, K. Kerr, and K. Yoon, *J. Am. Chem. Soc.*, *111*, 4982–4984 (1989).

155. T. Jüstel, J. Bendix, N. Metzler-Nolte, T. Weyhermüller, B. Nuber, and K. Wieghardt, *Inorg. Chem.*, *37*, 35 (1998).

156. C. Pierpont, *Prog. Inorg. Chem.*, *41*, 331 (1998).

157. K. R. Barnard, A. G. Wedd, and E. Tiekink, *Inorg. Chem.*, *29*, 29 (1990).

158. (a) J. Perkinson, S. Brodie, K. Yoon, K. Mosny, P. J. Carroll, and S. J. N. Burgmayer, *Inorg. Chem.*, *30*, 719–727 (1991); (b) S. J. N. Burgmayer, K. M. Everett, M. R. Arkin, and K. Mosny, *J. Inorg. Biochem.*, *43*, 581 (1991); (c) Y. Funhashi, T. Kohzuma, A. Odani, and O. Yamauchi, *Chem. Lett.*, 385–388 (1994); (d) O. Yamauchi, *Pure Appl. Chem.*, *67*, 297 (1995).

159. S. J. N. Burgmayer and E. I. Stiefel, *Inorg. Chem.*, *27*, 4059–4065 (1988).

160. (a) A. Schäfer, B. Fischer, H. Paul, R. Bosshard, M. Hesse, and M. Viscontini, *Helv. Chim. Acta*, *75*, 1955–1964 (1992); (b) A. Schäfer, H. Paul, B. Fischer, M. Hesse, and M. Viscontini, *Helv. Chim. Acta*, *78*, 1763–1776 (1995).

161. S. J. N. Burgmayer, M. R. Arkin, L. Bostick, S. Dempster, K. M. Everett, H. L. Layton, K. E. Paul, C. Rogge, and A. L. Rheingold, *J. Am. Chem. Soc.*, *117*, 5812–5823 (1995).

162. B. Fischer, H. Schmalle, E. Dubler, A. Schäfer, and M. Viscontini, *Inorg. Chem.*, *34*, 5726–5734 (1995).

163. H. L. Kaufmann, L. L. Liable-Sands, A. L. Rheingold, and S. J. N. Burgmayer, *Inorg. Chem.*, *38*, 2592–2599 (1999).

164. B. Fischer, H. Schmalle, M. Baumgartner, and M. Viscontini, *Helv. Chim. Acta*, *80*, 103–110 (1997).

165. H. L. Kaufmann, P. J. Carroll, and S. J. N. Burgmayer, *Inorg. Chem.*, *38*, 2600–2606 (1999).

166. S. J. Burgmayer, H. L. Kaufmann, G. Fortunato, P. Hug, and B. Fischer, *Inorg. Chem.*, *38*, 2607–2613 (1999).

167. (a) P. Hemmerich, F. Müller, and A. Ehrenberg, in *Oxidases and Related Redox Systems*, Vol. 1 (T. E. King, H. S. Mason, and M. Morrison, eds.), Wiley, New York, 1965, p 157; (b) P. Hemmerich and J. Lauterwein, in *Bioinorganic Chemistry* (G. L. Eichhorn, ed.), Elsevier, Amsterdam, 1973, pp. 1168–1190.

168. S. J. N. Burgmayer and E. I. Stiefel, *J. Am. Chem. Soc.*, *108*, 8310–8311 (1986).

169. F. M. Hornung, O. Heilmann, W. Kaim, S. Zalis, and J. Fiedler, *Inorg. Chem.*, *39*, 4052 (2000).

170. D.-H. Lee, N. N. Murthy, Y. Lin, N. S. Nasir, and K. D. Karlin, *Inorg. Chem.*, *36*, 6328 (1997).

171. J. L. Johnson, B. E. Hainline, K. V. Rajagopalan, and B. H. Arison, *J. Biol. Chem.*, *259*, 5414–5422 (1984).

172. W. Koschara, *Hoppe-Seyler's Z. Physiol. Chem.*, *263*, 78–79 (1940).

173. A. Sakurai and M. Goto, *J. Biochem.*, *65*, 755–757 (1969).

174. E. C. Taylor and L. A. Reiter, *J. Am. Chem. Soc.*, *111*, 285–291 (1989).

175. E. C. Taylor, P. S. Ray, and I. S. Darwish, *J. Am. Chem. Soc*, *111*, 7664–7665 (1989).

176. E. C. Taylor and P. S. Ray, *J. Org. Chem.*, *52*, 3997–4000 (1987).

177. S. P. Kramer, J. L. Johnson, A. A. Riberio, D. S. Millington, and K. V. Rajagopalan, *J. Biol. Chem.*, *262*, 16357–16363 (1987).

178. S. J. N. Burgmayer and A. Somogyi, manuscript in preparation.

179. B. Bradshaw, D. Collison, C. D. Garner, and J. A. Joule, *Chem. Commun.*, 123–124 (2001).

180. J. L. Johnson and S. K. Wadman, in *Metabolic Basis of Inherited Disease*, McGraw-Hill, New York, 1989, pp. 1463–1475.

181. W. Pfleiderer, *J. Inherit. Metab. Dis. 1*, 54–60 (1978).

182. J. L. Johnson, *Handbook of Nutritionally Essential Elements*, Marcel Dekker, New York, 1997.

183. C. Rieder, W. Eisenreich, J. O'Brien, G. Richter, E. Götze, P. Boyle, S. Blachard, A. Bacher, and H. Simon, *Eur. J. Biochem.*, *255*, 24–36 (1988).

184. D. Guschin and B. Fischer (in preparation).

185. D. M. A. M. Luykx, J. A. Duine, and S. de Vries, *Biochemistry*, *37*, 11366–11375 (1998).

9
Biosynthesis and Molecular Biology of the Molybdenum Cofactor (Moco)

Ralf R. Mendel and Günter Schwarz

Botanical Institute, Technical University of Braunschweig,
D-38023 Braunschweig, Germany

1. INTRODUCTION

1.1. The Beginning of Moco Genetics

In algae, fungi, and higher plants, the molybdenum enzyme nitrate reductase (NR) catalyzes the key step in organic nitrogen assimilation. With the genetic analysis of NR-deficient mutants of the filamentous fungus *Aspergillus nidulans* by Pateman et al. [1], also the genetic analysis of molybdenum metabolism had begun. Cove and Pateman [2] had isolated NR-deficient mutants that revealed a novel phenotype, namely, the simultaneous loss of the two Mo-dependent enzymes NR and xanthine dehydrogenase (XDH). Since Mo was the only common link between those two—otherwise very different—enzymes, the authors suggested that both enzymes should share a common Mo-related cofactor. Pateman et al. [1] described five genetic loci each of which exhibited that novel phenotype and designated them with the mnemonic *cnx* (cofactor for *n*itrate reductase and *x*anthine dehydrogenase). In the following years, *cnx*-like mutants were described for *Neurospora crassa* [3,4], *E. coli* [5], *Drosophila melanogaster* [6], and several higher plants [7].

1.2. The *"nit*-1 Assay" and the Discovery of Moco

In *Neurospora crassa*, Sorger and Giles [3] isolated the *nit*-1 mutant which was similar to the *Aspergillus cnx* mutants showing a pleiotropic loss of all Mo enzyme activities. Nason and coworkers [8–10] provided first biochemical evidence for the existence of a cofactor common to Mo enzymes. In crude extracts of *nit*-1 mutant cells, the inactive apoprotein of NR could be fully reconstituted by the addition of a low molecular weight fraction derived from denatured preparations of purified Mo enzymes of mammalian, plant, or bacterial origin. This fraction obviously included a Mo-containing cofactor that could be integrated into inactive apoNR of the fungal *nit*-1 mutant lacking the cofactor. Thus, it was suggested that there is a ubiquitous and universal Moco of identical structure that is able to associate with different apoenzymes to form the Mo holoenzymes. Nitrogenase, as another Mo-containing enzyme, did not release a *nit*-1 positive cofactor after denaturing of the enzyme [11]. Based on these data and functional analyses [12], it was shown that the dissociable cofactor of nitrogenase named Fe-Mo cofactor (see Chapter 5) is unique for this enzyme.

The observed complementation of *nit*-1 apoNR by Moco from different Mo enzyme sources served as a basis to develop a very sensitive biological in vitro assay referred to as the *"nit*-1 assay" for determination of the cofactor, which serves as a widely used tool for Moco analysis. Nason et al. [10,13] found that the reconstituting activity released from diverse Mo enzymes was very labile, with a lifetime of several minutes. This was a first indication for the labile nature of the cofactor in its isolated form. In the following time, a number of attempts were undertaken to stabilize the cofactor and to optimize the transfer of Moco to *nit*-1 apoNR. Different methods of cofactor release from Mo enzymes (acidification [10]; heat treatment [14]; detergent and organic solvent treatment [15,16]) were utilized to minimize the loss of cofactor due to its high sensitivity to air oxidation. Although an absolute quantification of Moco by the *nit*-1 assay was hard to achieve, it has been very useful for qualitative and semiquantitative detection of cofactor activity. Most of these in vitro complementations are dependent on an excess of molybdate (1–10 mM) in the reaction mixture [17–20], indicating a strong tendency of isolated Moco to lose Mo. Furthermore, it has been shown by using the *nit*-1 assay that different sulfhydryl protecting agents stabilize the isolated cofactor [14,18], while sulfhydryl-reactive inhibitors like *p*-hydroxymercuribenzoate totally abolished the reconstituting activity of the cofactor [19,21], thus indicating the involvement of free sulfhydryl groups in the *nit*-1 reconstitution process.

Since all attempts failed to use NR or other Mo enzymes from different organisms to get an assay for Moco detection working, the *nit-1* assay is still one of the easiest and simplest ways to detect Moco in biological systems. Only recently it was shown that human sulfite oxidase recombinantly expressed in *E. coli* can be complemented with exogenous Moco [22], which serves as a basis for developing a sensitive and fully defined in vitro system for quantitative Moco detection.

1.3. Discovery of the Chemical Nature of Moco

The elucidation of the chemical nature of Moco is based on the pioneering work of J. Johnson and K. V. Rajagopalan. Their final description of Moco was completely confirmed by crystal structures of Mo enzymes with the only exception that a third ring, a novel pyrano ring, is formed (Fig. 1). Due to the labile nature of Moco and its high sensitivity to air oxidation, most of the work was done by using degradation or oxidation products of the cofactor. By analysis of two oxidation products (form A and form B) the pterin nature of Moco and its C6 substitution with a unique four-carbon side chain were identified. Due to (1) the physiological link between urothione and form B [15], (2) the sensitivity of Moco to sulfhydryl reagents, and (3) its increased stability in the presence of reducing reagents, the presence of sulfur atoms in the cofactor was assumed. Later on, the coordination of Mo via a dithiolene function in the four-carbon side chain of the cofactor was demonstrated by carbamido methylation of Moco. The redox state of the pterin was assumed to be a dihydropterin, which later turned out to be formally right because

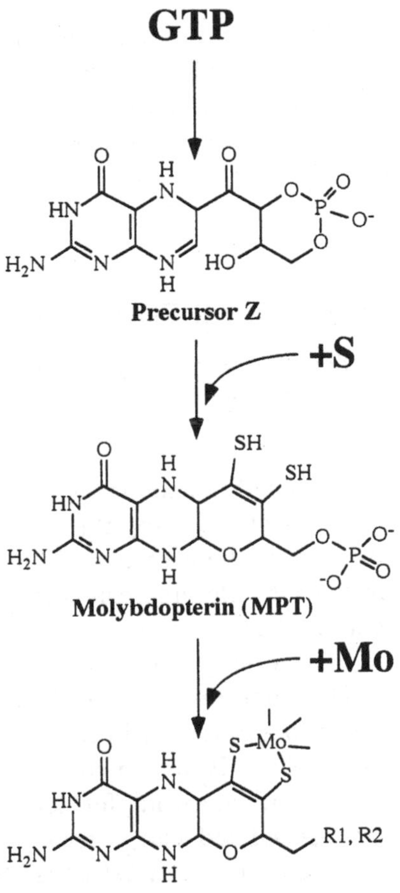

Molybdenum cofactor (Moco)

FIG. 1. General scheme of the biosynthesis of Moco in pro- and eukaryotes. A guanosine derivative (probably GTP) is converted to the sulfur-free precursor Z, already containing the four-carbon side chain and a cyclic phosphate. In the second step, the dithiolene function of MPT is generated under formation of MPT, which is followed by the transfer of Mo, thus forming Moco. Mo can be coordinated by one MPT molecule as it occurs in eukaryotes or by two pterins forming a bis-Mo-MPT cofactor as it can be found in bacteria. At the C4′ atom of the side chain (pyrano ring) a terminal phosphate (eukaryotes) or a nucleotide can be bound forming the dinucleotide form of the cofactor (bacteria).

the third pyrano ring is formed by ring closure of the OH group at C-3′ with C7 of the pterin in its 5,6-dihydro state. Once the pyrano ring is formed, Moco is fully reduced like tetrahydropterins. Because of the unique nature of the pterin in Moco, the metal-free form of the cofactor is called molybdopterin (MPT). The same pterin also coordinates tungsten in all W enzymes studied so far.

2. GENETICS OF THE MOLYBDENUM COFACTOR

2.1. Mutants and Deficiency Phenotypes in Bacteria, Plants, Insects, and Fungi

In parallel to the achievements of fungal biochemical genetics, also in *E. coli* Moco mutants had been isolated using the same selection principle: growth of mutagenized cells in the presence of high concentrations of chlorate. Mutants selected for chlorate resistance (*chl*) do not reduce chlorate to the toxic chlorite because they have lost chlorate reductase activity, which appears to be a nonphysiological catalytic activity of the Mo enzyme NR [5,23]. The chlorate-resistant phenotype either reflects the lack of NR activity due to a mutation in the NR structural genes in *E. coli* or is due to a loss of Moco. It turned out that the loci *chlA*, *chlB*, *chlD*, *chlE*, and *chlG* were all essential for Moco biosynthesis, and Stewart and MacGregor [24] isolated a large series of Mu-phage insertation mutants for all these loci. Finally, in 1992 the Moco-specific *chl* loci were renamed in *mo* loci [25]. Among eukaryotes, the molecular, biochemical, and genetic analysis of Moco mutants is most advanced in the filamentous fungus *Aspergillus nidulans* and in higher plants. Here the plant system will be discussed in greater detail. A mutational block of Moco biosynthesis leads to the loss of essential metabolic functions and can cause the death of the plant. Moco-deficient plant mutants show a pleiotropic loss of the activities of the Mo enzymes NR, XDH, AO, as well as the recently discovered SO [153]. Since these mutant plants can be kept alive (e.g., as in vitro plants) on media containing reduced nitrogen as N source, one can argue that the loss of NR activity is more dramatic for the plant than the lack of the other three Mo enzymes. Similar to fungi and bacteria, Moco-deficient plant mutants have been isolated as NR-deficient mutants by selection for chlorate resistance [26].

Moco mutants have been described in numerous higher plants, e.g., in tobacco [27,28], *Nicotiana plumbaginifolia* (reviewed in [7]) and barley (reviewed in [29]), and they were also found in green algae like in *Chlamydomonas reinhardtii* (reviewed in [30,31]). In the wild tobacco species *N. plumbaginifolia*, mutants in six Moco-specific genetic loci (*cnxA–cnxF*) were described, showing a similar morphology strongly deviating from that of the wild type: stunted growth, chlorosis of leaves, as well as small, narrow, and crinkled leaves [32], which is probably caused by the combined impairment of the plants to synthesize the phytohormones abscisic acid and indoleacetic acid due to the loss of AO activities [33,34]. For *Arabidopsis* five Moco loci (designated *chl*) were characterized [35]. The multitude of Moco-specific loci described in bacteria, fungi, algae, and higher plants led to the conclusion that Moco biosynthesis is a complex and probably ancient pathway involving several gene products [7,36–38].

In pregenomic times, the detailed mutant characterization contributed substantially to our understanding of the genetics and biochemistry of Moco in bacteria, plants, fungi, and insects. These analyses allowed a more precise description of the impairment caused by a given Moco mutation. By measuring intracellular amounts of Mo it could be ruled out that plant Moco mutants are defective in Mo uptake [18]. Subjecting plant Moco mutants to the *nit-1* assay demonstrated that in five of six mutants MPT was lacking. The mutants positive in the *nit-1* assay were also partially repairable by growing them on high-molybdate medium [17,18,32]. A similar result was found for the fungal Moco mutants: among five *A. nidulans cnx* loci, mutants in one locus were molybdate-repairable [39], and among four *Neurospora crassa nit* loci, again mutants in one locus were molybdate-repairable [40]. It was assumed, therefore, that the gene product of the molybdate-repairable locus should be involved in transferring or inserting Mo into the cofactor. Molecular analyses in plants proved this assumption to be correct (see below). Mutants in the loci nonrepairable by molybdate were interpreted to be defective in the biosynthesis of MPT itself. In *Drosophila melanogaster*, only two Moco-deficient mutant types were described [6,41] showing a pleiotropic loss of XO, AO, SO, and pyridoxal oxidase.

2.2. Human Moco Deficiency

In humans, a combined deficiency of SO and XDH was first described by Duran et al. [42]. To this end, more than 80 cases are known worldwide [43]. This disease is autosomal recessive and occurs in all racial groups [44]. Patients affected show neonatal seizures, severe neurological abnormalities, dislocated ocular lenses, feeding difficulties, and dysmorphic features of the brain and head, and they die in early childhood [45]. To this end, no therapy is available to cure the symptoms of this disease. The genetic analysis of cultured fibroblasts from different patients led to the description of two complementation groups [46]. Cocultivation experiments indicated that cells from group B patients excreted a relatively stable and diffusable precursor that could be converted to active Moco by cells from group A patients. Very recently, a third type of human mutant was identified that possesses MPT and thus is defective in the insertion of Mo into MPT [47]. The analysis of mammalian Mo proteins was highly important for deciphering the structure of Moco. Johnson et al. [15] found a structural relationship of Moco to urothione, a compound isolated from human urine [48]. Since human patients lacking active Moco [49] were devoid of urothione, a metabolic link to Moco was proposed.

3. BIOSYNTHESIS OF THE MOLYBDENUM COFACTOR

The greater part of our present knowledge about Moco biosynthesis has been obtained from studies of Moco mutants in *E. coli* where five Moco-specific operons, designed as *moa*, *mob*, *mod*, *moe*, and *mog*, are known comprising more than 15 genes. Comprehensive analyses of these mutants involving molecular, genetic, and biochemical studies by several laboratories led to a picture of Moco biosynthesis consisting of four stages in *E. coli* (summarized by Rajagopalan and coworkers [38,50], who are the major contributors to this model). In brief, during the first stage a guanosine-X-phosphate derivative (probably GTP) is transformed into a sulfur-free pterin compound, the precursor Z, possessing already the Moco-typical four-carbon side chain (Fig. 1). The chemical nature of precursor Z is not completely clear because Wuebbens and Rajagopalan [51] have described the intermediate as a chinoid form of the pterin with an enol function at the C2' atom of the side chain, which does not fit to the spectral properties of isolated precursor Z. It is most likely that precursor Z is already synthesized as pyranopterin (B. Fischer, unpublished results) maintaining the chirality of C6 and increasing its resistance against air oxidation. In the second stage, sulfur is transferred to precursor Z and the precursor is converted to MPT. Precursor Z and MPT are the only known intermediates of the Moco pathway. In the third stage, Mo has to be transferred to MPT in order to form Moco, linking the high-affinity molybdate uptake system to the MPT pathway. In bacteria, there is often a fourth stage in which a ribonucleotide has to be attached to form the dinucleotide form of Moco. In the following sections, we will discuss these steps in greater detail. Bacterial (mostly archaea) tungsten cofactor biosynthesis has not been studied yet in detail, but it is reasonable to assume that it proceeds in a similar way as Moco synthesis because all genes identified for Moco biosynthesis in *E. coli* have conterparts in archaea.

The MPT structure of Moco is conserved in all organisms. Therefore, it was tempting to conclude that perhaps also (part of) the biosynthetic pathway for Moco could be similar in all other organisms [36]. Several approaches have been successfully used to clone eukaryotic genes involved in Moco biosynthesis (functional complementation of *E. coli* Moco mutants, generation of Moco-defective mutants by tagging, cloning via "expressed sequence tags" by searching for homologies on protein level, exploiting protein–protein interactions by using the yeast two-hybrid system). On a DNA level most of these genes show only negligible homologies in the *E. coli* genes; however, on an amino acid level significant homologies (30–40% identity) do emerge between bacterial and eukaroytic proteins. It turned out that all *E. coli* Moco proteins (except MoaB and MobAB) have conterparts in plants, fungi, and humans, as shown in Table 1. However, beside these homologies to the bacterial proteins, there are also differences typical for the eukaryotes. In the following, we will discuss the single steps of Moco biosynthesis by presenting at first the data for bacteria (mostly *E. coli*) and then showing the similarities or differences as they occur in eukaryotes.

TABLE 1

Comparison of Proteins Involved in Moco Biosynthesis[a]

Bacteria *E. coli*	Fungi *A. nidulans*	Plants *A. thaliana*	Humans *H. sapiens*
MoaA [54]	CnxA [61]	Cnx2 [59]	MOCS1A [115]
MoaB [150]	—	—	—
MoaC [54,58]	CnxC [61]	Cnx3 [59]	MOCS1B [115]
MoaD [65,72]	CnxG [68]	Cnx7 [151]	MOCS2A [60]
MoaE [65,72]	CnxH [68]	Cnx6 [152]	MOCS2B [60]
MobA [92,97,98]	—	—	—
MobB [96]	—	—	—
MoeA [85]	CnxE (C) [113]	Cnx1 (N) [70]	Gephyrin (C) [81]
MoeB [50]	CnxF [74]	Cnx5 [75]	MOCS3 [76]
MogA [78,82]	CnxE (N) [113]	Cnx1 (C) [70]	Gephyrin (N) [81]

[a] For two-domain proteins, the position of the domain is given in brackets. N = N terminus; C = C terminus.

3.1. Conversion of GTP to Precursor Z

The pterin-based structure of Moco and the presence of a 6-alkyl side chain raised the possibility that its biosynthetic pathway could share steps or intermediates common to the biosynthesis of other pteridines. Several pteridines have three-carbon side chains, whereas Moco is unique in having a four-carbon side chain that forms the pyrano ring. Currently, two major pathways are known for the synthesis of pteridines and flavines involving the initial conversion of GTP by the enzymes cyclohydrolase I and II [52,53]. But none of them leads to a pteridine with a four-carbon side chain. Finally, detailed labeling studies in *E. coli* by Wuebbens and Rajagopalan [54] revealed evidence for a third route in pteridine synthesis that starts from a guanosine derivative, most likely GTP, with a novel and complex reaction sequence [51], and results in the formation of precursor Z (Fig. 2) as the first stable intermediate of Moco biosynthesis. This novel pathway is different from the other two pteridine biosynthetic pathways as the C8 of the purine is inserted between the 2' and 3' ribose carbon atoms, thus forming the four-carbon atoms of the pyrano ring that is typical for MPT. Using NMR with multiple labeled compounds, Rieder et al. [55] confirmed that each carbon atom of the ribose and the ring carbons of the guanine are incorporated into precursor Z; however, they suggested two possible reaction sequences, one of them being identical to that proposed by Wuebbens and Rajagopalan [54]. The latter model (Fig. 2) involves a ring opening reaction followed by the subsequent transfer of a formyl group (with C8) to the protein catalyzing precursor Z synthesis. The formyl group might be protein-bound via a thioester bond that might participate in a benzylic acid type of rearrangement of the side chain [55], resulting in the four-carbon side chain of precursor Z and the cyclic phosphate at C4' (Fig. 2). However, both suggested ways of converting guanosine to precursor Z are hypothetical multistep reactions, and further data are necessary to clarify the reaction mechanism.

In *E. coli*, the gene products MoaA and MoaC (Fig. 2) were identified to be essential for catalyzing the conversion of guanosine to precursor Z, and a mutation in *moaA* or *moaC* abolishes the formation of precurzor Z. MoaA contains an iron-sulfur cluster that is bound via highly conserved cysteine residues [56] and shows sequence similarities to a variety of proteins, including biotin synthase, pyruvate formate lyase, and anaerobic ribonucleotide reductase. The similarity to these proteins might indicate that the first committed step of Moco biosynthesis involves a radical-based reaction mechanism. Sofia et al. [57] propose a reductive cleavage of *S*-adenosylmethionine by an unusual Fe-S cluster in MoaA and a number of other homologous proteins that belong to a novel superfamily with more than 600 members. Interestingly, MoaA contains a C-terminal double-glycine motif that might be functionally important in the proposed transfer of the formyl group that could be bound via a thioester to the protein. The other protein important for precursor Z formation is MoaC, a homohexamer whose atomic structure has recently been solved [58]. The hypothetical active site of the protein is formed by residues of two MoaC monomers.

The plant counterparts to *E. coli* MoaA and MoaC are Cnx2 and Cnx3, respectively (Table 1, Fig. 2). These genes have been cloned by

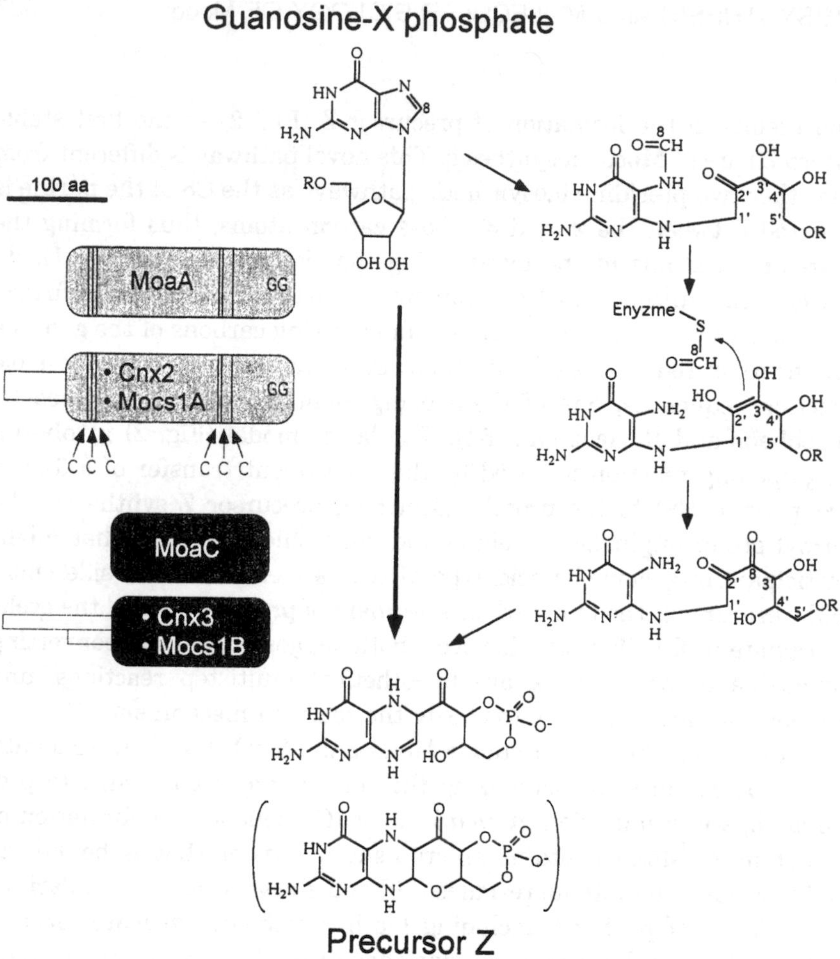

FIG. 2. Synthesis of precursor Z. The most poorly understood step in Moco biosynthesis is the formation of precursor Z by an alternative cyclohydrolase-like reaction. All carbon atoms of the guanosine are found within precursor Z, and the C8 atom transferred as formyl, is inserted between the C2′ and C3′ atoms of the ribose. Precursor Z is shown in its open form as 5,6-dihydropterin, which is very unstable. The proposed pyrano form of precursor Z is shown in brackets. In all organisms studied so far always two proteins catalyze the conversion of GTP to precursor Z. *E. coli* MoaA and homologous proteins are characterized by two cysteine clusters probably involved in Fe-S cluster binding. The conserved C-terminal double-glycine motif might participate in binding of the formyl group via a thioester. MoaC is smaller and thought to be involved in binding of the pterin because of its homologies to folate binding proteins [150]. Plant and human homologues of MoaA (Cnx2, Mocs1A) and MoaC (Cnx3, Mocs1B) show N-terminal extensions of yet unknown function.

328

functional complementation of the *E. coli* Moco mutants *moaA* and *moaC* [59], respectively, which indicates that the function of this pair of proteins has been strongly conserved during evolution. As in *E. coli* and plants, as well as in humans (MOCS1A and MOCS1B; [60]) and in the fungus *A. nidulans* (CnxA and CnxC; [61]), always two proteins are necessary to catalyze the first step in Moco biosynthesis, but nothing is known about their mechanism of function. In *Aspergillus*, it is not yet clear whether CnxA and CnxC are separate proteins [43] or whether they are domains of a fusion protein [61].

3.2. Conversion of Precursor Z to Molybdopterin

In order to form the dithiolene group in MPT, two sulfur atoms have to be incorporated into precursor Z during the second step of Moco biosynthesis. This reaction is catalyzed by the enzyme MPT synthase, a heterotetrameric complex of two small and two large subunits that stoichiometrically converts precursor Z to MPT. In a separate reaction, sulfur is transferred to the small subunit of MPT synthase by a sulfurase so as to reactivate the enzyme for the next reaction cycle of precursor Z conversion. In bacteria, fungi, and higher eukaryotes, these three proteins carry out the second step of Moco biosynthesis; however, some modifications occurred during evolution resulting in differences between bacteria and eukaryotes at this step.

3.2.1. *The Reaction of MPT Synthase and the Nature of the Dithiolene Sulfurs*

MPT synthase activity was first identified in *E. coli* by Johnson and Rajagopalan [62,63] who showed that *N. crassa nit-1* accumulates an MPT precursor that can be converted by an activity present in extracts of *E. coli moaA* (*chlA1*) mutants. Later, this activity (first called "converting factor") was purified. Several experiments indicated that it is a protein complex consisting of two different subunits that were identified as the small subunit MoaD (8.8 kDa) and the large subunit MoaE (17.0 kDa) [64].

Precursor Z and purified MPT synthase are sufficient to generate MPT in vitro [65]. Under conditions of precursor Z excess, the formation of MPT was stoichiometric with MPT synthase, indicating that a sulfur-regenerating system is needed to recharge the enzyme with activated sulfur. Pitterle and Rajagopalan [64] postulated that it is the small subunit of MPT synthase that carries the sulfur because a 16-Da difference was observed in the molecular weight between MoaD derived from active MPT synthase and MoaD derived from inactive MPT synthase purified from *moeB* (*chlN*) mutants. This difference was interpreted to be a replacement of an SH group for an OH group. MoaD shows sequence similarities to the functionally important C terminus of ubiquitin including the terminal double-glycine motif. In addition, the MPT synthase sulfurase MoeB that is essential for MoaD activation is homologous in its entire region to the N-terminal part of the ubiquitin-activating enzyme Uba1 from *Saccharomyces cerevisiae* [66,67]. Both similarities suggest a mechanism for sulfur transfer similar to ubiquitin activation by forming a thiocarboxylate at the C terminus of MoaD [50].

Among eukaryotes (cf. Table 1), genes for MPT synthases were identified in fungi [68], humans [60], and plants [154] showing the same high degree of conservation in the C terminus of the small subunit with the typical double-glycine motif. In plants, the cloning of the small subunit of MPT synthase was achieved in a yeast two-hybrid screen by making use of the assumed protein–protein interaction of the small subunit with the large subunit [154]. The result indicates that in plants both subunits of MPT synthase interact with each other to form a MPT synthase that might catalyze the same reaction of precursor Z conversion as the corresponding proteins in *E. coli*. However, there was no way to clone eukaryotic cDNAs encoding for MPT synthases by using the approach of functional reconstitution in *E. coli* as it worked successfully for the plant genes *cnx1* [70], *cnx2* and *cnx3* [59]. Even after isolation of the eukaryotic cDNAs was achieved, plant and human MPT synthases turned out to be incapable of complementing *E. coli* MPT synthase mutants [60,154], indicating functional differences between bacteria and eukaryotes at this step.

Besides the homology of MoaD and MoeB to the ubiquitin activation system, similarities between Moco biosynthesis and thiamin biosynthesis can also be seen in *E. coli*. The proteins ThiF, ThiS, and ThiI participate in the synthesis of the thiazole moiety. Begley and coworkers [71] have shown that ThiS is thiocarboxylated by ThiF (= homologous to MoeB) and ThiI (= sulfur transferase). Using an intein-based expression system, they generated in vitro large amounts of both carboxylated and thiocarboxylated ThiS, and the latter was capable of donating sulfur to thiamine biosynthesis. Using the same system, Gutzke et al. [69, 154] generated carboxylated and thiocarboxylated small subunits of MPT synthase from *E. coli* (MoaD) and plants (Cnx7). In both cases, only the thiocarboxylated small subunits and the corresponding large subunits were essential and sufficient for the assembly of active MPT synthase. This experiment shows that also in eukaryotic Moco biosynthesis the sulfur is transferred as thiocarboxylate from the C terminus of the small subunit to precursor Z, thus forming the dithiolene in MPT.

MPT synthase has to introduce two sulfurs to precursor Z in order to form the dithiolene group of MPT. This reaction is probably based on the heterotetrameric nature of *E. coli* MPT synthase. The crystal structure of *E. coli* MPT synthase (Fig. 3A) was recently solved [72], and also assembly experiments using the separated subunits confirm the heterotetrameric complex [69]. Interestingly, both forms (the inactive carboxylated and active thiocarboxylated MoaD) were able to generate a tetrameric protein complex with the corresponding large subunit, whereas the separated subunits themselves behaved exclusively monomerically on size exclusion chromatography. The crystal structure of MPT synthase revealed that the C terminus of the small subunit is inserted into the large subunit forming a heterodimer. The tetramer is formed by dimerization of two large subunits resulting in an elongated protein complex with two clearly separated active sites (Fig. 3A). A pocket on MoaE can be seen around the C terminus of MoaD that is formed by highly conserved residues of MoaE. A structural comparison of MoaD with ubiquitin shows a high degree of conservation that is accompanied by only 7% identity on amino acid sequence level. The finding that the monomeric large subunits dimerize only in the presence of small subunits [69] suggests that a conformational change occurs during insertion of MoaD.

In summary, it is justified to assume that the dithiolene group of MPT is generated by one tetrameric MPT synthase complex possessing two separate active sites each harboring one reactive thiocarboxylate. No proteins were found in databases having significant homologies on both the amino acid or structural level to the large subunit of MPT synthase. Hence, one may speculate that the large subunit was invented

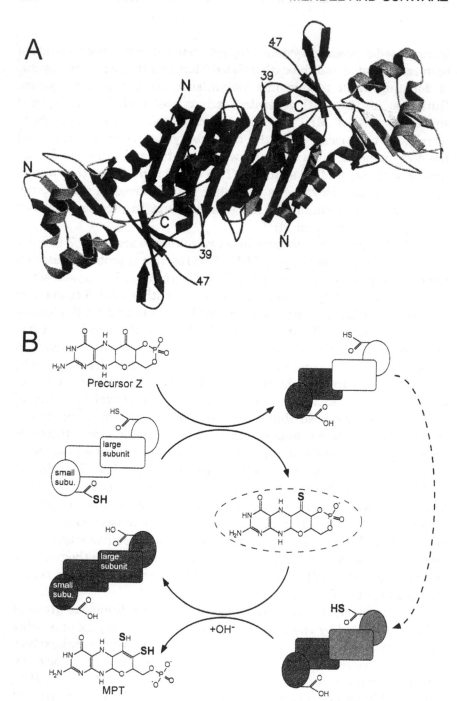

by nature to join two highly reactive thiocarboxylated small subunits, thereby facilitating the subsequent transfer of two sulfurs to precursor Z resulting in the formation of one dithiolene. Combining the present results and interpretations, we suggested [69] a model for a two-step mechanism of MPT synthase reaction (Fig. 3B) and propose a hypothetical intermediate (precursor G) carrying one sulfur atom as thione. Once the first small subunit has transferred the sulfur to precursor Z, the intermediate will be released causing a conformational change in the first active site (Fig. 3B, shaded in dark). In line with this hypothesis are experiments of Wuebbens and Rajagopalan [51] showing that compound Z (= the oxidation product of precursor Z) does not react with MPT synthase, which can be explained by the functional importance of the keto group at C1' that is chemically different in the oxidized form of precursor Z. Then an induced fit of the second active site (Fig. 3B, shaded in gray) could result in an increased affinity for the intermediate and a decreased affinity for precursor Z as compared with the starting enzyme. In the second half-reaction, the thiocarboxylate from the second small subunit opens the cyclic phosphate by generating the dithiolene and the terminal phosphate. Finally, MPT is released and a fully carboxylated MPT synthase is formed that can be resulfurated. Additional experiments are needed to elucidate the chemistry of the dithiolene synthesis in MPT.

FIG. 3.   Crystal structure of the heterotetrameric MPT synthase from *E. coli* (A), and a hypothetical reaction mechanism for the sequential transfer of two sulfur atoms to precursor Z forming the dithiolene (B). (A) Ribbon diagram of the high-resolution crystal structure of MPT synthase (provided by H. Schindelin, SUNY Stony Brook, USA [72]). The elongated heterotetramer consists of a dimer of two heterodimers where the large subunits (shaded in dark gray) mediate the interaction and the small subunits (shaded in light gray) are located at the opposite ends. (B) A model for the reaction of the thiocarboxylated heterotetrameric MPT synthase is shown. First the carboxyl group at C1' is attacked forming a thione (precursor G), resulting in the carboxylation of the first active site. In a second site, MPT is formed by opening the cyclic phosphate and incorporation of the second sulfur forming the dithiolene.

3.2.2. Sulfuration of MPT Synthase and the Evolutionary Relationship to Ubiquitin Activation

In *E. coli*, the MoeB protein has been interpreted to transfer a sulfur atom to MoaD, the small subunit of MPT synthase, since MoaD isolated from an *moeB* mutant was found to be sulfur-free [64]. As mentioned above, there are striking homologies of MoeB to the ubiquitin-activating enzyme E1 (Uba1) which adenylates ubiquitin and forms a thioester between the C terminus of ubiquitin and a thiol group in Uba1. ThiF, the MoeB homologous protein in thiamin biosynthesis initially forms an acyl adenylate at the C terminus of ThiS (= MoaD homologue), which is subsequently converted to the thiocarboxylate by the sulfur transferase ThiI. Assuming a similar mechanism for the generation of the thiocarboxylate at MoaD, one can conclude that MoeB is not a sulfur transferase but rather an activating enzyme similar to Uba1 and ThiF. However, there is no ThiI homologous protein known in Moco biosynthesis that cold supply the sulfur and mediate thiocarboxylation of the small subunit.

Recently, Leimkühler and Rajagopalan [73] found that MoeB itself is not sufficient to reactivate carboxylated MPT synthase in vitro. However, after adding a crude protein extract of an *moeB* mutant they observed activation of MPT synthase in an in vitro reaction. The crude extract could be replaced by its desalted protein fraction and cysteine or by sulfide itself, indicating that an enzyme providing sulfur from cysteine might be the real sulfur transferase generating the thiocarboxylate on the adenylated small subunit of MPT synthase. During extensive screenings for Moco mutants in *E. coli* and many other organisms, only MoeB was found to be exclusively essential for the activation of MPT synthase, while the novel sulfur transferase did not light up. Conclusively one can argue that this predicted sulfur transferase is probably involved in additional cellular processes different from Moco biosynthesis or that a number of sulfur transferases are capable of catalyzing this reaction (Fig. 4A).

Eukaryotic MPT synthase sulfurases (to be consistent, we will continue to use this terminology) were identified in fungi (*A. nidulans*, CnxF; [74]), plants (Cnx5; [75]), and humans (MOCS3; [76]). In all cases N- and C-terminal extensions were identified (Fig. 4B). Among the N-terminal extensions, no significant homologies were found and no motifs

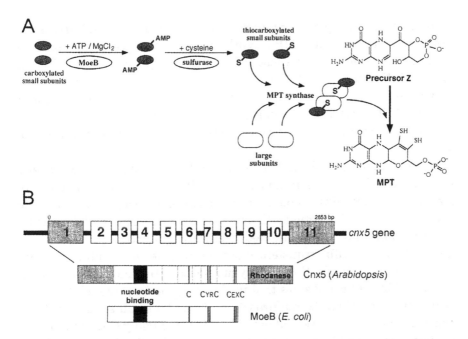

FIG. 4. Thiocarboxylation of the small subunit of MPT synthase. (A) Thiocarboxylation and assembly of the tetrameric MPT synthase is shown. In *E. coli*, the small subunit MoaD is first acyl-adenylated by the activity of MoeB (H. Schindelin, unpublished results) and subsequently sulfurated by a yet unknown sulfurase using cysteine as sulfur donor [73]. (B) Genomic structure and functional domains of *Arabidopsis* Cnx5 in comparison to *E. coli* MoeB. Exons 1 and 11 encode for N- and C-terminal extensions of Cnx5. The rhodanese-like domain is supposed to be involved in providing the sulfur for thiocarboxylation. The highly conserved cysteine clusters as well as the nucleotide binding site are indicated.

emerged during database searches. More interestingly, the C-terminal extensions form a rhodanese-like domain that is encoded by a single exon in Cnx5 [75], the plant homologue of MoeB. Rhodaneses are ubiquitous enzymes catalyzing the transfer of a sulfur atom, as defined in an in vitro system. In the active form, the sulfur is bound to a conserved cysteine residue as persulfide, and this cysteine is also conserved in the rhodanese-like domain of CnxF, Cnx5, and MOCS3, which might indicate that sulfur mobilization or activation differs between bacteria and eukaryotes. Support for this assumption comes from another line of experiments. Since plant and human MPT synthases and sulfurases are not able to reconstitute MPT synthesis in *E. coli*, we coexpressed all three proteins in an *E. coli moaD* mutant gaining a 10% reconstitution of *E. coli* NR activity (G. Gutzke and R. Mendel, unpublished result). Obviously, *E. coli* MPT synthase sulfurase is not able to activate plant or human MPT synthases, which is a hint that on the level of sulfur transfer and activation changes occurred during the evolution of Moco biosynthesis.

In eukaryotes, all MPT synthase sulfurases identified to this end have an MoeB-like domain and a second domain with striking homologies to rhodanese-like domains of other sulfur transferases. Thus, one can argue that eukaryotic MPT synthase sulfurase is a multifunctional protein combining the postulated adenylation function (= MoeB/ThiF domain) with a sulfur transferase domain in order to facilitate the subsequent formation of a thiocarboxylate. Besides the known homology of MoeB to ThiF we could show that in plants a protein–protein interaction occurs between Cnx5 (= sulfurase) and Cnx7 (= small subunit of MPT synthase) because we were able to identify Cnx7 using Cnx5 as a bait in a yeast two-hybrid screen (J. Nieder and R. Mendel, unpublished results). The direct protein interaction between the small subunit and the sulfurase was also shown by Schindelin and coworkers (unpublished results) who recently have solved the crystal structures of *E. coli* MoeB, of a complex of MoeB with MoaD, as well as the same complex soaked with ATP, yielding a structure of the bound substrate and a structure of adenylated MoaD. These new and exciting results demonstrate that MoeB adenylates the small subunit of MPT synthase as assumed by homologies to ubiquitin activation and thiamin biosynthesis.

3.3. Formation of Active Moco

After synthesis of the MPT moiety, the chemical backbone is built for binding and coordination of the Mo atom that catalyzes many diverse reactions once it is attached to its appropriate apoenzyme. Mo has to be taken up into the cell in the form of molybdate followed by the coordination to MPT. Finally, only in bacteria a nucleotide has to be attached via a pyrophosphate bond to MPT, resulting in the MPT dinucleotide cofactor. There is evidence that Mo insertion and dinucleotide formation are linked to each other, so that we will discuss both steps together as a third stage of Moco synthesis (Fig. 5).

3.3.1. Conversion of MPT to Moco in Bacteria

Analysis of the final step of Moco biosynthesis started with the identification of mutants exhibiting a so-called molybdate-repairable phenotype. These mutants were found in all organisms where Moco deficiency was studied and are characterized as mutants with partially or completely restored Mo enzyme activity after growth on unphysiologically high concentrations of molybdate (0.1–1 mM) [24]. For mutants in bacteria with total molybdate repair, a defect in the high-affinity uptake system for molybdate was identified, which is encoded by the *mod* operon [77]. Molybdate uptake and regulation of the *mod* operon are discussed in detail in Chapter 2. It is the key question of the last step of Moco synthesis whether it is molybdate that serves as donor for insertion of Mo into MPT or whether molybdate has to undergo intracellular processing prior to insertion. In the following we will discuss the proteins involved in this step.

3.3.1.1. MogA

In *E. coli*, two lines of evidence indicate that the MogA protein is involved in transfer of Mo to MPT: *E. coli mogA* mutants are the only molybdate-repairable mutants except the transporter mutants from the *mod* locus. They exhibit a measurable Mo enzyme background activity of 1–2% that goes up to 15% of wild-type level after growth on media containing 1 mM molybdate [24]. Accompanied with a defect in the last step of Moco synthesis is the ability of *mogA* mutants to synthesize MPT [78], resulting in a dramatic accumulation of MPT under Mo enzyme-inducing growth conditions [79]. A protein that acts as Mo insertase should bind both substrates, Mo and MPT. For plant and mammalian MogA homologous proteins, a tight binding of MPT was shown [80,81], but neither for MogA nor the other homologues could any binding of the molybdate ion be found, which is a first hint that molybdate itself might not be the physiological substrate for Mo insertion. Although MogA was the first protein of Moco biosynthesis for which the crystal structure was determined [82], it remains unclear how MPT is bound and which form of Mo is inserted. The structure of

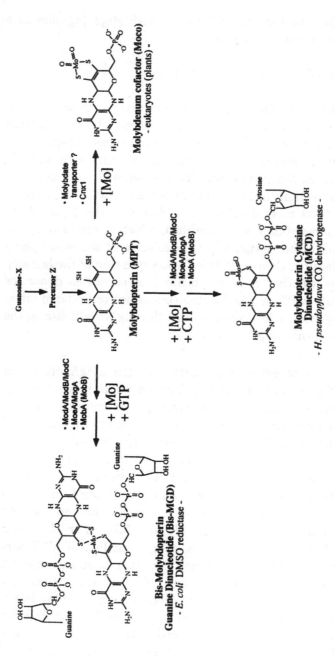

FIG. 5. The transfer of Mo to MPT in bacteria and eukaryotes and the required gene products. In *E. coli*, a bis-MPT guanine dinucleotide (bis-MGD) is formed. It is suggested that at first Mo is incorporated by the activities of MoeA and MogA forming a Mo-MPT which is then converted to bis-MGD by the activity of MobA. However, there are indications that metal incorporation proceeds after MGD synthesis. The CO dehydrogenase in *H. pseudoflava* contains an MPT cytosine dinucleotide (MCD) where Mo is coordinated in the mature enzyme by an oxygen and a sulfur atom. In eukaryotes, Mo is directly inserted into MPT by the activity of the two-domain proteins Cnx1 and gephyrin, thus forming the Moco.

MogA will later be discussed together with the structures of the homologous plant and human proteins.

3.3.1.2. MoaB

Within the *moa* operon the second open reading frame encodes for a protein (MoaB) with significant homologies to MogA but a yet unknown function because no mutants were found in the *moaB* gene resulting in Moco deficiency. Most bacteria have either a MogA-like protein or a MoaB-like protein; only a few organisms, such as *E. coli* and *Mycobacterium*, have both types. Deleting of *moaB* in *E. coli* did not cause deficiency in any of the analyzed Mo enzymes ([83]; D. Boxer and T. Palmer, unpublished results), and overexpression of MoaB was not sufficient for rescue of *mogA* mutants. However, a *mogA/moaB* deletion mutant lowered the residual activity of *mogA* mutants. When comparing MoaB and MogA sequences of diverse bacterial origin, one can observe that there are some residues in *E. coli* MoaB that differ from residues that are highly conserved in all MogA homologous proteins as well as MoaB proteins from other organisms. It is known from mutagenesis experiments [82] as well as functional studies of the homologous Cnx1 protein in plants [79] that in *E. coli* MogA aspartate 49 is functionally important. In MoaB, this aspartate is changed to a glutamate and the adjacent glutamate 50 highly conserved in MogA is replaced by glutamine in MoaB. Both residues were exchanged separately (D49 and E50) in MogA against the residue present in MoaB and transformed into *E. coli* *mogA* mutants. Whereas the D/E exchanges at position 49 did not affect the functional properties of MogA, the E/N exchange at position 50 resulted in a loss of activity [83]. The biological function of MoaB remains enigmatic. It might be that MoaB plays a role in the regulation of Moco biosynthesis by sensing active Moco. On the other hand, MoaB proteins could have a MogA-like function in the biosynthesis of W cofactors because in archaea predominantly MoaB homologous proteins were found, whereas to this end no MoaB and no W enzyme were identified in eukaryotes.

3.3.1.3. MoeA

Prior to identification of eukaryotic Moco synthetic proteins with domains having homologies to *E. coli* MogA and MoeA, no function was assigned to *E. coli* MoeA in the final step of Moco biosynthesis. The first *moeA* mutant, *chl*E5, was biochemically described [62,63] with low levels of MPT and the ability to synthesize a precursor similar to that accumulating in mutants defective in MPT synthase or MoeB. However, recent results (G. Schwarz, unpublished results) demonstrate that *E. coli moeA* mutants accumulated MPT in a similar way as *mogA* mutants do and no residual NR activity or any Mo repair of Mo enzyme activity could be observed. These results suggest that MPT itself is essential but not sufficient for incorporation of Mo into MPT. In contrast to *E. coli*, a *Rhodobacter capsulatus moeA* mutant was described that shows a molybdate repair for enzymes that contain the MPT-based cofactor (like that of eukaryotic enzymes), while activities of Mo enzymes with a dinucleotide-based cofactor, as occurring in eubacteria and archaebacteria, were not restored by molybdate [84]. Therefore, not only MogA but also MoeA seems to be essential for activating and transferring Mo to MPT, resulting in Moco formation. Furthermore, Hasona et al. [85] have found that *moeA* mutants can be complemented by supplementing the growth media with sulfide and molybdate, suggesting a role of MoeA in the activation of Mo by forming a yet unknown thiomolybdate compound. In summary, *E. coli* MoeA and MogA proteins are both essential for the incorporation of Mo into the cofactor, thus indicating a multistep reaction of Mo chelation.

3.3.2. Formation of the Dinucleotide Form of Moco

As first detected by Krüger and Meyer [86,87], bacteria can carry various substituents at the C4' atom of the side chain of Moco. Johnson et al. [88] showed that the Moco from DMSO reductase from *Rhodobacter sphaeroides* consists of an MPT guanine dinucleotide where a GMP is bound to the C4' atom via a pyrophosphate bond (Fig. 5). Other prokaryotic variants of the cofactor containing CMP, AMP, or IMP linked to the MPT were identified as well (reviewed in Rajagopalan and Johnson [38]). These dinucleotide forms were only found in prokaryotes. Bacteria may even contain both Moco forms because it was found that XDH from *R. capsulatus* and *Pseudomonas putida* contains the MPT form of the cofactor, whereas the other Mo enzymes (NR, DMSO reductase, chino-line oxidoreductase) contain the dinucleotide form [89,90].

The *mobAB* locus is responsible for the nucleotide attachment in *E. coli* Moco biosynthesis [91,92]. MobA catalyzes the conversion of MPT and GTP to MGD [93], whereas MobB is a GTP-binding protein with weak intrinsic GTPase activity [94]. Temple and Rajagopalan [95] have shown that in vitro only MobA, GTP, MgCl$_2$, and Mo-MPT are required for the assembly of active DMSO reductase from apoprotein. Under those conditions, the formation of Mo-MGD can proceed in the absence of any acceptor apoprotein and MobB seems to be most essential for MGD synthesis in vitro. Additional components are not necessary for in vitro synthesis of MGD, whereas in vivo chaperones like the NR-specific NarJ protein from *E. coli* may play an important role for this step [96].

Recently, the crystal structure of *E. coli* MobA was solved [97,98]. The protein has an α/β architecture with a nucleotide-binding Rossman fold formed by the N-terminal half of the protein. Lake et al. [97] have crystallized MobA in an octameric stage, but the native conformation of the protein seems to be monomeric because in solution most of the protein was found to be monomeric [97], and Stevenson et al. [98] have solved the structure of MobA as a monomer. The active site was defined by highly conserved residues as well as by cocrystallization of MobA with GTP which is bound in the N-terminal half [97]. The binding site of MPT, formed by another set of highly conserved residues, was modeled according to the position of GTP.

Interestingly, there are several indications that the conversion of MPT to its dinucleotide form is molybdate-dependent. Two lines of evidence argue for this conclusion: (1) Depletion of molybdate from the culture medium abolished MCD synthesis in *Hydrogenophaga pseudo-flava* [99], and a mutation in the *mod* locus of *E. coli* encoding for the high-affinity molybdate transport system resulted in the accumulation of MPT without detectable MGD levels (G. Schwarz, unpublished result). (2) When growing *E. coli* or *H. pseudoflava* on tungstate-rich media MGD or MCD synthesis was abolished and no MPT could be detected in DMSO reductase [100], NR [101], or CO dehydrogenase [99]. However, the corresponding nucleotides were found to be bound in the cofactor binding site of these enzymes, suggesting a tight binding of that portion of the mature cofactor to the apoenzyme. Based on these results one can argue that the MobA protein in *E. coli*, and the analogous MCD-synthesizing protein in *H. pseudoflava*, are probably specific for Mo-MPT and are unable to attach the guanine or cytosine nucleotide to Mo-free MPT.

However, recent findings show that *E. coli moeA* mutants accumulate MPT as well as MGD (G. Schwarz, unpublished results) without having any detectable Mo enzyme activity [63], suggesting the formation of Mo-free MGD in these mutants. Therefore, the observed Mo dependency of MGD synthesis could go back to the availability of molybdate itself instead of Mo-MPT. Further, crude extracts of *moeA* mutants grown on molybdate-rich media contain Mo-MPT (= eukaryotic Moco) as detected by molybdate-independent *nit-1* reconstitution. Therefore, it would also be conceivable that in bacteria at first the MPT dinucleotide is formed, and finally Mo is inserted forming the active Mo-MPT dinucleotide cofactor.

Another difference between bacterial and eukaryotic Moco biosynthesis is the formation of bis-MPT-based cofactors in bacteria where one Mo (or W) atom is coordinated by two dithiolenes of two MPT molecules (Fig. 5). If Mo insertion and bis-MPT formation were to take place prior to MGD formation, a binding site for bis-Mo-MPT would be formed on MobA. In summary, one has to state that although the final step of bacterial Moco biosynthesis still raises questions, the two processes of dinucleotide formation and of Mo insertion seem to be strongly linked. In all of the available crystal structures of Mo enzymes, Moco is buried deep in the protein, suggesting that cofactor insertion is coupled to the final steps of protein folding and Mo enzyme assembly.

3.3.3. Conversion of Molybdopterin to Moco in Eukaryotes— The Multidomain Plant Protein Cnx1

In eukaryotes, the final stage of Moco biosynthesis comprises transfer of Mo to MPT in order to form active Moco. This requires uptake of molybdate, but in eukaryotes information is limited about the way in which Mo is taken up into the cell. There are reports that sulfate as well as phosphate starvation of plants enhances Mo uptake up to 10-fold [102], which means that sulfate transporters and phosphate transporters might cotransport molybdate anions. The cotransport of molybdate by a sulfate transport system was also observed in filamentous fungi [103].

Also in *E. coli* where a high-affinity transport system is known, the sulfate transporter was shown to serve as a low-affinity molybdate transporter [104]. Very recently, mutant analysis in the green alga *Chlamydomonas reinhardtii* revealed for the first time a more detailed picture of molybdate uptake in eukaryotes [105]. This alga has two molybdate uptake systems: one is a high-affinity, low-capacity transporter that is insensitive to tungstate but can be inhibited by 0.3 mM sulfate. The other system is a bulk transporter (low-affinity, high-capacity) that can be inhibited by tungstate but not by sulfate.

Once having crossed the plasma membrane, Mo has to be inserted into MPT. Similar to *E. coli mogA* mutants, also among eukaryotes the type of molybdate-repairable mutants has been known for a long time: *Aspergillus cnxE* [39], *Neurospora nit-9ABC* [40,106], tobacco *cnxA* [17], and *Arabidopsis chl-6* [107]. The gene defective in the molybdate-repairable plant mutants is *cnx1*. It was isolated from *Arabidopsis* by functional complementation of *E. coli mogA* mutants [70]. The encoded protein Cnx1 consists of two domains (see Fig. 6 below), where the N terminus (E domain) is homologous to *E. coli* MoeA and the C terminus (G domain) to *E. coli* MogA. The high functional homology allows Cnx1 or its separately expressed G domain to replace the bacterial MogA function very efficiently [108]. In contrast, the N terminus of Cnx1 is not able to take over the task of its homologous bacterial protein MoeA.

In *A. thaliana*, *cnx1* was mapped to the molybdate-repairable *chl-6* locus on chromosome 5 [70]. Later on, the genomic structure of *cnx1* was solved, its functionality proven by complementation of *chl-6* mutants, and finally a point mutation was found in the Cnx1 E domain of *chl-6*, resulting in the exchange of the highly conserved glycine 107 residue to an aspartate [109]. In *N. plumbaginifolia*, *cnx1* represents the molybdate-repairable *cnxA* locus because transformation of *cnxA* mutants with *cnx1* fully reconstitutes the wild-type phenotype and restores activities of NR and XDH [109]. Amongst *cnxA* mutants, lines were found [32] with point mutations in the E domain and the G domain, as well as with a mutation affecting the entire Cnx1 protein (J. Schulze and R. Mendel, unpublished results), demonstrating that a defect in any of the two domains of *cnx1* results in a molybdate-repairable phenotype, whereas in *E. coli* only G domain-homologous *mogA* mutants feature this phenotype. However, in *moeA* mutants of *R. cap-*

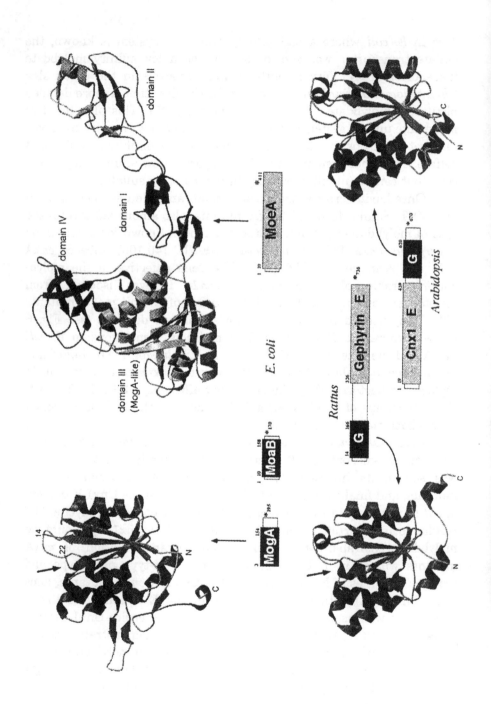

FIG. 6. Homologies of Cnx1 and gephyrin to *E. coli* MogA, MoaB, and MoeA in their primary and three-dimensional structures. Domains of Cnx1 and gephyrin homologous to *E. coli* MogA/MoaB as well as MoeA, respectively, are indicated. A ribbon diagram of the crystal structures of MoeA (provided by H. Schindelin, SUNY Stony Brooks; USA [110]), MogA (provided by H. Schindelin, SUNY Stony Brook; USA [82]), Cnx1 G domain and gephyrin G domain [155] are shown. MoeA is composed of four distinct domains, one of them showing high structural homologies to MogA. The structures of MogA and the eukaryotic G domains are highly homologous, with the exception of the extended hairpin between the last two α helices in MogA and the C terminus of MogA. The second α helix, which is in close proximity to the hypothetical active site (*arrow*), is disordered in MogA (residues 15–21).

sulatus [89], the XDH (MPT-based enzyme) was shown to be Mo-repairable as well. Together with the inability of Cnx1 to reconstitute *E. coli moeA* mutants one can argue that MoeA and the E domain of Cnx1 catalyze different reactions in Moco biosynthesis while both seem to be related to Mo insertion into MPT.

Biochemical characterization of Cnx1 and its domains revealed that in solution the G domain is a trimer whereas the E domain is a dimer [80], which is similar to the corresponding proteins in *E. coli* [82,110], indicating that the domain interaction in Cnx1 is conserved. However, for the holoprotein no distinct multimer form could be detected [80]. According to the characterization of *cnx1* mutants, the molybdate-repairable phenotype suggests that MPT and molybdate as hypothetical substrates should bind to Cnx1 in order to be converted to active Moco. MPT binding studies [80] revealed a high-affinity interaction between MPT and the G domain of Cnx1 ($K_d = 100$ nM). MPT binding to the G domain is equimolar and hyperbolic, indicating single independent binding sites on each monomer in the trimer. Interestingly, the E domain of Cnx1 also possesses the ability to bind MPT, but with a lower affinity than the G domain and in a cooperative manner.

Based on the observed MPT binding to both domains of Cnx1, a common substrate-binding fold can be assumed that is defined by three highly conserved amino acid motifs localized between residues 239 and 324 in the E domain and residues 512 and 600 in the G domain [109]. Recently, the crystal structure of *E. coli* MoeA (= E domain homologue) has been solved (Fig. 6) [110]. The dimeric protein is composed of very elongated monomers forming four distinct domains. In line with the previous assumption, the third domain of MoeA is structurally similar to *E. coli* MogA, the G domain homologue in *E. coli*. These data suggest a common binding fold for MPT in both domains of Cnx1, and the observed cooperative binding behavior of Cnx1E might be attributed to the fact that the hypothetical active site in MoeA/Cnx1E is formed by both subunits of the dimeric protein [110].

Several attempts were undertaken to show binding of molybdate to Cnx1 or its separated domains, but no binding was observed in the presence or absence of MPT. Based on the assumption that *E. coli* MoeA is involved in Mo activation [85], additional experiments are needed to show a binding of molybdate and/or synthesis of a thiomolybdate. It might be that Mo activation is nucleotide-dependent, since

Menendez et al. [111] have found a weak adenosine triphosphatase activity for MoeA from *Arthrobacter nicotinovorans*.

In order to identify functionally important residues within Cnx1 G domain a mutagenesis screen was performed followed by selection for loss of function using *E. coli mogA* complementation [79]. Several mutants were found, two of them similar (D515) to the already described D49 from *E. coli* [82]. These D515 mutants were still capable of binding MPT with high affinity, whereas the other mutants showed significant reduction in MPT binding. Hence one can assume that MPT binding is not the only function of Cnx1G. When purifying G domains expressed in *E. coli mogA* mutants, most of the MPT accumulated in these mutants (10–20 times of wild-type level) could be copurified with the G domain [79]. When using this copurified MPT for *nit-1* reconstitution to demonstrate Moco synthesis, only wild-type G domain but none of the mutant proteins could synthesize active Moco in the absence of external molybdate [79]. In the presence of high molybdate concentrations, MPT (free or protein-bound) can be converted noncatalytically to Moco [19]. Conclusively, one can say that D49 in *E. coli* and D515 in Cnx1 G domain seem to be essential for Mo insertion into MPT—and this function is distinct from MPT binding. Furthermore, when using only the protein fraction of the *nit-1* crude extract, the wild-type protein also lost the ability to generate active Moco, indicating that a low molecular weight compound from the crude extract, probably a thiomolybdate, is used by the G domain to generate Moco. In addition, when using MPT copurified with the Cnx1 holoprotein, molybdate-independent Moco synthesis could be observed in *nit-1* crude or gel-filtrated protein extracts, suggesting that the protein itself contains everything to generate Moco. These results indicate that within the Cnx1 holoprotein the E domain seems to be essential for generating that form of Mo that is incorporated by the G domain into the bound MPT. Finally, the G domain is also important for the stabilization of Moco [79].

The high-resolution crystal structure (1.5 Å) of *E. coli* MogA (Fig. 6) was the first X-ray structure solved for a Moco biosynthetic protein. Recently, the structure of the MogA-homologous G domain of Cnx1 was solved at 2.8 Å, the first structure of a eukaryotic Moco synthetic protein (Fig. 6). MogA and Cnx1-G show a trimeric arrangement in which each monomer contains a central, mostly parallel β sheet surrounded by α helices on either side (Fig. 6). Residues

15–21 are disordered in MogA, whereas in Cnx1-G they form a well-ordered short α helix. Another difference between the structures is found at the C-terminal part of the proteins. The last two α helices in MogA are interrupted by a β turn that is formed by residues 145–157. These residues are only conserved among bacterial homologues of MogA, but in eukaryotic proteins like Cnx1 both helices form a continuous helix. It might be that this β turn plays a role in mediating protein–protein interactions that are essential for cofactor biosynthesis in *E. coli*. The C-terminal end of MogA is not conserved and seems to be loosely folded. Removal of the last 42 residues of Cnx1 did not affect the functional properties of the protein. In all other aspects both structures are highly similar, underlining the evolutionary pressure to maintain the function of both proteins. The hypothetical active site of MogA and Cnx1G was assigned by the three-dimensional position of highly conserved residues as well as by residues that are functionally important (*E. coli* D49, Cnx1-G D515), by difference density features, and by conformational changes. All of these criteria point to a pocket in close proximity to the α helix that is disordered in MogA. It is assumed that this region of the protein is involved in MPT binding and Mo insertion [155].

 Cnx1-homologous proteins were also found in *Drosophila* (Cinnamon, [112]), in *A. nidulans* (CnxE, [113]), and in mammals (gephyrin [114]). All proteins show the same two-domain structure as Cnx1, but the order of the domains is inverted. The functional properties of the mammalian homologue gephyrin will be discussed in the following section.

3.4. Moco Biosynthesis in Humans

3.4.1. *Synthesis of MPT in Humans*

In humans, Moco biosynthesis seems to be very similar to the pathway described for higher plants. Table 1 shows that for each step of Moco biosynthesis there are human proteins (named MOCS) homologous to the plant Cnx proteins. Surprisingly, differences were found on the gene level. Like in all other organisms, two proteins are involved in the conversion of GTP to precursor Z (step 1); however, these two proteins are encoded by only one gene (*mocs1*) in humans [115]. The corresponding transcript is bicistronic with two consecutive reading frames separated by a stop codon. The first reading frame encodes for MOCS1A, the second one for MOCS1B, and for both proteins human patients were identified (summarized in [43]). Moreover, *mocs1* has already been used for prenatal diagnosis [116] of people with a family history in Moco deficiency. Recently, further transcripts of the *mocs1* gene were found [117] that are spliced in order to bypass the normal termination codon of *mocs1A*. These new transcripts encode for two-domain proteins embodying MOCS1A and MOCS1B. Nothing is known about the activity of these fusion proteins and their relative expression level with respect to the separated proteins. Nevertheless, it is a very rare case that a bicistronic mRNA is also alternatively spliced in order to produce different combinations of two proteins that catalyze one reaction step in a biosynthetic pathway.

The human system became even more fascinating when we found that the two subunits of human MPT synthase (step 2) were also encoded by only one gene, *mocs2* [60]. On the bicistronic messenger RNA, the first reading frame codes for the small subunit MOCS2A and the second one for the large subunit MOCS2B. Both reading frames overlap and exhibit a frameshift of +1 for *mocs2B*. Again here, human patients were identified for both proteins [118], thus confirming their functional role. In both cases of bicistronic expression, always the first of the two encoded proteins shows a typical double-glycine motif at its C terminus. For the small subunit of MPT synthase (MOCS2A) we know, that the C terminus is functionally essential. Therefore, the observed bicistronic expression of both MOCS2 proteins is a further indication of strong functional pressure for maintaining the free C terminus in MOCS2A, and possibly also in MOCS1A. However, it remains unclear whether or not a fusion of MOCS1A and MOCS1B affects the activity or functionality of MOCS1A.

Bicistronicity would ensure colinear expression and implicates the vicinity of the newly synthesized and interacting proteins. Such a microcompartmentalization is certainly advantageous for low substrate concentrations as in Moco biosynthesis. Yet it remains enigmatic why in the human Moco biosynthetic pathway two times the extremely rare case of bicistronicity is found whereas in higher plants the corresponding genes are widely separated and, as in the case of the step-1 proteins, are even located on different chromosomes. The plants are the ones that would need the vicinity of the newly synthesized and interacting proteins because here we encounter a strong cytoplasmic streaming.

Finally, with the identification of the human Moco genes *mocs1* and *mocs2* the previous biochemical characterization of mutant cell lines could be confirmed. Group A patients have a mutation in *mocs1* and are not able to synthesize precursor Z. Group B patients accumulate the precursor due to a mutation in *mocs2*, resulting in the loss of MPT synthase activity. Until now no mutant was found with a defect in *mocs3*, encoding for the MoeB homologous sulfurase MOCS3, suggesting for that protein additional functions different from Moco biosynthesis.

3.4.2. Incorporation of Mo into MPT: The Multifunctional Protein Gephyrin

The primary structures of Cnx1 as well as of the bacterial proteins MoeA and MogA show striking homologies to the mammalian protein gephyrin, which was first described as a neuroreceptor anchor protein linking glycine receptors in the postsynaptic membrane to the subcellular cytoskeleton. Gephyrin is thought to be an instructive molecule for the formation of glycinergic synapses [119], and its expression was shown to be essential for the postsynaptic aggregation of glycine receptors [120]. Based on the homology of gephyrin to Cnx1 an additional function of gephyrin in Moco biosynthesis was suggested, which was demonstrated by the following experiments. Recombinant gephyrin binds MPT with high affinity [81] and heterologous expression of gephyrin could restore Moco biosynthesis in *E. coli*, plants, and in the murine cell line L929 [81]. In addition, *gephyrin* knockout mice not only showed the expected absence of synaptic glycine receptor clustering but also developed symptoms identical to those of Moco deficiency [121] where no SO activity could be detected. Recently, identification of a *gephyrin* gene delection in a patient with symptoms typical of Moco deficiency was described [47]. Biochemical studies with the patient's fibroblasts demonstrated that gephyrin catalyzes the insertion of Mo into MPT, suggesting that this form of Moco deficiency might be curable with high doses of molybdate. It is obvious that gephyrin combines two different functions, such as (1) a biosynthetic activity in Moco formation and (2) a structural role in receptor clustering. The latter function is evolutionarily younger and must have been recruited from the older in primary metabolism.

Furthermore, gephyrin seems to have even more functional properties than neuroreceptor clustering and Moco synthesis because recent findings suggest that gephyrin may be implicated in the modulation of signaling pathways [122,123]. The observation that differentially spliced transcripts of gephyrin can be found not only in brain and spinal cord but also in liver, kidney, heart, and lung [124] raises the possibility that differential splicing of gephyrin results in the modulation of its functionality. To understand the molecular nature of this multifunctionality, the structure-function relationship has to be uncovered in order to assign protein regions that are functionally altered by alternative splicing. Recently, we have solved the high-resolution crystal structure of gephyrin G domain (Fig. 6; [155]), which is a first step in our understanding of the three-dimensional structure of gephyrin and the arrangements of both domains to one another. The overall folding of the G domain is extremely similar to the plant homologue Cnx1G, underlining once more the already observed functional similarity of both proteins. The C2 splice cassette in the G domain of gephyrin, which is essential for both Moco biosynthesis (G. Schwarz, unpublished results) and receptor clustering [125], is formed by α helices 1 and 2; the latter is in close proximity to the hypothetical active site. A splicing of cassette 2 would result in a new molecular surface of the protein that might generate a binding site for another interacting protein.

Not only gephyrin but also its plant homologue Cnx1 exhibits functional properties that are distinct from Moco biosynthesis. Based on the observed cytoskeleton binding of gephyrin, a binding of Cnx1 to actin filaments, exclusively mediated by the E domain, could be demonstrated [109]. Also, gephyrin binds to actin filaments via its E domain [126], and this domain also interacts with profilin, a major modulator of actin polymerization [126]. What could be the functional significance of cytoskeleton binding of Cnx1 in terms of Moco biosynthesis? As in the case of the bicistronic expression of mocs1 and mocs2 genes or the domain fusion in Cnx1 or gephyrin, one can conclude that for higher eukaryotic cells it becomes important to facilitate substrate-product flow, which could result in microcompartmentalization of a hypothetical Moco-biosynthetic multienzyme complex (Fig. 7). Therefore, anchoring to submembranous cellular structures like the

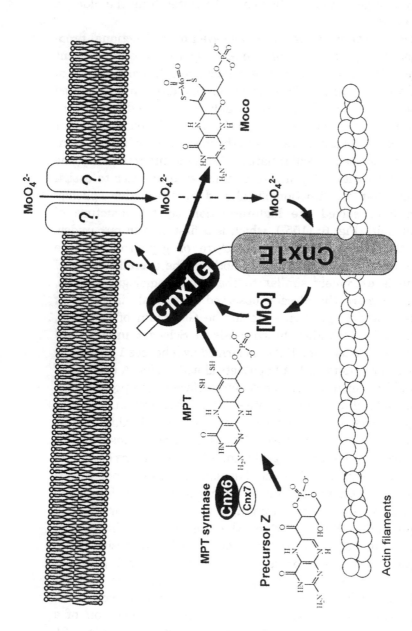

FIG. 7. Hypothetical model of the compartmentalization of Moco biosynthesis in a eukaryotic cell (plants) involving cytoskeleton binding as well as interaction with MPT synthase and an unidentified molybdate transport system [109]. The interaction of Cnx1 with an integral membrane protein is based on the function described for the homologous gephyrin protein in neuroreceptor anchoring, which is mediated by the G domain [125]. An interaction of Cnx1 with a molybdate anion transporter would facilitate substrate channeling to the E domain, which is thought to be involved in activating Mo prior its incorporation into MPT by the G domain.

cytoskeleton might help to organize such a biosynthetic machinery and bring it close to a yet unknown molybdate anion channel, providing the metal for Moco synthesis.

3.5. Terminal Sulfuration of the Mo Center in Eukaryotes

Figure 8 depicts the single steps of Mo enzyme formation in plants. (1) Moco is the end product of the Moco biosynthetic pathway. There are some indications that subsequent to Moco biosynthesis a Moco carrier/ storage protein could take over the task to buffer supply and demand of Moco. Such proteins were described for *Rhodospirillum rubrum* [127], *E. coli* [128], the green alga *Chlamydomonas reinhardtii* [129,130], and the field bean *Vicia faba* [131]. (2) Moco is distributed and incorporated into the apoproteins of Mo enzymes in an unknown way. (3) As a final step of maturation, XDH and AO as enzymes with a monooxo Mo center require the addition of a terminal inorganic sulfide to the Mo site. This sulfur ligand does not originate from the apoprotein nor does it come from the Moco moiety. For rat and fly, early work of Wahl et al. [132] demonstrated that in vitro this sulfur can be spontaneously lost or can be removed from AO/XDH by cyanide treatment generating an inactive enzyme. However, the reaction is reversible and the enzyme can be reactivated by sulfide treatment under reducing conditions.

In vivo, this terminal sulfur has to be added by a separate enzymatic reaction. There is good evidence for a Moco sulfurase (MCSU) catalyzing the activation of enzymes with a monooxo Mo center by insertion of the cyanolyzable sulfur into the active center (Fig. 8). MCSU activities were described for *D. melanogaster ma-l* [132]; the fungus *A. nidulans* (HxB [133]), cattle (MCSU [134]), and plants (ABA3 [135,136]). The N terminus of these MCSU proteins shares significant homologies to the bacterial sulfurase NifS. Furthermore, a highly conserved pyridoxal phosphate attachment site and a conserved cysteine residue are common to MCSU proteins and to NifS, so that a pyridoxal phosphate-dependent mechanism of (trans)sulfuration as described for NifS [137] was proposed for eukaryotes [136,138]: An MCSU-bound persulfide, resulting from the desulfuration of free L-cysteine to L-alanine, is likely to be transferred to Mo. The C-terminal

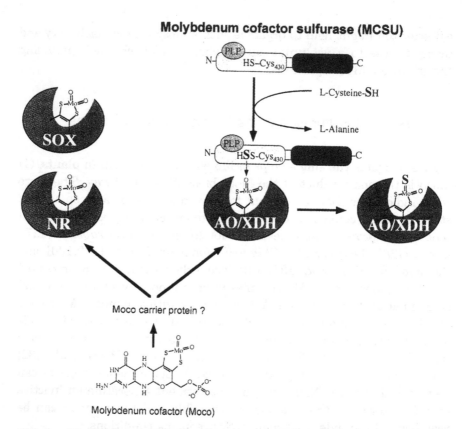

FIG. 8.   Working model for the function of the Moco sulfurase protein MCSU.
Moco as the end product of the Moco biosynthetic pathway is bound by an ill-
defined Moco storage protein and subsequently distributed and incorporated
into the apoproteins of Mo enzymes. As a final step of maturation, XDH and
AO—but not NR and SO—require the addition of a terminal inorganic sulfur
ligand to the Mo center. This reaction is catalyzed by MCSU encoded by *aba3* in
Arabidopsis. MCSU contains two functional domains; one is the NifS-like
domain with a PLP (pyridoxal phosphate) binding site that is thought to be
involved in desulfuration of L-cysteine [136].

domain non existent in NifS proteins but common to all MCSU proteins,
is probably responsible for mediating the contact between XDH/AO and
the transsulfurase (= NifS-like) domain of MCSU proteins [138].

Mutants defective in the terminal sulfurase MCSU were described
for fly [139], *A. nidulans* [133], and several plants [135,140,141].

They were lacking both XDH and AO activities but retained their activities of enzymes with a dioxo Mo center, like NR in plants [135] and SO in cattle and insects [132,134]. In both the fly *ma-l* and the fungal *hxB* genes, intragenic complementation was observed [138], demonstrating that the N-terminal NifS-like domain and the C-terminal domain function independently within the obviously oligomeric sulfurase protein. From this it could be assumed that in bacteria separate but interacting proteins were the predecessors of the two domains of the eukaryotic sulfurases. There are bacterial sequences in the databases homologous either to the N- or the C-terminal domain of the eukaryotic MCSU, but only functional assays in vivo and in vitro will reveal the bacterial counterparts.

3.6. Open Questions in Eukaryotes

Cnx4: There is one Moco-related gene [142] in plants that has no equivalent in *E. coli*. It represents the *Arabidopsis* Moco mutant *chl-7* [143], which is deregulated for the gene of γ-adaptin that is involved in vesicle transport to the plasma membrane, vacuole, and endosomes of eukaryotic cells. *Arabidopsis* plants expressing the γ-adaptin gene in antisense showed the typical Moco mutant phenotype, i.e., low MPT content and chlorate resistance [142]. The Cnx4 protein does not seem to be directly involved in MPT synthesis or Mo insertion; rather, it should have a pleiotropic effect. We favor the interpretation that a membrane protein essential for MPT synthesis is transported via the Golgi pathway to the plasma membrane.

Subcellular localization of Moco biosynthesis in eukaryotic cells is another open question. Here, the cytoskeleton association of the plant protein Cnx1 and mammalian gephyrin might be a clue for our understanding. Anchoring of a putative multienzyme complex for Moco biosynthesis on the cytoskeleton would ensure both the stabilization of such a complex and the fast and protected transfer of labile intermediates within the reaction sequence from GTP to Moco. Such a microcompartmentalization could be detected by advanced microscopic techniques using labeled antibodies, and the dynamics of formation and movement of a Moco biosynthesis complex could be followed in living cells by the noninvasive technique of time-lapse confocal laser scanning microscopy.

Moco carrier/storage proteins linking Moco biosynthesis with Moco users are still somewhat enigmatic. In fact, Moco-binding proteins were functionally characterized by their ability to reconstitute NR activity in the *nit-1* assay without prior heat release of the Moco-donating extract. However, their nature and detailed function within the cell is unknown.

Insertion of Moco into Mo enzymes as it occurs in the living eukaryotic cell is not understood. However, in a defined system of human SO and bacterial MPT synthase, Leimkühler and Rajagopalan [22] have shown that apoSO can directly take over MPT from MPT synthase and that molybdate is inserted into MPT as soon as the latter is bound to SO. Therefore, under in vitro conditions no additional proteins are required for conversion of MPT to Moco or for insertion of MPT into SO, provided that 5 mM molybdate is present in the mixture. In vivo, however, the concentrations of molybdate and of the target enzymes are orders of magnitude lower, so that mediating proteins such as MoeA/ MogA, Cnx1, and gephyrin are required for converting of MPT to Moco. For inserting of Moco into the target apoenzymes, either (still unknown) chaperones would be needed or the Moco carrier/storage proteins could become involved at this stage. For some bacterial Mo enzymes, system-specific chaperones are required for Moco insertion and protein folding, e.g., NarJ for *E. coli* NR [144] and XDHC for *Rhodobacter capsulatus* XDH [84].

4. REGULATION OF MOLYBDENUM COFACTOR
   SYNTHESIS

In *E. coli*, Moco biosynthesis is enhanced under anaerobiosis [145]. This regulation is mediated by the promoter of the *moa* operon. Also molybdate enhances Moco biosynthesis and the molybdate-binding protein ModE was found to be involved in this process [146]; for details compare Chapter 2 of this book. Once possessing sufficient amounts of active Moco, the bacterium represses Moco biosynthesis in a *moa* promoter-mediated way [147]. However, in the presence of tungstate this feedback inhibition of Moco biosynthesis is lost [147] so that one has to assume that there is a step in Moco formation that can discriminate between active Mo-MPT and functionally inactive W-MPT. The Mo-MPT dependent formation of the dinucleotide form of Moco is the most likely candidate for this regulatory link, however no mechanistic details were proposed yet. Recently Hasona et al. [148] analyzed the regulation of the *moe* operon and found that its expression is independent of genes coding for Mo transport and for MPT synthesis, rather anaerobic conditions as well as nitrate were stimulating *moe* expression. Earlier the authors had published that the product of the MoeA-catalyzed reaction is required for Mo-dependent control of genes coding for *E. coli* Mo enzymes [149]. Apparently the bacterium coordinates Moco biosynthesis with apoprotein synthesis at the level of *moe* operon transcription.

Little is known about the regulation of Moco genes in eukaryotes. In plants, their expression was found to be extremely low. As can be expected for house keeping genes, Moco genes should be expressed constitutively. *Arabidopsis* Cnx1 is constitutively synthesized in all organs throughout the plant and its amount does not depend on the nitrogen source of the plant [109]. By measuring RNA levels, the genes *cnx2* and *cnx3* were found to be expressed preferentially in the roots; they did not show regulation by nitrate [59].

The human *mocs* genes are expressed at very low level, and their mRNA could be detected in all organs although with varying abundance [60,115]. In particular muscle tissue and liver were rich in *mocs* expression. In the filamentous fungus *A. nidulans*, the expression of *cnx* genes was found to be constitutive and—with one exception—did not respond to the nitrogen source. The exception is the *cnxABC* locus (Table 1) involved in the first step of Moco biosynthesis, where a modest nitrate induction of transcription was observed [61].

The availability of sufficient amounts of Moco is essential for the cell to meet its changing demand for synthesizing Mo enzymes. In plants, the diurnal variation in the amount of NR protein requires a flexible regulation of Moco synthesis. Here the existence of Moco storage proteins would be a good means to buffer supply and demand of Moco. Downstream of Moco synthesis, insertion of the terminal sulfur ligand into the active Mo site of enzymes with a monooxo Mo center like AO and XDH could be another interesting regulatory switch point. The activity of this sulfurase could control the amount of functional AO/XDH molecules in the cell.

5. CONCLUSIONS

Moco is an ancient invention of nature. The structure of MPT as the organic moiety of Moco is highly conserved. In eubacteria, a nucleotide is attached to MPT and/or MPT can form dimers. Moco occurs ubiquitously—with the exception of the yeast *S. cerevisiae*, which possesses no Mo enzymes. During evolution of prokaryotes to eukaryotes the pterin core structure of Moco has not been changed. Accordingly its biosynthetic pathway is conserved and the diverse genome projects revealed a similar set of Moco genes in eubacteria, archaebacteria, fungi, fly, plants, and mammals. However, there are also differences between prokaryotes and eukaryotes to be mentioned. During evolution, the genes of single monofunctional prokaryotic proteins were fused in some cases to form two-domain multifunctional eukaryotic proteins like in *Arabidopsis* Cnx1 and animal gephyrin (comprising *E. coli* MogA and MoeA). Furthermore, the dinucleotide form of Moco has not been found in eukaryotes. Another difference resides in the complex morphology and size of a eukaryotic cell with its several compartments, thus making adaptations of the Moco biosynthetic pathway inevitable, as in the case of the cytoskeleton-binding function of Cnx1 protein or in the case of the Cnx4 protein that is part of the cellular vesicle transport.

It is important to note that proteins of the bacterial Moco biosynthetic pathway served also as evolutionary predecessors for very different eukaryotic pathways and functions: MoaD and MoeB are predecessors for the ubiquitin-dependent protein degradation pathway, and the fusion product of MogA and MoeA has—in addition to its Moco biosynthetic task—a neuroreceptor-anchoring function in mammals. For all organisms, Moco is essential, and a defective Moco has detrimental or lethal consequences due to the pleiotropic loss of all Mo enzymes. So the elucidation of Moco biosynthesis and regulation of this pathway as well as the steps of Moco assembly and insertion into the apoproteins are fascinating topics of ongoing research in several laboratories working with mammals, plants, insects, fungi, algae, and bacteria.

ACKNOWLEDGMENTS

We thank H. Schindelin (State University of New York at Stony Brook, USA), S. Leimkühler (Duke University, Durham, USA), D. Boxer (University of Dundee, Scotland, UK), T. Palmer (John Innes Center, Norwich, UK), and K. Shanmugam (University of Gainesville, Florida, USA) for providing unpublished results. We are grateful to P. Hänzelmann for critically reading the manuscript.

ABBREVIATIONS

AMP	adenosine 5'-monophosphate
AO	aldehyde oxidase
ATP	adenosine 5'-triphosphate
CMP	cytidine 5'-monophosphate
GTP	guanosine 5'-triphosphate
IMP	inosine 5'-monophosphate
MCD	molybdopterin cytosine dinucleotide
MCSU	molybdenum cofactor sulfurase
MGD	molybdopterin guanine dinucleotide
Moco	molybdenum cofactor
MPT	molybdopterin
NR	nitrate reductase
SO	sulfite oxidase
XDH	xanthine dehydrogenase
XO	xanthine oxidase

REFERENCES

1. J. A. Pateman, D. J. Cove, B. M. Rever, and D. B. Roberts, *Nature*, *201*, 58–60 (1964).

2. D. J. Cove and J. A. Pateman, *Nature*, *198*, 262–263 (1963).

3. G. J. Sorger and N. H. Giles, *Genetics*, *52*, 777–788 (1965).

4. A. B. Tomsett and R. H. Garrett, *Genetics*, *95*, 649–660 (1980).

5. J. H. Glaser and J. A. DeMoss, *J. Bacteriol*, *108*, 854–860 (1971).

6. C. K. Warner and V. Finnerty, *Mol. Gen. Genet*, *184*, 92–96 (1981).

7. A. J. Müller and R. R. Mendel, in *Molecular and Genetic Aspects of Nitrate Assimilation* (J. L. Wray and J. R. Kinghorn, eds.), Oxford University Press, Oxford, 1989, pp. 166–185.

8. P. A. Ketchum, H. Y. Cambier, W. A. Frazier, C. H. Madansky, and A. Naason, *Proc. Natl. Acad. Sci. USA*, *66*, 1016–1023 (1970).

9. A. Nason, A. D. Antoine, P. A. Ketchum, W. A. Frazier, and D. K. Lee, *Proc. Natl. Acad. Sci. USA*, *65*, 137–144 (1970).

10. A. Nason, K. Y. Lee, S. S. Pan, P. A. Ketchum, A. Lamberti, and J. DeVries, *Proc. Natl. Acad. Sci. USA*, *68*, 3242–3246 (1971).

11. P. T. Pienkos, V. K. Shah, and W. J. Brill, *Proc. Natl. Acad. Sci. USA*, *74*, 5468–5471 (1977).

12. V. K. Shah and W. J. Brill, *Proc. Natl. Acad. Sci. USA*, *74*, 3249–3253 (1977).

13. A. Nason, K. Y. Lee, S. S. Pan, and R. H. Ericson, *J. Less Common Metals*, *36*, 449–459 (1974).

14. R. R. Mendel, *Phytochemistry*, *22*, 817–819 (1983).

15. J. L. Johnson, B. E. Hainline, and K. V. Rajagopalan, *J. Biol. Chem.*, *255*, 1783–1786 (1980).

16. T. R. Hawkes and R. C. Bray, *Biochem J.*, *219*, 481–493 (1984).

17. R. R. Mendel, Z. A. Alikulov, N. P. Lvov, and A. J. Müller, *Mol. Gen. Genet.*, *181*, 395–399 (1981).

18. R. R. Mendel, R. J. Buchanan, and J. L. Wray, *Mol. Gen. Genet.*, *195*, 186–189 (1984).

19. R. C. Wahl, R. V. Hageman, and K. V. Rajagopalan, *Arch. Biochem. Biophys.*, *230*, 264–273 (1984).

20. R. V. Hagemann and K. V. Rajagopalan, *Meth. Enzymol.*, *122*, 399–412 (1986).

21. Z. A. Alikulov and R. R. Mendel, *Biochem. Physiol. Pflanzen.* *179*, 693–705 (1984).

22. S. Leimkühler and K. V. Rajagopalan, *J. Biol. Chem.*, *276*, 1837–1844 (2001).

23. J. H. Glaser and J. A. DeMoss, *Mol. Gen. Genet.*, *116*, 1–10 (1972).

24. V. Stewart and C. H. MacGregor, *J. Bacteriol.*, *151*, 788–799 (1982).

25. K. T. Shanmugam, V. Stewart, R. P. Gunsalus, D. H. Boxer, J. A. Cole, M. Chippaux, J. A. DeMoss, G. Giordano, E. C. Lin, and K. V. Rajagopalan, *Mol. Microbiol.*, *6*, 3452–3454 (1992).

26. A. J. Müller and R. Grafe, *Mol. Gen. Genet.*, *161*, 67–76 (1978).

27. R. R. Mendel and A. J. Müller, *Biochem. Physiol. Pflanzen*, *170*, 538–541 (1976).

28. R. R. Mendel and A. J. Müller, *Mol. Gen. Genet.*, *161*, 77–80 (1978).

29. A. Kleinhofs, R. L. Warner, J. M. Lawrence, J. M. Jeter, and D. A. Kudrna, in *Molecular and Genetic Aspects of Nitrate Assimilation* (J. L. Wray and J. R. Kinghorn, eds.), Oxford University Press, Oxford, 1989, pp. 197–211.

30. E. Fernandez and J. Cardenas, in *Molecular and Genetic Aspects of Ntirate Assimilation* (J. L. Wray and J. R. Kinghorn, eds.), Oxford University Press, Oxford, 1989, pp. 102–124.

31. M. R. Aguilar, J. Cardenas, and E. Fernandez, *Biochim. Biophys Acta*, *1073*, 463–469 (1991).

32. J. Gabard, F. Pelsy, A. Marion-Poll, M. Caboche, I. Saalbach, R. Grafe, and A. J. Müller, *Mol. Gen. Genet.*, *213*, 275–281 (1988).

33. M. Seo, S. Akaba, T. Oritani, M. Delarue, C. Bellini, M. Caboche, and T. Koshia, *Plant Physiol*, *116*, 687–693 (1998).

34. M. Seo, A. J. Peeters, H. Koiwai, T. Oritani, A. Marion-Poll, J. A. Zeevaart, M. Koornneef, Y. Kamiya, and T. Koshiba, *Proc. Natl. Acad. Sci. USA*, *97*, 12908–12913 (2000).

35. N. M. Crawford, in *Genetic Engineering: Principles and Methods*, Vol. 14 (J. K. Setlow, ed.), Plenum Press, New York, 1992, pp. 89–98.

36. R. R. Mendel, in *Plant Biotechnology and Development— Current Topics in Plant Molecular Biology*, Vol. 1 (P. M. Gresshoff, ed.), CRC Press, Boca Raton, FL, 1992, pp. 11–16.

37. R. R. Mendel, *Planta*, *203*, 399–405 (1997).

38. K. V. Rajagopalan and J. L. Johnson, *J. Biol. Chem.*, *267*, 10199–10202 (1992).

39. D. J. Cove, *Biol. Rev.*, *54*, 291–327 (1979).

40. N. S. Dunn-Coleman, *Curr. Genet.*, *8*, 581–588 (1984).

41. M. Stivaletta, C. K. Warner, S. Langley, and V. Finnerty, *Mol. Gen. Genet.*, *213*, 505–512 (1988).

42. M. Duran, F. A. Beemer, C. van de Heiden, J. Korteland, P. K. de Bree, M. Brink, S. K. Wadman, and I. Lombeck, *J. Inher. Metab. Dis.*, *1*, 175–178 (1978).

43. J. Reiss, *Hum. Genet.*, *106*, 157–163 (2000).

44. V. A. McKusick, *Mendelian Inheritance in Man. A Catalog of Human and Genetic Disorders*, Johns Hopkins University Press, Baltimore, 1994.

45. J. L. Johnson and S. K. Wadman, in *The Metabolic and Molecular Bases of Inherited Disease* (C. R. Scriver, A. L. Beaudet, W. S. Sly, and D. Valle, eds.), McGraw-Hill, New York, 1995, pp. 2271–2283.

46. J. L. Johnson, M. M. Wuebbens, R. Mandell, and V. E. Shih, *J. Clin. Invest.*, *83*, 897–903 (1989).

47. J. Reiss, S. Gross-Hardt, E. Christensen, P. Schmidt, R. R. Mendel, and G. Schwarz, *Am. J. Hum. Genet.*, *68*, 208–213 (2001).

48. M. Goto, A. Sakurai, K. Ohta, and H. Yamakami, *J. Biochem.*, *65*, 611–620 (1969).

49. J. L. Johnson, W. R. Waud, K. V. Rajagopalan, M. Duran, F. A. Beemer, and S. K. Wadman, *Proc. Natl. Acad. Sci. USA*, *77*, 3715–3719 (1980).

50. K. V. Rajagopalan, in *Escherichia coli and Salmonella typhimurium* (F. C. Neidhardt, ed.), ASM Press, Washington DC, 1996, pp. 674–679.

51. M. M. Wuebbens and K. V. Rajagopalan, *J. Biol. Chem.*, *268*, 13493–13498 (1993).

52. G. M. Brown, in *Folates and Pterins*, Vol. 2 (R. L. Blakley and S. J. Benkovics, eds.), Wiley, New York, 1985, pp. 299–419.

53. A. Bacher, in *Chemistry and Biochemistry of Flavoenzymes*, Vol. 1 (F. Müller, ed.), CRC Press, Boca Raton, FL, 1990, pp. 215–259.

54. M. M. Wuebbens and K. V. Rajagopalan, *J. Biol. Chem.*, *270*, 1082–1087 (1995).

55. C. Rieder, W. Eisenreich, J. O'Brien, G. Richter, E. Gotze, P. Boyle, S. Blanchard, A. Bacher, and H. Simon, *Eur. J. Biochem.*, *255*, 24–36 (1998).

56. C. Menendez, D. Siebert, and R. Brandsch, *FEBS Lett.*, *391*, 101–103 (1996).

57. H. J. Sofia, G. Chen, B. G. Hetzler, J. F. Reyes-Spindola, and N. E. Miller, *Nucleic Acids Res.*, *29*, 1097–1106 (2001).

58. M. M. Wuebbens, M. T. Liu, K. Rajagopalan, and H. Schindelin, *Structure Fold. Des.*, *8*, 707–718 (2000).

59. T. Hoff, K. M. Schnorr, C. Meyer, and M. Caboche, *J. Biol. Chem.*, *270*, 6100–6107 (1995).

60. B. Stallmeyer, G. Drugeon, J. Reiss, A. L. Haenni, and R. R. Mendel, *Am. J. Hum. Genet.*, *64*, 698–705 (1999).

61. S. E. Unkles, J. Smith, G. J. Kanan, L. J. Millar, I. S. Heck, D. H. Boxer, and J. R. Kinghorn, *J. Biol. Chem.*, *272*, 28381–28390 (1997).

62. M. E. Johnson and K. V. Rajagopalan, *J. Bacteriol.*, *169*, 110–116 (1987).

63. M. E. Johnson and K. V. Rajagopalan, *J. Bacteriol.*, *169*, 117–125 (1987).

64. D. M. Pitterle and K. V. Rajagopalan, *J. Biol. Chem.*, *268*, 13499–13505 (1993).

65. D. M. Pitterle, J. L. Johnson, and K. V. Rajagopalan, *J. Biol. Chem.*, *268*, 13506–13509 (1993).

66. J. P. McGrath, S. Jentsch, and A. Varshavsky, *EMBO J.*, *10*, 227–236 (1991).

67. A. Hershko and A. Ciechanover, *Annu. Rev. Biochem.*, *61*, 761–807 (1992).

68. S. E. Unkles, I. S. Heck, M. V. Appleyard, and J. R. Kinghorn, *J. Biol. Chem.*, *274*, 19286–19293 (1999).

69. G. Gutzke, J. Nieder, S. Orlich, M. Kuhnert, A. Sandmann, B. Riedel, R. R. Mendel, and G. Schwarz, *J. Biol. Chem.*, *276*, 36268–36274 (2001).

70. B. Stallmeyer, A. Nerlich, J. Schiemann, H. Brinkmann, and R. R. Mendel, *Plant J.*, *8*, 751–762 (1995).

71. T. P. Begley, D. M. Downs, S. E. Ealick, F. W. McLafferty, A. P. Van Loon, S. Taylor, N. Campobasso, H. J. Chiu, C. Kinsland, J. J. Reddick, and J. Xi, *Arch. Microbiol.*, 171, 293–300 (1999).

72. M. J. Rudolph, M. M. Wuebbens, K. V. Rajagopalan, and H. Schindelin, *Nat. Struct. Biol.*, 8, 42–46 (2001).

73. S. Leimkühler, M. M. Wuebbens, and K. V. Rajagopalan, *J Biol. Chem.*, 276, 34695–34701 (2001).

74. M. V. Appleyard, J. Sloan, G. J. Kana'n, I. S. Heck, J. R. Kinghorn, and S. E. Unkles, *J. Biol. Chem.*, 273, 14869–14876 (1998).

75. J. Nieder, B. Stallmeyer, H. Brinkmann, and R. R. Mendel, in *Sulphur Metabolism in Higher Plants* (W. J. Cram, L. J. De Kok, I. Stulen, C. Brunold, and H. Rennenberg, eds.), Backhuys, Leiden, 1997.

76. GeneBank, *mocs3*, accession number AF102544.

77. S. Rech, U. Deppenmeier, and R. P. Gunsalus, *J. Bacteriol.*, 177, 1023–1029 (1995).

78. M. S. Joshi, J. L. Johnson, and K. V. Rajagopalan, *J. Bacteriol.*, 178, 4310–4312 (1996).

79. J. Kuper, T. Palmer, R. R. Mendel, and G. Schwarz, *Proc. Natl. Acad. Sci. USA*, 97, 6475–6480 (2000).

80. G. Schwarz, D. H. Boxer, and R. R. Mendel, *J. Biol. Chem.*, 272, 26811–26814 (1997).

81. B. Stallmeyer, G. Schwarz, J. Schulze, A. Nerlich, J. Reiss, J. Kirsch, and R. R. Mendel, *Proc. Natl. Acad. Sci. USA*, 96, 1333–1338 (1999).

82. M. T. Liu, M. M. Wuebbens, K. V. Rajagopalan, and H. Schindelin, *J. Biol. Chem.*, 275, 1814–1822 (2000).

83. I. Lüke, J. Kuper, T. Palmer, T. Eilers, R. R. Mendel, and G. Schwarz (submitted).

84. S. Leimkühler, S. Angermüller, G. Schwarz, R. R. Mendel, and W. Klipp, *J. Bacteriol.*, 181, 5930–5939 (1999).

85. A. Hasona, R. M. Ray, and K. T. Shanmugam, *J. Bacteriol.*, 180, 1466–1472 (1998).

86. B. Krüger and O. Meyer, *Eur. J. Biochem.*, 157, 121–128 (1986).

87. B. Krüger and O. Meyer, *Biochim. Biophys. Acta.*, *912*, 357–364 (1987).

88. J. L. Johnson, N. R. Bastian, and K. V. Rajagopalan, *Proc. Natl. Acad. Sci. USA*, *87*, 3190–3194 (1990).

89. S. Leimkühler, M. Kern, P. S. Solomon, A. G. McEwan, G. Schwarz, R. R. Mendel, and W. Klipp, *Mol. Microbiol.*, *27*, 853–869 (1998).

90. D. Hettrich, B. Peschke, B. Tshisuaka, and F. Lingens, *Biol. Chem. Hoppe-Seyler*, *372*, 513–517 (1991).

91. J. L. Johnson, L. W. Indermaur, and K. V. Rajagopalan, *J. Biol. Chem.*, *266*, 12140–12145 (1991).

92. T. Palmer, A. Vasishta, P. W. Whitty, and D. H. Boxer, *Eur. J. Biochem.*, *222*, 687–692 (1994).

93. T. Palmer, I. P. Goodfellow, R. E. Sockett, A. G. McEwan, and D. H. Boxer, *Biochim. Biophys. Acta.*, *1395*, 135–140 (1998).

94. D. J. Eaves, T. Palmer, and D. H. Boxer, *Eur. J. Biochem.*, *246*, 690–697 (1997).

95. C. A. Temple and K. V. Rajagopalan, *J. Biol. Chem.*, *275*, 40202–40210 (2000).

96. T. Palmer, C. L. Santini, C. Iobbi-Nivol, D. J. Eaves, D. H. Boxer, and G. Giordano, *Mol. Microbiol.*, *20*, 875–884 (1996).

97. M. W. Lake, C. A. Temple, K. V. Rajagopalan, and H. Schindelin, *J. Biol. Chem.*, *275*, 40211-40217 (2000).

98. C. E. M. Stevenson, F. Sargent, G. Buchanan, T. Palmer, and D. M. Lawson, *Structure*, *8*, 1115–1125 (2000).

99. P. Hänzelmann and O. Meyer, *Eur. J. Biochem.*, *255*, 755–765 (1998).

100. R. A. Rothery, J. L. Grant, J. L. Johnson, K. V. Rajagopalan, and J. H. Weiner, *J. Bacteriol.*, *177*, 2057–2063 (1995).

101. R. A. Rothery, A. Magalon, G. Giordano, B. Guigliarelli, F. Blasco, and J. H. Weiner, *J. Biol. Chem.*, *273*, 7462–7469 (1998).

102. U. C. Gupta, *Molybdenum in Agriculture*, Cambridge University Press, Cambridge, UK, 1997.

103. J. W. Tweedie and I. H. Segel, *Biochim. Biophys. Acta*, *196*, 95–106 (1970).

104. J. K. Rosentel, F. Healy, J. A. Maupin-Furlow, J. H. Lee, and K. T. Shanmugam, *J. Bacteriol.*, *177*, 4857–4864 (1995).

105. A. Llamas, K. L. Kalakoutskii, and E. Fernandez, *Plant, Cell and Environment*, *23*, 1247–1255 (2000).

106. I. S. Heck and H. Ninnemann, *Photochem. Photobiol.*, *61*, 54–60 (1995).

107. F. J. Braaksma and W. J. Feenstra, *Theor. Appl. Genet.*, *64*, 83–90 (1982).

108. G. Schwarz, D. H. Boxer, and R. R. Mendel, in *Chemistry and Biology of Pteridines and Folates* (W. Pfleiderer and H. Rokos, eds.), Blackwell Scientific, Berlin, 1997, pp. 697–702.

109. G. Schwarz, J. Schulze, F. Bittner, T. Eilers, J. Kuper, G. Bollmann, A. Nerlich, H. Brinkmann, and R. R. Mendel, *Plant Cell*, *12*, 2455–2472 (2000).

110. S. Xiang, J. Nichols, K. V. Rajagopalan, and H. Schindelin, *Structure*, *9*, 299–310 (2001).

111. C. Menendez, A. Otto, G. Igloi, P. Nick, R. Brandsch, B. Schubach, B. Bottcher, and R. Brandsch, *Eur. J. Biochem.*, *250*, 524–531 (1997).

112. K. P. Kamdar, M. E. Shelton, and V. Finnerty, *Genetics*, *137*, 791–801 (1994).

113. L. J. Millar, I. S. Heck, J. Sloan, G. J. M. Kana'n, J. R. Kinhorn, and S. E. Unkles (submitted).

114. P. Prior, B. Schmitt, G. Grenningloh, I. Pribilla, G. Multhaup, K. Beyreuther, Y. Maulet, P. Werner, D. Langosch, J. Kirsch, and H. Betz, *Neuron*, *8*, 1161–1170 (1992).

115. J. Reiss, N. Cohen, C. Dorche, H. Mandel, R. R. Mendel, B. Stallmeyer, M. T. Zabot, and T. Dierks, *Nat. Genet.*, *20*, 51–53 (1998).

116. J. Reiss, E. Christensen, and C. Dorche, *Prenat. Diagn.*, *19*, 386–388 (1999).

117. T. A. Gray and R. D. Nicholls, *RNA*, *6*, 928–936 (2000).

118. J. Reiss, C. Dorche, B. Stallmeyer, R. R. Mendel, N. Cohen, and M. T. Zabot, *Am. J. Hum. Genet.*, *64*, 706–711 (1999).

119. J. Kirsch and H. Betz, *Nature*, *392*, 717–720 (1998).

120. J. Kirsch, I. Wolters, A. Triller, and H. Betz, *Nature*, *366*, 745–748 (1993).

121. G. Feng, H. Tintrup, J. Kirsch, M. C. Nichol, J. Kuhse, H. Betz, and J. R. Sanes, *Science*, *282*, 1321–1324 (1998).

122. D. M. Sabatini, R. K. Barrow, S. Blackshaw, P. E. Burnett, M. M. Lai, M. E. Field, B. A. Bahr, J. Kirsch, H. Betz, and S. H. Snyder, *Science*, *284*, 1161–1164 (1999).

123. S. Kins, H. Betz, and J. Kirsch, *Nat. Neurosci.*, *3*, 22–29 (2000).

124. M. Ramming, S. Kins, N. Werner, A. Hermann, H. Betz, and J. Kirsch, *Proc. Natl. Acad. Sci. USA*, *97*, 10266–10271 (2000).

125. J. Meier, M. De Chaldee, A. Triller, and C. Vannier, *Mol. Cell. Neurosci.*, *16*, 566–577 (2000).

126. T. Giesemann, G. Schwarz, M. Rothkegel, K. Schlüter, J. Paulukat, R. R. Mendel, and B. M. Jockusch (submitted).

127. P. A. Ketchum and C. L. Sevilla, *J. Bacteriol.*, *116*, 600–609 (1973).

128. N. K. Amy and K. V. Rajagopalan, *J. Bacteriol.*, *140*, 114–124 (1979).

129. M. Aguilar, K. Kalakoutskii, J. Cardenas, and E. Fernandez, *FEBS Lett.*, *307*, 162–163 (1992).

130. C. P. Witte, M. I. Igeno, R. Mendel, G. Schwarz, and E. Fernandez, *FEBS Lett.*, *431*, 205–209 (1998).

131. K. Kalakoutskii and E. Fernandez, *Planta*, *201*, 64–70 (1996).

132. R. C. Wahl, C. K. Warner, V. Finnerty, and K. V. Rajagopalan, *J. Biol. Chem.*, *257*, 3958–3962 (1982).

133. C. Scazzocchio, F. B. Holl, and A. I. Foguelman, *Eur. J. Biochem.*, *36*, 428–445 (1973).

134. T. Watanabe, N. Ihara, T. Itoh, T. Fujita, and Y. Sugimoto, *J. Biol. Chem.*, *275*, 21789–21792 (2000).

135. K. M. Leon-Klooserziel, M. A. Gil., G. J. Ruijs, S. E. Jacobsen, N. E. Olszewski, S. H. Schwartz, J. A. D. Zeevaart, and A Q Koorneef, *Plant J.*, *10*, 655–661 (1996).

136. F. Bittner, M. Oreb, and R. R. Mendel, *J. Biol. Chem.*, *276* (in press).

137. L. Zheng and D. R. Dean, *J. Biol. Chem.*, *269*, 18723–18726 (1994).

138. L. Amrani, J. Primus, A. Glatigny, L. Arcangeli, C. Scazzocchio, and V. Finnerty, *Mol. Microbiol.*, *38*, 114–125 (2000).

139. V. Finnerty, M. McCarron, and G. B. Johnson, *Mol. Gen. Genet.*, *172*, 37–43 (1979).

140. M.-T. Leydecker, T. Moureaux, Y. Kreapiel, K. Schnorr, and M. Caboche, *Plant Physiol.*, *107*, 1427–1431 (1995).

141. E. Marin and A. Marion-Poll, *Plant Physiol. Biochem.*, *35*, 369–372 (1997).

142. K. Schledzewski, PhD thesis, Technical University of Braunschweig, Germany, 1997.

143. S. T. LaBrie, J. Q. Wilkinson, Y. F. Tsay, K. A. Feldman, and N. M. Crawford, *Mol. Gen. Genet*, *233*, 169–176 (1992).

144. F. Blasco, J. P. Dos Santos, A. Magalon, C. Frixon, B. Guigliarelli, C. L. Santini, and G. Giordano, *Mol. Microbiol.*, *28*, 435–447 (1998).

145. K. P. Baker and D. H. Boxer, *Mol. Microbiol.*, *5*, 901–907 (1991).

146. P. M. McNicholas, S. A. Rech, and R. P. Gunsalus, *Mol. Microbiol.*, *23*, 515–524 (1997).

147. L. A. Anderson, E. McNairn, T. Leubke, R. N. Pau, and D. H. Boxer, *J. Bacteriol.*, *182*, 7035–7043 (2000).

148. A. Hasona, W. T. Self, and K. T. Shanmugam, *Arch. Microbiol.*, *175*, 178–188 (2001).

149. A. Hasona, W. T. Self, R. M. Ray, and K. T. Shanmugam, *FEMS Microbiol. Lett.*, *169*, 111–116 (1998).

150. S. L. Rivers, E. McNairn, F. Blasco, G. Giordano, and D. H. Boxer, *Mol. Microbiol.*, *8*, 1071–1081 (1993).

151. GeneBank, *cnx7*, accession number AF208343.

152. GeneBank, *cnx6*, accession number AJ133519.

153. T. Eilers, G. Schwarz, H. Brinkmann, C. Witt, T. Richter, J. Nieder, B. Koch, R. Hille, R. Hänsch, and R. R. Mendel, *J. Biol. Chem.* (in press).

154. G. Gutzke, J. Nieder, A. Freuer, A. Sandmann, S. Orlich, M. Kuhnert, B. Riedel, R. R. Mendel, G. Schwarz (submitted).

155. G. Schwarz, N. Schrader, R. R. Mendel, H. J. Hecht, and H. Schindelin, *J. Mol. Biol. 312*, 405–418 (2001).

10

Molybdenum in Nitrate Reductase and Nitrite Oxidoreductase[*]

Peter M. H. Kroneck and Dietmar J. Abt

Fachbereich Biologie, Universität Konstanz, D-78457 Konstanz,
Germany

---

[*] This work is dedicated to Professor Jack T. Spence on the occasion of his 70th birthday.

1. INTRODUCTION

The nitrogen cycle has received considerable attention over the past decades because of its ecological importance (Fig. 1). Nitrate reduction plays a key role and has important agricultural, environmental, and public health implications. Assimilatory nitrate reduction, performed by bacteria, fungi, algae, and higher plants, is one of the most fundamental biological processes. There is worldwide concern over the tremendous use of fertilizers in agricultural activities, leading to nitrate accumulation in groundwater. Consumption of drinking water with high levels of nitrate has been associated with various diseases due to formation of highly toxic N-nitroso compounds. Nitrogen oxides produced by denitrification are also linked with the greenhouse effect and the depletion of stratospheric ozone. Thus, the amount of dinitrogen monoxide (N_2O, nitrous oxide, laughing gas) is steadily increasing in the atmosphere where it can last 150 years [1].

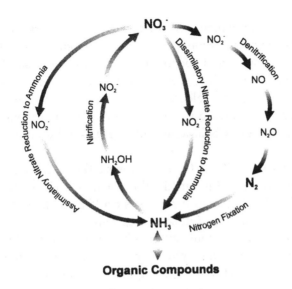

FIG. 1.   The biological nitrogen cycle.

Nitrogen is a basic element for life because it is a component of essential biomolecules, such as amino acids, proteins, and nucleic acids. Nitrogen exists in the biosphere in several oxidation states, ranging from N(V) to N(–III). Interconversions of these nitrogen species constitute the global biogeochemical nitrogen cycle which is sustained by biological processes, with bacteria playing a predominant role [2–7]. Inorganic nitrogen is converted to a biologically useful form by dinitrogen fixation ($N_2 \rightarrow NH_3$) or nitrate assimilation ($NO_3^- \rightarrow NH_3$) and the incorporation of ammonia into organic molecules. Inorganic nitrogen compounds are recycled from the environment by nitrification ($NH_3 \rightarrow NO_3^-$), the oxidative conversion of ammonia to nitrate, denitrification ($NO_3^- \rightarrow N_2$) whereby nitrate is successively transformed to nitrite, nitrogen monoxide (NO, nitric oxide), dinitrogen monoxide, and dinitrogen, and nitrate ammonification ($NO_3^- \rightarrow NH_3$), with nitrite as intermediate (Fig. 1).

Nitrate is a major source of inorganic nitrogen for plants, fungi, and many species of bacteria. Note that animals cannot assimilate nitrate. Assimilation involves three pathway-specific steps: uptake, reduction to nitrite, and further reduction to ammonia. Ammonia is then metabolized into central pathways. In recent years the genes for nitrate assimilation systems have been characterized in several bacterial species. The physiology and biochemistry of nitrate assimilation has been reviewed most recently [5,8,9]. In addition to its role as a nutrient, nitrate acts as an efficient electron acceptor for the conservation of metabolic energy, i.e., nitrate respiration, and the dissipation of excess reducing power for redox balancing, nitrate dissimilation, as discussed below.

It is thought that CO_2, CO, and N_2 were the main components by the time of the origin of life within the Earth's early atmosphere. In addition, NO, H_2, H_2O, and sulfuric gases could also have been present, and probably trace amounts of dioxygen as well [10]. Thus, it may not be too surprising that several molecular links exist between the evolution of denitrification enzymes and cytochrome oxidase, the terminal oxidase of aerobic respiration [11]. The most important is the homology between nitric oxide reductase and heme/copper cytochrome oxidases, and the presence of the mixed-valence $[Cu^{1.5+}(S_{Cys})_2 Cu^{1.5+}]$ Cu_A electron transfer center in nitrous oxide reductase and heme/copper cytochrome oxidases [12,13].

In this chapter we will focus on recent advances in the field of nitrate metabolism, with some emphasis on the pathway of ammonium nitrification in bacteria. Spectroscopic, structural, and mechanistic aspects of the molybdenum-dependent nitrate reductase will be presented, and the physiology, biochemistry, and molecular characteristics of the different types of nitrate reductase will be compared. In addition, biochemical and spectroscopic data of the molybdenum enzyme nitrite oxidoreductase from nitrifying bacteria will be discussed. Finally, environmental aspects, including the application of biosensors, will be addressed.

Because of its importance, research on nitrate metabolism has become a major issue as documented by a remarkable number of publications in biology, ecology, and chemistry. In view of the complexity of the topic, the following comprehensive reviews are recommended for further reading [2,4–7,11,14–17].

2. BACTERIAL RESPIRATORY CHAINS, METALLOENZYMES, AND BIOENERGETICS

Within the biogeochemical nitrogen cycle N_xO_y compounds serve as electron acceptors of anaerobic respiratory chains [17]. Prokaryotes carry out their energy conservation during reduction of nitrogen oxides and use complex transition metalloenzymes for their transformation [4] (Fig. 2). Dissimilatory nitrate reduction proceeds via two different pathways, i.e., denitrification and nitrate ammonification (Eqs. (1) and (2)].

$$^{V}NO_3^- \rightarrow {}^{III}NO_2^- \rightarrow {}^{II}NO \rightarrow {}^{I}N_2O \rightarrow {}^{0}N_2 \tag{1}$$

$$^{V}NO_3^- \rightarrow {}^{III}NO_2^- \rightarrow {}^{-III}NH_3 \tag{2}$$

$$^{-III}NH_3 \rightarrow {}^{-I}NH_2OH \rightarrow {}^{III}NO_2^- \tag{3}$$

Microbial nitrification is a two-step process carried out by two distinct groups of bacteria, placed together in the Nitrobacteriaceae [18]. The

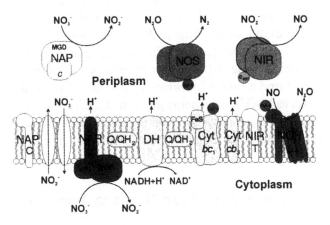

FIG. 2.   Molecular basis of denitrification in *Pseudomonas stutzeri*. The system comprises the respiratory nitrate reductase (NAR), the cd_1 nitrite reductase (NIR), the bc heme NO reductase (NOR), and the Cu-dependent nitrous oxide reductase (NOS). Evidence for the existence of the periplasmic nitrate reductase (NAP) is derived from DNA hybridization experiments. FeS, iron-sulfur centers; b, c, and d_1, heme b, heme c, and heme d_1. Components of the constitutive aerobic respiratory chain are NADH dehydrogenase complex (DH), quinone cycle (Q, QH_2), cytochrome bc_1 complex (Cyt bc_1), and the cytochrome cb_3 terminal oxidase complex (Cyt cb_3). (Courtesy of W. G. Zumft, unpublished).

first step, the oxidative conversion of ammonia to nitrite, is performed
by ammonia-oxidizing bacteria such as *Nitrosomonas europaea* [16,19].
The oxidation of ammonia to nitrite via hydroxylamine is catalyzed by
ammonia monooxygenase and the octaheme enzyme hydroxylamine oxidoreductase [Eq. (3)] [20,21]. The second step, conversion of nitrite to
nitrate, is performed by nitrite-oxidizing bacteria, such as *Nitrobacter*
species. In this case, a membrane-bound molybdenum-dependent nitrite
oxidoreductase is involved [22].

2.1. Denitrification—From Nitrate to Dinitrogen

In denitrification, NO_3^- is reduced via NO_2^- to the gaseous products
NO, N_2O, and N_2. The biology and microbial ecology of this pathway has been reviewed [4]. Denitrifying bacteria utilize the positive
redox potentials of nitrate $(E^{0'}(NO_3^-/NO_2^-) = +0.43$ V), nitrite
$[E^{0'}(NO_2^-/NO) = +0.35$ V], nitric oxide $[E^{0'} (NO/N_2O) = +1.175$ V],
and nitrous oxide $(E^{0'}(N_2O/N_2) = +1.355$ V] [23] for energy conservation by chemiosmotic coupling of ATP synthesis and electron transport.
The terminal oxidoreductases of denitrifying bacteria obtain the reducing equivalents via quinones from NADH and succinate dehydrogenase
[4]. The nitrite reductase is either a cytochrome cd_1 or a copper enzyme
carrying the type 1 and a type 2 Cu center [7,24–28]. The resulting NO is
further converted to N_2O by a *bc* heme protein [11,17,29] followed by the
reduction of N_2O to N_2, a strictly Cu-dependent process catalyzed by
nitrous oxide reductase carrying Cu_A and a tetranuclear cluster Cu_z
[12,30–32] (Fig. 2).

2.2. Nitrate Ammonification—From Nitrate to Ammonia

The reduction of nitrate to nitrite or ammonia $[E^{0'}/(NO_3^-/NH_4^+) =
+0.34$ V] was described for bacteria with a fermentative rather than a
respiratory metabolism [2]. However, growth of various nitrate-ammonifying bacteria by oxidation of nonfermentable substrates, such as formate linked to the reduction of nitrate or nitrite to ammonia,
demonstrated that nitrate ammonification may also function as a
respiratory energy-conserving process [3]. Nitrite is the only liberated

intermediate that is subsequently converted to ammonia in a six-electron step by a cytochrome c nitrite reductase [Eq. (2)] [33–35]. Dihydrogen and formate are the predominant electron donors [3], but sulfide can also serve as electron donor, thus connecting the biogeochemical cycles of nitrogen and sulfur [36]. In the sulfur- and sulfate-reducing species of the δ- and ε-proteobacteria, dissimilatory nitrate reduction to ammonia is the predominant metabolism during anaerobic growth on formate or dihydrogen. However, in γ-proteobacteria, such as *Escherichia coli*, the observed expression levels of nitrite reductase were considerably lower. The presence of a highly active nitrite reduction system might serve an important purpose, namely, the disposal of toxic nitrite anion.

2.2.1.   *From Nitrate to Nitrite*

A model for the organization of respiratory nitrate ammonification and direction of proton exchange associated with electron transfer from formate to nitrate in *E. coli* has been described earlier [3]. The reduction of nitrate to nitrite by the oxidation of formate is linked to the generation of a proton electrochemical potential. The membrane-bound nitrate reductase of *E. coli* forms a heterotrimeric complex (Fig. 2) [37]. Menaquinone and/or ubiquinone were shown to serve as electron mediator between the primary dehydrogenase and the nitrate reductase [38–43]. The formate dehydrogenase of *E. coli*, encoded by the *fdnGHJI* operon, has its hydrophilic active site subunit *fdnG* exposed to the periplasm [14,44,45]. A low-potential cytochrome b was coupled to the formate dehydrogenase of *E. coli* [38], *Wolinella succinogenes* [46,47], and *Sulfurospirillum deleyianum* [48]. The quinone involved in the oxidation of formate generated a transmembrane proton electrochemical potential through a redox loop mechanism. Reduction of nitrate resulted in the consumption of two protons from the cytoplasm and the formation of nitrite and water. Energy was conserved in a chemiosmotic mechanism by the transfer of four negative charges across the cytoplasmic membrane of *E. coli* during nitrate reduction to nitrite with formate, or H_2 [3]. Whether the topology of the nitrate reductase complex from *E. coli* also holds for other nitrate-ammonifying bacteria has to be shown. A cytochrome c-dependent periplasmic nitrate reductase was recently identified in *E. coli* [2]. *S. deleyianum* and *D. desulfuricans* also showed a membrane-associated or soluble nitrate reductase catalyzing the respiratory reduction of nitrate to nitrite [49–51]. A cytochrome c-dependent nitrate reductase was also isolated from *Geobacter metallireducens* [52]. Obviously, nitrate uptake and nitrite extrusion into the bulk phase is dispensable for periplasmic nitrate reductases. In addition, a periplasmic location of nitrate reductases will prevent the interaction of potentially toxic compounds, such as nitrite or NO, with cytoplasmic proteins.

2.2.2. From Nitrite to Ammonia

In nitrate-ammonifying bacteria, especially in enterobacteria, there may exist two independently regulated dissimilatory ways of nitrite reduction to ammonia with two different physiological functions [2]. A cytoplasmic siroheme-dependent nitrite reductase activity (NADH:nitrite oxidoreductase) [53,54] confers on *E. coli* the advantage of regenerating NAD^+ during anaerobic growth to maintain glycolysis [55]. Besides the supply of ammonia for biosynthesis, this nitrite reductase leads to the detoxification of intracellularly accumulated nitrite. The other nitrite reductase is a periplasmic cytochrome c that was isolated from *E. coli* [56,57]. A membrane-bound formate-nitrite oxidoreductase complex with a respiratory function was first described for *E. coli* [39,40].

A model of the respiratory chain for the transformation of nitrite was proposed for various nitrite-ammonifying bacteria [3]. A periplasmically oriented membrane-bound formate dehydrogenase, or a hydrogenase, was included in the model. Most likely, the primary dehydrogenases functioning in the reduction of nitrate to nitrite, and nitrate to ammonia, are identical entities [14]. The membrane-associated nitrite reductase should face the periplasm. The extracytoplasmic location is characteristic for c-type cytochromes due to their evolutionary advantage of a covalent heme binding [58–60].

Early on, there was good evidence that a quinone pool was mediating the electron flow between the formate dehydrogenase, or hydrogenase, and the cytochrome c nitrite reductase in *E. coli* and other microorganisms [3].

2.2.3. The Cytochrome c Nitrite Reductase Complex

The enzyme from *E. coli* was the first cytochrome *c* NiR that had been completely sequenced [61,62]. It is a heterooligomeric complex encoded by the *nrfABCDEFG* operon, with *nrfA* encoding for the periplasmic nitrite-reactive cytochrome *c* (M_r 51 kDa). Later on, sequences of cytochrome *c* NiR from other microorganisms, including *H. influenzae*, *S. deleyianum*, and *W. succinogenes*, became available in the GenBank/EBI sequence database, and high-resolution structures of cytochrome *c* NiR (soluble part, NrfA) of *S. deleyianum* and *W. succinogenes* were reported [34,35].

The functional nitrite reductase system of *W. succinogenes* and *S. deleyianum* was shown to be a complex of the soluble part NrfA and a membrane-anchored tetraheme *c*-type cytochrome NrfH, which acts as a quinol oxidase to receive electrons from the membranous quinone pool [63]. The current working hypothesis thus implies that the NrfA dimer is associated with the peripheral membrane proten NrfH as depicted in the model (Fig. 3).

Most recently, an enzyme complex that exhibited both nitrate and nitrite reductase activity was solubilized and partially purified from the membrane fraction of the dissimilatory iron-reducing bacterium *G. metallireducens*. The complex was composed of four different polypep-

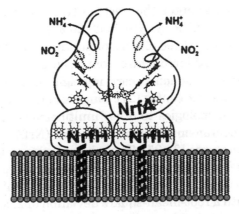

FIG. 3.   Schematic model of the formate-dependent cytochrome *c* nitrite reductase system for the *nrfHA* operon. (Reprinted from Ref. 35, with permission.)

tides, one of them a multiheme cytochrome c with a molecular mass of 62 kDa. Surprisingly, no molybdenum was detected in the preparation [64].

2.3. Ammonia Oxidation—From Ammonia to Hydroxylamine and Nitrite

Nitrifying bacteria connect the oxidized and reduced sides of the nitrogen cycle by nitrification, the conversion of ammonium to nitrogen oxyanions (Fig. 1) [16,65]. The most important microorganisms involved in the transformation of ammonia to nitrite, and nitrite to nitrate, are the lithoautotrophic ammonia- and nitrite-oxidizing bacteria placed in the family Nitrobacteraceae [65]. The obligate aerobic lithotrophic bacterium *Nitrosomonas europaea* derives energy from the oxidation of ammonia to nitrite. Thus, ammonia is oxidized to hydroxylamine by ammonia monooxygenase, which is converted to nitrite by hydroxylamine oxidoreductase [Eqs. (4) and (5)] [66,67].

$$^{-III}NH_3 + 2e^- + 2H^+ + O_2 \longrightarrow {}^{-1}NH_2OH + H_2O \qquad (4)$$
$$^{-1}NH_2OH + H_2O \longrightarrow {}^{III}NO_2^- + 4e^- + 5H^+ \qquad (5)$$

Two of the four electrons generated from hydroxylamine oxidation are used to support the oxidation of additional ammonia molecules, while the other two electrons enter the electron transfer chain and are used for CO_2 reduction and ATP biosynthesis [20,68].

Most recently, the three-dimensional structure (resolution 2.8 Å) of the hydroxylamine oxidoreductase from *N. europaea* revealed a homotrimeric unit, with each monomer carrying eight heme centers including the so-called P_{460} catalytic center [21]. Surprisingly, this enzyme showed obvious structural homologies to other multiheme c-type cytochromes, among them the pentaheme cytochrome c NiR (NrfA) from *S. deleyianum* and *W. succinogenes* [34,35]. These homologies mainly concern the arrangement of heme groups but not the surrounding protein. The five hemes of the cytochrome c NiR aligned to hemes 4–8 of hydroxylamine oxidoreductase with a root-mean-square deviation of 2.00 Å [35] (Fig. 4). However, while a protoporphyrin IX heme with a novel lysine coordination forms the active site of nitrite reductase, the hydroxylamine oxido-

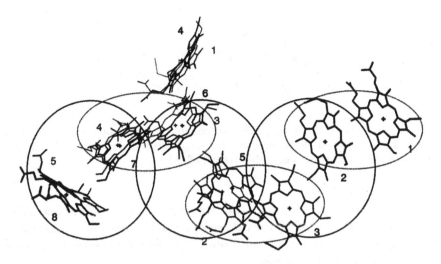

FIG. 4.   Heme-packing motifs. Superposition of the heme groups of cytochrome c nitrite reductase (*gray*) and hydroxylamine oxidoreductase (*black*) numbered according to their attachment to the protein chain. (Reprinted from Ref. 35, with permission.)

reductase features the P_{460}-type heme with a covalent link to a tyrosine residue. The iron atom of heme P460 appears to be five-coordinated, with a histidine as an axial protein ligand. The sixth position of the heme iron is not occupied and hydroxylamine can bind to this position [21].

3. NITRATE REDUCTASE

All eukaryotic and bacterial nitrate reductases are molybdenum-dependent enzymes and possess a unique cofactor, the so-called molybdopterin (Fig. 5) [69]. Nitrate-reducing archaea are also known, and putative molybdopterin-binding enzymes can be identified in the genome sequence of *Archaeoglobus fulgidus*, which suggests the presence of molybdopterin-dependent enzymes such as nitrate reductase, dimethyl sulfoxide (DMSO) reductase, or polysulfide reductase [17]. Most recently, two molybdenum-free nitrate-reducing systems have been reported. One was isolated from *Pseudomonas isachenkovii* and found

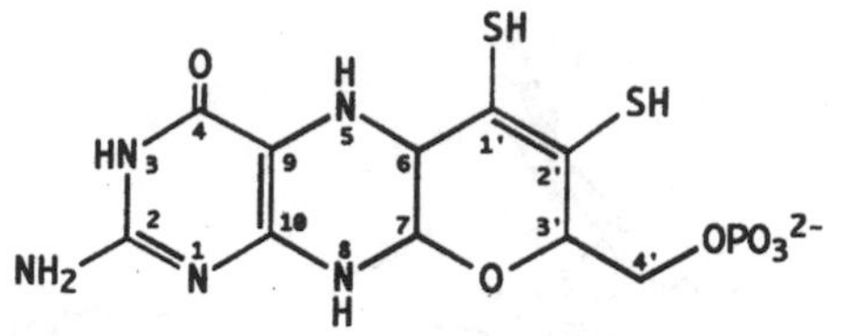

FIG. 5. Structure of the organic component of the molybdenum cofactor (Moco) in the tricyclic form as observed in all crystal structures of enzymes containing this cofactor. (Reprinted from Ref. 69.)

to contain vanadium [70]; the other was described as a heme c-containing enzyme complex of *G. metallireducens* [64]. Whether or not these nitrate-reducing enzymes depend on molybdenum will require further data on the purified enzymes.

The first assimilatory nitrate reductase was studied in the early 1950s [71]. In a classic experiment on the in vitro formation of assimilatory NADH-nitrate reductase from a *Neurospora* mutant using various acid-treated molybdenum enzymes from different phylogenetic sources, Nason and coworkers demonstrated the existence of a molybdenum-containing component that is shared by the known molybdenum enzymes from microbes, plants, and animals [72].

The molybdenum cofactor has been found in a variety of enzymes, most of which catalyze a net transfer of an oxygen atom to or from a substrate in a two-electron transfer reaction [69,73]. The basic structure of the eukaryotic cofactor is molybdopterin, a 6-alkyl pterin derivative with a phosphorylated C4 chain with two thiol groups [74,75]. The cofactor observed in bacterial nitrate reductase, and several other molybdenum enzymes, is the bismolybdopterin guanine dinucleotide (MGD) form [4,69]. MGD was found in a large variety of molybdenum enzymes, such as the formate dehydrogenases of bacteria and archaea [76,77].

The cofactor was initially believed to occur only in a molybdenum-containing form—hence the names molybdopterin and Moco—but later a tungsten-containing form of the cofactor was discovered [78,79]. In either case, the metal is ligated to the molybdopterin through the dithiolene sulfur atoms. The first crystal structure of a molybdopterin enzyme was the tungsten-dependent aldehyde ferredoxin oxidoreductase from the hyperthermophilic organism *Pyrococcus furiosus* [80]. Subsequently, several structures of *true* Moco-containing enzymes were determined by X-ray crystallography, i.e., aldehyde oxidoreductase from *Desulfovibrio gigas* [81,82], DMSO reductase from *Rhodobacter sphaeroides* [83] and *Rhodobacter capsulatus* [84,85], *E. coli* formate dehydrogenase H [86], chicken sulfite oxidase [87], and, most recently, the dissimilatory nitrate reductase from *Desulfovibrio desulfuricans* ATCC 27774 [88].

3.1. Location and Molecular Architecture

Four types of nitrate reductases catalyze the two-electron reduction of nitrate to nitrite: the eukaryotic assimilatory nitrate reductases and three distinct bacterial enzymes, comprising the cytoplasmic assimilatory (NAS), the membrane-bound (NAR), and the periplasmic dissimilatory (NAP) nitrate reductases [6]. Nitrite oxidoreductase, a membrane-bound enzyme from nitrifying bacteria, also exhibits nitrate reductase activity. This enzyme shows high sequence similarity to the membrane-bound NAR and catalyzes the nitrite oxidation to nitrate to allow chemoautotrophic growth [89].

Eukaryotic assimilatory nitrate reductases are cytosolic homodimeric enzymes, with pyridine nucleotides as electron donors. The monomeric unit (a 100- to 120-kDa polypeptide) carries a flavin adenine dinucleotide (FAD) and a *b*-type cytochrome as prosthetic groups in addition to Moco. The three cofactors are located in three functional domains that are highly conserved among eukaryotic species [8]. Note that eukaryotic and prokaryotic assimilatory nitrate reductases share no sequence similarity and have not much in common except their physiological function.

In addition to the assimilatory NAS enzyme, two types of dissimilatory nitrate reductases are found in bacteria, as observed in *Ralstonia eutropha* (formerly *Alcaligenes eutrophus*) [90]. These are the membrane-bound NAR, which generates a proton motive force for ATP synthesis, and the periplasmic NAP (Fig. 2). Two types of assimilatory nitrate reductases are present in bacteria, namely, the ferredoxin- or flavodoxin-dependent NAS, and the NADH-dependent NAS. Both contain the MGD cofactor and a [4Fe-4S] cluster at the N-terminal end, but no heme centers. The NADH-dependent NAS proteins from *Klebsiella pneumoniae* and *R. capsulatus* are heterodimers, composed of a 45-kDa FAD-containing diaphorase and a 95-kDa catalytic subunit carrying the Mo MGD and the Fe-S center. For further information on the various forms of NAS enzymes, refer to [6].

Membrane-bound nitrate reductases have been purified and characterized from denitrifying and nitrate-respiring bacteria, among them the thermophilic microorganism *Thermus thermophilus* [91]. NAR enzymes are composed of three subunits: the catalytic α subunit NarG carrying the MGD cofactor, the soluble β subunit NarH with one [3Fe-4S] and three [4Fe-4S] centers, and the membrane-bound biheme cytochrome *b* quinol-oxidizing γ subunit NarI. All NAR enzymes will reduce chlorate (ClO_3^-) and become inhibited by small ligands such as azide, chlorate, cyanide, or thiocyanate [6].

In *E. coli*, NAR proteins are synthesized during anaerobic growth, via the FNR protein, in the presence of nitrate or nitrite. The transcriptional regulator FNR plays a key role in anaerobic metabolism [92]. Disassembly of a labile Fe-S center was suggested for the O_2-dependent FNR inactivation [93]. Sequences with similarity to the *E. coli* FNR box have been observed upstream of anaerobic nitrate respiration and denitrification genes in many bacteria. FNR-like factors could be identified in both gram-positive and gram-negative bacteria [4].

Although periplasmic nitrate reductases were first discovered in denitrifying and phototrophic bacteria, they are widespread among gram-negative bacteria [17]. NAP enzymes, as a consequence of their location, do not directly contribute to the generation of a proton-motive force. It is assumed that the NAP enzyme is used for redox balancing, which can become important for optimal growth depending on the physiological conditions. Since dioxygen primarily inhibits denitrification at the level of nitrate transport [94], and since NAP does not require this step, several denitrifying bacteria perform aerobic denitrification by coupling the NAP enzyme to the nitrite and *N*-oxide reductases [14]. NAP enzymes have been investigated extensively in several bacteria, including *R. eutropha*, *Paracoccus pantotrophus* (formerly *Thiosphaera pantotropha*), *E. coli*, and *Rhodobacter* species. NAP has been isolated as a heterodimer, with a catalytic 90-kDa subunit NapA carrying the MGD cofactor, a [4Fe-4S] center, and a 15-kDa diheme *c*-type cytochrome NapB. NapB accepts electrons from NapC, a 25-kDa membrane-bound tetraheme cytochrome c [6]. NapC-type tetraheme proteins are involved in electron transfer between the membrane quinone pool and a series of periplasmic reductases, as described above for the cytochrome c nitrite reductase complex (Fig. 3).

3.2. Three-Dimensional Structure

So far, only one high-resolution structure (1.9 Å) of a nitrate reductase from the sulfate-reducing bacterium *D. desulfuricans* ATCC 27774 has been completed [88]. Crystals of the FAD domain of the assimilatory corn NADH:nitrate reductase expressed in *E. coli* had been obtained earlier, and the structure of that functional fragment was recently solved at 2.5 Å resolution [95–97]. The NAS enzyme is a dimer of two identical 100-kDa subunits carrying three different cofactors. The subunit can be cleaved by mild proteolysis and separated into three functional fragments. Each of the fragments contains one cofactor. The N-terminal fragment binds the molybdopterin moiety, the fragment of the middle part of the polypeptide chain binds the heme center, while the C-terminal fragment binds FAD and contains the binding site for NAD(P)H. The last two fragments share sequence homology with cytochrome b_5 and cytochrome b_5 reductase, respectively. The whole enzyme can be viewed as a group of subenzymes that are linked in one polypeptide chain to catalyze the transfer of the two electrons from NAD to FAD to cytochrome b to molybdopterin [8].

Based on sequence similarities, the Moco-containing enzymes have been currently divided into four different families [69], namely, the DMSO reductase, xanthine oxidase, sulfite oxidase, and aldehyde oxidoreductase families. Within each family sequence similarities are obvious, whereas no significant homologies can be detected between members of different families. This classification represents a unifying approach in terms of the overall enzyme structure, but at the same time it does not require the Mo/W-center to be coordinated in exactly the same way for members of the same family. The traditional classification of these enzymes into two families, one containing an oxothio Mo center and the other containing a dioxo Mo center, was based solely on the molybdenum environment and, with the knowledge of the crystal structures, proved to be inadequate for a general classification of these enzymes.

The features of the new classification scheme are illustrated best by DMSO reductase and formate dehydrogenase H, two members of the DMSO reductase family. Both enzymes share significant overall structural similarity, which is a consequence of their 23% identical amino acid sequence, but they differ in terms of their respective Mo ligand fields: DMSO reductase has at least one oxo ligand and a serine side chain coordinated to the Mo, whereas formate dehydrogenase H contains no oxo ligand and a selenocysteine side chain ligated to the metal. With the exception of *Rhodobacter* DMSO reductase, all enzymes contain at least one additional cofactor—either a heme, a Fe-S cluster, or a flavin—which is involved in intramolecular electron transfer to or from the Mo/W center.

Dissimilatory nitrate reductases belong into the DMSO reductase family. Members of this family are exclusively found in eubacteria and include, among others, DMSO reductase, the dissimilatory nitrate reductases, several formate dehydrogenases, and pyrogallol-phloroglucin transhydroxylase. With the exception of transhydroxylase, most of these enzymes serve as terminal reductases. On the other hand, assimilatory nitrate reductases, such as the enzymes from *Neurospora crassa*, *Chlorella vulgaris*, or *Spinacea oleracea*, were assigned to the sulfite oxidase family (Table 1).

The periplasmic NAP enzyme could be purified and crystallized from the sulfate-reducing bacterium *D. desulfuricans* [88]. It has one MGD cofactor and one [4Fe-4S] cluster in a single polypeptide chain of 723 amino acids (molecular mass approximately 80 kDa), heme centers and flavin were absent.

The protein is heart-shaped and its shorter dimension coincides with the direction of the cavity which leads to the active site. The NAP polypeptide fold is of the α/β type and is organized in four domains. All four domains are involved to a variable extent in cofactor binding and, while the [4Fe-4S] center is at the periphery of domain I, the MGD molybdenum cofactor is extended across the interior of the molecule and interacts with residues from all four domains (Fig. 6).

The molybdenum cofactor with the two MGD ligands extends along the interior of the NAP molecule. The pterin ring system is tricyclic, with a pyran ring fused to the pterin system. The pyran ring is tilted by 30–40° in relation to the pyrazine ring system. The molybdenum atom is located at the bottom of the deep crevice 12 Å away from

TABLE 1

Moco-Containing Enzyme Families and Selected Representatives

DMSO reductase family	
DMSO reductase	*R. sphaeroides/R. capsulatus/E. coli*
Nitrate reductase (dissimilatory)	*E. coli (narGHI)/E. coli (narZYW)*
Formate dehydrogenase	*E. coli (fdnGHI)/E. coli (fdoGHI)*
Pyrogallol-phloroglucinol transhydroxylase	*P. acidigallici*
Xanthine oxidase family	
Xanthine oxidase/dehydrogenase	*Bos taurus/Homo sapiens/Gallus gallus*
Aldehyde oxidoreductase	*D. gigas*
Sulfite oxidase family	
Sulfite oxidase	*Homo sapiens/Rattus norvegicus/ Gallus gallus/Thiobacillus novellis*
Nitrate reductase (assimilatory)	*Neurospora crassa/Chlorella vulgaris/Spinacea oleracea*
Aldehyde ferredoxin oxidoreductase family	
Aldehyde ferredoxin oxidoreductase	*Pyrococcus furiosus*
Formaldehyde ferredoxin oxidoreductase	*Pyrococcus furiosus*
Glyceraldehyde-3-phosphate ferredoxin oxidoreductase	*Pyrococcus furiosus*
Carboxylic acid reductase	*Clostridium formicoaceticum*
Aldehyde dehydrogenase	*Desulfovibrio gigas*
Unclassified at present	
Acetylene hydratase	*Pelobacter acetylenicus*

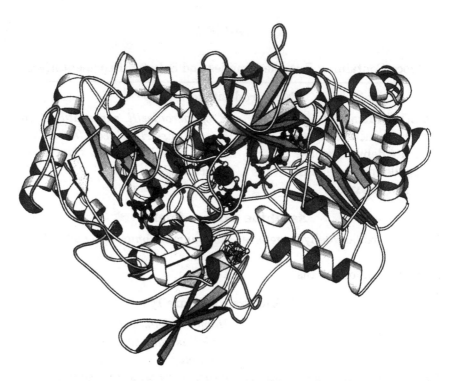

FIG. 6. The overall fold of the periplasmic nitrate reductase (NAP) from the sulfate-reducing bacterium *Desulfovibrio desulfuricans* ATCC 27774. Figure prepared with the program MOLSCRIPT (Brookhaven Protein Structure Database code 2nap, Ref. 88.)

the closest iron atom of the Fe-S cluster. In the structure, Mo is coordinated to four dithiolene sulfurs, a cysteine sulfur (Cys140), and a water molecule. In analogy to other molybdopterin-containing enzymes, there is a possible electron transfer pathway through bonds connecting the molybdenum and the iron centers. As in other redox proteins of known three-dimensional structure, in the NAP molecule the metal centers are properly aligned and close enough to allow efficient electron transfer within the protein and to external physiological redox partners.

The molybdenum active site is accessible from the surface through the funnel-like cavity, a structural feature also common to other molybdopterin-containing enzymes [69]. In the cavity, the nitrate anion is presumably fixed by interaction with arginine Arg354. The enzyme was crystallized in the Mo^{VI} state, with the metal coordinated by six ligands in a distorted trigonal prismatic geometry. Four ligands are provided by the two dithiolene sulfur atoms from the MGD molecules, the sulfur atoms approximately equidistant with Mo-S between 2.2 and 2.4 Å. A fifth ligand comes from the side chain of cysteine Cys140 (Mo-S ~ 2.5 Å). A hydroxo/water ligand (Mo$-OH_2$ ~ 2.2 Å) completes the coordination sphere around the molybdenum center (Fig. 7).

3.3. Spectroscopy and Oxidation-Reduction Potentials of the Molybdenum Centers

Electron paramagnetic resonance (EPR) and X-ray absorption spectroscopy were key to the identification of the ligands at the catalytic molybdenum site and to the in-depth understanding of its geometrical and electronic features [73].

3.3.1. Molybdenum(V) Electron Paramagnetic Resonance

The presence of EPR-active Mo^{V} in the nitrate reductase from *Pseudomonas aeruginosa* was reported as long ago as 1961 [98]. Further EPR studies (X-band, 80 K) on a homogeneous preparation of a nitrate reductase from *Micrococcus denitrificans* identified a signal attributable to Mo^{V} ($S = 1/2$), with typical resonances at g 1.985 and 2.045 [99]. Upon reaction with nitrate, the second g value shifted to 2.090, which appeared rather high for a pentavalent molybdenum center. Studies on the respiratory nitrate reductase from *E. coli* K12 gave EPR parameters comparable with those of other molybdopterin-dependent enzymes [100]. One of the signals showed hyperfine interaction of the Mo^{V} center with one solvent-exchangeable proton (A ^{1}H 0.9–1.2 mT). EPR spectra on the same enzyme under a variety of experimental conditions revealed well-resolved resonances that could be assigned to five different Mo^{V} species [101]. Oxidation-reduction potentials for the Mo^{VI}/Mo^{V} and the Mo^{V}/Mo^{IV} couple were determined by EPR to +220 ± 20 mV and +180 ± 20 mV at pH 7.14 [102]. In all molybdenum-containing enzymes, except nitrate reductase, i.e., sulfite oxidase, xanthine oxidase, or formate dehydrogenase, the oxidation-reduction potential of the half reaction catalyzed by the molybdenum center was rather low,

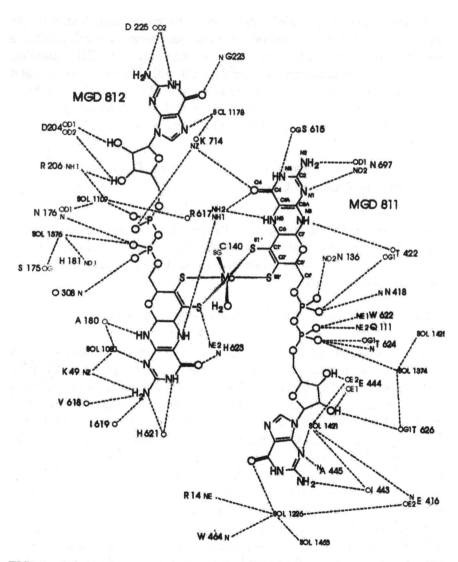

FIG. 7. Schematic representation of the hydrogen bond interactions of the MGD_2Mo cofactor with surrounding protein residues or buried water molecules in the periplasmic nitrate reductase (NAP) from the sulfate-reducing bacterium *Desulfovibrio desulfuricans* ATCC 27774. (Reprinted from Ref. 88, with permission.)

whereas for the respiratory nitrate reductase from *E. coli* K12 it was extremely high.

A study of the periplasmic respiratory nitrate reductase of denitrifying *Pa. pantotrophus* revealed that the molybdenum center of this enzyme had EPR characteristics very similar to those of the NAS enzyme from *Azotobacter vinelandii* but were slightly different from those of the NAR enzyme from species such as *E. coli* or *Paracoccus denitrificans*. The g values of the various Mo^V EPR signals observed in *Pa. pantotrophus* were typical for bacterial molybdenum-containing reductases. Incubation of the enzyme with an excess of the strong reductant dithionite (40 electrons/mol enzyme) resulted in two new Mo^V EPR signals (pseudorapid split and unsplit Mo^V signals). Examination of the parameters of these new signals revealed similarity to the so-called rapid signal observed in xanthine oxidase [103]. It was suggested that a change in oxidation state of the molybdopterin cofactor from a dihydropteridine to a tetrahydropteridine, which would change the conformation of the pterin cofactor [104], might affect the geometry at the molybdenum site and would explain the dithionite-induced changes of the EPR spectra. Under turnover conditions with nitrate, the so-called high-g-split Mo^V signal was observed with g_{av} at 1.9902 (g_1 1.9990, g_2 1.9905, g_3 1.9810) [103]. A Mo^V EPR investigation demonstrated the presence of this center in intact cells of *P. denitrificans* and suggested it as the physiologically relevant Mo^V site [105,106].

Reduction of purified assimilatory *Chlorella* nitrate reductase by NADH led to the appearance of a multiline EPR signal at $g_{av} = 1.977$, ascribed to a Mo^V center. The signal disappeared upon addition of the substrate nitrate or in the presence of a large excess of NADH [107,108]. At pH 7, this signal exhibited rhombic symmetry and a strong hyperfine coupling of a single $S = 1/2$ nucleus to the molybdenum center. The coupling component was identified as an exchangeable proton as described above for the respiratory nitrate reductase. Oxidation-reduction potentials for the Mo^{VI}/Mo^V couple and the Mo^V/Mo^{IV} were calculated from EPR monitored redox titrations to -34 mV and -54 mV at pH 7.0 [107].

3.3.2. Molybdenum X-ray Absorption Edge and Extended X-ray Absorption Fine Structure Spectroscopy

X-ray spectroscopic techniques were applied to characterize the molybdenum sites in several molybdenum enzymes including the assimilatory nitrate reductase from *Chlorella vulgaris* and the dissimilatory nitrate reductase from *E. coli* [109]. The molybdenum environment in the *Chlorella* nitrate reductase was found to strongly resemble that in sulfite reductase, with two terminal oxygens on molybdenum in the oxidized Mo^{VI} state. Upon reduction with NADH, one single terminal oxygen and a set of sulfurs were proposed for the fully reduced Mo^{IV} state. In contrast, the fully reduced molybdenum center in *E. coli* nitrate reductase seemed devoid of terminal oxygens, although an oxo species appeared upon reaction with nitrate. Related studies on the NAP enzyme from denitrifying *Pa. pantotrophus* gave an optimum fit of the EXAFS data for the Mo^{VI} site, using two terminal oxygens, three sulfurs, and either a fourth sulfur or a nitrogen ligand. The best fit to the Mo^{IV} EXAFS data (dithionite-reduced enzyme) was obtained by assuming one terminal oxygen, one oxygen and four sulfur/chloride ligands [110]. In the case of the NAP enzyme from *P. denitrificans*, the oxidized enzyme was best fitted as a dioxo Mo^{VI} species with five sulfur ligands (one at 2.82 Å, four at 2.43 Å) compared with the reduced Mo^{IV} species with one oxo and three sulfur ligands at 2.35 Å [106]. An X-ray spectroscopic investigation of a series of bis(dithiolene)molybdenum (IV,V,VI) and -tungsten (IV,V,VI) complexes, with accurately known three-dimensional structures, provided the opportunity to examine the effects of oxidation state, ligand types, and coordination geometries on absorption edge and EXAFS features. Systematic shifts of edge energies over the metal (IV,V,VI) oxidation state were observed, and features in the second derivative edge spectra were correlated with the number of oxo ligands, ranging from zero to two [111].

In the case of the periplasmic dissimilatory nitrate reductase from *D. desulfuricans*, with the exception of the oxygen ligand, the metal coordination found in the three-dimensional structure was reported to agree with the data obtained by EXAFS spectroscopy [88].

3.4. Reaction Mechanism

Early investigations on the reduction of nitrate to nitrite in the enzymes from *N. crassa* and *M. denitrificans* suggested the involvement of Mo^V and Mo^{VI} oxidation states [99,112]. Binding of nitrate to Mo^V in a bidentate manner was suggested on the basis of a biomimetic molybdenum complex. Facile electron transfer from an oxomolybdenum(V) center to nitrate would occur if the substrate anion would coordinate by way of one oxygen atom at a site cis to the oxo group [113]. Later it became well established that Mo^{IV} complexes were ideally suited to reduce the substrates of molybdenum oxotransferases by oxotransfer reactions [114,115]. In one Mo^{IV} model complex, nitrate was reduced to nitrite in a molybdenum-mediated oxotransfer reaction accompanied by oxidation of the $Mo^{IV}O$ site to the corresponding $Mo^{VI}O_2$ [116]. Using ^{18}O-enriched nitrate it could be demonstrated that the oxo group was transferred to the terminal oxo position of the catalytically active molybdenum species [117].

On the basis of the three-dimensional structure of the NAP nitrate reductase from *D. desulfuricans*, and using the results from biomimetic studies, a reaction mechanism was proposed where nitrate is coordinated to the Mo^{IV} state of the enzyme [88] (Fig. 8). Nitrate was coordinated to the $Mo^{IV}S_{Cys}(MGD)_2$ center in a unidentate manner. Electron transfer and oxotransfer to molybdenum produced the oxo $Mo^{VI}S_{Cys}(MGD)_2$ center accompanied by the release of nitrite. In an additional step, two protons had to be transferred, forming $Mo^{VI}(H_2 2O)S_{Cys}(MGD)_2$.

4. NITRITE OXIDOREDUCTASE

Nitrite oxidoreductase is a membrane-bound enzyme found in *Nitrobacter* species that shows a high sequence similarity to the respiratory NAR [118]. The enzyme exhibits nitrate reductase activity and catalyzes the oxidation of nitrite to nitrate [89]. Chemolithotrophic bacteria such as *Nitrobacter* species seem to derive their energy for growth by the oxidation of nitrite to nitrate. The lithotrophic growth is slow and inefficient because nitrite oxidation is thermodynamically unfavorable

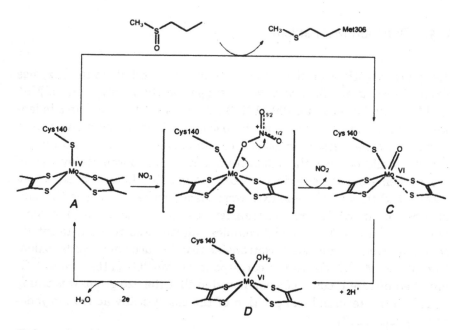

FIG. 8.   Proposed reaction mechanism for the reduction of nitrate to nitrite by the periplasmic nitrate reductase (NAP) from the sulfate-reducing bacterium *Desulfovibrio desulfuricans* ATCC 27774. The oxidation of methionine Met306 is regarded as a nonproductive side reaction. (Reprinted from Ref. 88, with permission.)

[119]. The primary product was shown to be NADH, which is used for ATP synthesis [120,121]. At present, it remains unclear how energy conservation proceeds in these bacteria.

4.1.   Location, Molecular Architecture, and Cofactors

Nitrite oxidoreductase was purified from the membrane fraction of *Nitrobacter hamburgensis* by heat treatment. The catalytically active enzyme consisted of two subunits with 115 kDa and 65 kDa molecular mass [22]. Most recently, a nitrite-oxidizing system was isolated from heat-treated membranes of *Nitrospira moscoviensis* consisting of four major proteins with 130, 62, 46, and 29 kDa each. Employing electron microscopic immunocytochemistry and monoclonal antibodies, the enzyme system was shown to be located in the periplasmic space [122].

In the presence of detergents, cytochromes a_1 and c_1 were detected and attributed to the enzyme from $N.$ $hamburgensis$. Cytochrome c_1 was an integral membrane protein, with a molecular mass of 32 kDa, and considered to be the third subunit of nitrite oxidoreductase [89,123]. The deduced amino acid sequence of the β subunit contained four cysteine clusters with striking homology to those of Fe-S centers of bacterial ferredoxins. It shared sequence similarity with the β subunit of two dissimilatory nitrate reductases of $E.$ $coli$, whereas part of the α subunit of nitrite oxidoreductase shared sequence similarity with the α-subunit of $E.$ $coli$ nitrate reductase. In view of these similarities it appears reasonable to assume the existence of four Fe-S centers in the β subunit of nitrite oxidoreductase, most likely three [4Fe-4S] and one [3Fe-4S] cluster [118]. The enzyme from $N.$ $moscoviensis$ contained a b-type cytochrome on the β subunit [122].

The enzyme from $N.$ $hamburgensis$ contained the MGD cofactor, and Mo, Fe, Zn, and Cu were present in the dimeric form of the enzyme. Fe-S centers were demonstrated by EPR spectroscopy of the dithionite-reduced enzyme, whereas in the enzyme as isolated or under turnover conditions with nitrite, multiline signals from Mo^V could be observed [22,124]. Some of these EPR signals had been earlier described in spectra of the membrane fraction from $Nitrobacter$ $winogradski$ [125].

4.2. Structure

Currently, a three-dimensional structure is not at hand. A two-dimensional structure of the nitrite oxidoreductase from $N.$ $hamburgensis$ was obtained by negative stain electron microscopy and digital image processing. The projection map of the protein showed the size, shape, and ultrastructure of the enzyme in its native environment, with a final resolution of 2.9 nm [126]. A molecular mass of 186 ± 43 kDa was derived for one single $\alpha\beta$ heterodimer, in good agreement with earlier biochemical data [22].

5. ENVIRONMENTAL ASPECTS AND BIOSENSORS

Nitrogen fertilizers have revolutionized agriculture in most parts of the world. Nitrate from fertilizers finds its way into groundwater, rivers, lakes, and seas. Water percolating through soil tends to remove or leach nitrate. Too much nitrate in drinking water can result in diseases, such as a severe blood disorder called *blue-baby syndrome*, caused by bacteria converting nitrate to nitrite which then interacts with hemoglobin. As a result, the baby suffers severe respiratory failure [127].

Removal of inorganic nitrogen compounds is achieved in most wastewater treatment plants by biological processes involving nitrification and denitrification. Due to the slow growth of nitrifying bacteria it is often very difficult, even impossible, to keep the biological nitrificiation process efficient. Disturbances of nitrification in sewage plants can lead to high concentrations of either ammonium or nitrite ions, which both act as dioxygen-consuming and toxic agents in surface waters.

Consequently, the quantitative determination of nitrate and nitrite concentrations is of rapidly growing interest, especially for the supervision of the quality of drinking water sources, wastewater treatment, for the food industry, and for remediation processes. Next to standard photometric procedures, the use of ion-selective electrodes (amperometric enzyme electrodes) might be an alternative method for the determination of nitrate and nitrite [128].

Enzyme electrodes were developed for the determination of nitrate and nitrite, by applying reductases which sequentially reduced nitrate to nitrite and to ammonia in the presence of a redox-active compound, such as methyl viologen. The final product ammonia was then determined by a potentiometric electrode [129,130]. More recently, immobilized nitrite reductases, such as the cytochrome cd_1 NiR from denitrifying bacteria, or the multiheme cytochrome c NiR from *S. deleyianum* or *W. succinogenes*, were coupled to an amperometric electrode modified with a suitable redox dye [128]. Especially through the use of the six-electron transferring cytochrome c NiR, in combination with an electron donor such as phenosafranin, a selective and highly sensitive determination of nitrite in drinking water was achieved. At present, the major problems to overcome are related to the stability of the immobilized enzyme, the interference by dioxygen, and the relatively long response time of the biosensor.

By a different approach, using an immobilized mixed nitrifying culture as the microbial component coupled to a Clark-type oxygen electrode, a microbial sensor was developed for detecting inhibitors of nitrification in wastewater [131]. The bacterial metabolic activity is determined by measuring the dioxygen consumption of the microbial immobilizate.

ACKNOWLEDGMENTS

This work was supported by the Deutsche Forschungsgemeinschaft (P.K., DFG-SPP 1071). We thank Professors E. Bock, B. Friedrich, and W. G. Zumft for providing valuable publications on nitrate reductase and nitrite oxidoreductase. Figures 2 and 6 were generously provided by Prof. W. G. Zumft and Dr. O. Einsle, respectively.

ABBREVIATIONS AND DEFINITIONS

ATP	adenosine-5'-triphosphate
DMSO	dimethyl sulfoxide
$E^{0'}$	midpoint redox potential for a given couple, at pH 7.0
EPR	electron paramagnetic resonance
EXAFS	extended X-ray absorption fine structure
FAD	flavin adenine dinucleotide
FNR	fumarate and nitrate reductase regulator
MGD	bismolybdopterin guanine dinucleotide
Moco	molybdenum cofactor
NAD^+	nicotinamide adenine dinucleotide (oxidized form)
NADH	nicotinamide adenine dinucleotide (reduced form)
NADPH	nicotinamide adenine dinucleotide phosphate (oxidized form)
NAP	periplasmic dissimilatory nitrate reductase
NAR	membrane-bound nitrate reductase
NAS	cytoplasmic assimilatory nitrate reductase
NiR	nitrite reductase

REFERENCES

1. K.-R. Kim and H. Craig, *Science*, *262*, 1855–1857 (1993).
2. J. A. Cole, *FEMS Microbiol. Lett.*, *136*, 1–11 (1996).
3. W. Schumacher, U. Hole, and P. M. H. Kroneck, in *Transition Metals in Biology* (G. Winkelmann and C. J. Carrano, eds.), Harwood Academic, Amsterdam, 1997, pp. 329–356.
4. W. G. Zumft, *Microbiol. Mol. Biol. Rev.*, *61*, 533–616 (1997).
5. J. T. Lin and V. Stewart, *Adv. Microb. Physiol.*, *39*, 1–30 (1998).
6. C. Moreno-Vivián, P. Cabello, M. Martinez-Luque, R. Blasco, and F. Castillo, *J. Bacteriol.*, *181*, 6573–6584 (1999).
7. D. J. Richardson and N. J. Watmough, *Curr. Opin. Chem. Biol.*, *3*, 207–219 (1999).
8. W. H. Campbell and J. R Kinghorn, *Trends Biochem. Sci.*, *15*, 315–319 (1990).
9. W. H. Campbell, *Plant Physiol.*, *111*, 355–361 (1996).
10. F. J. Kasting, *Science*, *259*, 902–926 (1993).
11. J. Hendriks, U. Gohlke, and M. Saraste, *J. Bioenerg. Biomembr.*, *30*, 15–24 (1998).
12. W. G. Zumft and P. M. H. Kroneck, *Adv. Inorg. Biochem.*, *11*, 193–221 (1996).
13. J. Hendriks, A. Oubrie, J. Castresana, A. Urbani, S. Gemeinhardt, and M. Saraste, *Biochim. Biophys. Acta.*, *1459*, 266–273 (2000).
14. B. C. Berks, S. J. Ferguson, J. W. B. Moir, and D. J. Richardson, *Biochim. Biophys. Acta*, *1232*, 97–173 (1995).
15. B. A. Averill, *Chem. Rev.*, *96*, 2951–2964 (1996).
16. A. B. Hooper, T. Vannelli, D. J. Bergmann, and D. M. Arciero, *Antonie van Leeuwenhoek*, *71*, 59–67 (1996).
17. D. J. Richardson, *Microbiol.*, *146*, 551–571 (2000).
18. E. Bock, H. P. Koops, and H. Harms, in *Nitrification* (J. I. Prosser, ed.), Spec. Pub. Soc. Gen. Microbiol, Vol. 20, IRL Press, Oxford, 1986, pp. 17–38.
19. P. M. Wood, in *Nitrification* (J. I. Prosser, ed.), Spec. Pub. Soc. Gen. Microbiol, Vol. 20, IRL Press, Oxford, 1986, pp. 39–62.

20. S. A. Ensign, M. R. Hyman, and D. J. Arp, *J. Bacteriol.*, *175*, 1971–1980 (1993).

21. N. Igarashi, H. Moriyama, T. Fujiwara, Y. Fukumori, and N. Tanaka, *Nat. Struct. Biol.*, *4*, 276–284 (1997).

22. M. Meincke, E. Bock, D. Kastrau, and P. M. H. Kroneck, *Arch. Microbiol.*, *158*, 127–131 (1992).

23. R. K. Thauer, K. Jungermann, and K. Decker, *Bacteriol. Rev.*, *41*, 100–180 (1977).

24. E. T. Adman, J. W. Godden, and S. Turley, *J. Biol. Chem.*, *270*, 27458–27474 (1995).

25. M. E. P. Murphy, P. F. Lindley, and E. T. Adman, *Prot. Sci.*, *6*, 761–770 (1997).

26. P. A. Williams, V. Fülop, E. F. Garman, N. F. Saunders, S. J. Ferguson, and J. Hajdu, *Nature.*, *389*, 406–412 (1997).

27. F. E. Dodd, S. S. Hasnain, Z. H. L. Abraham, R. R. Eady, and B. E. Smith, *Acta Crystallogr.*, *D53*, 406–418 (1997).

28. H. B. Gray, B. G. Malmström, and R. J. P. Williams, *J. Biol. Inorg. Chem.*, *5*, 551–559 (2000).

29. M. R. Cheesman, S. J. Ferguson, J. W. Moir, D. J. Richardson, W. G. Zumft, and A. J. Thomson, *Biochemistry*, *36*, 16267–16276 (1997).

30. K. Brown, M. Tegoni, M. Prudêncio, A. S. Pereira, S. Besson, J. J. Moura, I. Moura, and C. Cambillau, *Nat. Struct. Biol.*, *7*, 191–195 (2000).

31. K. Brown, K. Djinovic-Craugo, T. Haltia, I. Cabrito, M. Saraste, J. J. G. Moura, I. Moura, M. Tegoni, and C. Cambillau, *J. Biol. Chem.*, *275*, 41133–41136 (2000).

32. T. Rasmussen, B. C. Berks, J. Sanders-Loehr, D. M. Dooley, W. G. Zumft, and A. J. Thomson, *Biochemistry.*, *39*, 12753–12756 (2000).

33. O. Einsle, W. Schumacher, E. Kurun, U. Nath, and P. M. H. Kroneck, in *Biological Electron Transfer Chains*: Genetics, Composition, and Mode of Operation (G. Canters and E. Vijgenboom, eds.), Kluwer Academic, Dordrecht, 1998, pp. 197–208.

34. O. Einsle, A. Messerschmidt, P. Stach, G. P. Bourenkov, H. D. Bartunik, R. Huber, and P. M. H. Kroneck, *Nature*, *400*, 476–480 (1999).

35. O. Einsle, P. Stach, A. Messerschmidt, J. Simon, A. Kröger, R. Huber, and P. M. H. Kroneck, *J. Biol. Chem.*, *275*, 39608–39616 (2000).

36. E. Eisenmann, J. Beuerle, K. Sulger, P. M. H. Kroneck, and W. Schumacher, *Arch. Microbiol.*, *164*, 180–185 (1995).

37. F. Blasco, C. Iobbi, G. Giordano, M. Chippaux, and V. Bonnefoy, *Mol. Gen. Genet.*, *218*, 249–256 (1989).

38. A. Abou-Jaoudé, M. Chippaux, and M.-C. Pascal, *Eur. J. Biochem.*, *95*, 309–314 (1979).

39. A. Abou-Jaoudé, M.-C. Pascal, and M. Chippaux, *Eur. J. Biochem.*, *95*, 315–321 (1979).

40. U. Wissenbach, A. Kröger, and G. Unden, *Arch. Microbiol.*, *154*, 60–66 (1990).

41. U. Wissenbach, D. Ternes, and G. Unden, *Arch. Microbiol.*, *158*, 68–73 (1992).

42. M. Dubourdieu M. and J. A. DeMoss, *J. Bacteriol.*, *174*, 867–872 (1992).

43. F. Brito, J. A. DeMoss, and M. Dubourdieu, *J. Bacteriol.*, *177*, 3728–3735 (1995).

44. A. B. Hooper and A. A. DiSpirito, *Microbiol. Rev.*, *49*, 140–157 (1985).

45. B. L. Berg, J. Li, J.Heider, and V. Stewart, *J. Biol. Chem.*, *266*, 22380–22385 (1991).

46. A. Kröger and A. Innerhofer, *Eur. J. Biochem.*, *69*, 497–506 (1976).

47. A. Kröger, E. Dorrer, and E. Winkler, *Biochim. Biophys. Acta.*, *589*, 118–136 (1980).

48. A. Zöphel, M. C. Kennedy, H. Beinert, and P. M. H. Kroneck, *Eur. J. Biochem.*, *195*, 849–856 (1991).

49. W. Schumacher, Ph.D. thesis, Universität Konstanz, Konstanz, Germany, 1993.

50. C. Costa, J. J. G. Moura, I. Moura, M.-Y. Liu, H. D. Peck, Jr., J. LeGall, Y. Wang, and B. H. Huynh, *J. Biol.Chem.*, *265*, 14382–14388 (1990).

51. I. A. C. Pereira, J. LeGall, A. V. Xavier, and M. Teixeira, *Biochim. Biophys. Acta, 1491*, 119–130 (2000).

52. R. R. Naik, F. M. Murillo, and J. F. Stolz, *FEMS Microbiol. Lett., 106*, 53–58 (1993).

53. R. H. Jackson, A. Cornish-Bowden, and J. A. Cole, *J. Biochem., 193*, 861–867 (1981).

54. N. R. Harborne, L. Griffiths, S. J. Busby, and J. A. Cole, *Mol. Microbiol., 6*, 2805–2813 (1992).

55. J. A. Cole and C. M. Brown, *FEMS Microbiol. Lett., 7*, 65–72 (1980).

56. J. A. Cole, *Biochim. Biophys. Acta, 162*, 356–368 (1968).

57. T. Fujita, *J. Biochem. (Tokyo), 60*, 204–215 (1966).

58. P. M. Wood, *FEBS Lett., 164*, 223–226 (1983).

59. S. J. Ferguson, in *Bacterial Energy Transduction* (C. Anthony, ed.), Academic Press, London, 1988, pp. 151–182.

60. M. Bott, D. Ritz, and H. Hennecke, *J. Bacteriol., 173*, 6766–6772 (1991).

61. A. Darwin, H. Hussain, L. Griffiths, J. Grove, Y. Sambongi, S. Busby, and J. A. Cole, *Mol. Microbiol., 9*, 1255–1265 (1993).

62. H. Hussain, J. Grove, L. Griffiths, S. Busby, and J. A. Cole, *Mol. Microbiol., 12*, 153–163 (1994).

63. J. Simon, R. Gross, O. Einsle, P. M. H. Kroneck, A. Kröger, and O. Klimmek, *Mol. Microbiol., 35*, 686–696 (2000).

64. F. Martínez Murillo, T. Gugliuzza, J. Senko, P. Basu, and J. F. Stolz, *Arch. Microbiol., 172*, 313–320 (1999).

65. P. M. H. Kroneck, J. Beuerle, and W. Schumacher, in *Metal Ions in Biological Systems*, Vol. 28 (H. Sigel and A. Sigel, eds.), Marcel Dekker, New York, 1992, pp. 455–505.

66. A. B. Hooper, M. Logan, D. M. Arciero, and H. McTavish, *Biochim. Biophys. Acta, 1058*, 13–16 (1991).

67. M. Whittaker, D. Bergmann, D. M. Arciero, and A. B. Hooper, *Biochim. Biophys. Acta, 1459*, 346–355 (2000).

68. M. J. Frijlink, T. Abee, H. J. Laanbroek, W. de Boer, and W. N. Konings, *Arch. Microbiol., 157*, 194–199 (1992).

69. C. Kisker, H. Schindelin, D. Baas, J. Rétey, R. U. Meckenstock, and P. M. H. Kroneck, *FEMS Microbiol. Rev.*, *22*, 503–521 (1999).

70. A. N. Anitpov, N. N. Lyalikova, T. V. Khijniak, and N. P. L'vov, *FEBS Lett.*, *441*, 257–260 (1998).

71. H. J. Evans and A. Nason, *Plant. Physiol.*, *28*, 233–254 (1953).

72. A. Nason, K.-Y. Lee, S.-S. Pan, P. A. Ketchum, A. Lamberti, and J. DeVries, *Proc. Natl. Acad. Sci. USA*, *68*, 3242–3246 (1971).

73. R. Hille, *Chem. Rev.*, *96*, 2757–2816 (1996).

74. K. V. Rajagopalan, *Adv. Enzymol.*, *64*, 215–290·(1991).

75. K. V. Rajagopalan and J. L. Johnson, *J. Biol. Chem.*, *267*, 10199–10202 (1992).

76. A. Jankielewicz, R. A. Schmitz, O. Klimmek, and A. Kröger, *Arch. Microbiol.*, *162*, 238–242 (1994).

77. J. L. Johnson, N. R. Bastian, N. L. Schauer, J. G. Ferry, and K. V. Rajagopalan, *FEMS Microbiol. Lett.*, *61*, 213–216 (1991).

78. J. L. Johnson, K. V. Rajagopalan, S. Mukund, and M. W. W. Adams, *J. Biol. Chem.*, *268*, 4848–4852 (1993).

79. J. L. Johnson, D. C. Rees, and M. W. W. Adams, *Chem. Rev.*, *96*, 2817–2839 (1996).

80. M. K. Chan, S. Mukund, A. Kletzin, M. W. W. Adams, and D. C. Rees, *Science*, *267*, 1463–1469 (1995).

81. R. Huber, P. Hof, R. O. Duarte, J. J. G. Moura, I. Moura, M.-Y. Liu, J. LeGall, R. Hille, M. Archer, and M. J. Romão, *Proc. Natl. Acad. Sci. USA*, *93*, 8846–8851 (1996).

82. M. J. Romão, M. Archer, I. Moura, J. J. G. Moura, J. LeGall, R. Engh, M. Schneider, P. Hof, and R. Huber, *Science*, *270*, 1170–1176 (1995).

83. H. Schindelin, C. Kisker, J. Hilton, K. V. Rajagopalan, and D. C. Rees, *Science*, *272*, 1615–1621 (1996).

84. A. S. McAlpine, A. G. McEwan, A. L. Shaw, and S. Bailey, *J. Biol. Inorg. Chem.*, *2*, 690–701 (1997).

85. F. Schneider, J. Löwe, R. Huber, H. Schindelin, C. Kisker, and J. Knäblein, *J. Mol. Biol.*, *263*, 53–69 (1996).

86. J. C. Boyington, V. N. Gladyshev, S. V. Khangulov, T. C. Stadtman, and P. D. Sun, *Science*, *275*, 1305–1308 (1997).

87. C. Kisker, H. Schindelin, A. Pacheco, W. A. Wehbi, R. M. Garrett, K. V. Rajagopalan, J. H. Enemark, and D. C. Rees, *Cell*, *91*, 973–983 (1997).

88. J. Dias, M. E. Than, A. Humm, R. Huber, G. P. Bourenkov, H. D. Bartunik, S. Bursakov, J. Calvete, J. Caldeira, C. Carneiro, J. J. G. Moura, I. Moura, and M. J. Romão, *Structure*, 7, 65–79 (1999).

89. H. Sundermeyer-Klinger, W. Meyer, B. Warninghoff, and E. Bock, *Arch. Microbiol.*, *149*, 153–158 (1984).

90. U. Warnecke-Eberz and B. Friedrich, *Arc Microbiol.*, *159*, 405–409 (1993).

91. S. Ramírez-Arcos, L. A. Fernández-Herrero, and J. Berenguer, *Biochim. Biophys. Acta*, *1396*, 215–227 (1998).

92. S. Spiro, *Antonie Van Leeuwenhoek*, *66*, 23–36 (1994).

93. N. Khoroshilova, C. Popescu, E. Münck, H. Beinert, and P. J. Kiley, *Proc. Natl. Acad. Sci. USA*, *94*, 6087–6092 (1997).

94. K. S. Denis, F. M. Dias, and J. J. Rowe, *J. Biol. Chem.*, *265*, 18095–18097 (1990).

95. G. Lu, W. Campbell, Y. Lindquist, and G. Schneider, *J. Mol. Biol.*, *224*, 277–279 (1992).

96. G. Lu, W. Campbell, Y. Lindquist, and G. Schneider, *Structure*, 2, 809–821 (1994).

97. G. Lu, Y. Lindqvist, G. Schneider, U. Dwivedi, and W. H. Campbell, *J. Mol. Biol.*, *248*, 931–948 (1995).

98. C. A. Fewson and D. J. D. Nicholas, *Biochim. Biophys. Acta*, *49*, 335–349 (1961).

99. P. Forget and D. V. Dervartanian, *Biochim. Biophys. Acta*, *256*, 600–606 (1972).

100. R. C. Bray, S. P. Vincent, D. J. Lowe, R. A. Clegg, and P. B. Garland, *Biochem. J.*, *155*, 201–203 (1976).

101. S. P. Vincent and R. C. Bray, *Biochem. J.*, *171*, 639–647 (1978).

102. S. P. Vincent, *Biochem. J.*, *177*, 757–759 (1979).

103. B. Bennett, B. C. Berks, S. J. Ferguson, A. J. Thomson, and D. J. Richardson, *Eur. J. Biochem.*, *226*, 789–798 (1994).

104. W. Pfleiderer, in *Folates and pteridines*, Vol. 2 (R. L. Blakeley and S. J. Benkovic, eds.), Wiley Interscience, New York, 1985, pp. 42–113.

105. H. J. Sears, B. Bennett, S. Spiro, A. J. Thomson, and D. J. Richardson, *Biochem. J.*, *310*, 311–314 (1995).

106. C. S. Butler, J. M. Charnock, B. Bennett, H. J. Sears, A. J. Reilly, S. J. Ferguson, C. D. Garner, D. J. Lowe, A. J. Thomson, B. C. Berks, and D. J. Richardson, *Biochemistry*, *38*, 9000–9012 (1999).

107. L. P. Solomonson, M. J. Barber, W. D. Howard, J. L. Johnson, and K. V. Rajagopalan, *J. Biol. Chem.*, *259*, 849–853 (1984).

108. C. J. Kay and M. J. Barber, *Biochemistry*, *28*, 5750–5758 (1989).

109. S. P. Cramer, L. P. Solomonson, M. W. W. Adams, and L. E. Mortenson, *J. Am. Chem. Soc.*, *106*, 1467–1471 (1984).

110. B. Bennett, J. M. Charnock, H. J. Sears, B. C. Berks, A. J. Thomson, S. J. Ferguson, C. D. Garner, and D. J. Richardson, *Biochem. J.*, *317*, 557–563 (1996).

111. K. B. Musgrave, J. P. Donahue, C. Lorber, R. H. Holm, B. Hedman, and K. O. Hodgson, *J. Am. Chem. Soc.*, *121*, 10297–10307 (1999).

112. D. J. D. Nicholas and H. M. Stevens, *Nature*, *176*, 1066–1067 (1955).

113. C. D. Garner, M. R. Hyde, F. E. Mabbs, and V. I. Routledge, *Nature*, *252*, 579–580 (1974).

114. E. W. Harlan, J. M. Berg, and R. H. Holm, *J. Am. Chem. Soc.*, *108*, 6992–7000 (1986).

115. R. H. Holm, *Coord. Chem. Rev.*, *100*, 183–221 (1990).

116. J. A. Craig and R. H. Holm, *J. Am. Chem. Soc.*, *111*, 2111–2115 (1989).

117. R. H. Holm, *Chem. Rev.*, *87*, 1401–1449 (1987).

118. K. Kirstein and E. Bock, *Arch. Microbiol.*, *160*, 447–453 (1993).

119. J. G. Cobley, *Biochem. J.*, *156*, 493–498 (1976).

120. H. Sundermeyer and E. Bock, *Arch. Microbiol.*, *130*, 250–254 (1981).

121. A. Freitag and E. Bock, *FEMS Microbiol. Lett.*, *66*, 157–162 (1990).

122. E. Spieck, S. Ehrich, J. Aamand, and E. Bock, *Arch. Microbiol.*, *169*, 225–230 (1998).

123. Y. Tanaka, Y. Fukumori, and T. Yamanaka, *Arch. Microbiol.*, *135*, 265–271 (1983).

124. B. Krüger, O. Meyer, M. Nagel, J. R. Andreesen, M. Meincke, and E. Bock, *FEMS Microbiol. Lett*, *84*, 225–227 (1987).

125. W. J. Ingledew and P. J. Halling, *FEBS Lett.*, *67*, 90–93 (1976).

126. E. Spieck, S. Müller, A. Engel, E. Mandelkow, H. Patel, and E. Bock, *J. Struct. Biol.*, *117*, 117–123 (1996).

127. M. Saull, *New Scientist*, *37*, 1–4 (1990).

128. B. Strehlitz, B. Gründig, W. Schumacher, P. M. H. Kroneck, K.-D. Vorlop, and H. Kotte, *Anal. Chem.*, *68*, 807–816 (1996).

129. C. H. Kiang, S. S. Kuan, G. G. Guilbault, *Anal. Chim. Acta*, *80*, 209–214 (1975).

130. C. H. Kiang, S. S. Kuan, G. G. Guilbault, *Anal. Chem.*, *50*, 1319–1322 (1978).

131. A König, K. Riedel, and J. W. Metzger, *Biosensors and Bioelectronics*, *13*, 869–874 (1998).

11

The Molybdenum-Containing Hydroxylases of Nicotinate, Isonicotinate, and Nicotine

Jan R. Andreesen[1] *and Susanne Fetzner*[2]

[1]Institut für Mikrobiologie, Universität Halle, Kurt-Mothes-Strasse 3, D-06120 Halle, Germany

[2]Mikrobiologie, Fachbereich 7, Carl von Ossietzky Universität Oldenburg, P.O. Box 2503, D-26111 Oldenburg, Germany

1. INTRODUCTION

1.1. Hydroxylation Reactions Involving Nicotinate, Isonicotinate, and Nicotine: An Overview

Hydroxylation reactions occur in the metabolic pathways of all living organisms. They may involve the hydratase-catalyzed addition of a water molecule to a double bond, or they may involve the substitution of the hydrogen by a hydroxy group. Substitutions of -H by -OH are catalyzed by oxygenases, which utilize dioxygen as the source of the oxygen incorporated into the substrate, or by molybdenum-containing hydroxylases, which use a water molecule as cosubstrate, according to the following stoichiometries:

$$\text{Monooxygenases}: \text{R-H} + O_2 + 2[e^-] + 2H^+ \longrightarrow \text{R-OH} + H_2O$$

$$\text{Mo-containing hydroxylases}: \text{R-H} + H_2O \longrightarrow \text{R-OH} + 2[e^-] + 2H^+$$

Typical substrates of oxygenase-catalyzed reactions are aromatic compounds that are susceptible to electrophilic substitution reactions. In contrast to, e.g., the benzene ring, the pyridine ring due to its nitrogen heteroatom is not reactive toward electrophiles, but it is susceptible to attack by nucleophilic agents, preferably at positions C2, C6, and C4 [1].

Consistent with N-heteroaromatic chemistry, the bacterial degradation of pyridine carboxylic acids and of the pyridine ring of nicotine starts with a nucleophilic attack by a water molecule adjacent to the N heteroatom. (However, note that the 2-,4-, or 6-hydroxypyridines formed exist preferentially in the tautomeric pyridone form [2].) Actually, the bacterial degradation of nicotinate by *Bacillus niacini* (Sec. 2.1), the degradation of isonicotinate by *Mycobacterium* sp. INA1 (Sec. 3), and the utilization of nicotine by *Arthrobacter nicotinovorans* (Sec. 4) even involve two such hydroxylation steps, yielding 2,6-dihydroxy-substituted pyridine derivatives. Each hydroxylation step is catalyzed by a distinct molybdenum-containing hydroxylase. The picolinate hydroxylase of *Arthrobacter picolinophilus* (now termed *Rhodococcus erythropolis*) seems also to be a molybdenum-containing enzyme [3]. A reduction of the pyridine ring seems also to be possible [4–7]. However, firm evidence for an initiating reduction is still lacking, but was recently proposed for a facultatively anaerobic organism reducing pyridine [8].

In contrast to pyridine and pyridine carboxylic acids, pyridones react with electrophiles in *ortho* and *para* position relative to the activating oxo group [1,2]. Catabolic 4-pyridone (4-hydroxypyridine) hydroxylation, for example, involves dioxygen as cosubstrate and is catalyzed by a 4-pyridone 3-monooxygenase in an *Agrobacterium* sp. [9]. On the other hand, for pyridine and 2-pyridone (2-hydroxypyridine), a successive hydroxylation to 2,6-dioxopyridone (dihydroxypyridine) is reported for *Arthrobacter crystallopoietes* and *Rhodococcus opacus* [10]. The ability of gram-positive soil bacteria to transform pyridine and its derivatives becomes more obvious [11].

2. HYDROXYLATIONS OF NICOTINATE AND DERIVATIVES

2.1. Aerobic, Regioselective Hydroxylation Reactions

Aerobic bacterial degradation of nicotinate may proceed via different pathways, as shown in Fig. 1. Common to all known pathways is an initial regiospecific hydroxylation step that generates 6-hydroxynicotinate. Whereas *Azorhizobium caulinodans* subsequently reduces 6-hydroxynicotinate to 6-oxo-1,4,5,6-tetrahydronicotinate, which undergoes hydrolytic degradation [12] (Fig. 1, A), 6-hydroxynicotinate degradation by *Pseudomonas fluorescens* N-9 [13], *Bacillus niacini* DSM 2923 (formerly *Bacillus* sp.) [14–17], and *Achromobacter xylosoxidans* LK1 proceeds via distinct branches of the maleamate pathway (Fig. 1, B, C). *Achromobacter xylosoxidans* LK1 actually is used in an industrial biotransformation process to manufacture 6-hydroxynicotinate, which is an important building block in the chemical synthesis of insecticides [18].

The bacterial enzymes catalyzing the initial hydroxylation of nicotinate appear to differ in their cellular localization, cofactor content, and

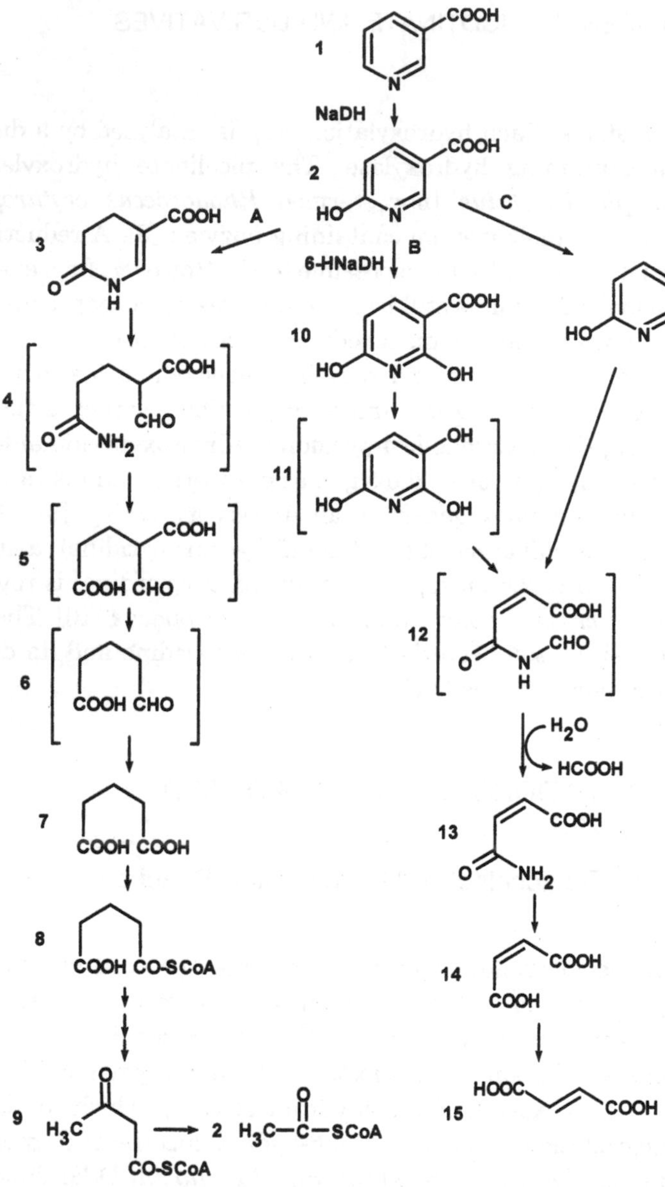

FIG. 1. Pathways of aerobic bacterial degradation of nicotinic acid. (A) *Azorhizobium caulinodans* [12], (B) *Bacillus niacini* [14,16,17], (C) *Pseudomonas fluorescens* [13]. **1**, nicotinic acid; **2**, 6-hydroxynicotinic acid; **3**, 6-oxo-1,4,5,6-tetrahydronicotinic acid; **4**, 2-formylglutaryl-5-amide; **5**, 2-formylglutaric acid; **6**, glutaric semialdehyde; **7**, glutaric acid; **8**, glutaryl-coenzyme A; **9**, acetoacetyl-coenzyme A; **10**, 2,6-dihydroxynicotinic acid; **11**, 2,3,6-trihydroxypyridine; **12**, *N*-formylmaleamic acid; **13**, maleamic acid; **14**, maleic acid; **15**, fumaric acid; **16**, 2,5-dihydroxypyridine.

structural properties. In *Pseudomonas fluorescens* N-9 (reclassified as *P. putida* [19]), the first enzyme of the nicotinate pathway is particulate and cytochrome-linked [13]. Nicotinate halogenated in 5-position was also transformed [20]. A preparation of partially purified particulate nicotinate hydroxylase from *P. fluorescens* KB1 contained flavin and cytochromes [21]. The hydroxyl group in 6-hydroxynicotinate became labeled only by $H_2^{18}O$, not by $^{18}O_2$ [22], indicating the action of a hydroxylase. Similar results were obtained for the enzyme from *Azorhizobium caulinodans* [23]. The membrane-bound, cytochrome-linked enzyme from *Pseudomonas ovalis* is presumably a nonheme iron protein [24]. Nicotinate dehydrogenase (NaDH) solubilized from the membrane fraction of *P. fluorescens* TN5 was obtained as homogenous preparation consisting of a monomeric 80-kDa protein with no absorbance in the region of 350–750 nm and, thus, is presumed to be devoid of any flavin or heme cofactor; however, it also is linked to the cytochrome respiratory chain [25]. This enzyme might contain only a molybdenum cofactor in analogy to the dimethyl sulfoxide reductase of *Rhodobacter sphaeroides* [26] or the cofactor and perhaps an iron/sulfur center as in certain dissimilatory 80-kDa nitrate reductases [27]. Nicotinate, pyrazine carboxylate, and 3-cyanopyridine serve as substrates [25].

NaDH as well as 6-hydroxynicotinate dehydrogenase (6-HNaDH) of *Bacillus niacini*, which are coordinately induced by 6-hydroxynicotinate [28,29], are readily soluble, immunologically related molybdo-iron/sulfur flavoproteins presumed to belong to the xanthine oxidase (XO) family [16]. Quite remarkable, HNaDH contains the Mo-MGD form of the molybdenum pyranopterin cofactor, whereas NaDH from the same *Bacillus* strain harbors Mo-MCD like many molybdenum enzymes involved in the bacterial degradation of N-heteroaromatic compounds (Table 1). In contrast to native NaDH, which is a dimer of heterotrimers, 6-HNaDH shows a subunit structure of ($\alpha\beta\gamma$) like nicotine dehydrogenase from *Arthrobacter oxidans* (now *A. nicotinovorans* or *A. ureafaciens*; see Sec. 4). NaDH and 6-HNaDH from *Bacillus niacini* are highly specific for their substrates; however, NaDH also catalyzes the hydroxylation of 6-hydroxynicotinate with up to 3% the activity observed for nicotinate [16].

In the eukaryote *Aspergillus nidulans*, nicotinate is the physiological substrate of purine hydroxylase II, encoded by the structural gene *hxnS*, whereas the similar purine hydroxylase I is a typical xanthine

TABLE 1

Comparison of Molybdenum Hydroxylases Involved in Bacterial Degradation of Nicotinate, Isonicotinate, Benzoisonicotinate, and Nicotine

Enzyme (Source)	Subunit structure	Subunit molecular masses (kDa)	Molybdenum cofactor	Fe/S	Flavin	Membrane association	Ref.
			- molar ratios per protomer -				
Nicotinate DH[a] *Eubacterium barkeri*	$(\alpha\beta\gamma\delta)_{1-2}$	50 37 33 23	Dinucleotide form of Mo-MPT 1 Mo	5 Fe Se[b]	1 FAD	No	[44,55]
Nicotinate DH *Bacillus niacini*	$(\alpha\beta\gamma)_{1-2}$	85 34 20	Mo-MCD[c] 0.8 Mo	4.2 Fe 0.8 S[d]	1.0 FAD	No	[16,62]
6-Hydroxynicotinate DH *Bacillus niacini*	$\alpha\beta\gamma$	85 34 15	Mo-MGD[e]	Fe S (n.d.)[f]	FAD	No	[16,62]
Isonicotinate DH *Mycobacterium* sp. INA1	$(\alpha\beta\gamma)_{1-2}$	83 31 19	Mo-MCD 0.9 Mo	4.1 Fe 4.1 S	1.0 FAD	"Partly"	[62]
2-Hydroxyisonicotinate DH *Mycobacterium* sp. INA1	$(\alpha\beta\gamma)_2$	97 31 17	Mo-MCD 0.8 Mo	3.8 Fe 3.7 S	0.9 FAD	No	[65]

Enzyme / Organism	Structure	Subunits (kDa)	Mo cofactor	Fe/S	Flavin / OR	Ref.
"Benzoisonicotinate DH"[a] Quinoline 4-carboxylate 2-OR[g] Agrobacterium sp. B1	$(\alpha\beta\gamma)_2$	85 35 21	Mo-MCD 1.1. Mo	4.2 Fe 3.6 S	0.9 FAD n.d.	[67]
Nicotine DH Arthrobacter nicotinovorans, Arthrobacter ureafaciens	$\alpha\beta\gamma$	82 (87.7)[h] 30 15	Dinucleotide form of Mo-MPT 0.9 Mo	4.0 Fe 2.0 S	0.9 FAD Yes	[75,76]
Ketone DH Arthrobacter nicotinovorans	$(\alpha\beta\gamma)_2$	89 27 18	n.d. - presumed molybdo-iron/sulfur-flavoprotein -[h]	n.d.	n.d. n.d.	[84]

[a]DH, dehydrogenase.
[b]Catalytically essential, but substoichiometric amounts observed.
[c] Mo-MCD, molybdenum pyranopterin cytosine dinucleotide (molybdopterin cytosine dinucleotide).
[d]S, acid-labile sulfur.
[e]Mo-MGD, molybdenum pyranopterin guanine dinucleotide (molybdopterin guanine dinucleotide).
[f] n.d., not determined.
[g] OR, oxidoreductase.
[h]As deduced from the nucleotide sequence.

411

dehydrogenase (XDH) [30]. Both enzymes require the HxB gene product for posttranslational activation by such a transsulfurylase [31]. Although both enzymes are quite similar in their structure and composition as NAD-dependent molybdo-iron/sulfur flavoproteins, they show no immunological relationship [32]. Nicotinate is no substrate for purine hydroxylase I in contrast to purine hydroxylase II, which converts it with about 15–21% of the rate observed for hypoxanthine; however, xanthine is no substrate for enzyme II [32,33]. The form II is more efficiently induced by its product 6-hydroxynicotinate than by nicotinate [31], thus emphasizing its physiological involvement in nicotinate metabolism. No sequence information is available up to now for purine hydroxylase II (HxnS) in contrast to form I (HxA) [34].

Whereas the initial hydroxylation step in bacterial nicotinic acid degradation usually involves regioselective oxygen incorporation at C6 (Fig. 1), a hydroxylation at C2 of nicotinate has been reported for the 6-methylnicotinate utilizing bacterial strains MCI3288, MCI3289 [35], and *Ralstonia/Burkholderia* sp. DSM 6920, which has been investigated in more detail [36,37]. Pyridine-3-carboxylic acid derivatives hydroxylated at C2 as well as C3-hydroxylated pyrazine-2-carboxylic acid derivatives are basic synthons for the production of a series of pharmaceuticals and herbicides [37]. 6-Methylnicotinate-grown resting cells of *Ralstonia/Burkholderia* sp. DSM 6920 indeed catalyze the regioselective hydroxylation of a variety of pyridine-3- and pyrazine-2-carboxylates, and may be used to produce 2-hydroxynicotinic acid (derivatives) or 3-hydroxy-substituted pyrazine-2-carboxylic acid derivatives. The hydroxylation at C2 occurs under both oxic and oxygen-limited conditions, suggesting oxygen incorporation from a water molecule into the substrates [36,37]. With respect to its broader substrate specificity, the putative (6-methyl)nicotinate 2-oxidoreductase markedly differs from NaDH and 6-HNaDH from *Bacillus niacini* and some other molybdenum hydroxylases involved in the hydroxylation of pyridine derivatives (see Sec. 3).

Several aerobic bacterial strains utilizing benzopicolinate (quinoline-2-carboxylate) and benzoisonicotinate (quinoline-4-carboxylate) have been isolated, and the molybdenum containing hydroxylases catalyzing the initial regiospecific hydroxylation at C4 and C2 of benzopicolinate and benzoisonicotinate, respectively, have been characterized (for a review, see [38]; see also Sec. 3). However, it is remarkable to note that bacterial strains utilizing benzonicotinate (quinoline-3-carboxylate) have not been described so far. This is quite interesting in the context of the potential biotransformation of quinoline antibiotics such as enrofloxacin and ciprofloxacin because the corresponding 2-position seems to be even resistant to an oxidative attack by fungi [39,40].

2.2. Anaerobic Biotransformation of Nicotinate by *Eubacterium barkeri*

Clostridium barkeri was isolated by its ability to ferment nicotinate under strictly anaerobic conditions [41]. Except for this unique property, it shares many similar properties with the acetogenic bacteria *Eubacterium limosum* and *Acetobacterium woodii* and, thus, was reclassified a *Eubacterium barkeri* [42]. Characteristic metabolites in nicotinate breakdown are 6-hydroxynicotinate, 1,4,5,6-tetrahydronicotinate, and methylene glutamate [43]. The NADP-dependent hydroxylation to 6-hydroxynicotinate is catalyzed by an NaDH of about 300 kDa, containing FAD (1.5 mol) and iron/sulfur (11.6 mol). The enzyme catalyzes also an NADPH-dependent oxidase and diaphorase reaction, presumably by these cofactors [44]. 6-Hydroxynicotinate is reduced (probably due to the pyridone form) by an unstable ferredoxin-dependent enzyme to 1,4,5,6-tetrahydronicotinate [45]; then the ring structure is hydrolytically cleaved (for further details of, e.g., B_{12}-coenzyme involvement and of stereospecific reactions, see [46,47]).

The reaction catalyzed by NaDH of *E. barkeri* suggested an involvement of molybdenum, as was tested by a possible antagonistic function of tungstate supplied to the nicotinate-containing media. Surprisingly, both metal ions exert no drastic effect in contrast to results obtained for purinolytic clostridia [48]. However, the presence of selenite increased the NaDH activity by an order of magnitude, but not the NADPH oxidase reaction or the diaphorase [49–51]. By purification of the NaDH, an incorporation of ^{75}Se into a labile form became evident [50], being in contrast to cotranslationally formed, TGA-encoded, selenocysteine-containing enzymes such as formate dehydrogenase [52]. Molybdenum, a fluorescent pterin cofactor, flavin, and selenium coeluted with NaDH activity, and the existence of a biologically active selenium-containing molybdocofactor was presumed [53] in accord with its specific effect on the hydroxylase reaction. This was confirmed by EPR studies showing a split Mo(V) signal by use of [77]Se, resulting from an interaction of selenium with molybdenum at the active site [54]. The purified NaDH loses ^{75}Se by SDS-PAGE. It consists of four subunits of 23, 33, 37, and 50 kDa (Table 1) containing substoichiometric amounts of Se (0.4–0.8 mol). The labile selenium moiety was of low molecular weight, presumably forming a series of polyselenides [55]. The NaDH consists of 160 kDa as the major form; however, the 400- and 120-kDa proteins contain also the same four subunits. Their N termini do not correspond to those of other molybdenum hydroxylases as evident by the deviating size of four instead of three subunits found to be present in most bacterial molybdenum-containing hydroxylases (Table 1; see also other chapters). The NaDH of *E. barkeri* is also unusual in its insensitivity to cyanide, but it is sensitive to selenide and sulfide. The extended EPR studies showed the usual signals in line with the cofactors present. Only some nicotinate derivatives or analogues change EPR signals, whereas the substrate spectrum resulting in hydroxylated compounds is even further restricted to nicotinate and pyrazine-2-carboxylate [55].

E. barkeri contains besides the NaDH a separate XDH being induced after growth on purines and quite related to a classical bacterial XDH except for its dodecameric structure and its substoichiometric content of molybdenum, tungsten, FAD, and labile selenium, the latter being similar to that of NaDH [51] (Table 1). The XDH is sensitive to cyanide but can later partially be reconstituted by selenide. An analogous binding of selenium was discussed as was derived from crystallographic studies of the carbon monoxide dehydrogenase of *Oligotropha carboxidovorans* [56], where the labile selenium was modeled to form a Mo-selanyl-cysteine bridge. However, further studies do not substantiate such a structure (O. Meyer, personal communication). Thus, the coordination of the labile selenium moiety—as also observed in XDH and NaDH of *E. barkeri*—is still unsettled. At least, a strict involvement of selenite, but not of molybdate, was also noticed for anaerobic nicotinate degradation by *Desulfurococcus niacini* [57]. Thus, selenium seems to play a more general role in anaerobic nicotinate hydroxylation.

3. CATABOLISM OF ISONICOTINATE AND DERIVATIVES

Isonicotinate utilization by a *Bacillus brevis* strain has been suggested to start with a reduction to 1,4-dihydroisonicotinic acid followed by oxygenative ring cleavage. A utilization of 2-hydroxyisonicotinate or citrazinate was not observed [58]. In other bacteria investigated, namely a presumed *Pseudomonas* sp. [59], a *Micrococcus* (formerly *Sarcina*) sp. [60], and *Mycobacterium* sp. INA1 [61,62], degradation of isonicotinate proceeds via 2-hydroxy- and 2,6-dihydroxyisonicotinate (citrazinate). These compounds also have been proposed as intermediates in *N*-methylisonicotinate degradation by the gram-positive strain 4C1. A subsequent oxygenolytic ring cleavage of either 2,6-dihydroxyisonicotinate or 2,6-dihydroxypyridine, similar to the ring cleavage reactions proposed in nicotinate degradation, was suggested to lead to intermediates of the "maleamate pathway" is strain 4C1 (cf. Sec. 2.1, Fig. 1) [63].

Whereas in the case of *Pseudomonas* sp. and *Micrococcus* sp., the metabolic steps involved in further degradation of 2,6-dihydroxyisonicotinate have not been identified, the pathway of isonicotinate degradation in *Mycobacterium* sp. strain INA1 has been established almost completely. This unique pathway involves coenzyme A activation of 2,6-dihydroxyisonicotinate, ring reduction, and hydrolysis of the cyclic imide function, and finally yields 2-oxoglutarate, as shown in Fig. 2. Presumably, the CoA esterification activates the heteroaromatic ring of 2-hydroxyisonicotinate for a nucleophilic attack of a hydride ion [62]. In analogy to benzoyl-CoA reduction to cyclohexa-1,5-diene-1-carboxyl-CoA by *Thauera aromatica*, the CoA ester might serve as entrance by facilitating two consecutive one-electron-reduction steps [64] of citrazyl-CoA to 2,6-dioxopiperidine-4-carboxyl-CoA yielding after hydrolysis propane-1,2,3-tricarboxylate as further intermediate [62].

The isonicotinate and 2-hydroxyisonicotinate hydroxylases from *Micrococcus* sp. have been enriched and shown to be highly specific iron- and flavin-containing enzymes that presumably incorporate the hydroxyl group from water into their substrates [60]; however, more detailed information about the enzymes involved in the first steps of isonicotinate degradation is restricted to *Mycobacterium* sp. INA1. Each hydroxylation step is catalyzed by a distinct molybdenum hydro-

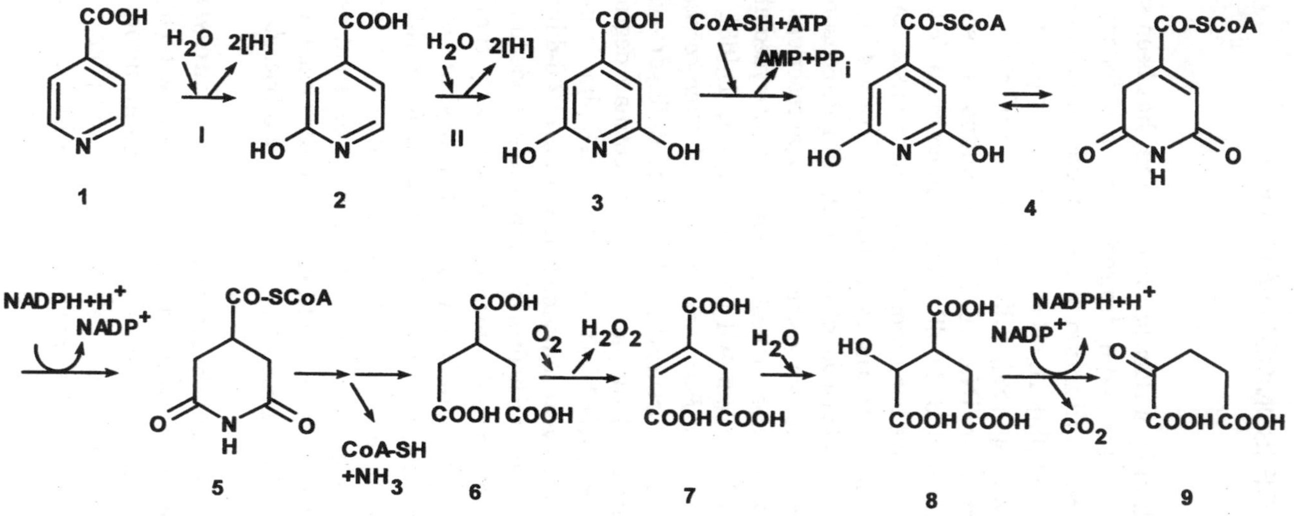

FIG. 2. Aerobic degradation of isonicotinic acid by *Mycobacterium* sp. INA1 [62]. 1, Isonicotinic acid; 2, 2-hydroxyisonicotinic acid; 3, 2,6-dihydroxyisonicotinic acid (citrazinic acid); 4, citrazyl-coenzyme A; 5, 2,6-dioxopiperidine-4-carboxyl-coenzyme A; 6, propane-1,2,3-tricarboxylic acid; 7, *cis*-aconitic acid; 8, isocitric acid; 9, 2-oxoglutaric acid. I, isonicotinate dehydrogenase; II, 2-hydroxyisonicotinate dehydrogenase.

xylase. Isonicotinate dehydrogenase (InaDH) oxidizes besides isonicotinate only quinoline-4-carboxylate (benzoisonicotinate), but is not active toward 2-hydroxyisonicotinate [62]. The second hydroxylation step is catalyzed by 2-hydroxyisonicotinate dehydrogenase (2-HInaDH), which likewise is highly specific for its substrate [65]. Whereas InaDH is a partly membrane-associated enzyme that exists as heterotrimer $(\alpha\beta\gamma)$ and also as a dimer of heterotrimers $[(\alpha\beta\gamma)_2]$ when analyzed by gel filtration, 2-HInaDH is a soluble enzyme with an $(\alpha\beta\gamma)_2$ structure. This subunit structure is rather common among the molybdenum hydroxylases; however the molecular mass of the large subunit of 2-HInaDH is somewhat higher than found in the other molybdenum hydroxylases [65] (Table 1). Like many molybdenum hydroxylases involved in the bacterial degradation of pyridine and quinoline derivatives (Table 1; see also Chap. 14), both enzymes contain the Mo-MCD form of the pyranopterin molybdenum cofactor, iron-sulfur centers, and noncovalently bound FAD [62,65].

Quinoline-4-carboxylate (benzoisonicotinate) utilization has been described for an *Agrobacterium* sp., a *Microbacterium* sp., and a *Pimelobacter simplex* strain [66]. Quinoline-4-carboxylate 2-oxidoreductase from *Agrobacterium* sp. 1B is another molybdo-iron/sulfur-flavoprotein harboring the Mo-MCD cofactor and showing an $(\alpha\beta\gamma)_2$ subunit structure [67] (Table 1). Besides quinoline-4-carboxylate, it catalyzes the hydroxylation at C2 of quinoline, 4-methylquinoline, and 4-chloroquinoline, but the analogue isonicotinate is not accepted as substrate [68]. Obviously, most enzymes involved in the hydroxylation of pyridine carboxylates and quinoline carboxylates are highly specific. Molybdenum hydroxylases involved in the degradation of apolar benzopyridines that show a somewhat relaxed substrate specificity, such as quinoline-2-oxidoreductase (Qor), or quinaldine-4-oxidase, do not catalyze the hydroxylation of carboxy-substituted benzopyridines (see Chap. 14).

4. NICOTINE CATABOLISM: ENZYMES AND GENES INVOLVED IN AEROBIC TRANSFORMATIONS

The aerobic soil bacteria *Arthrobacter nicotinovorans* (DSM 420) and *A. ureafaciens* (DSM 419, formerly strain P34) are able to utilize DL-nicotine as sole source of carbon, nitrogen, and energy. Both strains were formerly named *A. oxidans* but were shown to be different from the type strain of *A. oxidans* [69]. Early work of the groups of Decker and Rittenberg led to the elucidation of the reaction sequence of nicotine catabolism [70–72]. L- and D-nicotine are first converted to L- and D-6-hydroxynicotine. Oxidation of the 6-hydroxynicotines leads to the optically inactive 6-hydroxy-N-methylmyosmine that hydrolyzes spontaneously to [6-hydroxypyridyl-(3)]-(γ-N-methylaminopropyl)ketone. A subsequent second hydroxylation at C2 of the pyridine ring yields [2,6-dihydroxypyridyl-(3)]-(γ-N-methylaminopropyl)ketone (Fig. 3), which is degraded further via 2,6-dihydroxypyridine and maleamate [71]. The characteristic intensely blue-colored pigments (nicotine blue) formed from oxidation products were characterized to be two similar diazadipheno-quinones [73]. In contrast, *Pseudomonas convexa* degrades nicotine via 3-succinoylpyridine and 6-hydroxy-3-succinoylpyridine to 2,5-dihydroxypyridine, which is funneled into the maleamate pathway [74].

Nicotine dehydrogenases (Ndh) from *Arthrobacter nicotinovorans* and *A. ureafaciens* which catalyze the hydroxylation of L- and D-nicotine to L- and D-hydroxynicotine, respectively, are membrane-associated enzymes belonging to the family of molybdenum hydroxylases (XO family) and containing a molybdenum pyranopterin dinucleotide, iron-sulfur clusters, and FAD. The base in the pyranopterin dinucleotide of Ndh has not been determined yet [75,76]. Similar to carbon monoxide dehydrogenases (CO-DHs) of carboxidotrophic bacteria [56,77] and similar to a number of molybdenum hydroxylases involved in the degradation of other N-heteroaromatic compounds [38,78] (Table 1), Ndh from both sources (DSM 419 and DSM 420) consists of three subunits. However, the native Ndh enzyme like one of the active forms of isonicotinate dehydrogenase from *Mycobacterium* sp. INA1 and like 6-hydroxynicotinate dehydrogenase from *B. niacini* is a heterotrimer, whereas many other bacterial molybdenum hydroxylases are dimers of heterotrimers [78]. Its subunits show high amino acid sequence similarity to

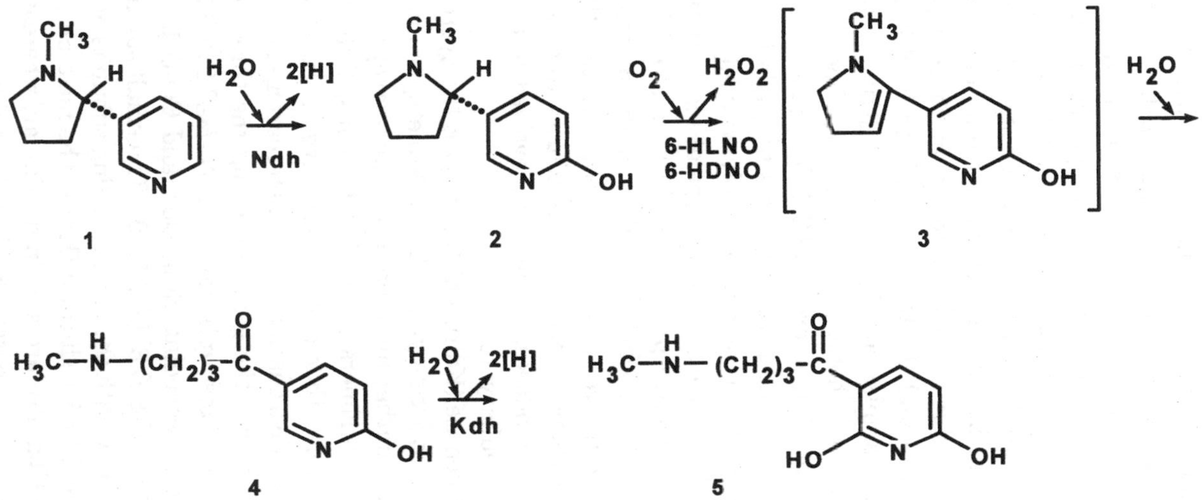

FIG. 3.   Early steps of nicotine degradation by *Arthrobacter nicotinovorans* and *A. ureafaciens* (both formerly *A. oxidans*) [71,72]. **1**, DL-nicotine; **2**, DL-6-hydroxynicotine; **3**, 6-hydroxy-*N*-methylmyosmine; **4**, [6-hydroxypyridyl-(3)]-(γ-*N*-methyl-aminopropyl)-ketone; **5**, [2,6-dihydroxypyridyl-(3)]-(γ-*N*-methylaminopropyl)-ketone.

Qor from *Pseudomonas putida* 86 (large subunit α: 41%, medium-sized subunit β: 38%, small subunit γ: 56% amino acid identity) [79].

Whereas Ndh catalyzes the hydroxylation of both L- and D-nicotine, 6-hydroxy-DL-nicotine is stereoselectively oxidized by a specific 6-hydroxy-L-nicotine oxidase (6-HLNO) and a specific 6-hydroxy-D-nicotine oxidase (6-HDNO) [72]. Since D-nicotine is not produced naturally, the evolution of a D-specific enzyme seems unexpected. However, 6-HDNO is instrumental in the catabolism not only of D-nicotine, but also of D-nornicotine, which has been found in tobacco plants [80]. Although both 6-HLNO and 6-HDNO are able to bind both enantiomeric substrates with similar affinity, their catalytic stereoselectivity is absolute. 6-HLNO is a homodimer containing one noncovalently bound FAD per subunit [81], whereas 6-HDNO is a monomeric protein containing a covalently bound FAD [80]. This flavinylation is an autocatalytic process that occurs posttranslationally without the action of a second enzyme [82]. Catalyzing the same type of reaction and forming an identical, optically inactive product, 6-HDNO and 6-HLNO nevertheless are unrelated at the DNA as well as at the protein level and thus are thought to be the result of convergent evolution [83,84]. The 6-HLNO gene, which shows considerable similarity to genes encoding eukaryotic monoamine oxidases, might have evolved after horizontal gene transfer from an eukaryote [85].

Ketone dehydrogenase (Kdh), which catalyzes the hydroxylation of the pyridine ring of [6-hydroxypyridyl-(3)]-(γ-*N*-methylaminopropyl)ketone at C2 (Fig. 3), is another molybdo-iron/sulfur-flavoprotein consisting of a dimer of heterotrimers [$(\alpha\beta\gamma)_2$] [84]. Thus, Kdh structurally resembles Ndh and some other molybdenum hydroxylases involved in the aerobic bacterial degradation of N-heteroaromatic compounds (Table 1) [38,78]. However, the genes encoding the three subunits of Kdh are arranged in the transcriptional order 5′*L-M-S* 3′(*kdhCAB*) [84], in contrast to the genes encoding Ndh, Qor, and CO-DHs, which are clustered in the order 5′*M-S-L* 3′ [76,77,79]. The small subunit of Kdh showed highest similarity with the corresponding subunits of Ndh, Qor, and a 4-hydroxybenzoyl-CoA reductase (Hba). The medium subunits of Kdh, Ndh, Qor, and Hba also are strikingly similar, whereas the large subunit of Kdh resembles the corresponding C-terminal region in XDH from chicken and other sources [84]. Based on amino acid sequence alignments, which allow the identification of putative cofactor binding sites, the small subunit of Kdh is assumed to bind two distinct [2Fe2S] clusters, whereas FAD and the molybdenum pyranopterin cofactor are thought to be located on the medium and the large subunit, respectively, as found in the CO-DHs from *Oligotropha carboxidovorans* and *Hydrogenophaga pseudoflava* [56,86,87], and as assumed for other heterotrimeric molybdenum hydroxylases.

The genes encoding Ndh, 6-hydroxynicotine dehydrogenases, Kdh, and the proteins involved in the molybdenum cofactor synthesis reside on the 160-kb plasmid pAO1 [76,84,88,89]. Cells of *A. nicotinovorans* grown on L-nicotine synthesize Ndh as well as 6-HLNO, whereas D- and DL-nicotine also induce 6-HDNO formation [72]. Kdh synthesis is coordinately induced with Ndh and 6-HLNO during the logarithmic phase of growth by DL-nicotine [90]. Induction of Ndh and 6-HLNO additionally depends on the presence of molybdate in the culture medium, whereas 6-HDNO synthesis is insensitive to molybdate [76]. Actually, the gene encoding 6-HDNO represents a separate transcription unit expressed only in stationary cells [91,92].

5. BIOTECHNOLOGICAL POTENTIALS AND MEDICAL IMPLICATIONS

Nicotinate and its amide are used per se as vitamins and food additives. In addition, the hydroxylated derivatives in positions 2 and 6 are important building blocks for further chemical synthesis. The strategy of certain commercial suppliers is the integration of chemistry and biotechnology [18]. The formation of 6-hydroxynicotinate by NaDH is an important reaction in the production of pesticides by *Achromobacter xylosoxidans*, *Serratia marcescens*, or *Pseudomonas fluorescens* yielding 40–65 g L^{-1} (up to 90% overall yield) [93–95]. Due to the high nicotinate concentration used in this process, the second enzyme, 6-HNaDH, is inhibited and its recovery is facilitated by the low solubility of the Mg salt of 6-hydroxynicotinate [18]. The unspecificity of a nitrilase and of NaDH of *Agrobacterium* sp. DSM 6336 is exploited for regiospecific biotransformation of 2-cyanopyrazine via pyrazinecarboxylate to 5-hydroxypyrazine-2-carboxylate (yielding about 280 mM), which is used for the development of new antitubercular drugs [96]. The use of a regioselective hydroxylation of nicotinate at position 2 and of pyrazine carboxylate at position 3 by *Ralstonia/Burkholderia* DSM 6920 will have a high industrial impact [37]. In contrast to this combined approach, a total chemical synthesis would require multiple steps in addition to a separation of differently substituted mixtures.

Isonicotinic acid hydrazide (isoniazid) and pyrazinamide are classical antitubercular drugs because they interfere specifically with the biosynthesis of mycolic acids, a key component of mycobacteria. [97]. The target of isoniazid seems to be the gene product InhA, an enoyl-acyl-carrier protein reductase, involved in the unusual fatty acid synthase II system, where the prodrug is transformed by a catalase/peroxidase (KatG) to a reactive isonicotinic acyl anion or radical that reacts with NAD at the active site forming a covalent attachment [98,99]. New drug candidates, such as N-heterocyclic compounds, which also require an activation step, e.g., by reduction, seem to be promising to fight the growing resistance of mycobacteria [100]. 5-Hydroxy-pyrazinecarboxylate, produced by an NaDH, is a building block of a drug being 1000 times more active than described drugs [93], pointing to the high industrial impact for such regioselective hydroxylations. The possible microbial transformations of isoniazid and derivatives [101] by, e.g., InaDH seem to be without influence.

Nicotine is a toxic alkaloid that leads to tobacco dependence by a transformation of nicotine to cotinine (the C5' is oxidized to the oxo form) as catalyzed by cytochrome P450 CYP2A6 or by the molybdenum-containing aldehyde oxidase [102]. An impaired nicotine metabolism protects a smoker against becoming dependent [103]. It seems that eukaryotes transform mainly the pyrrolidine ring, leaving the pyridine ring intact except forming an *N*-oxide [102]. However, (*S*)-nicotine and Ndh from certain bacteria are industrially used to form new analgesic compounds after combining the biotransformation with additional chemical steps [93].

6. CONCLUSIONS

In general, nicotinate, isonicotinate, and nicotine are initially transformed in bacteria by regioselective and substrate-specific hydroxylases being always molybdenum-iron/sulfur flavoproteins of similar structure ($\alpha\beta\gamma$) and molecular masses. A gene sequence is available only for two of these enzymes, Ndh and Kdh, involved in nicotine metabolism. They show a high similarity to other bacterial molybdenum-containing hydroxylases. No crystallographic X-ray structure is reported for these pyridine derivatives-converting bacterial enzymes in contrast to bacterial CO-DH and bovine XDH and XO [56,87,104]. A still unresolved, labile selenium moiety is additionally present in NaDH and XDH isolated from anaerobic bacteria that is essential for the hydroxylase function. All of these enzymes are biochemically not that well studied in comparison with other molybdoenzymes, such as XO, partly because of problems resulting from their instability. However, their pronounced substrate specificity is of high industrial importance as is now obvious for certain forms of NaDH or Ndh.

ACKNOWLEDGMENTS

Financial support of the authors' laboratories by the Deutsche Forschungsgemeinschaft, the Volkswagen-Stiftung, the former BMFT-Schwerpunkt "Bioprozesstechnik," and the Fonds der Chemischen Industrie is gratefully acknowledged.

ABBREVIATIONS

CoA	coenzyme A
CO-DH	carbon monoxide dehydrogenase
DH	dehydrogenase
EPR	electron paramagnetic resonance
FAD	flavin adenine dinucleotide
Hba	4-hydroxybenzoyl-coenzyme A reductase

2-HInaDH	2-hydroxynicotinate dehydrogenase
6-HDNO	6-hydroxy-D-nicotine oxidase
6-HLNO	6-hydroxy-L-nicotine oxidase
6-HNaDH	6-hydroxynicotinate dehydrogenase
InaDH	isonicotinate dehydrogenase
Kdh	ketone dehydrogenase
Mo-MCD	molybdenum pyranopterin cytosine dinucleotide ("molybdopterin cytosine dinucleotide")
Mo-MGD	molybdenum pyranopterin guanine dinucleotide ("molybdopterin guanine dinucleotide")
Mo-MPT	molybdopterin
NAD	nicotinamide adenine dinucleotide
NaDH	nicotinamide dehydrogenase
NADPH	nicotinamide adenine dinucleotide phosphate (reduced)
Ndh	nicotine dehydrogenase
Qor	quinoline 2-oxidoreductase
SDS-PAGE	sodium dodecyl sulfate polyacrylamide gel electrophoresis
XO	xanthine oxidase
XDH	xanthine dehydrogenase

REFERENCES

1. D. T. Davies, *Aromatic Heterocyclic Chemistry*, Oxford University Press, 1992.

2. T. L. Gilchrist, *Heterocyclic Chemistry*, Longman Scientific & Technical, Harlow, 1985.

3. I. Siegmund, K. Koenig, and J. R. Andreesen, *FEMS Microbiol. Lett.*, *67*, 281–284 (1990).

4. S. Fetzner, *Appl. Microbiol. Biotechnol.*, *49*, 237–250 (1998).

5. J. P. Kaiser, Y. Feng, and J. M. Bollag, *Microbiol. Rev.*, *60*, 483–498 (1996).

6. G. Schwarz and F. Lingens, in *Biochemistry of Microbial Degradation* (C. Ratledge, ed.), Kluwer Academic, Dordrecht, 1994, pp. 459–486.

7. O. P. Shukla, *J. Sci. Ind. Res.*, *43*, 98–116 (1984).

8. S. K. Rhee, G. M. Lee, J. H. Yoon, Y. H. Park, H. S. Bae, and S. T. Lee, *Appl. Environ. Microbiol.*, *63*, 2578–2585 (1997).

9. G. K. Watson, C. Houghton, and R. B. Cain, *Biochem. J.*, *140*, 265–276 (1974).

10. N. S. Zefirov, S. R. Agapova, P. B. Terentiev, I. M. Bulakhova, N. I. Vayukova, and L. V. Modyanova, *FEMS Microbiol. Lett.*, *118*, 71–74 (1994).

11. J. H. Yoon, S. S. Kang, Y. G. Cho, S. T. Lee, Y. H. Kho, C. J. Kim, and Y. H. Park, *Int. J. Syst. Evol. Microbiol.*, *50*, 2173–2180 (2000).

12. C. L. Kitts, J. P. Lapointe, V. T. Lam, and R. A. Ludwig, *J. Bacteriol.*, *174*, 7791–7797 (1992).

13. E. J. Behrman and R. Y. Stanier, *J. Biol. Chem.*, *228*, 923–945 (1957).

14. J. C. Ensign and S. C. Rittenberg, *J. Biol. Chem.*, *239*, 2285–2291 (1965).

15. M. Nagel and J. R. Andreesen, *FEMS Microbiol. Lett.*, *59*, 147–152 (1989).

16. M. Nagel and J. R. Andreesen, *Arch. Microbiol.*, *154*, 605–613 (1990).

17. M. Nagel and J. R. Andreesen, *Int. J. Syst. Bacteriol.*, *41*, 134–139 (1991).

18. H. G. Kulla, *Chimia*, *45*, 81–85 (1991).

19. J. J. Gauthier and S. C. Rittenberg, *J. Biol. Chem.*, *246*, 3737–3742 (1971).

20. E. J. Behrmann and R. Y. Stanier, *J. Biol. Chem.*, *228*, 947–953 (1957).

21. A. L. Hunt, *Biochem. J.*, *72*, 1–7 (1959).

22. A. L. Hunt, D. E. Hughes, and J. M. Lowenstein, *Biochem. J.*, *69*, 170–173 (1958).

23. C. L. Kitts, L. E. Schaechter, R. S. Rabin, and R. A. Ludwig, *J. Bacteriol. 171*, 3406–3411 (1989).

24. M. V. Jones, *FEBS Lett.*, *32*, 321–324 (1973).

25. B. Hurh, T. Yamane, and T. Nagasawa, *J. Ferment. Bioeng.*, *78*, 19–26 (1994).

26. J. C. Hilton, C. A. Temple, and R. V. Rajagopalan, *J. Biol. Chem.*, *274*, 8428–8436 (1999).

27. J. M. Dias, M. E. Than, A. Hummn, R. Huber, G. P. Bourenkov, H. D. Bartunik, S. Bursakov, J. Calvete, J. Caldeira, C. Carneiro, J. J. G. Moura, I. Moura, and M. J. Romão, *Structure*, *7*, 65–79 (1999).

28. R. Hirschberg and J. C. Ensign, *J. Bacteriol.*, *108*, 751–756 (1971).

29. R. Hirschberg and J. C. Ensign, *J. Bacteriol.*, *112*, 392–397 (1972).

30. C. Scazzocchio, in *Molybdenum and Molybdenum-containing Enzymes* (M. P. Coughlan, ed.), Pergamon Press, Oxford, 1980, pp. 487–515.

31. L. Amrani, G. Cecchetto, C. Scazzocchio, and A. Glatigny, *Mol. Microbiol.*, *31*, 1065–1073 (1999).

32. N. J. Lewis, P. Hurt, H. M. Sealy-Lewis, and C. Scazzocchio, *Eur. J. Biochem.*, *91*, 311–316 (1978).

33. R. K. Mehra and M. P. Coughlan, *Arch. Biochem. Biophys.*, *229*, 585–595 (1984).

34. A. Glatigny and C. Scazzocchio, *J. Biol. Chem.*, *270*, 3534–3550 (1995).

35. M. Ueda and R. Sashida, *J. Mol. Catal. B. Enzym.*, *4*, 199–204 (1998).

36. A. Tinschert, A. Kiener, K. Heinzmann, and A. Tschech, *Arch. Microbiol.*, *168*, 355–361 (1997).

37. A. Tinschert, A. Tschech, K. Heinzmann, and A. Kiener, *Appl. Microbiol. Biotechnol.*, *53*, 185–195 (2000).

38. S. Fetzner, B. Tshisuaka, F. Lingens, R. Kappl, and H. Hüttermann, *Angew. Chem. Int. Ed.*, *37*, 576–597 (1998).

39. H. G. Wetzstein, N. Schmeer, and W. Karl, *Appl. Environ. Microbiol.*, *63*, 4272–4281 (1997).

40. H. G. Wetzstein, M. Stadler, H. V. Tichy, A. Dahlhoff, and W. Karl, *Appl. Environ. Microbiol.*, *65*, 1556–1563 (1999).

41. E. R. Stadtman, T. C. Stadtman, I. Pastan, and L. D. S. Smith, *J. Bacteriol.*, *110*, 758–760 (1972).

42. M. D. Collins, P. A. Lawson, A. Willems, J. J. Cordoba, J. Fernandez-Garayzabal, P. Garcia, J. Cai, H. Hippe, and J. A. E. Farrow, *Int. J. Syst. Bacteriol.*, *44*, 812–826 (1994).

43. L. Tsai, I. Pastan, and E. R. Stadtman, *J. Biol. Chem.*, *241*, 1807–1813 (1966).

44. J. S. Holcenberg and E. R. Stadtman, *J. Biol. Chem.*, *244*, 1194–1203 (1969).

45. J. S. Holcenberg and L. Tsai, *J. Biol. Chem.*, *244*, 1204–1211 (1969).

46. D. H. Edwards, B. T. Golding, F. Kroll, B. Beatrix, G. Bröker, and W. Buckel, *J. Am. Chem Soc.*, *118*, 4192–4193 (1996).

47. H. Eggerer, *Cur. Top. Cell. Regul.*, *26*, 411–418 (1985).

48. R. Wagner and J. Andreesen, *Arch. Microbiol.*, *121*, 255–260 (1979).

49. D. Imhoff and J. R. Andreesen, *FEMS Microbiol. Lett.*, *5*, 155–158 (1979).

50. G. L. Dilworth, *Arch. Biochem. Biophys.*, *219*, 30–38 (1982).

51. T. Schräder, A. Rienhöfer, and J. R. Andreesen, *Eur. J. Biochem.*, *264*, 862–871 (1999).

52. F. Zinoni, A. Birkmann, W. Leinfelder, and A. Böck, *Proc. Natl. Acad. Sci. USA*, *84*, 3156–3160 (1987).

53. G. L. Dilworth, *Arch. Biochem. Biophys.*, *221*, 565–569 (1983).

54. V. N. Gladyshev, S. V. Khangulov, and T. C. Stadtman, *Proc. Natl. Acad. Sci. USA*, *91*, 232–236 (1994).

55. V. N. Gladyshev, S. V. Khangulov, and T. C. Stadtman, *Biochemistry*, *35*, 212–223 (1996).

56. H. Dobbek, L. Gremer, O. Myer, and R. Huber, *Proc. Natl. Acad. Sci. USA*, *96*, 8884–8889 (1999).

57. D. Imhoff-Stuckle and N. Pfennig, *Arch. Microbiol.*, *136*, 194–198 (1983).

58. R. P. Singh and O. P. Shukla, *J. Ferment. Technol.*, *64*, 109–117 (1986).

59. J. C. Ensign and S. C. Rittenberg, *Arch. Mikrobiol.*, *51*, 384–392 (1965).

60. R. C. Gupta and O. P. Shukla, *Ind. J. Biochem. Biophys.*, *16*, 72–75 (1979).

61. A. Kretzer and J. R. Andreesen, *J. Gen. Microbiol.*, *137*, 1073–1080 (1991).

62. A. Kretzer, K. Frunzke, and J. R. Andreesen, *J. Gen. Microbiol.*, *139*, 2763–2772 (1993).

63. C. G. Orpin, M. Knight, and W. C. Evans, *Biochem. J.*, *127*, 833–844 (1972).

64. W. Buckel and R. Keese, *Angew. Chem. Int. Ed.*, *34*, 1502–1506 (1995).

65. T. Schräder, C. Hillebrand, and J. R. Andreesen, *FEMS Microbiol. Lett.*, *164*, 311–316 (1998).

66. M. Schmidt, P. Röger, and F. Lingens, *Biol. Chem. Hoppe-Seyler.*, *372*, 1015–1020 (1991).

67. G. Bauer and F. Lingens, *Biol. Chem. Hoppe-Seyler*, *373*, 699–705 (1992).

68. G. Bauer, Ph.D. thesis, University of Hohenheim, Germany, 1992.

69. Y. Kodama, H. Yamamoto, N. amano, and T. Amachi, *Int. J. Syst. Bacteriol.*, *42*, 234–239 (1992).

70. F. A. Gries, K. Decker, H. Eberwein, and M. Brühmüller, *Biochem. Z.*, *335*, 285–302 (1961).

71. R. L. Gherna, S. H. Richardson, and S. C. Rittenberg, *J. Biol. Chem.*, *240*, 3669–3674 (1965).

72. K. Decker and H. Bleeg, *Biochim. Biophys. Acta.*, *105*, 313–334 (1965).

73. H. J. Knackmuss and W. Beckmann, *Arch. Microbiol.*, *90*, 167–169 (1973).

74. R. Thacker, O. Rørvig, P. Kahlon, and I. C. Gunsalus, *J. Bacteriol.*, *135*, 289–290 (1978).

75. W. Freudenberg, K. König, and J. R. Andreesen, *FEMS Microbiol. Lett.*, *52*, 13–18 (1988).

76. S. Grether-Beck, G. L. Igloi, S. Pust, E. Schilz, K. Decker, and R. Brandsch, *Mol. Microbiol.*, *13*, 929–936 (1994).

77. B. Santiago, U. Schübel, C. Egelseer, and O. Meyer, *Gene*, *236*, 115–124 (1999).

78. R. Hille, *Chem. Rev.*, *96*, 2757–2816 (1996).

79. M. Bläse, C. Bruntner, B. Tshisuaka, S. Fetzner, and F. Lingens *J. Biol. Chem.*, *271*, 23068–23079 (1996).

80. K. Decker and R. Brandsch, *BioFactors*, *3*, 69–81 (1991).

81. V. D. Dai, K. Decker, and H. Sund, *Eur. J. Biochem.*, *4*, 95–102 (1968).

82. R. Brandsch, V. Bichler, L. Mauch, and K. Decker, *J. Biol. Chem.*, *268*, 12724–12729 (1993).

83. R. Brandsch, A. E. Hinkkanen, L. Mauch, H. Nagursky, and K. Decker, *Eur. J. Biochem.*, *167*, 315–320 (1987).

84. S. Schenk, A. Hoelz, B. Krauss, and K. Decker, *J. Mol. Biol.*, *284*, 1323–1339 (1998).

85. S. Schenk and K. Decker, *J. Mol. Evol.*, *48*, 178–186 (1999).

86. L. Gremer, S. Kellner, H. Dobbek, R. Huber, and O. Meyer, *J. Biol. Chem.*, *275*, 1864–1872 (2000).

87. P. Hänzelmann, H. Dobbek, L. Gremer, R. Huber, and O. Meyer, *J. Mol. Biol.*, *301*, 1221–1235 (2000).

88. R. Brandsch and K. Decker, *Arch. Microbiol.*, *138*, 15–17 (1984).

89. R. Brandsch, W. Faller, and K. Schneider, *Mol. Gen. Genet.*, *202*, 96–101 (1986).

90. M. Gloger and K. Decker, *Naturforsch.*, *246*, 1016–1025 (1969).

91. L. Mauch, B. Krauss, and R. Brandsch, *Arch. Microbiol.*, *152*, 95–99 (1989).

92. L. Mauch, V. Bichler, and R. Brandsch, *Mol. Gen. Genet.*, *221*, 427–434 (1990).

93. A. Schmid, J. S. Dordick, B. Hauer, A. Kiener, M. Wubbolts, and B. Witholt, *Nature*, *409*, 258–268 (2001).

94. B. Hurh, M. Ohshima, T. Yamane, and T. Nagasawa, *J. Ferm. Bioeng.*, *77*, 382–385 (1994).

95. T. Nagasawa, B. Hurh, and T. Yamane, *Biosci. Biotech. Biochem.*, *58*, 665–668 (1994).

96. M. Wieser, K. Heinzmann, and A. Kiener, *Appl. Microbiol. Biotechnol.*, *48*, 174–176 (1997).

97. M. Daffe and P. Draper, *Adv. Microb. Physiol.*, *39*, 131–203 (1998).

98. D. A. Rozwarski, G. A. Grant, H. R. Barton, W. R. Jacobs, and J. C. Sacchettini, *Science*, *279*, 98–102 (1998).

99. D. A. Rozwarski, C. Vilcheze, M. Sugantino, R. Bittman, and J. C. Sacchettini, *J. Biol. Chem.*, *274*, 15582–15589 (1999).

100. C. K. Stover, P. Warrener, D. R. VanDevanter, D. R. Sherman, T. M. Arain, M. H. Langhorne, S. W. Anderson, J. A. Towell, Y. Yuan, D. N. McMurray, B. N. Kreiswirth, C. E. Barry, and W. R. Baker, *Nature.*, *405*, 962–966 (2000).

101. M. I. Sharma and O. P. Shukla, *Biol. Mem.*, *13*, 1–17 (1987).

102. G. A. Kyerematen and E. S. Vesell, *Drug. Metab. Rev.*, *23*, 3–41 (1991).

103. M. L. Pianezza, E. M. Sellers, and R. F. Tyndale, *Nature*, *393*, 750 (1998).

104. C. Enroth, B. T. Eger, K. Okamoto, T. Nishino, T. Nishino, and E. F. Pai, *Proc. Natl. Acad. Sci. USA*, *97*, 10723–10728 (2000).

12

The Molybdenum-Containing Xanthine Oxidoreductases and Picolinate Dehydrogenases

Emil F. Pai[1] and Takeshi Nishino[2]

[1]Departments of Biochemistry, Medical Biophysics, and Molecular & Medical Genetics, University of Toronto, 1 King's College Circle, Toronto, ON, M5S 1A8, and Division of Molecular & Structural Biology, Ontario Cancer Institute, University Health Network, 610 University Avenue, Toronto, ON, M5G 2M9, Canada

[2]Department of Biochemistry and Molecular Biology, Nippon Medical School, 1-1-5 Sendagi, Bunkyo-ku, Tokyo 113–8602, Japan

1.   INTRODUCTION

1.1.   History

In the late nineteenth century, a German pathologist was the first to recognize a uric acid generating catalytic activity in human tissue [1]. Although in 1902 Schardinger [2] described the enzyme causing this transformation as an aldehyde oxidase, it was later identified as xanthine oxidase (XO) [3]. More recently, it has become clear that the two catalytic activities listed as xanthine dehydrogenase (EC 1.1.1.204; XDH) and XO (E.C. 1.2.3.2) can be attributed to the same gene product [4–6]. The relative ratio between the two activities can change drastically during purification without dithiothreitol due to a loss of the enzyme's capacity to efficiently utilize NAD^+ as its substrate. In this chapter, the terms dehydrogenase and oxidase refer to the specific activities, respectively, whereas the term oxidoreductase (XOR) includes both.

Since its first purification three-quarters of a century ago [7], xanthine oxidoreductase has served as a benchmark for the whole class of complex metalloflavoproteins [8]. Given its prominent character, it is not surprising that this enzyme has been reviewed extensively in the past, including several quite recent reports [9–11]. In the absence of direct structural information, these publications often had to refer to several closely related and structurally known enzymes, e.g., aldehyde oxidase [12,13] and CO dehydrogenase [14,15], when discussing questions of substrate binding and catalysis. With the crystal structures of both the dehydrogenase and oxidase forms of bovine milk xanthine oxi-doreductase recently determined [16], this chapter will include some structural aspects, in addition to attempting a succinct description of the medical relevance and biochemical properties of this archetypal enzyme.

1.2.   Medical Relevance

1.2.1.   Gout and Hyperuricemia

In the purine degradation pathway, xanthine oxidoreductase catalyzes the oxidation of hypoxanthine and of xanthine, leading to the formation of uric acid [17]. In primates, uric acid is the ultimate product of purine catabolism, and its microcrystals are found deposited in the joints of gout patients. Under these circumstances, the enzyme is an obvious target of drugs directed against gout and hyperuricemia. One pyrazolo-pyrimidine derivative, allopurinol, has been used very effectively as a remedy for both hyperuricemia and gout [18]. Allopurinol is accepted as a substrate by the enzyme, and its oxidation product, oxypurinol, binds very tightly to the molybdenum center of the reduced enzyme [19]. However, the slow reoxidation of this complex by oxygen releases the inhibitor molecule [19]. Therefore, it is necessary to maintain a rela-tively high plasma level of the drug for efficient treatment.

1.2.2.   Xanthinuria

Xanthinuria is another disease involving abnormalities of xanthine oxi-doreductase [20]. It presents itself in three subtypes: classical xanthi-nuria type I, which is caused by XDH deficiency alone; classical xanthinuria type II, which represents both XDH and aldehyde oxidase deficiencies; and molybdenum cofactor deficiency, leading to the conco-mitant loss of function of three enzymes: XOR, aldehyde oxidase, and sulfite oxidase. Most patients with classical xanthinuria type I or type II have no symptoms, but some patients may develop urinary tract calculi, acute renal failure, or myositis due to tissue deposition of xanthine. On the other hand, molybdenum cofactor deficiency is associated with severe neurological disorders. These very serious symptoms are primarily linked to the absence of sulfite oxidase activity [21].

Analysis of the genes of patients presenting with classical xanthinuria type I identified defects in the XDH gene itself as the sole basis of this disease [22]. However, the cause of classical xanthinuria type II is mutation of the molybdenum cofactor sulfurase gene [23]. One of the features common to both XOR and aldehyde oxidase is their dependence on such an enzyme to place a sulfur atom as a ligand to the molybdenum ion, which in turn is bound to the pterin cofactor, as a prerequisite for becoming catalytically active [24]. In contrast, sulfite oxidase is fully functional without this addition of free sulfur, as one of its ligands is a cysteine residue.

In *Drosophila melanogaster*, mutations at the maroon-like locus (*ma-1*) are known to cause the inactivation of molybdenum hydroxylase, XDH, and aldehyde oxidase. The *ma-1* gene has recently been cloned and the protein it encodes is a candidate for the enzyme that sulfurates the common molybdenum cofactor [25]. Furthermore, changes in the homologous gene of cow (*MCSU*) show a close association with the appearance of the symptoms of bovine classical xanthinuria type II [26], and in the human homologue (*HMCS*), a nonsense mutation has been linked to the disease [23].

1.2.3. Postischemic Reperfusion Injury

As XO uses oxygen instead of NAD^+ as the terminal acceptor of electrons producing O_2^- as well as H_2O_2, the enzyme has been implicated in the pathogenesis of postischemic reperfusion injury [27]. During ischemia, purine nucleotide catabolism increases hypoxanthine concentrations [28] and XDH can be transformed to XO, e.g., by proteases set free by damaged cells. Upon reestablishment of circulation, XO can convert hypoxanthine and produce the highly toxic radicals [27]. The existence of the enzyme in endothelial cells is another factor supporting this hypothesis [29]. The initial evidence for the conversion of XDH to XO during ischemia was obtained in experiments using intestine [30], and its conclusions have since been extended to other organs, such as heart, kidney, lung, and brain [9]. However, the evidence for the conversion during ischemia is controversial [9], and other mechanisms for the production of oxyradical species by xanthine oxidoreductase have been proposed [31,32]. The situation seems to be more complex than originally envisaged, and firm proof for or against the hypothesis remains to be established.

2. XANTHINE DEHYDROGENASE/XANTHINE OXIDASE

2.1. Biochemistry

2.1.1. Characterization of Xanthine Oxidoreductases

Xanthine oxidoreductase enzymes have been isolated from a wide range of organisms, from bacteria to man. They catalyze the conversion of a large variety of substrates, such as purines, pteridines, pyrimidines, and aldehydes. All eukaryotic and most XORs from other sources have similar molecular weights and composition of redox centers [10,11]. XOR even seems to be the first example of a bacterial molybdopterin enzyme that does not contain the dinucleotide but the mononucleotide form of the cofactor [33].

XDHs from several bacterial organisms possess different subunit compositions. The enzyme from *Comamonas acidovorans* is a heterotetramer consisting of two copies each of a larger (90 kDa) and a smaller (60 kDa) subunit [33]; the enzyme from *Rhodobacter capsulatus* B10S [34] adopts the same subunit organization. Several $\alpha_4\beta_4$ structures have also been described [35–37], and *P. putida* 40 XDH seems to combine three subunits of 72 kDA into a homotrimeric assembly [38]. *Pseudomonas* species seem to be especially creative in combining various subunits in constructing their XDHs. Not only do they shuffle subunits of various sizes, but some of them have replaced the FAD cofactor with heme groups [36]. In the case of *P. putida* Ful, for example, a cytochrome *b* subunit is found instead [37].

Until recently, the *Veillonella atypica* enzyme was the only XDH to display the classical $(\alpha\beta\gamma)_2$ structure [39] found in many hydroxylating molybdenum-containing dehydrogenases [11]. Now it has been joined by two proteins with XDH activity and similar subunit arrangements, one purified from *Eubacterium barkeri* [40] and the other one from *Clostridium purinolyticum* [41]. Both of these enzymes display an

$\alpha_4\beta_4\gamma_4$ structure and require selenium for catalytic activity. While the cofactor content was not further characterized for the clostridial enzyme, it was shown that the *E. barkeri* XDH, in addition to selenium as already mentioned above, contains stoichiometric amounts of the dinucleotide form of molybdopterin, iron, acid-labile sulfur, FAD, and even tungsten.

2.1.2. Purification Protocol

Many methods of purification for a large selection of XORs from a variety of sources have been published [8,42,43]. In our quest to produce high-quality crystals of the enzyme from bovine milk, we found that the method of Ball [44] combined with affinity chromatography on a folate resin and ion exchange chromatography worked best for the preparation of the XO form of the enzyme [45]. In order to obtain the XDH form, however, pancreatin and butanol had to be replaced by protease-free lipase for solubilization of the protein from the fat globular membrane. Inclusion of the lipase treatment is essential for the removal of traces of contaminating lipids and leads to improved purity but more importantly allows for the growth of single crystals [46]. The enzyme isolated from regular sources contains an appreciable amount of enzyme whose activity has been abolished by the removal of the free sulfur ligand of the molybdenum ion and its replacement with an oxygen atom. Again, folate affinity chromatography has proven extremely useful in selecting only the enzymatically active component of the preparation [45].

2.1.3. Primary Sequences and Cofactor-Subunit Correlation

Due to the spectacular progress in obtaining nucleic acid sequences, we have witnessed in the last few years, the NCBI data bank now carries more than 350 entries of genes coding for XORs. Although many of these refer to closely related organisms (about 30 sequences from various *Drosophila* species alone), this dramatic increase enables meaningful comparisons. However, for the purpose of this chapter, we shall concentrate on the following sequences because they were the first ones determined and/or the knowledge gained from them was combined with the results of mutational studies as a valuable source of information in discussions about XOR's mechanism of catalysis. They comprise XDHs from *Aspergillus nidulans*, [47], *Drosophila melanogaster* [48,49], chicken [50], mouse [51], rat [52], cow [53], and *Homo sapiens* [54–56]. All these enzymes consist of approximately 1330 amino acids and are highly similar, with about 90% sequence identity among the mammalian enzymes, about 70% between mammalian and avian forms, and still a remarkable 52% between human and *Drosophila* enzymes.

The active forms of these enzymes are homodimers, with each of the 145-kDa monomers acting independently in catalysis [10,11]. Each subunit tightly binds one molybdopterin cofactor, two spectroscopically distinct [2Fe-2S] centers, and one FAD cofactor. Fragments of 20 kDa, 40 kDa, and 85 kDa, respectively, are the products of limited proteolysis of mammalian XDH with trypsin; cleavage is accompanied by the enzyme's irreversible conversion to the XO form [52]. However, similar cleavage with chicken enzyme does not cause the same change in substrate specificity [50]. A reversible conversion of XDH to XO of the bovine enzyme can be achieved by modification of two cysteine residues (Cys535 and Cys992) [57,58]. Comparative sequence alignment indicates that the two iron-sulfur centers are located in the N-terminal 20-kDa fragment, FAD in the intermediate 40-kDa fragment, and the molybdenum cofactor in the 85 kDa C-terminal fragment [10]. The schematic domain structure of the bovine milk enzyme is shown in Fig. 1, and the relative locations of the proteolytic cutting sites and of the modified cysteines are indicated.

2.2. Mechanistic Studies

2.2.1. Substrate Specificity

Studies with the enzymes from bovine milk or human liver have shown wide substrate specificity [42,43]. In addition to the physiological substrates xanthine and hypoxanthine or adenine, in some pathological cases [20], many other compounds, including aldehydes, pteridines, and other purines, including synthetic ones used in therapy like 6-mercaptopurine, can be oxidized by this oxidoreductase [20]. In contrast to most other hydroxylases, however, XDH utilizes a water molecule as the ultimate source of oxygen incorporated into the product [11]. The oxidation of these compounds takes place at the molybdopterin center, whereas the reduction of the natural oxidant substrate, NAD^+ in the

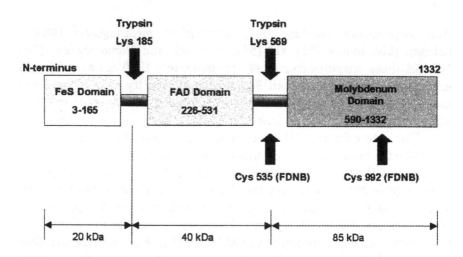

FIG. 1. Schematic description of the domain structure of bovine milk xanthine oxidoreductase. The N-terminal domain binds the two Fe_2S_2 clusters, the middle domain harbors the FAD binding site (also the site of NAD binding), and the C-terminal domain is the location of the Mo-pterin cofactor (binding site of xanthine). Long segments connect each of the three domains. Numbers in the boxes refer to the amino acid residues forming the respective domains of the bovine milk enzyme. Thick arrows indicate the sites of tryptic cleavage or FDNB modification. At bottom, the relative molecular weight of the domains is shown.

case of XDH and oxygen in XO, occurs through the isoalloxazine ring of the FAD cofactor [10,11]. Again NAD^+ or oxygen are not the only possible electron accepting substrates; many other synthetic compounds such as methylene blue, 2,6-dichlorophenol-indophenol, or phenazine methosulfate can serve in such a role, too, albeit some of them possibly via different sites of interaction [42,43].

2.2.2. Electron Transfer

The intramolecular distribution of electrons, once introduced into the enzyme, has been extensively probed under both thermodynamic as well as kinetic aspects [10,11]. The behavior of the enzyme during the reductive titration and stopped-flow studies has shown that the electron distribution within the various redox centers is very rapid and solely due to the relative levels of their redox potentials [59]. No evidence has been found that the redox potential of one center is influenced by the oxidation state of another center [59]. The reduction potentials of the XDH and XO forms of the bovine milk as well as chicken liver enzymes are shown in Table 1. The assignments of the cysteine ligands of the two iron sulfur centers by mutagenesis [60] and EPR studies [15,60] on the rat liver XO, and subsequently confirmed by the crystal structure of the bovine milk XDH, indicated the electron flow pathway during catalysis is as follows:

$$Mo\text{-pterin} \rightarrow Fe/S - I \rightarrow Fe/S - II \rightarrow FAD$$

The redox potentials for the two iron-sulfur clusters are different, but almost independent of whether the enzyme is in the dehydrogenase or oxidase form. The geometrical arrangements and redox potentials of these centers indicate that electrons are transferred from the molybdenum to the two iron-sulfur centers in a thermodynamically favorable process. However, the redox potential of the FAD/FADH$^{\bullet}$ couple is -270 mV, whereas that of the FADH$^{\bullet}$/FADH$_2$ couple is as low as -410 mV in bovine milk XDH [63] and -330 mV in chicken XDH [64]. The values for the XO forms are approximately -320 mV and -235 mV, respectively. Thus, the flavin semiquinone is thermodynamically much more stable in

TABLE 1

Midpoint Potentials of the Different Redox Centers of Bovine Milk XDH and XO as well as Chicken XDH (mV)

	Milk XO[a]	Milk XO[b]	Milk XDH[c]	Chicken XDH[d]
FADH$^{\bullet}$/FADH$_2$	-237	-234	-410	-330
FAD/FADH$^{\bullet}$	-310	-332	-270	-294
Fe/S II	-217	-255	-235	-275
Fe/S I	-310	-310	-310	-280
Mo^V/Mo^{IV}	-315	-377	nd	-337
Mo^{VI}/Mo^V	-345	-373	nd	-357

nd: not determined
[a]In 0.1 M potassium N, N'-bis(2-hydroxyethyl)glycine, pH 7.7, 25°C [61].
[b]In 50 mM bicine, 1 mM EDTA, pH 7.7, at 25°C [62].
[c]In 0.1 M pyrophosphate, 0.3 mM EDTA, pH 7.5, at 25°C [63].
[d]In 50 mM phosphate, pH 7.8, at 175 K and 25 K [64].

XDH than in XO. It had been reported that NAD^+ destabilizes the flavin semiquinone (increase in the potential of $FADH^\bullet/FADH_2$), whereas aminopyridine adenine dinucleotide binding has the opposite effect [65]. Although thermodynamically it is not favorable to reduce NAD^+ via a fully reduced flavin ring, the ratio of $NADH/NAD^+$ is high enough to overcome this problem under conditions as they are found in standard assays or in the cellular environment.

2.2.3. Biochemical Differences Between Xanthine Dehydrogenase and Xanthine Oxidase

Originally, XDH and XO were regarded as separate enzymes. This is reflected in their respective Enzyme Commission numbers. More recently, however, it has become clear that both catalytic activities are performed by different versions of the same gene product. Mammalian XORs are synthesized as the dehydrogenase form and exist mostly as such in the cell [9]. However, as pointed out above, they can be readily converted to the oxidase form by oxidation of sulfhydryl residues or by limited proteolysis [52]. XDH shows a preference for NAD^+ reduction at the FAD reaction site, while XO fails to react with the nicotinamide nucleotide and exclusively uses dioxygen as its substrate leading to the formation of superoxide anion and hydrogen peroxide [9–11]. It should be noted that both XDH and XO form H_2O_2 and O_2^- when only xanthine and molecular oxygen are provided as substrates. Under those conditions, XDH produces even more superoxide anion per mole oxygen utilized during turnover than XO. However, the presence of NAD^+ completely suppresses the oxidase reaction as the nucleotide outcompetes oxygen with very high efficiency [9–11]. The kinetic parameters of bovine milk and rat liver XDH and XO are shown in Table 2.

In parallel to these kinetic studies, it was also shown by binding experiments that XDH has an NAD binding site but XO does not [68,69]. Marked differences in the immediate environment of the flavin rings in the XDH and XO forms of bovine milk XOR have been deduced from experiments employing active site probes like artificial flavins that possess an ionizable group at the 6- or 8-position of the isoalloxazine ring [70–72]. As an example, the structures of 8-mercaptoflavin in its various ionic states are shown in Fig. 2. When the natural FAD of chicken liver [70], rat liver [71], and bovine milk [72] XDH was replaced

TABLE 2

Steady-State Kinetic Parameters for Mammalian Xanthine Oxidases and Xanthine Dehydrogenases (at 25°C)

	Bovine milk XO	Bovine milk XDH	Rat liver XO	Rat liver XDH
K_{cat} X-O$_2$	13 s^{-1}	2.1 s^{-1}	17.2 s^{-1}	4.5 s^{-1}
K_{cat} X-NAD	0.11 s^{-1}	6.3 s^{-1}	—	13.5 s^{-1}
K_m for xanthine	1 ~ 2× 10^{-6} M	~ 1× 10^{-6} M	1.8× 10^{-6} M	~ 2 × 10^{-6} M^a
K_m for O$_2$	5.3 × 10^{-5} M	6.5 × 10^{-6} M	4.6 × 10^{-5} M	2.6 × 10^{-4} M
K_m for NAD	6.9 × 10^{-6} M	7× 10^{-6} M	—	8.5 × 10^{-6} M
Ref.	66	67, 68	69	69

$^a K_m$ for xanthine is dependent on whether the activity of the xanthine-oxygen or xanthine-NAD reaction is measured.

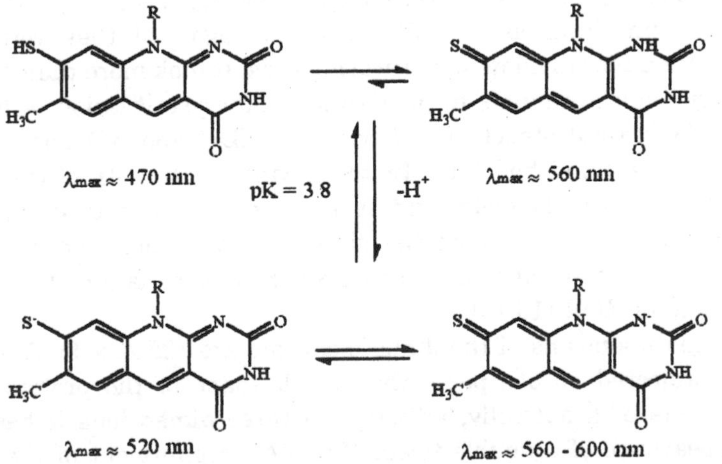

FIG. 2. Chemical formula of the analogue 8-mercaptoflavin in its neutral and anionic (partly quinoid) form. The absorption maxima for the various forms and the pK value for the transition are given. At pH 7.8, XDH largely stabilizes the neutral form of the analogue, in contrast to the prevalence of the anionic form in XO [71].

by 8-mercapto-FAD, the modified flavin was bound in its neutral form. On the other hand, XOs from rat liver and bovine milk bind the same flavin analogue in the benzoquinoid anionic form [70–72]. As the pK value of free 8-mercapto-FAD is 3.8 [73], XDH must perturb the pK of the ionizable substituent by more than four pH units. In the case of the milk enzyme, the individual pK values have been determined as 5.0 in XO and 9.0 in XDH [72]. Such relatively large shifts in pK suggest a strong negative charge in the flavin binding site of the XDH, which in turn should be absent in XO. A second series of experiments using another flavin analogue with a chemically reactive group at position 6 of the ring, 6-mercapto-FAD, uncovered evidence for conformational changes in the neighborhood of the isoalloxazine ring. The flavin ring seemed to be relatively open and easily accessible from the bulk solvent in XO but buried in XDH [70].

2.3. Crystal Structures

2.3.1. Xanthine Dehydrogenase

Although small XO crystals were obtained by chance during purification of the enzyme from bovine milk a long time ago [75], they were not of sufficient quality to allow structural analysis. It took more than 40 years and major improvements in purification and crystallization techniques until the crystal structures of both the XDH and XO forms of the enzyme were published [16]. In the meantime, the structures of the related enzymes aldehyde oxidase [12] and CO dehydrogenase [14] were determined and, given the absence of any such information on XOR, were used to infer structure-mechanism correlations for the Mo-pterin site of XOR [13,15].

The dimensions of the dimeric enzyme are 155 Å × 90 Å × 70 Å. When projected onto a page, the overall shape of the protein might remind one of a butterfly, although in three dimensions it has more the appearance of a double tower (Fig. 3A), reminiscent of the reconstruction obtained from single-molecule electron microscopy [74]. The three domains identified by limited proteolysis [10] correspond to three folding domains that are linked together by long, often mobile stretches of peptide chain. The closest intersubunit distance of cofactors is more

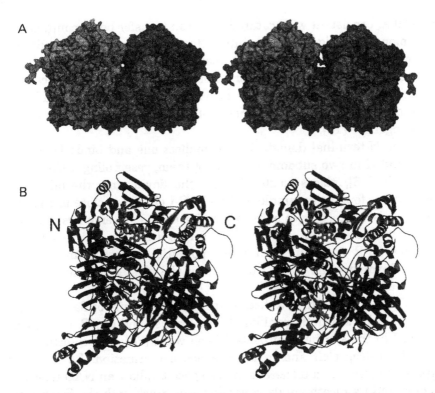

FIG. 3. Three-dimensional models of bovine milk xanthine dehydrogenase. (A) Parallel-view stereo of a surface representation reveals the "two-tower-like" shape of the catalytically active dimer strongly reminiscent of the single-molecule reconstruction resulting from electron microscopy techniques [74]. The overall dimensions are 155 Å × 90 Å × 70 Å. There is no experimental evidence of "electronic crosstalk" between the subunits. One remarkable feature of XOR is the flatness of one of its surfaces (the bottom one in this view). It is tempting to speculate that this property might reflect an interaction with the membrane of the fat globules in milk.(B) Parallel view stereo of a ribbon representation of one subunit of XDH (overall dimensions of 100 Å × 90 Å × 70 Å). The cofactors, Mo-pterin, two Fe_2S_2 clusters, and FAD (from bottom to top), together with a molecule of the inhibitor salicylate (in front of the Mo complex), are shown as stick models in a lighter shade of gray. Electrons would be transferred from the xanthine substrate via the Mo-pterin center consecutively to iron-sulfur cluster I and II and from there to FAD, whose isoalloxazine ring would pass them on to the nicotinamide ring of NAD^+ to form the reduced product NADH.

than 50 Å, consistent with catalytic electron transfer being contained to each monomer (Fig. 3B). The inter-cofactor distances of one subunit, however, are all shorter than 14 Å, making tunneling the most probable mechanism for electron transport. Despite only limited sequence conservation, the folds of the three domains of XOR are very similar to those of the corresponding domains of aldehyde oxidase [12] and CO dehydrogenase [14].

The N-terminal domain is the smallest one and binds Fe/S clusters I and II in two subdomains. One of them, resembling a plant-type ferredoxin [76], places cluster II near the flavin ring; the other one comprises a four-helix bundle and its Fe/S-I cluster is close to the Mo-pterin. It is interesting to note that mutating the cysteine ligand of one of the iron atoms in cluster II to serine leads to changes in the enzymes EPR spectrum and redox potential, whereas an identical change involving the other iron atom of the cluster has no effect on these features [77].

The FAD domain is a distant relative of a family of flavoproteins, whose members include vanillyl alcohol oxidase and MurB, an enzyme involved in the synthesis of bacterial cell walls [78]. The cofactor FAD binds in a deep cleft and in an extended conformation. Its *si* side is exposed to solvent, and there is ample space to allow an NAD molecule to bind and its nicotinamide ring could well stack with the flavin ring. That XDH, despite granting free access to its FAD cofactor, displays a relatively low oxygen reactivity has been explained by the stabilization of the neutral semiquinone flavin in the enzyme matrix [10]. On the *re* side of the isoalloxazine ring, the side chain of Phe337 undergoes a π-π interaction with the pyrimidine part of the flavin ring. The aromatic character of this residue is conserved in all XORs but it is changed to a leucine in bovine aldehyde oxidase, a constitutive oxidase. Another interesting feature is the very short distance of 3.6 Å between C6 of the flavin ring and the carboxylate of Asp429.

The C-terminal domain of XDH binds the Mo-pterin cofactor. In a new electron density map of the enzyme at 1.6 Å resolution [79], the two oxygen and three sulfur ligands of the pentavalent molybdenum complex are easily distinguished (Fig. 4). The arrangement corresponds to the one reported for the Mo cofactor of aldehyde oxidase [12] and is consistent with mechanistic features such as transfer to the purine substrate of a single-bonded oxygen originally supplied by solvent [80].

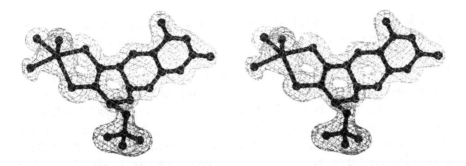

FIG. 4. The Mo-pterin cofactor of bovine milk xanthine dehydrogenase. Presented in parallel view stereo is the corresponding refined electron density at 1.6 Å resolution. The complex is shown as a black ball-and-stick model. The phosphate group is at center bottom and the molybdenum is at left. Due to its much stronger density, the apical ligand is easily identified as the sulfur atom known to be essential for catalytic activity

Sodium salicylate at 1 mM was included in the purification and crystallization of bovine XDH and XO to stabilize the proteins [46]. In the crystal structure, a salicylate molecule is found 6.5 Å from the Mo ion and assumes a position that overlaps with the binding site of purine substrates. It is kept in place by a variety of interactions with the protein matrix. The ring stacks with the side chain of Phe914 and undergoes an edge-on aromatic contact with the ring of Phe1009. The carboxylate group of salicylate interacts with Arg880 and the hydroxyl of Ser1010. In addition, it is bridged to Glu1261 via a water molecule. The inhibitor's hydroxyl group binds to both main chain amide and side chain of Ser1010. Several of these interactions, especially the ones involving Arg880 and Glu1261, confirm remarkably well the results of modeling [13] and mutational studies [81,82].

2.3.2. Xanthine Dehydrogenase to Oxidase Transition

The most obvious difference one realizes when looking at electron density maps of XDH and XO is the much longer stretches of polypeptide chain missing in XO [16]. The most pronounced one extends from residue 529 to residue 570 and adopts a short α helix, β strand, and a loop conformation in XDH. Together they span a distance of about 70 Å. As

XDH is composed of single chains of amino acids; electron density missing in its map must be due to mobility. For XO, given its proteolyzed
nature [46], no such argument can be made and the protein might have
lost parts of its polypeptide chain linking the three domains. The sites of
proteolysis by pancreatin, the enzyme used in the preparation of the XO
crystals [46], have been identified as Leu219 and Lys569 (Tomoko
Nishino, unpublished). They are both located in missing or mobile
stretches of the protein chain. Overall, the common parts of the XDH
and XO structures are very similar, consistent with kinetic studies,
which show no major difference in binding and catalysis of substrates
at the Mo center [10].

   When XDH is converted to XO, it is the immediate environment of
the flavin ring of FAD that shows the largest changes. Interestingly, all
of the residues playing a role in the transformation (via proteolysis or
via cysteine modification) are located at least 18 Å from the flavin ring
on the side of the molecule opposite to the opening that provides access
to the isoalloxazine ring. A direct influence of these amino acids on the
orientation of residues surrounding the FAD is therefore quite unlikely.
There is, however, a possible indirect route to couple events happening
on the surface of the protein to the buried flavin interior. In XDH, the
protein chain around Phe549 makes close contacts with the side chain of
Arg427. Upon removal of this interaction, the arginine becomes more
flexible and leads to a reorientation of Trp996 as well as to a very large
structural rearrangement (up to 20 Å) of a highly charged loop (Gln423
to Lys433) on the *si* side of the flavin ring. This replaces Asp429 with
Arg426 as the closest neighbor of the isolloxazine ring, drastically mod-
ifying its electrostatic environment. All major conformational changes
are restricted to the *si* side of the flavin and only a very minor shift of
Phe337 occurs, which still stacks with the ring's *re* side. The electrostatic effects of such a restructuring of the FAD site explain the different
chemical reactivities of active site probes [68,70] and the restriction in
access caused by the physical relocation of the loop underlie the inability
of NAD to compete successfully against the much smaller oxygen as a
substrate [10].

3. PICOLINATE DEHYDROGENASE

In stark contrast to the vast amounts of biochemical, biophysical, genetic, and medical information acquired on xanthine oxidoreductase, picolinic acid (or picolinate) dehydrogenase has almost completely escaped the scrutiny of scientific investigation. This enzyme's substrate, picolinic acid, is a product of the photolytic degradation of the herbicide Diquat (1,1'-dimethyl-4,4'-bipyridylium ion) [83], and this fact might have sparked some of the original interest.

In 1963, Dagley and Johnson [84] described an activity in cultures of *Aerococcus* QA and *Rhodotorula* MR that oxidized picolinic acid to 6-hydroxypicolinate. A cell-free *Rhodotorula* extract catalyzed the reaction in the presence of methylene blue and in the absence of oxygen. Nicotinamide nucleotides did not affect the rate. The next organism tested was a rod-shaped bacterium isolated from garden soil [85]. Its picolinic acid dehydrogenase was found to localize to the membrane and was released by ultrasonic treatment. As the *Rhodotorula* enzyme, it did not require oxygen and was active with dyes as electron acceptors. However, it also uses NAD^+ quite efficiently.

The first attempt to purify the enzyme responsible for the hydroxylation of picolinic acid focused on the bacterium *Arthrobacter picolinophilus* [86]. Again, the enzymatic activity proved to be particulate, probably membrane-bound, as it could be solubilized by detergents. Although the authors mention the enzyme's extreme instability in an oxygen atmosphere, they succeeded in preparing an enzyme that was 60-fold enriched. They determined its molecular weight as 230,000 and found that the enzyme required iron and flavin mononucleotide. More recently, Siegmund et al. [87] investigated the same enzyme together with the identical activity in a new bacterial isolate termed PQ1. Based on native gradient gel electrophoresis, they estimated its molecular weight as 130,000. They also presented convincing evidence that molybdenum was essential for its activity.

All published accounts are consistent in that the oxygen incorporated into the 6-position of the aromatic ring is derived from water molecules not molecular oxygen. Given the use of crude extracts in the early characterization experiments and the reported instability of some of the enzyme preparations, it is certainly a matter of discussion how conclusive are the results of testing the suitability of nicotinamide nucleotides as substrates or the nature of the flavin nucleotide leading to an increase in the reaction rate. The XDH to XO transition, which is difficult to avoid even with today's technology and which switches the enzyme's oxidative substrate from NAD^+ to oxygen, is an excellent example of such a "purification artifact." We would therefore argue that picolinic acid dehydrogenase is a regular member of the molybdenum/FeS/flavin family of hydroxylases.

4. CONCLUSIONS

In this chapter, we have tried to introduce two complex flavoproteins: xanthine oxidoreductase and picolinate dehydrogenase. The relative space devoted to their descriptions, though clearly dominated by XOR-related aspects, still does not do any justice to XOR, the archetype enzyme that has served as a benchmark for any scientist interested in complex flavoproteins.

Over the course of a century, investigators have probed it by chemical, biochemical, genetic, spectroscopic, and finally structural techniques making it one of the best understood enzymes. However, one major challenge remains as expression of fully functional recombinant XOR has only been achieved at relatively modest levels, most probably due to overload of the bacterial system that synthesizes and inserts the molybdenum cofactor [52,76,82]. Increasing the yield would open the door to studies of the effects mutations have on structure which, when combined with kinetic measurements possible at present, could provide finest details of XOR's catalytic mechanism.

Picolinate dehydrogenase, the "newcomer," has not yet entered the limelight and is still far removed from the prominent scientific exposure granted its famous "cousin." However, given its interesting talent to degrade a breakdown product of a herbicide, it might garner more interest in the future.

ACKNOWLEDGMENTS

The authors thank all of the former and present members of their laboratories who shared their interest in the "colorful" science of complex flavoproteins, especially xanthine oxidoreductase. They also want to thank many of their colleagues for countless stimulating discussions.

ABBREVIATIONS

EPR	electron paramagnetic resonance
FAD	flavin adenine dinucleotide
FDNB	fluoro-2,4-dinitrobenzene
Fe/S	cluster of two iron and two sulfur atoms
FMN	flavin mononucleotide
NAD^+	oxidized form of nicotinamide adenine dinucleotide
NADH	reduced form of nicotinamide adenine dinucleotide
NCBI	National Center for Biotechnology Information, http://www.ncbi.nlm.nih.gov
XDH	xanthine dehydrogenase
XO	xanthine oxidase
XOR	xanthine oxidoreductase (denoting both activities, dehydrogenase and oxidase)

REFERENCES

1. J. Horbaczewski, *Monatsh. Chem.* 12, 221–275 (1891).
2. F. Schardinger, *Z. Untersuch. Nahrungs Genussmittel*, 5, 1113–1121 (1902).
3. M. Dixon and S. Thurlow, *Biochem. J.*, *18*, 976 (1924).
4. E. Della Corte and F. Stirpe, *Biochem. J.*, *108*, 349–351 (1968).
5. F. Stirpe and E. Della Corte, *J. Biol. Chem.*, *244*, 3855–3863 (1969).
6. E. Della Corte and F. Stirpe, *Biochem. J.*, *126*, 739–745 (1972).

7. M. Dixon and K. Kodama, *Biochem. J.*, *20*, 1104 (1926).

8. V. Massey and C. M. Harris, *Biochem. Soc. Trans.*, *25*, 750–755 (1997).

9. T. Nishino, *J. Biochem. (Tokyo)*, *116*, 1–6 (1994).

10. R. Hille and T. Nishino, *FASEB J.*, *9*, 995–1003 (1995)

11. R. Hille, *Chem. Rev.*, *96*, 2757–2816 (1996).

12. M. J. Romão, M. Archer, I. Moura, J. J. Moura, J. LeGall, R. Engh, M. Schneider, P. Hof, and R. Huber, *Science*, *270*, 1170–1176 (1995).

13. R. Huber, P. Hof, R. O. Duarte, J. J. Moura, I. Moura, M. Y. Liu, J. LeGall, R. Hille, M. Archer, and M. J. Romão, *Proc. Natl. Acad. Sci. USA*, *93*, 8846–8851 (1996).

14. H. Dobbek, L. Gremer, O. Meyer, and R. Huber, *Proc. Natl. Acad. Sci. USA*, *96*, 8884–8889 (1999).

15. L. Gremer, S. Kellner, H. Dobbek, R. Huber, and O. Meyer, *J. Biol. Chem.*, *275*, 1864–1872 (2000).

16. C. Enroth, B. T. Eger, K. Okamoto, T. Nishino, T. Nishino, and E. F. Pai, *Proc. Natl. Acad. Sci. USA*, *97*, 10723–10728 (2000).

17. G. H. Hitchings, in *Handbook of Experimental Pharmacology*, Vol. 51, (W. N. Kelley and M. Weiner, eds.), Springer Verlag, New York, 1978, pp. 1–20.

18. G. B. Elion, *Science*, *244*, 41–47 (1989).

19. V. Massey, H. Komai, G. Palmer, and G. B. Elion, *J. Biol. Chem.*, *245*, 2837–2844 (1970).

20. H. A. Simmonds, S. Reiter, and T. Nishino, in *The Metabolic and Molecular Bases of Inherited Disease*, 7th ed., Vol. 2 (C. R. Scriver, A. L. Beaudet, W. S. Sly, and D. Valle, eds.), McGraw-Hill, New York, 1995, pp. 1781–1798.

21. J. L. Johnson and S. K. Wadman, in *The Metabolic and Molecular Bases of Inherited Disease*, 7th ed., Vol. 2, (C. R. Scriver, A. L. Beaudet, W. S. Sly, and D. Valle, eds.), McGraw-Hill, New York, 1995, pp. 2271–2283.

22. K. Ichida, Y. Amaya, N. Kamatani, T. Nishino, T. Hosoya, and O. Sakai, *J. Clin. Invest.*, *99*, 2391–2397 (1997).

23. K. Ichida, T. Matsumura, R. Sakuma, T. Hosoya, and T. Nishino, *Biochem. Biophys. Res. Commun.*, *282*, 1194–1200 (2001).

24. V. Massey and D. Edmondon, *J. Biol. Chem.*, *245*, 6595–6598 (1970).

25. L. Amrani, J. Primus, A. Glatigny, L. Arcangeli, C. Scazzocchio, and V. Finnerty, *Mol. Microbiol.*, *38*, 114–125 (2000).

26. T. Watanabe, N. Ihara, T. Itoh, T. Fujita, and Y. Sugimoto, *J. Biol., Chem.*, *275*, 21789–21792 (2000).

27. J. M. McCord, *N. Engl. J. Med.*, *312*, 159–163 (1985).

28. O. D. Saugstad, *Pediatr. Res.*, *23*, 143–150 (1988).

29. E.-D. Jarasch, C. Grund, G. Bruder, H. W. Heid, T. W. Keenan, and W. W. Fanke, *Cell*, *25*, 67–82 (1981).

30. R. S. Roy and J. M. McCord, *Can. J. Phys. Pharm.*, *60*, 1346–1352 (1982).

31. T. Nishino, S. Nakanishi, K. Okamoto, J. Mizushima, H. Hori, T. Iwasaki, T. Nishino, K. Ichimori, and H. Nakazawa, *Biochem. Soc. Trans.*, *25*, 783–786 (1997).

32. R. Harrison, *Biochem. Soc. Trans.*, *25*, 787–787 (1997).

33. Q. Xiang and D. E. Edmondson, *Biochemistry*, *35*, 5441–5450 (1996).

34. S. Leimkühler, M. Kern, P. S. Solomon, A. G. McEwan, G. Schwarz, R. R. Mendel, and W. Klipp, *Mol. Microbiol.*, *27*, 853–869 (1998).

35. D. Hettrich and F. Lingens, *Biol. Chem. Hoppe Seyler*, *372*, 203–211 (1991).

36. T. Sakai and H.-K. Jun, *Agric. Biol. Chem.*, *43*, 753–760 (1979).

37. K. Koenig and J. R. Andreesen, *J. Bacteriol.*, *172*, 5999–6009 (1990).

38. C. A. Woolfolk, *J. Bacteriol.*, *163*, 600–609 (1985).

39. L. Gremer and O. Meyer, *Eur. J. Biochem.*, *238*, 862–866 (1996).

40. T. Schrader, A. Rienhofer, and J. R. Andreesen, *Eur. J. Biochem.*, *264*, 862–871 (1999).

41. W. T. Self and T. C. Stadtman, *Proc. Natl. Acad. Sci. USA.*, *97*, 7208–7213 (2000).

42. R. C. Bray, in *The Enzymes*, 3rd ed., Vol. 12, Part B (P. D. Boyer, ed.), Academic Press, New York, 1975, pp. 299–419.

43. R. Hille and V. Massey, in *Molybdenum Enzymes*, Vol. 7 (T. G. Spiro, ed.), Wiley-Interscience, New York, 1985, pp. 443–518.

44. E. G. Ball, *J. Biol. Chem.*, *128*, 51–67 (1939).

45. T. Nishino, T. Nishino, and K. Tsushima, *FEBS Lett.*, *131*, 369–372 (1981).

46. B. T. Eger, K. Okamoto, C. Enroth, M. Sato, T. Nishino, E. F. Pai, and T. Nishino, *Acta Crystallogr.*, *D56*, 1656–1658 (2000).

47. A. Glatigny and C. Scazzocchio, *J. Biol. Chem.*, *270*, 3534–3550 (1995).

48. R. P. Keith, M. J. Riley, M. Kreitman, R. C. Lewontin, D. Curtis, and G. Chambers, *Genetics*, 116, 67–73 (1987).

49. C. S. Lee, D. Curtis, M. Mccarron, C. Love, M. Gray, W. Bender, and A. Chovnick, *Genetics*, *116*, 55–66 (1987).

50. A. Sato, T. Nishino, K. Noda, Y. Amaya, and T. Nishino, *J. Biol. Chem.*, *270*, 2818–2826 (1995).

51. M. Terao, G. Cazzaniga, P. Ghezzi, M. Bianchi, F. Falciani, P. Perani, and E. Garattini, *Biochem. J.*, *283*, 863–870 (1992).

52. Y. Amaya, K. Yamazaki, M. Sato, K. Noda, T. Nishino, and T. Nishino, *J. Biol. Chem.*, *265*, 14170–14175 (1990).

53. L. Berglund, J. T. Rasmussen, M. D. Andersen, M. S. Rasmussen, and T. E. Petersen, *J. Dairy Sci.*, *79*, 198–204 (1996).

54. K. Ichida, Y. Amaya, K. Noda, S. Minoshima, T. Hosoya, O. Sakai, N. Shimizu, and T. Nishino, *Gene*, *133*, 279–284 (1993).

55. P. Xu, T. P. Hueckstaedt, R. Harrison, and J. R. Hoidal, *Biochem. Biophys. Res. Commun.*, *199*, 998–1004 (1994); *Erratum, ibid*, *215*, 429 (1995).

56. M. Saksela and K. O. Raivio, *Biochem. J.*, *315*, 235–239 (1996).

57. T. Nishino and T. Nishino, *J. Biol. Chem.*, *272*, 29859–29864 (1997).

58. J. T. Rasmussen, M. S. Rasmussen, and T. E. Petersen, *J. Dairy Sci.*, *83*, 499–506 (2000).

59. J. S. Olson, D. P. Ballou, G. Palmer, and V. Massey, *J. Biol. Chem.*, *249*, 4363–4382 (1974).

60. T. Iwasaki, K. Okamoto, T. Nishino, J. Mizushima, H. Hori, and T. Nishino, *J. Biochem. (Tokyo)*, *127*, 771–778 (2000).

61. A. G. Porras and G. Palmer, *J. Biol. Chem.*, *275*, 11617–11626 (1982).

62. M. J. Barber and L. M. Siegel, *Biochemistry*, *21*, 1638–1647 (1982).

63. J. Hunt, V. Massey, W. R. Dunham, and R. H. Sands, *J. Biol. Chem.*, *268*, 18685–18691 (1993).

64. M. J. Barber, M. P. Coughlan, M. Kanda, and K. V. Rajagopalan, *Arch. Biochem. Biophys.*, *201*, 468–475 (1980).

65. L. M. Schopfer, V. Massey, and T. Nishino, *J. Biol. Chem.*, *263*, 13528–13538 (1988).

66. C. Harris and V. Massey, in *Flavins and Flavoproteins* (K. Yagi, ed.), Walters de Gruyter, New York, 1994, pp. 723–726.

67. C. Harris and V. Massey, *J. Biol. Chem.*, *272*, 8370–8379 (1997).

68. J. Hunt and V. Massey, *J. Biol. Chem.*, *267*, 5468–5473 (1992).

69. T. Saito and T. Nishino, *J. Biol. Chem.*, *264*, 10015–10022 (1989).

70. V. Massey, L. M. Schopfer, T. Nishino, and T. Nishino, *J. Biol. Chem.*, *264*, 10567–10578 (1988).

71. T. Saito, T. Nishino, and V. Massey, *J. Biol. Chem.*, *264*, 15930–15935 (1989).

72. J. Hunt and V. Massey, *J. Biol. Chem.*, *267*, 21479–21485 (1992).

73. E. G. Moore, S. Ghisla, and V. Massey, *J. Biol. Chem.*, *254*, 8173–8178 (1979).

74. D. R. Beniac, T. Iwasaki, and F. P. Ottensmeyer, (J. M. Corbett, ed.) ON, Canada, Microsc. Soc. Canada, Lethbridge, in *Proc 26th Ann. Meet. Microsc. Soc. Canada*, Guelph, Ontario, 1999, pp. 49–50.

75. P. G. Avis, F. Bergel, R. C. Bray, and K. V. Shooter, *Nature*, *173*, 1230–1231 (1954).

76. H. Sticht and P. Rösch, *Prog. Biophys. Mol. Biol.*, *70*, 95–136 (1998).

77. T. Nishino and K. Okamoto, *J. Inorg. Biochem.*, *82*, 43–49 (2000).

78. M. W. Fraaije and A. Mattevi, *Trends Biochem. Sci.*, *25*, 126–132 (2000).

79. B. T. Eger, K. Okamoto, C. Enroth, T. Nishino, T. Nishino, and E. F. Pai, to be published.

80. M. Xia, R. Dempski, and R. Hille, *J. Biol. Chem.*, *274*, 3323–3330 (1999).

81. A. Glatigny, P. Hof, M. J. Romão, R. Huber, and C. Scazzocchio, *J. Mol. Biol.*, *278*, 431–438 (1998).

82. W. A. Doyle, J. F. Burke, A. Chovnick, F. L. Dutton, J. R. White, and R. C. Bray, *Eur. J. Biochem.*, *239*, 782–795 (1996).

83. A. E. Smith and J. Grove, *J. Agric. Food Chem.*, *17*, 609–613 (1969).

84. S. Dagley and P. A. Johnson, *Biochem. Biophys. Acta.*, *78*, 577–587 (1963).

85. C. G. Orpin, M. Knight, and W. C. Evans, *Biochem. J.*, *127*, 819–831 (1972).

86. R. L. Tate and J. C. Ensign, *Can. J. Microbiol.*, *20*, 695–702 (1974).

87. I. Siegmund, K. Koenig, and J. R. Andreesen, *FEMS Microbiol. Lett.*, *67*, 281–284 (1990).

13

Enzymes of the Xanthine Oxidase Family: The Role of Molybdenum

David J. Lowe

Biological Chemistry Department, John Innes Centre,
Colney Lane, Norwich NR4 7UH, UK

I. INTRODUCTION

The xanthine oxidase family of molybdopterin-containing enzymes is widely distributed throughout the biosphere and has been thoroughly and carefully reviewed recently by Hille and colleagues [1–3]; reference should be made to Chapter 6 for an overview of the field by this author. I have taken Ref. 1 as a starting point for this chapter in which I will concentrate chiefly on recently published results on the molybdenum atom, its immediate environment, and its role in these proteins.

The distinguishing feature of the xanthine oxidase enzyme family is that the molybdenum atom(s) they all contain has a sulfido ligand; the molybdenum is also bound to molybdopterin via the latter's two dithiolene sulfur atoms and has an additional oxo ligand plus a fifth oxygen-containing ligand, which is probably a hydroxyl group. It is the sulfido ligand that gives these enzymes their characteristic of being irreversibly inhibited by cyanide ion, as the latter removes this sulfur as thiocyanate. The sulfido ligand is replaced by a second oxo group in the inactive form of the enzymes. Hille [1] gives a comprehensive list of family members known in 1996 that I will not repeat here. I will only comment that I would add 4-hydroxybenzoyl-CoA reductase from *Thauera aromatica* [4] to Hille's list, and perhaps also the pyrogallol transhydroxylase from *Pelobacter acidigallici* [5]. In addition, I would remove CO dehydrogenase since the recently published structure of this enzyme shows the molybdenum environment to be significantly different from that of other members of the family [6], as described in Chapter 7, especially since it lacks the crucial sulfido ligand; its inhibition by cyanide appears to be by a different mechanism from that of full family members.

There have been important advances in our understanding of these enzymes recently as a result of the publication of a number of X-ray crystallographically determined structures. Earlier results on the family member aldehyde oxidoreductase from *Desulfovibrio gigas* [7] in its inactive desulfo form at 2.0 Å have been supplemented by structure determinations on the same protein from *Desulfovibrio desulfuricans* [8] in the active sulfo form at 2.8 Å, and on the xanthine oxidase (to 2.5 Å) and dehydrogenase (to 2.1 Å) forms of this protein from bovine milk [9,10]. The latter papers concentrate chiefly on the details of the proteolytic conversion from dehydrogenase to oxidase; this is shown to be due to a significant rearrangement of residues close to the flavin after a proteolytic event. As expected there is no effect at the molybdenum center. All of the structures show that the coordination of the molybdenum is roughly square pyramidal, and its environment from the xanthine oxidase data is shown in Figure 1.

There are significant differences between the molybdenum-to-ligand distances in the xanthine oxidase and dehydrogenase structures although these may be due to poor resolution of the electron density and series truncation effects in the regions close to the molybdenum atom.

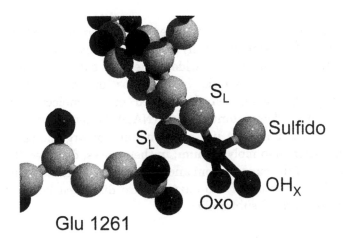

FIG. 1.   The environment of molybdenum in xanthine oxidase drawn from the PDB file deposited with Ref. 9 showing the dithiolene sulfurs (S_L) and Glu1261 which provides the proposed active site base.

Extended X-ray absorption fine structure spectroscopy (EXAFS) results will give more reliable distance estimates as discussed below. The dithiolene sulfurs, oxo, and putative hydroxyl ligands form the roughly square planar base of the pyramid with the molybdenum slightly above this plane and the sulfido at the apex. The coordination position trans to the sulfido is vacant. A glutamate residue, numbered 1261 in the xanthine oxidase/dehydrogenase and 869 in the *D. gigas* aldehyde oxidase sequences, has been implicated in mechanistic studies and is located close to this vacant position. The putative hydroxyl ligand is directed to the end of a solvent-containing pocket that extends some 15 Å or more in from the surface of the proteins; this pocket contains bound inhibitors in the various published structures and is presumably the channel by which substrates approach the molybdenum and by which products leave.

Enzymes of this family also contain either two-iron or four-iron-sulfur centers and in some cases an additional flavin center that are responsible for electron transfer away from the molybdenum to the oxidizing substrate. These will not be considered further here.

2. THE REACTIONS CATALYZED

Xanthine-oxidase-like enzymes are often known as molybdenum hydroxylases and characteristically transfer a hydroxyl group derived, uniquely, from water to a wide variety of substrates, generally at a carbon atom in aromatic heterocycles or an aldehyde carbonyl. Reducing equivalents can be passed on to molecular dioxygen or nicotinamide adenine dinucleotide (NAD). Other enzymes carrying out hydroxylation reactions use other metals, or organic centers such as flavin, rather than molybdenum; in these cases, dioxygen itself provides the source of the oxygen atom and they do not generate reducing equivalents. The activities of the enzymes are discussed below in relatively arbitrarily defined subfamilies.

2.1.   Eukaryotic Xanthine Oxidase/Dehydrogenase

Xanthine oxidase itself can be purified in large quantities from bovine milk and as a result of this its mechanism has been studied for many years. It has often been one of the first enzymes to which new physical techniques have been applied. The activity from which it gets its name is the hydroxylation of xanthine, at the C8 position, to generate uric acid, although it can also convert purine to hypoxanthine and then to xanthine, as well as hydroxylating other purines and pteridines, and converting many aldehydes to the corresponding carboxylic acid. The dehydrogenase form of the enzyme is found in particular animal tissues [1]—even, for example, in silkworms [11]—and there is a report of an apparently related xanthine dehydrogenase in the leaves of leguminous plants [12].

It is rare for enzymes of the xanthine oxidase family to possess high specificities, suggesting that such specificity that does exist must be conveyed by the side chains of the residues in the substrate-binding pocket, and the shape of the latter, which present substrates to the active site molybdenum in the appropriate orientation. Light has recently been shed on this by work on the fungal purine hydroxylase I, a xanthine dehydrogenase [13], and purine hydroxylase II, a nicotinate hydroxylase [14], from *Aspergillus nidulans* which are members of the eukaryotic enzyme subfamily. The principal activity of the first of these enzymes is to catalyze the oxidation of hypoxanthine (6-hydroxypurine) to xanthine (2,6-dihydroxypurine), which is then converted to uric acid (2,6,8-trihydroxypurine). However, the enzyme also converts 2-hydroxypurine to 2,8-dihydroxypurine, which is not a substrate. An interesting combination of information comes from random mutagenesis and X-ray crystallographic studies [15] on mutants that can convert 2-hydroxypurine to 2,6-dihydroxypurine, which is of course xanthine. Unfortunately, the mutant proteins are unable to utilize xanthine itself, but an alternative xanthine-metabolizing pathway exists so that these mutants are able to grow on 2-hydroxypurine. In addition, these mutants are not inhibited by allopurinol at levels far in excess of those required irreversibly to inhibit the wild-type protein. These mutants are all at Arg911 (*A. nidulans* sequence), which is changed to Gln or Gly and which is conserved among all xanthine dehydrogenases as well as the aldehyde oxidoreductase from *D. gigas*. Modeling of the various substrates and inhibitors into the crystallographically determined structure of the binding site of the *D. gigas* protein confirms that this arginine (501 in this structure) is situated so as to bind a hydroxyl group at a purine 6 position. Another conserved residue, a Glu that would be at position 425 in the *D. gigas* protein where it is actually a Phe, is positioned so that it could bind the purine N3 and N9 and help present the C8 of xanthine to the molybdenum. The models are consistent with the phenotypes of the various mutants.

2.2. Aldehyde Oxidoreductases

When sulfate-reducing prokaryotes, such as *D. gigas* and *D. desulfuricans*, are grown on aldehydes as the sole carbon source, large quantities of a broad-specificity aldehyde oxidoreductase, containing only molybdenum and iron-sulfur centers, are induced; this is used to generate reducing equivalents for the production of dihydrogen. The aldehydes are converted to the corresponding oxo acid. The broad specificity is again consistent with the suggestion that this does not lie in the molybdenum site itself but in the shape and charge distribution of residues in the substrate-binding pocket.

There are also several examples of the identification of molybdo-flavoprotein aldehyde oxidases from a number of eukaryotes that do not oxidize purines as their principal function. These include three distinct isozymes in mammals [16,17], where the physiological role is so far not clear, and four in plants [18–19, 20] where one has the important role of oxidizing abscisic aldehyde to abscisic acid in the last step of the synthesis of this plant hormone. Since the cDNAs of many of these proteins have recently been cloned, we can look forward to further structural and functional studies in this area.

2.3. Pyrimidine Oxidases

Allopurinol, a well-known irreversible suicide inhibitor of xanthine oxidase/dehydrogenase, is known to induce renal toxicity by impairing pyrimidine metabolism in mice. Suggestions that this is because it inhibits a xanthine oxidase family enzyme responsible for oxidizing pyrimidines can be discounted since other xanthine oxidase inhibitors do not have these effects [21]. Indeed the prime suspect for the enzyme affected is the one that catalyzes the first and rate-limiting step of pyrimidine catabolism, dihydropyrimidine dehydrogenase; this is now known to contain only iron-sulfur centers [22]. Some pyrimidines can be potent inhibitors of xanthine oxidase [23] although it is not clear that they bind at molybdenum.

2.4. Other Enzymes of the Family

Characteristics of some other family members may shed light on the reactions occurring at the molybdenum in these enzymes. An intriguing possibility has been introduced by the initial characterization of an enzyme from *Eubacterium barkeri* (formerly *Clostridium barkeri*), when grown on xanthine, that has xanthine dehydrogenase activity [24]. There is good evidence that this protein belongs to the xanthine oxidase family and contains a seleno group bound to molybdenum rather than the usual sulfido group. It is different from the nicotinic acid hydroxylase found in the same organism that is probably a member of the dimethyl sulfoxide (DMSO) reductase family of oxomolybdenum enzymes (see Chapter 6) also containing a seleno group bound to molybdenum. It is therefore necessary for models for the reactivity at the molybdenum to consider the effect of replacing sulfur by selenium in this position although the basic chemistry of the two systems is presumably the same.

There is strong evidence from sequence analysis that 4-hydroxybenzoyl-CoA reductase (dehydroxylating) from *T. aromatica* is also a member of the xanthine oxidase family of enzymes [4]. In addition to molybdenum, flavin, and two two-iron-sulfur centers, it also appears to contain a four-iron-sulfur center. It functions in the anaerobic metabolism of phenolic compounds. Interestingly, it catalyzes the reverse reaction to that of most other enzymes of this type in that it removes a hydroxyl group reductively rather than adding one oxidatively; again, this needs to be considered in any generalized mechanism. Finally, it may well be that the pyrogallol transhydroxylase from *P. acidigallici* also belongs to the group. This enzyme transfers a hydroxyl group from the 2-position of the pyrogallol (1,2,3-trihydroxybenzene) ring to the 5-position, with 1,2,3,5-tetrahydroxybenzene as a cosubstrate and coproduct [5]; thus, like 4-hydroxybenzoyl-CoA reductase, it is also capable of removing a hydroxyl group as well as having the usual activity of adding one.

3. SPECTROSCOPIC INVESTIGATIONS

The broad outline of the reactions taking place at molybdenum is generally agreed. In the hydroxylation reactions, substrate binds to the oxidized Mo(VI) in the first or second coordination sphere, and two electrons and a proton are transferred as hydride; oxygen is then added, leaving product bound at Mo(IV). This product then dissociates and molybdenum is reoxidized by electron transfer to the other cofactors. However, there remain a number of controversies about the details of these processes. Some recent results, using a number of techniques, and their consequences are discussed below.

3.1. UV-Visible Spectroscopy

UV/visible spectra of xanthine oxidase and its associated family are dominated by absorption bands from the iron-sulfur centers and flavin (where this is present) at all levels of reduction, so it requires considerable skill to detect the small effects due to molybdenum. Hille and coworkers recently succeeded in detecting changes in absorption and circular dichroism spectra during redox changes at the molybdenum using a double-difference technique [25]. They observed a marked pH dependence of these spectra indicating the presence of ionizable groups close to the molybdenum. I expect that observations of this kind will be exploited further in the future to enhance our knowledge of the system.

A related application of difference UV/visible spectroscopy has been to the observation of a catalytic cycle intermediate with product bound to molybdenum by magnetic circular dichroism (MCD) [26]. This intermediate is called the Very Rapid species from the kinetics of its characteristic EPR signal. Xanthine oxidase turns over very slowly with 2-hydroxy-6-methylpurine as substrate [27] and gives about 60% of the molybdenum as Very Rapid Mo(V) with much lower concentrations of other paramagnetic species. It was therefore possible to deconvolute the MCD spectrum of a rapidly frozen solution of this Mo(V) species and to compare its spectrum with a model compound. The comparison indicated that, in contrast with all the crystal structures, in the Very Rapid species the environment of molybdenum is altered such that the oxo ligand becomes cis to the dithiolene plane; this could be the result of a 90° rotation about the axis from the center of the dithiolene to the molybdenum, placing the oxo group in the vacant coordination site and the sulfido in the plane of the dithiolene. The authors suggested that the change in conformation takes place on the approach of substrate to molybdenum.

3.2. X-Ray Spectroscopies

Xanthine oxidase has been the subject of a number of molybdenum edge EXAFS studies (e.g. [28,29]) in order to define the molybdenum-to-ligand distances; the results are discussed elsewhere [1] and have not been extended recently. The conclusions clearly show the presence of an oxo ligand (at 1.67 Å) in all systems studied and a sulfido (at 2.15 Å) in oxidized enzyme that moves to a greater distance (2.39 Å), roughly the same as that of the dithiolene sulfurs, on reduction of the molybdenum with isolated enzyme and with inhibitors such as allopurinol or violapterin bound; this sulfur is presumed to be protonated in these states.

3.3. Electron Paramagnetic Resonance

Although Mo(V) is the only biologically accessible paramagnetic redox state of molybdenum, EPR has been used a great deal in the study of xanthine oxidase and its homologues. Many different signals can be elicited but only four will be considered here because of their importance in understanding how the enzymes work; they are shown in Fig. 2. The geometrical structures of these species have been deduced from EXAFS results together with isotopic substitution and detection of hyperfine couplings between the unpaired electron centered on molybdenum and

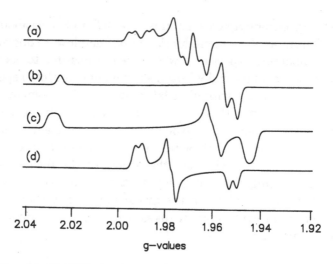

FIG. 2.   Four Mo(V) EPR signals given by xanthine oxidase and other members
of the family. The spectra are simulations at 9 GHz and are for the (a) Rapid
(Formamide Type 1), (b) Very Rapid, (c) Alloxanthine, and (d) Inhibited forms of
the enzyme. Parameters used in the simulations were taken from Refs. 30, 31,
and 38.

the magnetic nuclei introduced [32] by EPR and electron nuclear double
resonance (ENDOR). In all cases, the dithiolene sulfur and oxo ligand
core remains unaltered.

It is agreed that the Rapid signal, which really belongs to a family
of closely related signals, is given by enzyme without substrate or
product bound to molybdenum and appears during the catalytic cycle
when product dissociates from molybdenum before the latter is oxidized from Mo(IV) to Mo(V). Couplings to ^{1}H, ^{17}O, and ^{33}S are consistent with one hydroxyl and one sulfuryl ligand. Different members
of the family of Rapid signals may have product or substrate associated
with the enzyme but at a site somewhat remote from the molybdenum.
The Very Rapid species has product bound at molybdenum, in a form
under active debate [33,34], with the sulfur unprotonated; in this case,
the hyperfine interaction with ^{33}S at this site reveals that about 35% of
the electron spin density resides on the sulfur [35]. These signals are
named for their kinetics of appearance, with the Very Rapid signal
being a transient intermediate, maximal at about 10 ms with normal

substrates, and the Rapid signal appearing subsequently within the turnover time. Note that with isoquinoline-1-oxidoreductase in the presence of substrate and absence of oxidant the Very Rapid species is stable for hours [36]; however, it disappears very rapidly on addition of oxidant [37]. This enzyme is unusual in that it does not contain FAD and has much higher iron-sulfur center redox potentials than most of the related enzymes, so that in the two-electron-containing protein the redox equilibrium lies with Mo(VI), present at vanishingly small concentrations. A kinetic model [37] requires product to be released only from Mo(VI) after the sulfur-bound proton has been lost to generate the Very Rapid species.

The other two signals appear slowly in the presence of particular enzyme inhibitors. Alloxanthine and its dehydroxylated analogue allopurine have a nitrogen atom at the 8-position, which is the site of the carbon hydroxylated when xanthine is the substrate, and carbon at the 9-position which is nitrogen in xanthine; it forms a stable complex to give the alloxanthine signal with N8 bound close to molybdenum and the sulfur as sulfido. The Very Rapid and alloxanthine species have very similar EPR signals, and this has been used to imply that they have very similar molybdenum environments. There is, however, one important difference revealed by hyperfine interactions with ^{17}O in water and ^{14}N in alloxanthine and xanthine [38]. The Very Rapid signal shows a strong interaction with an oxygen derived from solvent water; this is the oxygen atom that is transferred to substrate and is assumed in all mechanistic models (see below) to be bound to molybdenum. In the alloxanthine signal there is no such interaction with oxygen, indicating that it is not present, but there is an interaction of equivalent magnitude (after allowing for differences in nuclear magnetic moments) with nitrogen; this has been assumed to be the N8 of alloxanthine. Thus, assuming that the interacting oxygen is bound to molybdenum, so is the nitrogen of alloxanthine [38]. Since this species is stable, it might be possible to prepare single crystals for X-ray analysis in order to confirm this structure; however, it may be that uncertainties due to the closeness of the heavy molybdenum atom would make it difficult to obtain accurate unrestrained distances by this method.

Finally, the Inhibited species forms slowly during turnover of aldehydes when substrate binds irreversibly in the incorrect orientation; again the precise geometry of the bound inhibitor is open to discussion.

The various structures described above as controversial bear considerably on the mechanism of the enzyme family and will therefore be discussed in some detail in the following section. The most important is that of the Very Rapid species since this is the only one that is catalytically competent.

3.4. ENDOR

ENDOR is a double-resonance technique that is used to measure the hyperfine interaction between unpaired electron(s) and magnetic nuclei by observing the change in the saturation of an EPR signal as a radio-frequency is scanned through the resonant frequency of the nuclei. It therefore only detects nuclei close to the paramagnetic center. To understand the current controversy on the structures at the molybdenum it is necessary briefly to describe the technique and how the results can be used. I will restrict the discussion to nuclei with $I = 1/2$ since only these are relevant here.

3.4.1. Theoretical Background

If the EPR signal of the center under study is anisotropic, as is the case for all Mo(V) signals being considered here, then specific orientations of the center can be selected by positioning the magnetic field at specific points on the EPR spectrum and the ENDOR spectrum measured at a number of these points. In this way single-crystal-like data can be obtained from a frozen solution and the hyperfine interaction between a magnetic nucleus and the unpaired electron resolved in three dimensions [39]. There is then the problem of how to make use of these data. Any hyperfine interaction can be broken down to an isotropic term, A_{iso}, with the same value at all orientations, and an anisotropic term, A_{aniso}, which sums to zero over all orientations.

The principal component of the isotropic term is generally taken to be the contact interaction between the electron and the nucleus, which is a measure of the contribution of the s orbital associated with the nucleus to the semioccupied molecular orbital (SOMO) containing the unpaired electron, and it indicates the extent to which this electron is delocalized onto this atom. It is possible to calculate the theoretical relationship between these quantities and the values have been tabulated [40].

Two main contributions to A_{aniso} are usually considered although there are additional contributions. These first two need to be described here in some detail as they could be used to derive geometrical structural information. The first contribution is the through-space dipolar interaction between the electronic and nuclear dipoles, and the second is the interaction between the nuclear moment and the unpaired electronic spin density in the p orbitals associated with the nucleus.

The first contribution, A_D, is the classical dipolar interaction and is given by:

$$A_D = D(3 \cos^2 \Phi - 1) \tag{1}$$

where $D = g_e \beta_e g_n \beta_n \rho / h r^3$ and Φ is the angle between the line joining the electron and the nucleus and the applied magnetic field. g_e and g_n are the electronic and nuclear g factors, β_e and β_n are the Bohr and nuclear magnetons, ρ is the spatial distribution of the unpaired electron, h is Plank's constant, and r is the distance between the electron and the nucleus. All of these values are either universal constants or are known, except for Φ, ρ, and r. If the electronic spin density is assumed to be centered on, in our case the molybdenum nucleus, then ρ can be taken as 1.0 if r is the distance from the molybdenum nucleus to the magnetic nucleus; thus, a complete analysis would give the distance and direction from the molybdenum to this nucleus. Even if the angular information is incompletely understood, the extreme values of A_D occur for $(3 \cos^2 \Phi - 1) = 2$ or -1, when Φ is $0°$ or $90°$, so that r can still be determined, with some sensitivity as it appears raised to the third power.

The second contribution introduces a complication to the deduction of distance information from the anisotropic component of the hyperfine interaction. If the contribution of the s and p orbitals of the distant atom to the SOMO are known as well as the geometrical relationships between these orbitals, those from the parent atom, and a full description of the **g** tensor, then in principal this second contribution to A_{aniso} can be fully taken into account. This is clearly impossible in the situations being considered here where the overall geometry is unknown. However, it is possible to estimate the size of the second contribution [41] using tabulated values of the anisotropic terms from

p orbitals [40] and assuming a particular hybridization of the SOMO at the distant atom. This argument indicates that if the estimated value of the p-orbital contribution is small compared with A_D, then the distance between the central and distant atoms can reasonably be estimated from A_{aniso}, if the two contributions are of the same order then such an estimate becomes very difficult if not impossible.

3.4.2.   Experimental Results

The most important ENDOR results on xanthine oxidase have been obtained on the Inhibited [42] and Very Rapid [43,43] signals. These have been discussed in a number of places [33,34,45]. ENDOR spectra with ^{13}C in $H^{13}CHO$ and 8-^{13}C xanthine, respectively, were determined at several points across the EPR spectra and the hyperfine interactions determined using simulations of the ENDOR spectra. For the Inhibited and Very Rapid species the results gave A_{aniso} values of 11.9 and 3.5 MHz, respectively, with A_{iso} values of 44.5 and 8.8 MHz. The two publications on the Very Rapid signal agree on the results to within experimental error. Ascribing the A_{aniso} values entirely to A_D for the Inhibited signal gives a molybdenum, with 100% of the unpaired electronic spin density, to aldehyde carbon distance of 1.7 Å; the correction for the carbon p-orbital contribution to the anisotropy, assuming delocalization into a single pure sp-hybridized orbital with no other orbitals involved, is a *small* contribution to A_{aniso} and gives a molybdenum-to-carbon distance of 1.9 Å [42]. This would be clearly within the range where there is a direct bonding interaction between these atoms and the existence of a sideways bound C=O group was postulated. The same calculation on the Very Rapid signal, with 65% spin density on molybdenum, since 35% is on the sulfur atom, gives an uncorrected molybdenum-to-carbon distance of 2.2 Å, which rises to 2.35 Å after again a *small* correction due to delocalization into a single-carbon orbitals with sp hybridization. This was taken to imply the existence of sideways-bound C=O also in the Very Rapid, catalytic intermediate, species [43]. A problem with this argument that has been discussed is that the theory must be confirmed using model compounds. As yet no such study has been published apart from the coupling of low-spin Fe(III) to ^{13}CN in a cyanide adduct of transferrin [41] where the hybridization clearly is expected to be sp; it is urgent that this be done for C=O bound side-on to Mo(V) although as yet there are no obvious model candidates.

The more recent publication [44], principally on the Very Rapid signal, performs the calculation for contributions from single sp_2 and sp_3 hybridized C8 orbitals and estimates that the molybdenum to carbon distances are greater, at 2.65 to 3.00 Å for the Very Rapid species, with a *large* correction due to the p-orbital contribution. The conclusion is that it is unnecessary to postulate the existence of a direct Mo-C bond in either the Inhibited or the Very Rapid species. The observed ^{13}C hyperfine tensor is explained using a Mo−O−C bonding system with the greater p-orbital contribution although it is acknowledged that such calculations are by their nature of limited value. In my view, the large size of the correction established [44] means that it is not possible to calculate the molybdenum-to-carbon distance accurately since, for example, any contribution to the SOMO from other sp_2 orbitals is likely to reduce the distance estimate; therefore, the ENDOR results cannot be used to give a reliable distance in cases such as this where the A_{iso} is so large. The authors of the earlier references [42,43] now agree that the ENDOR results described above do not indicate a direct Mo-C bond [46].

4. KINETIC STUDIES

The kinetics of interactions between the enzymes and a variety of substrates and inhibitors have added significantly to our understanding of the mechanism of action at the molybdenum.

4.1. Steady-State Kinetics

Steady-state kinetic studies on xanthine oxidase with substrates including xanthine [47] and lumazine (2,6-dihydroxypteridine) [48] indicate that the neutral form of substrates interacts with the enzyme. With the slow substrate 2-aminopteridine-6-aldehyde the formation of an early substrate-Mo(VI) complex can be detected, again with the neutral form of the subrate [49]. Further similar work together with UV/visible and resonance Raman spectroscopic characterization of stable enzyme-substrate complexes on violapterin (2,6,7-trihydroxypteridine) [50] are consistent with other work in showing that the first irreversible step during normal turnover is the formation of product bound to enzyme with molybdenum as Mo(IV) [1]. Decay of the violapterin complex is base-catalyzed, perhaps in common with other such intermediates as a result of deprotonation of the molybdenum-bound sulfuryl group [45], whereas that of the Very Rapid species is acid-catalyzed; the latter has been proposed to require protonation of the bound purine. Since different purines will have different pK_a values, it seems likely that these are important in defining the specificity of the enzymes and the extent to which different intermediates accumulate [1].

4.2. Pre-Steady-State Kinetics

Considerable work has been done both by EPR and UV/visible spectroscopies, as has been thoroughly reviewed by Hille [1]. The most extensive recent report was on isoquinoline-1-oxidoreductase [36]. Stopped-flow and rapid-freeze EPR data were fitted to a reaction scheme that was consistent with dissociation of product, after formation of the Very Rapid species, only from Mo(VI) and not from Mo(V).

5. THEORETICAL CALCULATIONS

Density functional theory (DFT) has advanced to a stage where it is now possible to do meaningful theoretical calculations on the energies of proposed catalytic intermediates, involving molybdenum and its immediate coordination sphere, and relate these to enzyme mechanisms. The method in essence involves constructing an electronic description of the target molecule derived from the atomic orbitals of its constituent nuclei, in which electron correlation is included via a function describing the electron density, and then looking for geometries that give energy minima. It has been applied to the molybdenum and its coordination sphere in xanthine oxidase and the related aldehyde oxidase [51–54], as well as to understanding the structure of substrates, especially the role of differently protonated tautomers in determining reactivity patterns [55,56]. All the work published so far had been done before the publication of the xanthine oxidase structure and used the aldehyde oxidoreductase structure as a basis for calculations. It should be noted, as stated by the various authors, that in general only the immediate molybdenum environment is taken into account in these studies, so that the effect of protein environment will influence structural details to an indeterminate extent.

One outcome of these studies has been the demonstration that a number of alternative structures are almost isoenergetic implying that, for example, protonation could occur at the sulfido, oxo, or even the metal at Mo(VI) and that transient intermediates of this nature should be considered. The oxo is unlikely to be protonated in the protein because of steric effects noted by the original authors [52], although significant molybdenum ligand movement [26] could alter the possibilities. The structure of the $-Mo-S-C=C-S-$ ring, especially the $S-S$ distance, the Mo-to-S distances, and the MoSCC dihedral angle, is very sensitive to other ligation changes; this fits with intuitive ideas about subtle ways in which the chemistry at molybdenum could be adjusted by changes in the dithiolene entity.

Calculations on the oxidation of formaldehyde [53] and formamide [54] have indicated that a feasible route for the reductive half-reaction is that substrate carbon approaches the molybdenum-bound hydroxyl, the proton of which is abstracted perhaps by the active site Glu as the $C-O$ bond forms by nucleophilic attack to generate a tetrahedral intermediate. Only subsequently to this does the hydride get transferred from the carbon to the sulfido group as the substrate becomes oxidized and the molybdenum reduced to Mo(IV). The bound product can then dissociate to be replaced by a new water-derived hydroxyl. The hydride transfer was calculated to be reversible, in line with the observation of substrate reduction instead of oxidation, by some of these enzymes and the spectroscopic results using violapterin [50]. Various transition states involving molybdenum-carbon bond formation were found to be too high in energy with difficult stereochemistries [54].

6. DISCUSSION

There have been two main advances during the last few years in understanding the functioning of the molybdenum in these enzymes. The first involved the question of which oxygen atom is transferred to substrate. The two contenders are the oxo and the hydroxyl (or water) ligands of molybdenum. Up to about 1996 almost all workers favored the oxo group (e.g. [47]) although there was already evidence, which was not sufficiently understood, that this is not the case [38,57,58]. Chemical precedence existed for either the oxo or the hydroxyl [59,60] group being the donor. Results of two experiments have allowed the conclusion that the oxo group is not involved: first, the observation that a slowly exchanging water could be detected both as ^{17}O by EPR in the Very Rapid signal and as ^{18}O in uric acid product by mass spectrometry, taken together with ENDOR and EPR evidence that there was only one class of exchangeable oxygen ligand to molybdenum, means that the oxygen atom transferred must be the hydroxyl/water ligand [43]; second, unlabeled deflavo xanthine oxidase in ^{17}O-labeled water exposed to reducing substrate (1-methyl xanthine) in the absence of oxidant can be detected as a strongly coupled oxygen by EPR on a resulting Rapid signal, which model compound work [60] indicates to be in the hydroxyl group.

During the last few years two mechanisms for the transfer of the oxygen from molybdenum to substrate carbon have been discussed in detail, and these are shown in Fig. 3. The first (Fig. 3a) is based on a nucleophilic attack by the hydroxyl molybdenum ligand on the carbon to be hydroxylated, in this case the C8 of xanthine, to generate a transient tetrahedral intermediate that breaks down by transferring its hydrogen to the sulfido ligand reducing Mo(VI) to Mo(IV) (e.g., [61,53]). Loss of the proton from the sulfur followed by internal electron transfer gives Mo(V) which shows the Very Rapid EPR signal and then Mo(VI) at which oxidation level product dissociates to be replaced by water or hydroxyl. The second (Fig. 3b) is derived from a proposal by Coucouvanis and coworkers [62] based on model chemistry that shows that unsaturated hydrocarbons can add across Mo=S bonds; such a reaction would result in a direct Mo−C bond, with the hydrogen transferred to the sulfur, followed by incorporation of the hydroxyl oxygen.

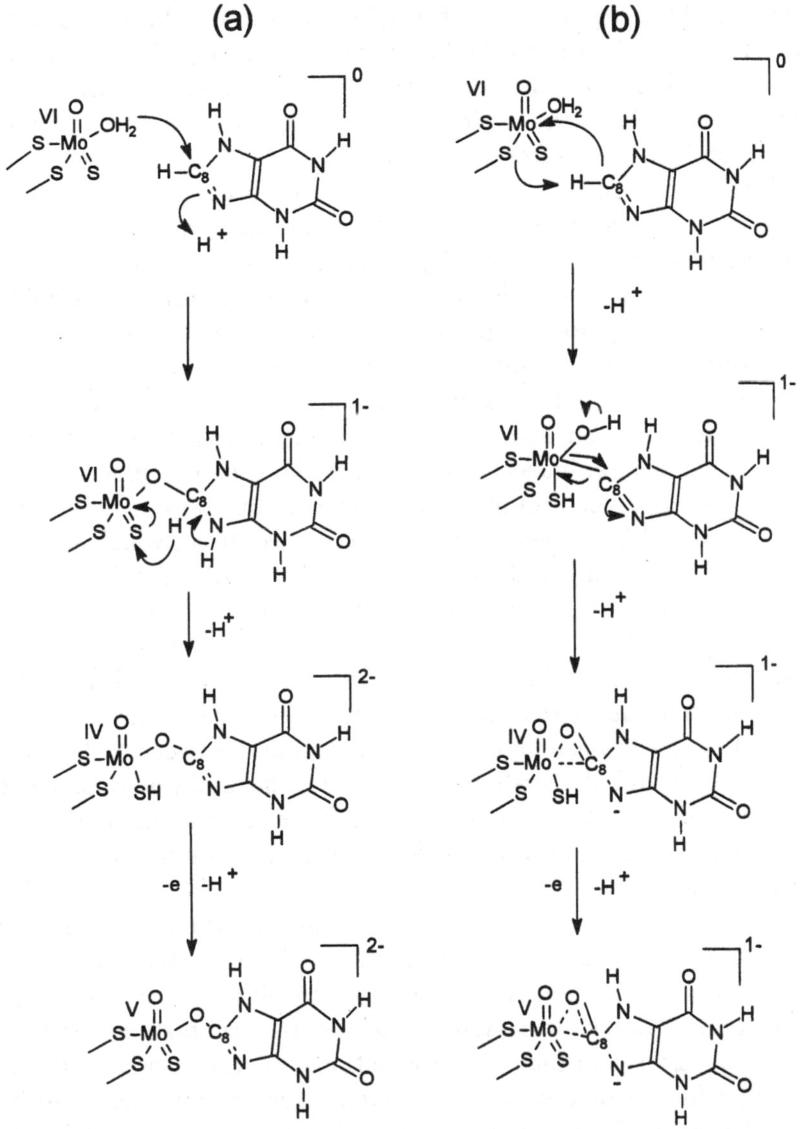

FIG. 3. Two proposed mechanisms for the activity of xanthine oxidase family enzymes involving (a) a tetrahedral intermediate and (b) a sideways-bound C=O group.

Again, loss of hydrogen from sulfur followed by internal electron transfer generates the Very Rapid Mo(V) species, this time with the retention of the Mo−C bond. Further oxidation to Mo(VI) results in the loss of electron density from the −Mo−C−O− ring, dissociation of product, and ligation by a new water molecule.

Evidence for the mechanism of Fig. 3a has been based on chemical precedence, the "reasonableness" of the chemistry, and density functional calculations, and that for Fig. 3b again on chemical precedence, on short molybdenum-to-carbon distances calculated from the ENDOR results on the Very Rapid species, and on the stability of this latter species when the Mo(VI) oxidation level is not populated. A reassessment of the ENDOR evidence indicates that this does not require short molybdenum-to-carbon distances so that these results are no longer useful in this debate; indeed, they can now be used to support Fig. 3a [44]. I would, however, still like to see ^{13}C hyperfine tensors determined for models of both proposed Very Rapid structures to see which gives the large isotropic coupling and associated anisotropy. Such large isotropic contributions require a large carbon s-orbital contribution to the SOMO and so are unlikely to result from coupling via bonding of a sideways-bound C=O with delocalization into this moiety's π system [44]; this bonding scheme therefore appears unlikely. In contrast, in work in progress we have so far been unable to detect coupling between Mo(V) and a ^{13}C-methoxy ligand of a trispyrazolylborate model and would have been able to do so if the isotropic coupling were of the order of that in the Very Rapid species [63]; the usefulness of this model compound remains to be evaluated.

Density functional calculations appear consistent with Fig. 3a rather than 3b because they give a reasonable energetic intermediate route for the former and find the latter to give unreasonably high-energy intermediates. It is fair to ask in this context whether the calculations give satisfactory explanations of the model reactions involving addition of unsaturated systems across molybdenum-sulfur double bonds [62], whether the significant rearrangement of the molybdenum environment that has been proposed to occur as substrate approaches [26] would affect the results of the calculations, and whether the apparent requirement of molybdenum to be oxidized to Mo(VI) before product can dissociate and be replaced by water/hydroxyl [37] is predicted.

7. CONCLUSIONS

Overall, we now agree on many aspects of the mechanism of action of xanthine oxidase family enzymes at their molybdenum atoms. The main area of contention has been the fine detail of the chemistry occurring during the formation of the molybdenum-product complex. In my view, the current balance of evidence appears to lie with the formation of the tetrahedral intermediate of Fig. 3a, although the molybdenum-carbon bond hypothesis of Fig. 3b cannot yet be definitely excluded.

ACKNOWLEDGMENTS

The author thanks the late Professor R. C. Bray, Professors R. L. Richard, and J. Hüttermann and Doctors B. D. Howes and R. Kappl for useful discussions, and Professors B. M. Hoffman and R. Hille for providing a copy of their work in press prior to publication.

ABBREVIATIONS

Coa	coenzyme A
DFT	density functional theory
DMSO	dimethyl sulfoxide
ENDOR	electron nuclear double resonance
EPR	electron paramagnetic resonance
EXAFS	extended absorption fine structure spectroscopy
FAD	flavin adenine dinucleotide
MCD	magnetic circular dichroism
NAD	nicotinamide adenine dinucleotide
SOMO	semioccupied molecular orbital

REFERENCES

1. R. Hille, *Chem. Rev.*, *96*, 2757–2816 (1996).
2. R. Hille, in *Essays in Biochemistry: Metalloproteins* (D. P. Ballou, ed.), Portland Press, London, 1999, pp. 125–137.
3. R. Hille, J. Rétey, U. Bartlewski-Hof, W. Reichenbecher, and B. Schink, *FEMS Microbiol. Rev.*, *22*, 489–501 (1999).
4. K. Breese and G. Fuchs, *Eur. J. Biochem.*, *251*, 916–923 (1998).
5. W. Reichenbecher, A. Rüdiger, P. M. H. Kroneck, and B. Schink, *Eur. J. Biochem.*, *237*, 406–413 (1996).
6. H. Dobbek, L. Gremer, O. Meyer, and R. Huber, *Proc. Natl. Acad. Sci. USA*, *96*, 8884–8889 (1999).
7. M. J. Romão, M. Archer, I. Moura, J. J. G. Moura, J. LeGall, R. Engh, M. Schneider, P. Hof, and R. Huber, *Science*, *270*, 1170–1176 (1995).
8. J. Rebelo, S. Macieira, J. K. Dias, R. Huber, C. S. Ascenso, F. Rusnak, J. J. G. Moura, I. Moura, and M. J. Romão, *J. Mol. Biol.*, *297*, 135–146 (2000).
9. C. Enroth, B. T. Eger, K. Okamoto, T. Nishino, T. Nishino, and E. Pai, *Proc. Natl. Acad. Sci. USA*, *97*, 10723–10728 (2000).
10. B. T. Eger, K. Okamoto, C. Enroth, M. Sato, T. Nishino, E. Pai, and T. Nishino, *Acta Cryst.*, *D56*, 1656–1658 (2000).
11. N. Komoto, K. Yukuhiro, and T. Tamura, *Insect Mol. Biol.*, *8*, 73–83 (1999).
12. P. Montalbini, *J. Plant Physiol.*, *156*, 3–16 (2000).
13. A. Glatigny and C. Scazzocchio, *J. Biol. Chem.*, *270*, 3534–3550 (1995).
14. R. K. Mehra and M. P. Coughlan, *Arch. Biochem. Biophys.*, *229*, 585–595 (1984).
15. A. Glatigny, P. Hof, M. J. Romão, R. Huber, and C. Scazzocchio, *J. Mol. Biol.*, *278*, 431–438 (1998).
16. M. Kurosaki, S. Demontis, M. M. Barzago, E. Garattini, and M. Terao, *Biochem. J.*, *341*, 71–80 (1999).
17. M. Terao, M. Kurosaki, G. Saltini, S. Demontis, M. Marini, M. Salmona, and E. Garattini, *J. Biol. Chem.*, *275*, 30690–30700 (2000).

18. H. Sekimoto, H. Seo, N. Dohmae, K. Takio, Y. Kamiya, and T. Koshiba, *J. Biol. Chem.*, *272*, 15280–15285 (1997).

19. S. Akaba, M. Seo, N. Dohmae, K. Takio, H. Sekimoto, Y. Kamiya, N. Furuya, T. Komano, and T. Koshiba, *J. Biochem.*, *126*, 395–401 (1999).

20. M. Seo, H. Koiwai, S. Akaba, T. Komano, T. Oritani, Y. Kamiya, and T. Koshiba, *Plant. J.*, *23*, 481–488 (2000).

21. H. Horiuchi, M. Ota, S. Nishimura, H. Kaneko, Y. Kasahara, T. Ohta, and K. Komoriya, *Life Sciences*, *66*, 2051–2070 (2000).

22. W. R. Hagen, M. A. Vanoni, K. Rosenbaum, and K. D. Schnackerz, *Eur. J. Biochem.*, *267*, 3640–3646 (2000).

23. T. Nagamatsu, T. Fujita, and K. Endo, *J. Chem. Soc. Perkin Trans. I*, 33–42 (2000).

24. T. Schräder, A. Rienhöfer, and J. R. Andreesen, *Eur. J. Biochem.*, *264*, 862–871 (1999).

25. M. G. Ryan, K. Ratnam, and R. Hille, *J. Biol. Chem.*, *270*, 19209–19212 (1995).

26. R. M. Jones, F. E. Inscore, R. Hille, and M. L. Kirk, *Inorg. Chem.*, *38*, 4963–4970 (1999).

27. R. C. Bray and T. Vänngård, *Biochem. J.*, *114*, 725–734 (1969).

28. N. A. Turner, R. C. Bray, and G. P. Diakun, *Biochem. J.*, *260*, 563–571 (1989).

29. R. Hille, G. N. George, M. K. Eidsness, and S. P. Cramer, *Inorg. Chem.*, *28*, 4018–4022 (1989).

30. F. F. Morpeth, G. N. George, and R. C. Bray, *Biochem. J.*, *220*, 235–242 (1984).

31. S. J. Tanner, R. C. Bray, and F. Bergmann, *Biochem. Soc. Trans.*, *6*, 1328–1330 (1978).

32. R. C. Bray and D. J. Lowe, *Biochem. Soc. Trans.*, *25*, 762–768 (1997).

33. D. J. Lowe, R. L. Richards, and R. C. Bray, *JBIC*, *3*, 557–558 (1998).

34. R. Hille, *J. Biol. Inorg. Chem.*, *3*, 559–560 (1998).

35. G. N. George and R. C. Bray, *Biochemistry*, *27*, 3603–3609 (1988).

36. C. Canne, I. Stephan, J. Finsterbusch, F. Lingens, R. Kappl, S. Fetzner, and J. Hüttermann, *Biochemistry*, *36*, 9780–9790 (1997).

37. C. Canne, D. J. Lowe, S. Fetzner, B. Adams, A. T. Smith, R. Kappl, R. C. Bray, and J. Hüttermann, *Biochemistry*, *38*, 14077–14087 (1999).

38. T. R. Hawkes, G. N. George, and R. C. Bray, *Biochem. J.*, *218*, 961–968 (1984).

39. B. M. Hoffman, R. A. Venters, J. E. Roberts, M. Nelson, and W. H. Orme-Johnson, *J. Am. Chem. Soc.*, *104*, 4711–4712 (1982).

40. J. R. Morton and K. F. Preston, *J. Magn. Reson.*, *30*, 577–582 (1978).

41. P. A. Snetsinger, N. D. Chasteen, and H. van Willigen, *J. Am. Chem. Soc.*, *112*, 8155–9160 (1990).

42. B. D. Howes, B. Bennett, R. C. Bray, R. L. Richards, and D. J. Lower, *J. Am. Chem. Soc.*, *116*, 11624–11625 (1994).

43. B. D. Howes, R. C. Bray, R. L. Richards, N. A. Turner, B. Bennett, and D. J. Lowe, *Biochemistry*, *35*, 1432–1443 (1996).

44. P. Manikandan, E.-Y. Choi, R. Hille, and B. M. Hoffmann, *J. Am. Chem. Soc. (in press)*.

45. D. J. Lowe, R. L. Richards, and R. C. Bray, *Biochem. Soc. Trans.*, *25*, 774–778 (1997).

46. B. D. Howes, R. C. Bray, R. L. Richards, B. Bennett, and D. J. Lowe, personal communication.

47. J. H. Kim, M. G. Ryan, H. Knaut, and R. Hille, *J. Biol. Chem.*, *271*, 6771–6780 (1996).

48. M. D. Davis, J. S. Olson, and G. Palmer, *J. Biol. Chem.*, *259*, 3526–3533 (1984).

49. M. Xia, P. Ilich, R. Dempski, and R. Hille, *Biochem. Soc. Trans.*, *25*, 768–773 (1997).

50. W. A. Oertling and R. Hille, *J. Biol. Chem.*, *265*, 17446–17450 (1990).

51. M. R. Bray and R. J. Deerth, *Inorg. Chem.*, *35*, 5720–5724 (1996).

52. A. V. Voityuk, K. Albert, S. Köstlmeier, V. A. Nasluzov, K. M. Neyman, P. Hof, R. Huber, M. J. Romão, and N. Rösch, *J. Am. Chem. Soc.*, *119*, 3159–3160 (1997).

53. A. V. Voityuk, K. Albert, M. J. Romão, R. Huber, and N. Rösch, *Inorg. Chem.*, *37*, 176–180 (1998).

54. P. Ilich and R. Hille, *J. Phys. Chem. B*, *103*, 5406-5412 (1999).

55. P. Ilich and R. Hille, *Inorg. Chim. Acta*, *263*, 87–93 (1997).

56. A. L. Michaud, J. A. Herrick, J. E. Duplain, J. L. Manson, C. Hemann, P. Ilich, R. J. Donohoe, R. Hille, and W. A. Oertling, *Biospectroscopy*, *4*, 235–256 (1998).

57. S. Gutteridge and R. C. Bray, *Biochem. J.*, *189*, 615–623 (1980).

58. N. A. Turner, R. C. Bray, and G. P. Diakun, *Biochem. J.*, *260*, 563–571 (1989).

59. R. H. Holm, *Coord. Chem. Rev.*, *100*, 183–221.

60. R. J. Greenwood, G. L. Wilson, J. R. Pilbrow, and A. G. Wedd, *J. Am. Chem. Soc.*, *115*, 5358–5392 (1993).

61. R. Huber, P. Hof, R. O. Duarte, J. J. G. Moura, I. Moura, M.-Y. Liu, J. LeGall, R. Hille, M. Archer, and M. Romão, *Proc. Natl. Acad. Sci. USA*, *93*, 8846–8851 (1996).

62. D. Coucouvanis, A. Toupadakis, J. D. Lane, S. M. Koo, C. G. Kim, and A. Hdajikyriacou, *J. Am. Chem. Soc.*, *113*, 5271–5282 (1991).

63. D. J. Lowe and R. L. Richards, unpublished.

14

The Molybdenum-Containing Hydroxylases of Quinoline, Isoquinoline, and Quinaldine

Reinhard Kappl[1], Jürgen Hüttermann[1], and Susanne Fetzner[2]

[1]Fachrichtung Biophysik und Physikalische Grundlagen der Medizin, Universität des Saarlandes, Klinikum Bau 76, D-66421 Homburg Saar, Germany

[2]Mikrobiologie, Fachbereich 7, Carl von Ossietzky Universität Oldenburg, P.O. Box 2503, D-26111 Oldenburg, Germany

1.　INTRODUCTION

The three prokaryotic enzymes quinoline 2-oxidoreductase (Qor), iso-quinoline 1-oxidoreductase (Ior), and quinaldine 4-oxidase (Qox) belong to the large family of molybdopterin-containing hydroxylases that introduce an oxygen atom originating from water into their substrates via cleavage of a CH bond. All of these proteins contain basically the same type of redox centers, a molybdenum (Mo) cofactor, two [2Fe2S] iron-sulfur clusters and, with a few exceptions, a flavin adenine dinucleotide (FAD), and they are found in all organisms from bacteria to man [1,2]. This family is named after its thoroughly investigated mammalian member xanthine oxidase (XO) from milk [2–6]. The xanthine oxidase family differs from the related sulfite oxidase and DMSO reductase families on account of the structures of their molybdenum cofactors. The molybdenum in the XO family is bound to the pyranopterin cofactor (MPT) moiety by a dithiolene group and comprises a monooxo-monosulfido ligation in the active enzyme. Mo-MPT as well as dinucleotide forms of the molybdenum pyranopterin cofactor have been identified in enzymes of this class [2]. In contrast to the other two families, only a single pterin moiety is present and no direct binding of molybdenum to the protein is found. The first X-ray structure of an enzyme of the XO family was obtained from aldehyde oxidoreductase (Aor) from *Desulfovibrio gigas*, which revealed the spatial arrangement of the molybdopterin cytosine dinucleotide (MCD) cofactor and the two [2Fe2S] iron-sulfur clusters as well as details of the interaction with surrounding protein moieties [7,8]. Whereas Aor is lacking the FAD cofactor, the recent X-ray structure determination of another member of the XO family, carbon monoxide dehydrogenase (Cox) from *Oligotropha carboxidovorans*, made structural details of the FAD site in the enzyme accessible. The Cox enzyme exhibits a special feature because it contains, besides FAD, the Mo-MCD cofactor, and two 2Fe2S clusters, a selenium and a Cu(I) ion. The cyanolysable sulfido group links the Mo and the Cu(I) ion, which is attached to the Sγ of Cys[388], thus forming a Mo-S-Cu-(S-Cys[388]) bridge [9–11]. Only very recently did the three-dimensional structure of milk XO in the oxidase and dehydrogenase form become available [12]. On the basis of such structural data several plausible models of the reaction mechanism for XO in particular and for enzymes of this family have been presented taking into account the wealth of spectroscopic, kinetic, and biochemical information available [2,8,13–22]. Within this framework we present here the biochemical and EPR-spectroscopic results for the three prokaryotic enzymes in comparison with the corresponding findings for XO and related systems.

2. BIOCHEMICAL AND GENETIC CHARACTERIZATION

2.1. Bacterial Catabolism of Quinoline, Isoquinoline, and Quinaldine

A number of aerobic bacterial strains that utilize quinoline or quinoline derivatives or isoquinoline as sole source of carbon and energy have been isolated by several groups (for a review, see [23]). Chemically, benzopyridines such as quinoline and isoquinoline are susceptible to nucleophilic substitution reactions at the pyridine ring and electrophilic substitutions at the benzene ring [24]. Consistent with N-heteroaromatic chemistry, the bacterial degradation of quinoline, isoquinoline, and quinaldine starts with a nucleophilic attack by a water molecule, yielding 1H-2-oxoquinoline, 2H-1-oxoisoquinoline, and 1H-4-oxoquinaldine, respectively (Scheme 1). These nucleophilic substitution reactions are catalyzed by distinct molybdenum hydroxylases, which will be discussed in the following sections. Whereas all known bacterial catabolic pathways involve the initial formation of such oxo compounds, the following degradative steps differ among the bacterial strains (for reviews, see

[23,25]). 1H-2-oxoquinoline, for example, is degraded via 8-hydroxycoumarin and 2,3-dihydroxyphenylpropionic acid by various *Pseudomonas* strains [26–28], whereas *Comamonas testosteroni* 63 forms the 5,6-*cis*-dihydrodiol of 1H-2-oxoquinoline, which is degraded further via 5,6-dihydroxy-1H-2-oxoquinoline and presumably trihydroxypyridine [29,30]. These examples already show that aerobic bacterial degradation pathways of heteroaromatic (as well as aromatic) compounds often involve several hydroxylation steps. In contrast to the hydroxylation reactions adjacent to or in para position relative to the N-heteroatom of the pyridine ring, which are catalyzed by molybdenum hydroxylases and thus involve water as cosubstrate, hydroxylations of the benzenoid ring of quinoline or related compounds usually are catalyzed by mono- or dioxygenases using molecular oxygen as electrophilic cosubstrate.

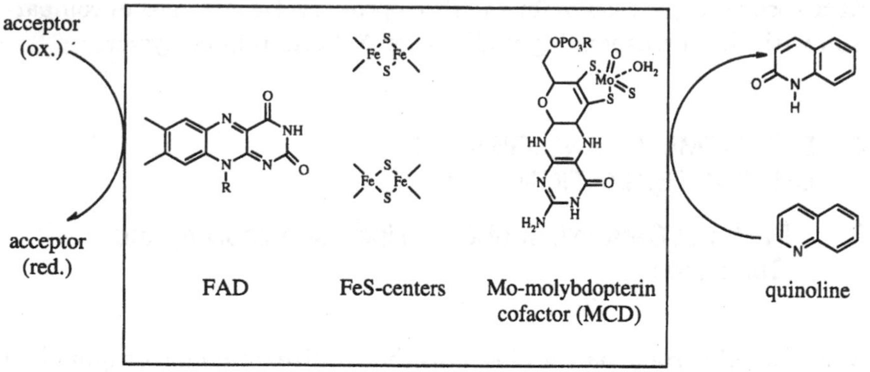

Scheme 1. Hydroxylation catalyzed by Qor is the first step in the degradation of quinoline.

Although the aerobic bacterial degradation of quinoline, various quinoline derivatives, and isoquinoline has been known for some years, most degradation pathways have not been elucidated in every detail, and apart from the initial hydroxylation and some oxygenase-catalyzed steps, many reactions of the pathways are not understood. This especially applies to the 8-hydroxycoumarin pathway of quinoline degradation [26–28], but also for the catabolism of isoquinoline. All known bacteria growing on isoquinoline initially form 2H-1-oxoisoquinoline. For *Brevundimonas diminuta* 7, further degradation has been suggested to proceed via phthalate, 4,5-dihydroxyphthalate, and proto-catechuate [31].

Quinoline, quinoline derivatives, and isoquinoline are degraded not only by aerobic but also by anaerobic bacteria (reviewed in [25]). Whereas the aerobic bacteria in oxygenase-catalyzed reactions may utilize dioxygen to modify and to cleave the aromatic rings of their (hetero)aromatic substrates, anaerobic bacteria have to use other strategies. *Desulfobacterium indolicum* after initial hydroxylation of quinoline to 1*H*-2-oxoquinoline forms 3,4-dihydro-1*H*-2-oxoquinoline, which is transformed into unidentified products [32]. The enzymes involved in anaerobic bacterial degradation of quinoline or its derivatives have not yet been investigated. Quinaldine (2-methylquinoline) is degraded aerobically by *Arthrobacter ilicis* Rü61a via the anthranilate pathway [33]. After initial hydroxylation at C4, catalyzed by quinaldine 4-oxidase, a monooxygenase mediates hydroxylation at C3. Ring cleavage of the *N*-heterocyclic ring of 1*H*-3-hydroxy-4-oxoquinaldine is catalyzed by a unique cofactor-free 2,4-dioxygenase, yielding carbon monoxide and *N*-acetylanthranilate [34–37]. The anthranilate formed from *N*-acetylanthranilate is converted to catechol, which undergoes dioxygenolytic intradiol ring cleavage [33]. In any case, catabolism eventually yields aliphatic metabolites that are funneled into central metabolic pathways to be used as carbon, nitrogen, and energy source.

2.2. Cofactors of Quinoline 2-Oxidoreductase, Isoquinoline 1-Oxidoreductase, and Quinaldine 4-Oxidase

The hydroxylases catalyzing the first step in the bacterial degradation of quinoline, isoquinoline, and quinaldine belong to the xanthine oxidase family of molybdenum enzymes, harboring monooxo-monosulfido-type molybdenum centers [2]. Quinoline 2-oxidoreductase and isoquinoline 1-oxidoreductase show significant amino acid sequence similarities to XOs/xanthine dehydrogenases (Xdh) (cf. Sec. 2.4). However, whereas the eukaryotic as well as almost all known prokaryotic Xdh compounds contain a pyranopterin derivative known as molybdopterin as an organic part of the molybdenum cofactor, Qor, Ior, and Qox, as well as aldehyde oxidoreductase (Aor) from *D. gigas* and the CO dehydrogenase (CODH) compounds belonging to the XO family, contain Mo-MPT that is modified by covalent attachment of cytidine monophosphate to its terminal phosphate group to form molybdenum molybdopterin cytosine dinucleotide [7–9,38–43]. The intramolecular electron transfer in Qor and Qox also involves two distinct iron-sulfur clusters [2Fe2S]I and [2Fe2S]II, and FAD. Ior, a heterodimeric enzyme, lacks the flavin cofactor. The cofactors of all three enzymes are listed in Table 1. The properties of these redox active centers will be discussed in Sec. 4.

2.3. Catalytic Activities of Quinoline 2-Oxidoreductase, Isoquinoline 1-Oxidoreductase, and Quinaldine 4-Oxidase

The molybdenum enzymes involved in the bacterial degradation of N-heteroaromatic compounds show strict regiospecificities of hydroxylation (Table 2). Quinoline, for example, is converted to $1H$-2-oxoquinoline by Qor, whereas Qox catalyzes the formation of $1H$-4-oxoquinoline [44]. A number of quinoline derivatives and some benzodiazines are accepted as substrates by Qor and Qox. Quite remarkably, the activity of Qox from A. ilicus Rü61a toward quinoline, isoquinoline, some chloro-substituted quinolines, and even some benzodiazines is much higher than toward quinaldine, i.e., the substrate used to grow the bacteria. Apart from isoquinoline Ior transforms phthalazine and quinazoline, but it is not able to oxidize quinoline derivatives. Xanthine or hypoxanthine, and quinoline carboxylic acids are not transformed by Qor, Qox, or Ior. The substrate specificity of the three enzymes is listed in Table 2.

The genes encoding Qox have not been sequenced yet; however, the amino acid sequence of Ior as deduced from the nucleotide sequence of the iorAB genes shows significant similarity with the sequence of aldehyde dehydrogenase (Adh) from Acetobacter polyoxogenes. Whereas aldehyde oxidoreductase activity of Qor is negligible, Ior as well as Qox catalyze the oxidation of aldehydes, especially aromatic ones, to the corresponding carboxylic acids (cf. Table 3). However, Ior is progressively inactivated during aldehyde oxidation.

A number of hydroxy-substituted quinolines, isoquinolines, and benzodiazines are important building blocks for the production of pharmaceuticals and agrochemicals. Due to the substrate and regiospecificity

TABLE 1

Cofactors of Quinoline 2-Oxidoreductase, Isoquinoline 1-Oxidoreductase, and Quinaldine 4-Oxidase

Enzyme (Source)	Subunit structure	Molybdenum cofactor	Iron-sulfur clusters	FAD	Organic phosphate	Ref.
			(contents in g atom, or mol per mol of enzyme)			
Quinoline 2-Oxidoreductase (*Pseudomonas putida* 86)	$(LMS)_2$	Mo-MCD 1.6 Mo 2.0 CMP	[2Fe2S]I[a] [2Fe2S]II[b] 6.8 Fe 8.5 S[c]	1.6	7.8	38 39, 57 58, 59
Quinaldine 4-Oxidase (*Arthrobacter ilicis* Rü61a)	$(LMS)_2$	Mo-MCD 1.6 Mo CMP[d]	[2Fe2S]I [2Fe2S]II 7.9 Fe 7.6 S	1.8	8.3	42 58
Isoquinoline 1-Oxidoreductase (*Brevundimonas diminuta* 7)	LS	Mo-MCD 0.85 Mo 1 CMP	[2Fe2S]II [2Fe2S]III 3.95 Fe 3.9 S	—	2.1	41, 46 58, 59

[a]EPR signals with smaller g-anisotropy.
[b]EPR signals with larger g-anisotropy.
[c]Acid-labile sulfur.
[d]Amount was not determined.

TABLE 2

Substrate Specificity of Quinoline 2-Oxidoreductase, Isoquinoline 1-Oxidoreductase, and Quinaldine 4-Oxidase Toward Some N-Heteroaromatic Compounds[a]

Substrate	Relative activity (%), and product formed by:		
	Qor	Ior	Qox
Quinoline	100, 1H-2-oxoquinoline	n.c.	301, 1H-4-oxoquinoline
Isoquinoline	n.c.	100, 2H-1-oxoisoquinoline	202, 2H-1-oxoisoquinoline
Quinaldine	n.c.	n.c.	100, 1H-4-oxoquinaldine
5-Hydroxyquinoline	50, 1H-5-hydroxy-2-oxoquinoline	n.d.	n.c.
6-Hydroxyquinoline	25, 1H-6-hydroxy-2-oxoquinoline	n.c.	n.c.
7-Hydroxyquinoline	17, 1H-7-hydroxy-2-oxoquinoline	n.c.	n.c.
8-Hydroxyquinoline	20, 1H-8-hydroxy-2-oxoquinoline	n.c.	n.c.
2-Chloroquinoline	n.c.	n.c.	514, n.d.

8-Chloroquinaldine	n.d.	n.d.	746, n.d.
Monocarboxyquinolines	n.c.	n.c.	n.c.
Hypoxanthine, xanthine	n.c.	n.c.	n.c
Phthalazine	n.c.	89, 2H-1-oxophthalazine	58, 2H-1-oxophthalazine
Quinazoline	10, 1H-2-oxoquinazoline	45, 3H-4-oxoquinazoline	206, 3H-4-oxoquinazoline
Cinnoline	n.c.	n.c.	132, 1H-4-oxocinnoline
Quinoxaline	45, 1H-2-oxoquinoxaline	n.c.	n.c.

aData were taken from [41,43,44].

Note: The specific activities of purified enzymes in the respective standard assays are: Qor, 19 U/mg (quinoline conversion); Ior, 14 U/mg (isoquinoline conversion); Qox, 3.3 U/mg (quinaldine conversion).

n.c., not converted; n.d., not determined

TABLE 3

Aldehyde Oxidation by Isoquinoline 1-Oxidoreductase
and Quinaldine 4-Oxidase[a]

Aldehyde	Relative activity (%) of	
	Ior	Qox
Acetaldehyde	n.c.	n.c.
Butyraldehyde	9[b]	1.5
Benzaldehyde	47[b]	70
Salicylaldehyde	52[b]	62
Vanillin	9[b]	5
Cinnamaldehyde	36[b]	9

[a]Data are from [44].
[b]Values estimated based on the initial reaction velocity; pro-
gessive decrease of activity.
Note: Activities of Ior and Qox with isoquinoline and quinaldine in the respective standard assay were defined as 100%.
n.c., not converted.

of the molybdenum hydroxylases, biotransformations using microbial
strains, or biocatalysis using bacterial enzymes might provide alternatives to chemical hydroxylation [23].

2.4. Comparative Sequence Analysis, and Expression Cloning of Genes Encoding Quinoline 2-Oxidoreductase and Isoquinoline 1-Oxidoreductase

The *qor* and *ior* genes from *P. putida* 86 and *B. diminuta* 7 are clustered
in the transcriptional order *qorMSL* (*M, S* and *L* for the gene encoding
the medium-sized, small, and large subunit, respectively), and *iorAB* (*A*
encoding the small subunit, *B* the large subunit) [45,46]. Since plasmids
have not been identified in these strains, a chromosomal location of the
structural genes is assumed. The alignment of the deduced amino acid
sequences of the subunits or Qor and Ior with the amino acid sequences
of the corresponding subunits or domains of pro- and eukaryotic

enzymes belonging to the XO family revealed significant similarities [45,46]. Qor, for example, shows high amino acid sequence similarity to nicotine dehydrogenase from *Arthrobacter nicotinovorans* (L: 41%, M: 38%, S: 56% amino acid identity) and to CODH from *Oligotropha carboxidovorans* (Cox; L: 32%, M: 34%, S: 49% amino acid identity). Identities of Qor and Xdh from eukaryotic sources are about 24%. Ior is most similar to aldehyde dehydrogenase from *A. polyoxogenes*. The amino acid motif, which in Aor and Cox are known to bind the iron-sulfur clusters and the pyranopterin molybdenum cofactor, are conserved in Qor and Ior (Fig. 1) [7–9,45,46]. However, the sequential arrangement of conserved peptide domains in IorB differs from the arrangement found in most other enzymes of the XO family. Whereas in CODHs, Aor, Qor, and Xdh, conserved domains of the large subunit or its corresponding segment are arranged in the order D-B-A-C (as defined in [46]), in IorB the domain A is located near the NH_2 terminus (arrangement A-D-B-C). Thus, the presumed Mo-MCD contacting segments MoCoIII and MoCoIV of IorB are located NH_2-terminal to MoCoI, II, and V (Fig. 1). Remarkably, Adh from *A. polyoxogenes* shows the same sequential order of domains as Ior. Amino acid segments similar to the FAD-binding motifs identified in CoxM [9] are found in the amino acid sequence of the medium-sized subunit of Qor (Fig. 1).

Hypotheses concerning the functions of various domains, motifs, and amino acid residues in cofactor binding and in substrate turnover ultimately have to be put to the test. A thorough study of the molybdenum hydroxylases requires the determination of the crystal structure of these enzymes, the construction of mutant proteins, and their structural, biochemical and spectroscopic characterization. However, a prerequisite for this approach is the availability of a suitable system for the regulated (over)expression of the genes encoding molybdenum hydroxylases and for the synthesis of functional enzymes so as to produce large amounts of protein for crystallization studies and to produce protein variants altered by site-directed mutagenesis. Despite many years of research on the genetics of molybdenum hydroxylases, heterologous functional expression of genes encoding a Mo-MCD-containing molybdenum hydroxylase and has not yet been achieved. Expression of CODH from *Pseudomonas thermocarboxydovorans* in *Escherichia coli*, for example, resulted in the synthesis of inactive protein [47]. However, in *E. coli* all known molybdoenzymes contain a molybdopterin guanine

A

	MoCoI	MoCoII	MoCoIII	MoCoIV	MoCoV
Aor	^{418}**GGTFGYK**^{424}	^{531}A**FR**-GYGAPQSM^{541}	^{653}H**GQG**^{656}	^{693}GPSG**GS**^{698}	^{867}VG**E**LPL^{872}
CoxL	^{268}**GGGFGNK**^{274}	^{385}AY**RC**SFRVTEAV^{396}	^{528}Q**GQG**^{531}	^{565}LGTYG**S**^{570}	^{761}VG**E**SPH^{766}
QorL	^{252}**GGGFGQK**^{258}	^{369}AY**R**-GVGFTAGQ^{379}	^{506}S**GQG**^{509}	^{542}FGAYA**S**^{547}	^{741}MG**E**SAM^{746}
IorB	^{444}**GGGFGRR**^{450}	^{563}YM**R**-GIGVAQVK^{573}	^{75}L**GQG**^{78}	^{119}QFTAA**S**^{124}	^{738}VA**E**LGL^{743}

B

Aor ^{99}**QCGFCS**PG**X**$_{26}$HR**NACRC**TGYKP^{144}
CoxS ^{101}**QCGYCT**PG**X**$_{26}$--**NLCRC**TGYQN^{144}
QorS ^{106}**QCGFCT**AG**X**$_{26}$--**NLCRC**TGYET^{149}
IorA ^{96}**QCGYCQ**SG**X**$_{26}$--**NLCRC**GTYQR^{139}

C

Aor ^{39}**GCEQGQCGACS**VX$_8$**ACVT**^{62}
CoxS ^{41}**GCDTSHCGACT**VX$_8$**SC**-**T**^{63}
QorS ^{47}**GCEQGVCGSCT**IX$_8$**SCLT**^{70}
IorA ^{38}**GCGLGLCGACT**VX$_8$**SCIT**^{61}

D

CoxM ^{32}**AGGHS**^{36} ^{111}**TIGG**^{114}
QorM ^{31}**AGGQS**^{35} ^{110}**TLGG**^{113}

FIG. 1.   Amino acid segments of Aor (*D. gigas*) and Cox (*O. carboxidovorans*) involved in cofactor binding [7–9], and similar motifs conserved in Qor and Ior [45,46]. (A) Mo-MCD contacting segments. C388 is present in the sequences of CODHs from different carboxidotrophic bacteria but is absent from other enzymes of the XO family. (B) Binding site of [2Fe2S]I, which in Aor and Cox is located proximal to the molybdenum cofactor. (C) [2Fe2S]II-contacting segment, resembling the motif of the plant-type [2Fe2S] ferredoxins. It is the distal FeS cluster in Aor, and located adjacent to the FAD residing on the medium-sized subunit in Cox. (D) FAD-contacting motifs, typical of the FAD binding site of the vanillyl alcohol oxidase family of flavoproteins.

dinucleotide (MGD) form of the molybdenum cofactor. Synthesis of Mo-MGD from Mo-MPT and Mg^{2+}-GTP is catalyzed by the MobA protein [48–52]. An enzyme catalyzing formation of Mo-MCD—presumably from Mo-MPT and (Mg^{2+})-CTP—so far has not been identified in *E. coli* or any other strain.

In recent work, we tested whether expression of structural genes and synthesis of functional Ior and Qor can be achieved in *Pseudomonas putida* recipient strains. *P. putida* KT2440 clones harboring the *qorMSL* genes from *P. putida* 86 in a *Pseudomonas* expression vector produced large amounts of Qor protein, as verified by Western blot analysis, when grown in the presence of a suitable effector to induce gene expression from the vector promoter. However, the specific activity of Qor protein purified from the *P. putida* KT2440 clone was about 80-fold lower than the specific activity of Qor purified from wild-type *P. putida* 86. Analysis of the cofactors of Qor purified from the *P. putida* KT2440 clone indicated the presence of the [2Fe2S] clusters and FAD; however, the recombinant Qor presumably contains only trace amounts of MCD [53].

Recombinant *P. putida* KT2440 clones harboring the *iorAB* genes in a *Pseudomonas* expression vector when grown in the presence of suitable effectors produced a protein that showed the same electrophoretic mobility as Ior from wild-type *Brevundimonas diminuta* 7. However, in crude extracts of the clones Ior activity was not detectable. When the quinoline-degrading strain *P. putida* 86 was transformed with the *iorAB*-containing expression plasmid, growth of this clone in the presence of effectors of succinate likewise did not produce functional Ior, whereas growth on quinoline led to the synthesis of catalytically active Ior. The specific activity of Ior in crude extracts of the *P. putida* 86 clone was about one-twentieth of the specific Ior activity found in crude extracts of fully induced *B. diminuta* 7. As outlined above (Sec. 2.2), *P. putida* 86 produces the Mo-MCD-, [2Fe2S]I-, [2Fe2S]II-, and FAD-containing enzyme Qor when utilizing quinoline. In *P. putida* 86 harboring *iorAB* and growing on quinoline, maturing Qor and Ior probably compete for the Mo-MCD cofactor (and maybe also for iron-sulfur cluster assembly). Expression of *iorAB* and synthesis of functional Ior protein may not be very efficient in wild-type *P. putida* 86, but in this recipient and under distinct physiological conditions, heterologous functional expression of genes encoding a Mo-MCD-containing molybdenum hydroxylase was achieved for the first time. A *qorMSL*-deficient mutant of *P. putida* 86 might be a suitable recipient for the expression of genes encoding Mo-MCD-containing molybdenum hydroxylases.

Since heterologous synthesis of functional Ior in *P. putida* 86 strictly required the presence of quinoline, we assume that accessory gene(s) encoding product(s) essential for the synthesis of catalytically competent Ior is(are) part of the quinoline regulon in *P. putida* 86. Whereas assembly of [2Fe2S] clusters and Mo-MPT biosynthesis involve ubiquitously conserved pathways, additional reactions that modify Mo-MPT appear to be restricted to certain organisms. Thus, expression of CODH genes from *P. thermocarboxydovorans* in *E. coli* [47], as well as expression of the *qorMSL* and *iorAB* genes in *P. putida* KT2440, and expression of *iorAB* in *P. putida* 86 in the absence of quinoline, might have resulted in the synthesis of inactive protein because the recipient strains did not catalyze the formation of the Mo-MCD cofactor and/or its insertion into the maturing protein. The gene product(s) whose synthesis is(are) coinduced by quinoline in *P. putida* 86 could be an enzyme catalyzing the formation of Mo-MCD from Mo-MPT and Mg^{2+}-CTP, or a Mo-MCD insertase, or a chaperone involved in folding of the molybdenum hydroxylase during or after insertion of Mo-MCD.

3. STRUCTURAL FEATURES OF RELATED ENZYMES

The aldehyde oxidoreductase from the sulfate-reducing anaerobic bacterium *D. gigas* was the first enzyme of the XO family whose three-dimensional structure was solved to a resolution of 1.8 Å [7]. It comprises a homodimeric (α_2) subunit composition and contains a Mo-MCD cofactor, deeply buried in the protein (15 Å away from the surface), and two [2Fe2S] clusters. One of the clusters is "proximal" to the Mo cofactor and has a unique binding motif; the "distal" cluster shows the typical plant-type cysteine-binding motif and is close to the surface in contact to solvent. Similar to Ior there is no FAD present in Aor, and both enzymes show significant amino acid sequence similarity [46]. From the spectroscopic viewpoint Aor may be considered as a closely related structural model for Ior.

The spatial arrangement of the redox centers of Aor is depicted in Fig. 2 (top). It was suggested that a water (or OH) is bound to the Mo

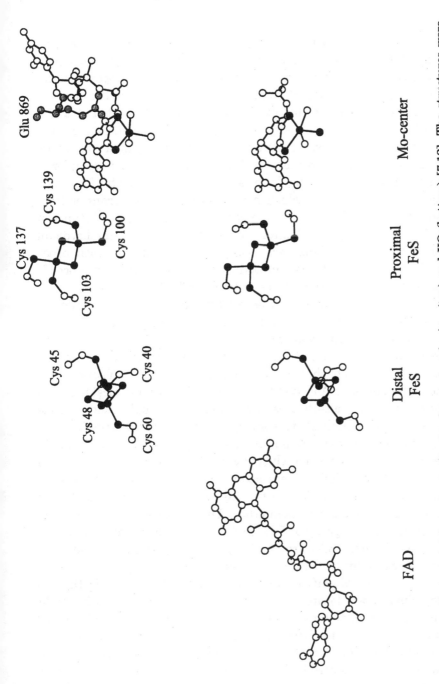

FIG. 2. Comparison of the arrangement of redox centers in Aor (top) and XO (bottom) [7,12]. The structures were extracted from the RCBS pdb-files 1alo and 1fiq. They are aligned such that the Mo, the dithiolene sulfurs, and the adjacent carbons are superimposable by a parallel shift. Molybdenum and the iron atoms are colored black, the sulfurs are represented by dark gray spheres, and glutamate is shaded in light gray.

(black sphere) in trans position to the dithiolene sulfurs (dark gray spheres) of the MCD cofactor. In the apical position the oxo ligand was replaced by sulfur in the resulfurated crystals. In the opposite apical direction a glutamate (Glu869, light gray in Fig. 2), which is conserved in other sequences of Mo-containing hydroxylases, is located at a distance of 3.5 Å to the molybdenum. It was suggested that this amino acid is interacting with the water bound to Mo and may play a role in the catalytic mechanism of enzymes of the class [8,19,20]. The cofactor is kinked by about 40° and extends toward the proximal cluster forming a hydrogen bond to Cys139, which may facilitate electron transfer from the Mo cofactor. The center of the proximal FeS cluster has a distance of about 16.2 Å to the molybdenum and is 13.4 Å away from the distal cluster. The [2Fe2S] planes are nearly perpendicular to each other (93°). Electron transfer between both FeS clusters probably proceeds via the hydrogen-bonding network in between [7,8,19,20].

For a direct comparison, the arrangement of the redox centers obtained from the X-ray structure of XO (bovine milk) [12] is given in Fig. 2 (bottom) in such a way that Mo, the dithiolene sulfurs, as well as the adjacent carbons of the pterin ring of both structures are superimposable by a parallel shift. In XO the Mo cofactor is MPT, also showing the distortion as in Aor. A sulfur was identified as an apical ligand to molybdenum, which also ligates two oxygens in a double and a single bond, respectively. Similarly to Aor a glutamate (Glu1261) is close to Mo in trans position to the apical sulfur (not shown). The distance from Mo to the center of the proximal cluster is about 16.1 Å and that between the centers of both FeS clusters about 13.7 Å. Both clusters show an arrangement nearly identical to that in Aor. The [2Fe2S] planes also are almost perpendicular to each other. The distances of the distal FeS cluster to the closest group of FAD and to the center of the ring system amount to 8.0 Å and 13.7 Å, respectively. Cox is composed as a dimer of heterotrimeric subunits $(LMS)_2$, with each subunit carrying a subset of redox centers. The Mo-MCD cofactor is found in the L subunit, the two [2Fe2S] clusters are located in the S and the FAD is located in the M subunit. In spite of this more complex subunit composition of Cox, as opposed to the homodimer subunit structure of Aor and XO, the relative positions of the redox centers extracted from the X-ray structure (not shown) are almost identical to those found for XO. Also, the intercenter distances are very similar.

It appears from this comparison that the arrangement of the redox centers can be expected to be very similar in other XO family enzymes for which a structure has not yet been obtained (as is the case for Ior, Qor, and Qox). Substrate specificity and regioselectivity of the hydroxylation reaction should thus mainly be governed by the composition of the binding pocket in the different enzymes. Similarly, the redox properties of the centers in various enzymes should be a matter of fine tuning by their environment. Therefore, the subsequent discussion of EPR results will refer to these structures.

4. EPR AND ENDOR CHARACTERIZATION OF REDOX CENTERS

The structural information obtained by X-ray crystallography of Aor and Cox as well as very recently of XO has directly corroborated the concept of spatial separation of the oxidative and reductive half cycles in Mo-containing hydroxylases [1,6,54,55]. The substrate is oxidized at the molybdenum center and electrons are then transferred via the two [2Fe2S] clusters to the flavin moiety, where the reduction of molecular oxygen takes place in the case of XO and Qox. For Ior, lacking the FAD, and Qor, which contains a FAD center but does not reduce oxygen, the electrons are transferred to unknown final electron acceptors. During the catalytic reaction the various redox centers cycle through several oxidation states, some of which are paramagnetic and, hence, are selected by EPR spectroscopy (cf. Scheme 2).

This selectivity is also maintained in case of fourfold or even sixfold reduced enzyme states encountered under conditions of excess of substrate (or generally of a reducing agent). For other spectroscopic methods like EXAFS or UV-Vis the superposition of multiple enzyme states often impedes a straight forward interpretation. In this respect EPR has become a valuable tool in the investigation of Mo hydroxylases. In conjunction with other spectroscopic techniques, e.g., EXAFS or resonance Raman, and rapid kinetic methods a very detailed picture on the structure of the molybdenum center, the intramolecular electron transfer between the redox centers and the models of the reaction mechanism have been developed for XO [1–6,54–56]. In the following sections we

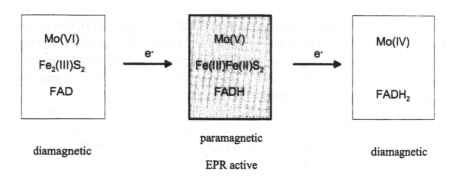

Scheme 2. The paramagnetic states appearing during the enzymatic cycle.

want to present the major EPR results obtained from the three bacterial enzymes Qor, Qox, and Ior giving reference mainly to XO, i.e., adopting the same nomenclature for corresponding species and their EPR signature [57–59].

4.1. EPR Signals of Redox Centers and Comparison

There are several different EPR signatures in XO that involve the Mo(V) oxidation state. Most of them are also found in Qor, Qox, and Ior. The so-called "*rapid*" species is readily induced by incubation with an excess of substrates, but also with dithionite. With substrate added in substoichiometric amounts no "*rapid*" signal was detected [57,58]. Generally, the "*rapid*" species exhibits a nearly axial g-tensor and a proton coupling, which is exchangeable in deuterated buffer (Fig. 3A, a,d). The EPR parameters of this species are listed in Table 4 together with those of XO for comparison. The exchangeable proton coupling in XO was associated with the presence of an SH group at the Mo(V) center [60,61]. In Qor the proton splitting is also resolved on top of the ^{95}Mo ($I = 5/2$) hyperfine split EPR signal obtained from an isotope-enriched sample (Fig. 3A, b,c,e,f). The EPR simulations yielded the parameters listed in Table 4 that require a rotation of the ^{95}Mo-hyperfine tensor of 25° around a common axis with the g-tensor [58]. Similar results have been reported for XO [62]. It is noted here that experiments on Mo centers substituted with paramagnetic nuclei (^{17}O, ^{33}S, ^{31}P) [63–69]

TABLE 4

EPR Parameters[a,b] of Mo(V) Centers, FAD Radicals, and FeS Centers

Center[c]	Qor	Qox	Ior	XO[d]
Rapid	g_1: 1.98801 g_2: 1.9679 g_3: 1.9650 ^{1}H: $A_\parallel$: 1.08 $A_\perp$: 1.36 ^{95}Mo: A_1: 6.57[e] A_2: 2.85[e] A_3: 2.70[e]	$g_\parallel$: 1.992 $g_\perp$: 1.966 nr	g_1: 1.9900 g_2: 1.9693 g_3: 1.9601 ^{1}H: A_1: 1.33 A_2: 1.75 A_3: 1.70	g_1: 1.9906 g_1: 1.9707 g_3: 1.9654 ^{1}H: A_1: 1.20 A_2: 1.30 A_3: 1.30
Very rapid	g_1: 2.0244 g_2: 1.9576 g_3: 1.9500	g_1: 2.0243 g_2: 1.945 g_3: 1.935	g_1: 2.0243 g_2: 1.9545 g_3: 1.9411	g_1: 2.0252 g_2: 1.9540 g_3: 1.9411
Resting 1, (2)	g_1: 1.991 g_2: 1.976(0) g_3: 1.954(63)	Not detected	Not detected	
Slow	g_1: 1.9663 g_2: 1.9656 g_3: 1.9581 ^{1}H: A_1: 1.60 A_2: 1.65 A_3: 1.70	Not investigated	Not detected	
FAD	g: 2.0034 LW: 1.91	g_1: 2.00407[f] g_2: 2.00353[f] g_1: 2.00205[f] LW: 1.60[f]	Not Present	g: 2.0035 LW: 1.9
FeSI	g_1: 2.035 g_2: 1.948 g_3: 1.898	$g_\parallel$: 2.021 $g_\perp$: 1.937	g_1: 2.004[g] g_2: 1.941 g_3: 1.914	g_1: 2.022 g_2: 1.935 g_3: 1.899
FeSII	g_1: 2.072 g_2: 1.973 g_3: 1.866	g_1: 2.075 g_2: 1.983 g_3: 1.874	g_1: 2.089 g_2: 1.980 g_3: 1.903	g_1: 2.12 g_2: 2.007 g_1: 1.91

[a]The parameters for Qor and Ior have been verified by simulation of the EPR spectra; the values for Qox are the apparent values.
[b]The hyperfine values A are given in mT, nr means "not resolved", LW is the linewidth in mT.
[c]The nomenclature for the centers was adopted from analogies to XO.
[d]Data from [4,60,65,79,97].
[e]g- and A-tensor are non-coaxial (25° between the maximal components).
[f]g-factors are obtained from multifrequency simulations, the line width is for X-band (9.5 GHz).
[g]Refers to the unsplit signal.

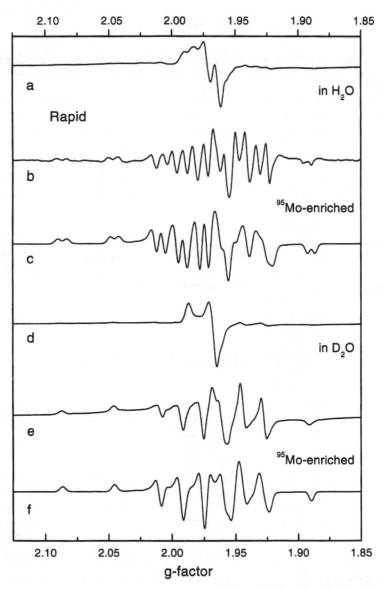

FIG. 3A.   The EPR spectra of the "*rapid*" species in Qor in H_2O (a) and D_2O (d) obtained at 77 K show the loss of an exchangeable proton coupling. The spectra of the ^{95}Mo-enriched samples as well as their simulations are given in traces b, c and e, f, respectively. The simulation parameters are compiled in Table 4.

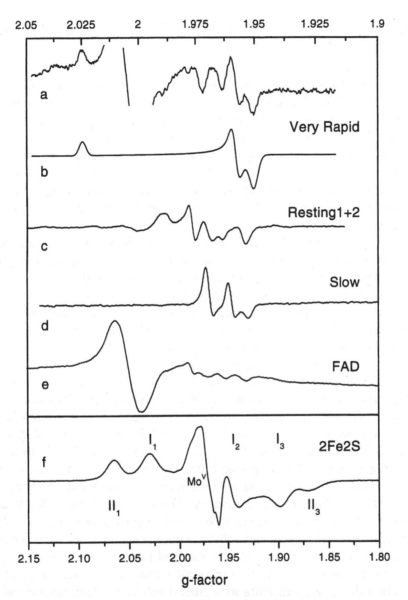

FIG. 3B.   In Qor the *"very rapid"* signal is observed at 77 K upon competitive reaction with quinoline and the nonsubstrate quinaldine in (a) after subtraction of the *"rapid"* signal contributions. Its simulation is shown in (b). Small amounts of inactive enzyme in the preparations give rise to the "resting" signals (c). The desulfo form of the enzyme causes the appearance of the "slow" signal (d). The semiquinone radical of FAD is characterized by a singlet line at $g = 2.004$ (e). At lower temperatures (< 60 K) the signals FeSI and FeSII of the [2Fe2S] clusters appear exhibiting different g-anisotropies and relaxation behavior (f). Their principal g-components are indicated. The spectrum also shows superposition with saturated Mo(V) signals.

as well as with isotope-labeled substrates (^{2}D, ^{13}C) [70,71] and on model compounds [72,73] disclosed details of ligation of the *"rapid"* species and other signal-giving species. In particular, the absence of a ^{13}C coupling of labeled substrate in the *"rapid"* species together with its appearance upon reduction with dithionite proved that no substrate is coordinated [70,71]. Hence, the signal-giving species for the *"rapid"* EPR spectrum appearing under excess of substrate is thought to be a paramagnetic Michaelis-Menten complex with substrate in the active site, but not directly bound [2].

The *"very rapid"* signal could hardly be detected in Qor and Qox after incubation with substrate and manual freezing. Taking advantage of the competitive inhibition of the nonsubstrate quinaldine and successive addition of quinoline to Qor allowed this species to accumulate sufficiently. The difference spectrum with the *"rapid"* species nearly completely subtracted shows the more anisotropic *"very rapid"* signal and its simulation in Fig. 3B, a,b. Typically, this species lacks a proton hyperfine interaction but shows only g-anisotropy [58]. In contrast, the *"very rapid"* species in Ior could easily be observed in comparable intensity as the *"rapid"* signal after addition of the substrate isoquinoline. In repeated freeze-thaw cycles the signal persisted for more than 1 hour, which was initially interpreted by the formation of an inhibited state of the enzyme similar to that observed for alloxanthine in XO [58,66]. The peculiar behavior of this species in Ior and the mechanistic and kinetic implications were subsequently clarified in fast kinetic experiments (see Sec. 4.3). The *"very rapid"* species in XO has been demonstrated to be a short-lived catalytic intermediate, which appears immediately after reaction with substrates and substrate analogues on time scales ranging from 10 ms to several seconds. No resolved proton hyperfine splitting is present in this species, its characteristic g-factors are given for comparison in Table 4. Experiments with labeled substrates have clearly shown that the substrate is bound to the Mo(V) center in this species [2,4,54,71,74–76]. In Qor a minor portion ($< 5\%$) of the enzyme consists of nonfunctional enzyme giving rise to the *"resting"* signals (Fig. 3B, c; Table 4), which are seen in untreated enzyme or in the desulfo form. Such preparation artefacts are also found in other Mo-containing hydroxylases, e.g., Aor [77]. They will not be considered further.

The *"slow"* signal (Fig. 3B, d) appears with some delay after inactivation of the enzyme with cyanide and subsequent reduction with dithionite. It has a clearly smaller g-anisotropy in comparison with, for example, the "*rapid*" signal and exhibits an exchangeable proton coupling of an OH group [57,58]. In the cyanide reaction the sulfur is replaced by an oxo group yielding the dioxo form, which is reversed to the monooxo-monosulfido form by addition of sulfides restoring the activity. This behavior is also found in other Mo-containing hydroxylases and represents a characteristic feature of this family of enzymes [54,78].

The chemical structures thought to be responsible for the different Mo(V) species are shown in Scheme 3. One should note that the mode of substrate binding in the catalytically competent "*very rapid*" species is currently under strong debate (for details, see Scheme 7 in Sec. 4.3) [2]. In the catalytic cycle the FAD cofactor adopts the radical semiquinone state producing the typical isotropic EPR signal at X-band frequency (9.5 GHz) close to the free electron g-factor (Fig. 3B, e). In Qor and Qox two different linewidths have been observed (Table 4) which can be assigned to the neutral and anionic FAD radicals, respectively [57,59,79]. In a multifrequency EPR study the intense FAD signal of Qox was investigated at several microwave frequencies ranging up to 284 GHz to determine g and hyperfine contributions to the linewidth. Only at 284 GHz could the intrinsic symmetry of the anionic FAD radical be resolved (Fig. 4). The parameters of the slightly rhombic g-tensor are listed in Table 4. Further high-frequency studies on various FAD species (or Qor, XO, etc.) may elucidate characteristic spectral differences, which should be related to specific properties of their protein environment.

Another feature common to all hydroxylases of the XO family is the presence of two distinct [2Fe2S] centers which give rise to two EPR

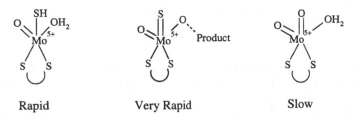

Scheme 3. Structures of the molybdenum center giving rise to the various Mo(V)-EPR species.

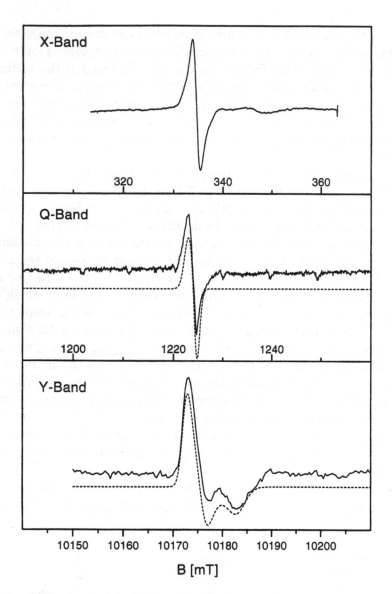

FIG. 4. EPR spectra of the FAD signal of Qox recorded at different frequencies: X-band, 9.5 GHz (top), Q-band, 34 GHz (middle), and Y-band at 284 GHz (bottom). Only at the highest frequency is the symmetry of the g-tensor of FAD resolved. The dashed lines represent simulations with the parameter set of Table 4. Spectra were recorded at about 10 K under nonsaturating conditions.

signals with different g-anisotropy and relaxation behavior (Table 4). The less anisotropic center is named FeSI, the other FeSII. In Qor, Qox, and Ior the FeSI center usually can be observed at temperatures around 60 K, whereas FeSII appears only below 45 K and shows an increased linewidth compared with FeSI. In the three proteins significant differences were found for the FeS clusters. While the two FeS centers of Qor (Fig. 3B, f) closely resemble those of XO, the FeSI cluster in Qox produces an axial EPR spectrum. In Ior a pronounced magnetic interaction between the FeS clusters was observed (see Sec. 4.4) [57–59]. In the X-ray structure of Aor and Cox two variant binding motifs of the clusters in the different locations were described [7,9]. For the one located close to the surface in Aor and, respectively, close to the subunit surface in Cox, a binding motif related to the spinach-type ferredoxin center was obtained. For the other cluster, which is near the Mo-MCD cofactor, a unique and unusual motif was found (see also Sec. 3). Until recently, the assignment of the FeSI and FeSII EPR signals to the two structural types of clusters was unresolved (see Secs. 4.3 and 4.4).

4.2. Redox Properties

In molybdenum-containing hydroxylases intramolecular electron transfer (and correlated proton transfer) occurs from the molybdenum center to the flavin or to another acceptor [80]. A fundamental property of each participating center is its corresponding redox potential, which is a determinant for the distribution of electrons and the kinetics of the system under consideration. (Short surveys of different methods to measure redox potentials are found, for example, in [81–83].) Due to its spectral selectivity for the various paramagnetic redox centers EPR has contributed significantly to the determination of redox potentials for, say, XO or Xdh [84–86] and has confirmed the rapid equilibrium model, in which electrons (or reducing equivalents) are distributed among the redox centers according to their relative redox potentials [87].

The three enzymes Qor, Qox, and Ior were subjected to a series of EPR redox titrations in order to obtain the individual potentials for a comparison with results from other hydroxylases. Some selected spectra for which the enzymes were poised at a certain redox potential, quickly frozen in liquid nitrogen, and then measured at 25 K are compiled in Fig. 5. At this temperature the two [2Fe2S] clusters were monitored. The spectra indicate immediately that the FeS clusters behave very differently in the three enzymes (the nomenclature I_i, II_i with $i = 1, 2, 3$ refers to the rhombic g-tensor components of center FeSI and FeSII of Ior and Qor, $i = $ ax1 or ax to the axial FeSI center of Qox). While for Qor both centers FeSI and FeSII are present for nearly all redox values (top), only center FeSII is visible at higher values in Qox (middle). On the other hand, in Ior at high redox potentials ($+60\,$mV) solely center FeSI is observed, which gradually develops a split feature (I_{1a}, I_{1b}) at more reductive potentials when center FeSII is growing in intensity (bottom). The splitting of component FeSI also appears in fully or partially reduced enzyme when the temperature is lowered from 70 K to 20 K where center FeSII becomes pronounced [58]. Both findings prove that a considerable magnetic interaction between FeSI and FeSII is prevailing in Ior. The redox data were fitted with the Nernst equation to extract the midpoint redox potentials of the clusters (Fig. 6), which are compiled in Table 5 for comparison with other hydroxylases. The midpoint potentials in Qox have the same order of values for FeSI and FeSII as in XO, but they are separated by 180 mV. In Qor and Ior the order is reversed, but there is only a small difference between the potentials of FeSI and FeSII. However, the midpoint potentials of Ior are unusually high in the positive range.

Comparable high values have not been found in any other Mo-containing hydroxylase. Although the same setup of two distinct FeS clusters is present in the enzymes of Table 5, the redox potentials are obviously tuned within each system. The determinants of this adjustment of redox potentials particularly for [2Fe2S] clusters are not well understood. The influence of solvent interaction, of hydrogen bonds to the cluster, as well as of electric dipoles caused by the protein backbone or charged side chains even at larger distances to the clusters may have to be considered. The rather large EPR linewidth encountered in studies of [2Fe2S] clusters in Mo-containing hydroxylases but also in classical plant or adrenodoxin-type ferredoxins is indicative of considerable structural flexibility which may be involved in the unusual potentiometric and spectroscopic properties of these moieties [88]. A detailed statistical analysis of the redox data of the FeS center of the three proteins showed that the fits of the Nernst curve yield significant deviations of the

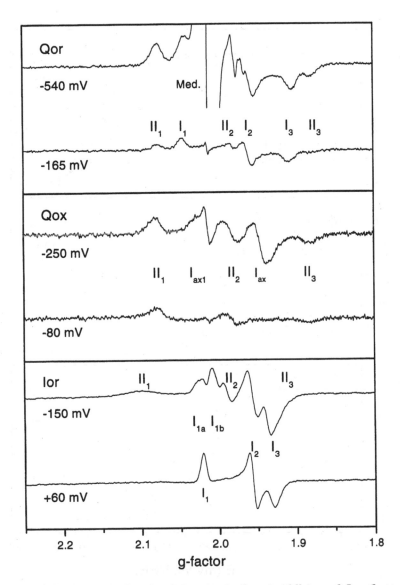

FIG. 5. EPR spectra (X-band) of Qor (top), Qox (middle), and Ior (bottom) poised at the potential indicated. For Qox and Ior a single FeS signal, FeSI for Ior and FeSII for Qox, can be isolated at certain redox potentials. For Ior the loss of the splitting (lines I_{1a}, I_{1b}) at high potentials (60 mV) becomes apparent when FeSII is in the oxidized state. In Qor the central part is overlapped by a strong signal of mediators (Med.) in radical form. All spectra were recorded at 25 K under nonsaturating conditions.

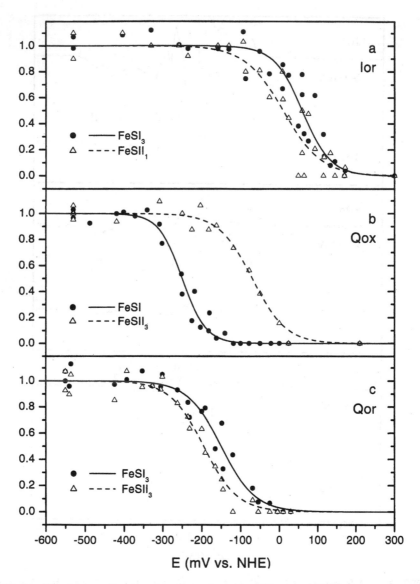

FIG. 6.   The degree of reduction of the FeSI (filled dots) and FeSII centers (open triangles) is plotted as a function of the redox potential with respect to standard hydrogen electrode for Ior (a), Qox (b), and Qor (c). The potentiometric titrations were performed in Tris-HCl buffer, pH 8. The data were fitted for the amount of reduced species formed in a single electron reduction process ($n = 1$) and a midpoint potential E_M (solid line, FeSI; dashed line, FeSII). The optimal fit values of E_M are compiled in Table 5.

TABLE 5

Comparison of Redox Potentials of Mo(V), FAD, and FeS Centers[a]

Redox center	Qor	Qox	Ior	XO[b]
Mo(VI/V)	−395	n.d.	n.d.	−345
Mo(V/IV)	−355			−315
FeSI	−155	−250	+65	−310
FeSII	−195	−70	+10	−217
FAD/H	−105	n.d.	—	−301
FADH/H$_2$	−105			−237

[a]The values are in mV and refer to SHE potential.
[b]The values are taken from [85].
n.d., not detected.

Nernst factor from the theoretical value of 59 mV for one-electron redox processes. For example, the fits for both centers of Qor give a Nernst factor of 100 mV or, in other words, the slopes of the Nernst curves are smaller than expected [58]. With a model that takes into account possible microheterogeneity, a standard deviation of the redox potential is defined which decreases the slope for an optimal fit to the experimental data. Assuming for example a standard deviation of 40 mV, the evaluation of the integral produces a curve that corresponds to a Nernst factor of 80 mV close to the experimental values [58]. Such a distribution of midpoint potentials might be modulated by a structural flexibility of protein domains in which the redox center is located and may play a functional role.

In an analogous manner, the redox titration of the molybdenum and FAD centers was performed and the EPR were spectra recorded at 77 K to avoid superposition with FeS signals. Because of the equilibrium between three oxidation states, intensity maxima for these centers appear during the variation of the potential of the solution. In an extension of our work [58], where the mean midpoint potential for the "*rapid*" signal of Qor was found to be −390 mV, the individual potentials of each redox pair E[Mo(VI)/Mo(V)] and E[Mo(V)/Mo(IV)] were derived from a quantitative analysis of the converted portion of Mo(V) with respect to the total concentration of Mo determined spectrophotometrically (Table 5). The potential of the second reduction step $E[Mo(V)/Mo(IV)] = -355\,mV$ is somewhat higher than that of the first $E[Mo(VI)/Mo(V)] = -395\,mV$. This implies that the Mo(V) state formed is slightly disproportionating to Mo(IV) and Mo(VI) in the presence of mediators [89]. A quantitative analysis of the redox equations shows that in case of equal potentials of both redox pairs maximally 33% of Mo is converted to the Mo(V) state [81,89]. In a situation where the second reduction step $E[Mo(V)/Mo(IV)]$ is higher than $E[Mo(VI)/Mo(V)]$, the Mo(V) state is formed in even lower relative concentrations. For Qor double integration of the Mo(V) signal yields about 20% of the total Mo concentration consistent with the order of the redox pairs. Apart from Aor, this order of potentials seems to be valid for most enzymes of the XO family (see Fig. 7 below).

In this context, the surprising result that nearly no Mo(V) signals could be detected in redox titrations of Ior and Qox [58] can be explained. Obviously, the potentials of the two redox pairs are even further apart as in Qor, which ultimately leads to very low concentrations of the Mo(V) state. For a difference of about 100 mV between the two potentials, less than 5% of Mo is converted to Mo(V). In Qox very small *"rapid"* signals in the range -300 to $-400\,mV$ could be detected, for which a quantitative evaluation was impossible because of the rather low total enzyme concentration available [89].

For the titration of the FADH radical signal a mediator composition was chosen to avoid formation of mediator signals which appear around $g = 2$ overlapping with the FAD signal. The quantitative determination yielded a redox potential for both redox couples of $-105\,mV$ (Table 5), which is significantly higher than that in XO. The measurements of the FAD potentials in Qox have not yet been completed. In addition, for the desulfo form of Qor the Mo(V) redox potential was measured with $-330\,mV$, which is even higher than that measured for the *"rapid"* signal. This indicates that the inactivity of the desulfo form is not caused by a decrease in its redox potential but by the modification of the Mo ligation to the dioxo form, which impedes the proton transfer to the sulfido ligand. This implies that the function of the sulfido ligand is not solely to modulate the redox potential of the Mo cofactor [80].

The potentials of all four redox centers of Qor are graphically compared with those of Aor, Xdh, and XO in Fig. 7. Similar to Aor [90], Qor provides potentials that are quite remote from each other

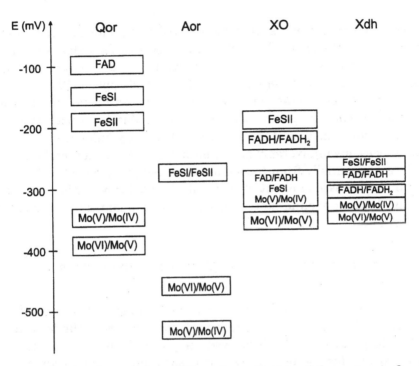

FIG. 7.   Graphical comparison of the redox potentials of the centers in Qor to those of Aor, XO (cow's milk), and Xdh (from chicken liver) showing the different spread in potential of the centers for the various enzymes. For Qor, Aor, and Xdh the redox potentials were determined by potentiometric titration and low-temperature EPR [58,86,90], for XO by room temperature CD spectroscopy [85].

and ordered in the direction of electron flow from the molybdenum center to FAD or to another electron acceptor. In this picture the exceptionally high values of the FeS centers in Ior must be considered, which again raises the question about the identity and properties of the final electron acceptor in this enzyme. For XO or Xdh the potentials are more closely correlated. The Mo potentials are also lying lowest in these cases, but the FAD potentials are partly overlapping with FeS potentials. The mode of reaction of XO is described by the formation of a rapid equilibrium under turnover conditions that finally is resolved by reoxidation of the enzyme [2,87]. For Qox similar behavior is observed, where under strict anaerobic conditions EPR signals of all redox centers can be detected. In the presence of the final electron acceptor oxygen the oxidative counterreaction is shifting the redox equilibrium such that mainly the FAD signal is left [58]. With respect to the redox potentials some caution is required, since significant differences, particularly for FeS clusters, are noted depending on the method applied [84,85].

It should be mentioned here that the product-bound Mo state of the *"very rapid"* species is not directly accessible by redox titration. However, complexation of the product will strongly influence its potential (as does any unfavorably bound mediator dye), which therefore may deviate substantially from the values measured in vitro for the *"rapid"* signal. An increase of about 90 mV and 180 mV for both Mo potentials relative to the unbound state was discussed in quantitative studies of XO [87].

4.3. Kinetic Investigations and Mechanistic Implications

The development of rapid-freeze quench techniques has allowed to resolve the first steps in the reactions of substrates with the Mo center in XO. In particular, the short-lived *"very rapid"* species was initially observed by this method [91–93] and subsequently characterized in detail by EPR of isotope-labeled ligands and substrates [63,65,67,68]. In analogy to XO, such kinetic studies were performed on Qor, Qox, and Ior in order to demonstrate that the reaction mechanism of XO is valid for Qor and Qox and also to clarify the origin and mechanism of the unusual longevity of the *"very rapid"* species found in manual freeze experiments of Ior (see Sec. 4.1). For that purpose, rapid-freeze samples under turnover conditions, i.e., in the presence of a final electron acceptor (ferricyanide was chosen), or under nonturnover conditions were produced for EPR measurements at 77 K and 30 K to select for Mo(V) and FeS signals [59]. The *"very rapid"* species of Qor together with the FADH signal could be trapped in considerable amounts within 10 ms after reaction with an excess of quinoline (Fig. 8a). At lower temperatures, contributions of both FeS centers I and II are visible indicating that the electrons are distributed over all redox centers. It is noted that a small splitting (0.9 mT) is observed on top of the g_2 component of the *"very rapid"* signal (Fig. 8b). Similarly, after 10 ms reaction time with quinaldine the *"very rapid"* species and FADH were found in Qox at 77 K. At 30 K only the features of the FeSII center were present (Fig. 8c,

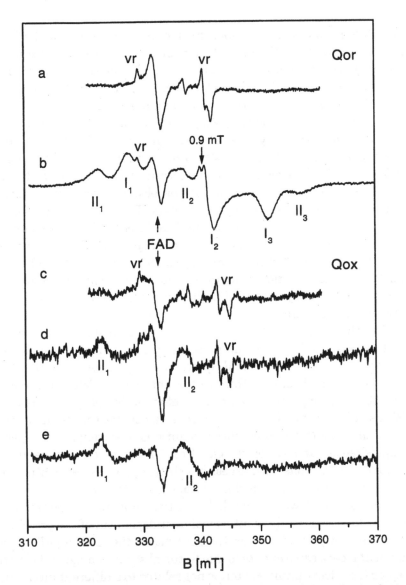

FIG. 8.   Rapid-freeze EPR spectra (X-band) of Qor and Qox obtained 10 ms after reaction with substrates (a: Qor + quinoline, at 80 K; b: as a but measured at 30K; c and d: Qox + quinaldine at 80 K and 30 K, respectively; e: Qox + isoquinoline measured at 20 K). The spectra at 80 K (a and c) show the "*very rapid*" species together with the FAD signals. At lower temperatures both FeS signals FeSI and FeSII are present in Qor (b) together with a splitting of the g_2 component of the "*very rapid*" signal, marked with an arrow. In Qox only FeSII is visible (d,e) with no indication of a splitting of the "*very rapid*" signal (d). This species does not appear in the isoquinoline-reduced sample (e).

d), which is in line with the much higher redox potential for the FeSII center of Qox, whereas the FeSI cluster remains in the oxidized state (cf. Table 5). Interestingly, after reaction with the substrate isoquinoline the *"very rapid"* signal could not be detected in Qox (Fig. 8e). This observation suggests that either different kinetic parameters are valid for isoquinoline or the resulting Mo-product complex has a considerably lower potential rendering the molybdenum in the hexavalent state. These experiments on Qor and Qox confirm that, in complete analogy to XO, a doubly reduced enzyme state with product bound to the Mo center E(2e)-P is prevailing in which the electron distribution follows the relative redox potentials of the centers. Due to the much smaller difference in potential (40 mV) for FeSI and FeSII of Qor as compared with Qox, both FeS clusters are reduced. A spectral comparison of the relative line intensities of the g_1 and g_3 components of FeSI and FeSII in a rapid freeze sample and a fully (dithionite) reduced sample shows that FeSII is diminished by one-third in the former, which is compatible with the potential difference.

The time course of the reaction of Ior with a slight excess of iso-quinoline was monitored in the presence of ferricyanide as an electron acceptor. The rapid-freeze EPR spectra at 80 K show that the *"very rapid"* species is formed within 10 ms and can be detected for more than 220 ms in appreciable amounts before it completely vanishes at 2000 ms (Fig. 9a, b). No *"rapid"* signal as observed in manual freeze experiments is formed under these conditions. For comparison, spectra of rapid-freeze samples produced after 10 ms reaction time in the absence of ferricyanide with excess of substrate (i.e., under nonturnover conditions) showed the immediate formation of the *"very rapid"* signal (Fig. 9c). Hence, it was concluded that the *"very rapid"*species in Ior behaves like a true catalytic intermediate independent of the presence of an electron acceptor. Moreover, the longevity of the *"very rapid"* species in manual freeze experiments is due to the absence of a natural electron acceptor (i.e., decay pathway) but is not related to a different mechanism of formation of the signal-giving species [59].

When the temperature of rapid-freeze samples quenched at 10 ms is lowered to 30 K, the EPR spectra are dominated by the FeSI signal with no or only minor traces of FeSII being visible (Fig. 10a, c). Due to the absence of the FeSII signal the g_1 component of FeSI remains unsplit, in contrast to dithionite-reduced samples with both FeS centers

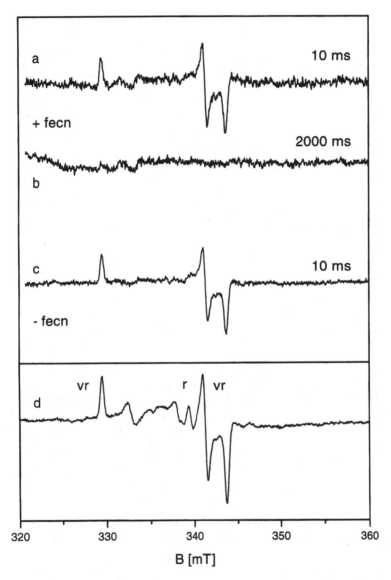

FIG. 9.   Rapid-freeze EPR spectra (X-band) of Ior obtained after reaction with 2.5-fold excess of substrate isoquinoline in the presence of the artificial electron acceptor potassium ferricyanide (fecn) at 10 ms (a) and 2000 ms (b) and in the absence of the electron acceptor (c) at 10 ms. For comparison the manual freeze spectrum (4 min reaction time) in the absence of an artificial electron acceptor shows both the *"very rapid"* (vr) and the *"rapid"* (r) species in spectrum (d). All spectra were recorded at 80 K.

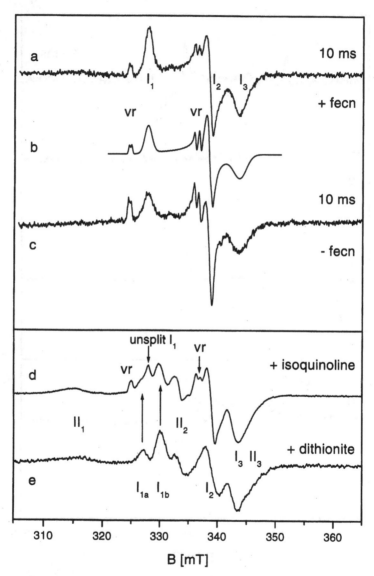

FIG. 10. Rapid-freeze EPR spectra (X-band) recorded at 30 K after reaction of Ior with substrate isoquinoline in the presence (a) and absence (c) of the artificial electron acceptor potassium ferricyanide (fecn). The FeSI species is present in sizable intensity, while for FeSII only marginal signals are detected. The g_2 component of the *"very rapid"* signal (vr) shows a dipolar splitting of 0.8 mT. The interaction spectrum of FeSI and the *"very rapid"* species is simulated in spectrum (b). A manual-freeze spectrum (d, 15 K) of Ior reacted with isoquinoline (4 min) is compared with the spectrum of a fully (dithionite) reduced sample (e, 20 K) to show the presence of unsplit FeSI and split *"very rapid"* signal (arrows).

reduced (Fig. 10e). On the other hand, a clearly resolved splitting of the g_2 component and a broadening of the g_1 line of the *"very rapid"* signal appear in the low-temperature spectra, indicative of a magnetic inter-action, which will be discussed in more detail in Sec. 4.4. The kinetic EPR experiments demonstrate that at 10 ms reaction time the enzyme has acquired two electrons from the substrate [E(2e) state] yielding the Mo(V) and FeSI signals. Only a small occupancy of FeSII is present (e.g., Fig. 10a), which in principle agrees with the lower redox potential of FeSII ($+10\,\text{mV}$ vs. $+65\,\text{mV}$ for FeSI).

For a quantitative description of the rapid-freeze experiments of Ior under turnover conditions, the following reaction scheme was analyzed.

$$ E + S \underset{k_{-1}}{\overset{k_{+1}}{\rightleftharpoons}} ES \xrightarrow{k_{+2}} E(2e)\text{-}P \xrightarrow{k_{+3}} E + P $$

Scheme 4. Model of the reaction mechanism of Ior turnover in the presence of ferricyanide as an artificial electron acceptor. E, enzyme; S, substrate; P, pro-duct; ES, Michaelis-Menten complex; E(2e)-P, species formed after transfer of two electrons from substrate to the enzyme, detectable as the *"very rapid"* signal.

The enzyme E and substrate S form a Michaelis-Menten complex ES, which reacts to the reduced enzyme state E(2e)-P. In this state the electrons are distributed thermodynamically controlled to the redox cen-ters which are manifest in EPR by the *"very rapid"* species and FeS signals. The enzymatic cycle is closed by a concerted electron transfer (to the acceptor) and product release. This last step of the kinetic scheme is rate limiting, so that most of the enzyme is present as E(2e)-P throughout the time range for which signals were observed during the turnover experiment. Based on the concentration of reac-tants and on the catalytic center activity of $6.8\,\text{s}^{-1}$ (k_{+3}) for the func-tional enzyme measured photometrically under identical conditions, substrate exhaustion would be expected after 290 ms (corresponding to about 2.5 turnovers), a value consistent with the data. The constants k_{-1} and k_{+1} are inversely related to the Michaelis-Menten constant of $K_M = 16\,\mu\text{M}$ [41] from which k_{-1} is calculated as $320\,\text{s}^{-1}$ by using a typical value of $k_{+1} = 20\,\mu\text{M}^{-1}\,\text{s}^{-1}$ for XO or Xdh [4]. The constant k_{+2} was chosen such that the reduced enzyme state E(2e)-P is formed within 10 ms, i.e., $2000\,s^{-1}$. The result of the time course simulation of the E(2e)-P state is shown in Fig. 11, which fits well to the data points. This simplified model does not take into account that ferricyanide only can pick up one electron at a time, so that two reaction steps are necessary to reoxidize the enzyme. It also requires that reoxidation and product release occur simultaneously.

Considering now the manual-freeze experiments, the spectra obtained after 4 min reaction time of isoquinoline and Ior in the absence of an electron acceptor can be fully explained. The EPR spectrum at 80 K shows the *"very rapid"* together with some *"rapid"* species (Fig. 9d), which is enhanced at higher pH values [59]. By lowering the temperature to 30 K a complex spectrum constituted of several species is recorded (Fig. 10d). Clearly, the FeSII center is formed at considerable concentrations. Its presence also leads to the magnetic splitting of the g_1 component of FeSI, which is compared by sticks to a fully (dithionite)

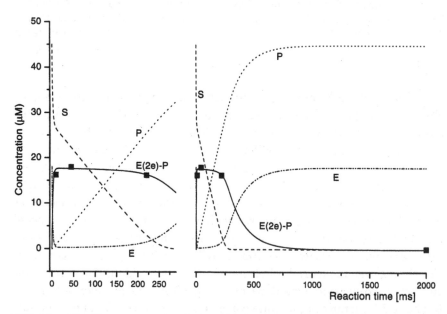

FIG. 11. The time course simulation of substrate conversion of Ior was derived for reaction Scheme 4. The reaction constants are given in the text. The concentration of the *"very rapid"* species (black squares, E(2e)-P) was normalized to the amount of functional enzyme (E) (18 μM). The initial substrate (S) concentration was 45 μM. The product formed is represented by curve P.

reduced sample displayed in Fig. 10e. In addition, an unsplit I_1 signal is visible arising from a population of molecules in which the FeSI center is reduced while FeSII remains in the oxidized diamagnetic state. The split signals I_{1a}, I_{1b} originate from enzyme molecules in which both FeS clusters are reduced and paramagnetic so that their dipolar interaction becomes apparent. Similar to the rapid-freeze spectra a splitting of the g_2 component of the *"very rapid"* species is less well resolved (0.8 mT, Fig. 10d).

The appearance and the origin of the signals observed in manual-freeze experiments in the absence of an electron acceptor is displayed graphically in Scheme 5. The unsplit FeSI species and the *"very rapid"* signal are associated with the product-bound E(2e)-P state, which decays very slowly leading ultimately to the persistence of the *"very rapid"* signal. The dissociation of product causes the appearance of the *"rapid"* signal with its characteristic proton hyperfine splitting. In this E(2e) state the two electrons may reside on Mo(V) and on one FeS center or, alternatively, on both FeS centers. In the latter case a split $FeSI_1$ component is produced. With excess substrate a new Michaelis-

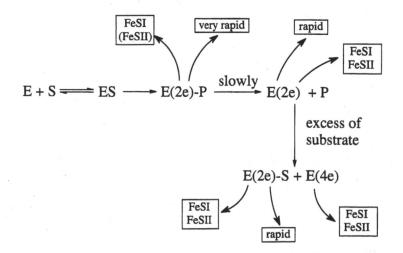

Scheme 5. Model for the reaction of Ior with substrate in the absence of an electron acceptor. The E(2e)-P species decays only slowly giving rise to the persisting *"very rapid"* signal. The EPR signals correlated to each step are indicated within the boxes.

Menten complex E(2e)-S may be formed, which yields the same signals
as the E(2e) state. When the reduced enzyme takes up two more electrons from the substrate a E(4e) state may result with two electrons
residing on a diamagnetic Mo(IV) state and both FeS centers reduced,
which contribute to the FeSII and split FeSI signals. It should be noted
that species with odd numbers of electrons, produced by disproportionation reactions or by reaction with the desulfo enzyme, may also contribute to a lesser extent to the spectra.

The restriction of the simplified reaction Scheme 4 mentioned
above did not allow for a direct application to stopped-flow experiments.
Hence, a more elaborate scheme had to be developed to fit the data,
particularly to account for an observed overshoot at the end of the
steady-state phase monitoring the absorption at 470 nm. To reproduce
the experimental absorbance changes, one- and three-electron-reduced
enzyme states E(1e), E(1e)-P, and E(3e)-P had to be included for a
satisfying simulation of the experiment. The refined model for enzyme
turnover in stopped-flow measurements is depicted in Scheme 6.

The reaction with ferricyanide opens up the possibility of an E(1e)-
P state, from which the product is released to yield E(1e). In a second
reaction with ferricyanide the original enzyme state E is restored or, by
reaction with a substrate molecule, the reduced states E(1e)-S and

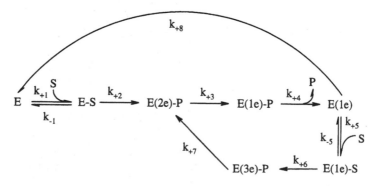

Scheme 6. Elaboration of Scheme 4 to account for stopped-flow data on enzyme
turnover. It incorporates separate reaction steps with two ferricyanide molecules, causing enzyme states with odd numbers of electrons [E(1e), E(1e)-P,
E(1e)-S, and E(3e)-P], of which E(1e)-P and E(3e)-P contribute to the "*very
rapid*" signal.

E(3e)-P are formed in an alternative subcycle. A detailed calculation of the electron distribution for the various states in Scheme 6 revealed that in E(1e)-P the electron is mainly on the Mo, while in E(1e) mostly on the FeS center [59,87]. The simulation of the time course was performed under the assumption of rapid equilibrium and statistical distribution of electrons controlled by the redox potentials of the individual centers. As the substrate becomes exhausted toward the end of the steady-state phase, the population of state E(1e) will transiently increase, which in turn is potentially responsible for the observed overshoot in the stopped-flow experiments [59]. As mentioned in Sec. 4.2, the redox potentials of the product bound state are not directly accessible but must be estimated or adjusted for calculations. The high quality of the fit of the stopped-flow data, however, indicates that the parameters used in the simulation are at least of the correct order of magnitude and that Scheme 6 is applicable to the action of Ior under turnover conditions.

An important piece of information deduced from the kinetic behavior of Ior is the finding that the decay of product-bound species is proportional to or related to the fraction of molybdenum in the hexavalent state. In addition, it was shown that the product does not, at least within 1 h, dissociate from the Mo(V) state of the enzyme but that it preferably dissociates from those molecules with the metal in the Mo(VI) state. Since it is generally assumed that all enzymes of the XO family act on their substrates by a common reaction mechanism, the implication of these findings of the two model mechanisms for XO currently debated in literature will be discussed.

The first model is based on a nucleophilic attack on the C8 atom of xanthine by a deprotonated water ligand of the molybdenum center followed by hydride transfer from the resulting tetrahedral intermediate to the sulfido ligand [2,14,19]. This results in a Mo(IV) species that converts to the "very rapid" intermediate after electron transfer to the FeS centers. The signal-giving species contains the product σ-bonded via the oxygen atom of the hydroxy group [Scheme 7 (1)]. The second mechanism postulates the addition of the $C8-H$ bond of xanthine across the Mo=S bond to give a Mo(VI) species with a Mo$-$C bond. The deprotonated water ligand of Mo subsequently attacks to give a side-on bound Mo(IV) carbonyl species which converts to the "very rapid" intermediate by electron transfer to the FeS centers [Scheme 7 (2)] [16,71,75]. It was stated that stable versions of such

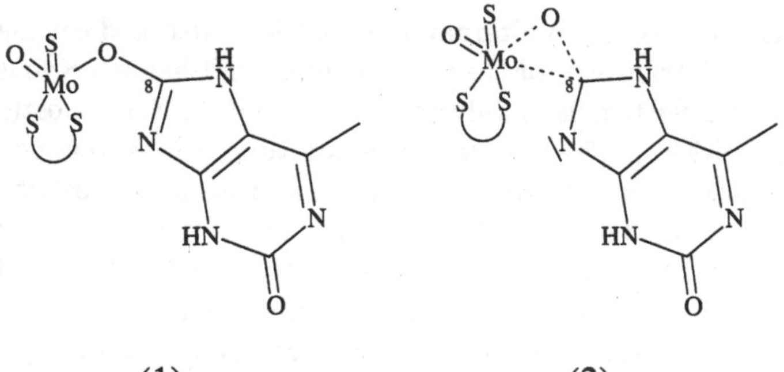

(1)   (2)

Scheme 7. Two models for the binding mode of xanthine to the molybdenum center in XO. In model 1, the product is bound via the oxygen atom of the hydroxy group. In model 2, it is bound side-on forming a carbon-molybdenum bond. The distance of C8 to molybdenum should be shorter than in the case of model 1.

side-on bound carbonyl species involve lower valence molybdenum centers. At the Mo(V) state the proposed side-on interaction can only involve a single electron and therefore must be very weak; at the Mo(VI) stage no back-bonding interaction is possible, so the product leaves. Thus, the second mechanism accounts very simply for the observation of a fast decay of the *"very rapid"* species of Ior, with product loss, on oxidation of Mo to Mo(VI). Conversely, the first mechanism provides no direct explanation for this phenomenon. Both models differ mainly in the distance between molybdenum and the substrates depending on the mode of binding. This question may be settled by a precise determination of the bond length in a product-bound (or inhibited) species employing spectroscopic or crystallographic methods [2,71].

4.4. Intercenter Interaction and Structural Aspects

Prior to the solution of the first three-dimensional structure of the XO family enzyme Aor [7] (see Sec. 3), some pieces of information on the relative arrangement of the redox centers had been provided by magnetic interactions observed in EPR spectroscopy. Generally, the magnetic interactions between two paramagnetic centers with spins S_A and

S_B consist of two contributions, the exchange and the dipolar interaction. The former is caused by a direct interaction of the orbitals of both centers and usually yields considerable, isotropic contributions for centers in close proximity. The dipolar part originates from the interaction of one center with the magnetic field induced by the other spin center and vice versa. It depends on the relative orientation of the individual magnetic moments (related to the g-tensor) with respect to the external magnetic field of the EPR experiment (angular dependence) and on the distance between the centers ($\sim r^{-3}$). A more refined local spin model has been developed for the description of the interaction between centers with a spatial distribution of spin population, which should be considered for adjacent centers [94]. When both the exchange contribution and the distribution of spin are negligible, the dipolar inter-center coupling is often sufficiently well described by the point-dipole approximation.

In the preceding sections several characteristic magnetic splittings of EPR signals have been mentioned that now will be analyzed with respect to structural aspects. The magnetic interaction between the Mo(V) *"very rapid"* signal and the FeSI center observed in rapid-freeze samples of Ior is of particular interest. It not only allows one to calculate the distance between the interaction partners but it proved unambiguously, for the first time, that the FeSI signal has to be associated with the FeS center proximal to the Mo cofactor [59]. The simulation of the interaction using a coaxial arrangement of g and an axial interaction tensor with 0.81 mT along g_2 is compared with the experimental pattern of the rapid-quench sample at 30 K in Fig. 10b. The *"very rapid"* signal shows the splitting at g_2, but it is not entirely resolved along the other principal components. Because of the much larger intrinsic linewidth of the FeSI signal (about 1.6 mT vs. 0.3 mT for *"very rapid"*), the interaction causes only a slight increase of 0.1–0.2 mT of the apparent linewidth in the FeSI pattern [59,89]. The distance between the Mo(V) center and FeSI in Ior is calculated with a value of 16.5 Å, which can be compared with the corresponding distance of 16.2 Å in Aor (Scheme 8) [7]. The almost exclusive presence of FeSI in the rapid-freeze samples (Fig. 10a, c) identifies the proximal FeS cluster of the crystal structure, e.g., in Aor, as the origin of the FeSI signal while the FeSII pattern has to be associated with the more remote distal FeS cluster. Additional support for this assignment was recently given by redox titration EPR experiments of Aor from *Desulfovibrio alaskensis* NCIMB 13491. Taking advantage of the gap between the redox potentials of both FeS clusters and the presence of a Mo(V) slow species, a population analysis was feasible that allowed assignment of the FeSI signal to the crystallographic FeS cluster proximal to the Mo-MCD in Aor [7,95]. The identical conclusion was drawn from the temperature dependence of the spin lattice relaxation of the FeSI and II EPR signals in Aor from *D. gigas* [96]. Thus it appears to be a general trait that the [2Fe2S] cluster giving rise to the FeSI EPR signal is proximal to the Mo cofactor.

In the low-temperature rapid-freeze spectra of Qor a very similar splitting (0.9 mT) as in Ior of the g_2 line of the *"very rapid"* signal was detected (Fig. 8b). However, in this case both FeS centers are present in the spectrum, so that a clear-cut assignment of the interaction is not possible. On the basis of spectral similarity to Ior it is suggested that the FeSI signal may correspond to the proximal FeS cluster at a distance of 15.9 Å from the Mo cofactor (Scheme 8). The same kind of plausibility argument can be used for the absence of a splitting of the *"very rapid"* signal in Qox (Fig. 8d) because here only the FeSII center is developed which was associated with the remote FeS cluster in Ior at a distance too far to induce a visible dipolar splitting [59,89]. A further dipolar splitting was observed for low-temperature spectra of the cyanide-treated *"slow"* species in Qor, which is characterized by an axial g-tensor and a proton splitting (see Sec. 4.1). A detailed simulation of the magnetic interaction in H_2O and D_2O yielded a maximal interaction value of 1 mT corresponding to a distance of 15.4 Å between the Mo(V) center and presumably the proximal FeS cluster giving rise to the FeSI signal. This value is close to the distance derived for the interaction between FeSI and the *"very rapid"* species (Scheme 8).

For the case of XO (from cow's milk), considerable evidence that the cluster causing the FeSI signal is also close to the Mo cofactor has been collected from analysis of dipolar interactions, temperature dependence, and saturation behavior of EPR signals [94,96–99]. For the splittings of the resting and Mo(V) signals limiting distances were estimated to be within 10 Å and 25 Å [97,98]. Considering long-range exchange effects a distance of 14 Å was obtained on the basis of line shape simulation for the crystalline case [100]. Applying the local spin model the distance between the Mo cofactor and the proximal cluster expanded to about 19 Å [94]. Site-directed mutagenesis and EPR of XO from rat

Distances from X-ray structures

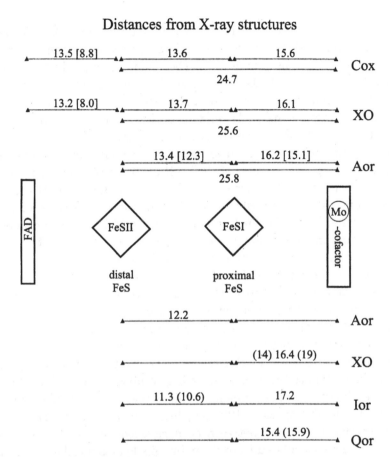

Distances from magnetic interactions

Scheme 8. Comparison of the intercenter distances derived from structural data (upper) of Aor, XO, and Cox and magnetic interactions observed in EPR of Aor, XO, Ior, and Qor (lower). The values are given in Ångstrom, the closest distances between the centers of the structures are set in square brackets. The values in parentheses represent distances either derived from interactions between different species or obtained by different calculation methods described in the text.

liver established the proximal cluster with its unusual binding motif as the origin of the FeSI signal, while the remote cluster with the plant-type binding motif was related to FeSII. The mutation Cys43Ser at an iron site of the FeSII cluster provoked changes in the EPR spectrum and its redox potential. According to the X-ray structure of XO, this iron shows the shortest distances to the FAD and to the proximal cluster [12,101]. The same assignment of the signal-giving clusters was achieved for Cox from FAD binding studies to the deflavo protein. Weak spectral changes affected the FeSII signal which therefore has to be located close to the FAD binding domain and remote from the Mo cofactor in agreement with the X-ray structure [9,10].

In fully (dithionite) reduced Ior lacking a paramagnetic Mo species a splitting of the g_1 component of the FeSI center was resolved at temperatures when the rather broad signals of the reduced FeSII center appeared (Fig. 10e). The absence of this splitting in the two-electron reduced enzyme state (E(2e)-P) produced in rapid-freeze experiments (Fig. 10a) and in the redox titration experiments (Fig. 5, bottom) has proven that the interaction is solely between FeSI and FeSII. Simulations of their interaction at X and Q band (10 and 34 GHz) yielded an axial dipolar tensor (D-tensor) with a maximal principal value of 3 mT assuming collinearity with the g_1 directions of the FeS centers [58]. From this value a distance of 10.6 Å is estimated with the point-dipole approximation, which is somewhat shorter than the distance of 12.3 Å measured in the related Aor crystal structure [7]. In this context it is noted that a g_1 splitting of about 2 mT of the FeSI center was observed in Aor, which primarily was assigned to an interaction with a Mo(V) species [102,103], but later on was clearly associated with the interaction between both reduced FeS clusters in Aor [20]. The distance calculated from the point-dipole model yields 12.2 Å in agreement with the crystallographic value (Scheme 8). This congruence can be taken as an indication that exchange interactions are indeed rather small, so that the distance between the FeS clusters determined for Ior may only be slightly overestimated.

Some novel aspects concerning the interaction of FeSI and FeSII in Ior can be derived from an analysis of manual-freeze spectra. Typically, after reaction with isoquinoline the low-temperature spectra show the persisting *"very rapid"* signal and the simultaneous presence of split and unsplit FeSI features (Fig. 10d). When such a sample is subsequently reduced with dithionite in order to induce full reduction the well-resolved pattern in Fig. 12a is observed at 15 K. Compared with the dithionite-reduced samples (Fig. 10e) the lines exhibit a smaller width, particularly for the signal II_1 and a pronounced difference in

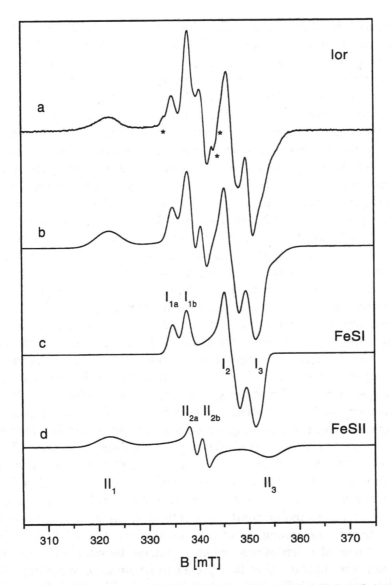

FIG. 12.   The EPR spectrum (20 K, X-band) of Ior reacted first with isoquino-line (4 min) and subsequently fully reduced with dithionite shows the magnetic interaction between FeSI and FeSII (a). Small signals of the *"very rapid"* species are indicated by asterisks. The relative arrangement of the FeS clusters in Aor [7] was combined with the orientation of the g-tensor of the related [2Fe2S] cluster of *A. spirulina* to simulate the interaction spectra of FeSI and FeSII (c and d). For FeSI the dipolar splitting occurs along g_1, for FeSII along g_2. The sum spectrum (b) reproduces the essential features of the experiment (a).

the intensities of the I_{1a} and I_{1b} components. Similar spectral differences have been observed for the FeS centers of Qor depending on the mode of production [57]. Also some traces of the *"very rapid"* and *"rapid"* species (marked with asterisks) remain visible under these conditions in Ior. The line patterns of the FeS species cannot be simulated properly when assuming collinear g_1 directions. On the other hand, there is no a priori knowledge of the relative arrangement of the individual g-tensors of the two FeS clusters. Moreover, due to the absence of any single-crystal EPR studies, the directions of the g-tensor axes with respect to any of the [2Fe2S] cluster frames were unknown until recently when we derived them from a detailed analysis of the proton ENDOR interactions of the [2Fe2S] cluster of *Arthrospira platensis* [104]. Combining the structural information from Aor with the knowledge about the orientation of the g-tensor could provide for an approach to resolve these uncertainties. It was shown for *A. platensis* that the g_1 direction is nearly parallel to the bond from the reduced iron to an adjacent cysteine sulfur. The g_2 and g_3 directions were found close to the [2Fe2S] cluster plane with g_2 oriented approximately collinear (16°) to the vector between the two irons. The structural congruence of the cluster of *A. platensis* [105,106] with the proximal and distal FeS clusters in Aor (and XO, Cox) suggests a transfer of the g-tensor assignment. Given the cluster arrangement in the Aor structure (see Fig. 2, top), a single possible combination requiring an angular deviation of less than 30° between tensor axes could be extracted. When g_1 of the proximal cluster (FeSI) is set close to the bond between Fe1 and Cys139 or between Fe2 and Cys103 with g_2, g_3 roughly in plane, then only the g_2 direction (Fe2-Fe1) in the distal cluster (FeSII) adopts an almost collinear alignment. For this setup a dipolar splitting is caused along g_1 of the proximal cluster (FeSI) and along g_2 of the distal cluster (FeSII). All other combinations of orientations result in angles between 40° and 60°, for which the dipolar effect is expected to be small or vanishing. The simulation for the individual clusters is given in Fig. 12c,d. The variation of the D-tensor yielded reasonable agreement to the experiment for values of 1.3 and 1.4 mT (corresponding to maximal dipolar values of 2.6 and 2.8 mT) upon summation of the individual spectra in a 1 : 1 ratio (Fig. 12b), for which the intensity pattern of the split features is closely reproduced.

A further improvement of the simulation is expected through the inclusion of deviations from the strict coaxiality of the g-tensor directions. From the maximal dipolar couplings a distance of 11–11.3 Å is estimated when applying the point dipole approximation (Scheme 8). An inspection of the corresponding EPR spectra of the FeS-cluster of Aor [102] shows the splitting of the $FeSI_1$ component ($\sim 2\,mT$), whereas for the g_2 component of FeSII a splitting is not resolved due to the smaller coupling and, partly, because it is obscured by a Mo(V) signal. The latter situation is also found in Qor, where FAD signals and Mo(V) species are superimposed on the $FeSII_2$ resonance. In this respect, Ior seems to represent a favorable case where the magnetic interaction is sufficiently large and the resolution (line width, superpositions) high enough to render a clear splitting. Whether this interpretation can be considered as a general rule for these types of enzymes, as is suggested by the similarities of the X-ray structures available so far, awaits verification in connection with related enzymes.

Of course, the analysis of proton ENDOR interactions in the clusters directly allows to identify the reducible iron in each cluster as well as to determine the g-tensor directions within the cluster frame as has been demonstrated for the [2Fe2S] ferredoxin of *A. platensis* [104]. This would give unequivocal evidence for the parameters necessary for simulation of the magnetic interactions. Two complete ENDOR datasets for the reduced clusters in Ior and Qor have been collected and are currently processed employing the available structural information. It is expected that the results will reveal additional intrinsic details and will help to assign the preferentially reduced irons of the proximal and distal FeS clusters. Moreover, it is hoped that ENDOR analysis will give a clue to the understanding of the different EPR g-anisotropies of the proximal (FeSI) and distal (FeSII) clusters.

The distances obtained from X-ray structures and the constraints derived from analysis of EPR spectra, particularly magnetic interactions, are compiled in Scheme 8. It is noted that the distances calculated from pure dipole approaches agree reasonably well with the crystallographic data.

5. CONCLUSIONS

It appears that the "structural" information gained from magnetic interactions now converges to a common picture with respect to the assignment of EPR signals to specific redox centers for enzymes of the XO family and that the distances obtained from various types of approaches generally coincide with the X-ray structures obtained for three members: Aor, Cox, and XO. The next level of information is to identify the reducible iron atoms of the iron-sulfur clusters by means of proton ENDOR. It may contribute to a more detailed understanding of the electronic properties and the somewhat variable redox potentials of these clusters encountered in the various enzymes of the XO family. In the near future, the availability of enzymes with specific mutations in the vicinity of the redox centers or the substrate binding site will disclose the determinants of substrate and site specificity as well as additional mechanistic details.

ACKNOWLEDGMENTS

We thank the many students who contributed to the work on molybdenum hydroxylases. We also thank C. Toia and A. L. Barra of the High Field Laboratory at CNRS in Grenoble. Generous financial support by the Deutsche Forschungsgemeinschaft, the Volkswagen-Stiftung, and the European Commission is gratefully acknowledged.

ABBREVIATIONS

Adh	aldehyde dehydrogenase
Aor	aldehyde oxidoreductase from *Desulfovibrio gigas*
CD	circular dichroism
CMP	cytidine monophosphate
CODH	carbon monoxide dehydrogenase
Cox	CODH from *Oligotropha carboxidovorans*
DMSO	dimethyl sulfoxide

ENDOR	electron nuclear double resonance
EPR	electron paramagnetic resonance
FAD	flavin adenine dinucleotide
Ior	isoquinoline 1-oxidoreductase
MCD	pyranopterin cytosine dinucleotide (molybdopterin cytosine dinucleotide)
MGD	pyranopterin guanine dinucleotide (molybdopterin guanine dinucleotide)
MoCo	molybdenum cofactor
Mop	Aor from *Desulfovibrio gigas*
MPT	pyranopterin cofactor (molybdopterin)
Qor	quinoline 2-oxidoreductase
Qox	quinaldine 4-oxidase
SHE	standard hydrogen electrode
UV-Vis	UV-visible
Xdh	xanthine dehydrogenase
XO	xanthine oxidase

REFERENCES

1. R. Hille, *Biochim. Biophys. Acta, 1184,* 143–169 (1994).

2. R. Hille, *Chem. Rev., 96,* 2757–2816 (1996).

3. R. C. Bray, B. G. Malmström, and T. Vänngård, *Biochem. J., 73,* 193–197 (1959).

4. R. C. Bray, *Adv. Enzymol. Relat Areas Mol. Biol., 51,* 107–165 (1980).

5. R. C. Bray, *Q. Rev. Biophys., 21,* 299–329 (1988).

6. R. Hille and V. Massey, in *Molybdenum Enzymes* (T. G. Spiro, ed.), John Wiley and Sons, New York, 1985, pp. 443–517.

7. M. J. Romão, M. Archer, I. Moura, J. J. G. Moura, J. LeGall, R. Engh, M. Schneider, P. Hof, and R. Huber, *Science, 270,* 1170–1176 (1995).

8. M. J. Romão and R. Huber, *Struct. Bonding, 90,* 69–95 (1998).

9. H. Dobbek, L. Gremer, O. Meyer, and R. Huber, *Proc. Natl. Acad. Sci. USA, 96,* 8884–8889 (1999).

10. L. Gremer, S. Kellner, H. Dobbek, R. Huber, and O. Meyer, *J. Biol. Chem.*, **275**, 1864–1872 (2000).

11. L. Gremer, V. Svetlitchnyi, O. Meyer, H. Dobbek, and R. Huber. Carbon monoxide dehydrogenases: Catalysis involving Ni, Mo, Cu, FeS, molybdopterin-cytosine dinucleotide and flavin. *Biospektrum Special Issue 3/2001* (VAAM2001 meeting), abstract KSP07.

12. C. Enroth, B. T. Eger, K. Okamoto, T. Nishino, T. Nishino, and E. F. Pai, *Proc. Natl. Acad. Sci. USA*, **97**, 10723–10728 (2000).

13. E. I. Stiefel, *J. Biol. Inorg. Chem.*, **2**, 772 (1997).

14. R. Hille, *J. Biol. Inorg. Chem.*, **2**, 804–809 (1997).

15. M. J. Romão, N. Rösch, and R. Huber, *J. Biol. Inorg. Chem.*, **2**, 782–785 (1997).

16. D. J. Lowe, R. L. Richards, and R. C. Bray, *Biochem. Soc. Trans.*, **25**, 774–778 (1997).

17. D. J. Lowe, R. L. Richards, and R. C. Bray, *J. Biol. Inorg. Chem.*, **3**, 557–558 (1998).

18. R. Hille, *J. Biol. Inorg. Chem.*, **3**, 559–560 (1998).

19. R. Huber, P. Hoff, R. O. Duarte, J. J. G. Moura, I. Moura, M. Y. Liu, J. LeGall, R. Hille, M. Archer, and M. J. Romão, *Proc. Natl. Acad. Sci. USA*, **93**, 8846–8851 (1996).

20. M. J. Romão, J. Knäblein, R. Huber, and J. J. G. Moura, *Prog. Biophys. Mol. Biol.*, **68**, 121–144 (1997).

21. M. J. Romão and R. Huber, *Biochem. Soc. Trans.*, **25**, 755–757 (1997).

22. M. J. Romão, *Rev. Port. Quim.*, **3**, 11–22 (1996).

23. S. Fetzner, B. Tshisuaka, F. Lingens, R. Kappl, and J. Hüttermann, *Angew. Chem. Int. Ed.*, **37**, 576–597 (1998).

24. D. T. Davies, *Aromatic Heterocyclic Chemistry*, Oxford University Press, Oxford, 1992.

25. S. Fetzner, *Appl. Microbiol. Biotechnol.*, **49**, 237–250 (1998).

26. O. P. Shukla, *Appl. Environ. Microbiol.*, **51**, 1332–1342 (1986).

27. O. P. Shukla, *Microbios 59*, 47–63 (1989).

28. G. Schwarz, R. Bauder, M. Speer, T. O. Rommel, and F. Lingens, *Biol. Chem. Hoppe-Seyler*, **370**, 1183–1189 (1989).

29. S. Schach, G. Schwarz, S. Fetzner, and F. Lingens, *Biol. Chem. Hoppe-Seyler, 374*, 175–181 (1993).

30. S. Schach, B. Tshisuaka, S. Fetzner, and F. Lingens, *Eur. J. Biochem., 232*, 536–544 (1995).

31. P. Röger, G. Bär, and F. Lingens, *FEMS Microbiol. Lett., 129*, 281–286 (1995).

32. S. S. Johansen, D. Licht, E. Arvin, H. Mosbaek, and A. B. Hansen, *Appl. Microbiol. Biotechnol., 47*, 292–300 (1997).

33. H. K. Hund, A. de Beyer, and F. Lingens, *Biol. Chem. Hoppe-Seyler 371*, 1005–1008 (1990).

34. I. Bauer, N. Max, S. Fetzner, and F. Lingens, *Eur. J. Biochem. 240*, 576–583 (1996).

35. F. Fischer, S. Künne, and S. Fetzner, *J. Bacteriol. 181*, 5725–5733 (1999).

36. S. Fetzner, *Naturwissenschaften, 87*, 59–69 (2000).

37. F. Fischer and S. Fetzner, *FEMS Microbiol. Lett., 190*, 21–27 (2000).

38. R. Bauder, B. Tshisuaka, and F. Lingens, *Biol. Chem. Hoppe-Seyler, 371*, 1137–1144 (1990).

39. D. Hettrich, B. Peschke, B. Tshisuaka, and F. Lingens, *Biol. Chem. Hoppe-Seyler, 372*, 513–517 (1991).

40. K. Parschat, C. Canne, J. Hüttermann, R. Kappl, and S. Fetzner, *Biochim. Biophys. Acta, 1544*, 151–165 (2001).

41. M. Lehmann, B. Tshisuaka, S. Fetzner, P. Röger, and F. Lingens, *J. Biol. Chem., 269*, 11254–11260 (1994).

42. A. de Beyer and F. Lingens, *Biol. Chem. Hoppe-Seyler, 374,* 101–110 (1993).

43. B. Tshisuaka, PhD thesis, University of Hohenheim, Germany (1992).

44. I. Stephan, B. Tshisuaka, S. Fetzner, and F. Lingens, *Eur. J. Biochem., 236*, 155–162 (1996).

45. M. Bläse, C. Bruntner, B. Tshisuaka, S. Fetzner, and F. Lingens, *J. Biol. Chem., 271*, 23068–23079 (1996).

46. M. Lehmann, B. Tshisuaka, S. Fetzner, and F. Lingens, *J. Biol. Chem., 270*, 14420–14429 (1995).

47. G. W. Black, C. M. Lyons, E. Williams, J. Colby, M. Kehoe, and C. O'Reilly, *FEMS Microbiol. Lett.*, *70*, 249–254 (1990).

48. J. J. Johnson, L. W. Indermaur, and K. V. Rajagopalan, *J. Biol. Chem.*, *266*, 12140–12145 (1991).

49. T. Palmer, A. Vasishta, P. W. Whitty, and D. H. Boxer, *Eur. J. Biochem.*, *222*, 687–692 (1994).

50. S. Leimkühler and W. Klipp, *FEMS Microbiol. Lett.*, *174*, 239–246 (1999).

51. M. W. Lake, C. A. Temple, K. V. Rajagopalan, and H. Schindelin, *J. Biol. Chem.*, *275*, 40211–40217 (2000).

52. C. A. Temple and K. V. Rajagopalan, *J. Biol. Chem.*, *275*, 40202–40210 (2000).

53. I. Ziemens, R. Kappl, J. Hüttermann, and S. Fetzner, unpublished results.

54. R. C. Bray, in *The Enzymes*, Vol. XII (P. D. Boyer, ed.) Academic Press, New York, 1975, pp. 300–419.

55. R. S. Pilato and E. I. Stiefel, in *Bioinorganic Catalysis* (J. Reedijk, ed.), Marcel Dekker, New York, 1993, pp. 131–188.

56. N. A. Turner, R. C. Bray, and G. P. Diakun, *Biochem. J.*, *260*, 563–571 (1989).

57. B. Tshisuaka, R. Kappl, J. Hüttermann, and F. Lingens, *Biochemistry*, *32*, 12928–12934 (1993).

58. C. Canne, I. Stephan, J. Finsterbusch, F. Lingens, R. Kappl, S. Fetzner, and J. Hüttermann, *Biochemistry*, *36*, 9780–9790 (1997).

59. C. Canne, D. J. Lowe, S. Fetzner, B. Adams, A. T. Smith, R. Kappl, R. C. Bray, and J. Hüttermann, *Biochemistry*, *38*, 14077–14087 (1999).

60. S. Gutteridge, S. J. Tanner, and R. C. Bray, *Biochem. J.*, *175*, 869–878 (1978).

61. G. N. George and R. C. Bray, *Biochemistry*, *22*, 5443–5452 (1983).

62. G. N. George and R. C. Bray, *Biochemistry*, *27*, 3603–3609 (1988).

63. R. C. Bray and S. Gutteridge, *Biochemistry*, *21*, 5992–5999 (1982).

64. S. P. Cramer, J. L. Johnson, K. V. Rajagopalan, and T. N. Sorrell, *Biochem. Biophys. Res. Commun.*, *91*, 434–439 (1979).

65. S. Gutteridge and R. C. Bray, *Biochem. J.*, *189*, 615–623 (1980).

66. T. R. Hawkes, G. N. George, and R. C. Bray, *Biochem. J.*, *218*, 961–968 (1984).

67. J. P. Malthouse and R. C. Bray, *Biochem. J.*, *191*, 265–267 (1980).

68. J. P. Malthouse, G. N. George, D. J. Lowe, and R. C. Bray, *Biochem. J.*, *199*, 629–637 (1981).

69. B. D. Howes, B. Bennett, A. Koppenhofer, D. J. Lowe, and R. C. Bray, *Biochemistry*, *30*, 3969–3975 (1991).

70. B. D. Howes, B. Bennett, R. C. Bray, R. L. Richards, and David J. Lowe, *J. Am. Chem. Soc.*, *116*, 11624–11625 (1994).

71. B. D. Howes, R. C. Bray, R. L. Richards, N. A. Turner, B. Bennett, and D. J. Lowe, *Biochemistry*, *35*, 1432–1443 (1996).

72. G. L. Wilson, R. J. Greenwood, J. R. Pilbrow, J. T. Spence, and A. G. Wedd, *J. Am. Chem. Soc.*, *113*, 6803–6812 (1991).

73. R. J. Greenwood, G. L. Wilson, J. R. Pilbrow, and A. G. Wedd, *J. Am. Chem. Soc.*, *115*, 5385–5392 (1993).

74. R. C. Bray and D. J. Lowe, *Biochem. Soc. Trans.*, *25*, 762–768 (1997).

75. D. J. Lowe, R. L. Richards, and R. C. Bray, *Biochem. Soc. Trans.*, *25*, 774–778 (1997).

76. G. A. Lorigan, R. D. Britt, J. H. Kim, and R. Hille, *Biochim. Biophys. Acta*, *1185*, 284–294 (1994).

77. N. Turner, B. Barata, R. C. Bray, J. Deistung, J. LeGall, and J. J. G. Moura, *Biochem. J.*, *243*, 755–761 (1987).

78. V. Massey and D. Edmondson, *J. Biol. Chem.*, *245*, 6595–6598 (1970).

79. G. Palmer, F. Müller, and V. Massey, in *Flavins and Flavoproteins* (H. Kamin, ed.), University Press, Baltimore, 1971, pp. 123–140.

80. R. S. Pilato and E. I. Stiefel, in *Bioinorganic Catalysis* (J. Reedijk, ed.), Marcel Dekker, New York, 1999, pp. 81–152.

81. R. Cammack, in *Bioenergetics: A Practical Approach* (G. C. Brown and C. E. Cooper, eds.), IRL Press, Oxford, 1980, pp. 85–109.

82. C. J. Kay and M. J. Barber, *Anal. Biochem.*, *184*, 11–15 (1990).

83. M. J. Barber and J. C. Salerno, in *Molybdenum and Molybdenum-containing Enzymes* (M. P. Coughlan, ed.), Pergamon Press, Oxford, 1980, pp. 543–568.

84. R. Cammack, M. J. Barber, and R. C. Bray, *Biochem. J.*, *157*, 469–478 (1976).

85. A. G. Porras and G. Palmer, *J. Biol. Chem.*, *257*, 11617–11626 (1982).

86. M. J. Barber, M. P. Coughlan, M. Kanda, and K. V. Rajagopalan, *Arch. Biochem. Biophys.*, *201*, 468–474 (1980).

87. J. S. Olson, D. P. Ballou, G. Palmer, and V. Massey, *J. Biol. Chem.*, *249*, 4363–4382 (1974).

88. B. Guigliarelli and P. Bertrand, in *Advances in Inorganic Chemistry*, Vol. 47, (A. G. Sykes and R. Cammack, eds.), Academic Press, San Diego, 1999, pp. 421–497.

89. C. Canne, PhD thesis, Universität des Saarlandes, Germany (1999).

90. B. A. Barata, J. LeGall, and J. J. Moura, *Biochemistry*, *32*, 11559–11568 (1993).

91. R. C. Bray and T. Vänngård, *Biochem. J.*, *114*, 725–734 (1969).

92. F. M. Pick and R. C. Bray, *Biochem. J.*, *114*, 735–742 (1969).

93. S. J. Tanner, R. C. Bray, and F. Bergmann, *Biochem. Soc. Trans. 6*, 1328–1330 (1978).

94. P. Bertrand, C. More, B. Guigliarelli, A. Fournel, B. Bennett, and B. Howes, *J. Am. Chem. Soc.*, *116*, 3078–3086 (1994).

95. S. L. Andrade, C. D. Brondino, M. J. Feio, I. Moura, and J. J. G. Moura, *Eur. J. Biochem.*, *267*, 2054–2061 (2000).

96. J. Caldeira, V. Belle, M. Asso, B. Guigliarelli, I. Moura, J. J. G. Moura, and P. Bertrand, *Biochemistry*, *39*, 2700–2707 (2000).

97. D. J. Lowe and R. C. Bray, *Biochem. J.*, *169*, 471–479 (1978).

98. D. J. Lowe, R. M. Lynden-Bell, and R. C. Bray, *Biochem. J.*, *130*, 239–249 (1972).

99. M. J. Barber, J. C. Salerno, and L. M. Siegel, *Biochemistry, 21*, 1648–1656 (1982).

100. R. E. Coffman and G. R. Buettner, *J. Phys. Chem., 83*, 2392–2400 (1979).

101. T. Iwasaki, K. Okamoto, T. Nishino, J. Mizushima, and H. Hori. *J. Biochem. Tokyo, 127*, 771–778 (2000).

102. J. J. Moura, A. V. Xavier, R. Cammack, D. O. Hall, M. Bruschi, and J. LeGall, *Biochem. J., 173*, 419–425 (1978).

103. R. C. Bray, N. A. Turner, J. Le Gall, B. A. Barata, and J. J. Moura, *Biochem. J., 280*, 817–820 (1991).

104. C. Canne, M. Ebelshäuser, E. Gay, R. Cammack, R. Kappl, and J. Hüttermann, *J. Biol. Inorg. Chem., 5*, 514–526 (2000).

105. K. Fukuyama, N. Ueki, H. Nakamura, T. Tsukihara, and H. Matsubara, *J. Biochem. Tokyo, 117*, 1017–1023 (1995).

106. T. Tsukihara, K. Homma, K. Fukuyama, Y. Katsube, T. Hase, H. Matsubara, N. Tanaka, and M. Kakudo, *J. Mol. Biol., 152*, 821–823 (1981).

15

Molybdenum Enzymes in Reactions Involving Aldehydes and Acids

Maria João Romão, Carlos A. Cunha, Carlos D. Brondino, and José J. G. Moura

Departamento de Química (and Centro de Química Fina e Biotecnologia), Faculdade de Ciências e Tecnologia, Universidade Nova de Lisboa, P-2825-114 Caparica, Portugal

1.   INTRODUCTION

The aldehyde oxidoreductases are members of the molybdenum hydroxylases, including xanthine oxidase (XO), which gives the name to this
family of enzymes, and the Mo-containing carbon monoxide dehydrogenases. These enzymes catalyze hydroxylation reactions of the type:

$$RH + H_2O \longrightarrow ROH + 2H^+ + 2e^-$$

Water is the ultimate oxygen atom source that is incorporated into the substrate. However, the group of enzymes that utilizes aldehydes shows a large diversity, since Mo and W bound to different pterins such as molybdopterin cytosine dinucleotide (MCD), molybdopterin guanine dinucleotide (MGD), and the monophosphate form of molybdopterin (MPT) can perform this task. From substrate to electron acceptors, the electron transfer chain may include Fe-S centers of different nuclearity, and flavin moieties are optional. A common feature observed is that no amino acid side chain participates as a Mo/W ligand.

2. ENZYMES OF THE XANTHINE OXIDASE FAMILY

2.1. Aldehyde Oxidoreductases from *Desulfovibrio gigas* and from *D. desulfuricans* ATCC 27774

Aldehyde oxidoreductases were purified and characterized in several sulfate reducers [1]. The enzymes from *Desulfovibrio gigas, D. desulfuricans* ATCC 27774, and *D. alaskensis* strain NCIMB 13491 [2–6] have been well characterized. The three-dimensional (3D) structures of the first two enzymes were solved as described and compared below. The enzymes catalyze the two-electron step oxidation of aldehydes to the carboxylic acid forms. The aldehyde oxidoreductases (AORs) are isolated as a homodimer (~ 100-kDa subunit) and each subunit contains one molybdenum atom associated with a pterin cofactor (MCD) and four iron atoms arranged as 2x[2Fe-2S] clusters. Flavin moieties are not found. The molybdenum cofactor was liberated from the enzyme from *D. gigas* and under appropriate conditions was transferred quantitatively to nitrate reductase in extracts of *Neurospora crassa* (nit-1 mutant) to yield active nitrate reductase [3].

The aldehyde oxidoreductases isolated from the sulfate-reducing bacteria *D. gigas* (MOP) and *D. desulfuricans* (MOD) belong to the sequence family of molybdenum hydroxylases [7–10] that includes XO [11], which gives the name to this family of enzymes, and the Mo-containing CO dehydrogenases (CODHs) [12].

MOP and MOD hydroxylate a large number of aldehydes but with very little specificity. However, glyceraldehyde is a substrate for MOP [2] and therefore its physiological role may be linked to the degradation of polyglucose, which is accumulated by *D. gigas* under normal growth conditions [13]. Both enzymes show almost identical spectroscopic properties [5] and their primary sequences exhibit 68% identity [14]. They are very closely related also in terms of their 3D structures (PDB codes 1ALO and 1DGJ for MOP and MOD, respectively), showing the same fold, overall architecture, and relative arrangement of the cofactors [14,15]. A comparison of MOP and MOD established essential features for their function, namely conserved residues in the active site, catalytically relevant water molecules, and recognition of the physiological electron acceptor docking site.

2.1.1. Three-Dimensional Structure

The 3D structure is organized into two major domains, one binding the two FeS clusters and a much larger one carrying the Moco. The FeS domain is subdivided into an N-terminal domain (residues 1–76) and a C-terminal domain (residues 84–156), each binding one of the distinct [2Fe-2S] clusters. The type I cluster is buried approximately 12 Å below the protein surface and the chain fold, a helix bundle with two central longer helices flanked by shorter ones, is new and typical of enzymes of this family (see Fig. 1a). Its binding motif C^{100}-X_2-C^{103}-X_{33}-C^{137}-X-C^{139} is also characteristic of the enzymes of the XO family [7]. The type II [2Fe-2S] cluster is solvent-exposed; its fold and binding motif C^{40}-X_4-C^{45}-X-C^{47}-X_{12}-C^{60} are typical of plant and cyanobacterial ferredoxins [16] and also characteristic of the Mo hydroxylases.

The Mo active site is located at the bottom of a 15-Å hydrophobic channel situated at the interface of two subdomains of the Moco domain. The Mo ion adopts a distorted square pyramidal coordination geometry and is bound by two sulfur atoms of the enedithiolate moiety of MCD and by three nonprotein oxygen ligands. An isopropanol (IPP) molecule, present in the second coordination sphere of the molybdenum, is an inhibitor of the enzyme [2] and provides the basis for the enzymatic mechanism [17]. Glu[869], which is very close to the Mo site and is totally conserved in the XO family (Fig. 2a), is involved in the catalytic cycle. It activates as a general base the coordinated water molecule and probably binds transiently to the metal when the carboxylic acid product is released, before the vacant coordination position is reoccupied by one water molecule from the chain or buried water molecules (Fig. 2b).

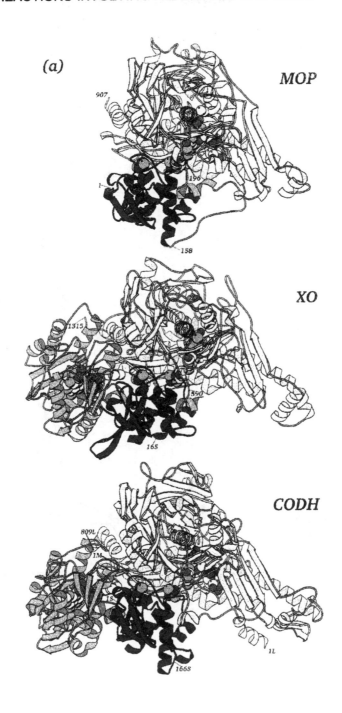

(b)

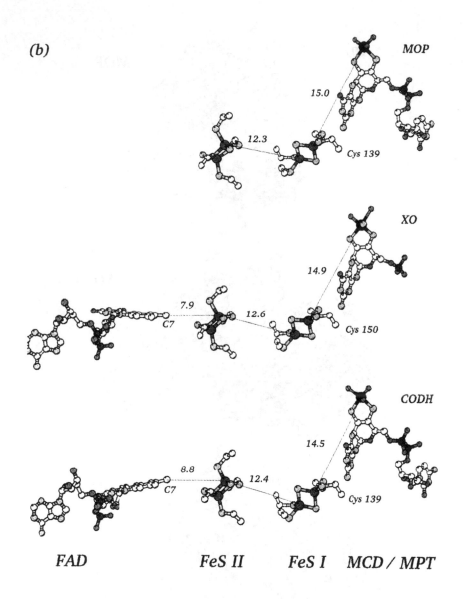

FAD   FeS II   FeS I   MCD / MPT

2.1.2. Structural Comparison with Other Members of the Family (XO and CODH)

Until recently, the crystal structure of MOP was the only one available for its enzymatic class, which includes XO and CODH. Therefore, due to the lack of structural information, the MOP protein structure was used as a basis for the proposal of a mechanism of the reductive half-reaction of XO, assuming the similarities of the respective active sites based in part in clearly conserved amino acid residues [7,15,17,18].

XO and xanthine dehydrogenase (XDH) are two forms of the same enzyme that catalyze the transformation of hypoxanthine to xanthine and xanthine to uric acid, in the latter steps of the catabolism of purines. Recently, the crystal structures of bovine milk XDH and XO, solved at 2.1 Å and 2.5 Å respectively, were reported [11] (PDB codes 1FO4 and 1FIQ, respectively). Several crystal structures were determined for another member of the family, CODH, which catalyzes the oxidation of CO with H_2O to produce CO_2. The structure from CODH isolated from *Oligotropha carboxidovorans* was solved and refined at 2.2 Å resolution [12] (PDB code 1QJ2). For the homologous enzyme from the carboxidotrophic eubacterium *Hydrogenophaga pseudoflava* structures were determined for two different forms of the enzyme synthesized at high and low intracellular molybdenum content [19]. These structures were solved at resolutions of 2.25 Å and 2.35 Å, respectively. While the two native structures are quite homologous, the structure of the catalytically inactive CODH grown under deficiency of intracellular Mo shows that, under these conditions, a nonfunctional cofactor was synthesized. In this form of CODH only the dinucleotide part of the

---

FIG. 1. (a) Crystal structures of MOP, bovine milk XO, and *O. carboxidovorans* CODH with the different domains represented in different tones of gray and all atoms of the cofactors represented as large spheres. (i) MOP:Fe,S domain (dark gray, residues 1–158) and Moco domain (light gray, residues 196–907). Residues 158–196 correspond to the polypeptide segment, which connects the two metal binding domains; (ii) XO:Fe,S domain (dark gray, residues 3–165), FAD domain (medium gray, residues 226–531), and Moco domain (lighter gray, 590–1315); (iii) CODH:Fe,S protein, subunit S (dark gray, 166 residues), FAD protein, subunit M (medium gray, 288 residues), and Moco protein, subunit L (lighter gray, 809 residues). (b) Arrangement of the prosthetic groups. The distances (in Ångstroms) between the cofactor metals are indicated, as well as the shortest distance between FAD and the Fe-S center II.

(a)

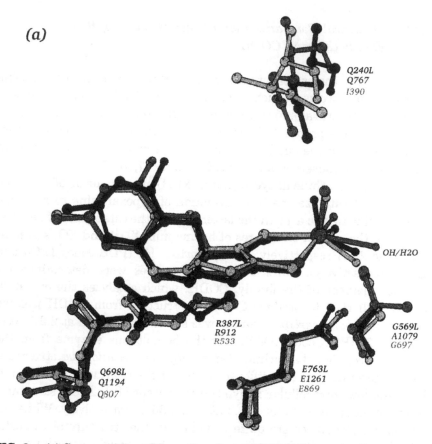

FIG. 2. (a) Superposition of the active sites of MOP (light gray), of bovine milk xanthine oxidase (medium gray), and of the L subunit of CODH (dark gray), including residues that are conserved in all three structures. (b) Superposition of the active sites of MOP (color-coded: carbon atoms in white, oxygen atoms in light gray, sulfurs in medium gray, and nitrogen atoms in dark gray) and of bovine milk xanthine oxidase (dark gray). Included are important conserved amino acid residues, internal ordered water molecules, and the inhibitors present in the second coordination sphere of Mo, isopropanol (IPP) for MOP and salicylate (SAL) for bovine milk XO.

cofactor (5′-CDP) is present, whereas the pyranopterin moiety and the Mo catalytic site are absent. However, this apo form of the enzyme does not involve conformational changes, which suggests that the 5′-CDP is integrated into the enzyme during protein folding, stabilizing the proper conformation and folding of the Moco binding site.

(b)

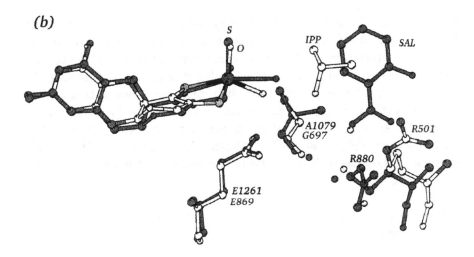

The overall fold and architecture of the three enzymes is remarkably similar as shown in Fig. 1a where the 3D structures are represented top/down in similar orientations. The domain organization corresponds to the binding of the cofactors. In MOP and XO the domains belong to a single polypeptide chain, whereas in CODH the domains correspond to three subunits: L (88.7 kDa), M (30.2 kDa), and S (17.8 kDa) [12]. The N-terminal ∼ 20-kDa domain (S subunit in CODH) binds the two spectroscopically distinct [2Fe-2S] centers, the 30- to 40-kDa FAD domain (M subunit in CODH and absent in MOP) harbors the FAD cofactor, and the C-terminal 80- to 90-kDa domain carries the Mo cofactor. This is an MCD in MOP and CODH and an MPT in XO.

2.1.3. Arrangement of the Cofactors and Electron Transfer Pathway

The redox centers are aligned in an identical fashion in all three enzymes defining a suitable intramolecular electron transfer pathway (Fig. 1b). The distances between the prosthetic groups are also very similar in all three structures. The geometrical alignment of the cofactors suggests the electron flow from the Mo site to the two spectroscopically distinct [2Fe-2S] centers I and II (see Sec. 2.1.5). The contact between the Moco and Fe/S I is through the exocyclic NH_2 of the pyranopterin and S^γ of one of the Fe/S cysteine ligands (Cys[139] in MOP, Cys[150] in XO and Cys[139] in CODH). The connection between the two [2Fe-2S] centers involves the main chain covalent bonds of the residue that precedes the third cysteine in the binding motif of center I and a hydrogen bond between the main chain nitrogen atom of this residue and the carbonyl group of the second cysteine in the binding motif of center II. An electron-tunneling mechanism may occur through this bond network [20]. The cluster [2Fe-2S] II is exposed to the protein surface in the case of MOP while in XO and CODH it contacts the C7 carbon of FAD.

2.1.4. Structure of the Molybdenum Site and Mechanistic Implications

The close relationship between MOP and the XO family of enzymes implies a similar mechanism of action. Essential structural features of MOP agree with a large number of experimental data for XO. On the basis of the MOP crystal structure obtained under different conditions [17], it was possible to have suitable models for the reaction mechanism of XO [7,18]. In this section, the active site of MOP will be compared with that of XO and of CODH highlighting conserved amino acid residues (Fig. 2).

In all three enzymes, the molybdenum atom adopts a square pyramidal geometry with no protein ligand binding to the metal. A glutamate (Glu[869] in MOP, Glu[1261] in XO, and Glu[763L] in CODH) is relatively close to the metal (3.2–3.5 Å) and may promote the nucleophilicity of the water ligand. In addition, this Glu residue may bind transiently to the Mo to maintain a five-fold coordination once the activated water ligand is transferred to the substrate during catalysis. In all three structures the water (or hydroxyl) ligand is within hydrogen bonding contact to the carboxylate of this conserved Glu as well as to the amide of Gly[697] in MOP, Ala[1079] in XO, or Gly[569L] in CODH. Other conserved residues included in the superposition of Fig. 2a are those involved in stabilizing the pyranopterin moiety of Moco, namely, Gln[807] in MOP, Gln[1194] in XO, and Gln[698L] in CODH, which define conserved hydrogen bonds to the pterin structure, as well as Arg[533] in MOP, Arg[912] in XO, and Arg[387L] in CODH. In the active site pocket there are additional conserved structural features in MOP and XO. In MOP, the location of the inhibitor isopropanol present in the second coordination sphere of Mo was proposed as the putative substrate binding site [17]. In the XO structure, its position is replaced by the competitive inhibitor salicylate, included during purification and crystallization of the enzyme [11]. The salicylate binding mode is proposed to mimic the binding of larger aromatic substrates. In both structures, well-ordered water molecules are present in the buried pocket (Fig. 2b).

Data on point mutations of the *Aspergillus nidulans* XDH allowed defining residues essential for substrate positioning [21]. Some of these mutations correspond to residues in the vicinity of the MOP substrate binding site and imply partial loss of function of *A. nidulans* XDH. Particularly interesting is Arg501 (Arg[501] in MOP and Arg[880] in bovine milk XO) whose mutation in *A. nidulans* XDH (Arg[911] to Gln or Gly) changes the hydroxylation position of 2-hydroxypurine from C8 to C6. This Arg is totally conserved within the XO/XDH family of enzymes and, on the basis of the crystallographic studies of MOP [17], was proposed to be essential to position purine substrates in the active site (see Fig. 2b).

2.1.5. Spectroscopic Characterization of the Metal Sites

A set of complementary spectroscopic tools have been used to probe the structural and magnetic properties of this group of enzymes. Diamagnetic and paramagnetic states can be obtained in native, substrate-reacted, and chemical reduced states. Visible, circular dichroism (CD), Raman, EPR, Mössbauer, and extended X-ray absorption line structure (EXAFS) spectroscopies have been applied and provided important pieces of information, complementing the structural data provided by the 3D structures now available. EPR at X and Q band revealed the presence of the Mo(V) rapid and slow signals after substrate reaction and chemical reduction, respectively. The different EPR signals that can be generated and observed in MOP (and in the others) reflect the reactivity and coordination versatility of the Mo site and enables placement of the enzyme in the group of the molybdenum hydroxylases. This variability of metal coordination can be discussed in conjunction with the 3D structural features of the Mo-pterin site in aldehyde oxidoreductases [4,8,15].

The UV/Vis absorption spectrum of the proteins is similar to those observed for the deflavo forms of xanthine and aldehyde oxidases

[7,22–26] and reminiscent of the one observed for plant-type ferredoxins. The CD spectrum is intense and indicative of the presence of [2Fe-2S] centers, again similar to the spectra of plant ferredoxins and XO [27].

The set of Mo(V) EPR signals detected in AORs isolated from sulfate reducers shows close homology with the ones from the molybdenum-containing hydroxylases group. Mössbauer and X/Q-band EPR spectroscopic studies complemented the UV/Vis and CD studies and the assignment of the [2Fe-2S] arrangement of the iron-sulfur cores (see below).

2.1.5.1. Electron Paramagnetic Resonance

(a) Molybdenum Center: The catalytically active form of the enzyme contains a sulfido ligand at the Mo site [7,17]. The conversion to an oxo ligand (desulfo form) results in the loss of catalytic activity. Molybdenum-containing hydroxylases are mixtures of inactive desulfo form Mo(V) slow type EPR signals that originate after long exposure to dithionite and active species (yielding rapid signals that are generated in the presence of substrates or by short time chemical reduction). Different catalytic competent forms have been detected. The Mo(V) EPR rapid type 2 signal shows two strongly coupled protons that can be exchanged. This signal, centered at $g_{av} = 1.9750$, is analogous to those observed for xanthine and aldehyde oxidases [3]. Mo(V) EPR rapid type 1 signal has one strong and one weakly coupled proton [3,24,28,29]. The rapid signals were not only obtained after a short time chemical reduction (with dithionite) but also in the presence of aldehydes [2,3]. AOR (as isolated) shows variable EPR signals (resting EPR signal) presumed to be due to Mo(V). The determined midpoint redox potentials associated with the slow signal are −415 mV (Mo(VI)/Mo(V)) and −530 mV (Mo(V)/Mo(IV)) [24]. Reaction of the enzyme with a high concentration of ethylene glycol for an extended period of time, in a sample where the slow signal develops, originates the appearance of a Mo(V) species designated as desulfo-inhibited [2]. The signal develops also in the presence of formaldehyde, methanol, and glycerol and persists on air oxidation. Inhibited EPR signals [1] are detected in MOP and D. alaskensis enzyme, mixed with rapid signals and clearly originated from a functional form.

(b) Iron-Sulfur Centers: EPR studies (complemented with Mössbauer studies) reveal the presence of two types of [2Fe-2S] cores, named spectroscopically as Fe-S I and Fe-S II signals [3,28]. The MOP data will be used because the g-values of Fe-S centers in the other enzymes are quite similar (see Table 1). X and Q band EPR studies established the presence of two EPR signals in equal amounts (Fe-S I and Fe-S II), as determined by computer simulation using the same set of parameters at the two different frequencies and double integration performed at different powers and temperatures. The Fe-S I center is observed at 77 K (g-values at 2.021, 1.938, and 1.919, $g_{av} = 1.959$). The Fe-S II center is only observable below 65 K (g-values at 2.057, 1.970, and 1.900, $g_{av} = 1.976$). Fe-S I EPR signal has parameters similar to those of spinach ferredoxin, whereas Fe/S II is unusual in having a larger g_{av} not being observable at temperatures above 40 K. The 3D structure indicates that one of the Fe-S centers is included in a protein domain with a fold similar to that of spinach ferredoxin (on the surface of the molecule), and the other, completely buried, is located in a domain with a unique fold. Variation of amplitude of Fe-S I and Fe-S II EPR signals as a function of redox potential revealed midpoint potentials of −260 and −285 mV, respectively, for MOP (see Table 2 for comparison with the other *Desulfovibrio* sp. enzymes).

2.1.5.2.   EXAFS Measurements

Two different samples of MOP were studied by EXAFS. These measurements reflect that the enzyme is purified as a variable mixture of sulfo and desulfo forms. One of the samples contained mainly the desulfo form [29]. In the oxidized form the molybdenum environment is found to contain two terminal oxo groups and two long (2.47 Å) Mo-S bonds. Evidence was also found for an oxygen or nitrogen ligand at 1.90 Å. The behavior of both oxidized and dithionite-reduced forms is similar to that observed previously with desulfo XO and resembles sulfite oxidase and nitrate reductase. A recent reanalysis of *D. gigas* and *D. desulfuricans* aldehyde oxidoreductases suggests that the fitting of the experimental data improve when a mixture of sulfo and oxo (desulfo) species is considered. The data were fitted with similar atomic distances (here indicated for the *D. gigas* enzyme): Mo−S 2.44 Å, Mo=O 1.67 Å, Mo=S (or O) 2.14 Å (the analysis was made keeping S + O = 2, in the proportion 0.7 Mo=S and 1.3 Mo=O) and a longer Mo−O distance 1.83 Å (assigned to a water or hydroxyl group) (G. George, personal communication).

TABLE 1

Mo(V) EPR Signals Detected in AORs and XO

Enzyme (mT)	Signal	g_1	g_2	g_3	^{1}H hyperfine interactions		
					A_1	A_2	A_3
D. gigas AOR	Slow (dith)	1.9705	1.9680	1.9580	1.68	1.60	1.44
D. alaskensis AOR	Slow (dith)	1.9716	1.9696	1.9596	1.60	1.60	1.60
XO	Slow (dith)	1.9719	1.9671	1.9551	1.50	1.48	1.52
D. gigas AOR	Rapid 2 (dith)	1.9882	1.9702	1.9643	1.15	1.66	1.26
					1.14	0.63	0.93
	Rapid 2 (ald)	1.9895	1.9715	1.9640	1.4	1.3	0.7
					0.35	0.5	1.3
XO	Rapid 2	1.9861	1.9695	1.9623	1.35	1.57	1.45
					1.35	0.74	0.99
	Rapid 1	1.9906	1.9707	1.9654	1.20	1.30	1.30
D. gigas AOR	Desulfo-inhib.	1.9787	1.9725	1.9673			
XO	Desulfo-inhib.	1.9784	1.9707	1.9641			
D. gigas AOR	Inhibited	1.9915	1.9795	1.9555	nd	nd	nd
XO	Inhibited	1.9911	1.9772	1.9513	0.50	0.34	0.54

nd, not determined.
Data from [1,49–52].

TABLE 2

EPR g-Values of [Fe-S] Centers in Aldehyde Oxidoreductases from *Desulfovibrio* Species: Comparison with Eukaryotic Systems

Enzyme (AOR)	g_1	g_2	g_3
Fe/S center I			
D. gigas	2.021	1.938	1.919
D. desulfuricans ATCC 27774	2.013	1.928	1.903
D. alaskensis NCIMB 13491	2.021	1.934	1.916
Eucaryontes	2.021 ± 0.004	1.933 ± 0.005	1.907 ± 0.01
Fe/S center II			
D. gigas	2.057	1.979	1.900
D. desulfuricans ATCC 27774	2.048	1.960	1.873
D. alaskensis NCIMB 13491	2.066	1.970	1.900
Eucaryontes	2.103 ± 0.005	2.002 ± 0.006	1.912 ± 0.01

Data from [1,3,5,6].

2.1.5.3. *Mössbauer Studies*

Spectra of the enzyme in oxidized, partially reduced, benzaldehyde-reacted, and fully reduced states were recorded at different temperatures and with variable externally applied magnetic fields [30]. In the oxidized enzyme, the clusters are diamagnetic ($\Delta E_Q = 0.62 \pm 0.02$ mm/s and $\delta = 0.27 \pm 0.01$ mm/s) typical of the +2 state (ferric-ferric). Reduced clusters also show characteristic parameters of the +1 state (ferric-ferrous). The spectra could be explained by the spin coupling model proposed for the [2Fe-2S] cluster, where the high-spin ferrous site ($S = 2$) is antiferromagnetically coupled to a high-spin ferric site ($S = 5/2$) to form an $S = 1/2$ system (localized valences). Two ferrous sites with different ΔE_Q (3.42 mm/s and 2.93 mm/s) are observed at 85 K, indicating the presence of two spectroscopically distinguishable [2Fe-2S] centers. A Mössbauer study of the protein reacted with benzaldehyde shows partial reduction of the iron-sulfur centers, indicating the involvement of the clusters in the process of substrate oxidation with rapid intramolecular electron transfer from the molybdenum to the iron-sulfur sites.

2.1.5.4. *Structural Spectroscopic Assignments of the Iron-Sulfur Centers/Magnetic Interactions*

EPR data accumulated in *D. gigas, D. desulfuricans,* and *D. alaskensis* aldehyde oxidoreductases and xanthine oxidase revealed that the Mo site and the Fe-S centers interact magnetically [3,6,31,32]. The 3D structures available indicate the precise location of the metal centers and their relative position, and give support to the interpretation of magnetic interactions observed in the EPR spectra. A few arguments can now be put forward in order to assign which Fe-S centers seen in the structure correspond to the spectroscopically distinct Fe/S signals. The Mo(V) slow signal is split at low temperature by a magnetic interaction with an iron-sulfur center. The center that serves as the origin for Fe-S EPR signal I is the one to be selected as interacting partner. The Fe-S signal I feature at g_{max} is clearly split below 40 K. The large splitting (about 20 G) must be due to the paramagnetic center that originates Fe-S EPR signal II; the low quantitation of Mo(V) paramagnetic species is also an argument to exclude the origin of this observation as being due to an interaction with the Mo site and Fe-S center I. The work was complemented by the study of the observable splittings and its temperature dependence, EPR saturation, and the effect of differential reduction of the Fe-S centers.

As indicated before, the two iron-sulfur clusters in MOP display distinct EPR signals: the less anisotropic one, called Fe-S I signal, is generally similar to the $g_{av} = 1.96$ type of signals given by ferredoxins, whereas the Fe S II signal often exhibits anomalous properties such as very large g values, broad lines, and very fast relaxation properties. A detailed comparison of the temperature dependence of the spin-lattice relaxation time and of the intensity of these signals in MOP suggests that the peculiar EPR properties of signal II arise from the presence of low-lying excited levels reflecting significant double-exchange interactions [31]. The issue raised by the assignment of Fe S I and II EPR signals to the [2Fe-2S] clusters in the reduced state was solved by using the EPR signal of the Mo(V) center as a probe. The temperature dependence of this signal could be quantitatively reproduced by assuming that the Mo(V) center is coupled to the cluster giving the Fe-S I EPR signal. This demonstrates unambiguously that in MOP and XO the Fe-S EPR signal I arises from the center, which is closest to the molybdenum cofactor (designated as "proximal"). Also, the two [2Fe-2S] clusters of *D. alaskensis* AOR are distinguishable by their relaxation behavior, g-values and midpoint redox potentials (−275 mV, Fe-S I and −325 mV, Fe-S II) (see Table 2). The detailed analysis of the EPR spectrum of a partially reduced sample made it possible, by using double EPR spectral integration and computer simulations, to assign the Fe-S EPR signal I as the proximal cluster to the Mo site [6]. A general consensus exists based on EPR arguments and by using 3D structural data that Mo is about 15 Å from the buried Fe-S (center that originates Fe-S EPR signal I) and about 27 Å from the solvent exposed Fe-S (plant ferredoxin like folding, Fe-S EPR signal II).

A magnetic interaction was also seen, in XO, between various Mo(V) EPR species and one of the Fe-S centers [32]. The striking structural similarities and the relative position of the metal sites and cofactors in MOP, XO and CODH [11,12,15] seem to be a conserved feature and a strong indication for the intramolecular electron transfer pathway.

2.1.5.5. Functionality of MOP

The persulfide coordination of the Mo site has been related to the active form of this group of enzymes. Its replacement by an oxo group leads to an inactive form (desulfo). The EPR and Mössbauer data analysis, together with the metal determination and observation of the effect of aldehydes in the bleaching of the chromophore absorption of the protein, indicated that MOP samples are a mixture of active (sulfo), inactive (desulfo), and demolybdo forms [14]. EPR indicates that the slow signals (related to the conversion to the desulfo form) increase after cyanide treatment with thiocyanide release (our unpublished results). The crystal structure of XO-related MOP was analyzed in desulfo, sulfo, oxidized, reduced, and alcohol-bound forms [17]. In the sulfo form the Mo-MCD cofactor has a dithiolene-bound fac-[Mo, $=O$, $= S$, $-H_2O$] substructure, which corroborates with the EXAFS analysis that deals with mixtures of species, making it more difficult to provide evidence for a short Mo-sulfido bond.

2.2. Aldehyde Oxidoreductase from *Sulfolobus acidocaldarius*

2.2.1. Molecular Properties and Catalytic Activity

Sulfolobus acidocaldarius is an extreme thermoacidophilic obligately aerobe. The purification and characterization of cytosolic molybdenum-containing aldehyde oxidoreductase was reported [33]. The enzyme has an apparent molecular mass of 177 kDa and consists of three subunits: 80.5 (α), 32 (β), and 19.5 (γ). Cofactor and metal analysis established the presence of one molybdenum and one MGD (identified after perchloric acid cleavage and thin-layer chromatography), one non-covalently bound extractable FAD, and iron atoms arranged as $2 \times$ [2Fe-2S] clusters (as suggested by EPR variable temperature measurements on reduced samples). It was suggested that the enzyme functions as a glyceraldehyde oxidoreductase in the course of the nonphosphorylated Entner-Doudoroff pathway. The enzyme shows a remarkable thermal stability.

The enzyme can be included in the Mo hydroxylases family, but it differs from *P. furiosus* aldehyde oxidoreductase, which contains two MGD pterins and tungsten, and from MOP with a single MCD cofactor and no flavin.

2.2.2. Spectroscopic Properties

The UV-Vis spectra of *Sulfolobus acidocaldarius* aldehyde oxidoreductase is similar to the one detected for comparable hydroxylases containing Mo, [2Fe-2S] centers, and a flavin moiety. EPR spectra of the aerobically prepared enzyme exhibit the so called "desulfo-inhibited" signal known from chemically modified forms of molybdenum-containing enzymes with g-values at 1.98, 1.97, and 1.96 ($g_{av} \sim 1.97$) [33]. The signal is detected without prior requirement of reducing agent or reaction with ethylene glycol, suggesting that previous modification already occurred at the molybdenum site. Anaerobically prepared samples show both active molybdenum EPR signals when reacted with substrate and from reduced [2Fe-2S] centers. One of the centers, observable at 70 K, has g-values at 2.02, 1.93, and 1.88 (equivalent to Fe-S EPR signal I in MOP). A detectable splitting occurs in the g_z line at 30–20 K due to the interaction with a second Fe-S center, and additional signals are observed at g-values 2.1, 1.99, and 1.91. An estimate of the redox potential of the Fe-S centers indicates −300 (Fe-S signal I) and −200 mV (Fe-S signal II, lower limit). The Mo(V) EPR signals were not studied in detail so far.

2.3. Aldehyde Oxidases from Mammalian Liver

2.3.1. Molecular Properties and Catalytic Activity

In mammals, three molybdoproteins have been identified and characterized: sulfite oxidase, xanthine dehydrogenase (XO/XDH), and aldehyde oxidase (AO) [34]. The last two enzymes are molybdenum hydroxylases using flavin and molybdenum cofactors to oxidize their substrates. XO/XDH and AO have a striking number of similarities and the corresponding genes seem to have a common origin [35]. AO is similar to XO, as observed by amino acid sequence alignment. AO has been studied in mammalian systems such as humans, cow, rabbit, and mouse. The rabbit liver enzyme is the most extensively studied. AO utilizes a wide range of aldehydes as substrates but not xanthine.

The active sites are identical to those of XO: one MPT, one flavin moiety, and two [2Fe-2S] centers. Like XO, these AOs are rendered nonfunctional upon reaction with cyanide (see Sec. 2.2.2). This reaction is faster in AO than in XO, indicating that the Mo site is more accessible to solvent.

Multiple variants of AO have been postulated in humans, and the existence of two benzaldehyde oxidase activities in mouse liver was reported [34,36]. Recently, the identification and characterization of two novel mouse cDNAs coding for putative molybdo flavoproteins named AOH1 and AOH2 (aldehyde oxidase homologues 1 and 2) was made, based on their striking resemblance with the molybdo proteins AO and XO/XDH. This was the first indication that, in the animal kingdom, the family of mammalian molybdenum proteins extends beyond sulfite oxidase, AO, and XO/XDH. No direct demonstration of Mo cofactors was made, but based on amino acid sequence comparisons, consensus sequences were observed for Mo-pterin, FAD, and the putative binding sites of 2 × [2Fe-2S] centers. The novel proteins are closer to

AO, due to the lack of a NAD binding site, always conserved in XOR. AOH1, AOH2 and AO are isoenzymes acting on different but overlapping substrates.

2.3.2. Spectroscopic Properties

The UV-Vis absorption spectra of native and deflavo forms of AO from rabbit liver are essentially identical to the ones from the corresponding forms obtained in XO. Low-intensity "resting"-type EPR signals are observed in certain preparations and considered to be due to enzyme molecules modified due to organic solvent manipulation (in general less than 5%). Rapid type 2 EPR signals have been generated in the rabbit liver enzyme, similar to XO, with g-values at 1.9895, 1.9071, and 1.9624, with coupled protons, a_{av} 14, and 8.3 G. Detailed measurements of proton coupling and product analysis suggest differences in orientation and coordination of substrates toward the Mo site when AO and XO are compared. Other typcial signals of molybdenum hydroxylases are generated under appropriate conditions (rapid type 1, slow and inhibited) ([7] and references therein).

The midpoint potentials of Mo(VI)/(V) and Mo(V)/(IV) transitions are similar to XO and the midpoint potentials of the Fe-S centers slightly different (−207 and −310 mV).

2.4. Retinal Oxidase from Mammalian Liver

2.4.1. Molecular Properties and Catalytic Activity

Retinal oxidase catalyzes the conversion of retinal to retinoic acid, an active metabolite of vitamin A or rethinol. Retinal oxidase obtained from rabbit liver cytosol is a homodimeric protein (135 kDa per monomer) that contains 1 FAD, 4 Fe, and 1 Mo per monomer (no pterin cofactor was reported) [37]. Retinal oxidase is able to catalyze a broad range of aldehydes such as all-*trans*-retinal, 9-*cis*-retinal, 13-*cis*-retinal, and N-methylnicotinamide. All these reactions are performed in aerobic conditions, the activity being highly suppressed under anaerobic conditions in the presence of electron donors (NAD$^+$ and NADP$^+$) and acceptors (NADH and NADPH). Comparison of the molecular, inmunological, and enzymatic properties of retinal oxidase and aldehyde oxidase from rabbit liver cytosol led to the conclusion that both enzymes are identical [38,39]. Therefore, it was suggested that a possible role for the mammalian liver aldehyde oxidase is the synthesis of retinoic acid from retinal.

Recently, direct evidence was provided that retinal oxidase is identical to aldehyde oxidase by cDNA cloning [40]. The cDNAs of rabbit and mouse retinal oxidases have a common sequence of similar size. Recently, a prokaryotic expression system for mouse retinal oxidase was constructed. The purified recombinant protein shows a typical spectrum of aldehyde oxidase. This is the first time an eukaryotic molybdenum-containing flavoprotein was expressed in an active form in a prokaryotic system.

3. ENZYMES OF THE ALDEHYDE OXIDOREDUCTASE FAMILY

Several tungsten enzymes have now been identified, mostly in thermophilic organisms (see also Chapter 19) [41,42]. These enzymes have been divided into two major classes: the aldehyde utilizers and those that activate CO. The former group belongs to the AOR family. Three members of this family were purified from *Pyrococcus furiosus*: AOR, FOR, and GAPOR (glyceraldehyde-3-phosphate ferredoxin oxidoreductase) ([43] and references therein). They all catalyze the oxidation of aldehydes to the corresponding carboxylic acids and use ferredoxin as electron donor. The purification and characterization of a tungsten-containing formate dehydrogenase from *D. gigas* [44] shows that W can also be used by mesophiles. A preliminary crystallographic structure of the enzyme was recently obtained by MAD methods [45] revealing the metal center architecture and the distribution of the centers in the two subunits.

3.1. Aldehyde Ferredoxin Oxidoreductase and the Formaldehyde Ferredoxin Oxidoreductase from *Pyrococcus furiosus*

3.1.1. Three-Dimensional Structures

AOR from *P. furiosus* was the first enzyme containing the molybdopterin cofactor to be characterized by crystallography (2.3 Å resolution: PDB code 1AOR) [46]. The crystal structure of FOR, another member of the family isolated from the same organism, was determined in the native form and complexed with the inhibitor glutarate at 1.85 Å and 2.4 Å resolution, respectively [43] (PDB codes 1B25 and 1B4N, respectively; see Fig. 3). Members of this family of enzymes have a single subunit of about 65 kDa, and contain one tungsten atom bound to two pyranopterin cofactors (MPT) and a [4Fe-4S] cluster. They share no sequence or structural similarities with other molybdopterin-containing enzymes (e.g., those described in Sec. 2.1).

While AOR oxidizes a large range of aromatic or aliphatic aldehydes derived from the catabolism of amino acids such as alanine, phenylalanine, or tryptophan, FOR can oxidize formaldehyde or C2-C4 aliphatic aldehydes. Both enzymes use the redox protein ferredoxin as the physiological electron acceptor. They show 40% amino acid sequence identity and 61% similarity. The 3D structures of AOR and FOR are very similar, showing an rmsd of 1.5 Å for 576 C^α atoms. Although AOR and FOR show different quaternary structures (AOR is a dimer and FOR a tetramer both in solution and in the crystal), both are folded into three domains: domain I includes residues 1–210, domain II residues 211–417, and domain III residues 418–605. In AOR, domains II and III define a tunnel at their interface, which gives access to the buried W active site. In FOR instead of an open channel, there is a large cavity close to the active site, which contacts the protein surface through a much narrower channel. In both enzymes, no protein side chain coordinates the W atom, which binds two pterins. An oxygen ligand is proposed as a fifth ligand in the FOR structure, in addition to the four sulfur atoms of the two dithiolene groups. The two pterins are linked via a magnesium ion, which bridges the respective phosphate groups.

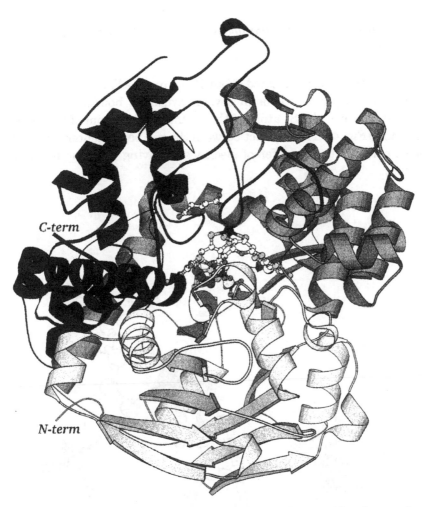

FIG. 3.   Crystal structure of the formaldehyde ferredoxin oxidoreductase from *Pyrococcus furiosus*; FOR with the three domains represented in different tones of gray and the two cofactors and the inhibitor glutarate represented in ball-and-stick mode (domain I: residues 1–208, light gray; domain II: residues 209–406, medium gray; domain III: residues 407–619, dark gray).

3.1.2.   Electron Transfer Pathways and Structure of the Metal Sites

The [4Fe-4S] cluster is located about 10 Å from the W atom and is linked to one of the pterins through several hydrogen bonds. One of these bonds is between the S^γ atom of Cys^{491} and the N8 atom of the pterinic structure (Fig. 4). This contact between the two redox centers provides a direct path for the transfer of electrons involving only one of the two pterins.

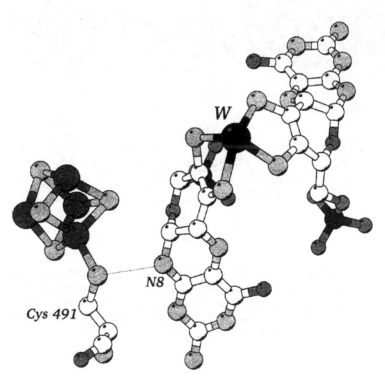

FIG. 4.   Electron transfer pathway defined through the redox centers of the formaldehyde ferredoxin oxidoreductase from *P. furiosus* (FOR) (or of the aldehyde ferredoxin oxidoreductase from *P. furiosus* (AOR)): W-bisMPT and [4Fe-4S] center.

3.1.3. Structural Comparison of the AOR and FOR Active Sites with Those from the D. gigas and D. desulfuricans Enzymes

As mentioned above, FOR and members of the XO family of enzymes are unrelated in terms of their amino acid sequences and 3D structures. However, on the basis of similarities of the active sites of MOP and FOR (Fig. 5) and since both enzymes catalyze similar oxidation reactions, similar reaction mechanisms were proposed that could also be relevant to other members of the AOR family [45]. As depicted in the superposition of Fig. 5, the substrate binding site occupied in MOP by an isopropanol molecule corresponds in FOR to the binding of the inhibitor glutarate. In addition, Glu^{308} corresponds to Glu^{869} of MOP and is also relatively close to the metal atom in a position trans to the apical ligand. In a similar fashion to MOP, Glu^{308} in FOR could probably play a similar role by activating a water molecule for the nucleophilic attack to the carbonyl carbon atom of the substrate. The apical ligand in MOP (a sulfur atom in XO) corresponds to one sulfur atom of the dithiolene group of the second pterin of FOR. Tyr^{416} of FOR was also proposed [43] to be involved in the mechanism, by polarizing the aldehyde group activating it for the nucleophilic attack (as Tyr^{622} in MOP [17]).

3.1.4. Spectroscopic Studies on the Tungsten-Containing Formaldehyde Ferredoxin Oxidoreductase from Thermococcus litoralis

The electronic and redox properties of this hyperthermophilic enzyme were recently studied combining EPR and magnetic circular dichroism [47]. The results revealed the presence of a [4Fe-4S] center that cycles between an oxidised $S = 0$ ground state and a reduced state that exhibits a mixture of $S = 1/2$ and $S = 3/2$ spin states that are temperature- and pH-dependent. Three distinct types of Mo(V) EPR signals are observed, that differ in redox potential and EPR g-values. Structural models were proposed for the W(V) species detected, being suggested that some species are derived from ligand-based oxidation of W(VI) species. It was proposed that the W(VI) species contains a terminal sulfido group.

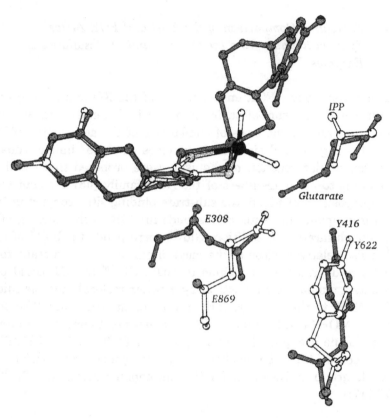

FIG. 5. Superposition of the active sites of formaldehyde ferredoxin oxidoreductase (FOR) from *P. furiosus* (dark gray) and of MOP (color coded: carbon atoms in white, oxygen atoms in light gray, sulfurs in medium gray, and nitrogen atoms in dark gray), showing the correspondence between residues of the active site (Glu[869] and Tyr[622] in MOP, and Glu[308] and Tyr[416] in FOR, respectively) as well as the bound inhibitors (isopropanol in MOP and glutarate in FOR). (Adapted from [46].)

3.2. Aldehyde Dehydrogenase from *D. gigas*

3.2.1. Molecular Properties and Catalytic Activity

Aldehyde dehydrogenase (ADH) is a tungsten-containing enzyme obtained from the sulfate-reducing bacterium *D. gigas* grown in a medium supplemented with W [48]. Like AOR from *D. gigas*, ADH catalyzes

the oxidation of a broad range of aldehydes, such as acetaldehyde, propionaldehyde, and benzaldehyde. However, contrary to MOP, ADH is
only expressed in growth mediums supplemented with W. ADH is a
homodimer (α_2) with a subunit of molecular weight 62 kDa, and the
metal contents is 0.68 ± 0.08 W, 4.8 Fe, and 3.2 ± 0.2 labile S per monomer. A pterin cofactor was also detected, though of unknown type. ADH
was obtained in anaerobic conditions in the presence of dithionite. The
as-purified form shows a fourfold overlapping EPR signal typical of W:
(1) $g_1 = 1.943$, $g_2 = 1.902$, $g_3 = 1.850$ (40%), (2) $g_1 = 1.974$, $g_2 = 1.932$,
$g_3 = 1.869$ (26%), (3) $g_1 = 1.949$, $g_2 = 1.889$, $g_3 = 1.844$ (24%), and (4)
$g_1 = 1.962$, $g_2 = 1.921$, $g_3 = 1.874$ (10%). The nature of this inhomo-
geneity at the W site is unknown, but it was suggested, on the basis of
similar EPR signals observed in the closely related *P. furiosus* W-AOR,
that this might be produced by the presence of glycerol in the purification buffer [43]. Signals 1 and 2 have similar g-values to W(V) species
detected in *P. furiosus* GAPOR, whereas signals 3 and 4 have g-values
close to the glycerol-inhibited AOR ([43] and references therein). W
signals are replaced by a broad EPR signal detected at 15 K (apparent
g-values 2.1-1.8) upon reduction of the samples with excess of dithionite
at pH = 7, which increases its intensity up to 0.2 spin per monomer
upon incubation with excess of dithionite at pH = 9.5. On the basis of
the g-values, fast relaxation behavior, spin quantitation, and metal contents, this signal was assigned to a Fe-S cluster of the [4Fe-4S] type,
having a redox potential less than -0.5 V. In addition, it was suggested
that the broadening of this signal is due to magnetic interactions with
another unknown paramagnet that does not change the redox state
upon reduction of the Fe-S cluster, presumably, a W(IV) (S = 1). An
alternative explanation considers that signals 3 and 4 are lost due to
reduction to W(IV), whereas signals 1 and 2 are lost because of the
magnetic interactions with the Fe-S cluster (S = 1/2) in the reduced
state).

4. CONCLUSIONS

The Mo (and W) containing enzymes have been classified according to
the Mo (and W) coordination site characteristics [7,9]:

Family 1—Xanthine oxidase
Family 2—DMSO reductase
Family 3—Sulfite oxidase
Family 4—*P. furiosus* aldehyde oxidoreductase (thermophiles)

The enzymes catalyzing reactions involving aldehydes belong to the first and last families indicated. Many differences are found in structural terms between families 1 and 4, in respect to pterin type and stoichiometry, Fe-S center types, domain composition and topology, and electron transfer pathways.

These differences are striking and surprising, and the homologies are mainly restricted to the vicinity of the active site. No amino acid side chain is used as ligands of the metal active site. Remarkably, as discussed, the conserved glutamate residue close to the Mo (or W) has a crucial role in the enzyme mechanism and the definition of the putative substrate binding site.

ACKNOWLEDGMENTS

Work supported by projects EC-TMR/HPRN-CT-1999-00084, EC-TMR/FMRXCT980204, PRAXIS-POCTI/1999/BME/35078, PRAXIS/BIA/11089/98 and PhD grant PRAXIS XXI/BD/15752/98 (to CC).

ABBREVIATIONS

ADH	aldehyde dehydrogenase
AOR	aldehyde ferredoxin oxidoreductase from *Pyrococcus furiosus*
CD	circular dichroism
5'-CDP	circular dichroism 5'-cytidine diphosphate
CODH	carbon monoxide dehydrogenase
EXAFS	extended X-ray absorption line structure
FAD	extended X-ray absorption line structure flavin adenine dinucleotide
FOR	formaldehyde ferredoxin oxidoreductase from *Pyrococcus furiosus*

GAPOR	glyceraldehyde-3-phosphate ferredoxin oxidoreductase
IPP	isopropanol
MAD	multiwavelength anomalous dispersion
MCD	molybdopterin cytosine dinucleotide
MGD	molybdopterin guanine dinucleotide
MOD	aldehyde oxidoreductase from *Desulfovibrio desulfuricans* ATCC 27774
MOP	aldehyde oxidoreductase from *Desulfovibrio gigas*
Moco	pyranopterin-ene-1,2-dithiolate cofactor or molybdopterin
MPT	monophosphate form of molybdopterin
rmsd	root mean square deviation
SAL	salicylate
XDH	xanthine dehydrogenase
XO	xanthine oxidase

REFERENCES

1. B. A. S. Barata and J. J. G. Moura, *Meth. Enzymol.*, *243*, 24–42 (1994).

2. B. A. S. Barata, J. LeGall, and J. J. G. Moura, *Biochemistry, 32*, 11559–11568 (1993).

3. N. A. Turner, B. A. S. Barata, R. C. Bray, J. Deistung, J. LeGall, and J. J. G. Moura, *Biochem. J.*, *243*, 755–761 (1987).

4. M. J. Romão and J. J. G. Moura, Aldehyde Oxidoreductase (MOP) in *Handbook of Metalloproteins* (K. Wieghardt, R. Huber, T. L. Poulos, and A. Messerschmidt, eds.), Wiley-VCH, New York, 2001, pp. 1037–1047.

5. R. O. Duarte, M. Archer, J. M. Dias, S. Bursakov, R. Huber, I. Moura, M. J. Romão, and J. J. G. Moura, *Biochem. Biophys. Res. Commun.*, *268*, 745–749 (2000).

6. S. L. A. Andrade, C. D. Brondino, M. J. Feio, I. Moura, and J. J. G. Moura, *Eur. J. Biochem.*, *267*, 2054–2061 (2000).

7. R. Hille, *Chem. Rev.*, *96*, 2757–2816 (1996).

8. M. J. Romão, N. Rösch, J. Knäblein, and R. Huber, *J. Biol. Inorg. Chem.*, *2*, 782–785 (1997).

9. M. J. Romão, J. Knäblein, R. Huber, and J. J. G. Moura, *Prog. Biophys. Mol. Biol.*, *68*, 121–144 (1997).

10. M. J. Romão and R. Huber, in *Metal Sites in Proteins and Models: Redox Centers, Structure and Bonding*, Vol. 90. (H. A. O. Hill, A. J. Sadler, and A. J. Thomson, eds.), Springer-Verlag, Berlin, 1998, pp. 69–96.

11. C. Enroth, B. T. Eger, K. Okamoto, T. Nishino, and E. F. Pai, *Proc. Natl. Acad. Sci. USA.*, *97*, 10723–10728 (2000).

12. H. Dobbek, L. Gremer, O. Meyer, and R. Huber, *Proc. Natl. Acad. Sci. USA*, *96*,, 8884–8889 (1999).

13. F. J. M. Stams, M. Veenhuis, G. H. Weenk, and T. A. Hansen, *Arch. Microbiol.*, *136*, 54–59 (1983).

14. J. Rebelo, S. Macieira, J. M. Dias, R. Huber, C. S. Ascenso, F. Rusnak, J. J. G. Moura, I. Moura, and M. J. Romão, *J. Mol. Biol.*, *297*, 135–146 (2000).

15. M. J. Romão, M. Archer, I. Moura, J. J. G. Moura, J. LeGall, R. Engh, M. Schneider, P. Hof, and R. Huber, *Science*, 1170–1176 (1995).

16. H. Sticht and P. Rosch, *Prog. Biophys. Mol. Biol.*, *70*, 95–136 (1998).

17. R. Huber, P. Hof, R. O. Duarte, J. J. G. Moura, I. Moura, M.-Y. Liu, J. LeGall, R. Hille, M. Archer, and M. J. Romão, *Proc. Natl. Acad. Sci. USA*, *93*, 8846–8851 (1996).

18. M. Xia, R. Dempski, and R. Hille, *J. Biol. Chem.*, *274*, 3323–3330 (1999).

19. P. Hanzelmann, H. Dobbek, L. Gremer, R. Huber, and O. Meyer, *J. Mol. Biol.*, *301*, 1221–1235 (2000).

20. D. N. Beratan, J. N. Betts, and J. N. Onuchic, *Science*, *252*, 1285–1288 (1991).

21. A. Glatigny, P. Hof, M. J. Romão, R. Huber, and C. Scazzocchio, *J. Mol. Biol.*, *278*, 431–438 (1998).

22. R. C. Bray, in *The Enzymes*, Vol. 12 (P. D. Boyer, ed.), 1975, pp. 299–419.

23. R. C. Bray, in *Adv. in Enzymol.*, *51*, 107–114 (1980).

24. J. J. G. Moura, A. V. Xavier, M. Bruschi, J. LeGall, D. O. Hall, and R. Cammack, *Biochem. Biophys. Res. Commun.*, 72, 782–789 (1976).

25. H. Komai, V. Massey, and G. Palmer, *J. Biol. Chem.*, *244*, 1692–1700 (1969).

26. K. V. Rajagopalan and P. Handler, *J. Biol. Chem.*, *239*, 2022 (1964).

27. K. Garbett, R. D. Gilard, P. F. Knowles, and J. E. Stangroom, *Nature*, *215*, 824–828 (1967).

28. J. J. G. Moura, A. V. Xavier, R. Cammack, D. O. Hall, M. Bruschi, and J. LeGall, *Biochem J.*, *173*, 419–425 (1978).

29. S. P. Cramer, J. J. G. Moura, A. V. Xavier, and J. LeGall, *J. Inorg. Biochem. 20*, 275 (1984).

30. B. A. S. Barata, J. Liang, I. Moura, J. LeGall, J. J. G. Moura, and B. H. Huynh, *Eur. J. Biochem.*, *204*, 773–778 (1992).

31. J. Caldeira, V. Belle, M. Asso, B. Guigliarelli, I. Moura, J. J. G. Moura, and P. Bertrand, *Biochemistry, 39*, 2700–2707 (2000).

32. D. J. Lowe, C. J. Mitchell, and R. C. Bray, *Biochem. Soc. Trans.*, *3*, 527S (1997).

33. S. Kardinahl, C. L. Schmidt, T. Hansen, S. Anemüller, A. Petersen, and G. Schäfer, *Eur. J. Biochem.*, *260*, 540–548 (1999).

34. M. Terao, M. Kurosaki, G. Saltini, S. Demontis, M. Marini, M. Salmona, and E. Garattini, *J. Biol. Chem.*, *275*, 30690–30700 (2000).

35. G. Cazzaniga, M. Terao, P. Lo Schiavo, F. Galbiati, F. Segalla, M. F. Seldin, and E. Garattini, *Genomics, 23*, 390–402 (1994).

36. R. S. Holmes, *Biochem. Genet.*, *17*, 517–528 (1979).

37. M. Tsujita, S. Tomita, S. Miura, and Y. Ichikawa, *Biochim. Biophys. Acta, 1204*, 108–116 (1994).

38. K. V. Rajagopalan, I. Fridovich, and P. Handler, *J. Biol. Chem.*, *237*, 922 (1962).

39. S. Tomita, M. Tsujita, and Y. Ichikawa, *FEBS Lett, 336*, 272–274 (1993).

40. D. Y. Huang, A. Furukawa, and Y. Ichikawa, *Arch. Biochem. Biophys.*, *364*, 264–272 (1999).

41. M. W. Adams, Tungsten enzymes, in *Encylopedia of Inorganic Chemistry* (R. B. King, ed.), John Wiley and Sons, Chichester, 1988, pp. 4284–4291.

42. M. K. Johnson, D. C. Rees, and M. W. Adams, *Chem. Rev.*, *96*, 2817–2839 (1996).

43. Y. Hu, S. Faham, R. Roy, M. W. Adams, and D. C. Rees, *J. Mol.. Biol.*, *286*, 899–914 (1999).

44. J. Almendra, C. D. Brondino, O. Gavel, A. S. Pereira, P. Tavares, S. Bursakov, R. O. Duarte, J. Caldeira, J. J. G. Moura, and I. Moura, *Biochemistry*, *38*, 16366–16372 (1999).

45. H. Raaijmakers, S. Teixeira, J. M. Dias, M. J. Almendra, C. D. Brondino, I. Moura, J. J. G. Moura, and M. J. Romão, *J. Biol. Inorg. Chem.*, 6(4), 398–404 (2001).

46. K. Chan, S. Mukund, A. Kletzin, M. W. Adams, and D. C. Rees, *Science*, *26*, 1463–1469 (1995).

47. I. K. Dhawan, R. Roy, B. P. Koehler, S. Mukund, M. W. Adams, and M. K. Johnson, *J. Biol. Inorg. Chem.*, *5*, 313–327 (2000).

48. C. M. H. Hensgens, W. R. Hagen, and T. O. Hansen, *J. Bacteriol.*, *177*, 6195–6200 (1995).

49. S. Gutterige, S. J. Tannerand, and R. C. Bray, *Biochem. J.*, *175*, 887–897 (1978).

50. R. C. Bray, in *Flavins and Flavoproteins* (R. C. Bray, P. C. Engels, and S. G. Mayhew, eds.), de Gruyter, Berlin, 1984, pp. 707–722.

51. M. J. Barber, M. P. Coughland, K. V. Rajagopalan, and L. M. Siegel, *Biochemistry*, *21*, 3561–3568 (1982).

52. R. C. Bray, G. N. George, S. Gutterige, L. Norlander, J. G. Stell, and C. Stubley, *Biochem. J.*, *203*, 263–267 (1982).

16

Molybdenum and Tungsten Enzymes in C1 Metabolism

Julia A. Vorholt and Rudolf K. Thauer

Max-Planck-Institut für terrestrische Mikrobiologie and Laboratorium für Mikrobiologie, Fachbereich Biologie, Philipps-Universität, Karl-von-Frisch-Strasse, D-35043 Marburg, Germany

1. INTRODUCTION

Four molybdenum or tungsten enzymes are known that catalyze the oxidoreduction of C1 compounds as expressed in reactions (1)–(4) (for the thermodynamic data, see [1]):

Formate dehydrogenase (FDH); reaction (1)

Formylmethanofuran dehydrogenase (FMD); reaction (2) [for the structures of methanofuran (MFR) and formyl-MFR, see Fig. 2 below]
Carbon Monoxide dehydrogenase (CODH); reaction (3)
(Form)aldehyde oxidoreductase (FODH); reaction (4)

$$HCOO^- + H^+ \rightleftharpoons CO_2 + 2[H] \qquad E^{0'} = -430\,mV^a \qquad (1)$$

$$Formyl\text{-}MFR + H_2O \rightleftharpoons CO_2 + 2[H] + MFR \qquad E^{0'} = -530\,mV^c \qquad (2)$$

$$CO + H_2O \rightleftharpoons CO_2 + 2[H] \qquad E^{0'} = -520\,mV^a \qquad (3)$$

$$H_2CO + H_2O \rightleftharpoons HCO_2^- + H^+ + 2[H] \qquad E^{0'} = -530\,mV\ (-580\,mV^b) \qquad (4)$$

The molybdenum formate dehydrogenases and the molybdenum formylmethanofuran dehydrogenases belong to the dimethy sulfoxide (DMSO) reductase family of mononuclear molybdenum enzymes [2]. The respective tungsten isoenzymes are structurally very similar [3]. The molybdenum carbon monoxide dehydrogenases and the molybdenum (form)aldehyde oxidoreductases belong to the xanthine oxidase family of mononuclear molybdenum enzymes [2]. The tungsten (form)aldehyde oxidoreductases form a family of their own [3].

Not all formate dehydrogenases, carbon monoxide dehydrogenases, and (form)aldehyde oxidoreductases contain molybdenum or tungsten. Thus, one of the NAD-dependent formate dehydrogenases in aerobic methylotrophs and the NAD-dependent formate dehydrogenase in methylotrophic yeast and plants and the formaldehyde dehydrogenase in most microorganisms are devoid of a redox-active transition metal and carbon monoxide dehydrogenases in anaerobic microorganisms are nickel proteins. Only formlymethanofuran dehydrogenases appear to be always molybdenum or tungsten enzymes. The molybdenum and tungsten enzymes and the respective non-molybdenum- and non-tungsten-

---

a Calculated for CO_2 in the gaseous state at 1 atm pressure and for formate at 1 molal activity.

b Calculated using a free energy of formation for formaldehyde in aqueous solution at 1 M of $-121.5\,kJ/mol$ [4] instead of $-130.54\,kJ/mol$ [1] (see [5]).

c Determined by Bertram and Thauer [6].

containing enzymes show no sequence similarity, indicating that they have evolved separately.

Xanthine dehydrogenases/oxidases catalyzing reactions (5) and (6) are molybdoenzymes which also have to be considered here. C2 and C8 of hypoxanthine are C1 units at the oxidation level of formate and C2 and C8 of uric acid are C1 units at the oxidation level of CO_2.

$$\text{Hypoxanthine} + H_2O \rightleftharpoons \text{xanthine} + 2[H] \qquad E^{0'} = -430\,\text{mV}^{\text{d}} \quad (5)$$

$$\text{Xanthine} + H_2O \rightleftharpoons \text{uric acid} + 2[H] \qquad E^{0'} = -300\,\text{mV}^{\text{d}} \quad (6)$$

For all tungsten enzymes molybdenum enzyme counterparts have been found but not vice versa. Formate dehydrogenases, formylmethanofuran dehydrogenases and formaldehyde oxidoreductases can be tungsten or molybdenum enzymes. Carbon monoxide dehydrogenase from aerobic microorganisms and the xanthine dehydrogenases/oxidases from all organisms appear to always contain molybdenum. In the case of xanthine dehydrogenase/oxidases it has been argued that the reason for this is that tungsten has a much more negative redox potential than molybdenum and can therefore not be involved in the catalysis of reactions with redox potential more positive than −400 mV. This argument has recently been proven wrong by showing that trimethylamine N-oxide reductase [8] and dimethyl sulfoxide reductase [9] are capable of catalysis with either molybdenum or tungsten at their active site (see also Chapter 20 of this book) despite the fact that the trimethylamine N-oxide/trimethylamine couple has a redox potential $E^{0'}$ of +130 mV and the DMSO/dimethylsulfide couple one of +160 mV ($E^{0'}$ data from [10]). In the case of the carbon monoxide dehydrogenase from aerobic organisms it has been argued that tungsten enzymes are restricted to anaerobic organisms, but there is evidence that formate dehydrogenases from aerobic methylotrophs can be tungsten enzymes [11–13]. It is probably the higher availability of tungsten in some anaerobic environ-

---

d Calculated from the free energies of formation of hypoxanthine (+89.5 kJ/mol) and uric acid (−356.9 kJ/mol) in aqueous solution at 1 M concentration and of xanthine in the crystalline solid state (−165.85 kJ/mol) [1] taking the solubility of xanthine in water at 20°C of 0.5 g/L and the pK_a value of uric acid (pK_{a1}=5.4) [7] into account.

ments which is the reason why most of the tungsten enzymes have been found in strictly anaerobic microorganisms (see Chapter 1).

The following overview concentrates on the structures and functions of formate dehydrogenases, formylmethanofuran dehydrogenases, carbon monoxide dehydrogenases, and formaldehyde dehydrogenases that contain molybdenum or tungsten. The properties of xanthine dehydrogenases and xanthine oxidases are dealt with in Chapters 12 and 13 and are therefore not considered here. At the end of each section some information on the respective enzymes not containing molybdenum or tungsten is also given.

With a few exceptions only literature published within the last 10 years will be cited.

2. FORMATE DEHYDROGENASE

Formate is an end product of many bacterial fermentations, e.g., the mixed-acid fermentations of enterobacteria, where it is formed in the pyruvate formate lyase reaction (reaction (7)). Formate does not, however, accumulate in the environment due to microorganisms that can grow at the expense of formate oxidation to CO_2 (reaction (1)) with either CO_2, sulfate, nitrate, fumarate, Mn(IV), Fe(III), polysulfide, or O_2 as electron acceptor.

$$\text{Pyruvate} + \text{CoA} \longrightarrow \text{acetyl-CoA} + \text{formate} \quad \Delta G^{0'} = -16.3\,\text{kJ/mol} \tag{7}$$

Formate is also an intermediate in the energy metabolism of some microorganisms. Thus, CO_2 reduction to acetic acid in acetogenic bacteria proceeds via formate as intermediate. These were just a few examples of the role of formate in energy metabolism (for more complete treaties, see [14–16]).

Besides being a product, substrate, or intermediate in the energy metabolism of microoganisms, formate is required or can be used by many cells for biosynthesis. Formate is incorporated into positions 2 and 8 of purines, the formyl group of formylmethionine, the hydroxymethyl group of serine, and the methyl group of methionine and thymine. The first step in these pathways is generally the formation of

N^{10}-formyltetrahydrofolate from formate and tetrahydrofolate (H_4F) catalyzed by N^{10}-formyltetrahydrofolate synthetase (reaction (8)) (for reviews, see [5,14]).

$$\text{Formate} + H_4F + ATP \longrightarrow N^{10}\text{-formyl-}H_4F + ADP + Pi$$

$$\Delta G^{0'} = +8.4\,\text{kJ/mol} \tag{8}$$

In many microorganisms incorporation of formate into C2 and C8 of purines can also proceed directly not involving N^{10}-formyl-H_4F as intermediate. In methanogenic- and sulfate-reducing archaea lacking H_4F, direct incorporation is the only pathway [5].

In strictly anaerobic bacteria with type 3 ribonucleotide reductase formate is required for ribonucleotide (NDP) reduction to deoxyribonucleotides (dNDP) (reaction (9)). Formate is a substrate of the glycine radical enzyme (for review, see [17]).

$$\text{Formate} + NDP \longrightarrow CO_2 + dNDP \tag{9}$$

Of the enzymes catalyzing reactions (6)–(9) none contain molybdenum or tungsten. Thus, in formate metabolism only the formate dehydrogenases are molybdenum or tungsten enzymes, and even that is not always the case because there are formate dehydrogenases that contain neither molybdenum nor tungsten (see Sec. 2.9).

The diversity of functions of formate dehydrogenases is reflected in differences in topology and electron acceptor specificity [16]. Formate dehydrogenases are known that are tightly associated with the cytoplasmic membrane facing either the cytoplasm or the periplasm and that involve membrane-associated electron carriers. Others are soluble cytoplasmic enzymes specific for NAD(P), coenzyme F_{420}, or ferredoxin. It is not surprising that the subunit and cofactor composition of formate dehydrogenases is diverse: one subunit or several different subunits; molybdenum or tungsten; a selenocysteine or a cysteine in the active site; one (?) or two molybdopterin guanine dinucleotides; one or several iron-sulfur clusters; zero, one, or two cytochrome b or cytochrome c; and FAD or FMN. However, all formate dehydrogenases appear to have in common that they catalyze the formation or reduction of CO_2 rather than of bicarbonate [18,19] (for literature, see [20]), indicating that formate oxidation is not associated with an oxygen transfer reaction catalyzed by many molybdenum and tungsten enzymes. They also have in common that they all appear to be inhibited by azide, which is a transition state analogue of formate [21].

In the following the properties and function of the three membrane-associated formate dehydrogenases from *Escherichia coli* will be discussed first since the crystal structure of one of the enzymes, of FDH-H, from this facultative organism has been solved. Sections 2.1–2.6 deal with formate dehydrogenases functioning in anaerobically growing microorganisms and Secs. 2.7–2.9 with formate dehydrogenases from aerobic microorganisms. The molybdate or tungsten formate dehydrogenases described in Secs. 2.1–2.8 are rapidly inactivated in the presence of O_2, whereas the formate dehydrogenases devoid of a prosthetic group (Sec. 2.9) are stable under these conditions. The oxygen lability may be the reason why in eukaryotes molybdenum or tungsten formate dehydrogenase, like nitrogenase, appears to be absent.

2.1. Membrane-Associated Formate Dehydrogenases from *Escherichia coli* (MoSe)

Escherichia coli belongs to the γ group of proteobacteria. It can grow aerobically and anaerobically, anaerobic growth being sustained either by mixed acid fermentation of carbohydrates or by anaerobic respiration with fumarate, nitrate, or DMSO as electron acceptor. *E. coli* has the capacity to synthesize three distinct formate dehydrogenase isoenzymes: FDH-H, FDH-N, and FDH-O = FDH-Z [22]. All three are membrane-associated proteins that are functional in the anaerobic metabolism of the organism. Only one of the formate dehydrogenases, FDH-O, is also synthesized in aerobic cells [23–25]. FDH-H is associated with hydrogenase 3 in the formate-hydrogen lyase pathway (FHL-1), which has recently been shown to be coupled with $2\,H^+/K^+$ exchange [26]. In conjunction with hydrogenase 4 it probably forms a respiratory-linked proton translocating formate-hydrogen lyase (FHL-2) [27]. FDH-N and FDH-O have a respiratory function during anaerobic growth of *E. coli* in the presence of formate and nitrate. The three formate dehydrogenases are molybdoselenocysteine proteins belonging to the DMSO reductase family of mononuclear molybdenum enzymes [2]. It has been proposed that FDH-N and FDH-O are oriented to the periplasm and FDH-H to the cytoplasm [22], which is supported by predictions from sequence analysis [28]. However, recent topological studies have provided evidence that the active site of FDH-O is probably located in the cytoplasm [25]. The same conclusion had previously been drawn from FDH-N by Graham and Boxer [29].

FDH-N is composed of an α, β, and γ subunit with molecular masses of 113 kDa, 32 kDa, and 21 kDa, respectively [30] encoded by the colinear genes *fdnG*, *fdnH*, and *fdnl*, respectively [31,32]. Per α, β, γ heterotrimer 1 mol of molybdenum, 1 selenocysteine, 2 hemes and 14 Fe/S were found. The Mo and Se and probably one [4Fe-4S] cluster are associated with the α subunit and the two hemes with the γ subunit, which is a diheme cytochrome *b*. The physiological electron acceptor is menaquinone. FdnG shows a twin arginine motif indicative of a periplasmic localization of the subunit [33–35].

FDH-O has the same composition as FDH-N, its three subunits α (107 kDa), β (34 kDa), and γ (22 kDa) being encoded by the colinear genes *fdoG*, *fdoH*, and *fdol*, respectively [36]. The sequences of the three subunits of FDH-O are very similar to those of the respective subunits of FDH-N, including the presence of a twin arginine motif in FdoG [37]. The physiological electron acceptor is also menaquinone [23,24].

FDH-H as isolated is composed of only one polypeptide of molecular mass 80 kDa [38] encoded by the *fdhF* gene which is monocistronic [39]. The enzyme contains one molybdenum, two molybdopterin guanine dinucleotides, one selenocysteine, and one [4Fe-4S] cluster [40]. The physiological electron acceptor is probably the *hycB* gene product which encodes for a ferredoxin-like protein. HycB shows convincing homology with the β subunit of formate dehydrogenase from *Wolinella succinogenes* [41]. In the formate hydrogen lyase reaction neither a cytochrome *b* nor menaquinone are involved. FDH-H can be homologously overproduced in *E. coli* [42].

The kinetic mechanism of FDH-H was found to be ping-pong Bi Bi [43]. With ^{13}C-labeled formate in ^{18}O-enriched water it was found that FDH-H produces [13]CO_2 gas that contains no [18]O label [19].

The crystal structure of FDH-H in the oxidized [Mo(VI), Fe_4S_{4ox}] and reduced [Mo(IV), Fe_4S_{4red}] form has been determined to 2.6 Å resolution [40,44]. The electron density revealed a four-domain α, β structure with the molybdenum directly coordinated in a square pyramidal geometry to the four equatorial dithiolene sulfur atoms from the pair of

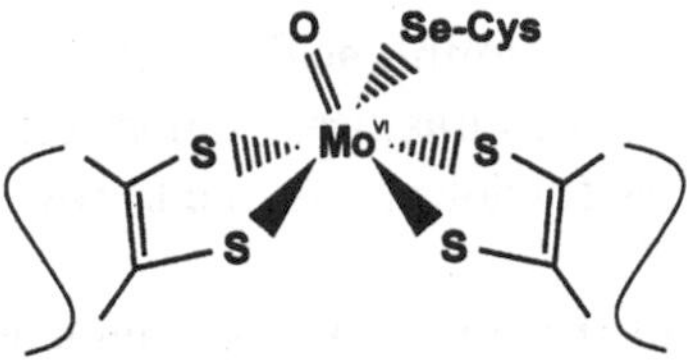

FIG. 1. The active site structure of formate dehydrogenase FDH-H from *Escherichia coli* [40]. The enzyme belongs to the DMSO reductase family of mononuclear molybdenum enzymes [2]. The molybdenum is coordinated to the dithiolene sulfur atoms of two molybdopterin guanine dinucleotides.

pterin cofactors and a Se atom of the selenocysteine[140] (Fig. 1). The structures suggest a reaction mechanism that directly involves SeCys[140] and His[141] in proton abstraction and molybdenum, molybdopterin, Lys[44] and the Fe_4S_4 cluster in electron transfer.

More detailed information on the active site structure was recently obtained by X-ray absorption spectroscopy [45]. From molybdenum K-edge data it was concluded that the oxidized and reduced enzyme are very similar, both containing a molybdenum site with four Mo-S ligands at 2.35 Å, probably one Mo-O at 2.1 Å and one Mo-Se ligand at 2.62 Å. Selenium K-edge EXAFS indicated the presence of Se-S ligation with a bond length of 2.19 Å, the sulfur originating from one of the pterin cofactor dithiolenes [45]. A direct coordination of selenium with Mo had previously already been shown by EPR spectroscopy via substitution of ^{77}Se for natural isotope abundance Se [19,46].

A formate dehydrogenase mutant with selenocysteine[140] replaced by cysteine was active albeit the catalytic efficiency of the mutant enzyme was two orders of magnitude lower [47]. This finding indicates that selenium is very important for the catalytic function of the enzyme from *E. coli*. In the molybdenum formate dehydrogenases from other organisms a cysteine naturally occurs in the position corresponding to selenocysteine[140]. These selenium-free formate dehydrogenases can be very active. The presence of selenium in the active site of formate dehydrogenase is therefore not a prerequisite for high catalytic efficiency. Why some enzymes contain selenium and others not is still not completely understood (see Chapter 18).

The biosynthesis of FDH-H from *E. coli* has extensively been studied in order to elucidate how selenocysteine is synthesized and incorporated selectively into the amino acid position 140 of the polypeptide chain. The mechanism, which has been almost completely resolved, will not be discussed here; the reader is referred to reviews [48,49] and to the original literature [50–53].

The biosynthesis of the formate dehydrogenases in *E. coli* is differentially regulated at the transcriptional level by various environmental factors. The expression of the *fdhF* gene encoding FDH-H proceeds only under anaerobic conditions and only in the presence of formate and molybdate [54–57]. FDH-N and nitrate reductase (NAR-A), which together catalyze formate oxidation coupled to nitrate reduction, are synthesized only in the absence of O_2 whereas synthesis of the isoenzymes FDH-O and NAR-Z appears not to be under the control of the two global regulatory proteins FNR and ArcA, which control genes for aerobic and anaerobic function [37]. For a comprehensive review on the control of formate dehydrogenase gene expression in *E. coli*, see [22].

2.2. Membrane-Associated Formate Dehydrogenases
   from *Wolinella succinogenes* (Mo)

Wolinella succinogenes belongs to the ε group of proteobacteria. The anaerobic organism grows on formate and fumarate and on formate and polysulfide (fore review, see [58]). It contains two membrane-associated formate dehydrogenases FDH-I and FDH-II with the encoding genes organized in two transcription units *fdhEABCD* [41,59]. The nucleotide sequences of *fdhI* and *fdhII* are almost identical, the sequences mainly being different in the promoter region. Available evidence indicates that *fdhI* expression is dependent only on the presence of formate whereas that of *fdhII* is also dependent on other factors [59]. Both enzymes are integrated in the cytoplasmic membrane with the catalytic subunit exposed to the periplasm [60]. The physiological electron acceptor for FDH is menaquinone when coupled with fumarate reductase [61]. When coupled with polysulfide reductase electron transport between formate dehydrogenase and polysulfide reductase is independent of the naphthoquinone [62]. FDH-I and FDH-II belong to the DMSO reductase family of mononuclear molybdenum enzymes [2].

FDH-I has been purified and characterized [63,64]. It is composed of three different subunits α, β, and γ of apparent molecular mass

110 kDa, 25 kDa, and 20 kDa, respectively. Per heterotrimer the enzyme was found to contain 1 mol Mo, 19 mol Fe/S, a pterin cofactor, and no selenium. The pterin cofactor was identified to be molybdopterin guanine dinucleotide [65]. The N-terminal half of FdhA is homologous to the largest subunit of the formate dehydrogenases from *E. coli*. It harbors a conserved cysteine cluster and two more domains, which in the enzyme in *E. coli* are involved in binding the molybdopterin cofactors. FdhB may represent an iron-sulfur protein, 12 cysteine residues of which are arranged in two clusters being typical for ligands of the iron-sulfur clusters of ferredoxins. FdhC is a hydrophobic protein with four predicted transmembrane segments that appear to be identical to the diheme cytochrome *b* present in the FDH. It forms the membrane anchor of the enzyme and reacts with bacterial menaquinone. Polypeptides corresponding to *fdhD* and *fdhE* were not detected in the enzyme preparation [41].

The genes *fdhA* and *fdhC* encode larger preproteins that differ from the corresponding mature proteins by N-terminal signal peptides [41,59]. The periplasmic orientation of FdhA is in agreement with this finding [60].

2.3. Periplasmic Formate Dehydrogenases from Sulfate-Reducing Proteobacteria (MoSe or W)

Formate dehydrogenases from three sulfate reducers belonging to the δ group of proteobacteria have been investigated. In the three anaerobic organisms, which can grow on formate and sulfate as energy sources, the enzyme has a periplasmic location and uses cytochrome *c* (553) as electron acceptor [66–69].

Formate dehydrogenase from *Desulfovibrio gigas* is a heterodimeric tungsten protein with a 92-kDa α subunit and a 29-kDa β subunit. Per mol heterodimer 1 mol tungsten, 8 mol Fe/S organized in two [4Fe-4S] clusters, and 1.3 mol molybdopterin guanidine dinucleotide were found but selenium was not [70]. The enzyme has been crystallized and an analysis of the crystal structure is available [71]. The enzyme is air-stable. It is induced during growth with formate as electron donor [72], growth with formate being dependent on the presence of tunstate in the growth medium [73].

Formate dehydrogenase from *D. vulgaris* is composed of three different subunits. The 83.5-kDa subunit is proposed to contain a molybdenum cofactor, the 27-kDa subunit to be a Fe/S protein, and the 14-kDa subunit to be a *c*-type cytochrome [74].

Formate dehydrogenase from *D. desulfuricans* is composed of three different subunits of apparent molecular masses 88 kDa, 29 kDa, and 16 kDa and contains three types of redox-active centers; a molybdopterin site, four *c*-type hemes, and two [4Fe-4S] centers. Selenium was chemically detected. The enzyme was air-stable [75].

2.4. Ferredoxin-Linked Formate Dehydrogenase from *Clostridium pasteurianum* (Mo)

Clostridium pasteurianum is a gram-positive anaerobic bacterium that ferments carbohydrates to acetate, butyrate, CO_2 and H_2. Also some formate is formed required by the organism for the biosynthesis of C1 units [14]. It has been shown that the formate is generated by CO_2 reduction to formate and that the formate dehydrogenase catalyzing this reaction is ferredoxin-dependent [18,76,77].

The enzyme has been purified and shown to be composed of two different subunits of apparent molecular masses 86 kDa and 34 kDa. For each mode of heterodimer, one or two molybdenums, a pterin cofactor, and 24 Fe/S were found. Selenium, tungsten, a flavin, and heme were not detected [78,79]. The redox-active centers of the enzyme have been characterized by EPR spectroscopy and redox potentiometry [80]. The enzyme exhibits no signal that can be attributed to molybdenum. For the 24 Fe/S, 20 can be extruded as [4Fe-4S] clusters. Yet only two EPR signals were found that might arise from the reduced form of such centers [80].

2.5. NAD(P)-Linked Formate Dehydrogenase from Acetogenic Bacteria (WSe)

The formate dehydrogenases from the acetogenic bacteria *Clostridium thermoaceticum* and *C. formicoaceticum* have been investigated. The function of the enzyme is to catalyze the reduction of CO_2 to formate in acetate formation from $2CO_2$ [81]. The enzyme from *C. thermoaceti-*

cum is NADP-specific and that from *C. formicoaceticum* NAD-specific. Both enzymes are soluble proteins that contain tungsten and selenium [82,83]. The formate dehydrogenase from *C. thermoaceticum* was the first enzyme shown to be a tungsten protein [84,85].

Only the NADP-specific enzyme from *C. thermoaceticum* has been purified and extensively characterized [86]. It is composed of two different subunits of apparent molecular mass 96 kDa and 76 kDa and contains per heterodimer one tungsten, a pterin cofactor, one selenium, and at least 18 Fe/S. The selenium resides in the α subunit. The presence of a flavin has not been reported but is very likely since it is required to transfer electrons from one-electron iron-sulfur centers to the obligate two-electron acceptor pyridine nucleotide. The enzyme exhibits EPR spectra characteristic for proteins with two [4Fe-4S] and two [2Fe-2S] clusters [87]. An EPR signal attributable to tungsten was not detected. The genes encoding the two subunits have been cloned and sequenced [88].

2.6. F_{420}-Linked Formate Dehydrogenases from Methanogenic Archaea (Mo or WSe)

The formate dehydrogenase from two methanogenic archaea have been analyzed in greater detail [16]. The function of the enzyme is to catalyze the oxidation of formate to CO_2 with coenzyme F_{420} as electron acceptor, reduced F_{420} being required for CO_2 reduction to CH_4, a pathway in which free formate is not an intermediate [89]. F_{420} is a 5-deazaflavin and as such an obligate two-electron acceptor. Methanogenic archaea do not appear to contain a membrane-associated formate dehydrogenase as revealed by the complete genome sequence of *Methanobacterium thermoautotrophicum* ΔH [90,91] and of *Methanococcus jannaschii* [92,93].

Formate dehydrogenase from *Methanobacterium formicicum* belongs to the DMSO reductase family of mononuclear molybdenum enzymes [2]. It is composed of two subunits of molecular mass 80 kDa and 30 kDa and contains molybdenum, iron, zinc, molybdopterin cofactor, and FAD but no selenium [94]. The genes encoding for the two polypeptides have been cloned and sequenced [95]. They are organized in the *fdhCAB* operon [96]. Inactive formate dehydrogenase was obtained when *M. formicicum* was grown on H_2/CO_2 in the presence of tungstate [97]. The pterin cofactor was identified to be molybdopterin guanine dinucleotide [98]. Together with F_{420}-reducing hydrogenase the F_{420}-dependent formate dehydrogenase forms a complex that catalyzes the formate-hydrogen lyase reaction [9]. The transcriptional regulation of the *fdhCAB* operon in *M. thermoformicicum* Z245 has been studied. A transcript was present in all growth stages in cells grown in formate but barely detectable during early exponential growth on H_2 plus CO_2. The level of the *fdh* transcript did, however, increase dramatically in cells grown on H_2 plus CO_2 when at the end of exponential growth the growth rate decreased [96].

A 105-kDa F_{420}-dependent formate dehydrogenase isolated from *Methanococcus vannielii* contains molybdenum, iron, and acid-labile sulfide but no selenium. The presence of flavin, though required as a $1e^-$/$2e^-$ switch for transfer of electrons from Fe/S clusters to F_{420}, was not reported. During growth of *M. vannielii* on selenite-supplemented medium a second higher molecular mass formate dehydrogenase is formed that appears to be a molybdenum-selenocysteine flavoprotein. This second high molecular mass enzyme is the predominant form in cells from medium additionally supplemented with tungstate. Under these conditions, almost complete (> 90%) replacement of molybdenum with tungsten appears to occur [100,101].

2.7. NAD-Linked Formate Dehydrogenase from *Ralstonia eutropha* (Mo)

R. eutropha (formerly *Alcaligenes eutropha*) belongs to the β group of proteobacteria. It can grow aerobically on formate as sole carbon and energy source. The aerobic organism contains a soluble NAD^+-linked formate dehydrogenase and a membrane-bound formate dehydrogenase [102]. The membrane-bound enzyme is apparently directly coupled to the respiratory chain but has remained uncharacterized so far.

The NAD-dependent formate dehydrogenase is a heterotetramer that is composed of four nonidentical subunits $\alpha\beta\gamma\delta$ and contains one molecule of molybdopterin guanine dinucleotide and FMN in addition to probably three [2Fe-2S] and four [4Fe-4S] centers as cofactors [103–105]. The gene locus encoding the enzyme has been identified [105]. It encompasses the four structural genes and an additional gene *fdsC* of unknown function forming the pentacistronic operon *fdsGBACD*. Sequence analysis have revealed that the α subunit (FdsA) represents the catalytic subunit of the enzyme complex and that the enzyme belongs to the DMSO reductase family of mononuclear molybdenum enzymes [2]. The β (FdsB) and the γ (FdsG) subunits, together with the NH_2-terminal domain of FdsA, constitute a functional entity corresponding to the NAD dehydrogenase fragment of respiratory chain complex I (NADH:ubiquinone oxidoreductases) and the diaphorase dimers of NAD(P)-dependent hydrogenases from various bacteria [106]. Transcription of the *fds* operon has been shown to be induced during growth of the organism on formate or oxalate. Because oxalate is metabolized to formate via oxalyl-CoA and formyl-CoA [102], the synthesis of the NAD-dependent formate dehydrogenase also appears to be induced by formate. A transcriptional regulatory gene *fdsR* was identified 150 bp upstream of the divergently oriented *fds* operon [106].

2.8.   NAD-Linked Formate Dehydrogenase from Methylotrophic Bacteria (Mo or W)

Methylotrophic bacteria that grow aerobically on methanol or methane as sole carbon and energy source usually contain an NAD-dependent formate dehydrogenase considered to serve as the terminal enzyme in C1 unit oxidation to CO_2, although this has recently been questioned [107]. When the organisms grow on medium containing molybdate they form an NAD-dependent formate dehydrogenase that is a molybdenum iron-sulfur protein. If the medium is supplemented with tungstate the synthesis of the molybdenum enzyme appears to be inhibited in most of these organisms with apparently one exception: In *Methylobacterium* sp. RXM grown in the presence of tungstate the specific activity of formate dehydrogenase was almost as high as in cells grown in the presence of molybdate [11–13]. When some methylotrophs grow on media low in molybdate they synthesize an NAD-dependent formate dehydrogenase devoid of a prosthetic group [108,109], the properties of which are described in Sec. 2.9.

The NAD-dependent formate dehydrogenase from *Methylosinus trichosporium* grown aerobically on molybdate-containing medium using methane as carbon and energy source was found to be composed of four different subunits of apparent molecular masses 98 kDa, 56 kDa, 20 kDa, and 11.5 kDa. The $\alpha_2\beta_2\gamma_2\delta_2$ hetero-octamer contained 1.5 ± 0.1 mol molybdenum, 1.8 ± 0.2 mol of an unusual flavin, and 46 ± 6 mol of Fe/S. EPR spectroscopy revealed the presence of a redox active Mo-pterin center [110]. The enzyme from the same organism was reported before to be composed of only two different subunits of molecular masses of 100 kDa and 53 kDa [111]. The enzyme was shown to be dependent on FMN and multiple iron-sulfur clusters for activity and to exhibit physical properties very similar to those of formate dehydrogenase from *Pseudomonas oxalaticus* that contains FMN and Fe/S clusters and therefore could also contain molybdenum, although this was never directly shown [112].

The NAD-dependent formate dehydrogenase from *Methylobacterium* sp. RXM grown aerobically on molybdate-containing minimal medium using methanol as carbon and energy source was partially purified. The molecular mass determined by SDS-PAGE was reported to be 75 kDa and by gel exclusion chromatography to be 300 kDa. The EPR signals assigned to iron-sulfur centers show a strong analogy with the aldehyde oxidoreductase from *Desulfovibrio gigas* known to contain a Mo-pterin and two [2Fe-2S] centers [113].

2.9. NAD-Linked Formate Dehydrogenase Not Containing Molybdenum or Tungsten

Some aerobic bacteria, when grown in the absence of molybdate, and some yeast and plants contain an NAD-specific formate dehydrogenase that is devoid of a prosthetic group (for reviews see [114,115]). Many of these organisms are methylotrophs. The enzyme is found, for example, in the bacteria *Pseudomonas* sp. 101 (formerly *Achromobacter parvulus*) [116], *Mycobacterium vaccae* N10 [117], and *Moraxella* sp. strain C1 [118] and the yeasts *Candida boidinii* [119–121], *Pichia augusta* [122], *Hansenula polymorpha* [123,124], *Candida methylica* [126,126] and *C. methanolica* [127], and *Neurospora crassa* [128]. The enzymes from the bacteria and the yeast show striking sequence similarity [128].

The enzyme from *Pseudomonas* sp. 101 has been studied in detail. It is composed of two identical subunits with a molecular mass of 46 kDa [129, 130]. The encoding gene has been cloned and heterologously expressed in *E. coli* [131] and the crystal structure has been determined with NAD and azide-bound [132,133], the latter being a transition state analogue of formate [21]. The azide molecule is located near the point of catalysis, the C4 atom of the nicotinamide moiety of NADP, and over-laps with the proposed formate binding site. The structure is consistent with the kinetic and chemical mechanism of yeast formate dehydrogen-ase determined by Blanchard and Cleland [21] and by Mesentsev et al. [134].

An NAD(P)-dependent formate dehydrogenase, which can be used to regenerate NADPH, has recently been described [135].

3. FORMYLMETHANOFURAN DEHYDROGENASE

N-Formylmethanofuran (formyl-MFR) (Fig. 2) is an intermediate in CO_2 reduction to methane and in methanol disproportionation to CO_2 and methane in methanogenic archaea [89]. It is an intermediate in lactate oxidation to $3CO_2$ and in autotrophic CO_2 fixation in sulfate-reducing archaea [136,137]. Genetic evidence indicates that formyl-MFR may also be an intermediate in formaldehyde oxidation to CO_2 in methylotrophic and methanotrophic bacteria [107,138,139].

Formyl-MFR is generated from CO_2 and MFR (Fig. 2) by reduction (reaction (2)) or by formyl transfer from N^5-formyltetrahydromethano-pterin (formyl-H_4MPT) to methanofuran (reaction (10)). Reaction (2)

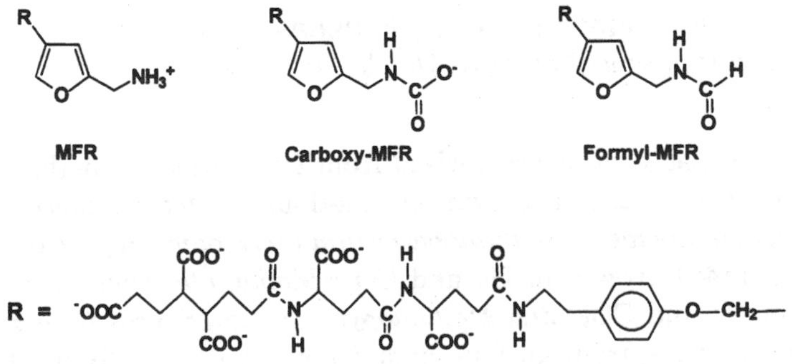

FIG. 2.   Structures of methanofuran (MFR), *N*-carboxymethanofuran (carboxy-MFR), and *N*-formylmethanofuran (formyl-MFR) [279].

proceeds via N-carboxymethanofuran (carboxy-MFR) (Fig. 2) as intermediate [20].

$$CO_2 + MFR \rightleftharpoons carboxy\text{-}MFR \qquad\qquad \Delta G^{0'} = +1.3\,kJ/mol \qquad (2.1)$$

$$Carboxy\text{-}MFR + 2[H] \rightleftharpoons formyl\text{-}MFR + H_2O \qquad E^{0'} = -520\,mV \qquad (2.2)$$

$$Formyl\text{-}MFR + H_4MPT \rightleftharpoons MFR + formyl\text{-}H_4MPT \quad \Delta G^{0'} = -4.4\,kJ/mol$$
$$(10)$$

Reaction (2.1) is spontaneous [140], reaction (2.2) is catalyzed by formylmethanofuran dehydrogenase (FMD), and reaction (10) by formylmethanofuran:tetrahydromethanopterin formyltransferase. The latter enzyme has been extensively characterized from *Methanopyrus kandleri* [141–144]. It is isolated from methanogenic archaea [145] and sulfate-reducing archaea [146] as a homotetrameric or homodimeric soluble enzyme with a subunit molecular mass of 35 kDa, which is devoid of a prosthetic group. The crystal structure of the enzyme from *Methanopyrus kandleri* [147] has been resolved. From *Methylobacterium extorquens* AM1 the formyltransferase is purified in complex with subunits of formylmethanofuran dehydrogenase (B. K. Pomper and J. A. Vorholt, unpublished results).

The formylmethanofuran dehydrogenases belong to the DMSO reductase family of mononuclear molybdenum enzymes [2]. They are molybdenum iron-sulfur proteins (Fig. 3A), tungsten iron-sulfur proteins (Fig. 3B), or tungsten iron-seleno proteins (Fig. 3C).

3.1. Formylmethanofuran Dehydrogenase from Methanogenic Archaea (Mo, W, or WSe)

The enzyme has been characterized from four different methanogens: *Methanosarcina barkeri* grown on methanol; *Methanothermobacter marburgensis* (formerly *Methanobacterium thermoautotrophicum* strain Marburg) [148] grown on H_2 and CO_2; *Methanothermobacter wolfeii* grown on H_2 and CO_2; and *Methanopyrus kandleri* grown on H_2 and CO_2. The enzyme from all four organisms is very rapidly inactivated under oxic conditions. The molybdenum formylmethanofuran dehydrogenases are inactivated by cyanide, which appears not to affect the tungsten isoenzymes [149]. Azide is not an inhibitor despite the fact

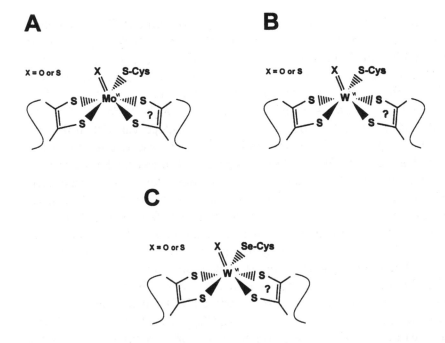

FIG. 3.   Proposed active site structure of formylmethanofuran dehydrogenases (FMD-M and FMD-W) from methanogenic and sulfate-reducing archaea. The enzymes probably belong to the DMSO reductase family of mononuclear molybdenum enzymes [2], although coordination of the molybdenum or tungsten to the dithiolene sulfur atoms of two molybdopterin guanine dinucleotides is not indicated by the primary structure of the catalytic subunit. (A) FMD-M from *Methanosarcina barkeri* and *Methanothermobacter marburgensis*; (B) FMD-W from *Methanothermobacter marburgensis, Methanopyrus kandleri*, and *Archaeoglobus fulgidus*; (C) seleno-FMD-W from *Methanopyrus kandleri*.

that some formylmethanofuran dehydrogenases can catalyze the dehydrogenation of formate albeit with very low catalytic efficiency [149].

The enzyme purified from *M. barkeri* is a soluble cytoplasmic protein composed of five different subunits of molecular masses of 65 kDa, 50 kDa, 37 kDa, 34 kDa, and 17 kDa. The heteropentamer contains per mol approximately one molybdenum, one molybdopterin guanine dinucleotide, 30 Fe/S, but no selenium or flavin [150–152]. The encoding genes have been cloned and sequenced. They are organized in a transcription unit *fmdEFACDB* [153]. FmdB (50 kDa) shows high sequence similarity to the molybdenum-harboring subunit of the formate dehydrogenases from *E. coli* and is thus the catalytic subunit. Besides molybdenum and the pterin cofactor it may also bind a [4Fe-4S] cluster as indicated by an N-terminal $CX_2CX_3CX_{18}CX_2G$ sequence motif. FmdA (65 kDa) shows sequence similarity to zinc-containing amidases [154] and is considered to provide part of the formylmethanofuran binding site. FmdF (37 kDa) is a polyferredoxin with eight [4Fe-4S] clusters. FmdC (34 kDa) and FmdE (17 kDa) are polypeptides with unknown function. A polypeptide corresponding to *fmdE* was not found in the purified enzyme. FmdE (22.8 kDa) is predicted to be an alkaline protein with a pI of 8.9. None of the polypeptides have domains characteristic for a membrane anchor. The purified enzyme contains a 29-kDa polypeptide that is not encoded within the *fmdEFACDB* operon. This polypeptide is therefore considered to be a contaminant. The physiological electron acceptor of the formylmethanofuran dehydrogenase is not known. Tungsten does not support growth and synthesis of active formylmethanofuran dehydrogenases in *M. barkeri*, indicating that the enzyme requires molybdenum to be active [155]. Formylmethanofuran dehydrogenase from *M. barkeri* ($V_{max}/K_m = 8800\,U\,mg^{-1}\,mM^{-1}$) can also catalyze the dehydrogenation *of* N-furfurylformamide ($0.1\,U\,mg^{-1}\,mM^{-1}$), *N*-methylformamide ($0.0001\,U\,mg^{-1}\,mM^{-1}$), formamide ($0.0001\,U\,mg^{-1}\,mM^{-1}$), and formate ($0.001\,U\,mg^{-1}\,mM^{-1}$) [149].

M. *marburgensis* contains two formylmethanofuran dehydrogenases, one of which is a tungsten enzyme (FMD-W) and the other a molybdenum enzyme (FMD-M). FMD-W is formed constitutively whereas FMD-M is only synthesized by the organism in the presence of molybdate. When grown on medium containing molybdate and tungstate, the molybdenum is also incorporated into FMD-W and tungsten into FMD-M yielding an active molybdenum-substituted tungsten enzyme and an active tungsten-substituted molybdenum enzyme, respectively [156–157]. The growth rates of the organism on medium containing molybdate or tungstate are identical [156].

Purified FMD-W from *M. marburgensis* is composed of four different subunits of molecular mass 65 kDa, 53 kDa, 31 kDa, and 15 kDa. Per heterotetramer 0.4 mol tungsten, 0.6 mol molybdopterin guanine dinucleotide, and approximately 8 mol Fe/S but no selenium or molybdenum were found [156]. The subunit encoding genes were cloned and found to form a transcription unit *fwdEFDGACB* [158]. The genes *fwdA*, *fwdB*, *fwdC*, and *fwdD* encode for the four polypeptides of the purified enzyme.

The other three open reading frames encode for proteins not present in the purified enzyme. From its sequence FwdB (53 kDa) is predicted to be the catalytic subunit harboring molybdenum, the pterin cofactor, and a [4Fe-4S] cluster. FwdA (65 kDa) might be involved in substrate binding [89,154]. FwdC (31 kDa) and FwdD (15 kDa) show sequence similarity to FmdC and FmdD of the enzyme from *M. barkeri* but not to any other proteins in the database. FwdF (38.6 kDa) is a polyferredoxin with eight [4Fe-4S] clusters, FwdG (8.6 kDa) a clostridial-type ferredoxin with two [4Fe-4S] clusters, and FwdE (17.8 kDa) an alkaline iron-sulfur protein with two [4Fe-4S] clusters.

Purified FMD-M from *M. marburgensis* is composed of three different subunits: one of molecular mass 65 kDa and two of molecular mass 45 kDa [150,159]. The heterotrimer contains per mol 0.6 mol molybdenum, molybdopterin guanine dinucleotide, molybdopterin adenine dinucleotide, and molybdopterin hypoxanthine dinucleotide in a ratio of 1:0.5:0.2 and 4 mol Fe/S [156,160]. The genes were cloned and sequenced and found to be organized in the transcription unit *fmd ECB* [159]. From the deduced amino acid sequence FmdB harbors the catalytic site with molybdenum, the pterin cofactor, and a [4Fe-4S] cluster. It is most closely related to FmdB of formylmethanofuran dehydrogenase from *M. barkeri*. The N-terminal part (approximately 230 amino acids) of FmdC (45 kDa) shows a sequence similarity to FwdC of the tungsten isoenzyme from *M. marburgensis* [158] and to FmdC of the molybdenum formylmethanofuran dehydrogenase from *M. barkeri*. The C-terminal part (approximately 170 amino acids) of FmdC shows 63.5% sequence similarity to FwdD of the tungsten isoenzyme from *M. marburgensis* and 39.5% similarity to FmdD of the molybdenum formylmethanofuran dehydrogenase from *M. barkeri*, FmdE is sequence-similar to FmdE of the enzyme from *M. barkeri*. In the purified molybdenum enzyme from both organisms this protein is absent. The operon encoding the tungsten isoenzyme in *M. marburgensis* does not contain an *fmdE* homologue; FmdE and FwdE are not sequence-similar. A gene coding for the 65-kDa subunit of FMD-M is not present in the transcription unit *fmdECB*. This protein is encoded by the *fwdA* gene of the *fwdHIGDACB* operon, which is constitutively transcribed allowing the two enzymes to share this subunit. As already indicated, FMD-M from *M. marburgensis* is only synthesized in *M. marburgensis* when the organism is grown in the presence of molybdate. A DNA-binding protein Tfx was identified and characterized that could be involved in transcriptional regulation of the *fmdEBD* operon [161].

M. wolfeii, which grows equally well in the presence of tungstate or molybdate, contains two formylmethanofuran dehydrogenases, a constitutively formed tungsten enzyme, and a molybdate-inducible molybdenum enzyme [162,163]. The two enzymes show essentially the same subunit composition and type of regulation by molybdate and tungstate as the two enzymes from *M. marburgensis* [157]. The tungsten-substituted molybdenum enzyme exhibits after air oxidation the EPR signal shown in Fig. 4. [164]. It was one of the first EPR spectra unambiguously assigned to tungsten V in an enzyme.

Methanopyrus kandleri is a hyperthermophilic archaeon with a growth temperature optimum of 98°C. Growth is dependent on the presence of tungstate in the medium. The methanogen contains two formylmethanofuran dehydrogenases, one of which is a tungsten selenoprotein (Fwu) and the other a tungsten enzyme (Fwc). The gene encoding for the catalytic subunit FwuB was found in the transcription unit *fwuGDB*; the *fwcB* gene appears to be transcribed monocistronically. Northern blot and primer extension analysis revealed that both genes are differentially transcribed. During growth of the methanogen on medium supplemented with selenium only *fwuGDB* and not *fwcB* were transcribed whereas transcription of both *fwuGDB* and *fwcB* was observed only on selenium-deprived medium indicating that transcription of *fwcB* is repressed in the presence of selenium [165].

3.2. Formylmethanofuran Dehydrogenase from *Archaeoglobus fulgidus* (W)

Archaeoglobus fulgidus is a hyperthermophilic sulfate-reducing archaeon that grows best on lactate and sulfate as energy sources, CO_2 and H_2S being the only end-products. Lactate is oxidized to $3CO_2$ via pyruvate, acetyl-CoA, N^5-methyltetrahydromethanopterin, N^5, N^{10}-methylenetetrahydromethanopterin, N^5, N^{10}-methenyltetrahydromethanopterin, N^5-formyltetrahydromethanopterin, and formylmethanofuran as intermediates (for review, see [136]). Formylmethanofuran dehydrogenase from *A. fulgidus* has been purified and shown to be composed of six different subunits of molecular masses 64 kDa,

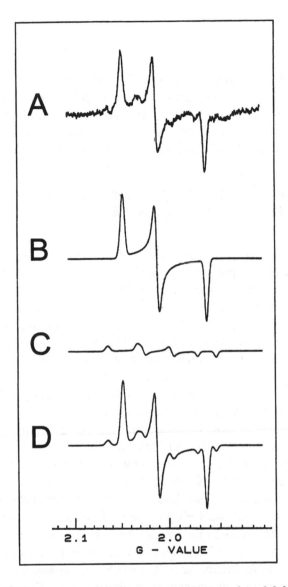

FIG. 4. (A) EPR spectrum of the tungsten-substituted molybdenum formyl-methanofuran dehydrogenase from *Methanothermobacter wolfii*, (B) computer simulation of the experimental rhombic spectrum (trace A) as an $S = \frac{1}{2}$ g_{xyz} = 2.0488, 2.0122, 1.9635) without a nuclear hyperfine interaction. (C) The same rhombic $S = \frac{1}{2}$ signal interacting with a nuclear spin of $I = \frac{1}{2}$. Signals B and C are plotted such that their double-integrated intensities are related in a ratio of 85.6 : 14.4. This ratio is derived from the natural abundance of the tungsten isotopes: $I = 0$: [180]W, 0.14%; [182]W, 26.4%; [184]W, 30.6%; and [186]W, 28.4%; and $I = \frac{1}{2}$: [183]W, 14.4%. (D) Sum of B plus C (from [164]). For EPR spectra of other tungsten enzymes, see [3].

593

52 kDa, 43 kDa, 32 kDa, 28 kDa, and 24 kDa [166]. The genome of *A. fulgidus* [167] harbors two putative transcription units, *fwdDBAC* and *fwdGBD*, predicted to encode for a tungsten formylmethanofuran dehydrogenase without a selenocysteine in the active site. Monocistronic genes encoding for FwdE and FwdF were also found.

3.3.   Formylmethanofuran Dehydrogenase from *Methylobacterium extorquens* (Mo?)

M. extorquens AM1 grows aerobically on methanol, which is oxidized to CO_2. It has been proposed that methanol oxidation proceeds via formaldehyde, N^5, N^{10}-methylenetetrahydromethanopterin, N^5, N^{10}-methenyltetrahydromethanopterin, N^5-formyltetrahydromethanopterin, and N-formylmethanofuran as intermediates [107]. The organism contains a formaldehyde-activating enzyme [168], an NAD-dependent and an NADP-dependent methylenetetrahydromethanopterin dehydrogenase [107,169], and an N^5, N^{10}-methenyltetrahydromethanopterin cyclohydrolase [170], which have been characterized. In addition, the formylmethanofuran dehydrogenase has been purified.

The dehydrogenase is composed of four different subunits with molecular masses of 65 kDa, 45 kDa, 35 kDa, and 20 kDa. The enzyme complex has formylmethanofuran:tetrahydromethanopterin formyltransferase activity (B. K. Pomper and J. A. Vorholt, unpublished results). From sequence comparisons it is deduced that the 35-kDa polypeptide is the subunit catalyzing the formyltransferase reaction. The 45-kDa polypeptide shows sequence similarity to FmdB from methanogens and sulfate-reducing archaea, and is therefore considered to be the catalytic subunit of formylmethanofuran dehydrogenase. The 65-kDa and 20-kDa polypeptides show sequence similarity to FmdA and FmdC, respectively, of formylmethanofuran dehydrogenases from methanogens. The four polypeptides are encoded by the genes *orf*2 (FmdB), *orf* 1 (FmdA), *ffsA* (formyltransferase), and *orf*3 (FmdC), which are located relative to one another such that they could form a transcription unit [138].

The presence of molybdenum or tungsten in this enzyme has not yet been shown. Also, the sequence identity of Orf2 with the catalytic subunit of formylmethanofuran dehydrogenase from methanogenic archaea is with 17% relatively low (the other subunits exhibit a sequence identity of more than 30%) [138]. Nor could a molybdopterin binding site be identified by sequence analysis. Therefore, it is uncertain whether this enzyme from *M. extorquens* AM1 falls within the class of molybdenum enzymes.

4.   CARBON MONOXIDE DEHYDROGENASE

Many aerobic organisms form CO, e.g., in the oxygenase reaction initiating heme degradation. But only bacteria and archaea are known that can catalyze the oxidation of CO to CO_2. This reaction is catalyzed by carbon monoxide dehydrogenase, which in aerobic microorganisms is a molybdoflavo iron-sulfur protein and in anaerobic microorganisms a nickel iron-sulfur protein [171,172]. Both enzymes have been shown to form CO_2 rather than HCO_3^- as product [173,174].

4.1.   Carbon Monoxide Dehydrogenase from Aerobic Bacteria (Mo, Se?)

Many bacteria can grow on CO and O_2 or nitrate as sole carbon and energy sources [175,176]. They are taxonomically diverse, encompassing more than 15 described species in 8 genera [177], examples are *Oligotropha carboxidovorans* [178] (formerly *Pseudomonas carboxidovorans*) [179], *Hydrogenophaga pseudoflava* [180], *Streptomyces thermoautotrophicus* [181], *Bradyrhizobium japonicum* [182], *Pseudomonas thermocarboxidovorans* [183], and *Acinetobacter* sp. [184,185]. The molybdenum carbon monoxide dehydrogenase belongs to the xanthine oxidase family of mononuclear molybdenum enzymes [2].

     O. carboxidovorans is the best studied representative; it carries the carbon monoxide dehydrogenase structural genes *coxMSL* on the 128-kb megaplasmid PHCG3 [186–188] and the expression of these genes in vivo is dependent on the presence of CO [189]. In contrast, the chromosomal *cutMLS* genes in *H. pseudoflava* [190] are expressed constitutively [180]. Dependent on the growth phase the enzyme in *O. carboxidovorans* occurs either associated with the inner aspect of the cytoplasmic membrane or soluble in the cytoplasm [191]. The physiological electron acceptor of the membrane-associated enzyme appears to be a cytochrome b_{561}. Unlike the membrane-associated carbon monoxide dehydrogenase, the cytoplasmic enzyme can transfer the electrons to O_2 thereby producing hydrogen peroxide and superoxide. Accumulation of the oxygen species results in the inactivation of the cytoplasmic carbon monoxide dehydrogenase. Purified carbon monoxide dehydrogenase inactivates rapidly in the presence of CO plus O_2 but remains completely active in the presence of CO or O_2 alone [172]. In vitro carbon monoxide dehydrogenase has the additional activity of uptake hydrogenases, which amounts to approximately 10% of the CO oxidizing activity [192].

Carbon monoxide dehydrogenase from all aerobic bacteria investigated to date appears to be phylogenetically related [193]. They are composed of three different subunits of apparent molecular masses 80–90 kDa (L), approximately 30 kDa (M) and approximately 20 kDa (S) occuring in heterohexameric ($L_2M_2S_2$) subunit composition. The enzymes are dimers of two heterotrimers. Per mole of heterotrimer the enzymes contains one bound molybdenum, one molybdopterin cytosine dinucleotide, probably one selenium not bound in selenocysteine, one FAD, and four Fe/S organized in two [2Fe-2S] clusters [194,195]. The molybdenum, the pterin cofactor, and the selenium are associated with the large subunit; the FAD with the medium-sized subunit; and the two Fe/S clusters with the small subunit [172,196,197].

The crystal structure of the enzyme from *O. carboxidovorans* [198] and that of *H. pseudoflava* [199] have been determined to 2.2 Å resolution. The molybdenum was found to be complexed by the enedithiolate group of the pyranopterin moiety and to carry two oxo groups and one sulfo group, which is in van der Waal's distance of a putative *S*-selanyl-cysteine possibly forming a selenoheterotrisulfide (-Mo-S-Se-S-Cys-) (Fig. 5). Since the selenium content of carbon monoxide dehydrogenase is substoichiometric, generally only 0.3 mol selenium per mole molybdenum, it is difficult to assign the electron density between the sulfo ligand and the cysteine sulfur unambiguously to Se. In this respect it is of interest that carbon monoxide dehydrogenase from *S. thermoautotrophicus* contains no selenium, although the enzyme displays a high specific activity and is fully reduced by CO [172]. The intramolecular electron transport chain in carbon monoxide dehydrogenase involves

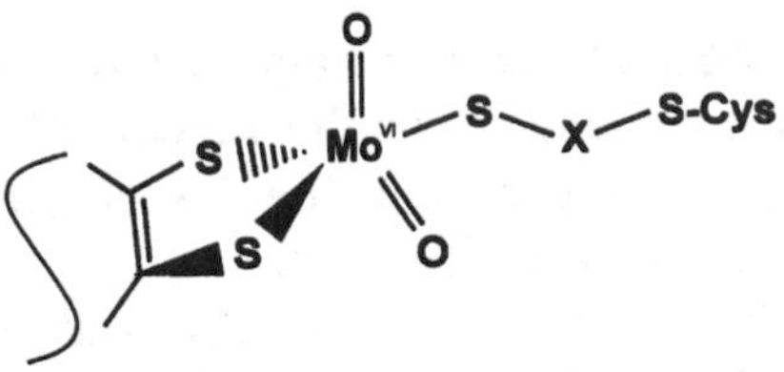

FIG. 5. Active site structure of molybdenum carbon monoxide dehydrogenase from *Oligotropha carboxidovorans* [198] and *Hydrogenophaga pseudoflava* [199]. The enzyme belongs o the xanthine oxidase family of mononuclear molybdenum enzymes [2]. The molybdenum is coordinated to the dithiolene sulfur atoms of one molybdopterin cytosine dinucleotide. It carries two oxo groups and one sulfo group which is in van der Waal's distance of a putative S-selanylcysteine possibly forming a selenoheterotrisulfide (Mo-S-Se-S-Cys) [172]. Note added in proof: In the meantime evidence is available that the active enzyme contains Cu instead of Se (Mo-S-Cu-S-Cys) (O. Meyer, personal communication).

the following groups and minimal distances: CO $\longrightarrow$ [Mo of the molybdenum cofactor]–14.6 Å–[2Fe-2S]–12.4 Å[2Fe-2S]–8.7 Å–[FAD] [172].

During growth of *H. pseudoflava* on media supplemented with tungstate or in media deficient in molybdate an inactive carbon monoxide dehydrogenase is formed that lacks molybdenum and that contains bound CDP rather than molybdopterin cytosine dinucleotide (MCD), indicating that molybdenum is involved in the biosynthesis and/or the insertion of Mo-MCD whereas protein synthesis, including integration of the Fe/S centers and FAD, is independent of Mo [199,200].

4.2. Carbon Monoxide Dehydrogenase from Anaerobic Bacteria and Archaea (Ni)

These are four groups of anaerobic microorganisms that contain a nickel carbon monoxide dehydrogenase: (1) bacteria such as *Rhodospirillum rubrum* [201,202] and *Carboxydothermus hydrogenoformans* [203–205] which can grow on CO forming CO_2 and H_2 as sole energy source [171]; (2) acetogenic and/or autotrophic bacteria, which synthesize acetyl-CoA from $2CO_2$ via the Wood-Ljungdahl pathway [81]; (3) acetate-oxidizing anaerobes without a tricarboxylic acid cycle, which grow, e.g., on acetate and sulfate as energy source [206,207]; (4) the acetoclastic methanogens, which ferment acetate to methane and CO_2 [208]. The carbon monoxide dehydrogenase from all of these organisms are phylogenetically related [209]. Therefore, the structural and catalytic properties of the few enzymes investigated in more detail can probably be generalized.

In *Rhodospirillum rubrum* carbon monoxide dehydrogenase is a membrane-associated cytoplasmic enzyme that can easily be detached from the membrane. In the acetate-forming or utilizing anaerobes (Groups 2–4), the carbon monoxide dehydrogenase is tightly associated with acetyl-CoA decarbonylase, which is also a nickel enzyme and which catalyzes the reversible formation of acetyl-CoA from the methyl group of a methylated corrinoid protein, CoA, and CO. The CO is channeled between the active sites and the two enzymes [210,211].

Carbon monoxide dehydrogenase from *R. rubrum* is a monomer of 61.8 kDa containing one nickel and two [4Fe-4S] clusters, one of these is in close vicinity to the nickel, which appears to form a binuclear [Fe-Ni] cluster [212,213], and the other is at a site that allows it to participate in electron transport to the physiological electron acceptor, which is a 22-kDa ferredoxin [214]. There is evidence for a ligand CO that is required for catalytic activity [215]. The enzyme from *Carboxydothermus hydrogenoformans* has been crystallized and the crystal structure has been refined. The structure revealed that carbon monoxide dehydrogenase is a functional dimer, with the two monomers sharing an additional bridging [4Fe-4S] cluster [281].

All carbon monoxide dehydogenases from anaerobes are nickel Fe/S proteins. That they contain nickel was first indicated by the finding that expression of carbon monoxide dehydrogenase activity in *Clostridium pasteurianum* was dependent on the availability of nickel in the growth medium [216].

The literature on nickel carbon monoxide dehydrogenase has recently been reviewed [217–220].

5.   FORMALDEHYDE DEHYDROGENASES

Formaldehyde is an intermediate in the energy metabolism of aerobic methylotrophic bacteria and yeast. In these organisms it is formed from methanol or methylated compounds, such as methylamine, by dehydrogenation. The formaldehyde thus generated is then oxidized to CO_2 and used for biosynthesis. The mechanism of formaldehyde oxidation differs in different organisms [221]. Only in a few organisms does the oxidation of formaldehyde appear to be catalyzed by a molybdenum or tungsten enzyme. In most organisms the reaction is catalyzed by an enzyme not containing these transition metals.

5.1. (Form)aldehyde Oxidoreductase (Mo or W)

In the gram-positive *Amycolatopsis methanolica* two different dye-linked (form)aldehyde oxidoreductases have been found that are molybdoflavo iron-sulfur proteins and that are induced 10-fold upon growth on methanol [222–224]. Both enzymes are composed of three different subunits of molecular mass 87 kDa, 35 kDa, and 17 kDa, and contain per heterotrimer probably one molybdenum, one FAD, and one [2Fe-2S] cluster as revealed by EPR spectroscopy [223]. Both enzymes have a broad substrate specificity, the catalytic efficiency (V_{max}/K_m) being more than 1000-fold higher for acetaldehyde oxidation than for formaldehyde oxidation. The relatively high K_m for formaldehyde of 3.6 mM and 63 mM, respectively, makes a function of the two enzymes in methanol metabolism of *A. methanolica* questionable.

Dye-linked (form)aldehyde oxidoreductases, which contain molybdenum or tungsten, have also been found in other microorganisms in which formaldehyde is not an apparent intermediate [225,226]; for example, in *Clostridium thermoaceticum* and *Clostridium formicoaceticum* [227–229], which are anaerobic gram-positive acetogens; in glucose-grown *Sulfolobus acidocaldarius* [230], which is a hyperthermophilic aerobic crenarchaeon; and in the peptide- and sugar-fermenting *Pyrococcus furiosus* [231–234] and *Thermococcus litorales* [231,235,236], which are hyperthermophilic euryarchaeota. These enzymes also have an unfavorable K_m for formaldehyde (> 10 mM) and show a much higher catalytic efficiency with other aldehydes. The molybdenum (form)aldehyde oxidoreductases appear to belong to the xanthine oxidase family of mononuclear molybdenum enzymes [2], whereas the tungsten (form)aldehyde oxidoreductases form a family of their own [3] (Fig. 6). The properties and function of the molybdenum aldehyde oxidoreductases and

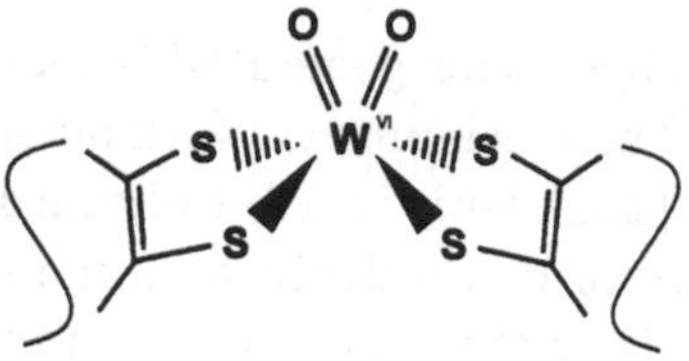

FIG. 6. Active site structure of the tungsten (form)aldehyde oxidoreductase from *Pyrococcus furiosus* [280]. The enzyme shows no sequence similarity to the three families of mononuclear molybdenum enzymes [3]. The tungsten is coordinated to the dithiolene sulfur atoms of two molybdopterin mononucleotides.

of the tungsten aldehyde oxidoreductases are described in Chapters 15 and 19, respectively.

5.2. Formaldehyde Dehydrogenase Not Containing Molybdenum or Tungsten

In *Paracoccus denitrificans* [237], *Rhodobacter sphaeroides* [238–240], and methylotrophic yeast [241,242], formaldehyde is oxidized to formate via a glutathione-dependent NAD-specific formaldehyde dehydrogenase [243,244]. This zinc class 3 type alcohol dehydrogenase is also found in many nonmethylotrophs [245] where it has a function in formaldehyde detoxification. Thus glutathione-dependent formaldehyde dehydrogenase is found in enterobacteriaceae [246–249], plants [250–254], insects [255], mouse and rat [256,257], and humans [258–262]. Its crystal structure has been resolved to 2.7 Å [263] and its kinetic mechanism studied in detail [262,265,265].

In the gram-positive *Amycolatopsis methanolica* [266] and *Rhodococcus erythropolis* [267], both belonging to the *Actinomycetes*, a mycothiol-dependent NAD-specific formaldehyde dehydrogenase is present [268] that shares sequence similarity with the glutathione-dependent enzymes [269].

In most of the gram-negative methylotrophic bacteria, formaldehyde probably reacts with tetrahydromethanopterin and/or tetrahydrofolate forming N^5, N^{10}-methylenetetrahydromethanopterin and

N^5, N^{10}-methylenetetrahydrofolate, respectively. The condensation reaction proceeds spontaneously but apparently not rapidly enough, so that a catalyst is required [168]. Bound to tetrahydromethanopterin or tetrahydrofolate the C1 unit is then oxidized to CO_2 [107]. In some methylotrophic bacteria formaldehyde is oxidized to CO_2 in a cyclic process starting with the aldol condensation of formaldehyde with ribulose-5-phosphate to hexulose-6-phosphate [270–272].

In *Methylococcus capsulatus* (Bath), formaldehyde is oxidized to formate via a glutathione-independent NAD-linked formaldehyde dehydrogenase that is dependent on an 8.8-kDa protein named modifine for formaldehyde specificity [273]. A glutathione-independent NAD-specific formaldehyde dehydrogenase is also present in *Pseudomonas putida* [274,275], *Arthrobacter* P1 [276], *Hyphomicrobium zarvarzinii* [277], and yeast [278].

All of these formaldehyde-consuming reactions are catalyzed by enzymes containing neither molybdenum nor tungsten. It thus appears that formaldehyde dehydrogenation in most methylotrophs is independent of these transition metals.

6. CONCLUSIONS

Molybdenum and tungsten enzymes play an important role in the C1 metabolism of bacteria and archaea but appear not to be involved in the C1 metabolism of eukarya:

Formate dehydrogenases in anaerobic bacteria and archaea are found to be always molybdenum or tungsten enzymes. These enzymes are involved both in the reduction of CO_2 to formate and in the oxidation of formate to CO_2.

Formate dehydrogenases in aerobic bacteria are generally molybdenum enzymes. But they can also be tungsten enzymes. In some aerobic methylotropic bacteria a metal-free formate dehydrogenase also occurs that appears to function only in formate oxidation rather than in CO_2 reduction.

In eukaryotic organisms only the metal free formate dehydrogenase is found.

Formylmethanofuran dehydrogenases in archaea are related to the formate dehydrogenases from anaerobic microorganisms. As these they are either molybdenum or tungsten enzymes and are functionally reversible.

Carbon monoxide dehydrogenase from aerobic bacteria appear always to be molybdenum enzymes and from anaerobic bacteria and archaea always to be nickel enzymes. These enzymes are not found in eukarya.

Formaldehyde dehydrogenases that contain molybdenum or tungsten are functionally probably aldehyde dehydrogenases that are not involved in C1 metabolism. Tungsten (form)aldehyde oxidoreductases appear to be restricted to anaerobic archaea and bacteria where they may function also as carboxylate reductases.

ACKNOWLEDGMENTS

This work was supported by the Max-Planck-Gesellschaft, by the Deutsche Forschungsgemeinschaft, and by the Fonds der Chemischen Industrie.

ABBREVIATIONS AND DEFINITIONS

ADP	adenosine 5′-diphosphate
ArcA	response regulator for "aerobic respiration control"
ATP	adenosine 5′-triphosphate
Bi Bi	indicates a ping-pong mechanism with two substrates and two products
CDP	cytosine 5′-diphosphate
CoA	coenzyme A
CODH	carbon monoxide dehydrogenase
DMSO	dimethyl sulfoxide
dNDP	deoxyribonucleotide
EPR	electron paramagnetic resonance
EXAFS	extended absorption fine structure spectroscopy

FAD	flavin adenine dinucleotide
FDH	formate dehydrogenase
FMD	formylmethanofuran dehydrogenase
FMN	flavin mononucleotide
FNR	response regulator for "fumarate and nitrate reduction"
FODH	formaldehyde dehydrogenase
H_4F	tetrahydrofolate
H_4MPT	tetrahydromethanopterin
MCD	molybdopterin cytosine dinucleotide
MFR	methanofuran
NAD	nicotinamide adenine dinucleotide
NADP	nicotinamide adenine dinucleotide phosphate
NAR	nitrate reductase
NDP	ribonucleotide
Pi	inorganic phosphate
SDS/PAGE	sodium dodecyl sulfate polyacrylamide gel electrophoresis

REFERENCES

1. R. K. Thauer, K. Jungermann, and K. Decker, *Bact. Rev.*, *41*, 100–180 (1977).

2. R. Hille, *Chem. Rev.*, *96*, 2757–2816 (1996).

3. M. K. Johnson, D. C. Rees, and M. W. W. Adams, *Chem. Rev.*, *96*, 2817–2839 (1996).

4. D. D. Wagman, W. H. Evans, V. B. Parker, R. H. Schumm, I. Hallow, S. M. Bailey, K. L. Churney, and R. L. Nutall, *J. Phys. Chem. Ref. Data.*, *11*, suppl 2, 2–83 (1982).

5. B. E. H. Maden, *Biochem. J.*, *350*, 609–629 (2000).

6. P. A. Bertram and R. K. Thauer, *Eur. J. Biochem.*, *226*, 811–818 (1994).

7. R. M. C. Dawson, D. C. Elliott, W. H. Elliott, and K. M. Jones (eds.), *Data for Biochemical Research*, Clarendon Press, Oxford, 1987.

8. J. Buc, C. L. Santini, R. Giordani, M. Czjzek, L. F. Wu, and G. Giordano, *Mol. Microbiol.*, *32*, 159–168 (1999).

9. L. J. Stewart, S. Bailey, B. Bennett, J. M. Charnock, C. D. Garner, and A. S. McAlpine, *J. Mol. Biol.*, *299* 593–600 (2000).

10. R. P. Gunsalus, *J. Bacteriol.*, *174*, 7069–7074 (1992).

11. F. M. Girio, J. C. Marcos, and M. T. Amaral-Collaco, *FEMS Microbiol. Lett.*, *97*, 161–166 (1992).

12. F. M. Girio, M. T. Amaral-Collaco, and M. M. Attwood, *Appl. Microbiol. Biotechnol.*, *40*, 898–903 (1994).

13. F. M. Girio, J. C. Roseiro, and A. I. Silva, *Curr. Microbiol.*, *36*, 337–340 (1998).

14. R. K. Thauer, G. Fuchs, and K. Jungermann, in *Iron-Sulfur Proteins*, Vol. 3 (W. Lovenberg, ed.), Academic Press, New York, 1977, pp. 121–156.

15. M. W. W. Adams and L. E. Mortenson, in *Molybdenum Enzymes* (T. G. Spiro, ed.), John Wiley & Sons, New York, 1985, pp. 519–593.

16. J. G. Ferry, *FEMS Microbiol. Rev.*, *7* 377–382 (1990).

17. J. A. Stubbe and W. A. van der Donk, *Chem. Rev.*, *98*, 705–762 (1998).

18. R. K. Thauer, B. Käufer, and G. Fuchs, *Eur. J. Biochem.*, *55*, 111–117 (1975).

19. S. V. Khangulov, V. N. Gladyshev, G. C. Dismukes, and T. C. Stadtman, *Biochemistry*, *37*, 3518–2528 (1998).

20. J. A. Vorholt and R. K. Thauer, *Eur. J. Biochem.*, *248*, 919–924 (1997).

21. J. S. Blanchard and W. W. Cleland, *Biochemistry*, *19*, 3543–3550 (1980).

22. G. Sawers, *A. v. Leeuwenhoek*, 66, 57–88 (1994).

23. G. Sawers, J. Heider, E. Zehelein, and A. Böck, *J. Bacteriol.*, *173*, 4983–4993 (1991).

24. J. Pommier, M. A. Mandrand, S. E. Holt, D. H. Boxer, and G. Giodano, *Biochim. Biophys. Acta.*, *1107*, 305–313 (1992).

25. S. Benoit, H. Abaibou, and M. A. Mandrand-Berthelot, *J. Bacteriol.*, *180*, 6625–6634 (1998).

26. A. A. Trchounian, K. A. Bagramyan, A. V. Vassilian, and A. A. Poladian, *Biologicheskie Membrany, 16,* 416–428 (1999).

27. S. C. Andrews, B. C. Berks, J. McClay, A. Ambler, M. A. Quail, P. Golby, and J. R. Guest, *Microbiology, 143,* 3633–3647 (1997).

28. G. Unden and J. Bongaerts, *Biochim. Biophys. Acta, 1320,* 217–234 (1997).

29. A. Graham and D. H. Boxer, *Biochem. J., 195,* 627–637 (1981).

30. H. G. Enoch and R. L. Lester, *J. Biol. Chem., 250,* 6693–6705 (1975).

31. B. L. Berg and V. Stewart, *Genetics, 125,* 691–702 (1990).

32. B. L. Berg, C. Baron, and V. Stewart, *J. Biol. Chem., 266,* 22386–22391 (1991).

33. B. C. Berks, *Mol. Microbiol., 22,* 393–404 (1996).

34. E. G. Bogsch, F. Sargent, N. R. Stanley, B. C. Berks, C. Robinson, and T. Palmer, *J. Biol. Chem., 273,* 18003–18006 (1998).

35. J. H. Weiner, P. T. Bilous, G. M. Shaw, S. P. Lubitz, L. Frost, G. H. Thomas, J. A. Cole, and R. J. Turner, *Cell, 93,* 93–101 (1998).

36. G. D. Plunkett, V. Burland, D. L. Daniels, and F. R. Blattner, *Nucleic Acids Res., 21,* 3391–3398 (1993).

37. H. Abaibou, J. Pommier, S. Benoit, G. Giordano, and M. A. Mandrand-Berthelot, *J. Bacteriol., 177,* 7141–7149 (1995).

38. M. J. Axley, D. A. Grahame, and T. C. Stadtman *J. Biol. Chem., 265,* 18213–18218 (1990)

39. F. Zinoni, A. Birkmann, T. C. Stadtman, and A. Böck, *Proc. Natl. Acad. Sci USA, 83,* 4650–4654 (1986).

40. J. C. Boyington, V. N. Gladyshev, S. V. Khangulov, T. C. Stadtman, and P. D. Sun, *Science, 275,* 1305–1308 (1997).

41. M. Bokranz, M. Gutmann, C. Kortner, E. Kojro, F. Fahrenholz, F. Lauterbach, and A. Kröger, *Arch. Microbiol., 156,* 119–128 (1991).

42. G. T. Chen, M. J. Axley, J. Hacia, and M. Inouye, *Mol. Microbiol., 6,* 781–785 (1992).

43. M. J. Axley and D. A. Grahame, *J. Biol. Chem., 266,* 13731–13736 (1991).

44. V. N. Gladyshev, J. C. Boyington, S. V. Khangulov, D. A. Grahame, T. C. Stadtman, and P. D. Sun, *J. Biol. Chem.*, *271*, 8095–8100 (1996).

45. G. N. George, C. M. Colangelo, J. Dong, R. A. Scott, S. V. Khangulov, V. N. Gladyshev, and T. C. Stadtman, *J. Am. Chem. Soc.*, *120*, 1267–1273 (1998).

46. V. N. Gladyshev, S. V. Khangulov, M. J. Axley, and T. C. Stadtman, *Proc. Natl. Acad. Sci. USA*, *91*, 7708–7711 (1994).

47. M. J. Axley, A. Böck, and T. C. Stadtman, *Proc. Natl. Acad. Sci. USA*, *88*, 8450–8454 (1991).

48. S. Commans and A. Böck, *FEMS Microbiol. Rev.*, *23*, 335–351 (1999).

49. A. Böck, *Biofactors*, *11*, 77–78 (2000).

50. M. Thanbichler, A. Böck, and R. S. Goody, *J. Biol. Chem.*, *275*, 20458–20466 (2000).

51. I. Chenier, F. Dube, and P. C. Hallenbeck, *Curr. Microbiol.*, *41*, 39–44 (2000).

52. H. Mihara, T. Kurihara, T. Yoshimura, and N. Esaki, *J. Biochem. (Tokyo)*, *127*, 559–567 (2000).

53. K. E. Sandman and C. J. Noren, *Nucleic Acids Res.*, *28*, 755–761 (2000).

54. R. Rossmann, G. Sawers, and A. Böck, *Mol. Microbiol.*, *5*, 2807–2814 (1991).

55. T. Maier, U. Binder, and A. Böck, *Arch. Microbiol.*, *165*, 333–341 (1996).

56. H. Abaibou, G. Giordano, and M. A. Mandrand-Berthelot, *Microbiology*, *143*, 2657–2664 (1997).

57. W. T. Self and K. T. Shanmugam, *FEMS Microbiol. Lett.*, *184*, 47–52 (2000).

58. R. Hedderich, O. Klimmek, A. Kröger, R. Dirmeier, M. Keller, and K. O. Stetter, *FEMS Microbiol. Rev.*, *22*, 353–381 (1999).

59. R. Lenger, U. Herrmann, R. Gross, J. Simon, and A. Kröger, *Eur. J. Biochem.*, *246*, 646–651 (1997).

60. A. Kröger, E. Dorrer, and E. Winkler, *Biochim. Biophys. Acta*, *589*, 118–136 (1980).

61. U. Geisler, R. Ullmann, and A. Kröger, *Biochim. Biophys. Acta*, *1184*, 219–226 (1994).

62. A. Jankielewicz, O. Klimmek, and A. Kröger, *Biochim. Biophys. Acta*, *1231*, 157–162 (1995).

63. A. Kröger, E. Winkler, A. Innerhofer, H. Hackenberg, and H. Schägger, *Eur. J. Biochem.*, *94*, 465–475 (1979).

64. G. Unden and A. Kröger, *Biochim. Biophys. Acta*, *725*, 325–331 (1983).

65. A. Jankielewicz, R. A. Schmitz, O. Klimmek, and A. Kröger, *Arch. Microbiol.*, *162* 238–242 (1994).

66. T. Yagi, *Biochim. Biophys. Acta*, *548*, 96–105 (1979).

67. C. Sebban-Kreuzer, A. Dolla, and F. Guerlesquin, *Eur. J. Biochem.*, *253*, 645–562 (1998).

68. C. Sebban-Kreuzer, M. Blackledge, A. Dolla, D. Marion, and F. Guerlesquin, *Biochemistry*, *37*, 8331–8340 (1998).

69. X. Morelli and F. Guerlesquin, *FEBS Lett.*, *460*, 77–80 (1999).

70. M. J. Almendra, C. D. Brondino, O. Gavel, A. S. Pereira, P. Tavares, S. Bursakov, R. Duarte, J. Caldeira, J. J. G. Moura, and I. Moura, *Biochemistry*, *38*, 16366–16372 (1999).

71. H. Raaijmakers, S. Teixeira, J. M. Dias, M. J. Almendra, C. D. Brondino, I. Moura, J. J. G. Moura, and M. J. Romão, *J. Biol. Inorg. Chem.*, *6*, 398–404 (2001).

72. T. S. Haynes, D. J. Klemm, J. J. Ruocco, and L. L. Barton, *Anaerobe*, *1*, 175–182 (1995).

73. C. M. H. Hensgens, M. E. Nienhuiskuiper, and T. A. Hansen, *Arch. Microbiol.*, *162*, 143–147 (1994).

74. C. Sebban, L. Blanchard, M. Bruschi, and F. Guerlesquin, *FEMS Microbiol. Lett.*, *133*, 143–149 (1995).

75. C. Costa, M. Teixeira, J. LeGall, J. J. G. Moura, and I. Moura, *J. Biol. Inorg. Chem.*, *2*, 198–208 (1997).

76. R. K. Thauer, G. Fuchs, and K. Jungermann, *J. Bacteriol.*, *118*, 758–760 (1974).

77. R. K. Thauer, G. Fuchs, and B. Käufer, *Hoppe Seyler's Z. Physiol. Chem.*, *356*, 653–662 (1975).

78. P. A. Scherer and R. K. Thauer, *Eur. J. Biochem.*, *85*, 125–135 (1978).

79. C. L. Liu and L. E. Mortenson, *J. Bacteriol.*, *159*, 375–380 (1984).

80. R. C. Prince, C.-L. Liu, T. V. Morgan, and L. E. Mortenson, *FEBS Lett.*, *189*, 263–266 (1985).

81. S. W. Ragsdale, *Biofactors.*, *6*, 3–11 (1997).

82. L. G. Ljungdahl and J. R. Andreesen, *FEBS Lett.*, *54*, 279–282 (1975).

83. U. Leonhardt and J. R. Andreesen, *Arch. Microbiol.*, *115*, 277–284 (1977).

84. J. R. Andreesen and L. G. Ljungdahl, *J. Bacteriol.*, *116*, 867–873 (1973).

85. J. R. Andreesen and L. G. Ljungdahl, *J. Bacteriol.*, *120*, 6–14 (1974).

86. I. Yamamoto, T. Saiki, S. M. Liu, and L. G. Ljungdahl, *J. Biol. Chem.*, *258*, 1826–1832 (1983).

87. J. C. Deaton, E. I. Solomon, G. D. Watt, P. J. Wetherbee, and C. N. Durfor, *Biochem. Biphys. Res. Commun.*, *149*, 424–430 (1987).

88. D. Gollin, X. L. Li, S. M. Liu, E. T. Davies, and L. G. Ljungdahl, *Adv. Chem. Conv. Mitig. Carb. Diox.*, *114*, 303–308 (1998).

89. R. K. Thauer, *Microbiology*, *144*, 2377–2406 (1998).

90. D. R. Smith, L. A. Doucette-Stamm, C. Deloughery, H. Lee, J. Dubois, T. Aldredge, R. Bashirzadeh, D. Blakely, R. Cook, K. Gilbert, D. Harrison, L. Hoang, P. Keagle, W. Lumm, B. Pothier, D. Qiu, R. Spadafora, R. Vicaire, Y. Wang, J. Wierzbowski, R. Gibson, N. Jiwani, A. Caruso, D. Bush, and J. N. Reeve, *J. Bacteriol.*, *179*, 7135–7155 (1997).

91. J. N. Reeve, J. Nölling, R. M. Morgan, and D. R. Smith, *J. Bacteriol.*, *179*, 5975–5986 (1997).

92. C. J. Bult, O. White, G. J. Olsen, L. Zhou, R. D. Fleischmann, G. G. Sutton, J. A. Blacke, L. M. FitzGerald, R. A. Clayton, J. D. Gocayne, A. R. Kerlavage, B. A. Dougherty, J. F. Tomb, M. D. Adams, C. I. Reich, R. Overbeek, E. F. Kirkness, K. G. Weinstock, J. M. Merrick, A. Glodek, J. L. Scott, N. S. M. Geoghagen, and J. C. Venter, *Science*, *273*, 1058–1073 (1996).

93. E. Selkov, N. Maltsev, G. J. Olsen, R. Overbeek, and W. B. Whitman, *Gene.*, *197*, GC11-26 (1997).

94. N. L. Schauer and J. G. Ferry, *J. Bacteriol.*, *150*, 1-7 (1982).

95. A. P. Shuber, E. C. Orr, M. A. Recny, P. F. Schendel, H. D. May, N. L. Schauer, and J. G. Ferry, *J. Biol. Chem.*, *261*, 12942-12947 (1986).

96. J. Nölling and J. N. Reeve, *J. Bacteriol.*, *179*, 899-908 (1997).

97. H. D. May, P. S. Patel, and J. G. Ferry, *J. Bacteriol.*, *170*, 3384-3389 (1988).

98. J. L. Johnson, N. R. Bastian, N. L. Schauer, J. G. Ferry, and K. V. Rajagopalan, *FEMS Microbiol. Lett.*, *61*, 213-216 (1991).

99. S. F. Baron and J. G. Ferry, *J. Bacteriol.*, *171*, 3854-3859 (1989).

100. J. B. Jones and T. C. Stadtman, *J. Biol. Chem.*, *255*, 1049-1053 (1980).

101. J. B. Jones and T. C. Stadtman, *J. Biol. Chem.*, *256*, 656-663 (1981).

102. C. G. Friedrich, B. Bowien, and B. Friedrich, *J. Gen. Microbiol.*, *115*, 185-192 (1979).

103. J. Friedebold and B. Bowien, *J. Bacteriol.*, *175*, 4719-4728 (1993).

104. J. Friedebold, F. Mayer, E. Bill, A. X. Trautwein, and B. Bowien, *Biol. Chem. Hoppe Seyler.*, *376*, 561-568 (1995).

105. J. I. Oh and B. Bowien, *J. Biol. Chem.*, *273*, 26349-26360 (1998).

106. J. I. Oh and B. Bowien, *Mol. Microbiol.*, *34*, 365-376 (1999).

107. J. A. Vorholt, L. Chistoserdova, M. E. Lidstrom, and R. K. Thauer, *J. Bacteriol.*, *180*, 5351-5356 (1998).

108. V. V. Karzanov, A. Bogatsky Yu, V. I. Tishkov, and A. M. Egorov, *FEMS Microbiol.*, *Lett.*, *60*, 197-200 (1989).

109. V. V. Karzanov, C. M. Correa, Y. G. Bogatsky, and A. I. Netrusov, *FEMS Microbiol. Lett.*, *81*, 95-99 (1991).

110. D. R. Jollie and J. D. Lipscomb, *J. Biol. Chem.*, *266*, 21853-21856 (1991).

111. D. C. Yoch, Y. P. Chen, and M. G. Hardin, *J. Bacteriol.*, *172*, 4456-4463 (1990).

112. U. Müller, P. Willnow, U. Ruschig, and T. Hopner, *Eur. J. Biochem., 83,* 485–498 (1978).

113. R. O. Duarte, A. R. Reis, F. Girio, I. Moura, J. J. Moura, and T. A. Collaco, *Biochem. Biphys. Res. Commun., 230,* 30–34 (1997).

114. N. Kato, *Meth. Enzymol., 188,* 459–462 (1990).

115. V. O. Popov and V. S. Lamzin, *Biochem. J., 301,* 625–643 (1994).

116. A. M. Rojkova, A. G. Galkin, L. B. Kulakova, A. E. Serov, P. A. Savitsky, V. V. Fedorchuk, and V. I. Tishkov, *FEBS Lett., 445,* 183–188 (1999).

117. A. Galkin, L. Kulakova, V. Tishkov, N. Esaki, and K. Soda, *Appl. Microbiol. Biotechnol., 44,* 479–483 (1995).

118. Y. Asano, T. Sekigawa, H. Inukai, and A. Nakazawa, *J. Bacteriol., 170,* 3189–3193 (1988).

119. H. Schütte, J. Flossdorf, H. Sahm, and M. R. Kula, *Eur. J. Biochem., 62,* 151–160 (1976).

120. Y. Sakai, A. P. Murdanoto, T. Konishi, A. Iwamatsu, and N. Kato, *J. Bacteriol., 179,* 4480–4485 (1997).

121. H. Slusarczyk, S. Felber, M. R. Kula, and M. Pohl, *Eur. J. Biochem., 267,* 1280–1289 (2000).

122. W. Neuhauser, M. Steininger, D. Haltrich, K. D. Kulbe, and B. Nidetzky, *Biotech. Techn., 12,* 565–568 (1998).

123. A. V. Mesentsev, T. B. Ustinnikova, T. V. Tikhonova, and V. O. Popov, *Appl. Biochem. Microbiol., 32,* 529–534 (1996).

124. O. V. Stasyk, G. P. Ksheminskaya, A. R. Kulachkovskii, and A. A. Sibirnyi, *Microbiology, 66,* 631–636 (1997).

125. T. V. Avilova, O. A. Egorova, L. S. Ioanesyan, and A. M. Egorov, *Eur. J. Biochem., 152,* 657–662 (1985).

126. S. J. Allen and J. J. Holbrook, *Gene, 162,* 99–104 (1995).

127. Y. Izumi, H. Kanzaki, S. Morita, H. Futazuka, and H. Yamada, *Eur. J. Biochem., 182,* 333–341 (1989).

128. C. M. Chow and U. L. RajBhandary, *J. Bacteriol., 175,* 3703–3709 (1993).

129. A. M. Egorov, T. V. Avilova, M. M. Dikov, V. O. Popov, Y. V. Rodionov, and I. V. Berezin, *Eur. J. Biochem., 99,* 569–576 (1979).

130. V. I. Tishkov, A. G. Galkin, G. N. Marchenko, O. A. Egorova, D. V. Sheluho, L. B. Kulakova, L. A. Dementieva, and A. M. Egorov, *Biochem. Biophys. Res. Commun., 192*, 976–981 (1993).

131. V. I. Tishkov, A. G. Galkin, G. N. Marchenko, Y. D. Tsygankov, and A. M. Egorov, *Biotechnol. Appl. Biochem., 18*, 201–207 (1993).

132. V. S. Lamzin, A. E. Aleshin, B. V. Strokopytov, M. G. Yukhnevich, V. O. Popov, E. H. Harutyunyan, and K. S. Wilson, *Eur. J. Biochem., 206*, 441–452 (1992).

133. V. S. Lamzin, Z. Dauter, V. O. Popov, E. H. Harutyunyan, and K. S. Wilson, *J. Mol. Biol., 236*, 759–785 (1994).

134. A. V. Mesentsev, V. S. Lamzin, V. I. Tishkov, T. B. Ustinnikova, and V. O. Popov, *Biochem. J., 321*, 475–480 (1997).

135. K. Seelbach, B. Riebel, W. Hummel, M. R. Kula, V. I. Tishkov, A. M. Egorov, C. Wandrey, and U. Kragl, *Tetrahedron Lett, 37*, 1377–1380 (1996).

136. R. K. Thauer and J. Kunow, in *Biotechnology Handbook* (N. Clark, ed.), Plenum Publishing, London, 1995, pp. 33–48.

137. J. A. Vorholt, D. Hafenbradl, K. O. Stetter, and R. K. Thauer, *Arch. Microbiol., 167*, 19–23 (1997).

138. L. Chistoserdova, J. A. Vorholt, R. K. Thauer, and M. E. Lidstrom, *Science, 281*, 9–102 (1998).

139. J. A. Vorholt, L. Chistoserdova, S. M. Stolyar, R. K. Thauer, and M. E. Lidstrom, *J. Bacteriol., 181*, 5750–5757 (1999).

140. S. Bartoschek, J. A. Vorholt, R. K. Thauer, B. H. Geierstanger, and C. Griesinger, *Eur. J. Biochem., 267*, 3130–3138 (2000).

141. S. Shima, D. S. Weiss, and R. K. Thauer, *Eur. J. Biochem., 230*, 906–913, 1995).

142. S. Shima, R. K. Thauer, H. Michel, and U. Ermler, *Proteins: Struct. Funct. Genet., 26*, 118–120 (1996).

143. S. Shima, C. Tziatzios, D. Schubert, H. Fukada, K. Takahashi, U. Ermler, and R. K. Thauer, *Eur. J. Biochem., 258*, 85–92 (1998).

144. S. Shima, R. K. Thauer, U. Ermler, H. Durchschlag, C. Tziatzios, and D. Schubert, *Eur. J. Biochem., 267*, 6619–6623 (2000).

145. J. Kunow, S. Shima, J. A. Vorholt, and R. K. Thauer, *Arch. Microbiol.*, *165*, 97–105 (1996).

146. B. Schwörer, J. Breitung, A. R. Klein, K. O. Stetter, and R. K. Thauer, *Arch. Microbiol.*, *159*, 225–232 (1993).

147. U. Ermler, M. C. Merckel, R. K. Thauer, and S. Shima, *Structure*, *5*, 635–646 (1997).

148. A. Wasserfallen, J. Nölling, P. Pfister, J. Reeve, and E. C. de Macario, *Int. J. System. Evol. Microbiol.*, *50*, 43–53 (2000).

149. P. A. Bertam, M. Karrasch, R. A. Schmitz, R. Böcher, S. P. J. Albracht, and R. K. Thauer, *Eur. J. Biochim.*, *220*, 477–484 (1994).

150. M. Karrasch, G. Börner, M. Enssle, and R. K. Thauer, *FEBS Lett.*, *253*, 226–230 (1989).

151. M. Karrasch, G. Börner, and R. K. Thauer, *FEBS Lett.*, *274*, 48–52 (1990).

152. M. Karrasch, G. Börner, M. Enssle, and R. K. Thauer, *Eur. J. Biochim.*, *194*, 367–372 (1990).

153. J. A. Vorholt, M. Vaupel, and R. K. Thauer, *Eur. J. Biochem.*, *236*, 309–317 (1996).

154. L. Holm and C. Sander, *Proteins*, *28*, 72–82 (1997).

155. R. A. Schmitz, P. A. Bertram, and R. K. Thauer, *Arch. Microbiol.*, *161*, 528–530 (1994).

156. P. A. Bertram, R. A. Schmitz, D. Linder, and R. K. Thauer, *Arch. Microbiol.*, *161*, 220–228 (1994).

157. A. Hochheimer, R. Hedderich, and R. Thauer, *Arch. Microbiol.*, *170*, 389–393 (1998).

158. A. Hochheimer, R. A. Schmitz, R. K. Thauer, and R. Hedderich, *Eur. J. Biochem.*, *234*, 910–920 (1995).

159. A. Hochheimer, D. Linder, R. K. Thauer, and R. Hedderich, *Eur. J. Biochem.*, *242*, 156–162 (1996).

160. G. Börner, M. Karrasch, and R. K. Thauer, *FEBS Lett.*, *290*, 31–34 (1991).

161. A. Hochheimer, R. Hedderich, and R. K. Thauer, *Mol. Microbiol.*, *31*, 641–650 (1999).

162. R. Schmitz, M. Richter, D. Linder, and R. K. Thauer, *Eur. J. Biochem.*, *207*, 559–565 (1992).

163. R. A. Schmitz, S. P. J. Albracht, and R. K. Thauer, *Eur. J. Biochem.*, *209*, 103–1018 (1992).

164. R. Schmitz, S. P. J. Albracht, and R. K. Thauer, *FEBS Lett.* *309*, 78–81 (1992).

165. J. A. Vorholt, M. Vaupel, and R. K. Thauer, *Mol. Microbiol.*, *23*, 1033–1042 (1997).

166. R. A. Schmitz, D. Linder, K. O. Stetter, and R. K. Thauer, *Arch. Microbiol.*, *156*, 427–434 (1991).

167. H. P. Klenk, R. A. Clayton, J. F. Tomb, O. White, K. E. Nelson, K. A. Ketchum, R. J. Dodson, M. Gwinn, E. K. Hickey, J. D. Peterson, D. L. Richardson, A. R. Kerlavage, D. E. Graham, N. C. Kyrpides, R. D. Fleischmann, J. Quackenbush, N. H. Lee, G. G. Sutton, S. Gill, E. F. Kirkness, B. A. Dougherty, K. McKenney, M. D. Adams, B. Loftus, and J. C. Venter, *Nature*, *390*, 364–370 (1997).

168. J. A. Vorholt, C. J. Marx, M. E. Lidstrom, and R. K. Thauer, *J. Bacteriol.*, *182*, 6645–6650 (2000).

169. C. H. Hagemeier, L. Chistoserdova, M. E. Lidstrom, R. K. Thauer, and J. A. Vorholt, *Eur. J. Biochem.*, *267*, 3762–3769 (2000).

170. B. K. Pomper, J. A. Vorholt, L. Chistoserdova, M. E. Lidstrom, and R. K. Thauer, *Eur. J. Biochem.*, *261*, 475–480 (1999).

171. J. G. Ferry, *Annu. Rev. Microbiol.*, *49*, 305–333 (1995).

172. O. Meyer, L. Gremer, R. Ferner, M. Ferner, H. Dobbek, M. Gnida, W. Meyer-Klaucke, and R. Huber, *Biol. Chem.*, *381*, 865–876 (2000).

173. S. Futo and O. Meyer, *Arch. Microbiol.*, *145*, 358–360 (1986).

174. M. Bott and R. K. Thauer, *Z. Naturforsch.*, *44c*, 392–396 (1989).

175. O. Meyer, K. Frunzke, D. Gadkari, S. Jacobitz, I. Hugendieck, and M. Kraut, *FEMS Microbiol. Rev.*, *87*, 253–260 (1990).

176. G. Mörsdorf, K. Frunzke, D. Gadkari, and O. Meyer, *Biodegradation*, *3*, 61–82 (1992).

177. O. Meyer, S. Jacobitz, and B. Krüger, *FEMS Microbiol. Rev.*, *39*, 161–179 (1986).

178. O. Meyer and H. G. Schlegel, *Arch. Microbiol.*, *118*, 35–43 (1978).

179. O. Meyer, E. Stackebrand, and G. Auling, *Syst. Appl. Microbiol.*, *16*, 390–395 (1993).

180. M. Kissling and O. Meyer, *FEMS Microbiol. Lett.*, 13, 333–338 (1982).

181. M. Ribbe, D. Gadkari, and O. Meyer, *J. Biol. Chem.*, *272*, 26627–26633 (1997).

182. M. J. Lorite, J. Tachil, J. Sanjuan, O. Meyer, and E. J. Bedmar, *Appl. Environ. Microbiol.*, *66*, 1871–1876 (2000).

183. D. M. Pearson, C. O'Reilly, J. Colby, and G. W. Black, *Biochim. Biophys. Acta*, *1188*, 432–438 (1994).

184. Y. J. Kim and Y. M. Kim, *FEMS Microbiol. Lett.*, *59*, 207–210 (1989).

185. Y. T. Ro and Y. M. Kim, *Kor. J. Microbiol.*, *31*, 214–217 (1993).

186. M. Kraut and O. Meyer, *Arch. Microbiol.*, *149*, 540–546 (1988).

187. I. Hugendieck and O. Meyer, *Arch. Microbiol.*, *157*, 301–304 (1992).

188. U. Schübel, M. Kraut, G. Mörsdorf, and O. Meyer, *J. Bacteriol.*, *177*, 2197–2203 (1995).

189. B. Santiago, U. Schubel, C. Egelseer, and O. Meyer, *Gene*, *236*, 115–124 (1999).

190. B. S. Kang and Y. M. Kim, *J. Bacteriol.*, *181*, 5581–5590 (1999).

191. S. Jacobitz and O. Meyer, *J. Bacteriol.*, *171*, 6294–6299 (1989).

192. B. Santiago and O. Meyer, *FEMS Microbiol. Lett.*, *136*, 157–162 (1996).

193. M. Kraut, I. Hugendieck, S. Herwig, and O. Meyer, *Arch. Microbiol.*, *152*, 335–341 (1989).

194. O. Meyer, *J. Biol. Chem.*, *257*, 1333–1341 (1982).

195. J. L. Johnson, K. V. Rajagopalan, and O. Meyer, *Arch. Biochem. Biophys.*, *283*, 542–545 (1990).

196. P. Hänzelmann, B. Hofmann, S. Meisen, and O. Meyer, *FEMS Microbiol. Lett.*, *176*, 139–145 (1999).

197. L. Gremer, S. Kellner, H. Dobbek, R. Huber, and O. Meyer, *J. Biol. Chem.*, *275*, 1864–1872 (2000).

198. H. Dobbek, L. Gremer, O. Meyer, and R. Huber, *Proc. Natl. Acad. Sci USA*, *96*, 8884–8889 (1999).

199. P. Hänzelmann, H. Dobbek, L. Gremer, R. Huber, and O. Meyer, *J. Mol. Biol.*, *301*, 1221–1235 (2000).

200. P. Hänzelmann and O. Meyer, *Eur. J. Biochem.*, *255*, 755–765 (1998).

201. R. L. Uffen, *Proc. Natl. Acad. Sci. USA*, *73*, 3298–3302 (1976).

202. R. L. Uffen, *J. Bacteriol.*, *155*, 956–965 (1983).

203. V. A. Svetlichny, T. G. Sokolova, M. Gerhardt, N. A. Kostrikina, and G. A. Zavarzin, *Microb. Ecol.*, *21*, 1–10 (1991).

204. V. A. Svetlichny, T. G. Sokolova, M. Gerhardt, M. Ringpfeil, N. A. Kostrikina, and G. A. Zavarzin, *Syst. Appl. Microbiol.*, *14*, 254–260 (1991).

205. J. M. Gonzáles and F. T. Robb, *FEMS Microbiol. Lett.*, *191*, 243–247 (2000).

206. R. K. Thauer, *Eur. J. Biochem.*, *176*, 497–508 (1988).

207. R. K. Thauer, D. Möller-Zinkhan, and A. Spormann, *Annu. Rev. Microbiol.*, *43*, 43–67 (1989).

208. J. G. Ferry, *FEMS Microbiol. Rev.*, *23*, 13–38 (1999).

209. P. A. Lindahl and B. Chang, *Origins for Life and Evolution of the Biosphere*, in press.

210. E. L. Maynard and P. A. Lindahl, *J. Am. Chem. Soc.*, *121*, 9221–9222 (1999).

211. J. Seravalli and S. W. Ragsdale, *Biochemistry*, *39*, 1274–1277 (2000).

212. C. R. Staples, J. Heo, N. J. Spangler, R. L. Kerby, G. P. Roberts, and P. W. Ludden, *J. Am. Chem. Soc.*, *121*, 11034–11044 (1999).

213. J. Heo, C. R. Staples, J. Telser, and P. W. Ludden, *J. Am. Chem. Soc.*, *121*, 11045–11057 (1999).

214. S. A. Ensign and P. W. Ludden, *J. Biol. Chem.*, *266*, 18395–18403 (1991).

215. J. Heo, C. R. Staples, C. M. Halbleib, and P. W. Ludden, *Biochemistry*, *39*, 7956–7963 (2000).

216. G. B. Diekert, E.-G. Graf, and R. K. Thauer, *Arch. Microbiol.*, *122*, 117–120 (1979).

217. U. Ermler, W. Grabarse, S. Shima, M. Goubeaud, and R. K. Thauer, *Curr. Opin. Struct. Biol.*, *8*, 749–758 (1998).

218. J. C. Fontecilla-Camps, and S. W. Ragsdale, *Adv. Inorg. Chem.*, *47*, 283–333 (1999).

219. M. J. Maroney, *Curr. Opin. Chem. Biol.*, *3*, 188–199 (1999).

220. S. W. Ragsdale, *Curr. Opin. Chem. Biol.*, *2*, 208–215 (1998).

221. L. Chistoserdova, L. Gomelsky, J. A. Vorholt, M. Gomelsky, Y. D. Tsygankov, and M. E. Lidstrom, *Microbiology*, *146*, 233–238 (2000).

222. P. W. van Ophem, J. Van Beeumen, and J. A. Duine, *Eur. J. Biochem.*, *206*, 511–518 (1992).

223. S. W. Kim, D. M. Luykx, S. de Vries, and J. A. Duine, *Arch. Biochem. Biophys.*, *325*, 1–7 (1996).

224. J. A. Duine, *Biofactors*, *10*, 201–206 (1999).

225. M. W. Adams and A. Kletzin, *Adv. Protein Chem.*, *48*, 101–180 (1996).

226. A. Kletzin and M. W. Adams, *FEMS Microbiol. Rev.*, *18*, 5–63 (1996).

227. H. White, R. Feicht, C. Huber, F. Lottspeich, and H. Simon, *Biol. Chem. Hoppe Seyler*, *372*, 999–1005 (1991).

228. H. White, C. Huber, R. Feicht, and H. Simon, *Arch. Microbiol.*, *159*, 244–249 (1993).

229. G. Strobl, R. Feicht, H. White, F. Lottspeich, and H. Simon, *Biol. Chem. Hoppe Seyler*, *373*, 123–132 (1992).

230. S. Kardinahl, C. L. Schmidt, T. Hansen, S. Anemuller, A. Petersen, and G. Schäfer, *Eur. J. Biochem.*, *260*, 540–548 (1999).

231. A. Kletzin, S. Mukund, T. L. Kelley-Crouse, M. K. Chan, D. C. Rees, and M. W. W. Adams, *J. Bacteriol.*, *177*, 4817–4819 (1999).

232. S. Mukund and M. W. W. Adams, *J. Bacteriol.*, *178*, 163–167 (1996).

233. Y. Hu, S. Faham, R. Roy, M. W. W. Adams, and D. C. Rees, *J. Mol. Biol.*, *286*, 899–914 (1999).

234. R. Roy, S. Mukund, G. J. Schut, D. M. Dunn, R. Weiss, and M. W. W. Adams, *J. Bacteriol.*, *181*, 1171–1180 (1999).

235. S. Mukund and M. W. W. Adams, *J. Biol. Chem.*, *268*, 13592–13600 (1993).

236. I. K. Dhawan, R. Roy, B. P. Koehler, S. Mukund, M. W. W. Adams, and M. K. Johnson, *J. Biol. Inorg. Chem.*, *5*, 313–327 (2000).

237. J. Ras, P. W. Van Ophem, W. N. Reijnders, R. J. Van Spanning, J. A. Duine, A. H. Stouthamer, and N. Harms, *J. Bacteriol.*, *177*, 247–251 (1995).

238. R. D. Barber, M. A. Rott, and T. J. Donohue, *J. Bacteriol.*, *178*, 1386–1393 (1996).

239. R. D. Barber and T. J. Donohue, *J. Mol. Biol.*, *280*, 775–784 (1998).

240. R. D. Barber and T. J. Donohue, *Biochemistry*, *37*, 530–537 (1998).

241. M. R. Fernandez, J. A. Biosca, A. Norin, H. Jornvall, and X. Pares, *FEBS Lett.*, *370*, 23–26 (1995).

242. G. Aggelis, S. Fakas, S. Melissis, and Y. D. Clonis, *J. Biotechnol.*, *72*, 127–139 (1999).

243. N. Kato, *Meth. Enzymol.*, *188*, 455–459 (1990).

244. P. W. Van Ophem and J. A. Duine, *FEMS Microbiol. Lett.*, *116*, 87–93 (1994).

245. M. Koivusalo, M. Baumann, and L. Uotila, *FEBS Lett.*, *257*, 105–109 (1989).

246. N. Kummerle, H. H. Feucht, and P. M. Kaulfers, *Antimicrob. Agents Chemother.*, *40*, 2276–2279 (1996).

247. W. G. Gutheil, B. Holmquist, and B. L. Vallee, *Biochemistry*, *31*, 475–481 (1992).

248. W. G. Gutheil, E. Kasimoglu, and P. C. Nicholson, *Biochem. Biophys. Res. Commun.*, *238*, 693–696 (1997).

249. P. M. Kaulfers and A. Marquardt, *FEMS Microbiol. Lett.*, *63*, 335–338 (1991).

250. U. Wippermann, J. Fliegmann, G. Bauw, C. Langebartels, K. Maier, and H. Sandermann, Jr., *Planta*, *208*, 12–18 (1999).

251. J. Fliegmann and H. Sandermann, Jr., *Plant. Mol. Biol.*, *34*, 843–854 (1997).

252. M. C. Martinez, H. Achkor, B. Persson, M. R. Fernandez, J. Shafqat, J. Farres, H. Jornvall, and X. Pares, *Eur. J. Biochem.*, *241*, 849–857 (1996).

253. R. Dolferus, J. C. Osterman, W. J. Peacock, and E. S. Dennis, *Genetics, 146,* 1131–1141 (1997).

254. J. Shafqat, M. El-Ahmad, O. Danielsson, M. C. Martinez, B. Persson, X. Pares, and H. Jornvall, *Proc. Natl. Acad. Sci. USA, 93,* 5595–5599 (1996).

255. T. Luque, S. Atrian, O. Danielsson, H. Jornvall, and R. Gonzalez-Duarte, *Eur. J. Biochem., 225,* 985–993 (1994).

256. S. Tsuboi, M. Kawase, A. Takada, M. Hiramatsu, Y. Wada, Y. Kawakami, M. Ikeda, and S. Ohmori, *J. Biochem. (Tokyo), 11,* 465-471 (1992).

257. M. H. Foglio and G. Duester, *Eur. J. Biochem., 237,* 496–504 (1996).

258. M. W. Hur and H. J. Edenberg, *Gene, 121,* 305–311 (1992).

259. B. Holmquist and B. L. Vallee, *Biochem. Biophys. Res. Commun., 178,* 1371–1377 (1991).

260. R. Kaiser, B. Holmquist, B. L. Vallee, and H. Jornvall, *J. Protein Chem., 10,* 69–73 (1991).

261. K. Engeland, J. O. Hoog, B. Holmquist, M. Estonius, H. Jornvall, and B. L. Vallee, *Proc. Natl. Acad. Sci. USA, 90,* 2491–2494 (1993).

262. M. Estonius, J. O. Hoog, O. Danielsson, and H. Jornvall, *Biochemistry, 33,* 15080–15085 (1994).

263. Z. N. Yang, W. F. Bosron, and T. D. Hurley, *J. Mol. Biol., 265,* 330–343 (1997).

264. P. C. Sanghani, C. L. Stone, B. D. Ray, E. V. Pindel, T. D. Hurley, and W. F. Bosron, *Biochemistry, 39,* 10720–10729 (2000).

265. M. R. Fernandez, J. A. Biosca, D. Torres, B. Crosas, and X. Pares, *J. Biol. Chem., 274,* 37869–37875 (1999).

266. P. W. Van Ophem, L. V. Bystrykh, and J. A. Duine, *Eur. J. Biochem., 206,* 519–525 (1992).

267. L. Eggeling and H. Sahm, *Eur. J. Biochem., 150,* 129–134 (1985).

268. M. Misset-Smits, P. W. van Ophem, S. Sakuda, and J. A. Duine, *FEBS Lett., 409,* 221–222 (1997).

269. A. Norin, P. W. Van Ophem, S. R. Piersma, B. Persson, J. A. Duine, and H. Jörnvall, *Eur. J. Biochem., 248,* 282–289 (1997).

270. C. Anthony, in *The Biochemistry of Methylotrophs*, Academic Press, London, 1982.

271. A. J. Beardsmore, P. N. G. Aperghis, and J. R. Quayle, *J. Gen. Microbiol., 128*, 1423–1439 (1982).

272. H. Yanase, K. Ikeyama, R. Mitsui, S. Ra, K. Kita, Y. Sakai, and N. Kato, *FEMS Microbiol. Lett., 135*, 201–205 (1996).

273. S. Tate and H. Dalton, *Microbiology, 145*, 159–167 (1999).

274. K. Ito, M. Takahashi, T. Yoshimoto, and D. Tsuru, *J. Bacteriol., 176*, 2483–2491 (1994).

275. D. Tsuru, N. Oda, Y. Matsuo, S. Ishikawa, K. Ito, and T. Yoshimoto, *Biosci. Biotechnol. Biochem., 61*, 1354–1357 (1997).

276. M. M. Attwood, N. Arfman, R. A. Weusthuis, and L. Dijkhuizen, *A. v. Leeuwenhoek, 62*, 201–207 (1992).

277. C. R. Klein, F. P. Kesseler, C. Perrei, J. Frank, J. A. Duine, and A. C. Schwartz, *Biochem. J., 301*, 289–295 (1994).

278. M. Grey, M. Schmidt, and M. Brendel, *Curr. Genet., 29*, 437–440 (1996).

279. A. A. DiMarco, T. A. Bobik, and R. S. Wolfe, *Annu. Rev. Biochem., 59*, 355–394 (1990).

280. M. K. Chan, S. Mukund, A. Kletzin, M. W. W. Adams, and D. C. Rees, *Science, 267*, 1463–1469 (1995).

281. H. Dobbek, V. Svetlitchnyi, L. Cremer, R. Huber, and O. Meyer, *Science, 293*, 1281–1285 (2001).

17
Molybdenum Enzymes and Sulfur Metabolism

John H. Enemark and Michele Mader Cosper

Department of Chemistry, University of Arizona,
Tucson, AZ 85721-0041, USA

1. INTRODUCTION

Molybdenum enzymes play important roles in sulfur metabolism of organisms ranging from bacteria to humans. Sulfite oxidase (SO) is essential for normal neurological development in neonatal children. Dimethyl sulfide (DMS) produced by the action of dimethyl sulfoxide (DMSO) reductase in certain bacteria is proposed to play a key role in cloud formation and the overall albedo of the earth [1,2].

The X-ray crystal structures of SO and DMSO reductase, respectively, provide the prototypical examples of the coordination of the Mo atom in the SO and DMSO reductase families of Mo enzymes [3,4]. A key feature of all Mo enzymes that catalyze two-electron oxidation-reduction reactions is coordination by the S atoms of the ene-dithiolate (also called dithiolene) fragment of the unusual pyranopterin whose basic structure is shown in **I**. The principal structural features of **I** were deduced by Rajagopalan and coworkers [5,6] from degradative and spectroscopic studies, and they named **I** "molybdopterin" (MPT). Variants that contained an appended dinucleotide instead of a phosphate group, such as guanine dinucleotide, were abbreviated MGD. Various systematic approaches to naming **I** and its derivatives have been discussed [7,8]. Here we will adopt the name "pyranopterindithiolate" for **I**. The abbreviation S_2pdt will be used for all forms of **I**; the abbreviation S_2dt will be used to designate generic dithiolate ligands in synthetic analogues. Due to space limitations, the scope of this chapter is restricted primarily to structural and spectroscopic characterization of SO and DMSO reductase and related bioinspired compounds. Recent results for biotin sulfoxide reductase and polysulfide reductase are briefly summarized.

I

2. SULFITE OXIDASE

2.1. Physiological Role in Sulfur Metabolism

SO, the most extensively studied member of the sulfite oxidase family [3,4], catalyzes the oxidation of sulfite (SO_3^{2-}) to sulfate (SO_4^{2-}) with the concomitant reduction of two equivalents of cytochrome c. The conversion of SO_3^{2-} to SO_4^{2-} is the final step in the degradation of the S-containing amino acids, cysteine and methionine. SO is common to all vertebrates, but the chicken, rat, and human enzymes have been most extensively studied. SO is located in the inner mitochondrial membrane space and is concentrated primarily in the liver, kidneys, and heart.

2.2. Sulfite Oxidase Deficiency

Sulfite oxidase deficiency is a genetic disease that was first described by Irreverre and coworkers [9] in 1967. The absence of SO activity in humans is characterized by dislocation of ocular lenses, mental retardation, and, in severe cases, attenuated growth of the brain and early death. These severe neurological symptoms result from the inability to properly produce the pyranopterindithiolate (**I**) [10,11], which results in simultaneous deficiencies in all mononuclear Mo enzymes, and from point mutations in the SO protein itself [11–14]. The biochemical basis of the pathology of sulfite oxidase deficiency is not yet known. Fatal brain damage may be due to the accumulation of a toxic metabolite (possibly SO_3^{2-}); alternatively, a deficiency in the reaction product (SO_4^{2-}) may disturb normal fetal and neonatal development of the brain [11].

2.3. Structural Characterization of Chicken Liver Sulfite Oxidase

To date, chicken liver SO [12] (Fig. 1) is the only member of the sulfite oxidase family whose crystal structure has been determined. SO is a α_2 dimer; each monomer consists of three domains. The N-terminal domain is structurally similar to bovine cytochrome b_5; the central domain surrounds the Mo center and has a fold that has not been observed previously; the C-terminal domain comprises much of the interface region between the subunits and has topology similar to the C2 subtype of the immunoglobin superfamily [15]. The Mo and heme domains are linked by a flexible loop with no secondary structure, and the Mo $\cdots$ Fe distance in the crystal is 32 Å. The active site of SO contains one S_2pdt per Mo atom, which is deeply buried beneath the protein surface and involved in extensive hydrogen bonding with side chains of highly conserved amino acid residues. The crystal structure revealed a novel five-coordinate, square pyramidal Mo center (Fig. 2) with a terminal oxo ligand in the apical position and thiolate S atoms in three of the equatorial positions. Two of the S atoms are from the S_2pdt ligand, and the third is the $S\gamma$ atom of Cys185, which is highly conserved in all SOs. Mutation of the

FIG. 1.   Ribbon diagram of the SO dimer viewed parallel to the pseudo-twofold axis. Unresolved portions of the flexible loop linking the Mo and heme domains are indicated by dotted lines. The Mo center and the heme center are shown in ball-and-stick representation. Coordinates from the Protein Data Bank, entry 1SOX.

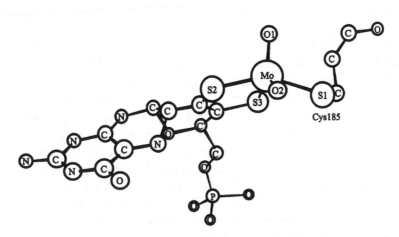

FIG. 2.   The Mo active site of chicken liver SO [12]. Bond distances (Å): Mo-O(1), 1.8; Mo-O(2), 2.3; Mo-S(1), 2.5; Mo-S(2), 2.4; Mo-S(3), 2.4. Bond angles (deg): S(1)-Mo-S(2), 93; S(1)-Mo-S(3), 147; S(2)-Mo-S(3), 84. Torsion angles (deg.): O(1)-Mo-S(1)-C, 88; O(1)-Mo-S(2)-C, 98; O(1)-Mo-S(3)-C, 117.

analogous cysteine residue in human SO severely attenuated activity [16] and produced an unusual trioxo Mo center [17]. The fourth equatorial position is occupied by an O atom that appears to be an OH^- or a H_2O ligand based on the long $Mo-O$ bond distance. However, SO has been shown to contain a $[Mo^{VI}O_2]^{2+}$ core in its fully oxidized state by X-ray absorption spectroscopy (XAS) studies on polycrystalline oxidized SO [18]. This result suggests that the Mo center may have become reduced to Mo(V) or Mo(IV) by synchrotron radiation during the X-ray crystal structure determination. Additionally, the tryptically cleaved human SO has been shown to possess a dioxo-Mo center in the oxidized form by resonance Raman spectroscopy [19]. The sixth coordination site, trans to the apical oxo ligand, is blocked by the protein backbone and is not accessible to substrate or other ligands.

2.4. Catalytic Mechanism

Investigation of the catalytic mechanism of SO is complicated by the multiple oxidation states that are accessible for the oxo-Mo center and the Fe of the b-type heme center. The six distinct combinations of oxidation states for each monomer of the dimeric protein are shown in Fig. 3. In the generally accepted catalytic cycle [20–23], the Mo(VI) center of the fully oxidized enzyme (1) reacts with SO_3^{2-} to produce SO_4^{2-}. The two-electron reduced enzyme, which is in the Mo(IV)/Fe(III) state (2), undergoes rapid intramolecular electron transfer (IET) to generate the Mo(V)/Fe(II) state (3) that can be detected by electron paramagnetic resonance (EPR) spectroscopy (Sec. 2.5). The Fe(II) of the b-type heme of 3 is then oxidized by exogenous cytochrome c to generate 4. A second IET step generates 5, which can react with a second molecule of cytochrome c to regenerate the enzyme in the fully oxidized resting state (1). States 2 and 3 are two-electron reduced forms; states 4 and 5 are one-electron reduced forms. Reaction with the strong reductant dithionite generates the three-electron reduced form of SO (6).

The intimate catalytic mechanism of the Mo center of SO (Scheme 1, bold numbers correspond to Fig. 3) is proposed to involve initial nucleophilic attack of SO_3^{2-} on the electrophilic equatorial oxo group of 1 in the solvent-exposed anion-binding pocket of SO to generate a transient oxo-Mo(IV)-SO_4 species [20–23]. Hydrolysis of SO_4^{2-} generates

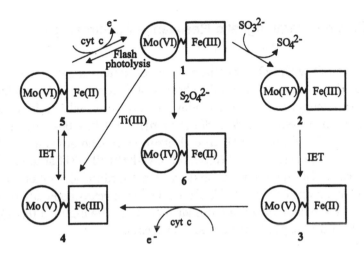

FIG. 3. Possible oxidation states for the Mo and heme centers of SO and the reactions that relate them to one another.

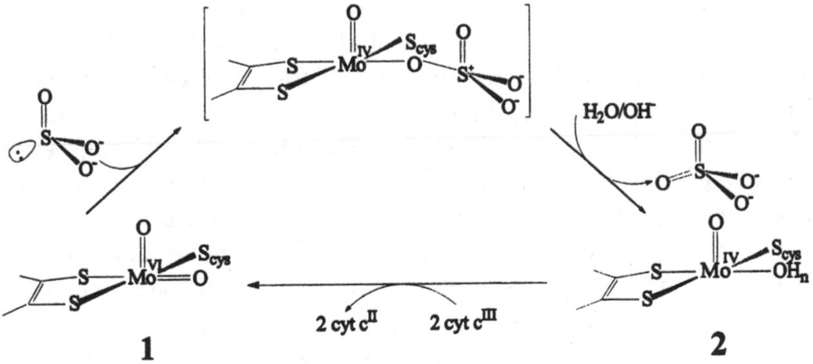

Scheme 1

2 and completes the oxygen atom transfer (OAT) phase of the reaction. Scheme 1 is also consistent with theoretical calculations on OAT reactions in model compounds [24,25]. Pilato and Stiefel have proposed an alternative mechanism in which a water molecule (or hydroxide ion) that is hydrogen-bonded to the equatorial oxo group of 1 becomes activated for incorporation into SO_3^{2-} to form SO_4^{2-} [26].

2.5. Electron Paramagnetic Resonance Spectroscopy

Reaction of SO with SO_3^{2-} in the absence of oxidants generates significant quantities of state **3** (Fig. 3) whose Mo(V) center can be studied by EPR spectroscopy. Cohen and Fridovich [27] observed the first Mo(V) EPR signal from SO in 1971. Subsequently, SO has been extensively examined by CW-EPR [28–31] and electron spin echo envelope modulation (ESEEM) spectroscopy [23,32,33]. CW-EPR [28–31] studies of the Mo(V) form of SO have shown that the Mo coordination sphere is very sensitive to its environment (Table 1). Three spectroscopically distinct forms that have been observed and identified are the high pH (hpH), low pH (lpH), and phosphate-inhibited (Pi) forms.

2.5.1. The Low-pH Form

The equilibrium between hpH and lpH forms of SO can be shifted to produce a majority of either form. The lpH form is obtained in low pH buffers (6.5–7.0) and in the absence of phosphate anions [28]. CW-EPR of the lpH form exhibits a large g_1 value (Table 1) and resolved doublets at the g_1 and g_2 positions, which collapse to singlets in D_2O buffer. Bray and coworkers [28] determined that the observed splitting resulted from a hyperfine interaction of Mo(V) with a single strongly coupled exchangeable proton. The EPR signal of the lpH form was postulated to arise from a Mo^V-OH species [28,32]. Two-pulse and four-pulse ESEEM and two-dimensional hyperfine sublevel correlation (HYSCORE) spectroscopies [32] were used to investigate the nature of the proposed Mo^V-OH lpH species and provided the first unambiguous evidence that the lpH form possesses a coordinated OH(D) group.

2.5.2 The High-pH Form

The hpH form is obtained in high-pH buffers (9–9.5) containing low concentrations of chloride and phosphate anions, and the CW-EPR spectrum shows no evidence for a hyperfine interaction of Mo(V) with nearby protons [28]. Recently, a refocused primary echo envelope modulation (RP ESEEM) technique was used to detect nearby protons in the hpH form for the first time [23]. Simulation of the RP ESEEM spectrum

TABLE 1

EPR Data for Chicken Liver Sulfite Oxidase and Mo(V) Model Compounds

Signal	g-values[a]				A (^{95,97}Mo)[b]				A(X)				Euler angles[c]		
	1	2	3	Av	1	2	3	Av	1	2	3	Av	α	β	γ
SO (low pH form)[d,e]	2.007	1.974	1.968	1.983	56.7	25.0	16.7	32.8	8.0[f]	7.4	11.9	9.1	0[g]	18	0
SO (high pH form)[d,e]	1.990	1.966	1.954	1.970	54.4	21.0	11.3	28.9					0[g]	14	22
SO (phosphate inhibited)[e]	1.992	1.969	1.961	1.978											
SO (sulfite form)[h]	2.000	1.972	1.963	1.978											
SO (arsenate inhibited)[i]	1.993	1.970	1.964	1.976					4.3[j]	8.3	5.3	6.0	49[k]	0	14
Models															
(L-N$_3$)Mo^{V}O(bdt)[d,e]	2.004	1.972	1.934	1.971	50.0	11.4	49.7	37.0					0[g]	0	0
(L-N$_2$S$_2$)Mo^{V}O(SCH$_2$Ph)[l]	2.022	1.963	1.956	1.980	58.4	23.7	22.3	34.8					0[g]	24	0
[MoO(edt)$_2$]$^{-m}$	2.017	1.979	1.979	1.990	52.6	23.0	23.0	31.5					0[g]	0	0
[MoO(edt)$_2$]$^{-n}$	2.052	1.983	1.979	2.005	49.4	18.7	20.7	29.6					0[g]	0	0

[a]Error ±0.001. [b]A (^{95,97}Mo), ×10^{-4} cm^{-1}, errors ±1 × 10^{-4} cm^{-1} [c]Errors ±2°. [d]Ref. [114,115]. [e]Ref. [28]. [f]A (^{1}H). [g]Angles between **g** and **A** (^{95,97}Mo). [h]Ref. [29]. [i]Ref. [34]. [j]A (^{75}As). [k]Angles between **g** and **A** (^{75}As). [l]Ref. [47]. [m]Ref. [116]. [n]Ref. [80].

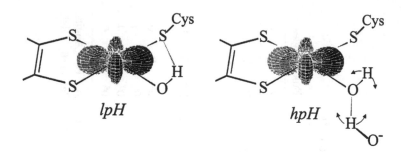

FIG. 4.   Proposed coordination environments of the Mo(V) centers for the *lpH* and *hpH* forms of SO. The *lpH* form has a single strongly coupled ordered proton that is ascribed to a Mo-OH $\cdots$ S$_{cys}$ hydrogen bonding interaction in the *XY* plane [32]. The *hpH* form appears to have two nearby protons in distributed orientations arising from hydrogen bonding between the Mo^V-OH group and a hydroxide ion (or water molecule) in the anion-binding pocket [23].

suggests a hyperfine interaction of Mo(V) with two nearby protons [23] (Fig. 4). The proton isotropic hyperfine constants (a_0) of the *hpH* and *lpH* forms differ significantly ($a_0 = 26$ MHz and ~ 0 MHz for the *lpH* and *hpH* forms, respectively). These differences in the value of a_0 are ascribed to variations in the orientation of the coordinated $-$OH ligand relative to the nodal plane of the single unpaired 4d electron of the Mo center. The ESEEM results for SO are consistent with the protein structure [12], which shows that the equatorial oxo group is exposed to solvent and therefore should be the one involved in OAT and coupled electron/proton transfer (CEPT) reactions of the active site, as well as interactions with exogenous anions. The EPR studies performed on chicken SO will serve as a prototype for EPR studies of the wild-type human SO, as well as the site-directed and clinical mutants.

2.5.3.   The Phosphate- and Arsenate-Inhibited Forms

The *Pi* form is generated in low pH buffers (6.5–7.0) in the presence of phosphate, and the CW-EPR spectrum exhibits no splitting due to a nearby proton [28]. Using ^{17}O-labeled phosphate, Bray and coworkers [31] demonstrated that phosphate coordinates to the Mo(V) center. Extensive two-pulse multifrequency ESEEM investigations of the *Pi* form revealed the ^{31}P interaction directly [33]. Comprehensive simulation of the Mo $\cdots$ P hyperfine interaction required a distribution of orientations of a monodentate phosphate group with a Mo$\cdots$P distance of 3.2–3.3 Å. This ESEEM study provided strong evidence that the Mo center of SO has a single anion binding site, which was borne out by the crystal structure [12]. When arsenate was added to the buffer instead of phosphate at low pH, K-edge Mo extended X-ray absorption fine structure (EXAFS) [34] showed a Mo$\cdots$As interaction at 3.20 Å, indicative of monodentate arsenate ligand binding. Bray and coworkers [29] also have presented evidence for coordination of SO_3^{2-} anions to the Mo(V) center of SO; the EPR parameters for this species were very similar to the *Pi* form.

2.6. Intramolecular Electron Transfer

Laser flash photolysis of SO in the presence of 5-deazariboflavin and a sacrificial donor (edta or semicarbazide) results in the rapid one-electron reduction of the Fe(III)-heme center of SO (conversion of **1** to **5** in Fig. 3), followed by IET between the Mo and Fe centers [eqs. (1)–(3)] [35–37].

$$dRF \xrightarrow[\text{edta}]{h\nu} dRFH^\bullet \tag{1}$$

$$Fe^{III}/Mo^{VI} + dRFH^\bullet \longrightarrow Fe^{II}/Mo^{VI} \tag{2}$$

$$Fe^{II}/Mo^{VI} \underset{k_r}{\overset{k_f}{\rightleftharpoons}} Fe^{III}/Mo^{V} \tag{3}$$

By analysis of the laser flash photolysis data as a function of pH and ionic strength, it has been possible to extract both the forward and reverse rate constants (k_f and k_r) of eq. (3), the IET reaction that interconverts the one-electron reduced forms of SO (**4** and **5** of Fig. 3) [35]. Combining this kinetic information with the crystal structure [12] and the ESEEM results [32,33] for the Mo(V) forms of the two-electron reduced state of SO (**3** in Fig. 3) has led to a self-consistent proposal for a CEPT reaction at the Mo center [35] [eq. (4)].

$$Fe^{II}/\overset{O}{\overset{\|}{Mo}}{}^{VI}{=}O \underset{-H^+}{\overset{+H^+}{\rightleftharpoons}} Fe^{III}/\overset{O}{\overset{\|}{Mo}}{}^{V}{-}OH \tag{4}$$

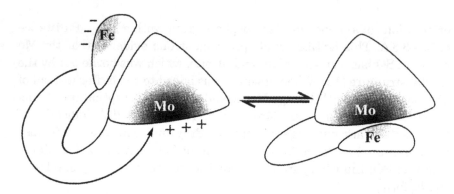

FIG. 5. Schematic view of the proposed conformational change that will move the positively charged Mo center and the negatively charged Fe center of SO closer together in solution, thereby favoring more rapid IET at low anion concentration. (Adapted with permission from [35]. Copyright 1999, Society of Biological Inorganic Chemistry.)

The wide range of IET rates ($20–1400$ s^{-1}) also led to the proposal that the flexible loop connecting the negatively charged heme center and the positively charged Mo center may allow SO to adopt more than one conformation in solution (Fig. 5), depending on the concentration of anions, and pH [35]. Conformations with shorter Mo$\cdots$Fe distances could exhibit substantially faster IET.

2.7. Bioinspired Molybdenum Compounds

The novel coordination about the Mo atom of SO has stimulated the investigation of the fundamental properties of the Mo$-$S bonding in well-characterized model compounds. Six-coordinate (L-N_3)MoO(S$_2$dt) complexes possess a cis-MoO(S$_2$dt) fragment that mimics the expected coordination of **I** to the Mo(V) state of SO (Figs. 2 and 4). The lowest energy electronic transitions are S p$\pi \longrightarrow$ Mo d$_{xy}$ charge transfer bands [38]. The resonance Raman spectrum of (L-N_3)MoO(bdt) [39] shows that the charge transfer absorption near $19,400$ cm^{-1}, strongly enhances vibrations at 362 and 393 cm^{-1}, which are assigned as symmetric Mo-S stretching and bending modes. These combined spectroscopic features

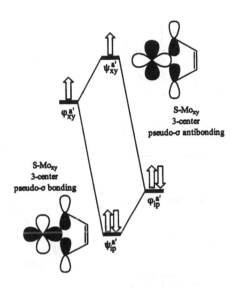

FIG. 6. Molecular orbital diagram showing the pseudo-σ bonding and antibonding interactions of the Mo d_{xy} orbital and the *in-plane* (ip) dithiolate S p orbitals [39]. (Reproduced with permission from [39]. Copyright 1999 American Chemical Society.)

have been associated with a unique three-center pseudo-σ bonding interaction (Fig. 6) that mixes substantial in-plane S p-character into the singly occupied Mo d_{xy} orbital and is proposed to provide a novel σ electron transfer pathway into the S_2pdt ligand system in SO.

The arrangements of three equatorial thiolate ligands about the Mo atom at the active site of SO (Fig. 2) is unprecedented for mononuclear Mo compounds [40–42] and has stimulated synthesis and spectroscopic characterization of new model compounds. Although several oxo-$Mo^V(S)_3$ species have been proposed from solution EPR studies [43–46], the Mo(V) complexes of the type (L-N_2S)MoO(SR) (Fig. 7), where L-$N_2S_2H_2$ is the tetradentate ligand N, N'-dimethyl-N, N'-bis(mercaptophenyl)ethylenediamine; R = C_2H_5, CH_2Ph, p-C_6H_4X (X = OCH_3, CF_3, Cl, Br, F, H, CH_3, C_2H_5) are the first to be structurally characterized [47]. These six-coordinate oxo-Mo(V) complexes possess three S donor atoms in the equatorial plane (Fig. 8). Although these models do not contain an ene-dithiolate ligand, the relative orientation of the pπ orbitals of the adjacent S atoms from the L-N_2S_2 and SR ligands (Fig. 8a)

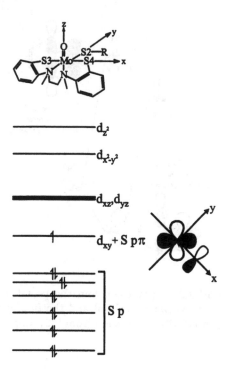

FIG. 7.   Energy level diagram for the *cis,trans*-(L-N_2S_2)Mo^{V}O(SR) compounds depicting the HOMO, which is composed of the Mo d_{xy} and the in-plane S4 p_π orbitals. (Reproduced with permission from [47]. Copyright 2000 American Chemical Society.)

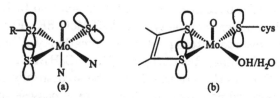

FIG. 8.   Coordination about the Mo centers of (a) *cis,trans*-L-N_2S_2)Mo^{V}O(SR) and (b) chicken liver SO, depicting the relative orientation of the S $p\pi$ orbitals. (Reproduced with permission from [47]. Copyright 2000 American Chemical Society.)

have the same spatial arrangement as those of an ene-dithiolate (Fig. 8b). Previous studies [38,39] have shown this relative orientation of the S $p\pi$ orbitals is the primary factor in determining spectroscopic properties of a $Mo^V O(S_2 dt)$ fragment [38]. The absorption spectra of the $cis,trans$-$(L$-$N_2 S_2)MoO(SR)$ compounds exhibit an intense broad band around 15,500 cm^{-1}, which has been ascribed to in-plane S $p\pi \longrightarrow Mo\, 4d_{xy}$ charge transfer from S4 (Figs. 7 and 8a). For the Mo(V)/Fe(II) state of SO (Fig. 3, state **3**) no MCD intensity is observed below $\sim 17,000$ cm^{-1} [48], consistent with the O1-Mo-S1-C torsional angle of $\sim 90°$ for Cys185 (Figs. 2 and 8b). This geometry precludes the efficient π bonding between the S $p\pi$ orbital and the Mo d$_{xy}$ orbital shown in Fig. 7. The MCD features anticipated near 19,000 cm^{-1} for the in-plane pseudo-σ interaction of the S$_2$pdt ligand (Fig. 6) are obscured by the intense temperature-independent A and B terms of the diamagnetic b-type Fe(II) heme center [48]. Very recently, Holm and coworkers synthesized a $Mo^{VI} O_2(S_2 dt)(SR)$ compound that mimics the active site structure of oxidized SO [49].

2.8 Sulfite Oxidase from Microorganisms and Plants

Various photo- and chemotrophic sulfur-oxidizing microorganisms effect the direct oxidation of SO$_3^{2-}$ to SO$_4^{2-}$ [117]. Recently, Dahl and coworkers [50] determined that *Thiobacillus novellus* carried out this reaction using a periplasmic $\alpha\beta$ heterodimeric SO. The α subunit (40.6 kDa) shows similarities to known vertebrate SOs and contains a Mo cofactor with one S$_2$pdt per Mo atom. The β subunit (8.8 kDa) possesses a mono-heme cytochrome c_{552} that does not appear to be closely related to any known c-type cytochromes. EPR spectroscopy revealed the presence of a SO$_3^{2-}$-inducible Mo(V) signal having g_{123} values of 1.9914, 1.9661, and 1.9541, which are very similar to the Mo(V) hpH EPR parameters observed for SO. In contrast with SO, no pH-dependent changes were observed in the EPR signal.

The existence of SO in plants has been controversial [51]. SO$_3^{2-}$ oxidizing activity that occurred in the dark was purified from spinach chloroplasts and found to be associated with the thylakoid membranes [52,53]. Gania et al. [54] reported the purification of SO activity from *Malva sylvestris*, but the enzyme had a molecular weight of only 27 kDa.

SO from *Arabidopsis thaliana* has been recently identified and characterized [118].

3. DIMETHYL SULFOXIDE REDUCTASE

3.1. Role in Sulfur Metabolism

DMSO reductase catalyzes the reduction of DMSO to DMS according to Eq. (5).

$$DMSO + 2H^+ \rightleftharpoons DMS + H_2O \qquad (5)$$

The possible ecological importance of this reaction and its ultimate relationship to the cloud cover and albedo of the earth have been discussed elsewhere [1,2]. DMSO reductase from photosynthetic bacteria of the genus *Rhodobacter* has been a key Mo enzyme for studies of structure-function relationships since its discovery [55,56] because its Mo center is the only prosthetic group in a protein of about 86 kDa. Thus, unlike most other Mo-containing enzymes, there are no other chromophoric groups to interfere with spectroscopic studies of the Mo center. Consequently, DMSO reductase from *Rhodobacter sphaeroides* and from *Rhodobacter capsulatus* have been extensively studied by X-ray crystallography [57–63], XAS [64–66], resonance Raman [67–70], and EPR spectroscopies [55,64,71].

3.2. Crystal Structures of the *R. sphaeroides* and *R. capsulatus* Enzymes

The multiple X-ray crystal structures of DMSO reductases [57–63] have produced a surprisingly complicated view of the active site that has stimulated considerable research on the enzymes. All of the X-ray crystal structures exhibit essentially identical arrangements of the protein matrix and two deeply buried MGD molecules that are extensively hydrogen-bonded to the protein. In addition, all structures concur that Ser147 is one of the ligands to the Mo atom. The structures differ primarily in the position of the Mo atom, which leads to concomitant

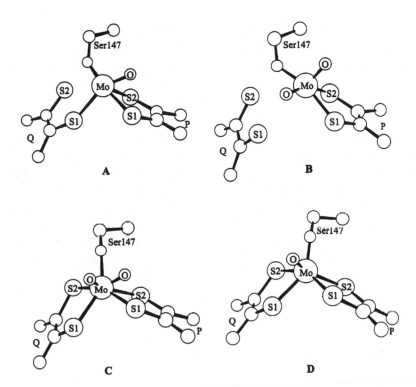

FIG. 9.   Active site structures proposed from X-ray crystallographic studies of DMSO reductase from *R. sphaeroides* and *R. capsulatus*, including the relative orientations of the two S_2pdt units (P and Q). Coordinates were taken from the Protein Data Bank using entries 1CXS, 1DMS, and 1DMR for structures **A–C** respectively. Plot **A** depicts the *R. sphaeroidies* oxidized enzyme of Schindelin et al. [57]. Plot **B** depicts the *R. capsulatus* oxidized enzyme of Schneider et al. [59]. Plot **C** depicts the *R. capsulatus* oxidized enzyme of Bailey et al. [61]. Plot **D** is a congruent structure in agreement with the recent reports [58,62] on the oxidized *R. sphaeroides* and *R. capsulatus* enzymes. Consensus bond distances (Å): 4 Mo−S, 2.4–2.5; Mo=O, 1.8; Mo−O$_{ser}$, 1.9.

differences in the number of oxo ligands to Mo and the nature of the coordination of the two S_2pdt moieties, as shown schematically in Fig. 9.

An inherent problem in the X-ray structures is the difficulty of determining the positions of coordinated O atoms in the presence of the much heavier Mo atom [72]. An additional problem for interpreting the original structures A and B (Fig. 9) is the lack of suitable chemical models for the proposed active site structures of DMSO reductase.

Mono-oxo-Mo(VI) centers, as proposed for A (Fig. 9), were then unknown in OAT chemistry. Partial and/or complete dissociation of an ene-dithiolate ligand, as implied by structures A and B (Fig. 9), had not been documented for coordination compounds. The dioxo-Mo(VI) structure C (Fig. 9) involving seven-coordination is also unprecedented in Mo coordination chemistry [73]. More recent biochemical, structural, and spectroscopic studies on DMSO reductase from *R. sphaeroides* and *R. capsulatus* in several laboratories [58,62,70] have revealed that the Mo coordination in DMSO reductase is highly dependent on the history of the sample and the buffers in the crystallization medium. The most recent X-ray structures for oxidized DMSO reductase from *R. sphaeroides* [58] and *R. capsulatus* [62] concur that D (Fig. 9) is the probable structure for the active enzyme, but crystallized samples often contain mixtures of more than one form. Earlier structures A–C (Fig. 9) most likely contain mixtures of B (Fig. 9) and D (Fig. 9) in various proportions. This view is supported by recent biochemical and resonance Raman data (see below) [58,62,70].

3.3. X-Ray Absorption Spectroscopy

Mo K-Edge XAS studies of DMSO reductase [64–66] have also played a major role in understanding the Mo coordination environment. An early study of the oxidized, reduced, and glycerol-inhibited forms of DMSO reductase from *R. sphaeroides* by George et al. [65] favored a novel mono-oxo-Mo(VI) center with approximately four equivalent thiolate ligands and an additional O (or N) donor ligand, analogous to structure D (Fig. 9). This result, which preceded any of the X-ray crystal structures, correctly predicted the novel coordination about the Mo in DMSO reductase. Of course, the EXAFS data could not determine the angular parameters about the Mo atom or that the additional oxygen-containing ligand was Ser147. The same study indicated that reduced DMSO reductase contained a des-oxo-Mo(IV) center with approximately four thiolate ligands and two O (or N) ligands, one of which was proposed to be OH$^-$ or H$_2$O. A particularly prescient suggestion by George et al. [65] was that one of the EPR signals observed under certain conditions for the *R. capsulatus* enzyme [71] was due to a Mo center that had lost one of its S$_2$pdt ligands and gained an additional oxogroup, e.g., structure B (Fig. 9). Subsequent XAS studies of recombinant *R. sphaeroides* DMSO reductase expressed in *E. coli* revealed that the enzyme initially has a dioxo-Mo(VI) structure with four equivalent Mo-S distances, but that the enzyme assumed the wild-type structure D (Fig. 9) after a cycle of reduction and reoxidation [64]. The reduced enzyme again was ascribed to a des-oxo-Mo(IV) center.

EXAFS studies on DMSO reductase from *R. capsulatus* also favored a structure with four equivalent Mo-S distances, consistent with both S_2pdt ligands being coordinated to the Mo atom. However, the data for the oxidized and reduced enzyme were interpreted as favoring dioxo-Mo(VI) and mono-oxo-Mo(IV)centers, respectively [66]. This interpretation has been questioned by George et al. [64].

3.4. Resonance Raman Spectroscopy

The electronic spectrum of oxidized DMSO reductase shows a unique low-energy band at about 720 nm with a molar absorptivity of about $2\,mM^{-1}$ [70,74,75]. Early resonance Raman studies by Spiro and cow-orkers [69] using 676.4 nm excitation revealed several vibrations in the 200–450 and 1300–1700 cm^{-1} regions whose frequencies were sensitive to ^{34}S enrichment, consistent with a coordinated ene-dithiolate. A subsequent detailed resonance Raman study by Johnson and colleagues [67,68] of oxidized and reduced DMSO reductase from *R. sphaeroides* confirmed the pattern of Mo-S vibrations. Moreover, the spectrum observed for the oxidized enzyme with 562-nm excitation showed a band at 826 cm^{-1} that downshifted by 42 cm^{-1} upon redox cycling with $DMS^{18}O$. This is only consistent with a mono-oxo Mo center because the coupling within a dioxo-Mo center would produce a smaller Mo=O frequency shift ($< 33\,cm^{-1}$) [76,77]. Reduction of the enzyme leads to loss of the 862 cm^{-1} band, consistent with a des-oxo-Mo(IV) center. The excitation profiles for the oxidized and reduced enzyme suggested two types of S_2pdt ligands, but all four S atoms were regarded as coordinated to Mo in both oxidation states. The spectra also provided evidence for a DMSO-bound Mo(IV) intermediate species [67,68]. Very recently, the resonance Raman spectra of DMSO reductase from *R. capsulatus* have been carefully investigated [70], taking into account the recent evidence for multiple forms of the Mo center that occur as a function of purification conditions and redox cycling [58,62,70]. "Redox-cycled" enzyme shows a single Mo=O band at 865 cm^{-1} upon excitation at 752 nm. This band downshifts by 46 cm^{-1} upon reaction with DMS^{18}O, clearly indicating a mono-oxo-Mo center. "As prepared" enzyme contains an extra band at 818 cm^{-1} that shifts by about 30 cm^{-1} in the presence of DMS^{18}O and that is assigned to the S=O vibration of coordinated DMSO. This band is absent in "redox-cycled" enzyme. An additional interesting feature of the resonance Raman spectra of DMSO reductase is a weak feature at 1210 cm^{-1} that shifts to 1161 cm^{-1} upon reaction with DMS^{18}O. This band has previously been assigned to a Mo=O and Mo-S combination band [67,68]. However, Bell et al. [70] suggest that this feature could arise from a S=O bond on an oxidized dithiolene. They note that residual electron density is found in the vicinity of one of the dithiolate sulfur atoms [58] and propose that a S=O group on the P-pterin might play a role in OAT during catalysis.

3.5. Electron Paramagnetic Resonance and Magnetic Circular Dichroism Spectroscopies

The Mo(V) state of DMSO reductase exhibits several EPR spectra, depending on sample history and treatment [55,64,71]. These multiple spectra can now be understood in part as resulting from the several coordination geometries that are possible for the Mo center (Fig. 9). The glycerol-inhibited Mo(V) form of DMSO reductase gives a rich magnetic circular dichroism (MCD) spectrum, but the multiple coordination geometries now known for the Mo atom (Fig. 9) preclude detailed reanalysis of this early data [74,75]. Recent MCD studies of model compounds with [Mo^{V}OS$_4$]$^-$ centers provide benchmarks for interpreting future MCD studies on DMSO reductase centers of controlled stoichiometry [78–80].

3.6. Bioinspired Molybdenum Complexes

The X-ray structural studies of DMSO reductase have led to renewed interest and emphasis on the chemistry of Mo complexes with two S$_2$dt ligands. Compounds of the type $[Mo^{VI}O_2(S_2dt)_2]^{2-}$ and $[Mo^{IV}O(S_2dt)_2]^{2-}$ were well known prior to the X-ray structure determinations for DMSO reductase [57–63], and it has been shown that such compounds can be interconverted by OAT reactions [81–87]. However, the initial structure report that DMSO reductase possessed a mono-oxo-Mo(VI) center, which carried out OAT chemistry to produce a des-oxo-Mo(IV) center [57], was without precedent in coordination chemistry of Mo dithiolate compounds. Thus, recent chemical research, particularly by Holm and collaborators [86,88–92], has focused on the synthesis, structure, reactivity, and spectroscopic characterization of Mo (and W) compounds in a variety of oxidation states that possess two S_2dt ligands and several types of coligands. Many of the compounds have been investigated in detail by Mo K-edge XAS and EXAFS [93,94] and provide important benchmark parameters for interpreting the XAS spectra of Mo- and W-enzyme centers. These studies have shown that the Mo-S distances are essentially independent of the nature of the substituents of the chelate skeleton of the dithiolate, validating the use of various synthetically convenient dithiolates in chemical approaches to mimicking the Mo centers of these enzymes. Mounting evidence from recent structural, spectroscopic, and biochemical investigations of DMSO reductase summarized above strongly supports the original suggestion [57] that catalysis involves mono-oxo-Mo(VI) and des-oxo-Mo(IV) centers. The formally des-oxo-M(IV) (M=Mo or W) species, $[M^{IV}(OPh)(S_2C_2Me_2)_2]^{1-}$, has been prepared by the reaction of $[M(CO)_2(S_2C_2Me_2)_2]$ [90,95] with one equivalent of NaOPh and NEt_4Cl. These five-coordinate square pyramidal $[M^{IV}(OPh)(S_2dt)_2]^-$ complexes (Fig. 10a, M=Mo and W) undergo OAT chemistry with DMSO and Me_3NO to give six-coordinate pseudo-octahedral compounds (Fig. 10). The second-order kinetics are relatively slow with a large negative entropy of activation, consistent with an associative transition state. $[M^{VI}O(OPh)O(S_2C_2Me_2)_2]^-$ (Fig. 10b), which has been structurally characterized for M=W, contains all of the chemical components of the active center of DMSO reductase. The analogous Mo(VI) compound is less stable, decomposing to the known square pyramidal oxo-Mo(V) center, $[Mo^VO(S_2dt)_2]^-$.

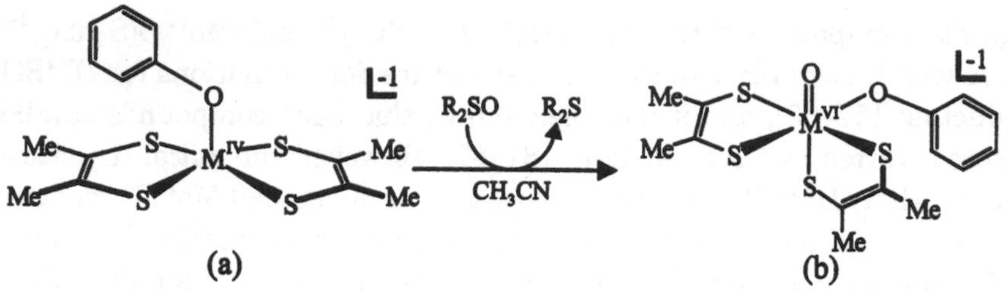

FIG. 10. Reaction of $[M^{IV}(OPh)(S_2C_2Me_2)_2]^{1-}$ (M=Mo or W) (a) with R_2SO to yield $[M^{VI}O(OPh)(S_2C_2Me_2)_2]^-$ (b) and R_2S. OAT transfer from Me_3NO to $[M^{VI}O(OPh)(S_2C_2Me_2)_2]^-$ was also reported. (Reproduced with permission from [89]. Copyright 2000 American Chemical Society.)

3.7. *E. coli* DMSO Reductase

In contrast to the soluble DMSO reductases isolated from *R. capsulatus* and *R. sphaeroides*, the *E. coli* DMSO reductase is a membrane-bound enzyme consisting of three subunits of 82.6, 23.6, and 22.7 kDa, which are the gene products of the *dmsABC* operon [96]. The Mo center is located in the DmsA subunit, and the S_2pdt is present as the MGD derivative [97,98]. The Mo(V) form has been characterized by EPR spectroscopy and has g_{123} of 1.987, 1.976, and 1.960 [99]. Protein film voltammetry of the complete DmsABC enzyme system of *E. coli* adsorbed at a graphite electrode reveals that the catalytic activity of this complex Mo-pterin/Fe-S enzyme is optimized within a narrow electrode potential window that approximately corresponds to the appearance of the Mo(V) EPR signal observed in potentiometric titrations. This result suggests that crucial stages of catalysis are facilitated while the active site is in the intermediate Mo(V) oxidation state [100].

4. BIOTIN SULFOXIDE REDUCTASE

4.1. Function

Biotin sulfoxide reductase (BSOR) is isolated from bacteria, such as *R. sphaeroides* and *E. coli*, and catalyzes the reduction of D-biotin D-sulfoxide (BSO) to D-biotin, as given in eq. (6). Although the exact role of

BSOR has not been determined, plausible roles could include scavenging and reducing D-biotin D-sulfoxide from the environment, protecting the cell from oxidative damage, and reducing bound intracellular biotin that has become oxidized [101]. BSOR is an 81-kDa enzyme consisting of 726 amino acid residues [102]. Enzymatic activity of this soluble protein requires a Mo cofactor and several auxiliary proteins, including a flavoprotein and a thioredoxin-like protein [103]. Until recently, extensive spectroscopic characterization of BSOR has been limited due to a low constitutive expression of BSOR in the bacterium, the requirement for accessory proteins, and the difficulty of detection [102]. Recent improvements in the heterologous expression of *R. sphaeroides* BSOR in *E. coli* have produced higher yields of recombinant BSOR [103,104].

$$+ 2H^+ + 2e^- \xrightarrow[\text{reductase}]{\text{BSOR}} + H_2O \qquad (6)$$

4.2. Physical Characteristics

BSOR is a member of the DMSO reductase family. A 50% amino acid sequence identity was found between DMSO reductase and BSOR, suggesting similar Mo coordination environments. Quantitative analysis and sequence homology suggest that a single Mo atom is coordinated by two MGD molecules and an O atom from a conserved serine residue [103,104], which has been determined to be necessary for enzymatic activity [105]. BSOR is unique among Mo enzymes because no additional auxiliary proteins or cofactors are required for the NADPH-dependent conversion of BSO to biotin [106].

4.3. Spectroscopic Characterization of the Molybdenum Active Site

The increased availability of purified BSOR has allowed investigation of the Mo active site by EPR [104], XAS [104], absorption [103,104], and

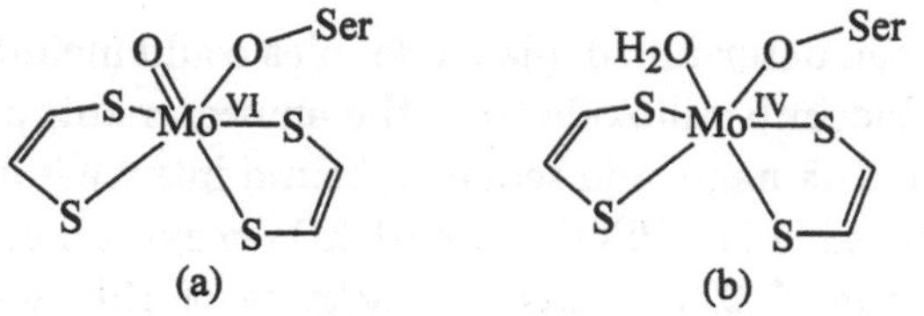

FIG. 11. Proposed active site structures of the oxidized (a) and reduced (b) forms of BSOR as deduced from EXAFS [104] and resonance Raman spectroscopy [107]. Bond lengths (Å). *Oxidized*: Mo=O, 1.70; Mo−S, 2.41; Mo−O$_{ser}$, 1.95. *Reduced*: Mo−S, 2.33; Mo−O/N, 2.19 and 1.94.

resonance Raman spectroscopies [107]. The similarity in the absorption spectra of oxidized BSOR and DMSO reductase are indicative of closely related Mo environments [103,104]. The EPR spectrum of the dithionite-reduced Mo(V) form of BSOR yields very broad lines, and simulation of the spectrum gives g_{123} values of 1.9962, 1.9804, and 1.9507 [104]. The spectrum lacks any resolved hyperfine splitting due to a coordinated hydroxyl or water [104].

XAS [104] at the Mo K-edge and resonance Raman spectroscopy [107] have been used to probe the Mo active site in the oxidized and reduced forms. The Mo site of the oxidized form (Fig. 11a) possesses a mono-oxo Mo(VI) center with four thiolate ligands, presumably from the two MGD molecules. EXAFS [104] reveals an additional O/N ligand, which is postulated to be the Oγ of the conserved serine residue. The dithionite-reduced form (Fig. 11b) possesses a des-oxo Mo(IV) center coordinated by four thiolate ligands, and two different O/N ligands, which have been assigned as an H$_2$O ligand and the Oγ of the conserved serine residue.

5.   POLYSULFIDE REDUCTASE

PSR is isolated from the bacterium *Wolinella succinogenes*, which gains energy in the form of ATP by S respiration [108–110]. PSR is a membrane-bound enzyme that catalyzes the reduction of polysulfide [111,112]. As depicted in eqs. (7) and (8), this reaction can be accomplished using either hydrogen or formate as the electron donor [111,113].

$$H_2 + S_n^{2-} \longrightarrow HS^- + S_{n-1}^{2-} + H^+ \tag{7}$$

$$HCO_2^- + S_n^{2-} + H_2O \longrightarrow HCO_3^- + HS^- + S_{n-1}^{2-} + H^+ \tag{8}$$

As predicted from the nucleotide sequence of the PSR operon, *psrABC*, the isolated enzyme consists of three subunits [108]. The PsrAB catalytic dimer faces into the periplasm, and PsrC serves as the membrane anchor [108]. PSR is placed in the DMSO reductase family based on the sequence homology of PsrA with mononuclear Mo enzymes, such as *E. coli* DMSO reductase and BSOR [108]. The largest subunit, PsrA, likely contains the Mo atom coordinated by two MGD [112].

6. CONCLUSIONS

In all of the enzymes described above, the Mo atom is coordinated by one or two S_2pdt ligands, and the reactions can be described as OAT processes followed by sequential one-electron steps to regenerate the active site. For SO and DMSO reductase, genetic, structural, spectroscopic, and model chemistry studies have greatly enhanced the understanding of their molecular mechanisms. PSR and BSOR have not been as extensively studied, but the methods that have been developed for studying other Mo enzymes should provide considerable insight into these enzymes as well.

ACKNOWLEDGMENTS

We thank Ms. Anne McElhaney, Dr. Frank Inscore, and Dr. Jon Cooney for helpful discussions and assistance with the figures. We thank D. Myers for technical assistance. Financial support by the NIGMS (GM 37773 to J.H.E.) is gratefully acknowledged.

ABBREVIATIONS AND DEFINITIONS

ATP	adenosine 5′-triphosphate
bdt	benzenedithiolate
BSO	biotin sulfoxide
BSOR	biotin sulfoxide reductase
CEPT	coupled electron/proton transfer
Cys	cysteine
CW-EPR	continuous-wave electron paramagnetic resonance
DMS	dimethyl sulfide
DMSO	dimethyl sulfoxide
D_2O	deuterium oxide
dRF	deazariboflavin
dRF·	deazariboflavin semiquinone
E. coli	*Escherichia coli*
edt	ethanedithiolate
edta	ethylenediaminetetraacetic acid
EPR	electron paramagnetic resonance
ESEEM	electron spin-echo envelope modulation
EXAFS	extended X-ray absorption fine structure
hpH	high pH form of sulfite oxidase
HYSCORE	hyperfine sublevel correlation
IET	intramolecular electron transfer
$L\text{-}N_2S_2H_2$	N,N'-dimethyl-N,N'-bis(mercaptophenyl)ethylene-diamine
$L\text{-}N_3$	hydrotris(3,5-dimethyl-1-pyrazolyl)borate
lpH	low pH form of sulfite oxidase
MCD	magnetic circular dichroism
MGD	molybdopterin guanine dinucleotide
MPT	molybdopterin
NADPH	nicotinamide adenine dinucleotide
OAT	oxygen atom transfer
Pi	phosphate-inhibited form of sulfite oxidase
RP ESEEM	refocused primary electron spin-echo envelope modulation
R. capsulatus	*Rhodobacter capsulatus*
R. sphaeroides	*Rhodobacter sphaeroides*

PSR	polysulfide reductase
S_2dt	dithiolate
S_2pdt	pyranopterindithiolate
Ser	serine
SO	sulfite oxidase
SO_3^{2-}	sulfite
SO_4^{2-}	sulfate
XAS	X-ray absorption spectroscopy

REFERENCES

1. E. I. Stiefel, *Science, 272,* 1599–1600 (1996).

2. E. I. Stiefel, Chapter 1 of this book.

3. R. Hille, *Chem. Rev., 96,* 2757–2816 (1996).

4. R. Hille, Chapter 6 of this book.

5. K. V. Rajagopalan, *Adv. Enzymol. Relat. Areas Mol. Biol., 64,* 215–290 (1991).

6. K. V. Rajagopalan and J. L. Johnson, *J. Biol. Chem., 267,* 10199–10202 (1992).

7. R. Hille, *J. Biol. Inorg. Chem., 2,* 804–809 (1997).

8. B. Fischer, J. H. Enemark, and P. Basu, *J. Inorg. Biochem., 72,* 13–21 (1998).

9. F. Irreverre, S. H. Mudd, W. D. Heizer, and L. Laster, *Biochem. Med., 1,* 187–217 (1967).

10. S. K. Wadman, M. Duran, F. A. Beemer, D. P. Cats, J. L. Johnson, K. V. Rajagopalan, J. M. Saudubray, H. Ogier, C. Charpentier, R. Berger, G. P. A. Smit, J. Wilson, and S. Krywawych, *J. Inher. Metab. Dis., 6,* 78–83 (1983).

11. J. L. Johnson and S. K. Wadman, in *The Metabolic and Molecular Bases of Inherited Disease* (C. R. Scriver, ed.), McGraw-Hill, New York, 1995, pp. 2271–2283.

12. C. Kisker, H. Schindelin, A. Paceco, W. A. Wehbi, R. M. Garrett, K. V. Rajagopalan, J. H. Enemark, and D. C. Rees, *Cell, 91,* 973–983 (1997).

13. C. A. Rupar, J. Gillert, B. A. Gordon, D. A. Ramsay, J. L. Johnson, R. M. Garrett, K. V. Rajagopalan, J. H. Jung, G. S. Bacheyie, and A. R. Sellers, *Neuropediatrics, 27*, 299–304 (1996).

14. R. M. Garrett, J. L. Johnson, T. N. Graf, A. Feigenbaum, and K. V. Rajagopalan, *Proc. Natl. Acad. Sci. USA, 95*, 6394–6398 (1998).

15. D. E. Vaughn and P. J. Bjorkman, *Neuron, 16*, 261–273 (1996).

16. R. M. Garrett and K. V. Rajagopalan, *J. Biol. Chem., 271*, 7387–7391 (1996).

17. G. N. George, R. M. Garrett, R. C. Prince, and K. V. Rajagopalan, *J. Am. Chem. Soc., 118*, 8588–8592 (1996).

18. G. N. George, I. J. Pickering, and C. Kisker, *Inorg. Chem., 38*, 2539–2540 (1999).

19. S. D. Garton, R. M. Garrett, K. V. Rajagopalan, and M. K. Johnson, *J. Am. Chem. Soc., 119*, 2590–2591 (1997).

20. M. S. Brody and R. Hille, *Biochemistry, 38*, 6668–6677 (1999).

21. R. Hille, J. Rétey, U. Bartlewski-Hof, W. Reichenbecher, and B. Schink, *FEMS Micro. Rev., 22*, 489–501 (1999).

22. K. Rajagopalan, in *Molybdenum and Molybdenum Containing Enzymes* (M. Coughlan, ed.), Pergamon Press, Oxford, 1980, pp. 241–272.

23. A. V. Astashkin, M. L. Mader, A. Pacheco, J. H. Enemark, and A. M. Raitsimring, *J. Am. Chem. Soc., 122*, 5294–5302 (2000).

24. M. A. Pietsch and M. B. Hall, *Inorg. Chem., 35*, 1273–1278 (1996).

25. Y. Izumi, K. Rose, T. Glaser, J. McMaster, P. Basu, J. H. Enemark, B. Hedman, K. O. Hodgson, and E. I. Solomon, *J. Am. Chem. Soc., 121*, 10035–10046 (1999).

26. R. S. Pilato and E. I. Stiefel, in *Bioinorganic Catalysis*, 2nd ed. (J. Reedijk and E. Bouwman, eds.), Marcel Dekker, New York, 1999, p. 139.

27. H. J. Cohen and I. Fridovich, *J. Biol. Chem., 246*, 359–366 (1971).

28. M. T. Lamy, S. Gutteridge, and R. C. Bray, *Biochem. J., 185*, 397–403 (1980).

29. R. C. Bray, T. Lamy, S. Gutteridge, and T. Wilkinson, *Biochem. J., 201*, 241–243 (1982).

30. R. C. Bray, in *Biological Magnetic Resonance*, J. Reuben and L. J. Berliner, eds.), Plenum Press, New York, 1980, pp. 45–84.

31. S. Gutteridge, M. T. Lamy, and R. C. Bray, *Biochem. J., 191*, 285–288 (1980).

32. A. M. Raitsimring, A. Pacheco, and J. H. Enemark, *J. Am. Chem. Soc., 120*, 11263–11278 (1998).

33. A. Pacheco, P. Basu, P. Borbat, A. M. Raitsimring, and J. H. Enemark, *Inorg. Chem., 35*, 7001–7008 (1996).

34. G. N. George, R. M. Garrett, T. Graf, R. C. Prince, and K. V. Rajagopalan, *J. Am. Chem. Soc., 120*, 4522–4523 (1998).

35. A. Pacheco, J. T. Hazzard, G. Tollin, and J. H. Enemark, *J. Biol. Inorg. Chem., 4*, 390–401 (1999).

36. C. A. Kipke, M. A. Cusanovich, G. Tollin, R. A. Sunde, and J. H. Enemark, *Biochemistry, 27*, 2918–2926 (1988).

37. E. P. Sullivan, Jr., J. T. Hazzard, G. Tollin, and J. H. Enemark, *Biochemistry, 32*, 12465–12470 (1993).

38. M. D. Carducci, C. Brown, E. I. Solomon, and J. H. Enemark, *J. Am. Chem. Soc., 116*, 11856–11868 (1994).

39. F. E. Inscore, R. McNaughton, B. L. Westcott, M. E. Helton, R. Jones, I. K. Dhawan, J. H. Enemark, and M. L. Kirk, *Inorg. Chem., 38*, 1401–1410 (1999).

40. E. I. Stiefel, in *Comprehensive Coordination Chemistry* (G. Wilkinson, R. D. Gillard, and J. A. McCleverty, eds.), Pergamon Press, New York, 1987, pp. 1375–1420.

41. R. S. Pilato and E. I. Stiefel, in *Bioinorganic Catalysis* (J. Reedijk, ed.), Marcel Dekker, New York, 1993, pp. 131–188.

42. J. H. Enemark and C. G. Young, *Adv. Inorg. Chem., 40*, 1–88 (1993).

43. K. R. Barnard, M. Bruck, S. Huber, C. Grittini, H. J. Enemark, R. W. Gable, and A. G. Wedd, *Inorg. Chem., 36*, 637–649 (1997).

44. D. Dowerah, J. T. Spence, R. Singh, A. G. Wedd, G. L. Wilson, F. Farchione, J. H. Enemark, J. Kristofski, and M. Bruck, *J. Am. Chem. Soc., 109*, 5655–5665 (1987).

45. R. Hahn, U, Küsthardt, and W. Scherer, *Inorg. Chim. Acta*, *210*, 177–182 (1993).

46. R. D. Guiles, MS thesis, San Francisco State University, San Francisco, CA (1983).

47. M. L. Mader, M. D. Carducci, and J. H. Enemark, *Inorg. Chem.*, *39*, 525–531 (2000).

48. M. E. Helton, A. Pacheco, J. McMaster, J. H. Enemark, and M. L. Kirk, *J. Inorg. Biochem.*, *80*, 227–233 (2000).

49. B. S. Lim, M. W. Willer, M. Miao, and R. H. Holm, *J. Am. Chem. Soc.*, *123*, 8343–8349 (2001).

50. U. Kappler, B. Bennettt, J. Rethmeier, G. Schwarz, R. Deutzmann, A. G. McEwan, and C. Dahl, *J. Biol. Chem.*, *275*, 13202–13212 (2000).

51. R. R. Mendel and G. Schwarz, *Crit. Rev. Plant Sci.*, *18*, 33–69 (1999).

52. P. Jolivet, E. Bergeron, A. Zimierski, and J.-C. Meunier, *Phytochemistry*, *38*, 9–14 (1995).

53. P. Jolivet, E. Bergeron, and J.-C. Meunier, *Phytochemistry*, *40*, 879–880 (1997).

54. B. A. Gania, A. Masood, and M. A. Baig, *Phytochemistry*, *45*, 879–880 (1997).

55. N. R. Bastian, C. J. Kay, M. J. Barber, and K. V. Rajagopalan, *J. Biol. Chem.*, *266*, 45–51 (1991).

56. T. Satoh and F. N. Kurihara, *J. Biochem (Toyko)*, *102*, 191 (1987).

57. H. Schindelin, C. Kisker, J. Huber, K. Rajagopalan, and D. Rees, *Science*, *272*, 1615–1621 (1996).

58. H. Li, C. Temple, K. V. Rajagopalan, and H. Schindelin, *J. Am. Chem. Soc.*, *122*, 7673–7680 (2000).

59. F. Schneider, J. Löwe, R. Huber, H. Schindelin, C. Kisker, and J. Knäblein, *J. Mol.. Biol.*, *263*, 53–69 (1996).

60. A. S. McAlpine, A. G. McEwan, and S. Bailey, *J. Mol. Biol.*, *275*, 613–623 (1998).

61. A. S. McAlpine, A. G. McEwan, A. L. Shaw, and S. Bailey, *J. Biol. Inorg. Chem.*, *2*, 690–701 (1997).

62. R. C. Bray, B. Adams, A. T. Smith, B. Bennett, and S. Bailey, *Biochemistry, 39*, 11258–11269 (2000).

63. L. J. Stewart, S. Bailey, B. Bennett, J. M. Charnock, C. D. Garner, and A. S. McAlpine, *J. Mol. Biol., 299*, 593–600 (2000).

64. G. N. George, J. Hilton, C. Temple, R. C. Prince, and K. V. Rajagopalan, *J. Am. Chem. Soc. 121*, 1256–1266 (1999).

65. G. N. George, J. Hilton, and K. V. Rajagopalan, *J. Am. Chem. Soc., 118*, 1113–1117 (1996).

66. P. E. Baugh, C. D. Garner, J. M. Charnock, D. Collison, E. S. Davies, A. S. McAlpine, S. Bailey, I. Lane, G. R. Hanson, and A. G. McEwan, *J. Biol. Inorg. Chem., 2*, 634–643 (1997).

67. S. D. Garton, H. James, H. Oku, B. R. Crouse, K. V. Rajagopalan, and M. K. Johnson, *J. Am. Chem. Soc., 119*, 12906–12916 (1997).

68. M. K. Johnson, S. D. Garton, and H. Oku, *J. Biol. Inorg. Chem., 2*, 797–803 (1997).

69. L. Kilpatrick, K. V. Rajagopalan, J. Hilton, N. R. Bastian, E. I. Stiefel, R. S. Pilato, and T. G. Spiro, *Biochemistry, 34*, 3032–3039 (1995).

70. A. F. Bell, X. He, J. P. Ridge, G. R. Hanson, A. G. McEwan, and P. J. Tonge, *Biochemistry, 40*, 440–448 (2001).

71. B. Bennett, N. Benson, A. G. McEwan, and R. C. Bray, *Eur. J. Biochem., 225*, 321–331 (1994).

72. H. Schindelin, C. Kisker, and D. C. Rees, *J. Biol. Inorg. Chem., 2*, 773–781 (1997).

73. J. H. Enemark and C. D. Garner, *J. Biol. Inorg. Chem., 2*, 817–822 (1997).

74. M. G. Finnegan, J. Hilton, K. V. Rajagopalan, and M. K. Johnson, *Inorg. Chem., 32*, 2616–2617 (1993).

75. N. Benson, J. A. Farrar, A. G. McEwan, and A. J. Thomson, *FEBS Lett., 307*, 169–172 (1992).

76. A. K. Smiemke, T. M. Loehr, and J. Sanders-Loehr, *J. Am. Chem. Soc., 108*, 2437–2443 (1986).

77. L. J. Willis and T. M. Loehr, *Spectrochim. Acta, 43A*, 51–58 (1987).

78. R. L. McNaughton, M. E. Helton, N. D. Rubie, and M. L. Kirk, *Inorg. Chem., 39*, 4386–4387 (2000).

79. R. L. McNaughton, A. A. Tipton, N. D. Rubie, R. R. Conry, and M. L. Kirk, *Inorg. Chem.*, *39*, 5697–5706 (2000).

80. J. McMaster, M. D. Carducci, Y. Yang, E. I. Solomon, and J. H. Enemark, *Inorg. Chem.*, *40*, 687–702 (2001).

81. H. Oku, N. Ueyama, M. Kondo, and A. Nakamura, *Inorg. Chem.*, *33*, 209–216 (1994).

82. H. Oku, N. Ueyama, and A. Nakamura, *Inorg. Chem.*, *34*, 3667–3676 (1995).

83. H. Oku, N. Ueyama, and A. Nakamura, *Chem. Lett.*, *8*, 621–622 (1995).

84. N. Ueyama, H. Oki, M. Kondo, T. Okamura, N. Yoshinaga, and A. Nakamura, *Inorg. Chem.*, *35*, 643–650 (1996).

85. G. C. Tucci, J. P. Donahue, and R. H. Holm, *Inorg. Chem.*, *37*, 1602–1608 (1998).

86. J. P. Donahue, C. R. Goldsmith, U. Nadiminti, and R. H. Holm, *J. Am. Chem. Soc.*, *120*, 12869–12881 (1998).

87. S. K. Das, P. K. Chaudhury, D. Biswas, and S. Sarkar, *J. Am. Chem. Soc.*, *116*, 9061–9070 (1994).

88. C. Lorber, J. P. Donahue, C. A. Goddard, E. Nordlander, and R. H. Holm, *J. Am. Chem. Soc.*, *120*, 8102–8012 (1998).

89. B. S. Lim, K. Sung, and R. H. Holm, *J. Am. Chem. Soc.*, *122*, 7410–7411 (2000).

90. B. S. Lim, J. P. Donahue, and R. H. Holm, *Inorg. Chem. 39*, 263–273 (2000).

91. K. Sung and R. H. Holm, *Inorg. Chem.*, *39*, 1275–1281 (2000).

92. D. V. Fomitchev, B. S. Lim, and R. H. Holm, *Inorg. Chem.*, *40*, 645–654 (2001).

93. K. B. Musgrave, J. P. Donahue, C. Lorber, R. H. Holm, B. Hedman, and K. O. Hodgson, *J. Am. Chem. Soc.*, *121*, 10297–10307 (1999).

94. K. B. Musgrave, B. S. Lim, K. Sung, R. H. Holm, B. Hedman, and K. O. Hodgson, *Inorg. Chem.*, *39*, 5238–5247 (2000).

95. C. A. Goddard and R. H. Holm, *Inorg. Chem.*, *38*, 5389–5398 (1999).

96. P. T. Bilous and J. H. Weiner, *J. Bacteriol.*, *170*, 1511–1518 (1988).

97. J. H. Weiner, D. P. MacIsaac, R. E. Bishop, and P. T. Bilous, *J. Bacteriol., 170*, 1505–1510 (1988).

98. R. Rothery, J. L. Simala Grant, J. L. Johnson, K. V. Rajagopalan, and J. H. Weiner, *J. Bacteriol., 177*, 2057 (1995).

99. J. H. Weiner and R. Cammack, *Biochemistry, 29*, 8410–8416 (1990).

100. K. Heffron, C. Léger, R. A. Rothery, J. H. Weiner, and F. A. Armstrong, *Biochemistry, 40*, 3117–3126 (2001).

101. D. E. Pierson and A. Campbell, *J. Bacteriol., 172*, 2194–2198 (1990).

102. V. V. Pollock and M. J. Barber, *Arch. Biochem. Biophys., 318*, 322–332 (1995).

103. V. V. Pollock and M. J. Barber, *J. Biol. Chem., 272*, 3355–3362 (1997).

104. C. A. Temple, G. N. George, J. C. Hilton, M. J. George, R. C. Prince, M. J. Barber, and K. V. Rajagopalan, *Biochemistry, 39*, 4046–4052 (2000).

105. V. V. Pollack and M. J. Barber, *J. Biol. Chem., 275*, 35086–35090 (2000).

106. V. V. Pollock and M. J. Barber, *Biochemistry, 40*, 1430–1440 (2001).

107. S. D. Garton, C. A. Temple, I. K. Dhawan, M. J. Barber, K. V. Rajagopalan, and M. K. Johnson, *J. Biol. Chem., 275*, 6798–6805 (2000).

108. T. Krafft, M. Bokranz, O. Klimmek, I. Schröder, F. Fahrenholz, E. Kojro, and A. Kröger, *Eur. J. Biochem., 206*, 503–510 (1992).

109. R. Hedderich, O. Klimmek, A. Kröger, R. Dirmeier, M. Keller, and K. O. Stetter, *FEMS Microbiol. Rev., 22*, 353–381 (1999).

110. J. Castresana and D. Moreira, *J. Mol. Evol., 49*, 453–460 (1999).

111. A. Kröger, V. Geisler, E. Lemme, F. Theis, and R. Lenger, *Arch. Microbiol., 158*, 311–314 (1992).

112. G. D. Fauque, O. Klimmek, and A. Kröger, *Meth. Enzymol., 243*, 367–383 (1994).

113. V. Geisler, R. Ullmann, and A. Kröger, *Biochim. Biophys. Acta, 1184*, 219–226 (1994).

114. I. K. Dhawan, A. Pacheco, and J. H. Enemark, *J. Am. Chem. Soc.*, *116*, 7911–7912 (1994).

115. L. K. Dhawan and J. H. Enemark, *Inorg. Chem.*, *35*, 4873–4882 (1996).

116. J. R. Bradbury, M. F. Mackay, and A. G. Wedd, *Aust. J. Chem.*, *31*, 2423–2430 (1978).

117. U. Kappler and C. Dahl, *FEMS Microbiol. Lett.*, *203*, 1–9 (2001).

118. T. Eilers, G. Schwarz, H. Brinkmann, C. Witt, T. Richter, J. Neider, B. Koch, R. Hille, R. Hänsch, and R. R. Mendel, *J. Biol. Chem.*, (2001) in press.

18

Comparison of Selenium-Containing Molybdoenzymes

Vadim N. Gladyshev

Department of Biochemistry, University of Nebraska,
Lincoln, NE 68588, USA

1. INTRODUCTION

Known molybdopterin-dependent enzymes are often grouped into three
principal families on the basis of the structure of the molybdenum center and amino acid homology of molybdopterin-binding subunits [1].
These groups include (1) hydroxylases, which typically contain
MoOS(OH) coordinated to a single molybdopterin molecule; (2) prokaryotic oxotransferases, which typically contain MoOX (X is oxygen, sulfur
or selenium in serine [Ser], cysteine [Cys] or selenocysteine [Sec],
respectively) coordinated to two molybdopterin molecules; and (3)
eukaryotic oxotransferases, which contain MoO_2X (X is S of Cys).
Structures and reaction mechanisms of representatives of these classes
of enzymes have been characterized in great detail. Interestingly, these
analyses revealed that several representatives of the first two groups of
molybdoenzymes contain a trace element selenium coordinated to Mo in
enzyme active sites (Table 1).

The catalysis by molybdenum-containing hydroxylases differs from
most other hydroxylase enzymes in that water is used as a source of
oxygen for substrate oxidation, even if these molybdoenzymes are O_2-
dependent. Perhaps the best studied representative of this enzyme
family is xanthine oxidase. Historically, studies on this enzyme provided
many important clues to the mechanism of molybdenum-containing
enzymes [2]. Se-containing enzymes of this group incorporate Se in
the form of a dissociable cofactor that is directly coordinated to Mo
and is essential for enzyme activity. For many years, the only such

protein identified was nicotinic acid hydroxylase (NAH) [3–7]. This enzyme was also the only known natural Se-containing enzyme in which Se was not present in the form of Sec. But later, through analogies to NAH and the observations that Se is essential to purine degradation in certain clostridia, two additional selenomolybdoenzymes, xanthine dehydrogenase (XDH) [8,9] and purine hydroxylase (PH) [9], were found to contain Se. Finally, recent structural analysis of carbon monoxide dehydrogenase (CODH) revealed the presence of Se in the enzyme active center, which allowed classifying this enzyme as a selenoprotein [10–12].

The family of bacterial oxotransferases includes diverse enzymes [1], in which the Mo atom is coordinated to one of three amino acid residues—Ser, Cys, or Sec—that occupy equivalent positions in homologous polypeptides. Coordination of these amino acids to Mo was first demonstrated using EPR spectroscopy [13] and later was confirmed by X-ray crystallography [14]. Many, but not all, enzymes of this group utilize water as a source of oxygen. Formate dehydrogenase (FDH) is a notable exception because no oxygen is transferred during the reaction catalyzed by this enzyme [15]. Se-containing members of this family are FDH and formylmethanofuran dehydrogenase (FMDH). A potential additional member of this family is perchlorate reductase, which was reported to contain 1 Mo and 1 Se per enzyme subunit [16].

The family of eukaryotic oxotransferases is exemplified by nitrate reductase and sulfite oxidase. Again, independently of the dependency on the molecular oxygen, these enzymes employ water as an oxygen source or generate water as a product of the reaction [1]. No selenium-containing enzymes have been described that belong to this group of molybdoenzymes.

The six known Se-Mo-containing enzymes (Table 1) represent approximately half of all previously characterized prokaryotic selenoproteins [17]. Such abundance of selenoproteins among molybdoenzymes points to an important role Se plays in catalysis by this class of enzymes. Besides Se-Mo hydroxylases, no other selenoproteins are known that contain Se in the form of a cofactor, suggesting that molybdoenzymes acquired a highly specialized pathway of Se utilization for their catalytic functions. The purpose of this chapter is to provide an overview of selenium- and molybdenum-containing proteins with the emphasis on

TABLE 1

Selenium- and Molybdenum-Containing Proteins

Hydroxylases containing dissociable Mo-coordinated Se moiety
    Nicotinic acid hydroxylase
    Xanthine dehydrogenase
    Proline hydroxylase
    Carbon monoxide dehydrogenase
Prokaryotic dehydrogenases containing Mo-coordinated Sec residue
    Formate dehydrogenase
    Formylmethanofuran dehydrogenase

---

chemical identity, evolution, and function of the Se moiety and its role in catalysis.

2. OVERVIEW OF SELENIUM-CONTAINING MOLYBDOENZYMES

2.1. Nicotinic Acid Hydroxylase

Nicotinic acid hydroxylase from *Clostridium barkeri* was first recognized as a Se- and Mo-containing protein by Dilworth approximately two decades ago [4,5]. His studies were carried out several years after the findings of the first natural Se-containing proteins, mammalian glutathione peroxidase and selenoprotein A of the clostridial glycine reductase complex. In these first known selenoproteins, Se was found to occur in the form of a Sec residue [17] and thus the Se moiety was relatively stable to protein denaturation. In contrast, the Se moiety of NAH was extremely labile and was lost during protein isolation or denaturation with chaotropic agents [4–7]. This unexpected finding generated further interest in characterizing this enzyme and the role of its Se atom.

It was hypothesized that Se could be involved in redox interaction with one of the redox centers in the protein, which include FAD, two iron-sulfur clusters, and Mo. Since these cofactors could be monitored by EPR spectroscopy, this technique was used to determine the presence of

Se in the vicinity of these redox groups in the enzyme [6]. As isolated, NAH exhibited a Mo(V) signal, which disappeared upon reduction, while the reduction also generated iron-sulfur cluster and flavin radical signals. Substitution of ^{77}Se for the normal selenium isotope abundance resulted in splitting of the Mo(V) EPR signal of the native protein without affecting the signals of the FeS clusters. These results indicated a coordination of Se to Mo in the enzyme [6].

A second important finding was the ability to isolate NAH from cells grown under Se deficiency. The fact that NAH was synthesized by these cells argued against cotranslational incorporation of Se into protein (as occurs in the case of Sec), confirming the novelty of the Se moiety in the protein [6]. Moreover, Se-deficient NAH was inactive in oxidation of nicotinic acid, indicating the essentiality of Se for the catalytic function of the enzyme. In addition, isolated Se-deficient NAH showed no Mo(V) signals, and the Mo signal that was generated upon reduction of the protein with a substrate was distinct from that of the Se-containing NAH. Thus, the dissociable selenium moiety of NAH was found to be coordinated directly with molybdenum in the molybdopterin cofactor, and this selenium was essential for nicotinic acid hydroxylase activity of the enzyme.

Due to extreme lability of the Se-Mo cofactor of NAH, this enzyme was rapidly inactivated during isolation. Parallel losses of Se and catalytic activity were also observed during purification and storage of the enzyme [7]. Nevertheless, a procedure has been developed that allowed isolation of the protein in milligram amounts. Characterization of NAH revealed that the protein consists of four dissimilar subunits and is most stable at alkaline pH and in the presence of glycerol. There were 5–7 Fe, 1 FAD, 1 molybdopterin cytosine dinucleotide, 1 Se, and 1 Mo per 160-kDa protein protomer. NAH exhibits some unusual kinetic and spectroscopic properties that differentiate the enzyme from other Mo-containing hydroxylases [7].

2.2. Xanthine Dehydrogenase

The presence of a clostridial Se-containing XDH with properties similar to those of NAH was recognized early on [18], but this enzyme was only recently characterized in sufficient detail to classify it as selenoenzyme

[8,9]. Besides NADP$^+$-dependent xanthine oxidation, XDH showed NADPH-dependent oxidase and diaphorase activities [8]. The enzyme contained Fe, acid-labile sulfur, Mo, tungsten, Se, and FAD at molar ratios of 17.5, 18.4, 2.3, 1.1, 0.95, and 2.8 per mol of a 540-kDa native enzyme. XDH was inactivated upon incubation with cyanide, and the activity could be partially restored by addition of selenide or selenite, indicating that selenium likely occurred in a form similar to that found in NAH rather than in the form of Sec [8]. Selenium-containing XDH was also purified from *Clostridium purinolyticum*. The enzyme is insensitive to the presence of oxygen during purification. Xanthine was a preferred substrate, and Se was required for the xanthine dehydrogenase activity of the enzyme [9].

2.3. Purine Hydroxylase

PH was discovered during purification of Se-dependent XDH from *C. purinolyticum* and was distinct from the latter enzyme [9]. However, it also contained selenium and exhibited similar spectral properties to those of NAH and XDH. PH used purine, 2-OH-purine, and hypoxanthine as substrates, which was the reason the enzyme was designated as PH. A concomitant release of selenium from the enzyme and loss of catalytic activity on the treatment with cyanide indicated that selenium was essential for PH activity [9]. These biochemical data as well as characterization of microorganisms employing Se- and Mo-containing enzymes revealed an important role of Se in pathways of purine fermentation.

2.4. Carbon Monoxide Dehydrogenase

The first indication that CODH from *Oligotropha carboxidovorans* might be a selenoprotein came from the unexpected observation in the early 1980s that incubation of the isolated enzyme with selenite activated the enzyme [10]. However, this enzyme was only recently shown to be a selenoprotein, when its structure was solved by X-ray crystallography [11]. CODH is a dimer of heterotrimers (molybdoprotein, flavoprotein, and an iron-sulfur protein) and catalyzes oxidation of CO according to the following equation: $CO + H_2O \longrightarrow CO_2 + 2e^- + 2H^+$. The site for CO oxidation is the Mo-Se center coordinated to an essential Cys residue and a single molybdopterin cytosine dinucleotide. The Se moiety occurs in the form of S-selanylcysteine; that is, Se is coordinated to both Mo and the S of Cys [11]. Because NAH, XDH, and PH are hydroxylases with properties similar to those of CODH, the occurrence of the Se moiety in CODH as Mo-coordinated S-selanylcysteine suggests a similar organization of catalytic centers in these Mo- and Se-containing enzymes.

2.5. Formate Dehydrogenase

FDH is perhaps the best studied Se- and Mo-containing enzyme, which was extensively characterized by kinetic, spectroscopic, and structural techniques. This enzyme had served as a model for studies on Sec incorporation into bacterial selenoproteins [19]. FDH catalyzes the following reaction: $HCOO^- \longrightarrow CO_2 + H^+ + 2e^-$, which differs from those catalyzed by other molybdoenzymes in that water is not a substrate or product of the reaction [15].

Se- and Mo-containing FDHs occur in methane-producing bacteria, clostridia, and several facultative enterobacteria when grown under anaerobic conditions. In addition, certain archaea have been reported to contain Se-dependent FDHs, which incorporate Mo or W [20]. The genome of *E. coli* contains three selenoprotein FDH genes encoding FDH_H, FDH_N, and FDH_O [21]. Of these enzymes, *E. coli* FDH_H has been characterized in great detail. This enzyme is a component of the formate-hydrogen lyase complex that detoxifies formic acid to carbon dioxide and molecular hydrogen when cells are grown anaerobically on glucose and in the absence of nitrate.

FDH_H is an 80-kDa polypeptide containing a single molybdenum cofactor (Mo and 2 molybdopterin cytosine dinucleotides), an Fe_4S_4 cluster, and a Sec residue that was shown to be coordinated to Mo through its Se atom [13,14,22–27]. The protein is extremely sensitive to oxygen, especially in a reduced state. Its purification requires strictly anaerobic conditions.

2.6. Formylmethanofuran Dehydrogenase

Formylmethanofuran dehydrogenase (FMDH) occurs in certain archaea, such as *Methanopyrus kandleri* and *Methanococcus jannaschii* [28,29]. This enzyme shows homology to FDHs from various sources, including conservation of a Sec residue at the enzyme active site. Some FMDHs incorporate W in place of Mo, which is predicted to coordinate Se. FMDH has not been characterized in detail structurally or mechanistically, but many elements of its mechanism likely resemble the FDH_H mechanism.

3. INCORPORATION OF SELENIUM INTO MOLYBDOENZYMES

3.1. Selenocysteine is Encoded by UGA

Natural Se-containing proteins occur in all three major domains of life: bacteria, archaea, and eukaryotes [17,30]. In these proteins, Sec is the major form of Se, and this amino acid is the 21st amino acid in protein. Sec is encoded by UGA codon (Fig. 1) in every organism that has natural Sec-containing proteins. Incorporation of Sec into polypeptides is cotranslational and requires a specific Sec insertion system composed of selenophosphate synthetase that generates a selenium donor compound, selenophosphate, selenocysteine tRNA [17,30,31], selenocysteine synthase, and the Sec-specific elongation factor EFsec [35] (also designated as SELB [19]). The presence of in-frame Sec-encoding UGA codons makes it difficult to interpret protein products of selenoprotein genes. As a result, the majority of these proteins are incorrectly annotated in completely sequenced prokaryotic genomes. No algorithms are currently available that correctly identify Sec UGA codons.

3.2. SECIS Elements

Besides encoding Sec, UGA codons also serve as signals for the cessation of protein synthesis. To distinguish between these two translation pos-

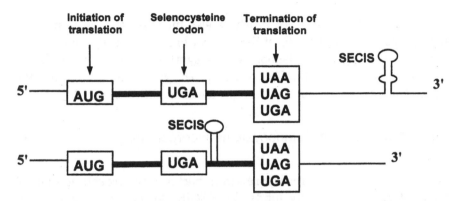

FIG. 1.   Structure of genes for selenocysteine-containing molybdoenzymes in archaea (top) and bacteria (bottom). Selenocysteines are encoded by UGA codons located in open reading frames. The stem-loop structure (designated as the SECIS element) is necessary for Sec insertion and also prevents termination of protein synthesis at UGA codons. It is present in the 3′-untranslated regions in archaea and immediately downstream of the UGA selenocysteine codon in bacteria.

sibilities, selenoprotein genes contain a stem-loop structure, designated Sec insertion sequence (SECIS) element [31]. SECIS elements are located immediately downstream of UGA codons (within coding regions) in bacterial selenoprotein genes, and in 3′-untranslated regions in archaeal and eukaryotic selenoprotein genes (Fig. 1). Bacterial and archaeal SECIS elements have neither sequence homology nor common structural features [29–31]. Searches for SECIS elements in completely sequenced genomes allow identification of selenoprotein genes, but such approaches have been only developed for eukaryotic sequences [32,33].

3.3.   Biosynthesis of Selenocysteine

An interesting feature in Sec biosynthesis is that this amino acid is synthesized on tRNA [34]. Sec tRNA is first aminoacylated with Ser by seryl-tRNA synthetase [19]. In bacteria, a pyridoxal phosphate-dependent Sec synthase converts the Ser moiety on the tRNA to an aminoacrylyl intermediate by removal of water. Selenophosphate that is synthesized from selenite and ATP by selenophosphate synthetase [17] is the donor of selenium to the aminoacrylyl intermediate to yield selenocysteyl-tRNA, and Sec synthase also mediates this later step. In eukaryotes, the biosynthesis of Sec has not been completely established, although two separate selenophosphate synthetases have been identified. Interestingly, Sec tRNAs in both prokaryotes and eukaryotes are longer than other tRNAs and contain fewer modified bases [34].

3.4. Mechanism of Selenocysteine Incorporation

In the first stage of Sec incorporation in bacteria, the translation factor SELB (EFsec) specifically binds the SECIS element in selenoprotein genes, which in turn recruits selenocysteyl-tRNA allowing insertion of Sec at in-frame UGA codons. While bacterial SELB is composed of two domains (i.e., the SECIS binding and elongation factor domains), its archaeal counterpart lacks the SECIS binding domain [36]. In archaea and mammals, separate factors are utilized for Sec insertion into protein. One factor specifically binds to the SECIS element and designated as the SECIS-binding protein (SBP2) [37], while a second factor, EFsec, binds to SBP2 and selenocysteyl-tRNA [35,38].

The biological mechanism for insertion of Se into Mo-containing hydroxylases as S-selenylcysteine is currently not known. Although selenophosphate could not activate Se-deficient NAH in vitro [7], it may serve as a potential Se donor compound for Se-Mo hydroxylases in vivo.

4. SELENIUM VERSUS SULFUR IN CATALYSIS

4.1. Why Is Selenium Used in Molybdoenzymes?

In all previously characterized Se-containing proteins except selenophosphate synthetase (SPS), the Se moiety was found to be located at enzyme active centers [17,30,34]. Molybdenum- and selenium-containing enzymes conform to this rule. For example, replacement of Sec with Cys in FDH$_H$ by site-directed mutagenesis led to an approximately 300-fold decrease in the rate of formate oxidation [23], and the Ser-for-Sec mutant was inactive. Although Sec, Cys, and Ser differ by only a single chalcogen atom (Se vs. S or O), these amino acids have important differences. Under physiological pH, Sec ($pK_a \sim 5.5$) is fully ionized, whereas most cysteines ($pK_a \sim 8$) are protonated and all serines are fully protonated [17,30,34]. In addition, Sec is a better nucleophile and has a lower redox potential than Cys. These characteristics allow utilization of Sec for redox reactions that involve electron and/or proton transfer and for fine-tuning molybdenum centers, adapting them for catalyzing a specific chemical reaction.

4.2.   Role of Selenocysteine in Formate Dehydrogenase H

A role for Sec in formate dehydrogenases has been established through X-ray crystallographic [14,26], spectroscopic (EPR and XAS) [13,15,27], and mechanistic studies [15,23–25] of *E. coli* FDH$_H$. These studies revealed one of the most detailed structure-based mechanisms known for Mo-containing enzymes. In FDH$_H$, Mo(VI) is coordinated to four sulfurs of two molybdopterin guanine dinucleotides, Se of Sec140 and a hydroxyl group (Fig. 2). In the initial stage of the reaction, the latter ligand is replaced with formate such that formate binds Mo through its ionized carboxyl group and is stabilized through interaction with Arg333. A subsequent transfer of two electrons to Mo generates Mo(IV) and is linked to the transfer of a formate-derived proton to a nearby His141 residue, which subsequently forms a hydrogen bond with the Mo-coordinated Sec. His141 is conserved in most FDHs, but a few enzymes contain Gln instead (Fig. 3).

In the reduced state, the five-coordinate Mo(IV) and dithiolene ligands from two molybdopterins are located in a planar orientation. Both oxidized Mo(VI) (with and without substrate analogues) and reduced Mo(IV) forms have been characterized by X-ray crystallography [14]. In the next step, one electron is transferred to an Fe_4S_4 cluster through the backbone of one of the molybdopterins, a water molecule and Lys44 (Fig. 2). The resulting Mo(V) center was extensively characterized by EPR spectroscopy [15], which revealed that at this step of the reaction the formate-derived proton was still located in the vicinity of the Mo center. After the Fe_4S_4 cluster is oxidized with benzylviologen (in vitro), a second electron is transferred from Mo(V) to the iron-sulur cluster forming the initial Mo(VI) center.

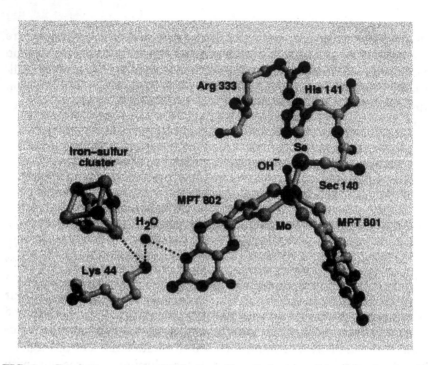

FIG. 2.   Catalytic center in FDH$_H$ from *E. coli*. Amino acids and cofactors that directly participate in the catalyzed reaction are shown in the figure. The FDH$_H$ is shown in a fully oxidized state, in which Mo(VI) is coordinated to dithiolene groups of two molybdopterins (MPT801 and MPT802). The hydroxyl is replaced with formate prior to electron and proton transfers. The predicted pathway for electron transfer within the enzyme is Mo → MPT802 → H$_2$O → Lys44 → Fe$_4$S$_4$. The roles of Arg333, His141, and Sec140 are discussed in the text.

From this mechanism it was suggested that the role of Se in FDH$_H$ is to fine-tune the Mo center and to influence substrate binding [14]. In addition, Se is directly involved in the transfer of the substrate-derived proton and in the subsequent phased release of the proton from the active center.

5. EVOLUTION OF SELENOCYSTEINE-CONTAINING MOLYBDOENZYMES

Homologues of most selenoproteins, including Mo-containing proteins, are known that contain Cys in place of Sec. For example, approximately one-third of known FDHs have Sec in the active center, whereas the remaining FDHs incorporate Cys (Fig. 3). Cys and Sec have similar chemical properties, suggesting a close evolutionary relationship between these two amino acids. On the other hand, it appears that Cys can only partially compensate for the loss of Sec [17,30,34], whereas replacement of Cys with Sec may achieve a superior catalyst [39].

```
 1 E.coli              AE005355      FLYQLFAR-EYGSNN FPDCSNMCHEPTSVG LAASIGVGKGTVLL EDFEKCDLVICIGHN
 2 R.eutropha          REU60056      YLYQLFAR-EYGTNN FPDCSNMCHEPTSVG LPRSIGIGKGTVSL EDFDTCELIISIGHN
 3 S.coelicolor        SC7H2         FLYQLFAR-ELGTNN LPDCSNMCHESSGSA LSETIGVGKGSVLL EDLYKSDLIIVAGQN
 4 P.abyssi            CNSPAX03      YLLQKIAR-LLGTNN IDNCARLCHEASVHA LKMTLGAGVQTNPY SDLENFKAIMIWGYN
 5 P.horikoshii        AP000006      YLIQKIAR-LLGTNN VDNCARLCHESSVHA LKLTLGDGVQTNPY SDLERFGAIMIWGYN
 6 T.litoralis         AF039208      YLFQKLAR-NIGTNN VDNTSHLCHGVSVRA ILDANFERNWAT-Y DDIEESNVIILWGAN
 7 B.halodurans        AP001515      YTAAKVAR-FLGTNN IDNASRICHSPSKTA LKRSLGIGASSCNY QDWIGTDVLVFWGSV
 8 S.solfataricus      SSOLP2N05     YSFMKLAR-ALGTNN VDSCARVCHEPSAMA LKELVGIGASSVTV SEILNARNIVISGES
 9 C.jejuni            CJ11168X5     YYIRKFAA-FFGTNN VDHQARIUHSATVAG VANTFGYGAMTNHL GDIQRSKCIIIIGAN
10 W.succinogenes      WSFDHABCD     YYFRKFAA-FFGTNN LDTIARICHAPTVAG VSNTLGYGGMTNHL ADMMHSKAIFIIGGN
11 V.cholerae          AE004229      YLFRKMAS-LWGTNN VDHQARICHSTTVAG VANTWGYGAMTNSF NDMHNCKSMLFIGSN
12 B.subtilis          AF015825      YLMQKLARGVIGTNN VDNCSRYCQSPATAG LFRTVGYGGDSGSI TDIAQADLVLIIGSN
13 T.volcanium         AP000991_2    YLMQKLARQVFGTNN VDNSSRFCQAPATTG LWRTVGYGGDSGSI QDIYMADLVIAIGTN
14 T.acidophilum       TACID2        YLVQKLARQVFGTNN VDNSSRFCQAPATTG LWRTVGYGGDAGSI SDLYVSDLILAVGTN
15 B.subtilis          BSU93874      YVIQKLARQVFETNN VDNCSRYCQSPATDG LFRTVGMGGDAGTI KDIAKAGLVIVAGSN
16 S.solfataricus      SSOLP2N18     YLLQKLARAIIGTNN VDNSARYCQSPATVG LWRTVGIGADSGTI RDIENANLIVIVGHN
17 S.meliloti          AF298190      FLVQKLVRAGFGNNN VDTCARVCHSPTGYG LNQTFGTSAGTQDF DSVEHTDVAVIIGAN
18 M.loti              AP003006      YLVQKLVRQGFRNNN VDTCARVCHSPTGYG LGQTYGTSAGTQDF DSVEFTDVAVVIGAN
19 R.eutropha          RAAJ3295      YLVQKLVRARFGNNN VDTCARVCHSPTGYG LKQTLGESAGTQTF KSVEKADVIMVIGAN
20 T.acidophilum       TACID1        YLLQKIAR-MIGTNN VDHCARSCHSSTVAG LIRTLGTAAATGSI KSLKSTQTFFVIGSN
21 T.volcanium         AP000991_1    YLMQKIAR-MFGTNN IDHCARSCHSSTVAG LIKTIGTAAATGSI KSLKSTKTYFIIGSN
22 M.thermoautotrop    AE000915      YILQKFTRAVMLSGN IDHCARLCHAPSVRS LSMTLGSGAMTNSI AELEDSACILAVGTN
23 M.thermoautotrop    MTDNAFLPF     YLLQKFTRAVMGSGN IDHCARLCHAPSLTG LRMSLGSGAMTNSI SELGAAGCILAVGTN
24 M.thermoformic.     MTU52681      YLLQKFARAVIGTQN VDHCARLCHGPSVAG LAKTFGSGAMTNSI SDIEESSCIFIIGSN
25 M.thermoacetica     MTU73807      YLLQKLARGVLGTNN VDHCARLUHSSTVAG LATTFGSGAMTNSI ADIASADCIFVIGSN
26 M.jannaschii        U67575        YILQKFARVALKTNN VDHCARLUHSATVTG MSACFGSGAMTNSI EDIELADCILIIGSN
27 M. formicicum       MBFFDHAB      YVNQKFARIVVGTHN IDHCARLCHGPTVAG LAASFGSGAMTNSY ASFEDADLIFSIGAN
28 S.coelicolor        SC4B5         YVAQKFARVVMGTHN VDSCNRTCHAPSVAG LSAAFGSGGGTSSY EEIEHTDVIVMWGSN
29 E.coli FdhO         AE000464      YLTQKFSRA-LGMLA VDNQARVUHGPTVAS LAPTFGRGAMTNHW VDIKNANLVVVMGGN
30 Y.enterocolitica    YESODAGEN     YLTQKFSRA-LGMLA VDNQARVUHGPTVAS LAPTFGRGAMTNHW VDIKNADLIIVMGGN
31 E.coli FdhN         AE005358      MLTQKFARS-LGMLA VDNQARVUHGPTVAS LAPTFGRGAMTNHW VDIKNANVVMMGGN
32 P. aeruginosa       AE004894      YITHKVMRS-LGILG FDNQARVUHGPTVAS LAPTFGRGAMTNHW TDIKNADLVLIMGGN
33 H. influenzae       U32686        LLTQKWIRM-LGMVP VCNQANTUHGPTVAS LAPSFGRGAMTNHW VDIKNANLIIVQGGN
34 A. aeolicus         AE000720      WLMVKIGIA-LGLSA RETQATIUHAPTVAS LAPTFGRGAMTNHW VDISNSDLVFVMGGN
35 S.typhimurium       AF146729      YVMQKFARAVIGTNN VDCCARVUHGPSVAG LHQSVGNGAMSNAI NEIDNTDLVFVFGYN
36 E.coli FDHH         AE000481      YVMQKFARAVIGTNN VDCCARVUHGPSVAG LHQSVGNGAMSNAI NEIDNTDLVFVFGYN
```

FIG. 3. Multiple alignment of formate dehydrogenase Sec-flanking sequences and their Cys-flanking homologues. Sec (U) and Cys (C) residues that coordinate Mo are shown in bold. Also bolded are His residues that were predicted to accept the formate-derived proton. In several FDHs, the His is replaced with Gln. Accession numbers and organisms which encode sequences are shown on the left.

Indeed, kinetic analyses of Sec-containing FDHs and functionally related Cys-containing homologues demonstrated that Sec is preferred over Cys to achieve superior catalytic efficiency [17].

These observations suggest that the use of Sec in molybdoenzymes may represent an evolutionary advancement. An alternative hypothesis on the evolution of Se- and Mo-containing proteins suggests that Sec is a relic of the primordial world and that the use of Sec decreased during evolution [19]. If the use of Sec represents a survival of the ancient genetic code, then Sec was likely replaced in archaic proteins with more stable and metabolically available Cys. However, if the use of Sec was a novelty in the primordial genetic code, certain molybdoproteins could have benefited from the replacement of Cys or Ser with Sec.

These hypotheses may be reconciled if Sec/Cys pairs in homologous molybdopterin-dependent sequences are viewed both as an evolutionary advancement of inserting Sec in place of Cys or Ser at active centers of certain molybdoproteins and as a disadvantage of using Sec. The disadvantage of incorporating Sec is that it is metabolically expensive due to the scarcity of selenium in the environment and the necessity to maintain a multicomponent Sec incorporation system for insertion of just a few amino acids or even a single amino acid. Thus, it is possible that in addition to an evolutionary advantage of utilizing Sec in protein, organisms may also be under selective pressure to inactivate the Sec insertion system.

6. CONCLUSIONS

The frequent use of Se in molybdoenzymes illustrates advantages that this trace element provides in fine-tuning active centers of prokaryotic molybdoenzymes. It appears that for certain reactions Se is preferred over sulfur due to its unique ionization and redox properties. Two groups of Mo-containing proteins evolved that utilize these features. In the group that includes FDH and FMDH, Sec, the 21st natural amino acid in proteins, is employed. It is inserted cotranslationally and requires a unique multicomponent system for its biosynthesis and insertion. Remarkably, in certain bacteria, such as *E. coli*, FDHs are the only selenoproteins, and thus the Sec system is maintained exclusively for these enzymes. Sec-encoding UGA codons are difficult to identify unless homologous sequences are available or SECIS elements can be identified downstream of UGA codons. Another group of Se-Mo-enzymes includes NAH, XDH, CODH, and PH, which employ Se in the form of a cofactor. The use of Se as a natural dissociable cofactor has not been described in any other known proteins.

ACKNOWLEDGMENTS

I am grateful to my Mo-Se collaborators. In particular, I thank Sergey Khangulov for a wonderful collaboration, help and for being a dear colleague and friend; Thressa Stadtman for support and encouragement; and Peter Sun, Jeffrey Boyington, and David Grahame for fruitful and enjoyable collaborations. I also thank Dolph Hatfield for his helpful comments and Jeffrey Boyington for providing Figure 2.

ABBREVIATIONS

CODH	carbon monoxide dehydrogenase
Cys	cysteine
EPR	electron paramagnetic resonance
FAD	flavin adenine dinucleotide
FDH	formate dehydrogenase
FMDH	formylmethanofuran dehydrogenase
MPT	molybdopterin
NADP$^+$	nicotinamide adenine dinucleotide phosphate
NADPH	nicotinamide adenine dinucleotide phosphate (reduced)
NAH	nicotinic acid hydroxylase
PH	purine hydroxylase
Sec	selenocysteine
SECIS	Sec insertion sequence
Ser	serine
SPS	selenophosphate synthetase
XDH	xanthine dehydrogenase

REFERENCES

1. R. Hille, *Essays Biochem.*, *34*, 125–137 (1999).
2. R. C. Bray, *Q. Rev. Biophys.*, *21*, 299–329 (1988).
3. J. S. Holcenberg and E. R. Stadtman, *J. Biol. Chem.*, *244*, 1194–1203 (1969).
4. G. L. Dilworth, *Arch. Biochem. Biophys.*, *219*,. 30–38 (1982).
5. G. L. Dilworth, *Arch. Biochem. Biophys.*, *221*, 565–569 (1983).
6. V. N. Gladyshev, S. V. Khangulov, and T. C. Stadtman, *Proc. Natl. Acad. Sci. USA, 91*, 232–236 (1994).
7. V. N. Gladyshev, S. V. Khangulov, and T. C. Stadtman, *Biochemistry, 35*, 212–223 (1996).
8. T. Schrader, A. Rienhofer, and J. R. Andreesen, *Eur. J. Biochem.*, *264*, 862–871 (1999).
9. W. T. Self and T. C. Stadtman, *Proc. Natl. Acad. Sci. USA, 97*, 7208–7213 (2000).
10. O. Meyer, and K. V. Rajagopalan, *J. Biol. Chem.*, *259*, 5612–5617 (1984).
11. H. Dobbek, L. Gremer, O. Meyer, and R. Huber, *Proc. Natl. Acad. Sci. USA, 96*, 8884–8889 (1999).
12. O. Meyer, L. Gremer, R. Ferner, M. Ferner, H. Dobbek, M. Gnida, W. Meyer-Klaucke, and R. Huber, *Biol. Chem.*, *381*, 865–876 (2000).
13. V. N. Gladyshev, S. V. Khangulov, M. J. Axley, and T. C. Stadtman, *Proc. Natl. Acad. Sci. USA, 91*, 7708–7711 (1994).
14. J. C. Boyington, V. N. Gladyshev, S. V. Khangulov, T. C. Stadtman, and P. D. Sun, *Science, 275*, 1305–1308 (1997).
15. S. V. Khangulov, V. N. Gladyshev, G. C. Dismukes, and T. C. Stadtman, *Biochemistry, 37*, 3518–3528 (1998).
16. S. W. Kengen, G. B. Rikken, W. R. Hagen, C. G. van Ginkel, and A. J. Stams, *J. Bacteriol.*, *181*, 6706–6711 (1999).
17. T. C. Stadtman, *Annu. Rev. Biochem. 65*, 83–100 (1996).
18. R. Wagner and J. R. Andreesen, *Arch. Microbiol.*, *121*, 255–260 (1979).
19. A. Bock, K. Forchhammer, J. Heider, W. Leinfelder, G. Sawers, B. Veprek, and F. Zinoni, Mol. Microbiol., 5, 515–520 (1991).

20. I. Yamamoto, T. Saiki, S. M. Liu, and L. G. Ljundahl, *J. Biol. Chem., 258*, 1826–1832 (1983).

21. J. Heider and A. Bock, *Adv. Microb. Physiol., 35*, 71–109 (1993).

22. M. J. Axley, D. A. Grahame, and T. C. Stadtman, *J. Biol. Chem., 265*, 18213–18218 (1990).

23. M. J. Axley, A. Bock, and T. C. Stadtman, *Proc. Natl. Acad. Sci. USA, 88*, 8450–8454 (1991).

24. M. J. Axley and D. A. Grahame, *J. Biol. Chem. 266*, 13731–13736 (1991).

25. V. N. Gladyshev and P. Lecchi, *Biofactors, 5*, 93–97 (1995).

26. V. N. Gladyshev, J. C. Boyington, S. V. Khangulov, D. A. Grahame, T. C. Stadtman, and P. D. Sun, *J. Biol. Chem., 271*, 8095–8100 (1996).

27. G. N. George, C. M. Colangelo, J. Dong, R. A. Scott, S. V. Khangulov, V. N. Gladyshev, and T. C. Stadtman, *J. Am. Chem. Soc., 120*, 1267–1273 (1998).

28. J. A. Vorholt, M. Vaupel, and R. K. Thauer, *Mol. Microbiol., 23*, 1033–1042 (1997).

29. R. Wilting, S. Schorling, B. C. Persson, and A. Bock, *J. Mol. Biol., 266*, 637–641 (1997).

30. V. N. Gladyshev and D. L. Hatfield, *Curr. Protocols Protein Sci.*, 3.8.1–3.8.19 (2000).

31. S. C. Low and M. J. Berry, *Trends Biochem. Sci., 21* 203–208 (1996).

32. G. V. Kryukov, V. M. Kryukov, and V. N. Gladyshev, *J. Biol.. Chem., 274*, 3388–33897 (1999).

33. A. Lescure, D. Gautheret, P. Carbon, and A. Krol, *J. Biol. Chem., 274*, 38147–38154 (1999).

34. D. L. Hatfield, V. N. Gladyshev, J. Park, S. I. Park, H. S. Chittum, H. J. Baek, B. A. Carlson, E. S. Yang, M. E. Moustafa, and B. J. Lee, *Comp. Nat. Prod. Chem., 4*, 353–380 (1999).

35. R. M. Tujebajeva, P. R. Copeland, X. M. Xu, B. A. Carlson, J. W. Harney, D. M. Driscoll, D. L. Hatfield, and M. J. Berry, *EMBO Rep., 1*, 158–163 (2000).

36. M. Kromayer, R. Wilting, P. Tormay, and A. Bock, *J. Mol. Biol., 262*, 413–420 (1996).

37. P. R. Copeland, J. E. Fletcher, B. A. Carlson, D. L. Hatfield, and D. M. Driscoll, *EMBO J.*, *19*, 306–314 (2000).

38. D. Fagegaltier, N. Hubert, K. Yamada, T. Mizutani, P. Carbon, and A. Krol, *EMBO J.*, *19*, 4796–4805 (2000).

39. S. Hazebrouck, L. Camoin, Z. Faltin, A. D. Strosberg, and Y. Eshdat, *J. Biol. Chem.*, *275*, 28715–28721 (2000).

19

Tungsten-Dependent Aldehyde Oxidoreductase: A New Family of Enzymes Containing the Pterin Cofactor

*Roopali Roy and Michael W. W. Adams**

Department of Biochemistry and Molecular Biology, and Center for Metalloenzyme Studies, University of Georgia, Athens, GA 30602, USA

---

* Correspondence should be addressed to this author.

1. INTRODUCTION

As described elsewhere in this volume, the essential role of molybdenum in biological systems has been recognized for many decades. Molybdoenzymes are widespread and have been extensively studied from all three of the recognized domains of life, namely, bacteria, archaea, and higher organisms or eukarya. In comparison, a positive role in biology for the analogous metal tungsten has emerged very recently. Historically, W has been regarded as an antagonist of the biological function of Mo. Early attempts to substitute W into active sites of molybdoenzymes resulted in inactive metal-free enzymes or W-substituted enzymes with little or no activity [1]. Clearly, the chemical properties of the two metals are sufficiently different that organisms can distinguish between them, they do not readily substitute one for the other, either at the level of uptake and/or incorporation into various enzymes, and molybdenum is obviously the preferred element.

However, in the early 1970s the stimulatory effect of W on the growth of certain microorganisms was reported [2]. Subsequently, the first tungsten-containing enzyme (tungstoenzyme) was purified and characterized, in this case from an acetogen, a bacterium that grows by producing acetate [3]. To date, more than a dozen tungstoenzymes have been isolated and characterized (Table 1). However, so far a biological role for tungsten has been limited to bacteria and archaea, and a tungstoenzyme has yet to be isolated from a higher organism. Indeed, tungstoenzymes have been purified from a wide variety of microorganisms. These include hyperthermophilic archaea, which grow at temperatures near 100°C, methanogens, gram-positive and negative bacteria, and acetylene-utilizing and sulfate-reducing bacteria. Of these, only the hyperthermophilic archaea appear to be obligately dependent on tungsten for growth. All other organisms either have active Mo isoforms of their "tungstoenzymes" or express active Mo-substituted counterparts of their tungstoenzymes [1].

The known tungstoenzymes can be classified into three functionally and phylogenetically distinct families termed AOR, F(M)DH, and AH. The AOR family is represented by aldehyde ferredoxin oxidoreductase (AOR) [4,5], formaldehyde ferredoxin oxidoreductase (FOR), [6,7] and glyceraldehyde-3-phosphate ferredoxin oxidoreductase (GAPOR) [8] from hyperthermophilic archaea, carboxylic acid reductase (CAR) from certain clostridia [9,10], and aldehyde dehydrogenase (ADH) from sulfate reducing-bacterium *Desulfovibrio gigas* [11]. The F(M)DH family includes formate dehydrogenase (FDH) from *Clostridium* species [3,12], *Methanococcus vannielii* [13], and *Desulfovibrio gigas* [14], and formyl methanofuran dehydrogenase (FMDH) from the thermophilic methanogens *Methanobacterium thermautotrophicum* [15] and *Methanobacterium wolfei* [16]. So far the AH family includes only acetylene hydratase from the acetylene-utilizing anaerobic bacterium *Pelobacter acetylenicus* [17]. This chapter gives a brief description of the various types of tungstoenzymes followed by a detailed discussion on the AOR family of W enzymes from the hyperthermophilic archaea. All of the known tungstoenzymes listed in Table 1 have been isolated and characterized from anaerobic organisms, and most are extremely sensitive to inactivation by oxygen. The enzymes are located in the cytoplasm of cells in which they are found—none has been shown to be membrane-associated.

TABLE 1

Molecular Properties of Tungsten-Containing Enzymes

Organism/enzyme[a]	Holoenzyme M_r (kDa)	Subunits/M_r (kDa)	W content[b]	FeS or cluster content[c]	Pterin cofactor[d]
AOR family					
Pf AOR	136	$\alpha_2(67)$	2	$2[Fe_4S_4] + 1Fe$	Nonnuc
Pf FOR	280	α_4 (69)	4	$4[Fe_4S_4]$	Nonnuc
Pf GAPOR	73	α (73)	1	$1[Fe_4S_4]$	Nonnuc
Ct CAR (form I)	86	$\alpha\beta$ (64,14)	1	~ 29 Fe, ~ 25 S	Nonnuc
Ct CAR (form II)	300	$\alpha_3\beta_3\gamma$ (64,14,43)	3	~ 82 Fe, ~ 54 S	Nonnuc
Cf CAR	134	α_2 (67)	2	~ 11 Fe, ~ 16 S	Nonnuc
Dg ADH	132	α_2 (62)	2	$2[Fe_4S_4]$	NR
F(M)DH family					
Ct FDH	340	$\alpha_2\beta_2$ (96,76)	2	20–40 Fe, 2 Se	NR
Mw FMDH	130	$\alpha\beta\gamma$ (64,51,35)	1	2–5 Fe	Nuc[e]
AH family					
Pa AH	73	α (73)	1	$1[Fe_4S_4]$	Nuc[e]

[a]The abbreviations and sources of data are: Pf (*P. furiosus*) AOR [4,24]; Pf FOR [7,36]; Pf GAPOR [8,38]; Ct (*C. thermoaceticum*) CAR [9,10]; Cf (*C. formicoaceticum*) CAR [10]; Dg (*D. gigas*) ADH [11]; Ct FDH [3,12]; Mw (*M. wolfei*) FMDH [16]; Pa (*Pe. acetylenicus*) AH [17,22]. Modified from Adams and Kletzin [40].

[b]Expressed as integer value per mole of holoenzyme.

[c]Cluster content expressed per mole of holoenzyme.

[d]Indicates whether the pterin is with (Nuc) or without (Nonnuc) an appended nucleotide. NR, not reported.

[e]Appended nucleotide is GMP.

2. CLASSIFICATION OF TUNGSTOENZYMES

2.1. The AOR Family

The majority of the known tungstoenzymes belong to this family (Table 1). Named after the most extensively studied example of a tungstoenzyme, aldehyde ferredoxin oxidoreductase (AOR) from the hyperthermophilic archaeon *Pyrococcus furiosus*, this family has additional members from both hyperthermophilic (FOR, GAPOR) and mesophilic microorganisms (CAR and ADH). The enzymes in this family are related phylogenetically and display high sequence similarity at the amino acid level. AOR, FOR, and GAPOR are most closely related. For example, AOR and FOR have 40% sequence identity with each other, whereas GAPOR is more distantly related with 23% identity with AOR or FOR [7]. However, none of the enzymes of the AOR family shows any sequence similarity to any known molybdoenzyme, showing that they are phylogenetically distinct.

As the name implies, the enzymes of the AOR family catalyze the oxidation of various types of aldehyde to the corresponding acid according to Eq. (1), where RCHO represents the substrate with the aldehyde functional group, and Fd(ox) and Fd(red) represent oxidized and reduced forms of the electron acceptor ferredoxin (Fd), respectively.

$$RCHO + H_2O + 2Fd(ox) \longrightarrow RCOOH + 2H^+ + 2Fd(red) \tag{1}$$

Aldehyde oxidation is a two electron reaction and in vivo the redox protein Fd serves as the electron acceptor for AOR, FOR, and GAPOR. The physiological electron mediator for ADH and CAR are not known. Although the enzymes in this family catalyze the same type of reaction, they differ in their substrate specificities. In fact, CAR was first isolated based on its ability to catalyze the reductive activation of carboxylic acids (the reverse of the reaction shown in Eq. (1)) although it can also carry out aldehyde oxidation. The acid-aldehyde couple has a very low reduction potential, one of the lowest of any biological system. For instance, the E_0' value for the acetaldehyde/acetate couple is $-580 \, \text{mV}$ (SHE). Consequently, under biological conditions aldehyde oxidation is much more thermodynamically favorable than acid reduction [1].

With the exception of CAR, the enzymes of the AOR family are composed of a single type of subunit that contains a mononuclear W site and one or more FeS cluster, and the only other known cofactor is a monomeric Fe site (found in the AOR of *P. furiosus*). On the other hand, CAR is a more complex enzyme with two or more types of subunit, a higher Fe content, and it contains flavin as an additional cofactor (Table 1). The other AOR-type enzymes differ in their quaternary structures, e.g., GAPOR is a monomer, AOR and ADH are dimers, and FOR is a tetramer. The W atom in all of these tungstoenzymes is coordinated by the same cofactor that binds the Mo at the active site of molybdoenzymes, the organic moiety termed *pterin*. This cofactor consists of a tricyclic ring structure with side chains containing dithiolene and phosphate groups. The metal (Mo or W) is coordinated to the pterin through dithiolene sulfur atoms. In molybdoenzymes from bacteria the pterin is usually modified with a mononucleotide (AMP, CMP, GMP, or IMP) attached to the terminal phosphate. Amongst the tungstoenzymes, AOR, FOR, and CAR have the nonnucleotide form of pterin, whereas the F(M)DH and AH enzyme families have the modified dinucleotide form.

2.2. The F(M)DH Family

A more detailed discussion of this family is given in Chapter 16. Herein some of its features are described for comparison with the AOR type of tungstoenzymes. In contrast to the AOR group, both members of the F(M)DH family use CO_2 as substrate. FDH catalyzes the reversible reduction of CO_2 to formate with the concomitant transfer of two electrons from the appropriate electron carrier, usually NADPH, according to Eq. (2).

$$CO_2 + NADPH + H^+ \rightarrow HCOOH + NADP^+ \tag{2}$$

Typically, FDH obtained from anaerobic organisms is a molybdoenzyme, but in recent years growth studies and biochemical analyses indicate that several types of bacteria and methanogens may have W-containing FDHs [18,19]. Indeed, the first tungstoenzyme to be isolated and characterized was the W-containing (W-)FDH from the thermophilic acetogenic bacterium *Clostridium thermaceticum* [3]. However, this organism also contains a Mo-containing FDH. W-FDH from *C. thermo-aceticum* is a heterotetramer with two distinct types of subunits (Table 1). In addition to the W and FeS centers, this enzyme contains the unusual amino acid selenocysteine. Both Mo- and W-containing FDHs are reported to be extremely sensitive to oxygen inactivation, rendering them difficult to isolate and characterize. However, recently the isolation of an air-stable FDH has been reported from the sulfate-reducing bacterium *D. gigas* [14]. Like the enzyme from *C. thermoaceticum*, this FDH has multiple types of subunits, is heterodimeric, and has mononuclear W and two [4Fe-4S] clusters per protein molecule. The enzyme was isolated in the presence of oxygen, although activity could be measured only under strictly anoxic conditions. Also, unlike clostridial FDH, Se was not detected in the *D. gigas* enzyme. The W atom at the active site of the enzyme is coordinated by two pterin cofactors, which are of the dinucleotide form (MGD).

The other member of this family of tungstoenzymes is FMDH. Found only in methanogens, FMDH is involved in the first step of the conversion of CO_2 to methane. It catalyzes the reductive addition of CO_2 to the organic cofactor methanofuran (MFR) in a two-electron reaction, the physiological electron donor for which is not known (Eq. (3)).

$$CO_2 + MFR + 2H^+ \rightarrow CHO\text{-}MFR + H_2O \qquad (3)$$

Like FDH, FMDHs are typically molybdoenzymes. In fact, *M. thermoautotrophicum* and *M. wolfei* each contain two isoforms of FMDH, one Mo- and one W-containing [20]. These are complex multisubunit enzymes (Table 1) that are highly sensitive to oxygen and intrinsically labile, making it difficult to correctly estimate their cofactor contents. Complete amino acid sequences are available for both W-containing FDH and FMDH. However, they show no sequence similarity to the AOR family of tungstoenzymes and instead appear to be closely related to the DMSO family of molybdoenzymes [20,21].

2.3. The AH Family

This family of tungstoenzymes is represented by a single member. Isolated from the acetylene-utilizing anaerobic bacterium *Pelobacter acetylenicus*, AH consists of a single type of subunit that contains one

W atom and a single FeS cluster (Table 1) [17,22]. The enzyme catalyzes the hydration of acetylene to acetaldehyde, according to Eq. (4).

$$C_2H_2 + H_2O \longrightarrow CH_3CHO \tag{4}$$

Therefore, unlike the other tungstoenzymes, AH does not seem to carry out an overall oxidation or reduction reaction but it is active only in the presence of a strong reducing agent. Thus, it seems likely that the tungsten atom and FeS center carry out consecutive reduction, hydration, and oxidation reactions. Furthermore, a Mo-containing form of AH has been isolated from *P. acetylenicus* grown in the presence of Mo (see Chapter 20 for details). The Mo-containing enzyme is fully active indicating that both W and Mo can catalyze the hydration reaction. An active AH has also been found in an aerobic acetylene-degrading bacterium [23]. However, the expression of the enzyme is molybdenum-dependent, and activity does not require addition of a strong reductant. Therefore, this type of AH appears to be biochemically distinct from the W-containing AH isolated from anaerobic bacteria.

3. ALDEHYDE FERREDOXIN OXIDOREDUCTASE

3.1. Molecular Properties

AOR, first purified from the hyperthermophilic archaeon *Pyrococcus furiosus*, is the most extensively studied example among all the tungstoenzymes [4,24]. The enzyme has also been purified and characterized from two other hyperthermophilic archaea, *Pyrococcus* strain ES4 [25] and *Thermococcus* strain ES1 [5]. So far AOR has not been found in any mesophilic microorganism. *P. furiosus* is an anaerobic heterotroph that grows on both simple and complex carbohydrates as well as peptides (yeast extract, peptone, casein) at temperatures near 100°C. It metabolizes carbohydrates by a fermentative pathway producing acetate, CO_2 and H_2 as end-products. *Thermococcus* strain ES1 and *Pyrococcus* strain ES4 only use peptides as a carbon source and require elemental sulfur for growth, which is reduced to H_2S. *P. furiosus* also requires sulfur for growth on peptides although it grows to high cell densities without sulfur on carbohydrates.

The crystal structure of *P. furiosus* AOR was the first for a tungsten-containing protein and for a pterin-containing protein (and the first for an enzyme from an organism that can grow at 100°C [24]). AOR is a homodimer where each subunit (M_r ~ 66 kDa) contains one W atom. The enzyme is extremely oxygen-sensitive and makes up a significant portion of the total protein in the cell. A survey of the genome databases reveals the presence of homologues of AOR in various hyperthermophilic archaea, including *P. horikoshii* [26], *Archaeoglobus (A.) fulgidus* [27], *P. abyssi* (www.genoscope.cns.fr/cgi-bin/Pab.cgi), and *Pyrobaculum (Pm) aerophilum* [28]. However, this homology is based solely on the amino acid sequence, and whether the enzymes encoded by the predicted ORFs in these organisms really belong to the AOR family of tungstoenzymes remains to be seen. The high sequence similarity between AOR and the hypothetical AORs in the other organisms suggests that the latter are true homologues [7].

3.2. Catalytic Properties and Physiological Role

AOR oxidizes a broad range of both aliphatic and aromatic aldehydes to their corresponding acid. This is a two-electron oxidation, and under physiological conditions Fd serves as the electron acceptor. The Fd of *P. furiosus* contains a single [4Fe-4S] cluster and this undergoes a one-electron redox reaction. Consequently, assuming each W-containing subunit of AOR (also FOR and GAPOR; see below) functions independently, one catalytic turnover per subunit requires the reduction of two molecules of Fd. It is proposed that during the reaction each Fd molecule is reduced sequentially rather than at the same time. Some evidence for this comes from the crystal structure of FOR (see below).

Although AOR has the ability to oxidize a wide range of aldehyde substrates, the enzyme shows the highest catalytic efficiency (R_{cat}/K_m) with acetaldehyde, isovaleraldehyde, indoleacetaldehyde, and phenylacetaldehyde. The K_m values are less than 100 μM (Table 2) [4,5], although higher concentrations (in the mM range) inhibit the enzyme. These aldehyde substrates are derivatives of the common amino acids such as alanine, valine, tryptophan, and phenylalanine. They are produced when *P. furiosus* grows on peptides and are byproducts of keto acid oxidoreductases which oxidize pyruvate, branched-chain 2-keto

TABLE 2

Substrate Specificity of AOR, FOR, and ADH

Substrate[a]	Apparent K_m (mM)		
	AOR	FOR	ADH
Formaldehyde	1.42	25.0	NA
Acetaldehyde	0.02	60.0	0.01
Propionaldehyde	0.15	62.0	0.01
Crotonaldehyde	0.14	ND	NA
Benzaldehyde	0.06	ND	0.02
Isovaleraldehyde	0.03	ND	NA
Phenylacetaldehyde	0.08	ND	NA
Phenylpropionaldehyde	NA	15.0	NA
Indoleacetaldehyde	0.05	25.0	NA
Succinic semialdehyde	NA	8.00	NA
Glutaric dialdehyde	NA	0.80	NA
Furfural	NA	ND	0.03

[a]For AOR (from *Thermococcus* strain ES-1 [5]) and FOR (from *P. furiosus* [7]), K_m values were determined at 80°C in 100 mM EPPS buffer (pH 8.4) with benzylviologen (2.5 mM) as the electron acceptor. For ADH (from *D. gigas* [11]), reactions were carried out at 30°C in 50 mM potassium phosphate buffer (pH 7.5) with benzylviologen (2 mM) as the electron acceptor.
ND, activity not detectable; NA, data not available.

acids, and aromatic 2-keto acids to their coenzyme A derivative. A fraction of the keto acid substrates are converted by nonoxidative reaction to the corresponding aldehyde [25]. Inside the cell, it is thought that AOR functions to oxidize the aldehydes derived from pyruvate (when *P. furiosus* grows on carbohydrates) or aldehydes generated during amino acid oxidation in peptide metabolism (in proteolytic organisms).

3.3. Structure and Mechanism

The crystal structure of *P. furiosus* AOR was resolved at 2.3 Å resolution [24]. It showed that AOR exists as a homodimer, where each subunit contains a mononuclear W atom and a [4Fe-4S] cluster in close proximity. Another mononuclear metal center, most likely Fe, is positioned at the dimer interface (Fig. 1). However, this site is situated more than 20 Å from the W and probably has a purely structural function. Each subunit of AOR has three domains, with binding sites for the tungstopterin cofactor and [4Fe-4S] cluster located at the interface of these domains. The W atom in AOR is coordinated by dithiolene side chains of the pterin cofactor. This cofactor was first identified by Johnson and Rajagopalan [29], who analyzed its bis(carboxyamidomethyl) derivative from molybdoenzymes and proposed that it had a bicyclic pterin ring structure with a side chain that contained hydroxyl, phosphate, and *cis*-dithiolene groups. The AOR crystal structure confirmed the proposed structure, except that the cofactor contained a third ring formed by the closure of the hydroxyl side chain with the pterin ring to form a nonplanar ring system that binds the W/Mo atom via the dithiolene sulfurs. Subsequently, similar tricyclic pterin cofactors have been found to be present in all W/Mo-containing enzyme structures [30].

As the first crystal structure for a pterin-containing enzyme, one surprising result was the revelation that the W atom is coordinated by

FIG. 1.   Schematic representation of the AOR dimer (PDB set 1 AOR) [24]. The two subunits are shown with the associated metal centers. The mononuclear Fe site is indicated at the dimer interface.

two dithiolene side chains from not one but two pterin cofactors, giving rise to the term bispterin cofactor. The two pterins are linked to each other through their terminal phosphate groups which coordinate the same Mg^{2+} ion. The tungstopterin site is buried deep within the protein and located approximately 10 Å away from the [4Fe-4S] cluster (Fig. 2). The cluster is closer to the protein surface, serving as an intermediary in the electron transfer between the W active site and the physiological electron acceptor, Fd. The [4Fe-4S] cluster is coordinated by the protein via four cysteine ligands. One of these Cys forms a hydrogen bond with a pterin ring nitrogen, indicating that the pterin might have an active role in the redox chemistry of the enzyme. The coordination sphere around the active site W in AOR includes four dithiolene sulfur atoms and an oxo ligand. However, it is difficult to establish the exact nature of the

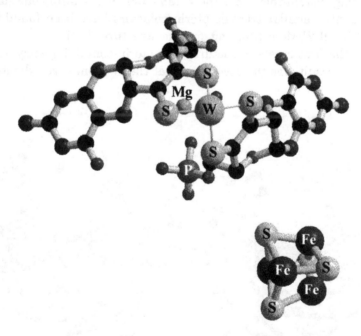

FIG. 2. Three-dimensional structure of the bispterin-tungsten cofactor and [4Fe-4S] cluster in *P. furiosus* AOR. The ball-and-stick figure depicts tungsten coordinated by dithiolene sulfurs from two pterins. The Mg^{2+} ion and phosphate side chains that link the two pterins are also indicated. Modified from Chan et al. [24].

ligands near the W atom due to either heterogeneity of the AOR active site and/or crystallographic problems associated with locating light atoms in the vicinity of heavy metals at moderate resolution [31]. X-ray absorption spectroscopy studies on the reduced active form of AOR reveal a single W=O, 4 or 5 W−S, and possibly an additional W−O/N coordination. The oxidized, catalytically inactive form of AOR, in contrast, has 2 W=O, 3 W−S, and possibly a W−O/N [32,33]. This suggests that in the oxidized form the enzyme exists in the W(VI) oxidation state. No protein ligands are coordinated to the tungsten. A hydrophobic channel of about 15 Å in length leads from the surface of the protein to the active site tungsten, which is at the bottom of the cavity. The channel is large enough to accommodate both small and large substrates, consistent with the ability of AOR to oxidize both aliphatic and aromatic aldehydes.

4.   CARBOXYLIC ACID REDUCTASE AND ALDEHYDE DEHYDROGENASE

4.1.   Molecular Properties

In addition to the tungstoenzymes isolated from hyperthermophilic archaea, the AOR family has two other members: carboxylic acid reductase (CAR) found in certain moderately thermophilic (optimum growth temperature around 40°C) acetogenic bacteria and aldehyde dehydrogenase (ADH) from mesophilic sulfate-reducing bacteria.

   CAR was discovered in acetate-producing bacteria by its ability to catalyze the reduction of nonactivated carboxylic acids (see Sec. 1). The enzyme also catalyzes the reverse reaction, aldehyde oxidation. The molecular properties of CAR vary depending on its source. CAR isolated from *C. formicoaceticum* has properties very similar to those of AOR. These include subunit size and composition (Table 1), metal cofactor content, pterin type, and N-terminal sequence. The complete amino acid sequence for CAR is not known. However, the related species *Clostridium thermoaceticum* has two forms of CAR (Table 1). Compared with the other enzymes in the AOR family these two CAR enzymes are more complex, with multiple subunits and a much higher Fe content. In addition, one of the forms contains a flavin group (FAD). Nevertheless, it is important to note that the catalytic subunit of both forms of CAR show N-terminal sequence similarity with CAR from *C. formicoaceticum* and the other enzymes in the AOR family.

Aldehyde dehydrogenase has been isolated and characterized from the sulfate-reducing *Desulfovibrio gigas* [11]. Its properties resemble those of AOR very closely. ADH is a homodimer ($M_r \sim 126\,$kDa) where each subunit contains a tungstobispterin cofactor and one [4Fe-4S] cluster. The enzyme is sensitive to inactivation by oxygen. It should be mentioned that when *D. gigas* is grown in the absence of tungsten (with Mo present in the growth medium) it produces a Mo-containing, aldehyde-oxidizing enzyme termed aldehyde oxidoreductase (AOX, sometimes referred to as Mop for molybdenum protein). AOX is distinct from W-ADH. It is a homodimeric enzyme, where each subunit has one Mo atom and two [2Fe-2S] clusters, and the enzyme uses flavodoxin as a physiological acceptor [1]. Although the complete sequence of ADH is not known, it is unlikely to have any similarity to that of AOX, as these represent the phylogenetically distinct AOR and molybdoenzyme families, respectively.

4.2. Catalytic Properties and Physiological Role

Like AOR, CAR displays a broad substrate specificity, and the enzyme can reduce a wide range of aliphatic and aromatic carboxylic acids [10]. However, neither its physiological electron carrier nor its function inside the cell is known, despite the fact that CAR is expressed at high levels ($\sim 4\%$ of the total cellular protein in *C. formicoaceticum*) [1]. *D. gigas* ADH also oxidizes a wide variety of aldehyde substrates. The enzyme efficiently oxidizes C_3-C_4 aldehydes like acetaldehyde and propionaldehyde, with low apparent K_m values (μM range) (Table 2) although high concentrations of aldehydes do not seem to cause inhibition. The electron carrier for this enzyme inside the cell is not known. W-dependent aldehyde dehydrogenase activity has been reported in the cell-free extract of another sulfate-reducing bacterium, *Desulfovibrio simplex* [34]. This enzyme oxidizes both aliphatic and aromatic aldehydes and can use flavins (FMN or FAD) as electron carriers. The activity of ADH in cell-free extracts of *D. simplex* was reported to increase (threefold) when the growth medium was supplemented with tungsten. However, so far the enzyme has not been purified and characterized, and it remains to be seen if it is closely related to the *D. gigas* ADH and belongs to the AOR family of tungstoenzymes.

5.   FORMALDEHYDE FERREDOXIN OXIDOREDUCTASE

5.1.   Molecular Properties

Like AOR, FOR has been found only in hyperthermophilic archaea. So far, the enzyme has been isolated and characterized from two species, from *Thermococcus litoralis* [6] and *P. furiosus* [7]. *T. litoralis*, like *P. furiosus*, is an obligate heterotroph and can grow on complex protein sources, as well as simple and complex carbohydrates both with and without sulfur. Based on similarity at the amino acid level, homologues of FOR are present in the genomes of other archaea such as *P. horikoshii* [26], *P. abyssi* (www.genoscope.cns.fr/cgi-bin/Pab.cgi), and *Pm aerophilum* [28]. Whether the encoded proteins are biochemically and structurally similar to FOR remains to be seen. Like the other members of the AOR family, FOR is extremely oxygen-sensitive, rapidly losing activity when exposed to air. This loss can be reversed by treating the enzyme with sulfide under highly reducing conditions, a process termed *sulfide activation* [35]. FOR exists in solution as a tetramer ($M_r \sim 270$ kDa) of identical subunits. Each subunit contains a mononuclear tungsten atom and a single [4Fe-4S] cluster.

5.2.   Catalytic Properties and Physiological Role

Originally isolated based on its ability to oxidize formaldehyde, FOR can also use other short-chain (C_1-C_4) aldehydes as substrates. However, the K_m values for such substrates are very high (> 10 mM), especially for formaldehyde (25 mM), indicating that these are not unlikely to be physiological substrates (Table 2). The lack of significant activity with longer chain aldehydes ($\geq$ C5) or with aromatic aldehydes indicates that the catalytic site of FOR is probably accessible to only short-chain aldehydes. On the other hand, C_4-C_6 acid-substituted aldehydes and dialdehydes appear to be good substrates for FOR. For example, FOR can rapidly oxidize succinic semialdehyde (C_4) and glutaric dialdehyde (C_5), whereas similar sized unsubstituted aldehydes are very poor substrates. Some C_4-C_6 semialdehydes are involved in the metabolism of certain amino acids, such as Arg, Pro, and Lys. Since the organisms in which FOR has been found, such as *P. furiosus* and *T. litoralis*, grow using proteins as a primary carbon source, FOR may have a role in amino acid metabolism, although this has yet to be established [7].

5.3. Structural Properties

The crystal structure of FOR from *P. furiosus* has been solved at 1.8 Å resolution [36]. Based on the high degree of sequence similarity between AOR and FOR, it was expected that the two enzymes would be structurally related, and this turned out to be the case. Indeed, the overall folding of the FOR subunit is virtually superimposable on that of AOR [36]. Nevertheless, their subunits interact with each other very differently, as reflected in their quaternary structures: AOR is a dimer, FOR forms a tetramer. The four subunits of FOR are arranged around a channel of about 27 Å diameter that passes through the center of the molecule, the function of which is not known. Furthermore, unlike AOR, FOR does not have the mononuclear Fe site at the subunit interface. Each subunit in FOR has one tungsten center coordinated by the dithiolene sulfurs of two pterin molecules and a single [4Fe-4S] cluster located about 10 Å away. The distance between two W atoms is about 50 Å and the subunits are thought to be catalytically independent. As in AOR, the two pterins are linked to each other by their phosphate groups via a magnesium ion. A calcium ion is located close to one of the pterin cofactors and coordinates a ring keto group, but this is proposed to have a structural rather than a catalytic role. Cocrystals of FOR with its physiological electron acceptor, *P. furiosus* Fd, show that the Fd binds close to the [4Fe-4S] cluster of the FOR subunit [36]. The result is that the [4Fe-4S] cluster is located halfway between the tungsten atom and the [4Fe-4S] cluster of Fd, clearly indicating a possible electron transfer pathway between these two centers during the reaction.

In FOR, a cavity (rather than an open channel) provides access for the substrate to the W active site. The cavity has two distinctive parts: a large chamber at the bottom and a narrower channel leading to the protein surface. The bottom chamber is linked with amino acids with bulky side chains such as tyrosine, arginine, leucine, and valine, whereas the channel is lined with hydrophobic residues. As a result, the active site cavity is much smaller in FOR than in AOR, contributing to the difference in substrate specificity between the two enzymes. FOR crystals were soaked with the dicarboxylic acid, glutarate, which is the product of glutaric dialdehyde oxidation [36]. Glutaric dialdehyde was chosen since the K_m value is the lowest for any of the substrates characterized so far (Table 2). In the crystal structure, the glutarate molecule can be clearly seen within the active site of FOR (Fig. 3). One carboxylate group of the glutarate is located near the W site, stabilized by hydrogen bond interactions with a side chain carboxylate group of Glu308, Tyr416, and His437. The second carboxylate group of glutarate

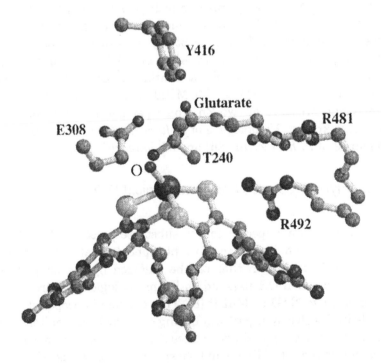

FIG. 3. Active site structure of *P. furiosus* FOR (PDB set 1B4N) [36]. The bound glutarate molecule is indicated along with nearby residues that are proposed to determine the substrate specificity for the enzyme.

is bound to the protein through electrostatic interactions with the side chains of Arg 481 and Arg 492. This structure lends credence to the proposal that a C_4- to C_6- substituted aldehyde might be the physiological substrate for FOR, although precisely what it is remains to be seen.

6. GLYCERALDEHYDE-3-PHOSPHATE FERREDOXIN OXIDOREDUCTASE

6.1. Molecular Properties

GAPOR is the third tungstoenzyme to be isolated and characterized from *P. furiosus* [8]. It is a monomeric protein of 73 kDa and is extremely oxygen-sensitive. It is the least characterized of the three tungstoenzymes that have been purified from *P. furiosus*, in part because its crystal structure has not been determined. However, many of the amino acid residues involved in binding the pterins as well as the four cysteines coordinating the FeS clusters in AOR and FOR are conserved in GAPOR, indicating that GAPOR also contains a tungstobispterin cofactor and a single [4Fe-4S] cluster [7]. Metal analyses of the pure GAPOR also show the presence of two Zn atoms per subunit, the function of which is not known. Homologues of *P. furiosus* GAPOR are present in the genomes of *P. horikoshii* [26] and *M. janaschii* [37].

6.2. Catalytic Properties and Physiological Role

GAPOR is absolutely specific for its substrate glyceraldehyde-3-phosphate (GAP), which it oxidizes to 3-phosphoglycerate. So far this is the only substrate known that can be oxidized by the enzyme. Like AOR and FOR, GAPOR uses Fd as its physiological electron carrier and does not use NAD or NADP [8]. This enzyme is proposed to have a key role in the glycolytic pathway during carbohydrate (maltose) metabolism in *P. furiosus* wherein it replaces the more conventional enzymes GAP dehydrogenase (GAPDH) and phosphoglycerate kinase (PGK), and converts GAP directly to 3-phosphoglycerate, bypassing the intermediate 1,3-bisphosphoglycerate. Consistent with this proposal, the activities of GAPDH and PGK are very low in maltose-grown *P. furiosus* [8].

Conversely, the activity of GAPOR has been shown to be about fivefold higher in cells grown on carbohydrate (cellobiose) than in cells grown on pyruvate. Expression of the gene encoding GAPOR is significantly induced after the addition of cellobiose to pyruvate-grown cultures of *P. furiosus* [38].

7. HYPOTHETICAL TUNGSTOENZYMES—WOR4 AND WOR5

7.1. Phylogeny and Molecular Properties

A search of the complete genome sequence of *P. furiosus* with the amino acid sequences of AOR and FOR revealed the presence of two previously uncharacterized open reading frames (ORFs) that show significant similarity. These two ORFs appear to encode for two additional tungstoenzymes in *P. furiosus* [7]. Tentatively termed *wor4* and *wor5* (to indicate genes encoding tungsten-containing oxidoreductases), these genes encode 622 and 582 codons corresponding to proteins WOR4 ($M_r \sim 69\,kDa$) and WOR5 ($M_r \sim 65\,kDa$), respectively. At the amino acid sequence level, both WOR4 and WOR5 are more closely related ($\sim 58\%$ sequence similarity) to AOR and FOR than they are to GAPOR ($\sim 49\%$ similarity). A survey of the database reveals that the genomes of other hyperthermophilic archaea, such as *P. horikoshii* [26], *Pm. aerophilum* [28], and *A. fulgidus* [27], also contain genes that might encode tungstoenzymes that do not fall into the AOR, FOR, or GAPOR class. Whether any or all of these genes actually encode tungstoproteins, and what their functions are, remains to be determined.

The two putative tungstoenzymes in *P. furiosus*, WOR4 and WOR5, would have subunit molecular weights comparable to those of the other members of the AOR family. Furthermore, the amino acid sequence of these putative proteins each contains the same four Cys residues that bind a single [4Fe-4S] cluster in AOR and FOR, as well as the motifs that bind the bispterin cofactor (Fig. 4). Therefore, it appears that the two putative tungsten oxidoreductases in *P. furiosus* are closely related both in structure and metal cofactor content to the more extensively characterized tungstoenzymes. It will be intriguing to determine under what conditions these genes are expressed.

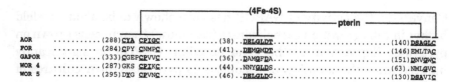

FIG. 4. Alignment of the cofactor-binding motifs of FOR, AOR, GAPOR, and the two putative tungstoenzymes WOR4 and WOR5 from *P. furiosus*. Based on the three-dimensional structures of AOR and FOR, the CxxCxxxC motif is known to coordinate a single [4Fe-4S] cluster, whereas two DxxGxD/C motifs bind the bispterin cofactor. Numbers in parenthesis indicate the numbers of residues between the indicated motifs. (Modified from [7].)

8. TUNGSTEN VERSUS MOLYBDENUM IN THE AOR FAMILY

Once considered merely as an antagonist to Mo uptake in organisms, a role for tungsten has slowly emerged in biological catalysis. Although tungstoenzymes have been isolated from diverse sources, they all appear to catalyze reactions involving extremely low-potential chemistry that require anaeobic conditions and in some cases temperatures close to 100°C. So, the question is, why do these microorganisms use W instead of the analogous metal Mo? In fact, many of the W-utilizing microorganisms produce functional Mo-containing isoforms of their tungstoenzymes when they are grown in the presence of Mo rather than W. For example, a Mo form of CAR has been isolated from *C. formicoaceticum*, and *D. gigas* expresses Mo-AOX that resembles the W-ADH in function. However, replacement of tungstoenzymes with Mo analogues appears not to be the case for hyperthermophilic archaea like *P. furiosus*. The W in AOR, FOR, and GAPOR is not substituted with Mo when the organism is grown with Mo in the absence of added W [39]. An explanation for this might lie in the chemical properties of Mo and W and their availability in different environments. Comparing the properties of mononuclear W and Mo synthetic complexes [1], it appears that W complexes are much more oxygen-sensitive and display greater thermal stability than the Mo complexes. Furthermore, W complexes display lower redox potentials in comparison with their Mo analogues.

These data suggest that W complexes might be better suited for catalyzing low-potential redox reactions under anoxic conditions at higher temperatures. On the other hand, Mo complexes should catalyze similar reactions at lower temperatures. The distribution of these metals may also play a critical role in their utilization. For example, in freshwater and marine environments the concentration of Mo is higher than that of W, whereas in deep-sea hydrothermal vents (where hyperthermophilic archaea are found) W is more abundant than Mo [40]. These factors may collectively account for the obligate W-dependence of the hyperthermophilic archaea. Such organisms carry out chemical reactions near the limits of biology, at extremely low potentials and at high temperatures. It remains to be determined if catalysis under such extreme conditions is a feat that can be accomplished by W and not by Mo.

9. CONCLUSIONS

The fact that W can play an active role in biological systems has become clear in recent decades. W is relatively scarce on this planet, but is enriched in certain ecological niches such as hot springs, brine lakes, and hydrothermal vent fluids [40]. At present, all of the known tungstoenzymes catalyze reactions involving oxo atom transfer and coupled electron proton transfer, very similar to the reactions catalyzed by molybdoenzymes. While several microbial species synthesize genetically distinct molybdoenzymes and tungstoenzymes that catalyze the same reaction, the hyperthermophilic archaea are an exception in that the enzymes of the AOR family appear to function only with W and the organism contains no functional Mo homologues. Thus members of this family are referred to as "true" tungstoenzymes and are phylogenetically distinct from the major classes of molybdoenzyme. A survey of the genomic databases shows that W-containing enzymes might be more prevalent than previously anticipated, at least among the hyperthermophilic archaea. However, these putative W enzymes have yet to be characterized. Similarly, very little is known about the uptake, transport, and storage of W in hyperthermophilic organisms. For example, it is not known if the mechanisms involved are analogous to those used for Mo utilization in more conventional organisms. Clearly, much is still to be understood about the AOR family, a unique group of what are, so far at least, W-only enzymes.

ACKNOWLEDGMENT

Research in the author's laboratory was supported by grants from the U.S. Department of Energy.

ABBREVIATIONS

ADH	aldehyde dehydrogenase
AH	acetylene hydratase
AMP	adenosine 5′-monophosphate
AOR	aldehyde ferredoxin oxidoreductase
AOX	aldehyde oxidoreductase
CAR	carboxylic acid reductase
DMSO	dimethylsulfoxide
EPPS	(N-[2-hydroxyethyl]piperazine-N'-[3-propanesulfonic acid])
FAD	flavin adenine dinucleotide
Fd	ferredoxin
FDH	formate dehdyrogenase
FMDH	formyl methanofuran dehydrogenase
FMN	flavin mononucleotide
FOR	formaldehyde ferredoxin oxidoreductase
GAP	glyceraldehyde-3-phosphate
GAPDH	glyceraldehyde-3-phosphate dehydrogenase
GAPOR	glyceraldehyde-3-phosphate ferredoxin oxidoreductase
GMP	guanosine 5′-monophosphate
IMP	inosine 5′-monophosphate
MFR	methanofuran
MGD	molybdopterin guanine dinucleotide
NAD	nicotinamide adenine dinucleotide
NADP	nicotinamide adenine dinucleotide phosphate

NADPH	nicotinamide adenine dinucleotide phosphate (reduced)
ORF	open reading frame
PGK	phosphoglycerate kinase
SHE	standard hydrogen electrode

REFERENCES

1. M. K. Johnson, D. C. Rees, and M. W. W. Adams, *Chem. Rev.,* *96*, 2817–2840 (1996).

2. L. G. Ljungdahl, *Trends Biochem. Sci.,* *1*, 63–65 (1976).

3. I. Yamamoto, T. Saiki, S.-M. Liu, and L. G. Ljungdahl, *J. Biol. Chem.,* *258*, 1826–1832 (1983).

4. S. Mukund and M. W. W. Adams, *J. Biol. Chem.,* *266*, 14208–14216 (1991).

5. J. Heider, K. Ma, and M. W. W. Adams, *J. Bacterial.,* *177*, 4757–4764 (1995).

6. S. Mukund and M. W. W. Adams, *J. Biol. Chem.,* *268*, 13592–13600 (1993).

7. R. Roy, S. Mukund, G. J. Schut, D. M. Dunn, R. Weiss, and M. W. W. Adams, *J. Bacteriol.,* *181*, 1171–1180 (1999).

8. S. Mukund and M. W. W. Adams, *J. Biol. Chem.,* *270*, 8389–8392 (1995).

9. H. White, G. Strobl, R. Feicht, and H. Simon, *Eur. J. Biochem.,* *184*, 89–96 (1989).

10. H. White, R. Feicht, C. Huber, F. Lottspeich, and H. Simon, *Biol. Chem. Hoppe-Seyler,* *372*, 999–1005 (1991).

11. C. M. H. Hensgens, W. R. Hagen, and T. H. Hansen, *J. Bacteriol.,* *177*, 6195–6200 (1995).

12. J. R. Andreesen and L. G. Ljungdahl, *J. Bacteriol.,* *116*, 867–873 (1973).

13. J. B. Jones and T. C. Stadtman, *J. Bacteriol.,* *130*, 1404–1406 (1977).

14. M. J. Almendra, C. D. Brondino, O. Gavel, A. S. Pereira, P. Tavares, S. Bursakov, R. Duarte, J. Caldeira, J. J. G. Moura, and I. Moura, *Biochemistry,* *38*, 16366–16372 (1999).

15. P. A. Bertram, R. A. Schmitz, D. Linder, and R. K. Thauer, *Arch. Microbiol. 161*, 220–228 (1994).

16. R. A. Schmitz, M. Richter, D. Linder, and R. K. Thauer, *Eur. J. Biochem., 207*, 559–565 (1992).

17. B. Rosner and B. Schink, *J. Bacteriol., 177*, 5767–5772 (1995).

18. K. A. Burke, K. Calder, and J. Lascelles, *Arch. Microbiol., 126*, 155–159 (1980).

19. F. M. Girio, J. C. Roseiro, and A. L. Silva, *Curr. Microbiol., 36*, 337–340 (1998).

20. A. Hochheimer, R. Hedderich, and R. K. Thauer, *Arch. Microbiol., 170*, 389–393 (1998).

21. D. Gollin, X. L. Li, S. M. Liu, E. T. Davies, and I. G. Ljungdahl, *Advs. Chem. Conversions, 114*, 303–308 (1998).

22. R. U. Meckenstock, R. Kreiger, S. Ensign, P. M. H. Kroneck, and B. Schink, *Eur. J. Biochem., 264*, 176–182 (1999).

23. B. M. Rosner, F. A. Rainey, R. M. Kroppenstedt, and B. Schink, *FEMS Microbiol. Lett., 148*, 175–180 (1997).

24. M. K. Chan, S. Mukund, A. Kletzin, M. W. W. Adam, and D. C. Rees, *Science, 267*, 1463–1469 (1995).

25. K. Ma, A. Hutchins, S.-H. S. Sung, and M. W. W. Adams, *PNAS, 94*, 9608–9613 (1997).

26. Y. Kawarabayasi, M. Sawada, H. Horikawa, Y. Haikawa, Y. Hino, S. Yamamoto, M. Sekine, S. Baba, H. Kosugi, A. Hosoyama, Y. Nagai, M. Sakai, K. Ogura, R. Otsuka, H. Nakazawa, M. Takamiya, Y. Ohfuku, T. Funahashi, T. Tanaka, Y. Kudoh, J. Yamazaki, N. Kushida, A. Oguchi, K. Aoki, and H. Kikuchi, *DNA Res. 5*, 147–155 (1998).

27. H. P. Klenk, R. A. Clayton, J. F. Tomb, O. White, K. E. Nelson, K. A. Ketchum, R. J. Dodson, M. Gwinn, E. K. Hickey, J. D. Peterson, D. L. Richardson, A. R. Kerlavage, D. E. Graham, N. C. Krypides, R. D. Fleischmann, J. Quackenbush, N. H. Lee, G. G. Sutton, S. Gill, E. F. Kirkness, B. A. Dougherty, K. McKenney, M. D. Adams, B. Loftus, S. Peterson, C. I. Reich, L. K. McNeil, J. H. Badger, A. Glodek, L. Zhou, R. Overbeek, J. D. Gocayne, J. F. Weidman, L. McDonald, T. Utterback, M. D. Cotton, T. Spriggs, P. Artiach, B. P. Kaine, S. M. Sykes, P. W. Sadow, K. P. D'Andrea, C. Bowman, C. Fujii, S. A. Garland, T.

M. Mason, G. J. Olsen, C. M. Fraser, H. O. Smith, C. R. Woese, and J. C. Venter, *Nature, 390*, 364–370 (1997).

28. S. Fitz-Gibbon, A. J. Choi, J. H. Miller, K. O. Stetter, M. I. Simon, R. Swanson,and U-J. Kim, *Extremophiles, 1*, 36–51 (1997).

29. J. L. Johnson and K. V. Rajagopalan, *Proc. Natl. Acad. Sci. USA, 79*, 6856–6859 (1982).

30. C. Kisker, H. Schindelin, D. Bass, J. Retey, R. U. Meckenstock, and P. M. Kroneck, *FEMS Microbiol. Rev., 22*, 503–521 (1998).

31. H. Schindelin, C. Kisker, and D. C. Rees, *J. Biol. Inorg. Chem., 2*, 773–781 (1997).

32. G. N. George, R. C. Prince, S. Mukund, and M. W. W. Adams, *J. Am. Chem. Soc., 114*, 3521–3523 (1992).

33. S. Mukund and M. W. W. Adams, *J. Biol. Chem., 265*, 11508–11516 (1990).

34. G. Zellner and A. Jargon, *Arch. Microbiol., 168*, 480–485 (1997).

35. R. Roy, A. L. Menon, and M. W. W. Adams, *Meth. Enzymol., 331*, 132–144 (2001).

36. Y. Hu, S. Faham, R. Roy, M. W. W. Adams, and D. C. Rees, *J. Mol. Biol., 286*, 899–914 (1999).

37. C. J. Bult, O. White, G. J. Olsen, L. Zhou, R. D. Fleischmann, G. G. Sutton, J. A. Blake, L. M. Fitzgerald, R. A. Clayton, J. D. Gocayne, A. R. Kerlavage, B. A. Dougherty, J. F. Tomb, M. D. Adams, C. I. Reich, R. Overbeek, E. F. Kirkness, K. G. Weinstock, J. M. Merrick, A. Glodeck, J. L. Scott, N. S. M. Geoghagen, J. F. Weidman, J. L. Fhurmann, D. Nguyen, T. R. Utterback, J. M. Kelley, J. D. Peterson, P. W. Sadow, M. C. Hanna, M. D. Cotton, K. M. Roberts, M. A. Hurst, B. P. Kaine, M. Borodovsky, H. P. Klenk, C. M. Fraser, H. O. Smith, C. R. Woese, and J. C. Venter, *Science, 273*, 1058–1073 (1996).

38. J. van der Oost, G. Schut, S. W. M. Kengen, W. R. Hagen, M. Thomm, and W. M. de Vos, *J. Biol. Chem., 273*, 28149–28154 (1998).

39. S. Mukund and M. W. W. Adams, *J. Bacteriol., 178*, 163–167 (1996).

40. A. Kletzin and M. W. W. Adams, *FEMS Microbiol. Rev., 18*, 5–63 (1996).

20
Tungsten-Substituted Molybdenum Enzymes

C. David Garner and Lisa J. Stewart

School of Chemistry, University of Nottingham,
Nottingham NG7 2RD, UK

1. INTRODUCTION

The group 6 metals molybdenum and tungsten are unique in that they are the only 4d (Mo) and 5d (W) transition metals that are required for the normal metabolism of biological systems. Molybdenum enzymes are found in all types of living systems [1–3]. Furthermore, molybdenum plays a vital role in the nitrogen cycle, as it is essential for both routes to fixed nitrogen, via the nitrogenases (see Chapters 3 and 5) and the nitrate reductases. The latter enzymes belong to a large class of enzymes that catalyze key reactions in the metabolism of C, N, and S in bacteria, plants, animals, and humans (see Chapters 6, 10–18). Each conversion involves the net transfer of an oxygen atom, either to or from the substrate, with the metal cycling between the oxidation states Mo^{VI} and Mo^{IV}. Examples of these enzymes (and the conversion accomplished) include the nitrate reductases ($NO_3^- \rightarrow NO_2^-$); the sulfite oxidases ($SO_3^{2-} \rightarrow SO_4^{2-}$); the aldehyde oxidases ($RCHO \rightarrow RCO_2H$); the xanthine oxidases (xanthine $\rightarrow$ uric acid); and the dimethyl sulfoxide (DMSO) reductases (DMSO $\rightarrow$ DMS) [2,3]. Recently, a series of protein crystallographic studies have determined the molecular structure of several of these enzymes and defined the nature of their catalytic centers (see Chapter 7). In each enzyme, the molybdenum is coordinated to one or two molecules of a special pterin "molybdopterin" (MPT) (see Chapters 6, 7, and 9) via the ene-1,2-dithiolate (or dithiolene) group.

In contrast to molybdenum, evidence for the involvement of tungsten in biological systems (see Chapter 19) has only been obtained relatively recently. The initial studies showed that the production and activity of formate dehydrogenase, which catalyzes the first step in the reduction of CO_2 to acetate, was significantly higher in the acetogens

Clostridium thermoaceticum and *C. formicoaceticum* if the bacteria were grown in the presence of $[WO_4]^{2-}$ rather than $[MoO_4]^{2-}$ [4]. The first naturally occurring tungsten enzyme was purified in 1983 [5]. Major progress in the demonstration of the biological significance of tungsten was made in the 1990s. A special aspect of the role of tungsten in biology is its involvement in the metabolism of hyperthermophilic archaea that live in volcanic vents on the sea bed at > 100°C [6]. The first structural characterization of MPT was achieved by the determination of the protein crystal structure of the tungsten enzyme aldehyde oxidoreductase from the thermophile *Pyrococcus furiosus* [7]. All tungsten enzymes that have been identified to date are MPT-dependent, and it appears that each has two MPTs bound to the metal, i.e., they have a catalytic center of DMSO reductase-type in the Hille classification of the MPT-dependent molybdenum enzymes [2]. Furthermore, with the possible exception of acetylene hydratase [8], tungsten centers in enzymes, like their molybdenum counterparts, catalyze oxygen atom transfer reactions.

This chapter will consider the progress made in the substitution of tungsten for molybdenum in both the MPT-dependent enzymes and the nitrogenases; in the latter respect, contrary to initial reports [9], the substrate reduction pattern of the enzyme is unchanged by tungsten substitution [10].

2. MOLYBDENUM AND TUNGSTEN CHEMISTRY: SIMILARITIES AND DIFFERENCES

The atomic and ionic radii and the chemical properties of molybdenum and tungsten are very similar, although "there are differences between them in various types of compounds that are not easy to explain" [11]; *the same can be said of the MPT-dependent enzymes of these two metals!* The coordination chemistries of molybdenum and tungsten are comparable and both elements form complexes in oxidation states −II to +VI, inclusively, of which the +IV, +V, and +VI states are accessible to biology [12]. Specific investigations of molybdenum and tungsten chemistry, inspired by the involvement of these metals in biological systems, commenced in the 1970s and have resulted in a significant number of new and important developments in the chemistry of these elements [13,14].

With respect to the chemistry relevant to the known roles of these metals in the MPT-dependent enzymes, it is important to note that the primary mechanisms for the uptake and transport of both molybdenum and tungsten involves the $[MoO_4]^{2-}$ tetrahedral anions (see Chapter 2). The proteins responsible for the transport of these metals bind the $[MO_4]^{2-}$ anion in a cavity of a suitable size, locating it by a series of hydrogen bonds from the polypeptide chain. As $[MoO_4]^{2-}$ and $[WO_4]^{2-}$ are essentially of the same size, the transport proteins do not discriminate significantly between the two $[MoO_4]^{2-}$ anions, but they will not bind the much smaller anion $[SO_4]^{2-}$ [15].

Oxo groups are ubiquitous in the complexes formed by molybdenum and tungsten in their higher oxidation states [13,14,16,17]. Furthermore, and crucial to the roles of these metals in the MPT-dependent enzymes, a change in oxidation state from M^{VI} to M^{IV} (or M^V) invariably involves the loss of an oxo group, and vice versa. This concomitant change in the metal's oxidation state and the number of oxo groups bound, is well developed in the chemistry of these metals bound to S-donor ligands, including dithiolenes. Dithiolenes were introduced as analytical reagents for metals in the 1930s [18]. An extensive series of investigations of d-transition metal complexes of these ligands were accomplished in the 1960s, stimulated by the facile redox chemistry of these systems and the novel trigonal prismatic geometry of $[M(dithiolene)_3]^{n-}$ complexes [19]. An extensive range of dithiolene complexes has been characterized for both molybdenum and tungsten, including examples of $[M(dithiolene)_3]^{n-}$ ($n = 0, 1,$ or 2), $[MO(dithiolene)_2]^{n-}$ ($n = 1$ or 2), and $[MO_2(dithiolene)_2]^{2-}$ systems. Two particular aspects of the chemistry of these systems have been observed for corresponding complexes of molybdenum and tungsten, each of which may be relevant to the differences in the behavior of molybdenum and tungsten centers of MPT-dependent enzymes.

2.1 Differences in Redox Potentials

The $E_{1/2}$ values for corresponding couples of molybdenum and tungsten complexes are invariably in the sense that Mo $\gg$ W, i.e., a tungsten complex is significantly more stable in the higher oxidation state than its molybdenum counterpart; e.g., for the $[MF_6]^-/[MF_6]^{2-}$ couple the difference is about 987 mV. Such behavior is also manifest in dithiolene complexes; however, the dithiolene ligands modulate the inherent differences between these metals so that, for example, the $E_{1/2}$ values for corresponding $[MO(dithiolene)_2]^-/[MO(dithiolene)_2]^{2-}$ couples are greater for the molybdenum than the tungsten systems by only about 225 mV [20,21].

2.2   Relative Rates of Reactions Involving Cleavage of M=O Bonds

$[MO(dithiolene)_2]^{2-}$ and $[MO_2(dithiolene)_2]^{2-}$ complexes are related by the oxygen atom transfer reactions (1) [22,23]; these reactions are analogous to those catalyzed by molybdenum and tungsten enzymes of the DMSO reductase family.

$$[MO(dithiolene)_2]^{2-} + Me_3NO \rightarrow [MO_2(dithiolene)_2]^{2-} + Me_3N$$

$$(1a)$$

$$[MO_2(dithiolene)_2]^{2-} + Ph_3P \rightarrow [MO(dithiolene)_2]^{2-} + Ph_3PO$$

$$(1b)$$

M = Mo or W

The relative rates of oxygen atom transfer from the metal to the phosphorus in the second-order reactions (2) has been determined and, at all of the temperatures examined, $k_{Mo}/k_w \sim 10^2$–10^3 [24].

$$[MO_2(mnt)_2]^{2-} + (RO)_{3-n}R_n'P \rightarrow [MO(mnt)_2]^{2-}$$
$$+ (RO)_{3-n}R_n'PO \qquad (2)$$

(M = Mo or W; mnt = 1, 2-dicyanoethylenedithiolate;

R = Me, Et, or Ph; $n = 0$ or 1)

Thus, the ease of reducibility of the metal center, Mo > W, and the strength of the M=O bonds, Mo=O < W=O, are two important factors that increase the activation energy for oxygen atom transfer occurring at a tungsten center, as compared with a corresponding molybdenum center.

3. MOLYBDENUM AND TUNGSTEN GEOCHEMISTRY AND LINK TO BIOAVAILABILITY

The different roles that have evolved for molybdenum and tungsten in biological systems require some comment, which inevitably involves some speculation. For a more comprehensive account of the "biogeochemistry" of molybdenum and tungsten, the reader is referred to Chapter 1.

Starting from the premise that life evolved on this planet, it is relevant to observe that both molybdenum and tungsten are present in the Earth's crust at a concentration of 1.5 ppm. Thus, both of these elements are reasonably abundant, but significantly less so than copper (55 ppm), zinc (70 ppm), and iron (50 000 ppm), three metals that are widely employed in biological systems. Despite their similar chemistries, the geochemistry of molybdenum and tungsten is quite different. Thus, molybdenum occurs in the Earth's crust primarily as molybdenite, MoS_2, whereas tungsten occurs as scheelite, $CaWO_4$, and wolframite, $(Fe,Mn)WO_4$ [25].

Given that molybdenum and tungsten are generally taken up by organisms as the $[MO_4]^{2-}$ ion, the relative availability of $[MoO_4]^{2-}$ and $[WO_4]^{2-}$ will be an important factor in determining whether molybdenum or tungsten is incorporated by an organism. This aspect is clearly demonstrated by studies of the DMSO reductase of *Rhodobacter capsulatus* (Sec. 5.3) [26,27]. Thus, a physiologically competent enzyme in which tungsten is substituted for molybdenum is produced under $[WO_4]^{2-}$-rich, $[MoO_4]^{2-}$-depleted, conditions. However, in a growth medium containing equal concentrations of $[MoO_4]^{2-}$ and $[WO_4]^{2-}$, the bacterium *R. capsulatus* shows an only modest preference for incorporating molybdenum over tungsten. One important caveat to emerge from this study was that a trace of molybdenum was essential for cell growth [26], *indicating that there is a valid role that molybdenum is required to perform in the metabolism of this organism that cannot be fulfilled by tungsten.*

The significantly greater presence of molybdenum over tungsten at the catalytic center of enzymes can be understood by noting the relative concentrations of $[MoO_4]^{2-}$ and $[WO_4]^{2-}$ in seawater that is found at the present time. $[MoO_4]^{2-}$ is present at the relatively high concentration of 1×10^{-2} mgL^{-1}, about 100 times the concentration of $[WO_4]^{2-}$. Thus, for metal incorporation by an organism from an aqueous medium where this concentration difference applies, $[MoO_4]^{2-}$ is expected to be incorporated in preference to $[WO_4]^{2-}$. However, the 100-fold excess of $[MoO_4]^{2-}$ over $[WO_4]^{2-}$ in seawater has (probably) not always been the case. A significant aspect of this argument is the fact that MoS_2, the predominant form of molybdenum in the Earth's crust, is insoluble in water. Therefore, it would appear that molybdenum was relatively unavailable for involvement in the early stages of evolution of life on Earth. However, in an oxidizing atmosphere (i.e., post-photosynthesis, about 10^9 years ago, when the Earth's atmosphere gradually gained oxygen) MoS_2 is converted to $[MoO_4]^{2-}$ (3).

$$2MoS_2 + 7O_2 + 2H_2O \longrightarrow 2[MoO_4]^{2-} + 4SO_2 + 4H^+ \qquad (3)$$

$[WO_4]^{2-}$, arising from the dissolution of tungsten oxide sources, has probably always been available in seawater.

These observations offer some rationalization for the association of tungsten with primitive organisms [6] and the significantly greater involvement of molybdenum in biological systems of the present biosphere [2].

4. EARLY ATTEMPTS TO SUBSTITUTE TUNGSTEN FOR MOLYBDENUM

Early attempts to substitute tungsten for molybdenum in MPT-dependent enzymes failed because the organism was incapable of growing on the tungstate-containing medium. This failure could be due, at least in part, to the total absence of molybdenum, as was found for the investigations of tungsten incorporation into *R. capsulatus* [26]. Also, the failure could be due to the pathway used for growth requiring a molybdenum enzyme that was inactivated by the presence of tungsten. In several instances, once an alternative growth medium was established to provide a suitable alternative pathway, growth was achieved and the effect of tungsten on the activity of the enzyme could be examined. For example, *Chlorella* cell growth is inhibited by $[WO_4]^{2-}$ in media with nitrate as the nitrogen source, but use of ammonia allowed cell growth. Subsequent removal of ammonia from the *Chlorella* growth medium derepressed nitrate reductase synthesis, and this could be done in the presence of $[MoO_4]^{2-}$ or $[WO_4]^{2-}$. Radiolabeled tungsten (^{185}W) was used in the derepression step to establish whether tungsten was actually incorporated into the enzyme or merely hindered the incorporation of molybdenum. The former was found to be the case, although the tungsten was bound by the enzyme more weakly than molybdenum. Interestingly, while the nitrate reductase activity was lost, NADH-diaphorase activity (i.e., the ability to reduce Fe^{III} to Fe^{II} in heme) was retained by the tungsten-substituted enzyme [28].

A second example of tungsten's ability to substitute for molybdenum in a regulatory role, enabling enzyme synthesis but producing an inactive tungsten-substituted enzyme, is provided by the nitrate reductase from *Paracoccus denitrificans*. Suspensions of cells grown without molybdenum or tungsten were incubated with and without molybdenum, tungsten, and azide, and the induced nitrate reductase activity was assessed. With molybdenum and azide, the nitrate reductase activity was found to be associated with a membrane protein of M_r 150,000 Da. This polypeptide was also present after incubation with tungsten and azide, but no nitrate reductase activity was detected under such conditions. The polypeptide was absent after incubation with azide alone, proving that although tungsten prevents development of nitrate reductase activity, it plays a role in enzyme regulation and stimulates production of nitrate reductase [29].

Two methanogenic archaea, *Methanobacterium thermoautotrophicum* and *M. wolfei*, each contain two separate formylmethanofuran dehydrogenases (FMDHs). These enzymes belong to the DMSO reductase family [6]. In each organism, one FMDH contains molybdenum ("FMDH I") and the other tungsten ("FMDH II") and the two enzymes exhibit different chromatographic properties. The optimal activity for FMDH I was observed at pH 7.9 and this enzyme showed no activity at pH 6.5. FMDH II has maximal activity at pH 7.4; this activity is reduced by 70% at pH 6.5. The N-terminal sequences and molecular weights of two subunits (α and β) of the two enzymes are identical, but the nature of the third subunit (γ) is different: the molecular weights of this subunit are 31 kDa for FMDH I and 35 kDa for FMDH II [30]. The two organisms react differently to change in their growth medium. Thus, *M. wolfei* expresses FMDH I if grown on medium containing molybdenum and not tungsten, but produces both enzymes if grown on medium containing tungsten but not molybdenum (i.e., molybdenum cannot substitute for tungsten in FMDH I, but tungsten can replace molybdenum in FMDH I) [30]. *M. thermoautotrophicum* shows the opposite response and can produce molybdenum-substituted FMDH II but not tungsten-substituted FMDH I [31].

Tungsten-substituted FMDH I of *M. wolfei* is an active enzyme and the catalytic activity with the physiological substrate formylmethanofuran is identical to that of the natural molybdenum enzyme [32]. However, the activity with the alternative substrate *N*-furylformamide was found to be increased for the tungsten-substituted enzyme. The optimum pH for catalytic activity is also affected by such a substitution, changing from 7.9 for the molybdenum enzyme to 7.4 for the tungsten-substituted enzyme. This suggests that the difference in response to pH change between FMDH I and II is *only* a consequence of the nature of the metal center (molybdenum or tungsten). Electron paramagnetic resonance (EPR) spectroscopy confirmed the presence of tungsten and a W^V EPR signal was observed only after air oxidation of the purified enzyme, suggesting that the enzyme was purified in the W^{IV} state [32].

5. TUNGSTEN-SUBSTITUTED MOLYBDENUM ENZYMES

5.1. Xanthine Oxidase and Sulfite Oxidase

The activity of the molybdenum enzymes xanthine oxidase and sulfite oxidase in rats has been shown to be lowered by addition of $[WO_4]^{2-}$ to the drinking water of rats on a molybdenum-depleted diet [33]. Under these conditions, xanthine oxidase was produced as the apoprotein [34], and sulfite oxidase was produced as a mixture of apoprotein and tungsten-substituted protein [35]. The molybdenum-free proteins (both xanthine oxidase and sulfite oxidase) were found to be inactive.

EPR studies on the inactive tungsten-substituted sulfite oxidase suggested that the tungsten had an identical coordination to that of molybdenum in the native enzyme. Therefore, this enzyme provides a direct comparison of the W^V and Mo^V EPR spectra recorded for MPT-dependent enzymes. The g values of the W^V are lower than those of the corresponding Mo^V center; the g-value anisotropy of both centers is pH-dependent and is affected by the presence of anions such as phosphate and fluoride. The two enzymes also show very similar proton hyperfine interactions and doublets in the spectra appear as singlets when D_2O is used as solvent [36].

One interesting and informative difference between the molybdenum- and tungsten-containing forms of sulfite oxidase is the ease of reduction at different pHs. While the molybdenum enzyme is readily reduced to Mo^V in the pH range 6–9, the W^V EPR signal is more easily generated at low pH, and the tungsten enzyme cannot be fully reduced at pH 9 [36]. This behavior is consistent with the normal preference for the adoption of a higher oxidation state, W > Mo, with reduction of these centers involving coupled proton-electron transfer [37].

5.2. Trimethylamine Oxide Reductase

Tungsten substitution for molybdenum in the periplasmic trimethyl-amineoxide (TMAO) reductase from *Escherichia coli* has been achieved in a mutant lacking the usual molybdate uptake (*mod*) pathway. In this case, molybdenum or tungsten was imported via the sulfate uptake pathway and the amount of tungsten-containing TMAO reductase (W-TMAOR) was about 15% of the amount of Mo-TMAOR produced. Analysis of *moa* and *mob* mutants, which lack particular stages in the biosynthetic pathway of MPT, revealed that $[WO_4]^{2-}$ is capable of substituting for $[MoO_4]^{2-}$ in the activation of TMAOR and its translocation from the cytoplasm [38]. The tungsten-substituted enzyme has been purified, characterized, and shown to be enzymatically active [39]. W-TMAOR is capable of catalyzing not only the reduction of TMAO to trimethylamine (TMA) but also, and in contrast to the natural (Mo-TMAOR) enzyme, the reduction of DMSO to dimethyl sulfide (DMS). Thus, studies involving a variety of substrates revealed that while Mo-TMAOR can catalyze the reduction of various *N*-oxides but not sulfoxides, W-TMAOR is capable of catalyzing the reduction of both N-oxides and sulfoxides. Interestingly, tungsten was found to be incapable of substituting for molybdenum in the membrane-bound DMSO reductase from $E.$ $coli$ [39]. This last observation is consistent with a previous observation that $E.$ $coli$ cannot grow on a minimal medium containing tungsten and DMSO, although it does grow satisfactorily on the same medium containing molybdenum in place of tungsten [40].

The kinetics of TMAO reduction by the tungsten-substituted enzyme have been probed by use of a benzylviologen-coupled assay [39]. This approach is analogous to that used to assess the activity of $Rhodobacter$ $capsulatus$ DMSO reductase (see Sec. 5.3). The catalytic efficiency (k_{cat}/k_m, where k_{cat} is the turnover number and k_m the Michaelis-Menten constant) was greater for the tungsten enzyme than the molybdenum enzyme, although the specific activity (k_{cat}) of the former was lower than that of the latter by about 50%. However, this lower value is probably due, at least in part, to the measurement of the activity of the tungsten enzyme being accomplished at a significantly lower temperature than that required for optimal activity.

Substitution of tungsten for molybdenum in TMAOR affects the activity profile of the enzyme. While both forms of the enzyme exhibit maximum activity at pH 5, the activity at pH 5.5 is significantly decreased for W-TMAOR but little altered for Mo-TMAOR. Maximum activity for Mo-TMAOR is reached at 60°C and remains unchanged at 80°C, but the activity of W-TMAOR increases with temperature up to 80°C. Incubation at 80°C for an extended time dramatically reduces the activity of Mo-TMAOR (50% activity after 4 min incubation; 3% after 90 min). However, W-TMAOR is more stable to high temperatures and retains 50% of its initial activity after 90 min incubation. Similarly, TMAOR is more resistant to an increase in the ionic strength; thus, Mo-TMAOR retains only 4% of its activity in the presence of 2 M NaCl, compared with 15% for W-TMAOR [39].

5.3. Dimethyl Sulfoxide Reductase

The periplasmic DMSO reductases (DMSORs) of the photosynthetic bacteria $Rhodobacter$ $capsulatus$ and $Rhodobacter$ $sphaeroides$ function in a respiratory chain with DMSO as the terminal electron acceptor.

These enzymes catalyze the environmentally important reaction (4) [41,42] that involves the direct transfer of an oxygen atom from DMSO to Mo^{IV}, producing DMS [43]. As expected, these enzymes have a high affinity for DMSO and also catalyze the reduction of TMAO to TMA [44]

$$DMSO + 2e^- + 2H^+ \rightleftharpoons DMS + H_2O \qquad (4)$$

The DMSORs from *R. capsulatus* and *R. sphaeroides* are important as two of the simplest known molybdenum enzymes: they have a relatively low molecular weight ($M_r \sim 85,000$) and contain only one redox-active center [42]. This center involves molybdenum bound to two molybdopterin guanine dinucleotides (MGDs). Several crystallographic characterizations of these systems have been reported [45–49], and the nature of the active site in the oxidized state of the enzyme has been defined, although this has been complicated by the cocrystallization of more than one form of the enzyme [49]. An important result was obtained when DMS was added to crystals of the oxidized enzyme and shown to bind to an oxo-group [48], thereby generating DMSO and corroborating the direct nature of oxygen atom transfer accomplished by this enzyme [43].

The possible substitution of molybdenum by tungsten in DMSOR of *R. capsulatus* has been explored, together with the nature and catalytic activity of the resultant metalloprotein [26,27]. The impact of Na_2WO_4 on the growth of *R. capsulatus* strain H123 has been investigated by monitoring the cell density over a period of several days (Fig. 1). Too high a concentration of Na_2WO_4 prevented all cell growth, but the cells grew well with $3\,\mu M$ Na_2WO_4. *However, the presence of a low concentration (6 nM) of Na_2MoO_4 was found to be essential for cell growth.* The metal content of the purified enzyme was measured by inductively coupled plasma mass spectrometry and the W/Mo ratio found to be more than $99:1$ in each case. In a separate experiment, when the growth medium contained equal quantities of the two metals ($3\,\mu M$ Na_2MoO_4 plus $3\,\mu M$ Na_2WO_4), the Mo/W ratio in the isolated DMSOR was about $1.5:1$. Thus, the processes of metal uptake, delivery, and/or incorporation lead to a slight but significant preference for the binding of molybdenum over tungsten in DMSOR of *R. capsulatus*.

The structure of oxidized W-DMSOR, with respect to the polypeptide, the two MGDs, *and* the nature of the tungsten center, in each case

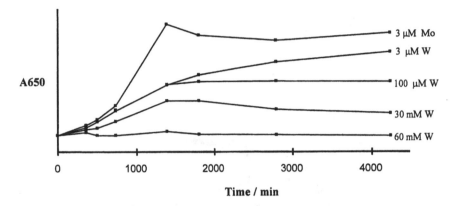

FIG. 1. Growth curves for *Rhodobacter capsulatus* grown phototrophically with various concentrations of Na$_2$WO$_4$; all media contained Na$_2$MoO$_4$ (6 nM), kanamycin (25 mg mL^{-1}), and DMSO (45 mM) [26].

was found to be very similar to the corresponding details of Mo-DMSOR [45–49]. However, a complete interpretation of the nature of the tungsten site was complicated by poorly defined positive difference density in the electron density distribution about the metal. Nevertheless, the tungsten in W-DMSOR is ligated by four dithiolene sulfurs—two from each of the MGDs (P & Q) with W-S distances in the range 2.4–2.5 Å—and the Oγ of serine147 with a W-O distance of 1.9 Å (Fig. 2). Whether, in addition to the serine oxygen and the four dithiolene sulfur atoms, one or two oxo groups are bound to the molybdenum of oxidized Mo-DMSOR has been debated [45–49]. In the case of W-DMSOR, it is not clear whether there is a third oxygen ligand (equivalent to O1 in Mo-DMSOR [47]) since significant positive difference density was observed close to one of the sulfur atoms of the P-MGD ligand. However, this density may be attributed to ripples in the electron density due to series termination errors from the heavy atom [50] or to a degree of heterogeneity in the tungsten environment. Nevertheless, it is clear that the coordination sphere of the metal in W-DMSOR is very similar to that in Mo-DMSOR, and this is corroborated by X-ray absorption spectroscopic studies of these centers [26,51].

W-DMSOR, as isolated, exhibited no EPR signal. Incubation of W-DMSOR with dithionite for 1 to 10 min produces an EPR signal that corresponds to $\leq 15 \pm 2\%$ of the tungsten content. Further incubation

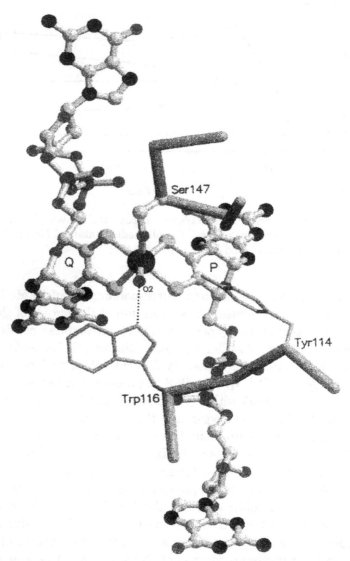

FIG. 2.   The nature of the tungsten center of W-DMSOR from *Rhodobacter capsulatus*; the tungsten is bound to the dithiolene group of two molybdopterin guanine dinucleotides (P and Q), the oxygen of serine147, and an oxo group [26].

with dithionite extinguishes the EPR signal. The addition of $K_3[Fe(CN)_6]$ to a solution of as-isolated W-DMSOR elicits no detectable EPR signal at 110 K. These results are consistent with the majority of the tungsten centers in W-DMSOR, as isolated, being present as W^{VI} and dithionite reduction producing W^V, then W^{IV}. The W^V EPR signal is similar to those reported for W^V centers in enzymes [6,52] and involves an $I = \frac{1}{2}$, W^V-^{1}H superhyperfine interaction, very suggestive of a W^V-OH moiety. The rhombicity $[(g_1 - g_2)/(g_1 - g_3)]$ and the orientation of the W^V g values are very similar to those of the Mo^V "high-g split" signal of Mo-DMSOR [53,54]. Furthermore, both the magnitude and orientation of the metal-proton superhyperfine coupling tensors of the two centers are very similar. These observations indicate that the Mo^V and the W^V centers of the DMSO reductase from *R. capsulatus* experience essentially the same ligand field.

The UV/visible spectra of Mo-DMSOR and W-DMSOR as isolated are shown in Fig. 3. The spectra are distinctive and that of Mo-DMSOR is in good agreement with spectra reported previously [42,48,53]; that of

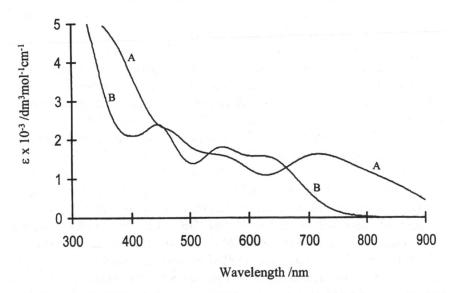

FIG. 3.   The UV/Vis absorption spectra, recorded at room temperature for (A) Mo-DMSOR (5 mg mL^{-1} in 50 mM Hepes, pH 7.5) and (B) W-DMSOR (5 mg mL^{-1} in 50 mM Tris pH 8.0) [26].

W-DMSOR has a similar profile to its molybdenum counterpart with the λ_{max} values blue-shifted by about 150 nm (i.e., 3000–5000 cm^{-1}). The absorptions are considered to arise from ligand-to-metal charge transfer transitions, from sulfur-based orbitals to a d^0 metal center. Consistent with this view, the blue shift observed from molybdenum to tungsten is similar to that (about 4350 cm^{-1}) for the two lowest energy transitions of the [MS$_4$]$^{2-}$ (M = Mo or W) anions [55]. Furthermore, this shift is in the sense (see Sec. 2) that it is more difficult to reduce a tungsten center than its molybdenum counterpart.

The redox potentials of W-DMSOR reductase have been measured by dye-mediated EPR potentiometric titrations and compared with the redox potentials of Mo-DMSOR. The results obtained for the latter enzyme are in good agreement with those reported by Bastian et al. [53], with midpoint potentials (vs. SHE) of +141 mV for the Mo^{VI}/Mo^V couple and +200 mV for the Mo^V/Mo^{IV} couple, at pH 7.0 and 173 K. The values for W-DMSOR, under the corresponding conditions, are −203 mV for the W^{VI}/W^V couple and −105 mV for the W^V/W^{IV} couple [56]. Thus, as compared with Mo-DMSOR, W-DMSOR favors the adoption of the higher oxidation states by about 325 mV. However, both systems are poised to undergo a *two*-electron redox process, as is required for oxygen atom transfer.

The ability of W-DMSOR from *R. capsulatus* to catalyze the reduction of DMSO has been examined and compared with that of Mo-DMSOR. This was accomplished as previously described for Mo-DMSOR [57], using dithionite-reduced methylviologen (MV) as the electron donor. The steady state rate of MV oxidation is $52.8 \pm 1.6\,\mathrm{s}^{-1}$ for Mo-DMSOR but $936 \pm 20\,\mathrm{s}^{-1}$ for W-DMSOR. In the DMS/phenazine ethosulfate (PES)/2,6-dichlorophenolindophenol (DCIP) oxidoreduction assay [57], the activity of Mo-DMSOR and W-DMSOR is $8.5 \pm 0.1\,\mathrm{s}^{-1}$ and $\leq 0.05\,\mathrm{s}^{-1}$ molecules of DCIP reduced (= DMS oxidized), respectively. The latter value could correspond to the presence of a small amount of Mo-DMSOR in the W-DMSOR. Thus, W-DMSOR reduces DMSO about 17 times faster than Mo-DMSOR. This result is consistent with Sung and Holm's analysis [58] of information obtained for molybdenum and tungsten isoenzymes, that oxygen atom transfer from the substrate to the reduced metal center is faster for tungsten than molybdenum. However, and in stark contrast to Mo-DMSOR, W-DMSOR does not appear to catalyze the oxidation of DMS [26].

However, the in vitro studies are only a guide to the in vivo situation. The reduction of DMSOR by ubiquinol in vivo is mediated by the pentaheme c-type cytochrome DorC, which has midpoint potentials of −34, −128, −184, −185 and −276 mV (vs. SHE) [59]. Therefore, DorC should be capable of reducing oxidized W-DMSOR, allowing this protein to turn over inside a cell. As an alternative to spectrophotometric assays of the enzyme activity (by use of a redox-coupled dye indicator on isolated enzymes in the presence of nonphysiological reductants), ^{1}H NMR spectroscopy has been used to follow the activity of both Mo-DMSOR and W-DMSOR in vivo [27]. This procedure employed the approach of King et al. [60] and Richardson et al. [61], who monitored the turnover of DMSO and TMAO by R. capsulatus cells. In the investigations of Stewart et al. [27], the molybdenum- and tungsten-grown cells contained the same amount of DMSOR, suggesting that R. capsulatus uses the same pathway to incorporate molybdenum or tungsten from the corresponding $[MoO_4]^{2-}$ anion. This result contrasts with that of Santini et al. [38], who observed that the amount of tungsten-substituted TMAO reductase produced from E. coli by genetic manipulation of the pathway for metal uptake was about 15% of the level found with molybdenum.

The ^{1}H NMR studies accomplished [27] assessed the ability of Mo-DMSOR and W-DMSOR to catalyze oxo transfer from DMSO or TMAO in intact R. capsulatus cells. Each of the four species of principal interest exhibits a single ^{1}H NMR spectrum with a distinctive chemical shift: 2.73 ppm (DMSO), 2.11 ppm (DMS), 3.27 ppm (TMAO), and 2.90 ppm (TMA) with the DMS signal slightly overlapping the propionate methine signal (2.19 ppm). In each experiment, the initial ^{1}H NMR spectrum was dominated by the DMSO (or TMAO) singlet, which decreased steadily in amplitude over time with concomitant growth of the DMS (or TMA) signal. As noted by King et al. [60], the reduction of both DMSO and TMAO produced a second product (X) in parallel with the main reduction; X for TMAO reduction is dimethylamine (DMA). The time course for the reduction of DMSO and TMAO with turnover of the R. capsulatus cells is readily followed by ^{1}H NMR spectroscopy. Plots of the concentration of substrate or product vs. time (Fig. 4) were essentially linear, indicating a zeroth-order process, and the rates of reactions were readily obtained. DMSO was reduced at about 21% of the rate of TMAO reduction for the molybdenum-grown

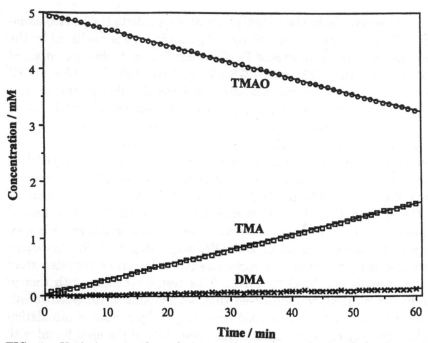

FIG. 4. Variations in the relative concentrations of trimethylamine oxide (TMAO), trimethylamine (TMA), and dimethylamine (DMA) with time, as observed by ^{1}H NMR spectra, recorded at 1-min intervals for the reduction of TMAO by *Rhodobacter capsulatus* cells in a medium containing Na_2WO_4. Solid lines show the results of linear regression [27].

cells and about 8% for the tungsten-grown cells. The tungsten-grown cells reduce both DMSO and TMAO at a significantly slower rate than the molybdenum-grown cells; the relative rates are about 9% and 22%, respectively. Nevertheless, the tungsten-grown cells are capable of turnover with DMSO or TMAO as the terminal electron acceptor.

Tungsten-substituted *R. capsulatus* cells turn over at a slower rate than their molybdenum counterparts, even though the reduction of the substrate in the isolated enzyme proceeds at a significantly faster (about 17 times) rate for W-DMSOR than Mo-DMSOR [26]. Thus, reduction of the oxidized state (M^{VI}) to the reduced state (M^{IV}) appears to contain the rate-determining step of the catalytic cycle employed by these enzymes (Fig. 5). As noted above, reduction of Mo^{VI} to Mo^{IV} in DMSOR involves the loss of an oxo group; a similar process is expected to occur for the corresponding step in the catalytic cycle of W-DMSOR. This reduction is

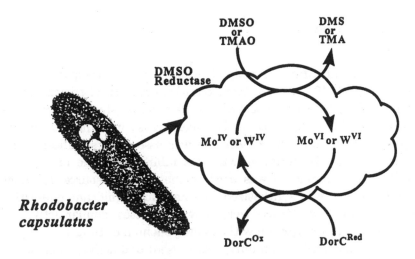

FIG. 5.   Diagrammatic representation of the catalytic cycle of DMSO reductase in *Rhodobacter capsulatus* with either molybdenum or tungsten at the active site and DMSO or TMAO as the electron acceptor; reduction of the oxidized enzyme by ubiquinol is mediated by the pentaheme *c*-type cytochrome DorC [27].

expected to proceed at a slower rate for W-DMSOR than for Mo-DMSOR because:

1. The redox potentials observed for these enzymes [26] (see above) show that this W^{VI} center is more difficult to reduce than its Mo^{VI} counterpart; and

2. As argued by Tucci et al. [62], more energy is required to deform a W-oxo bond than an equivalent Mo-oxo bond.

   This characterization of W-DMSOR is the first demonstration that tungsten can directly replace molybdenum in an MPT-dependent enzyme and produce an active enzyme. As-isolated W-DMSOR is in the W^{VI} oxidation state and the enzyme is capable of accessing the same range of oxidation states (VI, V, and IV) as Mo-DMSOR [54,63]. The UV/Vis and EPR spectra recorded for W-DMSOR indicate that the electronic structure of the W^{VI} and W^{V} centers are similar to their molybdenum counterparts. MPT-dependent tungsten enzymes catalyze oxygen atom transfer at a carbon site [6,52], but the studies of Stewart et al. [27] and Buc et al. [39] show that reduction at a sulfur center can also be catalyzed by a tungsten-containing enzyme.

5.4. Nitrogenase

Tungsten can substitute for molybdenum in nitrogenase and, contrary to initial reports [9], the substrate reduction pattern of the enzyme is unchanged [10]. The first report of the effect of tungsten on nitrogenase showed that this element could be incorporated into the nitrogenase complex of *Azotobacter vinelandii* and was associated specifically with component I [9]. Bacterial growth was inhibited by the presence of $[WO_4]^{2-}$, unless ammonia was used as the nitrogen source, in which case nitrogenase was synthesized in small but significant quantities. The tungsten-grown extracts were shown to possess similar activity to extracts from cultures grown without molybdenum or tungsten; component II, which contains iron only, was present and active, whereas the activity of component I was low. A subsequent report examined the effect of tungsten on nitrogenase activity in intact cells of *Rhodospirillum rubrum* [64]. These results suggested a true nitrogenase activity in the tungsten-grown cells and provided the first instance of active substitution, although the level of enzyme activity was lower than that of the molybdenum-grown cells.

It was reported in 1974 [65] that *A. vinelandii* was unable to grow on tungsten-containing medium where N_2 was the sole source of nitrogen. However, the cells were capable of growth in the absence of either molybdenum or tungsten, presumably due to low-level molybdenum contamination, the uptake of which was competitively inhibited by the presence of tungsten. Addition of tungsten or molybdenum to cells previously grown without these metals induced the synthesis of nitrogenase. Components I and II were assayed independently and component II exhibited a consistently high activity. The activity of component I was low when tungsten or no metal was added to the cells and high when molybdenum was added, demonstrating that tungsten cannot replace molybdenum in the active protein.

The amount of component I present was found to be 10 times greater than the amount predicted from the specific activity. Thus, while tungsten can stimulate synthesis of component I (comparable experiments performed with vanadium in place of tungsten produced very low levels of component I), the protein produced is inactive. The inactive protein might contain tungsten, but because activity was restored upon the addition of molybdate, apoprotein seemed to be a more likely product [65].

Investigations of wild-type A. *vinelandii* and a mutant tungsten-resistant strain revealed that both produced an active FeMo protein that contained molybdenum and tungsten in an approximately 1:1 ratio [66]. The concentration of molybdenum in the growth medium was about 0.01% of the tungsten concentration. Thus, the 1:1 ratio observed in the protein demonstrated that the bacteria have a significantly higher affinity for molybdenum than tungsten. The EPR spectrum of component I of the nitrogenases in the presence of dithionite contains a signal that is characteristic of the iron-molybdenum cofactor (FeMoco). The EPR spectrum of the protein sample that contains both molybdenum and tungsten possesses a signal characteristic of FeMoco and another signal assigned to an analogous iron-tungsten cofactor, FeWco. The intensity of only the molybdenum-derived signal was diminished by enzyme activity; thus, it would appear that turnover of the enzyme does not lead to reduction of FeWco [66]. The addition of purified FeMoco to the mixed enzyme had no effect on its activity. This result supports the suggestion (see above) that the enzyme isolated by Nagatani and Brill [65], the activity of which *did* increase on such an addition, is probably apoprotein rather than a tungsten-substituted protein.

As the role of molybdenum in the nitrogenases has yet to be determined (see Chapter 3), it is difficult to comment on the changes that are introduced by the change from FeMoco to FeWco.

6.   MOLYBDENUM-SUBSTITUTED TUNGSTEN ACETYLENE HYDRATASE

Acetylene hydratase (AH) from *Pelobacter acetylenicus* has been characterized as an iron-sulfur, tungsten enzyme, as purified from cells transferred repeatedly into molybdenum-free media [67]. This is an unusual enzyme with no sequence homology to any other known molybdenum or tungsten enzyme. Also, the nature of the catalytic action is quite distinct in that the enzyme apparently catalyzes a non-redox reaction, the hydration of acetylene to acetaldehyde. AH has been purified as a monomer about 83 kDa in weight that contains one [4Fe-4S] cluster and one tungsten atom coordinated by two MGD ligands [67]. A fully active AH, with identical molecular mass and N-terminal sequence to the tungsten enzyme, has been purified from cells grown in the absence of tungsten [68]. The enzyme contains molybdenum and exhibits different redox properties from those of the tungsten enzyme, being inactive under the strongly reducing conditions necessary for activity of the tungsten enzyme. However, the molybdenum enzyme was shown to have some activity upon careful reduction, which was lost upon further reduction.

It remains unclear whether AH is a genuine tungsten enzyme—and hence the above study would represent a unique example of a molybdenum-substituted tungsten enzyme—or rather a versatile enzyme capable of using either metal. Given that the tungsten content appears to be dependent on multiple transfers into a medium containing $[WO_4]^{2-}$, native AH may be a molybdenum enzyme. Clearly, further studies are required to establish the physiologically preferred metal.

It should also be noted that *Pyrococcus furiosus* has been grown on a medium containing molybdenum in an attempt to replace the tungsten of the aldehyde ferredoxin oxidoreductase (AOR), formaldehyde ferredoxin oxidoreductase (FOR), and glyceraldehyde-3-phosphate ferredoxin oxidoreductase (GAPOR) [69]. However, molybdenum does not appear to become incorporated into any of these enzymes, and the resultant low activities correlate with the residual tungsten content. Thus, *P. furiosus*, unlike many other organisms that use tungsten, does not seem to be capable of producing catalytically active molybdenum isoenzymes when molybdenum and not tungsten is provided in the nutrient medium.

7. CONCLUSIONS

We are still in the early stages of understanding the roles of molybdenum and tungsten in the MPT-dependent enzymes and the nitrogenases. We need to explain why tungsten MPT-dependent enzymes often occur in thermophilic organisms and generally catalyze reactions that have a low redox potential [70], whereas molybdenum MPT-dependent enzymes operate at a higher redox potential and are employed by a wide range of organisms that exist at ambient temperature [2].

The challenges presented by these questions clearly include the development of an understanding of the differences in the nature and the rate of the processes whereby these two chemically very similar metals are recognized, incorporated, and used by biological systems. The information obtained to date suggests that the differences between the behavior of molybdenum and tungsten will be subtle and a satisfactory interpretation will require detailed structural, spectroscopic, redox, and kinetic information. In these respects, the Mo- and W-DMSOR enzymes represent the best characterized systems [26,27]. Furthermore, to develop a clear perspective of the similarities and differences of these molybdenum and tungsten isoenzymes, it was necessary to investigate their properties as isolated entities and in vivo.

ABBREVIATIONS

AH	acetylene hydratase
AOR	aldehyde ferredoxin oxidoreductase
DCIP	2,6-dichlorophenolindophenol
DMA	dimethylamine
DMS	dimethyl sulfide
DMSO	dimethyl sulfoxide
DMSOR	dimethyl sulfoxide reductase
EPR	electron paramagnetic resonance
Et	ethyl
FeMoco	iron-molybdenum cofactor
FeWco	iron-tungsten cofactor
FMDH	formylmethanofuran dehydrogenase
FOR	formaldehyde ferredoxin oxidoreductase
GAPOR	glyceraldehyde-3-phosphate ferredoxin oxidoreductase
Hepes	N-(2-hydroxyethyl)piperazine-N'-[2-ethanesulfonic acid)
Me	methyl
MGD	molybdopterin guanine dinucleotide
mnt	1,2-dicyanoethylenedithiolate
MPT	molybdopterin

MV	methylviologen
NMR	nuclear magnetic resonance
PES	phenazine ethosulfate
Ph	phenyl
SHE	standard hydrogen electrode
TMA	trimethylamine
TMAO	trimethylamine oxide
TMAOR	trimethylamine oxide reductase
Tris	tris(hydroxymethyl)aminomethane
UV	ultraviolet

REFERENCES

1. C. G. Young and A. G. Wedd, in *Molybdenum: Molybopterin-Containing Enzymes, Encyclopedia of Inorganic Chemistry*, Vol. 4 (R. B. King, ed.), John Wiley & Sons, New York, 1994, pp. 2330–2346.

2. R. Hille, *Chem. Rev.*, *96*, 2757–2816 (1996) and references therein.

3. C. D. Garner, R. Banham, S. J. Cooper, E. S. Davies, and L. J. Stewart, in *Enzymes and Proteins Containing Molybdenum or Tungsten* (I. Bertini, A. Sigel, and H. Sigel, eds.), *Handbook on Metalloproteins*, Marcel Dekker, New York, 2001, pp. 1023–1090.

4. J. R. Andreesen and L. G. Ljungdahl, *J. Bacteriol.*, *116*, 867–873 (1973); J. R. Andreesen and E. El Ghazzawi, *Arch. Microbiol.*, *96*, 103–110 (1974).

5. I. Yamamoto, T. Saiki, S.-M. Liu, and L. G. Ljungdahl, *J. Biol. Chem.*, *258*, 1826–1832 (1983).

6. M. K. Johnson, D. C. Rees, and M. W. W. Adams, *Chem. Rev.*, *96*, 2817–2839 (1996) and references therein.

7. M. K. Chan, S. Mukund, A. Kletzin, M. W. W. Adams, and D. C. Rees, *Science, 267*, 1463–1469 (1995).

8. B. M. Rosner and B. Schink, *J. Bacteriol.*, *177*, 5767–5772 (1995).

9. H. H. Nagatani and W. J. Brill, *Biochim. Biophys. Acta, 362,* 160 (1974).

10. B. J. Hales and E. E. Case, *J. Biol. Chem., 262,* 16205–16211 (1987).

11. F. A. Cotton, G. Wilkinson, C. A. Murillo, and M. Bochmann, *Advanced Inorganic Chemistry,* 6th ed., John Wiley & Sons, New York, 1999, pp. 920–974.

12. A. Kletzin and M. W. W. Adams, *FEMS Microbiol. Rev., 18,* 5–63 (1996).

13. R. H. Holm and E. D. Simhon, in *Molybdenum Enzymes* (T. G. Spiro, ed.), *Metal Ions in Biology,* John Wiley & Sons, New York, 7, 1985, pp. 1–87.

14. C. D. Garner and S. Bristow, in *Molybdenum Enzymes* (T. G. Spiro, ed.), *Metal Ions in Biology,* John Wiley & Sons, New York, 7, 1985, pp. 343–410.

15. D. M. Lawson, C. E. Williams, D. J. White, A. P. Choay, L. A. Mitchenhall, and R. N. Pau, *J. Chem. Soc., Dalton Trans.,* 3981–3984 (1997); D. M. Lawson, C. E. M. Williams, L. A. Mitchenhall, and R. N. Pau, *Structure, 6,* 1529–1539 (1998).

16. A. G. Sykes, G. J. Leigh, R. L. Richards, C. D. Garner, J. M. Charnock, and E. I. Stiefel, in *Comprehensive Coordination Chemistry,* Vol. 3 (G. Wilkinson, R. D. Gillard, and J. A. McCleverty, eds.), Pergamon Press, Oxford, 1987, pp. 1229–1444.

17. Z. Dori, in *Comprehensive Coordination Chemistry,* Vol. 3 (G. Wilkinson, R. D. Gillard, and J. A. McCleverty, eds.), Pergamon Press, Oxford, 1987, pp. 973–1022.

18. W. H. Mills and R. E. D. Clark, *J. Chem. Soc.,* 175–181 (1936).

19. J. A. McCleverty, *Prog. Inorg. Chem., 10,* 49–221 (1968).

20. E. S. Davies, R. L. Beddoes, D. Collison, A. Dinsmore, A. Docrat, J. A. Joule, C. R. Wilson, and C. D. Garner, *J. Chem. Soc., Dalton Trans.,* 3985–3996 (1997).

21. E. S. Davies, G. M. Aston, R. L. Beddoes, D. Collison, A. Dinsmore, A. Docrat, J. A. Joule, C. R. Wilson, and C. D. Garner, *J. Chem. Soc., Dalton Trans.,* 3647–3656 (1998).

22. S. K. Das, P. K. Chaudhury, D. Biswas, and S. Sarker, *J. Am. Chem. Soc., 116,* 9061–9070 (1994); S. K. Das, D. Biswas, R.

Maiti, and S. Sarker, *J. Am. Chem. Soc.*, *118*, 1387–1397 (1996); J. Yadav, S. K. Das, and S. Sarker, *J. Am. Chem. Soc.*, *119*, 4315–4316 (1997).

23. H. Oku, N. Ueyama, and A. Nakamura, *Bull. Chem. Soc. Japan*, *69*, 3139–3150 (1996); N. Ueyama, H. Oku, M. Kondo, T. O. Kamura, N. Yoshinago, and A. Nakamura, *Inorg. Chem.*, *35*, 643–650 (1996).

24. G. C. Tucci, J. P. Donahue and R. H. Holm, *Inorg. Chem.*, *37*, 1602–1608 (1998).

25. N. N. Greenwood and A. Earnshaw, *The Chemistry of the Elements*, Pergamon Press, Oxford, UK, 1984, p. 1170.

26. L. J. Stewart, S. Bailey, B. Bennett, J. M. Charnock, C. D. Garner, and A. S. McAlpine, *J. Mol. Biol.*, *299*, 593–600 (2000).

27. L. J. Stewart, S. Bailey, D. Collison, G. A. Morris, I. Preece, and C. D. Garner, *ChemBioChem.*, *2*, 703–706 (2001).

28. A. Paneque, J. M. Vega, J. Cádenas, J. Herrera, P. J. Aparicio, and M. Losada, *Plant Cell Physiol.*, *13*, 175–178 (1972).

29. K. A. Burke, K. Calder, and J. Lascelles, *Arch. Microbiol.*, *126*, 155–159 (1980).

30. R. A. Schmitz, S. P. J. Albracht, and R. K. Thauer, *Eur. J. Biochem.*, *209*, 1013–1018 (1992).

31. P. A. Bertram, R. A. Schmitz, D. Linder, and R. K. Thauer, *Arch. Microbiol.* *161*, 220–228 (1994).

32. R. A. Schmitz, S. P. J. Albracht, and R. K. Thauer, *FEBS Lett.*, *309*, 78–81 (1992).

33. J. L. Johnson and K. V. Rajagopalan, *J. Biol. Chem.*, *249*, 859–866 (1974).

34. J. L. Johnson, W. R. Waud, H. J. Cohen, and K. V. Rajagopalan, *J. Biol. Chem.*, *249*, 5056–5061 (1974).

35. J. L. Johnson, H. J. Cohen, and K. V. Rajagopalan, *J. Biol. Chem.*, *249*, 5046–5055 (1974).

36. J. L. Johnson and K. V. Rajagopalan, *J. Biol. Chem.*, *251*, 5505–5511 (1976).

37. E. I. Stiefel, *Proc. Natl. Acad. Sci.*, *70*, 988–992 (1973).

38. C.-L. Santini, B. Ize, A. Chanal, M. Müller, G. Giordano, and L.-F. Wu, *EMBO J.*, *17*, 101–112 (1998).

39. J. Buc, C.-L. Santini, R. Giordano, M. Czjek, L.-F. Wu, and G. Giordano, *Mol. Microbiol., 32* 159–168 (1999).

40. R. A. Rothery, J. L. Simala-Grant, J. L. Johnson, K. V. Rajagopalan, and J. H. Weiner, *J. Bacteriol, 177,* 2057–2063 (1995).

41. N. R. Bastian, C. J. Kay, M. J. Barber, and K. V. Rajagopalan, *J. Biol. Chem., 266,* 45–51 (1991).

42. A. G. McEwan, S. J. Ferguson, and J. B. Jackson, *Biochem. J., 274,* 305–307 (1991).

43. B. E. Schultz, R. Hille, and R. H. Holm, *J. Am. Chem. Soc., 117,* 827–828 (1995).

44. T. Satoh and F. N. Kurihara, *J. Biochem (Tokyo), 102,* 191–197 (1987).

45. H. Schindelin, C. Kisker, J. Hilton, K. V. Rajagopalan, and D. C. Rees, *Science, 272,* 1615–1621 (1996).

46. F. Schneider, J. Löwe, R. Huber, H. Schindelin, C. Kisker, and J. Knäblein, *J. Mol. Biol., 263,* 53–69 (1996).

47. A. S. McAlpine, A. G. McEwan, A. L. Shaw, and S. Bailey, *JBIC, 2,* 690–701 (1997).

48. A. S. McAlpine, A. G. McEwan, and S. Bailey, *J. Mol. Biol., 275,* 613–623 (1998).

49. H.-K. Li, C. Temple, K. V. Rajagopalan, and H. Schindelin, *J. Am. Chem. Soc., 122,* 7673–7680 (2000).

50. H. Schindelin, C. Kisker, and D. C. Rees, *JBIC, 2,* 773–781 (1997).

51. G. N. George, J. Hilton, and K. V. Rajagopalan, *J. Am. Chem. Soc., 118,* 1113–1117 (1996); G. N. George, *JBIC, 2,* 790–796 (1997); G. N. George, J. Hilton, C. Temple, R. C. Prince, and K. V. Rajagopalan, *J. Am. Chem. Soc., 121,* 1256–1266 (1999).

52. W. R. Hagen and A. F. Arendson, *Struct. Bond., 90,* 161–192 (1998).

53. N. R. Bastian, C. J. Kay, M. J. Barber, and K. V. Rajagopalan, *J. Biol. Chem., 266,* 45–51 (1991).

54. B. Bennett, N. Benson, A. G. McEwan, and R. C. Bray, *Eur. J. Biochem., 225,* 321–331 (1994).

55. E. Diemann and A. Müller, *Coord. Chem. Rev., 10,* 79–122 (1973).

56. P. L. Hagedoorn, W. R. Hagen, L. J. Stewart, A. Docrat, S. Bailey, and C. D. Garner, to be published.

57. B. Adams, A. T. Smith, S. Bailey, A. G. McEwan, and R. C. Bray, *Biochemistry*, *38*, 8501–8511 (1999).

58. K.-M. Sung and R. H. Holm, *J. Am. Chem. Soc.*, *123*, 1931–1943 (2001).

59. A. L. Shaw, A. Hochkoeppler, P. Bonora, D. Zannoni, G. R. Hanson, and A. G. McEwan, *J. Biol. Chem.*, *274*, 9911–9914 (1999).

60. G. F. King, D. J. Richardson, J. B. Jackson, and S. J. Ferguson, *Arch. Microbiol.*, *149*, 47–51 (1987).

61. D. J. Richardson, G. F. King, D. J. Kelly, A. G. McEwan, S. J. Ferguson, and J. B. Jackson, *Arch. Microbiol.*, *150*, 131–137 (1988).

62. G. C. Tucci, J. P. Donahue, and R. H. Holm, *Inorg. Chem.*, *37*, 1602–1608 (1998) and references therein.

63. P. E. Baugh, C. D. Garner, J. M. Charnock, D. Collison, E. S. Davies, A. S. McAlpine, S. Bailey, I. Lane, G. R. Hanson, and A. G. McEwan, *JBIC*, *2*, 634–643 (1997).

64. H. Paschinger, *Arch. Microbiol.*, *101*, 379–389 (1974).

65. H. H. Nagatani and W. J. Brill, *Biochim. Biophys. Acta*, *362*, 160–166 (1974)

66. B. J. Hales and E. E. Case, *J. Biol. Chem.*, *262*, 16205–16211 (1987).

67. R. U. Meckenstock, R. Kriger, S. Ensign, P. M. H. Kroneck, and B. Schink, *Eur. J. Biochem.*, *264*, 176–182 (1999).

68. C. Kisker, H. Schindelin, D. Baas, J. Rétey, R. U. Meckenstock, R. Kriger, and P. M. H. Kroneck, *FEMS Microbiol. Rev.*, *22*, 503–521 (1999).

69. S. Mukund and M. W. W. Adams, *J. Bacteriol.*, *178*, 163–167 (1996).

70. A. Kletzin and M. W. W. Adams, *FEMS Microbiol. Rev.*, *18*, 5–63 (1996).

21
Molybdenum Metabolism and Requirements in Humans

Judith R. Turnlund

USDA/ARS/Western Human Nutrition Research Center, University of
California at Davis, One Shields Avenue, Davis, CA 95616, USA

1. INTRODUCTION

Molybdenum has been known to be present in plants since 1900 and by 1930 was shown to have a role in nitrogen metabolism [1]. In 1953 it was found to have a role in animal nutrition as a constituent of xanthine oxidase [2], and in 1954 it was found to be a constituent of aldehyde oxidase [3]. It was not known to be essential for humans in 1964 when the first study of the metabolic fate of molybdenum, using the radio-isotope ^{99}Mo, was conducted in terminally ill patients [4]. In 1967 a genetic deficiency of sulfite oxidase was reported in a child [5] and in 1971 it was discovered that sulfite oxidase was a molybdenum enzyme [6], establishing a role for molybdenum in human health.

2. ESSENTIALITY OF MOLYBDENUM IN HUMANS

The essentiality of molybdenum for humans was established relatively recently, based on its presence in the molybdenum enzyme sulfite oxidase [6–8]. Molybdenum-containing enzymes occur in plants, microorganisms, and animals. Of the 11 or more of these enzymes that occur in nature [1], three are found in animals. These enzymes—sulfite oxidase, xanthine oxidase, and aldehyde oxidase—all contain the molybdenum cofactor, a complex of molybdenum, and an organic component, molybdopterin [9]. A detailed discussion of these enzymes and the molybdopterin cofactor can be found in other chapters in this volume and in a review by Johnson [10]. Of the three enzymes, sulfite oxidase is critical to human health. More than 100 patients have been identified who lack this functioning enzyme [10]. These patients either have a defect in the gene coding for the sulfite oxidase enzyme or have a genetic deficiency in the molybdenum cofactor.

3. DIETARY INTAKE AND BIOAVAILABILITY OF MOLYBDENUM

3.1. Dietary Molybdenum

Estimates of the dietary intake of molybdenum vary widely. This is due in part to analytical difficulties in the determination of molybdenum and also because the molybdenum contents of food crops vary with the amount of molybdenum in the soil [11].

There is no simple analytical method for measuring molybdenum; therefore, limited data are available on the molybdenum content of foods and diets. Older estimates of dietary intake in the United States are considerably higher than more recent estimates. In 1970, the usual intake was estimated to be 0.1–0.5 mg/day [12]. A 1980 report estimated the average intake ranged from 120 to 240 μg/day, with an average of 180 μg/day [13]. A still more recent study reported the usual intake in the United states is 76 μg/day for women and 109 μg/day for men [14].

Estimates of dietary molybdenum intake from countries other than the United States vary widely. Again, this is likely due to a combination of analytical techniques and variations in the molybdenum content of the soil. Data from several countries have been summarized, and reported intakes are as high as 1 mg/day [15]. Recent estimates agree with the lower, more recent estimates from the United States [15,16].

3.2. Food Sources of Molybdenum

Good sources of molybdenum include legumes, grain products, and nuts. Animal products, fruits, and many vegetables tend to be low in molybdenum [14,17].

Human milk has much less molybdenum than cow's milk and infant formulas. Intake of infants fed human milk and formulas has been measured in a number of studies [18–22]. It ranges from 1 to 15 μg/L and decreases rapidly during the first month postpartum.

3.3. Dietary Recommendations

The first dietary molybdenum recommendations for the United states were made in 1980 in the form of an Estimated Safe and Adequate Daily Dietary Intakes (ESADDI) [23]. Inadequate information was available at that time to provide a Recommended Dietary Allowance (RDA) and an ESADDI of 0.15–0.5 mg/day was based on early balance studies and dietary intake data available then. The dietary recommendation was revised downward in 1989 to 75–250 µg/day, based on more recent estimates of dietary intake [24]. Considerably more is now known about molybdenum metabolism based on controlled human studies. An estimate of the minimal requirement for young men of about 25 µg/day was based on controlled studies in young men [25]. This agrees with estimates based on extrapolation from nonruminant animal studies [26]. The new Dietary Reference Intake (DRIs) [27] indicate, after including a factor for bioavailability, that an Estimated Adequate Requirement is 34 µg/day for adults and the RDA is 45 µg/day. This is about half of the usual dietary intake in the United States, discussed above.

3.4. Bioavailability

Molybdenum is very well absorbed from the diet over a broad range in dietary intakes [28]. Absorption ranged from 88% to 93% with dietary intakes from 22 to 1490 µg/day. The mechanism of absorption is not known, but the efficient absorption over a range of intakes suggests a passive process.

Little is known about the bioavailability of molybdenum from different food sources. Bioavailability studies in men and women suggest that the molybdenum in some foods is less available than the molybdenum in others. Molybdenum absorption from soy was only 57%, while absorption of molybdenum from kale and molybdenum added to the diet was similar, i.e., 88% and 86%, respectively [29]. The molybdenum concentration in soy is high, and some of it may be bound to components that are not digested and absorbed, making it unavailable.

4.   MOLYBDENUM DEFICIENCY AND TOXICITY

4.1.   Dietary Deficiency

There has been only one documented case of dietary molybdenum deficiency. This was in a patient with Crohn's disease who developed short bowel syndrome due to multiple small bowel resections. He was maintained on total parenteral nutrition (TPN) for 18 months [7,30]. In the last 6 months of TPN, he developed episodes of a syndrome that included tachycardia, tachypnea, severe bifrontal headaches, night blindness, nausea, vomiting, central scotomas, generalized edema, lethargy, disorientation, and coma. The symptoms were associated with a number of biochemical abnormalities. Blood plasma methionine levels were elevated. Urinary total sulfur and inorganic sulfate excretions were low; sulfite was elevated in the blood. Sulfite, normally not detected in the urine, was elevated in the patient's urine. Serum uric acid and urinary uric acid excretion were low. Urinary hypoxanthine and xanthine excretions were very high. The administration of commercial amino acid preparations precipitated the syndrome. Supplementation with ammonium molybdate (300 μg/day) reversed these abnormalities and the patient could tolerate TPN administration. Abumrad observed elevated molybdenum losses in ileostomy fluids of patients with Crohn's disease [7], which may have contributed to the molybdenum deficiency observed in the molybdenum-deficient patient. Many of his symptoms were very similar to those observed in those with metabolic defects in sulfite oxidase or molybdenum cofactor deficiency.

4.2.   Metabolic Defects in Molybdenum Metabolism

Isolated sulfite oxidase deficiency has been identified in at least 20 cases [10]. It usually appears shortly after birth and results in death at an early age. Symptoms include seizures, mental retardation, dislocated lenses, and brain atrophy and lesions. Biochemical abnormalities include increased sulfite and thiosulfate and decreased sulfate. S-sulfocysteine, taurine, xanthine, hypoxanthine are increased, and uric acid is low.

A more prevalent inborn error of molybdenum metabolism is molybdenum cofactor deficiency [10]. This condition affects the activities of the three molybdoenzymes due to lack of functional molybdopterin. The lack of sulfite oxidase is responsible for the serious symptoms of the disease. Individuals have been identified with xanthinuria due to lack of xanthine dehydrogenase and with deficiency in aldehyde oxidase, but they have mild symptoms, confirming that the lack of sulfite oxidase in molybdenum cofactor deficiency causes the severe illness and early death associated with the disease [10].

4.3. Molybdenum Toxicity

Molybdenum toxicity data in humans is limited. Ruminants are more sensitive to molybdenum toxicity than monogastric animals [17]. In ruminants, the amounts of copper and sulfur in the diet also affect the toxicity, but this is not the case in monogastric animals. Impaired reproduction and growth has been observed in monogastric laboratory animals [9,17].

Results of molybdenum toxicity studies in humans have produced conflicting results. High serum molybdenum and hyperuricemia accompanied by decreases in blood copper levels were observed in Armenians who consumed 10–15 mg/day of molybdenum [31]. In contrast, another study found decreased uric acid in serum and urine with high molybdenum intakes [32]. Investigators found no change in uric acid levels at 1.5 mg/day from sorghum, but increased urinary copper was observed [33]. No changes were observed in urinary copper or copper status in a controlled human study with an intake of 1.5 mg/day of molybdenum [34].

Data from humans are insufficient to establish an upper intake limit for humans. However, a Tolerable Upper Intake Level for humans of 2 mg/day was recently established by the Institute of Medicine, Food and Nutrition Board in the United States, based on extrapolation from animal studies.

5. STABLE ISOTOPE STUDIES OF MOLYBDENUM METABOLISM

5.1. Isotopic Tracers

Isotopic tracers are valuable tools for studies of mineral metabolism. Rosoff and Spencer first used a radioisotope of molybdenum to study its metabolism in humans [4]. By following an injected dose of ^{99}Mo, they found that the main route of molybdenum excretion was via the kidney and that fecal excretion was low. This was followed by other metabolic studies in mice [35].

Stable isotopes have a number of advantages over radiosotopes and recent research in humans has employed these tools [36]. They are safe and their administration results in no exposure to radioactivity. Therefore, they can be used safely in pregnant women, infants, and children. They do not decay, so that samples can be stored for long periods of time and long-term studies can be conducted. The advantages and limitations of stable isotopes for studies of mineral metabolism have been reviewed [37].

5.2. Stable Isotope Tracers

Molybdenum has seven stable isotopes, permitting multiple isotopes to be used simultaneously for studies of its metabolism. The isotopes and their natural abundances are 92–14.8%, 94–9.247%, 95–15.92%, 96–16.68%, 97–9.555%, 98–24.13%, and 100–9.634% [38]. All have abundances sufficiently low that they can be used as tracers. Thus, multiple treatments can be tested simultaneously. In addition, they can be used with stable isotopes of other trace elements to evaluate interactions. We have used three molybdenum isotopes simultaneously (96, 97, and 100), each in a different dietary component, and used a fourth isotope (94) as an isotopic diluent to quantify molybdenum [29]. An isotope can also be used for normalization of mass fractionation [39]. We have also administrated stable isotopes of zinc, copper, and iron along with the stable isotopes of molybdenum [40].

Two methods have been used for analysis of stable isotopes for most of the studies in humans. These are neutron activation analysis

(NAA) [41] and thermal ionization mass spectrometry (TIMS) [39]. Recently, methods have been developed using inductively coupled plasma mass spectrometry (ICPMS) [42]. We are now using ICPMS in most of our molybdenum-related work.

5.3. Human Studies of Molybdenum Metabolism Using Stable Isotopes

Recently, investigators have made considerable use of stable isotopes of molybdenum to study its metabolism. This work has advanced the understanding of molybdenum metabolism in humans.

Studies were conducted in our metabolic research unit to determine the minimal dietary requirement for molybdenum [25], to assess the effect of the level of molybdenum intake on molybdenum metabolism [28], to develop a kinetic model of molybdenum metabolism [43,44], and to evaluate the effect of varying food source on the bioavailability of molybdenum [29].

A study was conducted on a metabolic research unit in an attempt to determine the minimal dietary requirement for molybdenum [25]. The diet contained 22 μg of molybdenum a day for 102 days, then 467 μg a day for 18 days. To achieve the low level of 22 μg of molybdenum, it was necessary to remove the molybdenum in the salts used in the diet by recrystallizing them. Stable isotope tracers were fed and infused at both intake levels. It was possible for the men to achieve molybdenum balance after adaptation to the low molybdenum intake. Absorption was very efficient, and molybdenum turnover decreased with the low dietary intake. In addition, oral load tests were done, feeding compounds that would stress the molybdenum enzymes and following their metabolism [24,45,46]. The data on the metabolites of these compounds suggested that, after 100 days of a diet with only 22 μg/day molybdenum, there may have been small changes in the activity of sulfite oxidase and xanthine oxidase. However, no signs of molybdenum deficiency were observed. Based on these data, a minimum dietary requirement of about 25 μg/day or possibly less was suggested.

In another study, young men were fed five levels of dietary molybdenum ranging from 22 to 1490 μg/day for 24 days at each intake [28].

Stable isotopes were again fed at each dietary intake and infused at three of the intake levels. Molybdenum was very efficiently absorbed at all levels of intake, and the fraction of absorbed molybdenum and the total amount of molybdenum increased in the urine. Molybdenum turnover was slow when intake was low and increased as molybdenum intake increased. These results suggested that molybdenum retention is regulated by urinary excretion. No adverse effects were observed at any of the molybdenum intakes.

Studies on the kinetics of molybdenum metabolism are made possible with the combination of stable isotopes and SAAM/CONSAM computer modeling programs [40,47]. One stable isotope of molybdenum, [100]Mo, was fed and another, [97]Mo, was infused into an arm vein. A kinetic model was developed based on appearance and disappearance of the isotopes in urine and stools. The nine-component model was used to predict residence time, turnover rates, and compartmental masses at varying controlled intakes of dietary molybdenum [44], and during molybdenum depletion [43]. The model demonstrated that there are efficient homeostatic mechanisms of molybdenum metabolism over a range of intakes.

Cantone and coworkers [41,48] have also used stable isotopes of molybdenum for biokinetic studies of molybdenum metabolism and to estimate absorption. They used the stable isotopes [95]Mo and [96]Mo. They followed plasma enrichment to develop a four-compartment model of metabolism and to estimate absorption.

Stable isotopes were used to label plants foods intrinsically and these foods were used in human studies [29]. Soy and kale were grown hydroponically in nutrient solutions containing [97]Mo and [100]Mo. It was possible to label sufficient quantities of these plants to be used in human studies. The studies, conducted in young men and young women, demonstrated that the molybdenum in kale is absorbed as efficiently as the molybdenum added to the diet as an extrinsic label. However, the molybdenum in soy was less well absorbed. Once absorbed, the utilization of the different sources of molybdenum was similar. Other studies demonstrated that molybdenum is poorly absorbed from tea and is less well absorbed from a composite meal than from an aqueous solution [49].

6. CONCLUSIONS

Molybdenum is essential in the diet of humans, but little has been known about the metabolism of this trace element until recently. The usual intake of molybdenum is well above the required amount, so that molybdenum deficiency is unlikely. It has not been seen except under unusual circumstances. Most cases of deficiency are associated with metabolic defects in one or more molybdenum enzymes. The understanding of molybdenum metabolism in humans has increased greatly during the last 10 years. New approaches, such as use of stable isotope tracers and computer modeling, have enabled researchers to make significant advances in the understanding of the roles for molybdenum in human nutrition.

ABBREVIATIONS

DRI	Dietary Reference Intakes
ESADDI	Estimated Safe and Adequate Daily Dietary Intakes
ICPMS	inductively coupled plasma mass spectrometry
NAA	neutron activation analysis
RDA	Recommended Dietary Allowance
TIMS	thermal ionization mass spectrometry
TPN	total parenteral nutrition

REFERENCES

1. K. V. Rajagopalan, in *Biochemistry of the Essential Ultratrace Elements* (E. Frieden, ed.), Plenum Press, New York, 1984, pp. 149–174.

2. D. A. Richert and W. W. Westerfeld, *J. Biol. Chem., 203*, 915–923 (1953).

3. H. R. Mahler, B. Mackler, D. E. Green, and R. M. Bock, *J. Biol. Chem., 210*, 480 (1954).

4. B. Rosoff and H. Spencer, *Nature, 202*, 410–411 (1964).

5. S. H. Mudd, F. Irreverre, and L. Laster, *Science, 156*, 1599–1602 (1967).

6. H. J. Cohen, T. Fridovich, and K. V. Rajagopalan, *J. Biol. Chem., 246*, 374–382 (1971).

7. N. N. Abumrad, *Bul. N. Y. Acad. Med., 60*, 163–170 (1984).

8. K. V. Rajagopalan, *Nutr. Rev., 45*, 321–328 (1987).

9. K. V. Rajagopalan, *Annu. Rev. Nutr., 8*, 401–427 (1988).

10. J. L. Johnson, in *Clinical Nutrition in Health and Disease: Handbook of Nutritionally Essential Mineral Elements* (B. L. O'Dell and R. A. Sunde, eds.), Marcel Dekker, New York, 1997, pp. 413–438.

11. *Traced Elements in Human Nutrition and Health*, World Health Organization, Geneva, 1996.

12. H. A. Schroeder, J. J. Balassa, and I. H. Tipton, *J. Chron. Dis., 23*, 481–499 (1970).

13. T. A. Tsongas, R. R. Meglen, P. A. Walravens, and W. R. Chappell, *Am. J. Clin. Nutr., 33*, 1103–1107 (1980).

14. J. A. T. Pennington and J. W. Jones, *J. Am. Diet. Assoc., 87*, 1644–1650 (1987).

15. R. Van Cauwenbergh, P. Hendrix, H. Robberecht, and H. Deelstra, *Z. Lebensm. Unters. Forsch. A., 205*, 1–4 (1997).

16. S. Holzinger, M. Anke, B. Rohig, and D. Gonzales, *analyst, 123*, 447–450 (1998).

17. C. F. Mills and G. K. Davis, in *Trace Elements in Human and Animal Nutrition* (W. Mertz, ed.), Academic Press, San Diego, 1987, pp. 429–463.

18. J. K. Friel, A. C. MacDonald, C. N. Mercer, S. L. Belkhode, G. Downton, P. G. Kwa, K. Aziz, and W. L. Andres, *J. Parenteral Enteral Nutr., 23*, 155–159 (1999).

19. D. Bougle, F. Bureau, P. Foucault, J.-F. Duhamel, G. Muller, and M. Drosdowsky, *Am. J. Clin. Nutr., 48*, 652–654 (1998).

20. C. E. Casey and M. C. Neville, *Am. J. Clin. Nutr., 45*, 921–926 (1987).

21. E. Rossipal and M. Krachler, *Nutr. Res., 18*, 11–24 (1998).

22. G. H. Biego, M. Joyeax, P. Hartemann, and G. Debry, *Food Add. Contam., 15*, 775–781 (1998).

23. National Research Council, *Recommended Dietary Allowances*, National Academy of Sciences, Washington, DC, 1980.

24. National Research Council, *Recommended Dietary Allowances*, National Academy Press, Washington, DC, 1989.

25. J. R. Turnlund, W. R. Keys, G. L. Peiffer, and G. Chiang, *Am. J. Clin. Nutr.*, *61*, 1102–1109 (1995).

26. M. Anke, B. Groppel, H. Kronemann, and M. Grun, *Nutr. Res.*, *1*, S-180–S-186 (1985).

27. Institute of Medicine. *Dietary Reference Intakes: Vitamin A, Vitamin K, Arsenic, Boron, Chromium, Copper, Iodine, Iron, Manganese, Molybdenum, Nickel, Silicon, Vanadium, and Zinc.* Food and Nutrition Board. National Academy Press, Washington, DC, 2001.

28. J. R. Turnlund, W. R. Keyes, and G. L. Peiffer, *Am. J. Clin. Nutr.*, *62*, 790–796 (1995).

29. J. R. Turnlund, C. M. Weaver, S. K. Kim, W. R. Keyes, Y. Gizaw, K. H. Thompson, and G. L. Peiffer, *Am. J. Clin. Nutr.*, *69*, 1217–1223 (1999).

30. N. N. Abumrad, A. J. Schneider, D. Steel, and L. S. Rogers, *Am. J. Clin. Nutr.*, *34*, 2551–2559 (1981).

31. V. V. Kovalsky, G. A. Yarovaya, and D. M. Shmavonyan, *Zh. Obshch. Biol.*, *22*, 179–191 (1961).

32. W. R. Chappell, R. R. Meglen, R. Moure-Eraso, C. C. Solomons, T. A. Tsongas, P. A. Walravens, and P. W. Winston, *Human Health Effects of Molybdenum in Drinking Water.* EPA-600/1-79-006, 1-91. 1979. Cincinnati, OH, U.S. Environmental Protection Agency, Health Effects Research Laboratory.

33. Y. G. Deosthale and C. Gopalan, *Br. J. Nutr.*, *31*, 351–355 (1974).

34. J. R. Turnlund and W. R. Keyes, in *Trace Elements in Man and Animals 10*, (A. M. Roussel, R. A. Anderson, and A. E. Favrier, eds.), Kluwer Academic/Plenum Publishers, New York, 2000, pp. 951–953.

35. B. Rosoff and H. Spencer, *Health Physics, 25*, 173–175 (1973).

36. J. R. Turnlund, *J. Nutr., 199*, 7–14 (1989).

37. J. R. Turnlund, in *New Techniques in Nutritional Research* (R. G. Whitehead, ed.), Academic Press, San Diego, 1991, pp. 113–130.

38. N. E. Holden, R. L. Martin, and I. L. Barnes, *Pure Appl. Chem.*, *55*, 1119–1136 (1983).

39. J. R. Turnlund, W. R. Keyes, and G. L. Peiffer, *Anal. Chem.*, *65*, 1717–1722 (1993).

40. J. R. Turnlund, in *Kinetic Models of Trace Element and Mineral Metabolism During Development* (K. N. S. Subramanian and M. E. Wastney, eds.), CRC Press, Boca Raton, FL, 1995, pp. 133–143.

41. M. C. Cantone, D. de Bartolo, G. Gambarini, A. Giussani, A. Ottolenghi, and L. Pirola, *Med. Phys.*, *22*, 1293–1298 (1995).

42. E. T. Luong, R. S. Houk, and R. E. Serfass, *J. Anal. Atom. Spectrosc.*, *12*, 703–708 (1997).

43. K. H. Thompson and J. R. Turnlund, *J. Nutr.*, *126*, 963–972 (1996).

44. K. H. Thompson, K. C. Scott, and J. R. Turnlund, *J. Appl. Physiol.*, *81*, 1404–1409 (1996).

45. G. Chiang, *Studies of Biochemical Markers Indicating Molybdenum Status in Human Subjects Fed Diets Varying in Molybdenum Content*, Ph.D. thesis, University of California, Los Angeles, 1991.

46. J. R. Turnlund, W. R. Keyes, and G. L. Peiffer, in *Trace Elements in Man and Animals, Eighth International Symposium* (M. Anke, D. Meissner, and C. F. Mills, eds.), Verlag Media Touristik, Gersdorf, 1994, pp. 189–193.

47. M. Berman, W. F. Beltz, P. C. Greif, R. Chabay, and R. O. Boston, *CONSAM User's Guide*, National Institutes of Health, Bethesday, MD, 1983.

48. M. C. Cantone, D. de Bartolo, A. Giussani, A. Ottolenghi, L. Pirola, Ch. Hansen, P. Roth, and E. Werne, *Appl. Radiat. Isot.*, *48*, 333–338 (1997).

49. E. Werner, A. Giussani, U. Heinrichs, P. Roth, And D. Gertz, *Isotopes Environ. Health Stud.*, *34*, 297–301 (1998).

22

Metabolism and Toxicity of Tungsten in Humans and Animals

Florence Lagarde and Maurice Leroy

Laboratoire de Chimie Analytique et Minérale-UMR 7512, ECPM,
25 rue Becquerel F-67087 Strasbourg Cédex 02, France

1. INTRODUCTION

Tungsten has been used for many years as a component of high-speed and hot-worked steels, sintered hard metals, and cast hard as well as highly heat-resistant special alloys. The industrial, medical, and military applications of W and its alloys are still expanding which could result, within the next few years, in a substantial increase of tungsten levels in the environment. It is thus very important to understand the metabolism and evaluate the toxicity of this metal and its compounds in humans and animals.

2. METABOLISM OF TUNGSTEN

Until now, the metabolism of tungsten has been principally studied in animals and in workers occupationally exposed to hard-metal dust. The following paragraphs summarize the most significant results obtained and describe a recent model proposed for the distribution and retention of tungsten in the human body.

2.1. Data from Animal Studies

*2.1.1. Absorption, Distribution, Retention, and Excretion of
         Tungsten*

Numerous studies have been carried out on dogs, cows, sheep, goats, and rodents, using different concentration and forms of tungsten as well as different routes of administration [1–17]. Experimental conditions used in some of these studies are given in Table 1.

2.1.1.1. Absorption

Aamodt studied the fate of tungsten in beagle dogs exposed to radioactive $^{181}WO_3$ by inhalation [1]. In vivo γ-ray measurements indicate that 60% of the inhaled activity is rapidly deposited in the respiratory tract, half of it in the lower portion of the tracheobronchial compartment and in the pulmonary compartment. About 33% of the deposited activity enters the systemic circulation, chiefly during the 10 days following inhalation. The remaining activity is cleared from the lung via the ciliary escalatory system.

Different studies have also been carried out by instillating calcium tungstate in the dog lung [2,3]. In 1990, Grande et al. [2] showed that the transport of tungsten particles from the alveolus to the lymph nodes is slow and is mediated by pulmonary macrophages. W particles were also found inside alveolar macrophages in mice [4]. The absorption of tungsten following oral administration of tungsten is rapid, taking, for example, 1–2 h for dogs and rats that ingested a solution of sodium tungstate (25 or 50 mg/kg) [5]. The uptake of orally administered tungsten in beagle dogs was found to lie in the range of 57–74% when using sodium tungstate [5], whereas the level of absorption is only 25% after administration of a weakly acidic aqueous solution of tungstic oxide (4% HCl) [1]. In experiments on rats, similar results are obtained (40–92% when tungsten is administered as tungstate [5–8] and only 1% with tungstic acid [6]). Dairy cows, on their part, absorb only 25% of ^{181}W-labeled tungstate [9].

Bell and Sneed evaluated the metabolism of ^{185}W as ammonium tungstate in sheep and swine [10]. It appears that growing swine absorb at least 75% of orally administered tungsten, whereas sheep absorb 44% at the same conditions. When tungsten is introduced in the sheep directly into the abomasum, the uptake increases and reaches more than 65%.

Sodium[^{181}W]tungstate given to goats by gastric intubation was weakly absorbed (about 5% of the administered dose), but this result might be explained by the high content of roughage of the diet and absorption of tungsten to feed particles [11].

TABLE 1

Summary of Experimental Conditions Used in Studies on Absorption, Distribution, and Excretion of Tungsten in Animals

| | Animals | | | Administration | | | |
	Type	Number of specimen	Age or weight	Amount of tungsten	Form administered	Route	Duration	Ref.
Cows	Dairy Holsteins	12	3–12 yr	22 μg in a single dose	$Na_2^{187}WO_4$	Capsule	1 d	[9]
		4		58 μg in 14 equal doses	$Na_2^{181}WO_4$	Capsule	7 d	
		4		22 μg	$Na_2^{187}WO_4$	Intravenous		
	Nonlactating Holstein	1	3–12 yr	67 μg in 6 equal doses	$Na_2^{181}WO_4$	Capsule	3 d	
Goats	—	4	—	—	$Na_2^{181}WO_4$	Oral (gavage)		[11]
Dogs	Beagles	6	—	—	$^{181}WO_3$	Inhalation		[1]
					$^{181}WO_3$	Oral (gavage)		[1]
					$Na_2^{181}WO_4$	Intravenous		[12]
	Mongrels	10	Adults	3.2 g	$CaWO_4$	Instillation		[2]
	Mongrels	4	Adults	1.3 g	$CaWO_4$	Instillation		[3]
	Beagles (males)	6	11.5 ± 0.24 kg	16–32 mg/kg	Na_2WO_4	Intravenous		[5]
						Oral (gavage)		

Species	Strain / description	N	Age / weight	Compound	Dose	Route	Ref.
Rats	Wistars, males and females	70	Adults	Tungstate, pH 8	2 μg	Oral (gavage)	[8]
	Wistars, females	24	Adults	Tungstate, pH 8	3.4 and 5 μg	Oral (gavage)	[15]
	Domryu rats	—	150–200 g	$Na_2{}^{181}WO_4$	3.3 μg	Intravenous	[5]
	Sprague-Dawley rats	216	10 weeks (316–532 g)	Na_2WO_4	8.97 mg/kg	Intravenous	
					35.9 and 107.7 mg/kg	Oral (gavage)	
	Albino rats	—	Adults	Na_2WO_4	—	Intramuscular	[13]
		—	—	$K_2{}^{185}WO_4$	15 mg/kg	Intraperitoneal	[16]
				$Na_2{}^{185}WO_4$	120 μg/kg	Intravenous	[17]
Mouse	Pregnant mice (albino of the NMRI strain)	3	Day 8 of gestation		20 mg/kg		
		1	day 12 of gestation		120 μg/kg		
		16			20 mg/kg		
		9	day 17 of gestation		20 mg/kg		
		10	gestation		20 mg/kg		
	Male NMRI mice	3	—	$Na_2{}^{185}WO_4$	120 μg/kg	Intravenous	[4]
	Female mouse	1	—	$Na_2{}^{185}WO_4$	120 μg/kg	Intravenous	
	CD-1 mice			$CaWO_4$	160 μg	Instillation	[14]
Sheep	Texel wethers	5	66–71 kg	$Na_2{}^{185}WS_3O$	0.5 mg	Intravenous	
				$Na_2{}^{185}WS_4$	0.5 mg	Intravenous	
				$Na_2{}^{185}WO_4$	0.5 mg	Intravenous	
				$Na_2{}^{185}WS_3O$	50 mg/day	Intravenous	
				$Na_2{}^{185}WS_4$	50 mg/day	Intravenous	

— unknown.

2.1.1.2. Distribution and Retention

(a) *Circulating Tungsten.* In vivo experiments in animals [5,6,9,11–15] yield variable residence times of tungsten in blood but generally indicate that a large part of the administered tungsten (whatever the animal, the route of administration, and the form of tungsten) is rapidly removed from blood. The remaining part (a few tenths of a percent) is more slowly evacuated, this persistence being due to a return of tungsten from extravascular spaces to blood and to its binding to nondiffusible constituents such as proteins or red blood cells. As an example, it has been observed, after intravenous injection of radiolabeled tungstate into dogs, that about 70% is removed from blood with a biological half-life (B.H.L.) of 35 min, 25% with a B.H.L. of 70 min, and most of the remainder with a B.H.L. of 5 h [12].

The distribution of tungsten between plasma and red blood cells (RBCs) is variable. In beagle dogs receiving radiotungsten intravenously, the ratio of the concentration in plasma to that in RBCs is about 3 during the first 24 h [12]. Higher ratios have been found in rodents (9 and 14 in rats and mice, respectively [8,16], at day 3 after administration). 6 h after intravenous injection in goats, the RBCs contain 10% of [181]W in blood [11].

(b) *Tissues and Organs.* [185]W-tungstate, injected into male and pregnant female mice, accumulates in skeleton, kidneys, liver and spleen before being rapidly excreted to urine and intestinal contents [17]. Relatively high concentrations are also observed in thyroid, adrenal medulla, and pituary as well as in the seminal vesicle of males and in the follicules of female ovaries. Furthermore, W is easily transferred from mother to fetus, particularly in late gestation.

Other studies carried out in animals [1,5,7–8,14–15] confirm that the major site of long-term retention of injected and ingested tungsten is generally bone. The quantity accumulated represents a few tenths of a percent of absorbed tungsten, and the retention is greater in growing than in mature bone. According to Kaye, for example, about 0.4% of a tungstate dose administered to adult rats by gavage was retained with a biological half-life of 1100 days for the slowest component of a three-component elimination curve [8].

Target organs are lung and kidney when [181]WO$_3$ is inhaled by dogs [1]. It was found that 69% of the activity in lung is removed with a

B.H.L. of 4 h, 23% with a B.H.L. of 20 h, 4.6% with a B.H.L. of 6.3 days, and 3% with a B.H.L. of 100 days.

The levels of tungsten detected in the liver of animals following injection, ingestion, or inhalation of this metal [1,6–8,11,13,15] are generally higher than in other soft tissues, which may be explained by the ability of W to replace Mo in certain liver enzymes (see Section 2.1.2).

Results concerning the retention in spleen are more variable but W levels, observed a long time after administration, are generally higher than the average for all other soft tissues [1,6–9,11,17]. In goats, the level of tungsten in spleen [11] is slightly less than in liver during the first few days after exposure to radiotungstate. In dogs [1], the concentration in spleen is very close to that in liver 165 days after the inhalation of tungstic oxide.

The other soft tissues seem to accumulate a substantial portion of deposited tungsten soon after introduction into blood but lose most of the initial burden within a few hours.

2.1.1.3. Elimination

Experiments on animals indicate that a large portion of tungsten is rapidly eliminated via urine or feces when injected or orally administered as tungstate. A total of about 80–95% is excreted during the first day following administration in rats [8,13,15], goats [11], dogs [1,12], and sheep [14]. Lower rates of elimination are observed for lactating cows (only 50% of tungsten is eliminated 30 h after oral administration and 68% 92 h after intravenous injection of [187]W as tungstate). The use of trithiotungstate or tetrathiotungstate instead of tungstate also decreases the rate of excretion. About 50% of activity is found in urine and feces of sheep 24 h after intravenous injection of [[185]W] trithiotungstate, and it is necessary to wait 120 h to reach 40% of excretion with [[185]W] tetrathiotungstate.

Elimination via urine seems to be the major excretion pathway for absorbed and injected tungsten. Only a small fraction of the tungsten filtered is deposited in the kidneys and remains a few hours or days in the renal tubules. Urinary-to-fecal excretion ratios are generally greater than 10 in the first days following the administration [7,11,12]. As an exception, a ratio of 0.23 was found in lactating cows after oral intake of tungstate.

2.1.2. Mode of Action

Tungsten and molybdenum belong to the same subgroup of the periodic table. Their atomic and ionic radii are almost identical and their chemical properties are similar. Different studies, carried out on chickens [18], rats [19], goats and cows [20], have shown that tungsten acts antagonistically toward Mo, significantly decreasing the activity of sulfite and xanthine oxidase (two liver molybdenum enzymes) as well as hepatic Mo concentration. The most likely mechanism is the replacement of molybdenum by tungsten in the enzyme, although other modes of action are possible. As an example, tungstate administered to rats prevents the incorporation of Mo in xanthine oxidase without W being incorporated into the enzyme [21]. It also inhibits the transport of sulfate, molybdate, sulfite, and thiosulfate in the gut [22]. Dietary tungstate rapidly decreases the activities of xanthine dehydrogenase and sulfite oxidase in the liver, kidneys, and intestine of the rat [23].

Sodium tungstate stimulates the production of adenylate cyclase in ovarian homogenates of rats. The activation is rapid and takes place also in brain, heart, lung, kidneys, and liver. Hwang and Ryan's explanation is based on a phosphate transfer mechanism [24]. Tungstate, as molybdate, can also replace phosphate in bone [7]. Ammonium di- and tetra-thiotungstates, added to the diet of weanling rats [25], induce signs of copper deficiency, including a progressive decline in plasma ceruloplasmin oxidase activity. Similar effects, observed for thiomolybdates, were attributed to the creation of new Cu binding sites, which alters the distribution of systemic copper. Copper metabolism is also distributed by intravenous injection of thiotungstates into sheep [13] or cattle [26].

It has also been demonstrated that tungstate, orally administered to diabetic rats, is capable of decreasing glycemia to levels similar to those observed in healthy animals [27]. In insulin-dependent animals, an increase of glucokinase, 6-phosphofructo-2-kinase, and pyruvate activity as well as a decrease of phosphoenolpyruvate carboxykinase expression was observed. Glucose production and disposal was thus improved. The mechanism is slightly different for non-insulin-dependent diabetic rats, where tungstate administration causes a normalization of glycemia through the restoration of islet function.

2.2. Data Collected for Humans

The fate of tungsten in humans has been principally studied on workers chronically exposed to hard-metal dust. Results are based on the determination of tungsten concentrations in blood, tissues, and excreta in contaminated subjects [28–32]. Sabbioni et al. have carried to the largest epidemiological study, involving 251 individuals from which 23 workers, representing the highest number of cases ever monitored, have been diagnosed as diseased subjects [32]. Four different groups of workers and seven individual cases, employed in different factories in northern Italy, have been examined. The data collected [concentration of tungsten in the urine, blood, pubic hair, and toe nails as well as in biopsy and bronchoalveolar lavage (BAL) of the patients] are indicated in Table 2, together with kidney and liver contents obtained by Brune et al. [28] in another work. These values are compared to the levels found in nonexposed people.

In all cases, the values observed in exposed workers exceed the normal levels. The corresponding high standard deviations are indicative of wide individual fluctuations. One can see that concentrations are particularly high in pubic hair, toe nails, and lung, but this approach cannot provide much information about the biokinetics of tungsten in the human body.

2.3. Proposed Model for Tungsten Distribution and Retention in Humans

In order to be able to predict the behavior of tungsten in the human body, a model was proposed in 1981 by the International Commission on Radiological Protection (ICRP) [35]. It is based on exponential curve fits to data obtained in dogs, goats, and rats but lacks biological reality. Recently, Leggett [36] has started to build a more realistic model, taking into account the data on the biokinetics in laboratory animals and information on the kinetics of molybdenum and other physiological analogues of tungsten in humans and animals. The compartmental model structure is the same as that developed for uranium and described in IRCP Publication 69 [37]. Transport of tungsten between compartments

TABLE 2

Tungsten Concentration in Occupationally Exposed Subjects and in Controls

Specimen	Occupationally exposed people				Diseased workers		Controls	Ref.
	Nondiseased workers				From groups 1 and 2	Ref.		
	Group 1 (mold fillers, grinding...)	Group 2 (production of tools)	Group 3 (production of diamond wheels)	Group 4 (hard metal manufacturing)				
Blood	1.2 ± 1.6 (0.04–6.5); n=43	1.29 ± 2.7 (0.11–10); n=16	—	—	—	[32]	4.2 (n = 3); 0.4 (n = 6); ~0.2	[30] [29] [33]
Urine	6.7 ± 19.4 (0.11–230); n = 78	9.32 ± 6.5 (1.1–25); n = 16	2.29 ± 2.79 (0.25–12.5); n = 23	12.8 ± 14.7 (0.35–55); n = 24	—	[32]	0.21 ± 0.09 (0.1–0.32); n = 14; 0.4 (n = 3); 0.7 (n = 6); ~0.7	[34] [30] [29] [33]
Pubic hair	2147 ± 5151 (25–59,000); n = 75	7018 ± 16570 (227–76,000); n = 20	—	9585 ± 9159 (340–4000); n = 24	—	[32]	~15 (n = 3); ~100	[30] [33]
Toe nails	3056 ± 10760 (27–105,000); n = 82	17298 ± 32470 (1070–127,000); n = 23	—	—	—	[32]	18 (n = 3); ~17	[30] [33]

Open lung biopsy	—	—	—	52,000; 107,000 [32]	1.5 (n = 6) [29]	
Transbronchial biopsy	—	—	—	78980 ± 39316 [32] (33900–134000); n = 5		
BAL	448 ± 602 n = 24	—	—	[32]	1.15 ± 0.9 (n = 9) [31] 1.5 (n = 6) [29]	

Workers from a refinery and a smeltery

Kidney	< 3–18 (n = 19)	[28]	< 3–5 (n = 8) [28]
Liver	< 3– 14 (n = 19)	[28]	< 3–36 (n = 8) [28]

— unknown.
Mean Results ± standard deviation expressed in μg/L for blood, urine, and BAL; Results in ng/g wet weight for hair, nails, lung biopsy, kidney, and liver).

is assumed to follow first-order kinetics and the transfer rates are divided into groups comprising (1) circulating tungsten, (2) uptake and retention in the kidneys and urinary excretion, (3) fecal excretion, (4) uptake and retention in the liver, (5) uptake and retention in the spleen, (6) uptake and retention in the remaining soft tissues, and (7) uptake and retention in the skeleton. This model was compared with experimental data and the tungsten IRCP model [35], which allowed Legett to make suggestions for further research regarding the biokinetics of tungsten.

3. TOXICITY OF TUNGSTEN

3.1. Acute Toxicity

3.1.1. Lethal Doses

Tungstate toxicity has been studied in different animals using several modes of administration. By subcutaneous injection the LD_{50} has been estimated to be 71 mg W/kg in rabbits [38], whereas the lethal dose is 450 mg tungstate/kg in guinea pig [39]. 80 mg W/kg and 112 mg W/kg are the LD_{50} values obtained by intraperitoneal injection in mice and rat, respectively [40]. Intravenously, the following lethal doses have been found: 463 mg W/kg for dog, 272 mg W/kg for chicken, 154 mg W/kg for rat, 128 mg W/kg for pigeon, 79 mg W/kg for cat, and only 59 mg W/kg for rabbit [41]. Orally, Karantassis found a lethal dose of 550 mg tungstate in guinea pig [39].

3.1.2. Symptoms of Intoxication

Pham-Huu and Som report that the principle signs of acute intoxication by sodium tungstate injected intraperitoneally in rats and mice are asthenia, adynamia, prostration, coma, and finally death [41]. Similar signs were described by Caujolle et al. [40]. Oral administration or injection of tungstate in guinea pigs produces anorexia, colic, disorganized movements, trembling, and dyspnea [39]. Lewin and Pouchet observed vomiting in dogs following the ingestion of tungstate [42]. WC-Co powder (84% W) instilled in female rats produces a severe alveolitis and fatal pulmonary edema [43].

3.2. Short-Term Subacute or Chronic Toxicity

As early as 1890, Berstein-Kohan, quoted by Kinard and Van de Erve [44], studied the effect of 17 mg W given to rabbits in food during 45 days. Only diarrhea was observed. In 1941, Kinard and Van de Erve measured the effects of dietary tungsten (0.1–0.5 wt% per day during 70 days) as tungstic oxide, sodium tungstate, and ammonium paratungstate in growing rats. At a concentration of 0.5% tungstic oxide was responsible for 9 deaths among 11 rats, sodium tungstate killed 7 animals, and paratungstate had no effect. A decrease of xanthine oxidase activity and a slower growth rate was also observed following the introduction of tungsten in the diet of chickens [18,45], goats and cows [20].

3.3. Fetal Toxicity

In 1978, Nadeeko et al. observed that sodium tungstate may be toxic to rat fetus at concentrations that usually induce no effect on the maternal organism [46]. For his part, Wide found that tungstate, when administered to mice in a single dose at early organogenesis, does not induce any fetal malformation but leads to a high frequency of resorptions [47]. These adverse effects on embryonic survival may be due to toxic effects directly on the embryo or to a disturbance of certain maternal functions. Toxicity to embryonic cells in vitro was observed at concentrations close to those found in vivo after administration of a fetotoxic dose of tungstate. Cohen et al. have also shown that tungstate, administered to pregnant rats inhibits the production of xanthine oxidase and high doses may be fatal to fetus [48]. Accumulation of tungsten is observed in maternal hypophisis and ovaries and W added to the ovarian homogenates modifies the activity of adenylate cyclase.

3.4. Long-Term Chronic Toxicity

3.4.1. Toxicity to Lungs

Experimental studies carried out in rats and guinea pigs indicate that W metal [49–52], tungsten carbide [49,50,53], and tungsten carbide-carbone [54] alloys are not very toxic. On the other hand, inhalation of a tungsten carbide-cobalt alloy (CW/Co = 91/9) by guinea pigs induces an inflammatory reaction accompanied by peribronchal fibrosis and bronchial mucosa hypertrophy [50,55].

It has been observed that workers daily exposed to tungsten carbide/cobalt dusts (WC > 80%) may suffer from the so-called hard-metal lung disease which is mainly characterized by an interstitial fibrosis, occupational asthma, and sometimes a syndrome that resembles extrinsic allergic alveolitis. A detailed description of pathological symptoms and mechanisms is given in [56]. Symptoms appear after a variable period (1 month to more than 2 years), and it seems that the evolution of the disease does not depend on the age but mostly on the sensitivity of the subjects. This disease is unusual among workers exposed to pure cobalt dust, and in vitro and in vivo experiments have shown that the biological response to the mixture (WC-Co) is different from that of the single components. Nevertheless, the mechanisms of interstitial fibrosis and cancer are not completely understood. A mechanism, based on the interaction between tungsten carbide and cobalt metal to produce activated oxygen species, has been proposed. In this process, which is still in debate [57,58], the source of critical reactions that lead to cytotoxicity is considered to be the oxygen species and not the Co ions that are produced. Recent in vivo experiments [59] indicate that the pulmonary response evoked in the lung by inhalation of high levels of WC-Co particles is due to alterations in the F-actin microfilaments of lung epithelial cells.

3.4.2. Cutaneous Toxicity

In hard-metal manufactures, workers' skin is directly exposed to metals, mainly tungsten and cobalt. W has a primary irritant effect. It has been shown that allergic eczemas observed are exclusively due to Co, whereas W is responsible for contact eczema, pruritis, folliculitis, and neurodermatitis [60,61].

3.4.3. Tungsten and Cancer

Tungsten was proved to accelerate the development of mammary cancer in rats [62]. This was attributed to a decrease of hepathic molybdenum, which severely disturbs the function of liver, the site of estrogen metabolism.

It has also been shown that a WC-Co mixture produces a higher level of DNA damage than Co alone, which may be responsible for the increased incidence of lung cancers observed in the population of hard-metal workers. The results are consistent with the implication of an increased production of hydroxyl radicals and also suggest that WC modifies the structure of chromatin [63].

A recent epidemiological study has confirmed that the excess mortality from lung cancer among hard-metal production workers is not due to smoking. This excess occurs mostly in subjects to unsintered hard-metal dust [64].

4. CONCLUSIONS

Tungsten has been the subject of numerous in vivo and in vitro studies in view of determining its metabolism and toxicity in humans and animals. It seems that W metal, tungsten carbide, and tungsten carbide-carbon are not very toxic and that the severe pneumoconosis often observed among the hard-metal workers is due to the simultaneous presence of cobalt and tungsten. A model for the retention and distribution in humans has been proposed, but some experimental results are still lacking. In particular, better information concerning the accumulation of tungsten in soft tissues other than liver, kidney and spleen is needed. Data for large and presumably more "human-like" laboratory animals, such as swine, are also required.

ABBREVIATIONS AND DEFINITIONS

BAL	bronchoalveolar lavage
B.H.L.	biological half-life

IRCP   International Commission on Radiological Protection
LD_{50} lethal dose of a substance that leads, in a single dose,
                to the death of 50% of a population
RBC   red blood cell
wt   weight

REFERENCES

1. R. L. Aamodt, *Health Physics, 28*, 733–742 (1975).
2. N. R. Grande, C. Moreira de Sa, A. P. Aguas, E. Carvahlo, and M. Soares, *Lymphology, 23*, 171–182 (1990).
3. A. De Souza Pereira, N. R. Grande, E. Carvahlo, and A. Ribeiro, *Acta Anatomica, 145*, 416–419 (1992).
4. M. N. D. Peao, A. P. Aguas, C. M. Moreira de Sa, and N. R. Grande, *Lung, 171*, 187–201 (1993).
5. S. Le Lamer, P. Poucheret, G. Cros, R. Kiesgen de Richter, P. A. Bonnet, and F. Bressolle, *J. Pharmacol. Exp. Ther. 294(2)*, 714–721 (2000).
6. J. E. Ballou, Metabolism of ^{185}W in the rat, *AEC Res. Dev. Rep.*, HW-64112, 1960.
7. D. Fleshman, S. Krokz and A. Silva, The metabolism of elements of high atomic number, University of California Radiation Laboratory, 14739, 69–86 (1966).
8. S. V. Kaye, *Health Physics, 15*, 399–418 (1968).
9. A. L. Mullen, E. W. Bretthauer, and R. E. Stanley, *Health Physics, 31*(Nov), 417–424 (1976).
10. M. C. Bell and N. N. Sneed, in *Trace Element Metabolism in Animals* (C. F. Mills, ed.), E. and S. Livingstone, Edinburgh, 1970, pp. 70–72.
11. L. Ekman, H. D. Figueiras, B. E. V. Jones, and S. Myamoto, Metabolism of 181W-labeled sodium tungstate in goats, *FOA Rep.*, C-40070-A3, 1977.
12. R. L. Aamodt, *Health Physics, 24*, 519–524 (1973).
13. P. W. Durbin, *Health Physics, 2*, 225–238 (1960).

14. J. Mason, G. Mulryan, M. Lamand, and C. Lafarge, *J. Inorg. Biochem., 35*, 115–126 (1989).

15. A. Ando, J. Ando, T. K. Hiraki, and K. Hisada, *Nucl. Med. Biol., 16*, 57–80 (1989).

16. A. W. Wase, *Arch. Biochem. Biophys., 61*, 272–277 (1956).

17. E. M. Wide, B. R. G. Danielsson, and L. Dencker, *Environ. Res., 40*, 487–498 (1986).

18. S. Higgins, D. A. Richert, and W. N. Westerfeld, *J. Nutr., 59*, 539–559 1956).

19. J. L. Johnson and K. V. Rajagopalan, *J. Biol. Chem., 249*, 859–866 (1974).

20. L. Hart, E. C. Owen, and R. Proudfoot, *Br. J. Nutr., 21(3)*, 617–630 (1967).

21. J. L. Johnson, W. R. Wau, H. J. Cohen, and K. V. Rajagopalan, *J. Biol. Chem., 249*, 5056–5061 (1974).

22. C. J. Cardin and J. Mason, *Biochim. Biophys. Acta. 394*, 46–54 (1975).

23. G. Kazantsis, in *Handbook of the Toxicology of Metals*, L. Friberg, G. F. Nordberg, and V. B. Vouk, eds.), Elsevier, Amsterdam, 1979, pp. 637–646.

24. P. L. Hwang and R. J. Ryan, *Endocrinology, 108*, 435–439 (1981).

25. B. W. Young, I. Bremmer, and C. F. Mills, *J. Inorg. Biochem., 16*, 121 (1982).

26. J. Mason, A. Smith, and Z. Y. Wang, *J. Comp. Pathol., 98*, 375–379 (1988).

27. A. Barbera, J. Fernandez-Alvarez, A. Truc, and R. Gomis and J.-J. Guinovart, *Diabetologia, 40*, 143–149 (1997).

28. D. Brune, G. Nordberg, and P. O. Wester, *Sci. Total Environ., 16*, 13–35 (1980).

29. G. Rizzato, S. Lo Cicero, M. Barberis, M. Torre, R. Pietra, and E. Sabbioni, *Chest, 90(1)*, 101–106 (1986).

30. G. Nicolaou, R. Pietra, E. Sabbioni, G. Mosconi, G. Cassina, and P. Seghizzi, *J. Trace Elem. Electrolytes Health Dis., 1(2)*, 73–77 (1987).

31. E. Sabbioni, R. Pietra, F. Mousty, F. Colombo, G. Rizzato, and G. Scansetti, *J. Radioanalyt. Nuclear Chem.*, *110*(2), 595–601 (1987).

32. E. Sabbioni, C. Minoia, R. Pietra, G. Mosconi, A. Forni, and G. Scansetti, *Sci. Total Environ.*, *150*, 41–54 (1994).

33. G. V. Iyengar, M. E. Kollmer, and H. J. M. Bowen (eds.), *The Elemental Composition of Human Tissues and Body Fluids*, Verlag Chemie, New York, 1978.

34. P. Schramel, I. Wendler, and J. Angerer, *Int. Arch. Occup. Environ. Health*, *69*(3), 219–223 (1997).

35. International Commission on Radiological Protection, *Limits for Intakes of Radionuclides by Workers*, IRCP Publication 30, Part 3, Pergamon Press, Oxford, 1981.

36. R. W. Legett, *Sci. Total Environ.*, *206*, 147–165 (1997).

37. International Commission on Radiological Protection, *Age-Dependent Doses to Members of the Public from Intake of Radionuclides*, IRCP Publication 69, Part 3, Pergamon Press, Oxford, 1995.

38. L. M. Lusky, H. A. Braun, and E. P. Laug, *J. Ind. Hyg.*, *31*, 301–308 (1949).

39. T. Karantassis, *Bull. Sci. Pharm.*, *11*, 561–567 (1924).

40. F. Caujolle, J. C. Godfrain, D. Meynier, and C. Pham Huu, *Compte Rendu Acad. Sci. 248*, 2667–2669 (1959).

41. C. Pham-Huu and C. Som, *Agressologie, 8*(5), 43–439 (1968).

42. L. Lewin and G. Pouchet, *Traité de Toxicologie*, Octave Douin, Paris, 1903, p. 372.

43. G. Lasfargues, D. Lison, P. Maldague, and R. Lauweris, *Toxicol. Appl. Pharmacol, 112*, 41–50 (1992).

44. F. W. Kinard and J. Van de Erve, *J. Pharm. Exp. Therm.*, *72*, 196–201 (1941).

45. R. A. Teekel and A. B. Watts, *Poult. Sci.*, *38*, 791 (1959).

46. V. G. Nadeeko, V. Lenchenko, S. B. Genkina, and T. Arkhipenko, *Famakol. Toksikol.*, *41*, 620–623 (1978).

47. M. Wide, *Environ. Res.*, *33*, 47–53 (1984).

48. H. J. Cohen, J. L. Johnson and K. V. Rajagopalan, *Arch. Biochem. Biophys.*, *164*, 440–446 (1974).

49. H. E. Harding, *Br. J. Ind. Med.*, 7, 76–78 (1950).
50. A. Delahant, *Arch. Ind. Health*, 12, 116–120 (1955).
51. G. W. H. Schepers, *Arc. Ind. Health*, 12, 134–136 (1955).
52. Z. S. Kaplun and N. W. Mezencewa, *J. Hyg. Epidemiol. Microbiol. Immun.*, 4, 390–399 (1960).
53. C. W. Miller, M. M. Davis, A. Goldman, and J. P. Wyatt, *Arch. Ind. Hyg.*, 8, 453–465 (1953).
54. G. W. H. Schepers, *Arch. Ind. Health*, 12, 137–140 (1955).
55. G. W. H. Schepers, *Arch. Ind. Health*, 12, 140–146 (1955).
56. D. Lison, *Crit. Rev. Toxicol.*, 26(6), 585–616 (1996).
57. G. Roesems, P. H. M. Hoet, D. Dinsdale, M. Demedts, and B. Nemery, Toxicol. Appl. Pharmacol., 162, 2–9 (2000).
58. D. Lison, *Toxicol. Appl. Pharmacol.*, 168, 2–9 (2000).
59. J. M. Antonini, K. Starks, J. R. Roberts, L. Millecchia, H. M. Yang, and K. M. Rao, *In Vitr. Mol. Toxicol.*, 13(1), 5–16 (2000).
60. E. Skog, *Ind. Med. Surg.*, 32(7), 266–268 (1963).
61. Rystedt, T. Fischer, and B. Lagerholm, *Contact Dermatitis, 9*, 69–73 (1983).
62. H. J. Wei, X. M. Luo, and S. P. Yang, *J. Natl. Cancer Inst.*, 74(2), 469–473 (1985).
63. D. Anard, M. Kirsch-Volders, A. Elhajouji, K. Belpaeme, and D. Lison, *Carcinogenesis*, 18(1), 177–184 (1997).
64. P. Wild, A. Perdrix, S. Romazini, J. J. Moulin, and F. Pellet, *Occup. Environ. Med.*, 57(8), 568–573 (2000).

Subject Index

A

A.

caulinodans, see Azorhizobium
crystallopoietes, see Arthrobacter
eutrophus, see Alcaligenes
faecalis, see Alcaligenes
fulgidus, see Archaeoglobus
ilicis, see Arthrobacter
methanolica, see Amycolatopsis
nicotinovorans, see Arthrobacter
nidulans, see Aspergillus
oxidans, see Arthrobacter
parvulus, see Achromobacter
picolinophilus, see Arthrobacter
platensis, see Arthrospira
polyoxygenes, see Acetobacter
spirulina, 527
thaliana, see Arabidopsis
ureafaciens, see Arthrobacter
vinelandii, see Azotobacter

[A.]
woodii, see Acetobacterium
xylosoxidans, see Achromobacter
ABC transporters, 40, 41, 43, 47, 64
Abundance of
  copper, 704
  iron, 704
  molybdenum, 3–6, 704
  tungsten, 3–6, 704
  zinc, 704
Acetaldehyde, 490, 565, 677, 680
  formation, 680, 719
  oxidation, 599, 686
Acetobacter polyoxygenes, 486, 491
Acetobacterium woodii, 413
Acetylacetone, 143, 148
Acetyl coenzyme A, 177, 575, 592
  biosynthesis, 597
Acetyl-coenzyme A decarbonylase,
  598
  nickel in, 598

761

T